MATH 1411 CALCULUS I

Larson/Edwards

University of Texas at El Paso

Australia • Brazil • Japan • Korea • Mexico • Singapore • Spain • United Kingdom • United States

Math 1411 Calculus I,
University of Texas at El Paso

Calculus, Eleventh Edition
Ron Larson, Bruce Edwards

For product information and technology assistance, contact us at
Cengage Learning Customer & Sales Support, 1-800-354-9706
For permission to use material from this text or product,
submit all requests online at **cengage.com/permissions**
Further permissions questions can be emailed to
permissionrequest@cengage.com

This book contains select works from existing Cengage Learning resources and was produced by Cengage Learning Custom Solutions for collegiate use. As such, those adopting and/or contributing to this work are responsible for editorial content accuracy, continuity and completeness.

ISBN: 978-1-337-69720-0

Cengage Learning
20 Channel Center Street
Boston, MA 02210
USA

Cengage Learning is a leading provider of customized learning solutions with office locations around the globe, including Singapore, the United Kingdom, Australia, Mexico, Brazil, and Japan. Locate your local office at: **www.international.cengage.com/region.**

Cengage Learning products are represented in Canada by Nelson Education, Ltd.

For your lifelong learning solutions, visit **www.cengage.com/custom.**

Visit our corporate website at **www.cengage.com.**

Printed at CLDPC, USA, 05-19

Contents

*Available at the text-specific website *www.cengagebrain.com*

Preface

Welcome to *Calculus*, Eleventh Edition. We are excited to offer you a new edition with even more resources that will help you understand and master calculus. This textbook includes features and resources that continue to make *Calculus* a valuable learning tool for students and a trustworthy teaching tool for instructors.

Calculus provides the clear instruction, precise mathematics, and thorough coverage that you expect for your course. Additionally, this new edition provides you with **free** access to three companion websites:

- **CalcView.com**—video solutions to selected exercises
- **CalcChat.com**—worked-out solutions to odd-numbered exercises and access to online tutors
- **LarsonCalculus.com**—companion website with resources to supplement your learning

These websites will help enhance and reinforce your understanding of the material presented in this text and prepare you for future mathematics courses. CalcView® and CalcChat® are also available as free mobile apps.

Features

NEW CalcView®

The website *CalcView.com* contains video solutions of selected exercises. Watch instructors progress step-by-step through solutions, providing guidance to help you solve the exercises. The CalcView mobile app is available for free at the Apple® App Store® or Google Play™ store. The app features an embedded QR Code® reader that can be used to scan the on-page codes and go directly to the videos. You can also access the videos at CalcView.com.

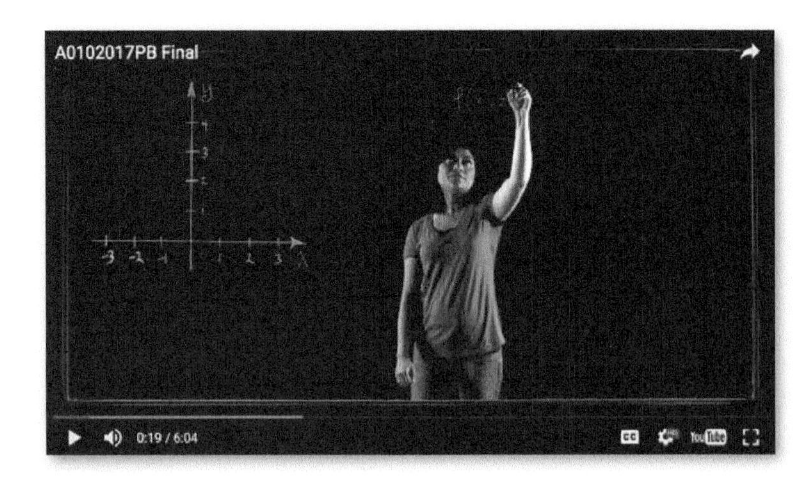

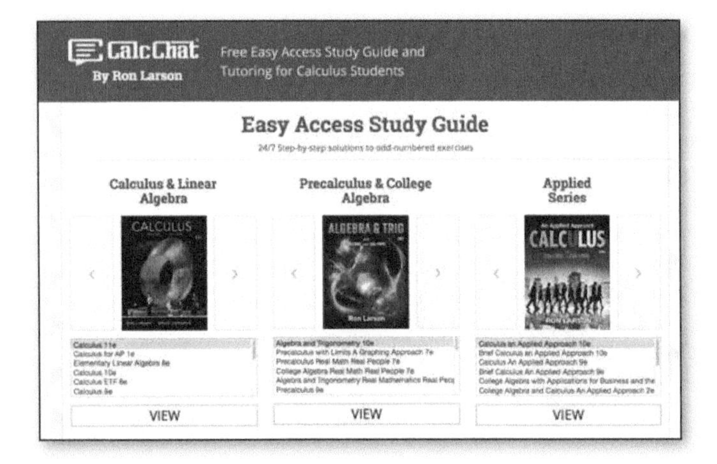

UPDATED CalcChat®

In each exercise set, be sure to notice the reference to *CalcChat.com*. This website provides free step-by-step solutions to all odd-numbered exercises in many of our textbooks. Additionally, you can chat with a tutor, at no charge, during the hours posted at the site. For over 14 years, hundreds of thousands of students have visited this site for help. The CalcChat mobile app is also available as a free download at the Apple® App Store® or Google Play™ store and features an embedded QR Code® reader.

App Store is a service mark of Apple Inc. Google Play is a trademark of Google Inc.
QR Code is a registered trademark of Denso Wave Incorporated.

REVISED LarsonCalculus.com

All companion website features have been updated based on this revision. Watch videos explaining concepts or proofs from the book, explore examples, view three-dimensional graphs, download articles from math journals, and much more.

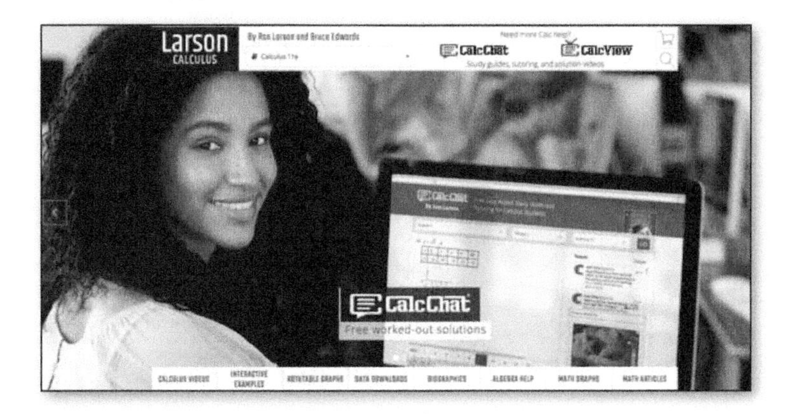

NEW Conceptual Exercises

The *Concept Check* exercises and *Exploring Concepts* exercises appear in each section. These exercises will help you develop a deeper and clearer knowledge of calculus. Work through these exercises to build and strengthen your understanding of the calculus concepts and to prepare you for the rest of the section exercises.

REVISED Exercise Sets

The exercise sets have been carefully and extensively examined to ensure they are rigorous and relevant and to include topics our users have suggested. The exercises are organized and titled so you can better see the connections between examples and exercises. Multi-step, real-life exercises reinforce problem-solving skills and mastery of concepts by giving you the opportunity to apply the concepts in real-life situations.

REVISED Section Projects

Projects appear in selected sections and encourage you to explore applications related to the topics you are studying. We have added new projects, revised others, and kept some of our favorites. All of these projects provide an interesting and engaging way for you and other students to work and investigate ideas collaboratively.

Table of Contents Changes

Based on market research and feedback from users, we have made several changes to the table of contents.

- We added a review of trigonometric functions (Section P.4) to Chapter P.
- To cut back on the length of the text, we moved previous Section P.4 *Fitting Models to Data* (now Appendix G in the Eleventh Edition) to the text-specific website at *CengageBrain.com.*
- To provide more flexibility to the order of coverage of calculus topics, Section 3.5 *Limits at Infinity* was revised so that it can be covered after Section 1.5 *Infinite Limits*. As a result of this revision, some exercises moved from Section 3.5 to Section 3.6 *A Summary of Curve Sketching.*
- We moved Section 4.6 *Numerical Integration* to Section 8.6.
- We moved Section 8.7 *Indeterminate Forms and L'Hôpital's Rule* to Section 5.6.

Chapter Opener

Each Chapter Opener highlights real-life applications used in the examples and exercises.

Section Objectives

A bulleted list of learning objectives provides you with the opportunity to preview what will be presented in the upcoming section.

Theorems

Theorems provide the conceptual framework for calculus. Theorems are clearly stated and separated from the rest of the text by boxes for quick visual reference. Key proofs often follow the theorem and can be found at *LarsonCalculus.com.*

Definitions

As with theorems, definitions are clearly stated using precise, formal wording and are separated from the text by boxes for quick visual reference.

Explorations

Explorations provide unique challenges to study concepts that have not yet been formally covered in the text. They allow you to learn by discovery and introduce topics related to ones presently being studied. Exploring topics in this way encourages you to think outside the box.

Remarks

These hints and tips reinforce or expand upon concepts, help you learn how to study mathematics, caution you about common errors, address special cases, or show alternative or additional steps to a solution of an example.

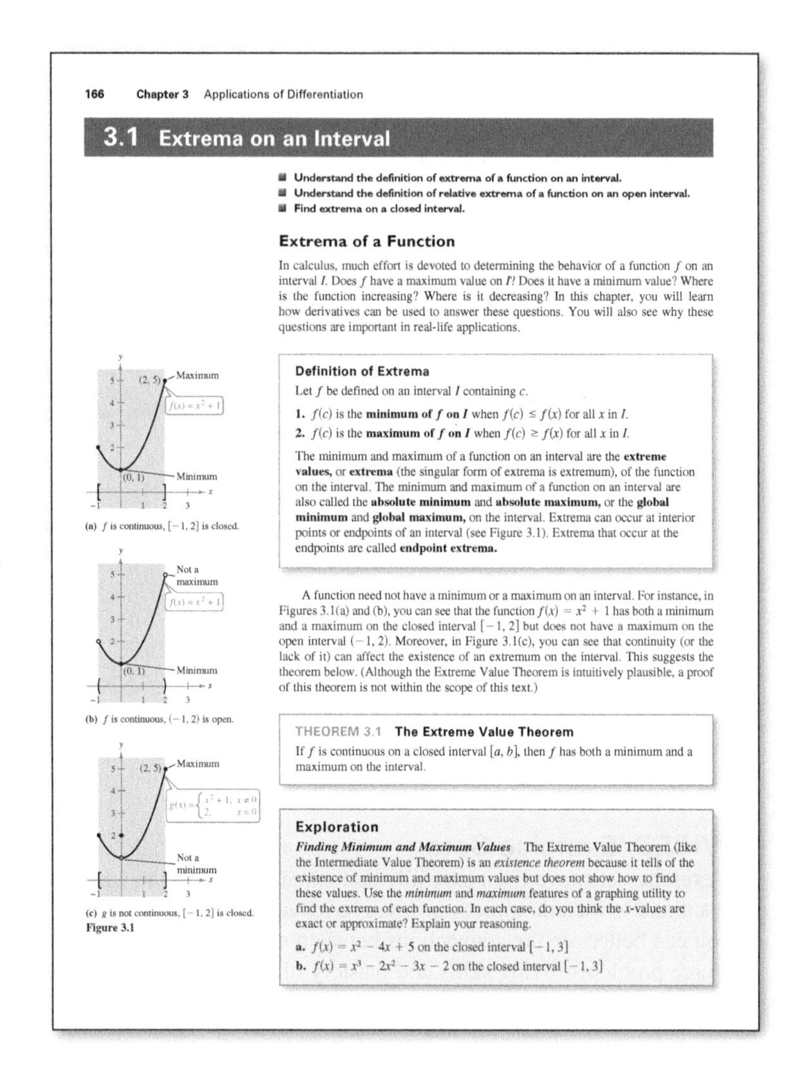

166 **Chapter 3** Applications of Differentiation

3.1 Extrema on an Interval

- **Understand the definition of extrema of a function on an interval.**
- **Understand the definition of relative extrema of a function on an open interval.**
- **Find extrema on a closed interval.**

Extrema of a Function

In calculus, much effort is devoted to determining the behavior of a function f on an interval I. Does f have a maximum value on I? Does it have a minimum value? Where is the function increasing? Where is it decreasing? In this chapter, you will learn how derivatives can be used to answer these questions. You will also see why these questions are important in real-life applications.

Definition of Extrema

Let f be defined on an interval I containing c.

1. $f(c)$ is the **minimum of f on I** when $f(c) \le f(x)$ for all x in I.
2. $f(c)$ is the **maximum of f on I** when $f(c) \ge f(x)$ for all x in I.

The minimum and maximum of a function on an interval are the **extreme values,** or **extrema** (the singular form of extrema is extremum), of the function on the interval. The minimum and maximum of a function on an interval are also called the **absolute minimum** and **absolute maximum,** or the **global minimum** and **global maximum,** on the interval. Extrema can occur at interior points or endpoints of an interval (see Figure 3.1). Extrema that occur at the endpoints are called **endpoint extrema.**

(a) f is continuous, $[-1, 2]$ is closed.

(b) f is continuous, $(-1, 2)$ is open.

(c) g is not continuous, $[-1, 2]$ is closed.

Figure 3.1

A function need not have a minimum or a maximum on an interval. For instance, in Figures 3.1(a) and (b), you can see that the function $f(x) = x^2 + 1$ has both a minimum and a maximum on the closed interval $[-1, 2]$ but does not have a maximum on the open interval $(-1, 2)$. Moreover, in Figure 3.1(c), you can see that continuity (or the lack of it) can affect the existence of an extremum on the interval. This suggests the theorem below. (Although the Extreme Value Theorem is intuitively plausible, a proof of this theorem is not within the scope of this text.)

THEOREM 3.1 **The Extreme Value Theorem**

If f is continuous on a closed interval $[a, b]$, then f has both a minimum and a maximum on the interval.

Exploration

Finding Minimum and Maximum Values The Extreme Value Theorem (like the Intermediate Value Theorem) is an *existence theorem* because it tells of the existence of minimum and maximum values but does not show how to find these values. Use the *minimum* and *maximum* features of a graphing utility to find the extrema of each function. In each case, do you think the x-values are exact or approximate? Explain your reasoning.

a. $f(x) = x^2 - 4x + 5$ on the closed interval $[-1, 3]$

b. $f(x) = x^3 - 2x^2 - 3x - 2$ on the closed interval $[-1, 3]$

How Do You See It? Exercise

The How Do You See It? exercise in each section presents a problem that you will solve by visual inspection using the concepts learned in the lesson. This exercise is excellent for classroom discussion or test preparation.

Applications

Carefully chosen applied exercises and examples are included throughout to address the question, "When will I use this?" These applications are pulled from diverse sources, such as current events, world data, industry trends, and more, and relate to a wide range of interests. Understanding where calculus is (or can be) used promotes fuller understanding of the material.

Historical Notes and Biographies

Historical Notes provide you with background information on the foundations of calculus. The Biographies introduce you to the people who created and contributed to calculus.

Technology

Throughout the book, technology boxes show you how to use technology to solve problems and explore concepts of calculus. These tips also point out some pitfalls of using technology.

Putnam Exam Challenges

Putnam Exam questions appear in selected sections. These actual Putnam Exam questions will challenge you and push the limits of your understanding of calculus.

Student Resources

Student Solutions Manual for Calculus of a Single Variable
ISBN-13: 978-1-337-27538-5

Student Solutions Manual for Multivariable Calculus
ISBN-13: 978-1-337-27539-2

Need a leg up on your homework or help to prepare for an exam? The *Student Solutions Manuals* contain worked-out solutions for all odd-numbered exercises in *Calculus of a Single Variable* 11e (Chapters P–10 of *Calculus* 11e) and *Multivariable Calculus* 11e (Chapters 11–16 of *Calculus* 11e). These manuals are great resources to help you understand how to solve those tough problems.

CengageBrain.com
To access additional course materials, please visit *www.cengagebrain.com.* At the *CengageBrain.com* home page, search for the ISBN of your title (from the back cover of your book) using the search box at the top of the page. This will take you to the product page where these resources can be found.

MindTap for Mathematics
MindTap® provides you with the tools you need to better manage your limited time—you can complete assignments whenever and wherever you are ready to learn with course material specifically customized for you by your instructor and streamlined in one proven, easy-to-use interface. With an array of tools and apps—from note taking to flashcards—you'll get a true understanding of course concepts, helping you to achieve better grades and setting the groundwork for your future courses. This access code entitles you to 3 terms of usage.

Enhanced WebAssign® 

Enhanced WebAssign (assigned by the instructor) provides you with instant feedback on homework assignments. This online homework system is easy to use and includes helpful links to textbook sections, video examples, and problem-specific tutorials.

Instructor Resources

Complete Solutions Manual for Calculus of a Single Variable, Vol. 1
ISBN-13: 978-1-337-27540-8

Complete Solutions Manual for Calculus of a Single Variable, Vol. 2
ISBN-13: 978-1-337-27541-5

Complete Solutions Manual for Multivariable Calculus
ISBN-13: 978-1-337-27542-2

The *Complete Solutions Manuals* contain worked-out solutions to all exercises in the text. They are posted on the instructor companion website.

Instructor's Resource Guide (on instructor companion site)
This robust manual contains an abundance of instructor resources keyed to the textbook at the section and chapter level, including section objectives, teaching tips, and chapter projects.

Cengage Learning Testing Powered by Cognero (login.cengage.com)
CLT is a flexible online system that allows you to author, edit, and manage test bank content; create multiple test versions in an instant; and deliver tests from your LMS, your classroom, or wherever you want. This is available online via *www.cengage.com/login.*

Instructor Companion Site
Everything you need for your course in one place! This collection of book-specific lecture and class tools is available online via *www.cengage.com/login*. Access and download PowerPoint® presentations, images, instructor's manual, and more.

Test Bank (on instructor companion site)
The Test Bank contains text-specific multiple-choice and free-response test forms.

MindTap for Mathematics
MindTap® is the digital learning solution that helps you engage and transform today's students into critical thinkers. Through paths of dynamic assignments and applications that you can personalize, real-time course analytics, and an accessible reader, MindTap helps you turn cookie cutter into cutting edge, apathy into engagement, and memorizers into higher-level thinkers.

Enhanced WebAssign® ENHANCED WebAssign
Exclusively from Cengage Learning, Enhanced WebAssign combines the exceptional mathematics content that you know and love with the most powerful online homework solution, WebAssign. Enhanced WebAssign engages students with immediate feedback, rich tutorial content, and interactive, fully customizable e-books (YouBook), helping students to develop a deeper conceptual understanding of their subject matter. Quick Prep and Just In Time exercises provide opportunities for students to review prerequisite skills and content, both at the start of the course and at the beginning of each section. Flexible assignment options give instructors the ability to release assignments conditionally on the basis of students' prerequisite assignment scores. Visit us at **www.cengage.com/ewa** to learn more.

Acknowledgments

We would like to thank the many people who have helped us at various stages of *Calculus* over the last 43 years. Their encouragement, criticisms, and suggestions have been invaluable.

Reviewers

Stan Adamski, *Owens Community College;* Tilak de Alwis; Darry Andrews; Alexander Arhangelskii, *Ohio University;* Seth G. Armstrong, *Southern Utah University;* Jim Ball, *Indiana State University;* Denis Bell, *University of Northern Florida;* Marcelle Bessman, *Jacksonville University;* Abraham Biggs, *Broward Community College;* Jesse Blosser, *Eastern Mennonite School;* Linda A. Bolte, *Eastern Washington University;* James Braselton, *Georgia Southern University;* Harvey Braverman, *Middlesex County College;* Mark Brittenham, *University of Nebraska;* Tim Chappell, *Penn Valley Community College;* Mingxiang Chen, *North Carolina A&T State University;* Oiyin Pauline Chow, *Harrisburg Area Community College;* Julie M. Clark, *Hollins University;* P.S. Crooke, *Vanderbilt University;* Jim Dotzler, *Nassau Community College;* Murray Eisenberg, *University of Massachusetts at Amherst;* Donna Flint, *South Dakota State University;* Michael Frantz, *University of La Verne;* David French, *Tidewater Community College;* Sudhir Goel, *Valdosta State University;* Arek Goetz, *San Francisco State University;* Donna J. Gorton, *Butler County Community College;* John Gosselin, *University of Georgia;* Arran Hamm; Shahryar Heydari, *Piedmont College;* Guy Hogan, *Norfolk State University;* Marcia Kleinz, *Atlantic Cape Community College;* Ashok Kumar, *Valdosta State University;* Kevin J. Leith, *Albuquerque Community College;* Maxine Lifshitz, *Friends Academy;* Douglas B. Meade, *University of South Carolina;* Bill Meisel, *Florida State College at Jacksonville;* Shahrooz Moosavizadeh; Teri Murphy, *University of Oklahoma;* Darren Narayan, *Rochester Institute of Technology;* Susan A. Natale, *The Ursuline School, NY;* Martha Nega, *Georgia Perimeter College;* Sam Pearsall, *Los Angeles Pierce College;* Terence H. Perciante, *Wheaton College;* James Pommersheim, *Reed College;* Laura Ritter, *Southern Polytechnic State University;* Leland E. Rogers, *Pepperdine University;* Paul Seeburger, *Monroe Community College;* Edith A. Silver, *Mercer County Community College;* Howard Speier, *Chandler-Gilbert Community College;* Desmond Stephens, *Florida A&M University;* Jianzhong Su, *University of Texas at Arlington;* Patrick Ward, *Illinois Central College;* Chia-Lin Wu, *Richard Stockton College of New Jersey;* Diane M. Zych, *Erie Community College*

Many thanks to Robert Hostetler, The Behrend College, The Pennsylvania State University, and David Heyd, The Behrend College, The Pennsylvania State University, for their significant contributions to previous editions of this text.

We would also like to thank the staff at Larson Texts, Inc., who assisted in preparing the manuscript, rendering the art package, typesetting, and proofreading the pages and supplements.

On a personal level, we are grateful to our wives, Deanna Gilbert Larson and Consuelo Edwards, for their love, patience, and support. Also, a special note of thanks goes out to R. Scott O'Neil.

If you have suggestions for improving this text, please feel free to write to us. Over the years we have received many useful comments from both instructors and students, and we value these very much.

Ron Larson

Bruce Edwards

1 Limits and Their Properties

Average Speed (Exercise 62, p. 93)

Charles's Law and Absolute Zero (Example 5, p. 78)

Free-Falling Object (Exercises 101 and 102, p. 73)

Sports (Exercise 68, p. 61)

Bicyclist (Exercise 5, p. 51)

1.1 A Preview of Calculus

- **Understand what calculus is and how it compares with precalculus.**
- **Understand that the tangent line problem is basic to calculus.**
- **Understand that the area problem is also basic to calculus.**

REMARK As you progress through this course, remember that learning calculus is just one of your goals. Your most important goal is to learn how to use calculus to model and solve real-life problems. Here are a few problem-solving strategies that may help you.

- Be sure you understand the question. What is given? What are you asked to find?
- Outline a plan. There are many approaches you could use: look for a pattern, solve a simpler problem, work backwards, draw a diagram, use technology, or any of many other approaches.
- Complete your plan. Be sure to answer the question. Verbalize your answer. For example, rather than writing the answer as $x = 4.6$, it would be better to write the answer as, "The area of the region is 4.6 square meters."
- Look back at your work. Does your answer make sense? Is there a way you can check the reasonableness of your answer?

What Is Calculus?

Calculus is the mathematics of change. For instance, calculus is the mathematics of velocities, accelerations, tangent lines, slopes, areas, volumes, arc lengths, centroids, curvatures, and a variety of other concepts that have enabled scientists, engineers, and economists to model real-life situations.

Although precalculus mathematics also deals with velocities, accelerations, tangent lines, slopes, and so on, there is a fundamental difference between precalculus mathematics and calculus. Precalculus mathematics is more static, whereas calculus is more dynamic. Here are some examples.

- An object traveling at a constant velocity can be analyzed with precalculus mathematics. To analyze the velocity of an accelerating object, you need calculus.
- The slope of a line can be analyzed with precalculus mathematics. To analyze the slope of a curve, you need calculus.
- The curvature of a circle is constant and can be analyzed with precalculus mathematics. To analyze the variable curvature of a general curve, you need calculus.
- The area of a rectangle can be analyzed with precalculus mathematics. To analyze the area under a general curve, you need calculus.

Each of these situations involves the same general strategy—the reformulation of precalculus mathematics through the use of a limit process. So, one way to answer the question "What is calculus?" is to say that calculus is a "limit machine" that involves three stages. The first stage is precalculus mathematics, such as the slope of a line or the area of a rectangle. The second stage is the limit process, and the third stage is a new calculus formulation, such as a derivative or integral.

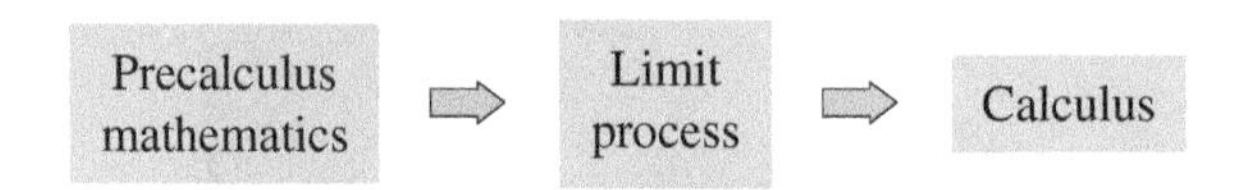

Some students try to learn calculus as if it were simply a collection of new formulas. This is unfortunate. If you reduce calculus to the memorization of differentiation and integration formulas, you will miss a great deal of understanding, self-confidence, and satisfaction.

On the next two pages are listed some familiar precalculus concepts coupled with their calculus counterparts. Throughout the text, your goal should be to learn how precalculus formulas and techniques are used as building blocks to produce the more general calculus formulas and techniques. Do not worry if you are unfamiliar with some of the "old formulas" listed on the next two pages—you will be reviewing all of them.

As you proceed through this text, come back to this discussion repeatedly. Try to keep track of where you are relative to the three stages involved in the study of calculus. For instance, note how these chapters relate to the three stages.

Chapter P: Preparation for Calculus	Precalculus
Chapter 1: Limits and Their Properties	Limit process
Chapter 2: Differentiation	Calculus

This cycle is repeated many times on a smaller scale throughout the text.

Without Calculus		With Differential Calculus	
Value of $f(x)$ when $x = c$	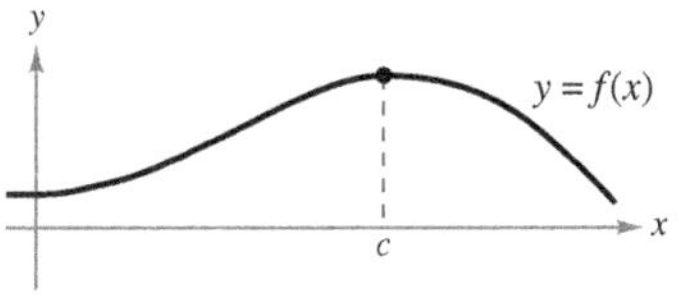	Limit of $f(x)$ as x approaches c	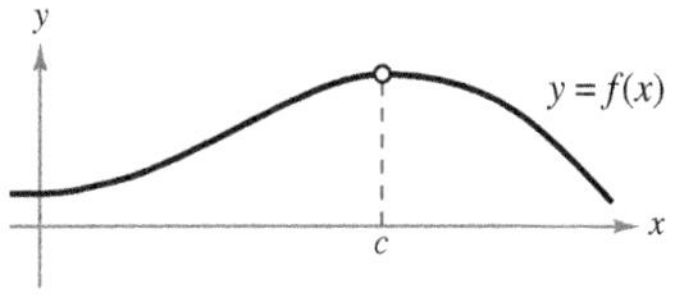
Slope of a line	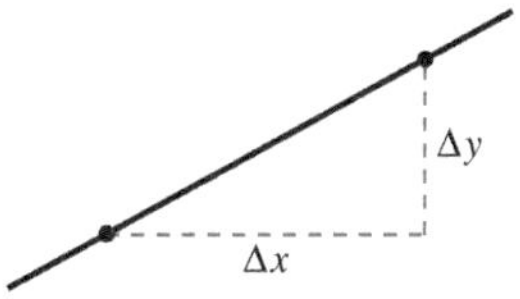	Slope of a curve	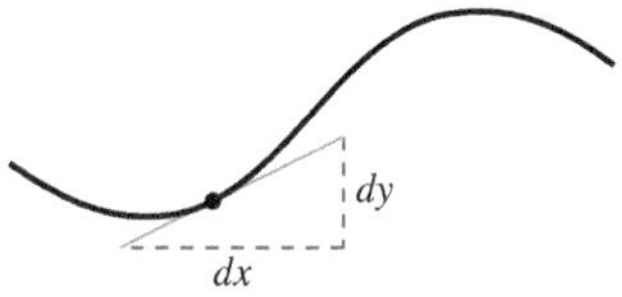
Secant line to a curve	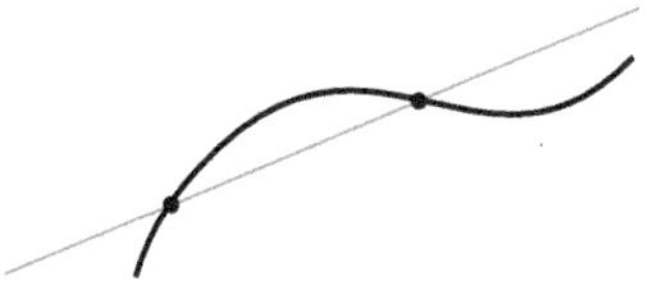	Tangent line to a curve	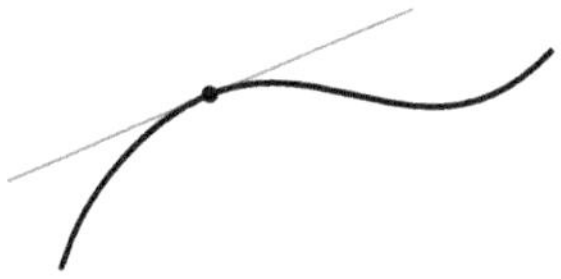
Average rate of change between $t = a$ and $t = b$		Instantaneous rate of change at $t = c$	
Curvature of a circle	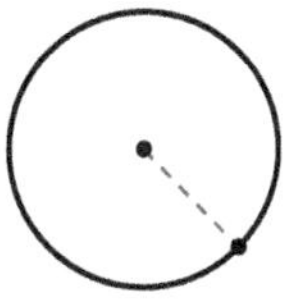	Curvature of a curve	
Height of a curve when $x = c$	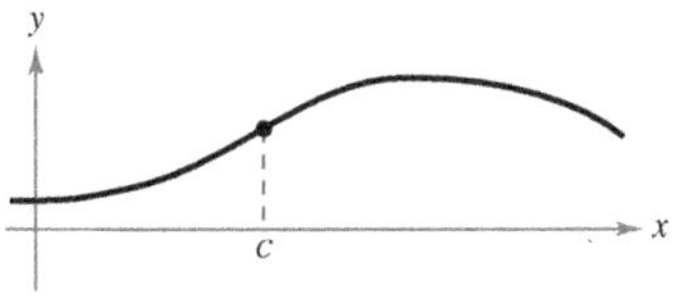	Maximum height of a curve on an interval	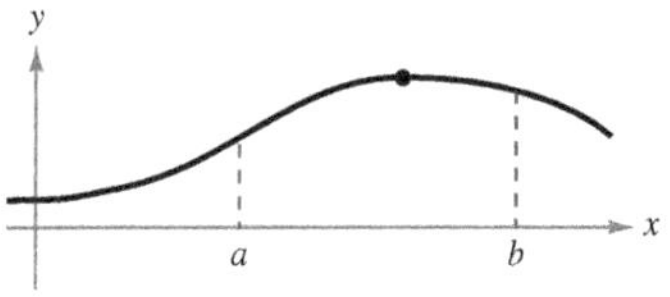
Tangent plane to a sphere	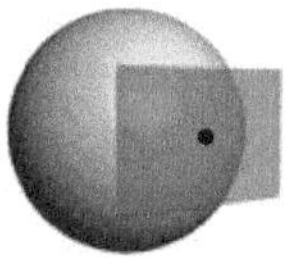	Tangent plane to a surface	
Direction of motion along a line		Direction of motion along a curve	

Without Calculus		With Integral Calculus	
Area of a rectangle		Area under a curve	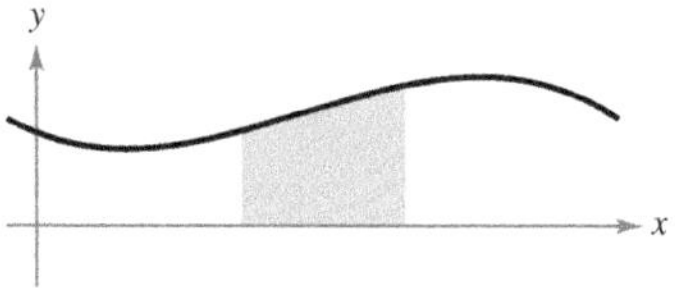
Work done by a constant force	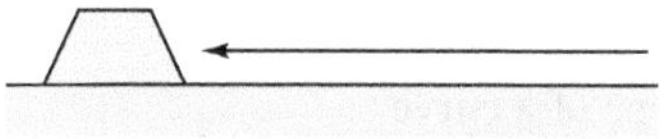	Work done by a variable force	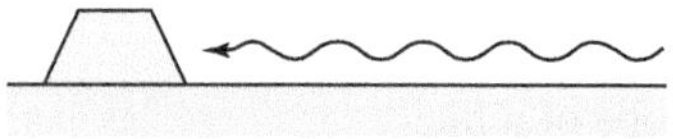
Center of a rectangle	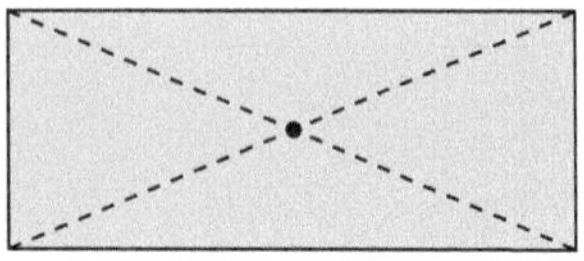	Centroid of a region	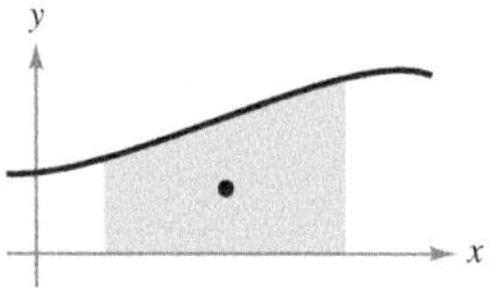
Length of a line segment		Length of an arc	
Surface area of a cylinder		Surface area of a solid of revolution	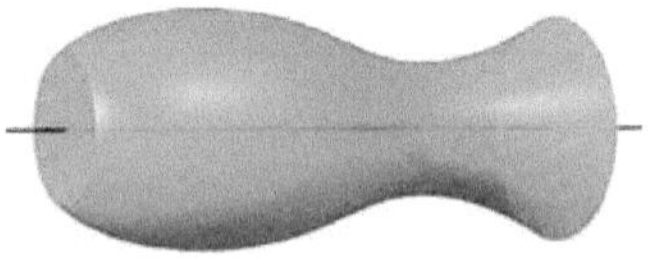
Mass of a solid of constant density	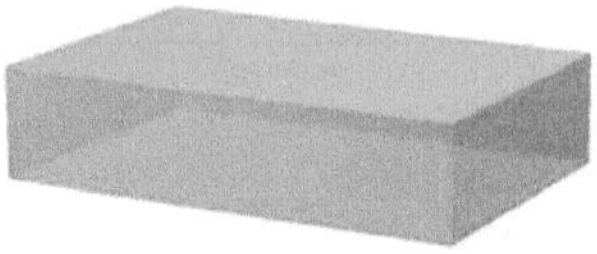	Mass of a solid of variable density	
Volume of a rectangular solid		Volume of a region under a surface	
Sum of a finite number of terms	$a_1 + a_2 + \cdots + a_n = S$	Sum of an infinite number of terms	$a_1 + a_2 + a_3 + \cdots = S$

The Tangent Line Problem

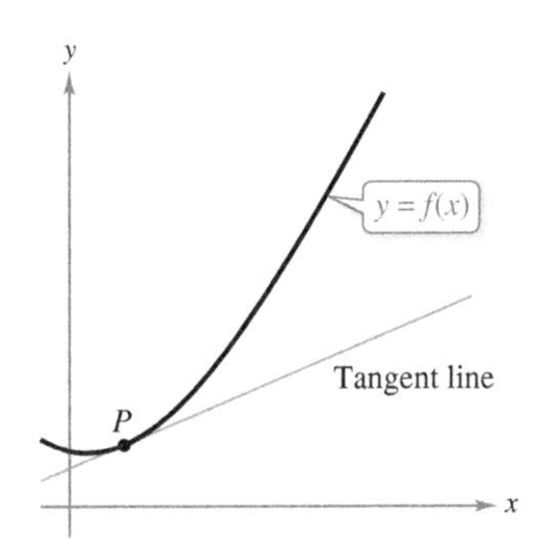

The tangent line to the graph of f at P
Figure 1.1

The notion of a limit is fundamental to the study of calculus. The following brief descriptions of two classic problems in calculus—*the tangent line problem* and *the area problem*—should give you some idea of the way limits are used in calculus.

In the tangent line problem, you are given a function f and a point P on its graph and are asked to find an equation of the tangent line to the graph at point P, as shown in Figure 1.1.

Except for cases involving a vertical tangent line, the problem of finding the **tangent line** at a point P is equivalent to finding the *slope* of the tangent line at P. You can approximate this slope by using a line through the point of tangency and a second point on the curve, as shown in Figure 1.2(a). Such a line is called a **secant line.** If $P(c, f(c))$ is the point of tangency and

$$Q(c + \Delta x, f(c + \Delta x))$$

is a second point on the graph of f, then the slope of the secant line through these two points can be found using precalculus and is

$$m_{sec} = \frac{f(c + \Delta x) - f(c)}{c + \Delta x - c} = \frac{f(c + \Delta x) - f(c)}{\Delta x}.$$

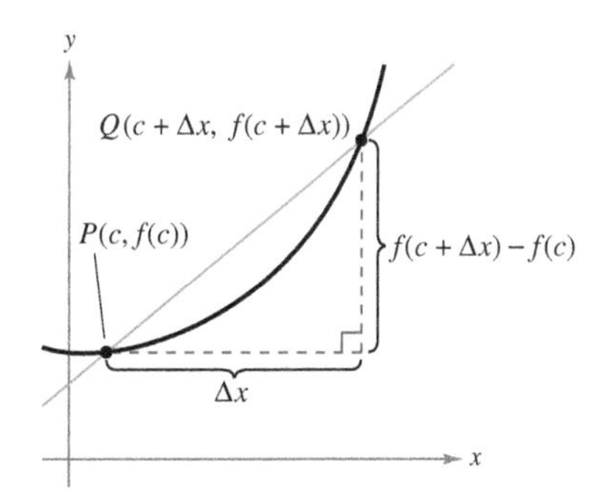

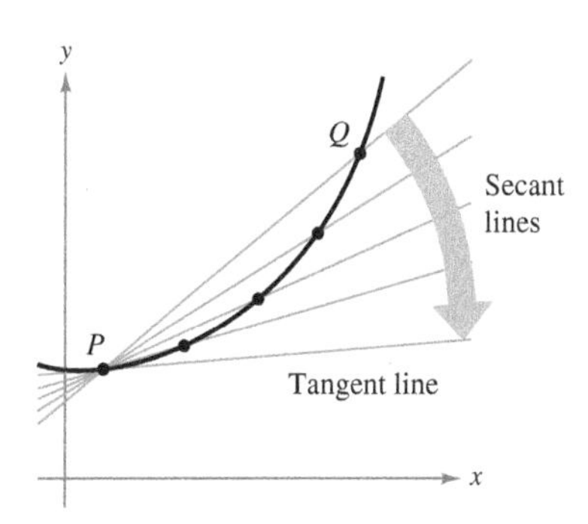

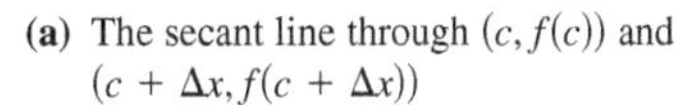
(a) The secant line through $(c, f(c))$ and $(c + \Delta x, f(c + \Delta x))$

(b) As Q approaches P, the secant lines approach the tangent line.

Figure 1.2

As point Q approaches point P, the slopes of the secant lines approach the slope of the tangent line, as shown in Figure 1.2(b). When such a "limiting position" exists, the slope of the tangent line is said to be the **limit** of the slopes of the secant lines. (Much more will be said about this important calculus concept in Chapter 2.)

GRACE CHISHOLM YOUNG (1868–1944)

Grace Chisholm Young received her degree in mathematics from Girton College in Cambridge, England. Her early work was published under the name of William Young, her husband. Between 1914 and 1916, Grace Young published work on the foundations of calculus that won her the Gamble Prize from Girton College.

Exploration

The following points lie on the graph of $f(x) = x^2$.

$Q_1(1.5, f(1.5))$, $Q_2(1.1, f(1.1))$, $Q_3(1.01, f(1.01))$,
$Q_4(1.001, f(1.001))$, $Q_5(1.0001, f(1.0001))$

Each successive point gets closer to the point $P(1, 1)$. Find the slopes of the secant lines through Q_1 and P, Q_2 and P, and so on. Graph these secant lines on a graphing utility. Then use your results to estimate the slope of the tangent line to the graph of f at the point P.

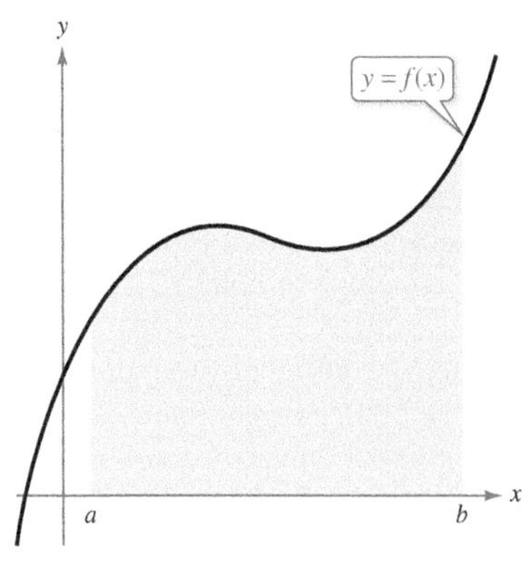

Area under a curve
Figure 1.3

The Area Problem

In the tangent line problem, you saw how the limit process can be applied to the slope of a line to find the slope of a general curve. A second classic problem in calculus is finding the area of a plane region that is bounded by the graphs of functions. This problem can also be solved with a limit process. In this case, the limit process is applied to the area of a rectangle to find the area of a general region.

As a simple example, consider the region bounded by the graph of the function $y = f(x)$, the x-axis, and the vertical lines $x = a$ and $x = b$, as shown in Figure 1.3. You can approximate the area of the region with several rectangular regions, as shown in Figure 1.4. As you increase the number of rectangles, the approximation tends to become better and better because the amount of area missed by the rectangles decreases. Your goal is to determine the limit of the sum of the areas of the rectangles as the number of rectangles increases without bound.

HISTORICAL NOTE

In one of the most astounding events ever to occur in mathematics, it was discovered that the tangent line problem and the area problem are closely related. This discovery led to the birth of calculus. You will learn about the relationship between these two problems when you study the Fundamental Theorem of Calculus in Chapter 4.

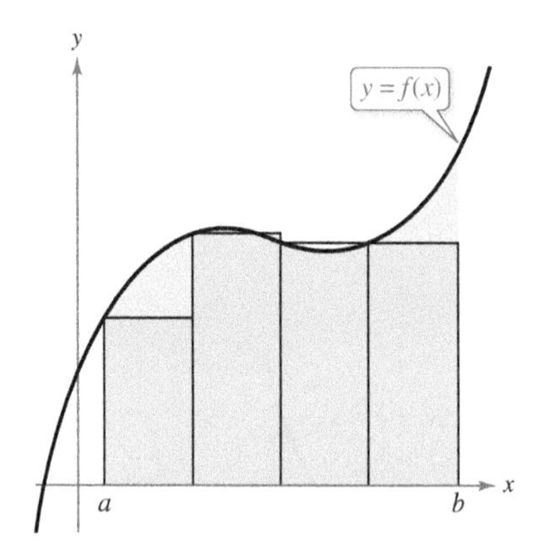

Approximation using four rectangles

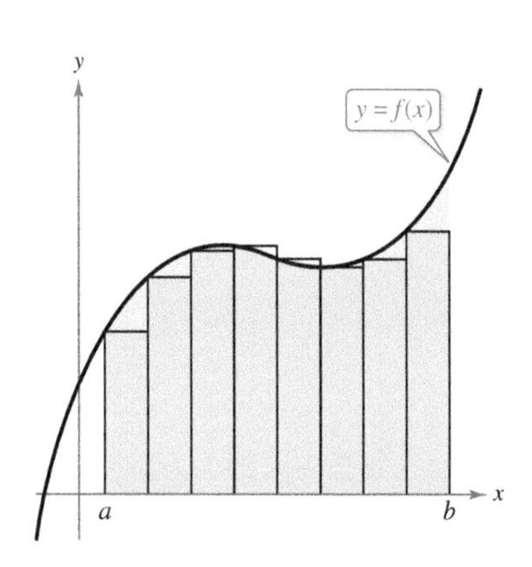

Approximation using eight rectangles

Figure 1.4

Exploration

Consider the region bounded by the graphs of

$$f(x) = x^2, \quad y = 0, \quad \text{and} \quad x = 1$$

as shown in part (a) of the figure. The area of the region can be approximated by two sets of rectangles—one set inscribed within the region and the other set circumscribed over the region, as shown in parts (b) and (c). Find the sum of the areas of each set of rectangles. Then use your results to approximate the area of the region.

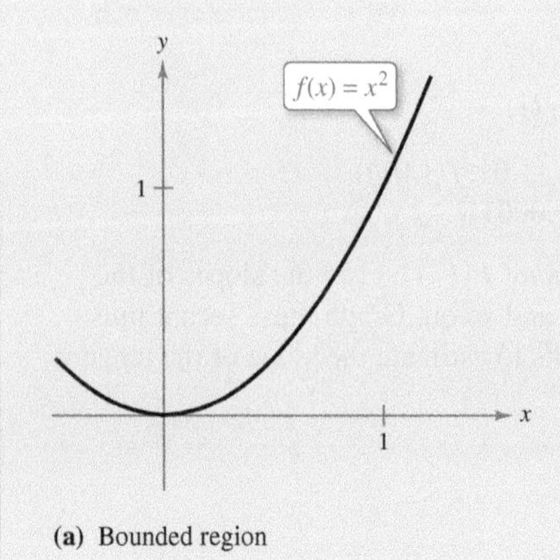

(a) Bounded region

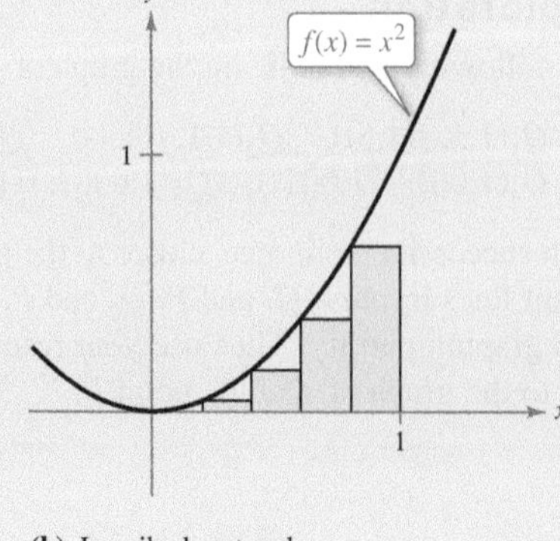

(b) Inscribed rectangles

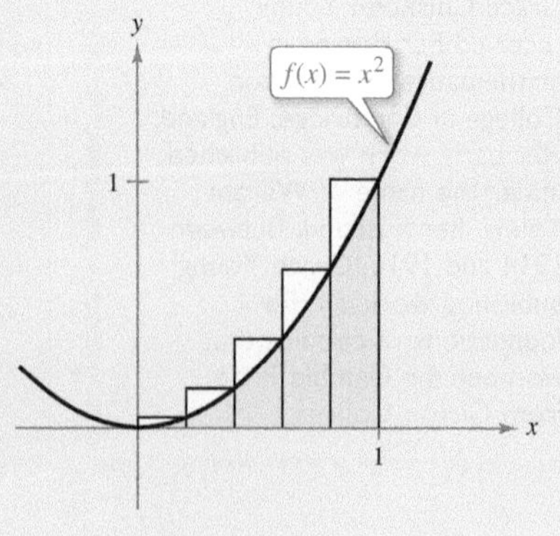

(c) Circumscribed rectangles

1.1 Exercises

See CalcChat.com for tutorial help and worked-out solutions to odd-numbered exercises.

CONCEPT CHECK

1. **Precalculus and Calculus** Describe the relationship between precalculus and calculus. List three precalculus concepts and their corresponding calculus counterparts.
2. **Secant and Tangent Lines** Discuss the relationship between secant lines through a fixed point and a corresponding tangent line at that fixed point.

Precalculus or Calculus **In Exercises 3–6, decide whether the problem can be solved using precalculus or whether calculus is required. If the problem can be solved using precalculus, solve it. If the problem seems to require calculus, explain your reasoning and use a graphical or numerical approach to estimate the solution.**

3. Find the distance traveled in 15 seconds by an object traveling at a constant velocity of 20 feet per second.
4. Find the distance traveled in 15 seconds by an object moving with a velocity of $v(t) = 20 + 7\cos t$ feet per second.
5. **Rate of Change**

A bicyclist is riding on a path modeled by the function $f(x) = 0.04(8x - x^2)$, where x and $f(x)$ are measured in miles (see figure). Find the rate of change of elevation at $x = 2$.

6. A bicyclist is riding on a path modeled by the function $f(x) = 0.08x$, where x and $f(x)$ are measured in miles (see figure). Find the rate of change of elevation at $x = 2$.

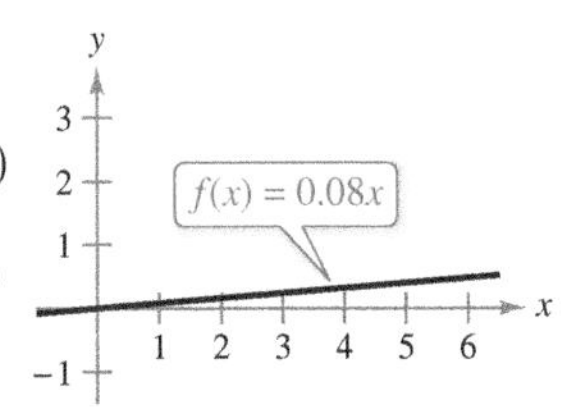

7. **Secant Lines** Consider the function $f(x) = \sqrt{x}$ and the point $P(4, 2)$ on the graph of f.
   (a) Graph f and the secant lines passing through $P(4, 2)$ and $Q(x, f(x))$ for x-values of 1, 3, and 5.
   (b) Find the slope of each secant line.
   (c) Use the results of part (b) to estimate the slope of the tangent line to the graph of f at $P(4, 2)$. Describe how to improve your approximation of the slope.

The symbol and a red exercise number indicates that a video solution can be seen at *CalcView.com*.

Raphael Christinat/Shutterstock.com

8. **Secant Lines** Consider the function $f(x) = 6x - x^2$ and the point $P(2, 8)$ on the graph of f.
   (a) Graph f and the secant lines passing through $P(2, 8)$ and $Q(x, f(x))$ for x-values of 3, 2.5, and 1.5.
   (b) Find the slope of each secant line.
   (c) Use the results of part (b) to estimate the slope of the tangent line to the graph of f at $P(2, 8)$. Describe how to improve your approximation of the slope.
9. **Approximating Area** Use the rectangles in each graph to approximate the area of the region bounded by $y = 5/x$, $y = 0$, $x = 1$, and $x = 5$. Describe how you could continue this process to obtain a more accurate approximation of the area.

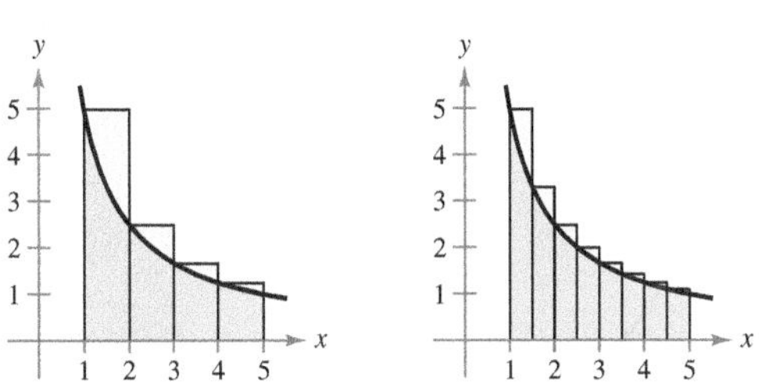

10. **HOW DO YOU SEE IT?** How would you describe the instantaneous rate of change of an automobile's position on a highway?

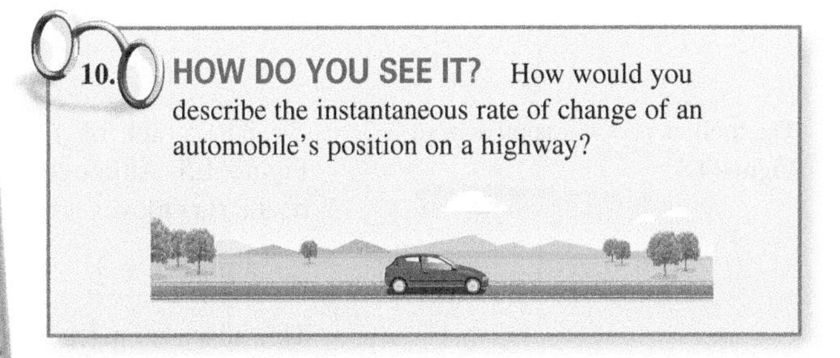

EXPLORING CONCEPTS

11. **Length of a Curve** Consider the length of the graph of $f(x) = 5/x$ from $(1, 5)$ to $(5, 1)$.

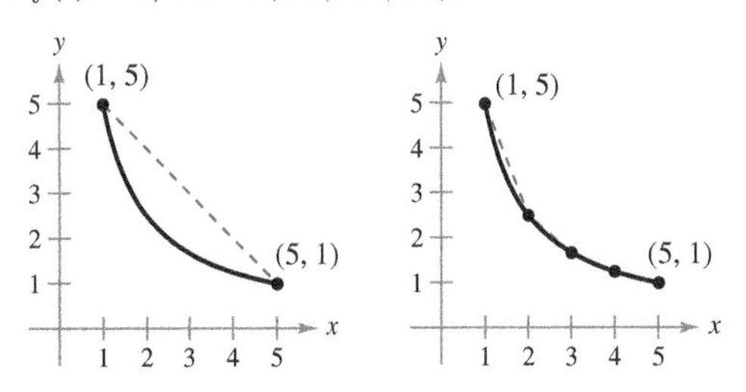

   (a) Approximate the length of the curve by finding the distance between its two endpoints, as shown in the first figure.
   (b) Approximate the length of the curve by finding the sum of the lengths of four line segments, as shown in the second figure.
   (c) Describe how you could continue this process to obtain a more accurate approximation of the length of the curve.

1.2 Finding Limits Graphically and Numerically

- **Estimate a limit using a numerical or graphical approach.**
- **Learn different ways that a limit can fail to exist.**
- **Study and use a formal definition of limit.**

An Introduction to Limits

To sketch the graph of the function

$$f(x) = \frac{x^3 - 1}{x - 1}$$

for values other than $x = 1$, you can use standard curve-sketching techniques. At $x = 1$, however, it is not clear what to expect. To get an idea of the behavior of the graph of f near $x = 1$, you can use two sets of x-values—one set that approaches 1 from the left and one set that approaches 1 from the right, as shown in the table.

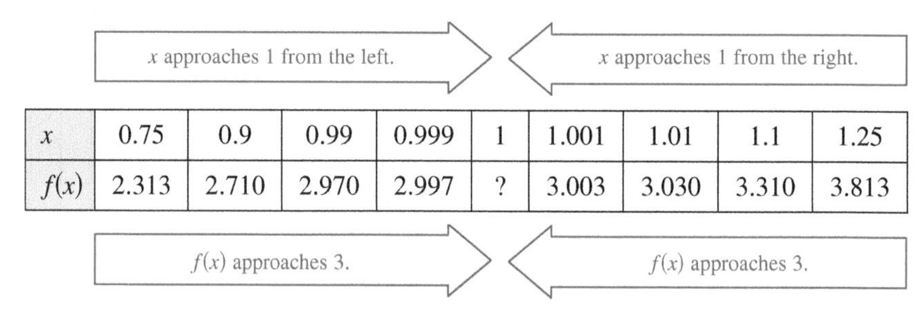

x	0.75	0.9	0.99	0.999	1	1.001	1.01	1.1	1.25
$f(x)$	2.313	2.710	2.970	2.997	?	3.003	3.030	3.310	3.813

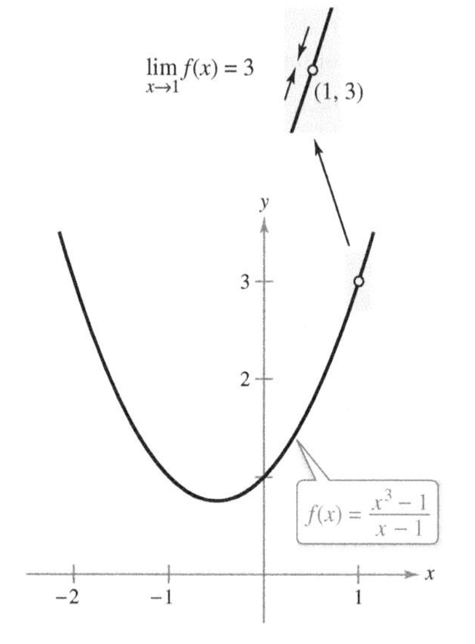

The limit of $f(x)$ as x approaches 1 is 3.
Figure 1.5

The graph of f is a parabola that has a hole at the point $(1, 3)$, as shown in Figure 1.5. Although x cannot equal 1, you can move arbitrarily close to 1, and as a result $f(x)$ moves arbitrarily close to 3. Using limit notation, you can write

$$\lim_{x \to 1} f(x) = 3.$$ This is read as "the limit of $f(x)$ as x approaches 1 is 3."

This discussion leads to an informal definition of limit. If $f(x)$ becomes arbitrarily close to a single number L as x approaches c from either side, then the **limit** of $f(x)$ as x approaches c is L. This limit is written as

$$\lim_{x \to c} f(x) = L.$$

Exploration

The discussion above gives an example of how you can estimate a limit *numerically* by constructing a table and *graphically* by drawing a graph. Estimate the following limit numerically by completing the table.

$$\lim_{x \to 2} \frac{x^2 - 3x + 2}{x - 2}$$

x	1.75	1.9	1.99	1.999	2	2.001	2.01	2.1	2.25
$f(x)$	?	?	?	?	?	?	?	?	?

Then use a graphing utility to estimate the limit graphically.

EXAMPLE 1 Estimating a Limit Numerically

Evaluate the function $f(x) = x/\left(\sqrt{x+1} - 1\right)$ at several x-values near 0 and use the results to estimate the limit

$$\lim_{x\to 0} \frac{x}{\sqrt{x+1} - 1}.$$

Solution The table lists the values of $f(x)$ for several x-values near 0.

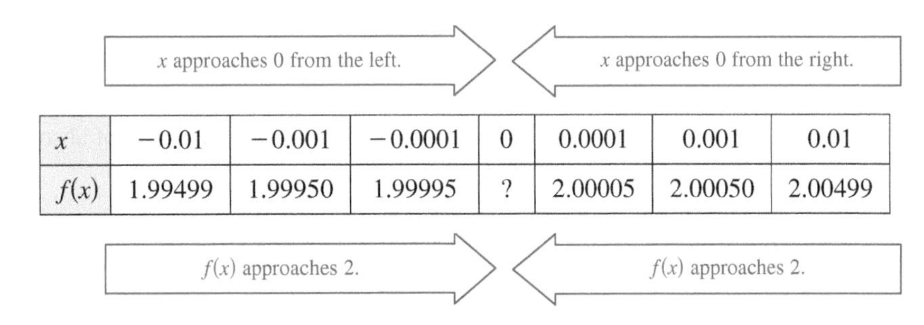

x approaches 0 from the left. x approaches 0 from the right.

x	−0.01	−0.001	−0.0001	0	0.0001	0.001	0.01
$f(x)$	1.99499	1.99950	1.99995	?	2.00005	2.00050	2.00499

$f(x)$ approaches 2. $f(x)$ approaches 2.

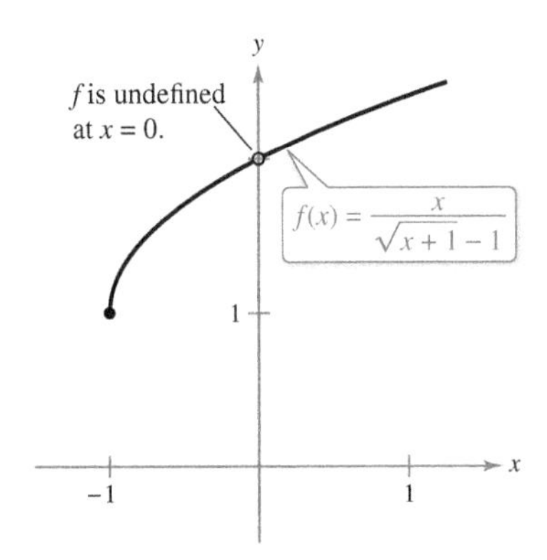

The limit of $f(x)$ as x approaches 0 is 2.
Figure 1.6

From the results shown in the table, you can estimate the limit to be 2. This limit is reinforced by the graph of f shown in Figure 1.6.

In Example 1, note that the function is undefined at $x = 0$, and yet $f(x)$ appears to be approaching a limit as x approaches 0. This often happens, and it is important to realize that *the existence or nonexistence of* $f(x)$ *at* $x = c$ *has no bearing on the existence of the limit of* $f(x)$ *as* x *approaches* c.

EXAMPLE 2 Finding a Limit

Find the limit of $f(x)$ as x approaches 2, where

$$f(x) = \begin{cases} 1, & x \neq 2 \\ 0, & x = 2 \end{cases}.$$

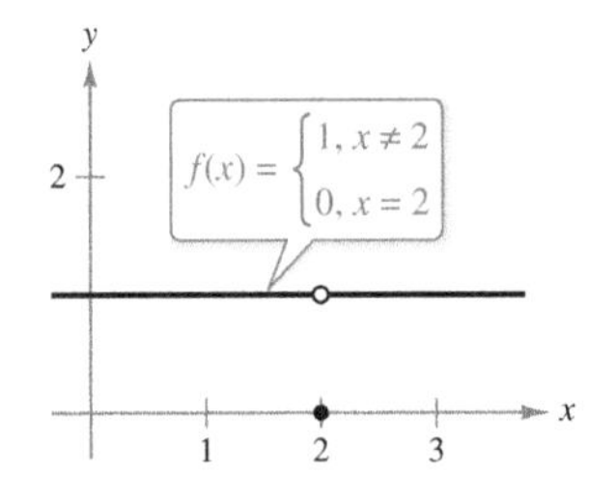

The limit of $f(x)$ as x approaches 2 is 1.
Figure 1.7

Solution Because $f(x) = 1$ for all x other than $x = 2$, you can estimate that the limit is 1, as shown in Figure 1.7. So, you can write

$$\lim_{x\to 2} f(x) = 1.$$

The fact that $f(2) = 0$ has no bearing on the existence or value of the limit as x approaches 2. For instance, as x approaches 2, the function

$$g(x) = \begin{cases} 1, & x \neq 2 \\ 2, & x = 2 \end{cases}$$

has the same limit as f.

So far in this section, you have been estimating limits numerically and graphically. Each of these approaches produces an estimate of the limit. In Section 1.3, you will study analytic techniques for evaluating limits. Throughout the course, try to develop a habit of using this three-pronged approach to problem solving.

1. Numerical approach — Construct a table of values.
2. Graphical approach — Draw a graph by hand or using technology.
3. Analytic approach — Use algebra or calculus.

Limits That Fail to Exist

In the next three examples, you will examine some limits that fail to exist.

EXAMPLE 3 Different Right and Left Behavior

Show that the limit $\lim_{x\to 0} \frac{|x|}{x}$ does not exist.

Solution Consider the graph of the function

$$f(x) = \frac{|x|}{x}.$$

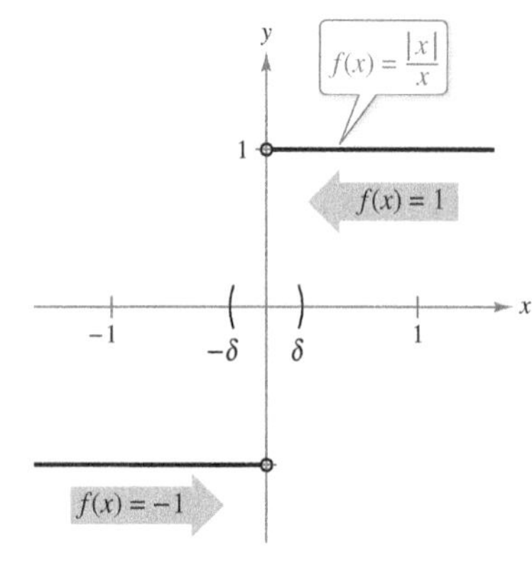

$\lim_{x\to 0} f(x)$ does not exist.
Figure 1.8

In Figure 1.8 and from the definition of absolute value,

$$|x| = \begin{cases} x, & x \geq 0 \\ -x, & x < 0 \end{cases}$$ Definition of absolute value

you can see that

$$\frac{|x|}{x} = \begin{cases} 1, & x > 0 \\ -1, & x < 0 \end{cases}.$$

So, no matter how close x gets to 0, there will be both positive and negative x-values that yield $f(x) = 1$ or $f(x) = -1$. Specifically, if δ (the lowercase Greek letter delta) is a positive number, then for x-values satisfying the inequality $0 < |x| < \delta$, you can classify the values of $|x|/x$ as -1 or 1 on the intervals

$(-\delta, 0)$ or $(0, \delta)$.

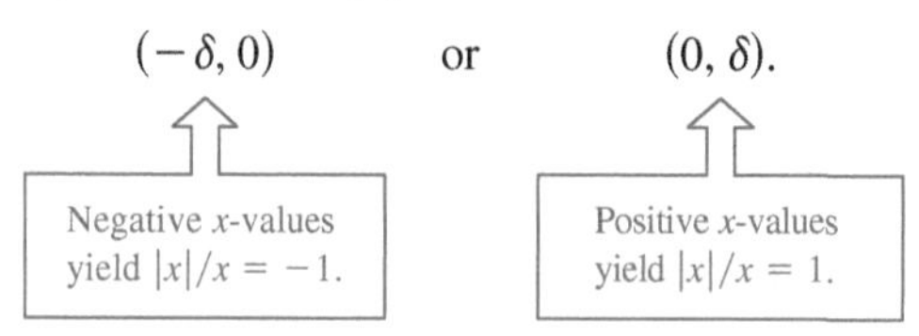

Because $|x|/x$ approaches a different number from the right side of 0 than it approaches from the left side, the limit $\lim_{x\to 0} (|x|/x)$ does not exist.

EXAMPLE 4 Unbounded Behavior

Discuss the existence of the limit $\lim_{x\to 0} \frac{1}{x^2}$.

Solution Consider the graph of the function

$$f(x) = \frac{1}{x^2}.$$

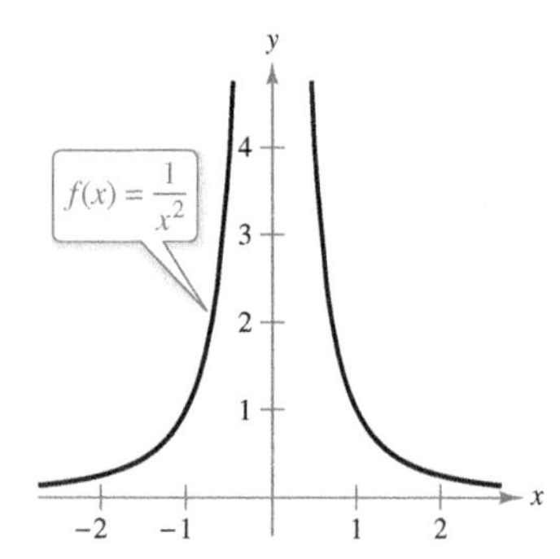

$\lim_{x\to 0} f(x)$ does not exist.
Figure 1.9

In Figure 1.9, you can see that as x approaches 0 from either the right or the left, $f(x)$ increases without bound. This means that by choosing x close enough to 0, you can force $f(x)$ to be as large as you want. For instance, $f(x)$ will be greater than 100 when you choose x within $\frac{1}{10}$ of 0. That is,

$$0 < |x| < \frac{1}{10} \implies f(x) = \frac{1}{x^2} > 100.$$

Similarly, you can force $f(x)$ to be greater than 1,000,000, as shown.

$$0 < |x| < \frac{1}{1000} \implies f(x) = \frac{1}{x^2} > 1{,}000{,}000$$

Because $f(x)$ does not become arbitrarily close to a single number L as x approaches 0, you can conclude that the limit does not exist. ∎

EXAMPLE 5 Oscillating Behavior

See LarsonCalculus.com for an interactive version of this type of example.

Discuss the existence of the limit $\lim_{x \to 0} \sin \frac{1}{x}$.

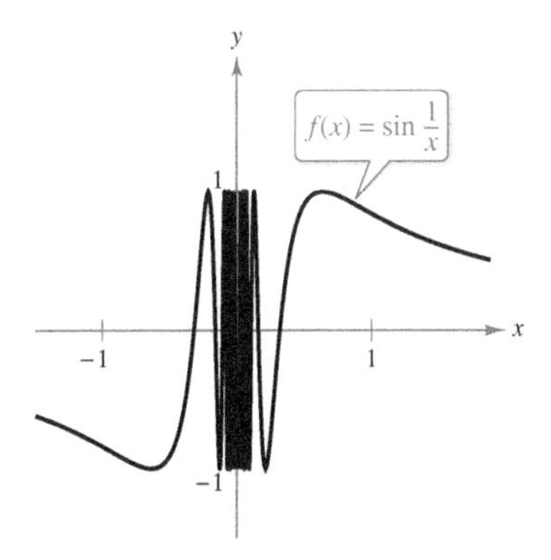

$\lim_{x \to 0} f(x)$ does not exist.
Figure 1.10

Solution Let $f(x) = \sin(1/x)$. In Figure 1.10, you can see that as x approaches 0, $f(x)$ oscillates between -1 and 1. So, the limit does not exist because no matter how small you choose δ, it is possible to choose x_1 and x_2 within δ units of 0 such that $\sin(1/x_1) = 1$ and $\sin(1/x_2) = -1$, as shown in the table.

x	$\frac{2}{\pi}$	$\frac{2}{3\pi}$	$\frac{2}{5\pi}$	$\frac{2}{7\pi}$	$\frac{2}{9\pi}$	$\frac{2}{11\pi}$	$x \to 0$
$\sin \frac{1}{x}$	1	-1	1	-1	1	-1	Limit does not exist.

Common Types of Behavior Associated with Nonexistence of a Limit

1. $f(x)$ approaches a different number from the right side of c than it approaches from the left side.
2. $f(x)$ increases or decreases without bound as x approaches c.
3. $f(x)$ oscillates between two fixed values as x approaches c.

PETER GUSTAV DIRICHLET (1805–1859)

In the early development of calculus, the definition of a function was much more restricted than it is today, and "functions" such as the Dirichlet function would not have been considered. The modern definition of function is attributed to the German mathematician Peter Gustav Dirichlet.

See LarsonCalculus.com to read more of this biography.

In addition to $f(x) = \sin(1/x)$, there are many other interesting functions that have unusual limit behavior. An often cited one is the *Dirichlet function*

$$f(x) = \begin{cases} 0, & \text{if } x \text{ is rational} \\ 1, & \text{if } x \text{ is irrational} \end{cases}.$$

Because this function has *no limit* at any real number c, it is *not continuous* at any real number c. You will study continuity more closely in Section 1.4.

TECHNOLOGY PITFALL When you use a graphing utility to investigate the behavior of a function near the x-value at which you are trying to evaluate a limit, remember that you cannot always trust the graphs that graphing utilities draw. When you use a graphing utility to graph the function in Example 5 over an interval containing 0, you will most likely obtain an incorrect graph such as that shown in Figure 1.11. The reason that a graphing utility cannot show the correct graph is that the graph has infinitely many oscillations over any interval that contains 0.

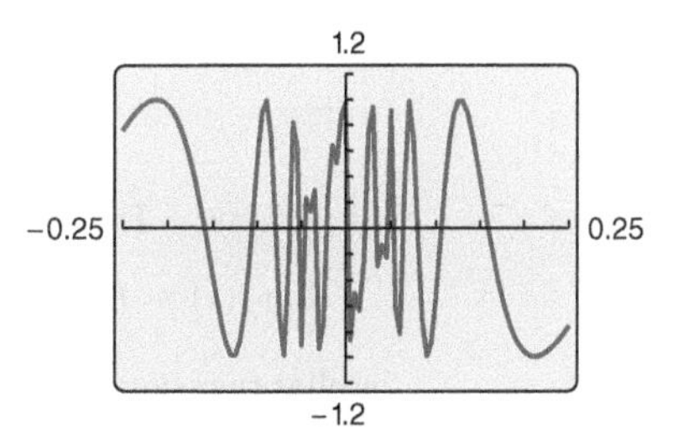

Incorrect graph of $f(x) = \sin(1/x)$
Figure 1.11

■ **FOR FURTHER INFORMATION** For more on the introduction of rigor to calculus, see "Who Gave You the Epsilon? Cauchy and the Origins of Rigorous Calculus" by Judith V. Grabiner in *The American Mathematical Monthly*. To view this article, go to *MathArticles.com*.

A Formal Definition of Limit

Consider again the informal definition of limit. If $f(x)$ becomes arbitrarily close to a single number L as x approaches c from either side, then the limit of $f(x)$ as x approaches c is L, written as

$$\lim_{x \to c} f(x) = L.$$

At first glance, this definition looks fairly technical. Even so, it is informal because exact meanings have not yet been given to the two phrases

"$f(x)$ becomes arbitrarily close to L"

and

"x approaches c."

The first person to assign mathematically rigorous meanings to these two phrases was Augustin-Louis Cauchy. His **ε-δ definition of limit** is the standard used today.

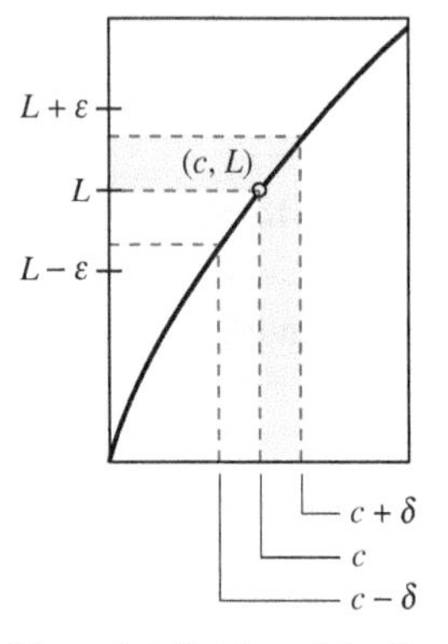

The ε-δ definition of the limit of $f(x)$ as x approaches c
Figure 1.12

In Figure 1.12, let ε (the lowercase Greek letter epsilon) represent a (small) positive number. Then the phrase "$f(x)$ becomes arbitrarily close to L" means that $f(x)$ lies in the interval $(L - \varepsilon, L + \varepsilon)$. Using absolute value, you can write this as

$$|f(x) - L| < \varepsilon.$$

Similarly, the phrase "x approaches c" means that there exists a positive number δ such that x lies in either the interval $(c - \delta, c)$ or the interval $(c, c + \delta)$. This fact can be concisely expressed by the double inequality

$$0 < |x - c| < \delta.$$

The first inequality

$$0 < |x - c|$$ The distance between x and c is more than 0.

expresses the fact that $x \neq c$. The second inequality

$$|x - c| < \delta$$ x is within δ units of c.

says that x is within a distance δ of c.

Definition of Limit

Let f be a function defined on an open interval containing c (except possibly at c), and let L be a real number. The statement

$$\lim_{x \to c} f(x) = L$$

means that for each $\varepsilon > 0$ there exists a $\delta > 0$ such that if

$$0 < |x - c| < \delta$$

then

$$|f(x) - L| < \varepsilon.$$

REMARK Throughout this text, the expression

$$\lim_{x \to c} f(x) = L$$

implies two statements—the limit exists *and* the limit is L.

Some functions do not have limits as x approaches c, but those that do cannot have two different limits as x approaches c. That is, *if the limit of a function exists, then the limit is unique* (see Exercise 81).

The next three examples should help you develop a better understanding of the ε-δ definition of limit.

EXAMPLE 6 Finding a δ for a Given ε

Given the limit

$$\lim_{x \to 3} (2x - 5) = 1$$

find δ such that

$$|(2x - 5) - 1| < 0.01$$

whenever

$$0 < |x - 3| < \delta.$$

Solution In this problem, you are working with a given value of ε—namely, $\varepsilon = 0.01$. To find an appropriate δ, try to establish a connection between the absolute values

$$|(2x - 5) - 1| \quad \text{and} \quad |x - 3|.$$

Notice that

$$|(2x - 5) - 1| = |2x - 6| = 2|x - 3|.$$

Because the inequality $|(2x - 5) - 1| < 0.01$ is equivalent to $2|x - 3| < 0.01$, you can choose

$$\delta = \tfrac{1}{2}(0.01) = 0.005.$$

This choice works because

$$0 < |x - 3| < 0.005$$

implies that

$$|(2x - 5) - 1| = 2|x - 3| < 2(0.005) = 0.01.$$

As you can see in Figure 1.13, for x-values within 0.005 of 3 ($x \neq 3$), the values of $f(x)$ are within 0.01 of 1.

REMARK In Example 6, note that 0.005 is the *largest* value of δ that will guarantee

$$|(2x - 5) - 1| < 0.01$$

whenever

$$0 < |x - 3| < \delta.$$

Any *smaller* positive value of δ would also work.

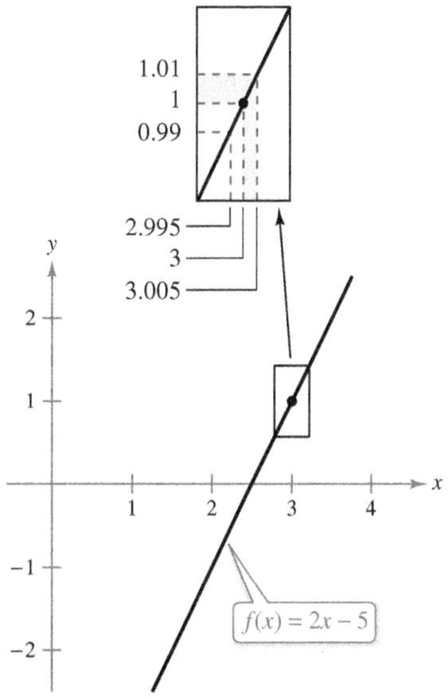

The limit of $f(x)$ as x approaches 3 is 1.
Figure 1.13

In Example 6, you found a δ-value for a *given* ε. This does not prove the existence of the limit. To do that, you must prove that you can find a δ for *any* ε, as shown in the next example.

EXAMPLE 7 Using the ε-δ Definition of Limit

Use the ε-δ definition of limit to prove that

$$\lim_{x\to 2}(3x - 2) = 4.$$

Solution You must show that for each $\varepsilon > 0$, there exists a $\delta > 0$ such that

$$|(3x - 2) - 4| < \varepsilon$$

whenever

$$0 < |x - 2| < \delta.$$

Because your choice of δ depends on ε, you need to establish a connection between the absolute values $|(3x - 2) - 4|$ and $|x - 2|$.

$$|(3x - 2) - 4| = |3x - 6| = 3|x - 2|$$

So, for a given $\varepsilon > 0$, you can choose $\delta = \varepsilon/3$. This choice works because

$$0 < |x - 2| < \delta = \frac{\varepsilon}{3}$$

implies that

$$|(3x - 2) - 4| = 3|x - 2| < 3\left(\frac{\varepsilon}{3}\right) = \varepsilon.$$

As you can see in Figure 1.14, for x-values within δ of 2 ($x \neq 2$), the values of $f(x)$ are within ε of 4.

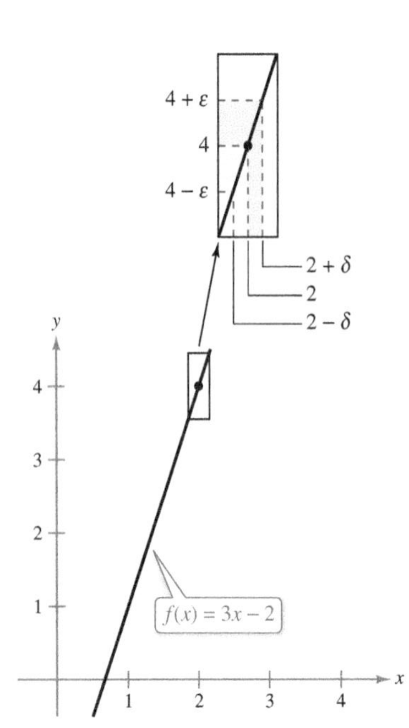

The limit of $f(x)$ as x approaches 2 is 4.
Figure 1.14

EXAMPLE 8 Using the ε-δ Definition of Limit

Use the ε-δ definition of limit to prove that $\lim_{x\to 2} x^2 = 4$.

Solution You must show that for each $\varepsilon > 0$, there exists a $\delta > 0$ such that

$$|x^2 - 4| < \varepsilon$$

whenever

$$0 < |x - 2| < \delta.$$

To find an appropriate δ, begin by writing $|x^2 - 4| = |x - 2||x + 2|$. You are interested in values of x close to 2, so choose x in the interval $(1, 3)$. To satisfy this restriction, let $\delta < 1$. Furthermore, for all x in the interval $(1, 3)$, $x + 2 < 5$ and thus $|x + 2| < 5$. So, letting δ be the minimum of $\varepsilon/5$ and 1, it follows that, whenever $0 < |x - 2| < \delta$, you have

$$|x^2 - 4| = |x - 2||x + 2| < \left(\frac{\varepsilon}{5}\right)(5) = \varepsilon.$$

As you can see in Figure 1.15, for x-values within δ of 2 ($x \neq 2$), the values of $f(x)$ are within ε of 4. ■

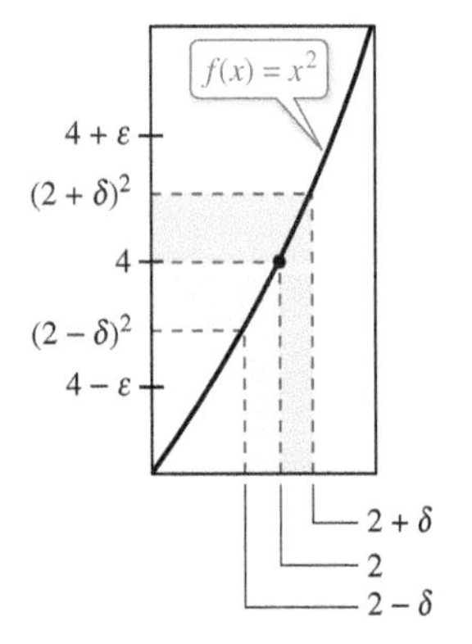

The limit of $f(x)$ as x approaches 2 is 4.
Figure 1.15

Throughout this chapter, you will use the ε-δ definition of limit primarily to prove theorems about limits and to establish the existence or nonexistence of particular types of limits. For *finding* limits, you will learn techniques that are easier to use than the ε-δ definition of limit.

1.2 Exercises

See CalcChat.com for tutorial help and worked-out solutions to odd-numbered exercises.

CONCEPT CHECK

1. **Describing Notation** Write a brief description of the meaning of the notation $\lim_{x \to 8} f(x) = 25$.
2. **Limits That Fail to Exist** Identify three types of behavior associated with the nonexistence of a limit. Illustrate each type with a graph of a function.
3. **Formal Definition of Limit** Given the limit

   $\lim_{x \to 2} (2x + 1) = 5$

   use a sketch to show the meaning of the phrase "$0 < |x - 2| < 0.25$ implies $|(2x + 1) - 5| < 0.5$."
4. **Functions and Limits** Is the limit of $f(x)$ as x approaches c always equal to $f(c)$? Why or why not?

Estimating a Limit Numerically **In Exercises 5–10, complete the table and use the result to estimate the limit. Use a graphing utility to graph the function to confirm your result.**

5. $\lim_{x \to 4} \dfrac{x - 4}{x^2 - 5x + 4}$

x	3.9	3.99	3.999	4	4.001	4.01	4.1
$f(x)$				?			

6. $\lim_{x \to 3} \dfrac{x - 3}{x^2 - 9}$

x	2.9	2.99	2.999	3	3.001	3.01	3.1
$f(x)$				?			

7. $\lim_{x \to 0} \dfrac{\sqrt{x + 1} - 1}{x}$

x	−0.1	−0.01	−0.001	0	0.001	0.01	0.1
$f(x)$				?			

8. $\lim_{x \to 3} \dfrac{[1/(x + 1)] - (1/4)}{x - 3}$

x	2.9	2.99	2.999	3	3.001	3.01	3.1
$f(x)$				?			

9. $\lim_{x \to 0} \dfrac{\sin x}{x}$

x	−0.1	−0.01	−0.001	0	0.001	0.01	0.1
$f(x)$				?			

10. $\lim_{x \to 0} \dfrac{\cos x - 1}{x}$

x	−0.1	−0.01	−0.001	0	0.001	0.01	0.1
$f(x)$				?			

Estimating a Limit Numerically **In Exercises 11–18, create a table of values for the function and use the result to estimate the limit. Use a graphing utility to graph the function to confirm your result.**

11. $\lim_{x \to 1} \dfrac{x - 2}{x^2 + x - 6}$
12. $\lim_{x \to -4} \dfrac{x + 4}{x^2 + 9x + 20}$
13. $\lim_{x \to 1} \dfrac{x^4 - 1}{x^6 - 1}$
14. $\lim_{x \to -3} \dfrac{x^3 + 27}{x + 3}$
15. $\lim_{x \to -6} \dfrac{\sqrt{10 - x} - 4}{x + 6}$
16. $\lim_{x \to 2} \dfrac{[x/(x + 1)] - (2/3)}{x - 2}$
17. $\lim_{x \to 0} \dfrac{\sin 2x}{x}$
18. $\lim_{x \to 0} \dfrac{\tan x}{\tan 2x}$

Limits That Fail to Exist **In Exercises 19 and 20, create a table of values for the function and use the result to explain why the limit does not exist.**

19. $\lim_{x \to 0} \dfrac{2}{x^3}$
20. $\lim_{x \to 0} \dfrac{3|x|}{x^2}$

Finding a Limit Graphically **In Exercises 21–28, use the graph to find the limit (if it exists). If the limit does not exist, explain why.**

21. $\lim_{x \to 3} (4 - x)$

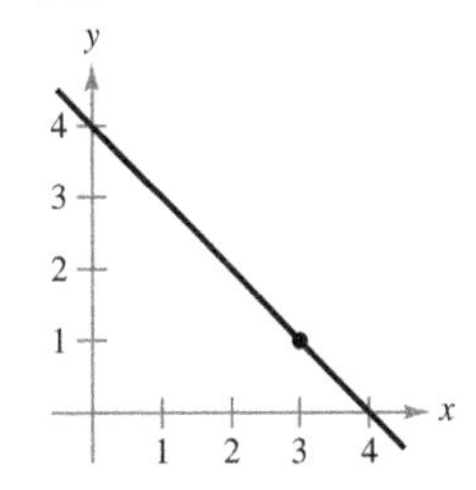

22. $\lim_{x \to 0} \sec x$

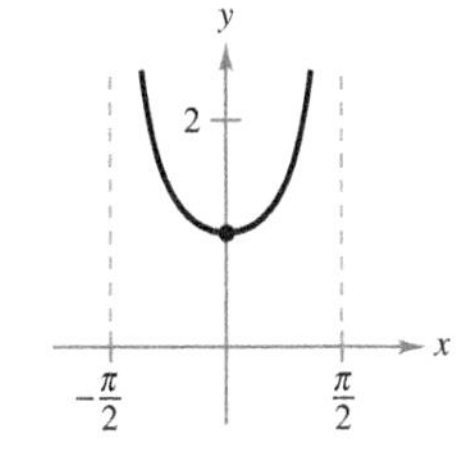

23. $\lim_{x \to 2} f(x)$

$$f(x) = \begin{cases} 4 - x, & x \neq 2 \\ 0, & x = 2 \end{cases}$$

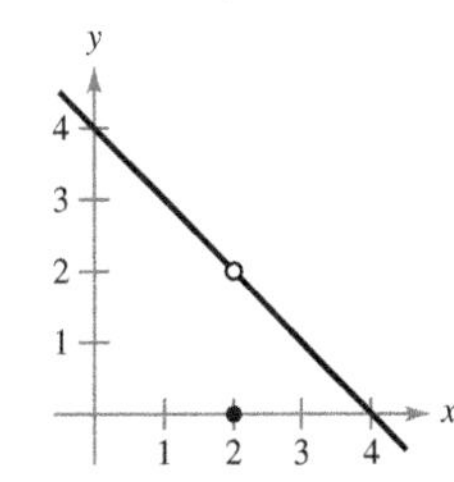

24. $\lim_{x \to 1} f(x)$

$$f(x) = \begin{cases} x^2 + 3, & x \neq 1 \\ 2, & x = 1 \end{cases}$$

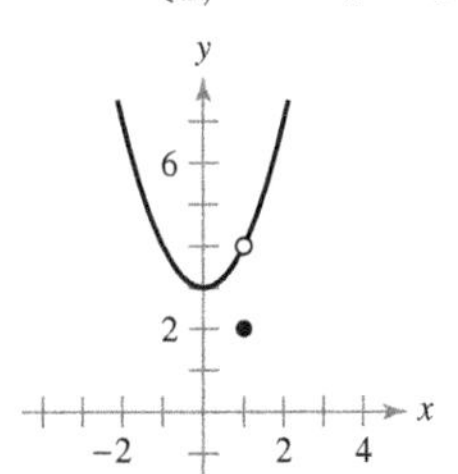

25. $\lim_{x \to 2} \frac{|x-2|}{x-2}$

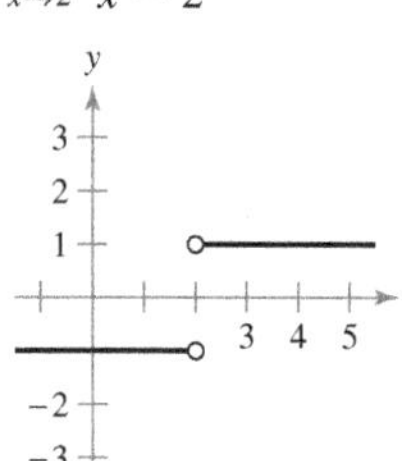

26. $\lim_{x \to 5} \frac{2}{x-5}$

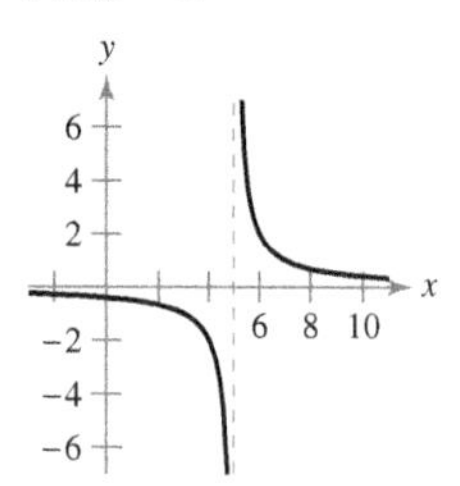

27. $\lim_{x \to 0} \cos \frac{1}{x}$

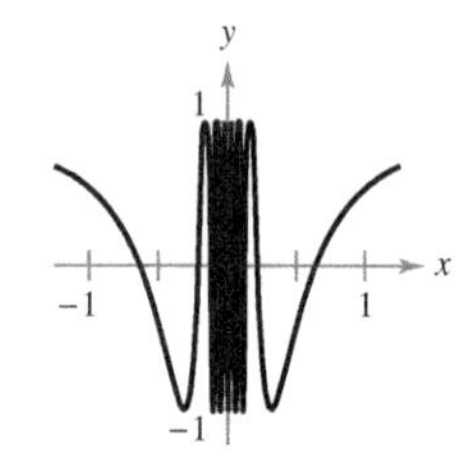

28. $\lim_{x \to \pi/2} \tan x$

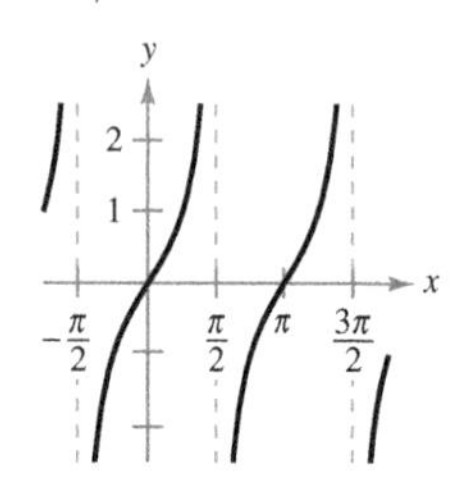

Graphical Reasoning **In Exercises 29 and 30, use the graph of the function f to decide whether the value of the given quantity exists. If it does, find it. If not, explain why.**

29. (a) $f(1)$

(b) $\lim_{x \to 1} f(x)$

(c) $f(4)$

(d) $\lim_{x \to 4} f(x)$

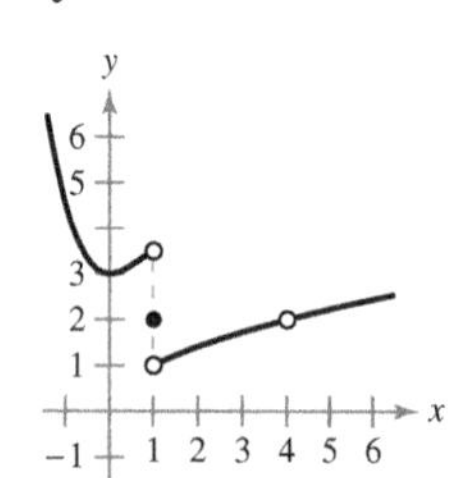

30. (a) $f(-2)$

(b) $\lim_{x \to -2} f(x)$

(c) $f(0)$

(d) $\lim_{x \to 0} f(x)$

(e) $f(2)$

(f) $\lim_{x \to 2} f(x)$

(g) $f(4)$

(h) $\lim_{x \to 4} f(x)$

Limits of a Piecewise Function **In Exercises 31 and 32, sketch the graph of f. Then identify the values of c for which $\lim_{x \to c} f(x)$ exists.**

31. $f(x) = \begin{cases} x^2, & x \le 2 \\ 8 - 2x, & 2 < x < 4 \\ 4, & x \ge 4 \end{cases}$

32. $f(x) = \begin{cases} \sin x, & x < 0 \\ 1 - \cos x, & 0 \le x \le \pi \\ \cos x, & x > \pi \end{cases}$

Sketching a Graph **In Exercises 33 and 34, sketch a graph of a function f that satisfies the given values. (There are many correct answers.)**

33. $f(0)$ is undefined.

$\lim_{x \to 0} f(x) = 4$

$f(2) = 6$

$\lim_{x \to 2} f(x) = 3$

34. $f(-2) = 0$

$f(2) = 0$

$\lim_{x \to -2} f(x) = 0$

$\lim_{x \to 2} f(x)$ does not exist.

35. **Finding a δ for a Given ε** The graph of $f(x) = x + 1$ is shown in the figure. Find δ such that if $0 < |x - 2| < \delta$, then $|f(x) - 3| < 0.4$.

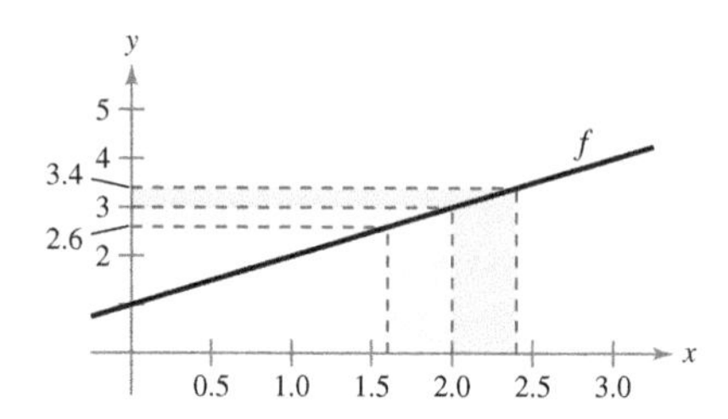

36. **Finding a δ for a Given ε** The graph of

$$f(x) = \frac{1}{x-1}$$

is shown in the figure. Find δ such that if $0 < |x - 2| < \delta$, then $|f(x) - 1| < 0.01$.

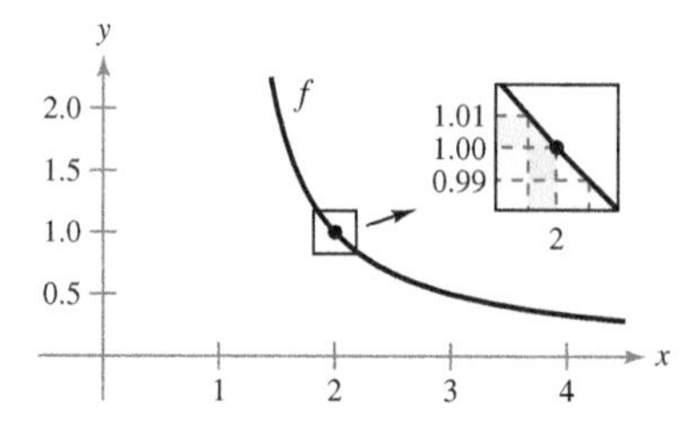

37. **Finding a δ for a Given ε** The graph of

$$f(x) = 2 - \frac{1}{x}$$

is shown in the figure. Find δ such that if $0 < |x - 1| < \delta$, then $|f(x) - 1| < 0.1$.

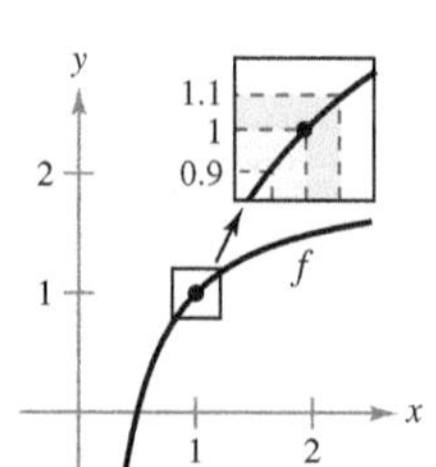

38. **Finding a δ for a Given ε** Repeat Exercise 37 for $\varepsilon = 0.05$, 0.01, and 0.005. What happens to the value of δ as the value of ε gets smaller?

Finding a δ for a Given ε **In Exercises 39–44, find the limit L. Then find δ such that $|f(x) - L| < \varepsilon$ whenever $0 < |x - c| < \delta$ for (a) $\varepsilon = 0.01$ and (b) $\varepsilon = 0.005$.**

39. $\lim_{x\to 2} (3x + 2)$

40. $\lim_{x\to 6} \left(6 - \frac{x}{3}\right)$

41. $\lim_{x\to 2} (x^2 - 3)$

42. $\lim_{x\to 4} (x^2 + 6)$

43. $\lim_{x\to 4} (x^2 - x)$

44. $\lim_{x\to 3} x^2$

Using the ε-δ Definition of Limit **In Exercises 45–56, find the limit L. Then use the ε-δ definition to prove that the limit is L.**

45. $\lim_{x\to 4} (x + 2)$

46. $\lim_{x\to -2} (4x + 5)$

47. $\lim_{x\to -4} \left(\frac{1}{2}x - 1\right)$

48. $\lim_{x\to 3} \left(\frac{3}{4}x + 1\right)$

49. $\lim_{x\to 6} 3$

50. $\lim_{x\to 2} (-1)$

51. $\lim_{x\to 0} \sqrt[3]{x}$

52. $\lim_{x\to 4} \sqrt{x}$

53. $\lim_{x\to -5} |x - 5|$

54. $\lim_{x\to 3} |x - 3|$

55. $\lim_{x\to 1} (x^2 + 1)$

56. $\lim_{x\to -4} (x^2 + 4x)$

57. Finding a Limit What is the limit of $f(x) = 4$ as x approaches π?

58. Finding a Limit What is the limit of $g(x) = x$ as x approaches π?

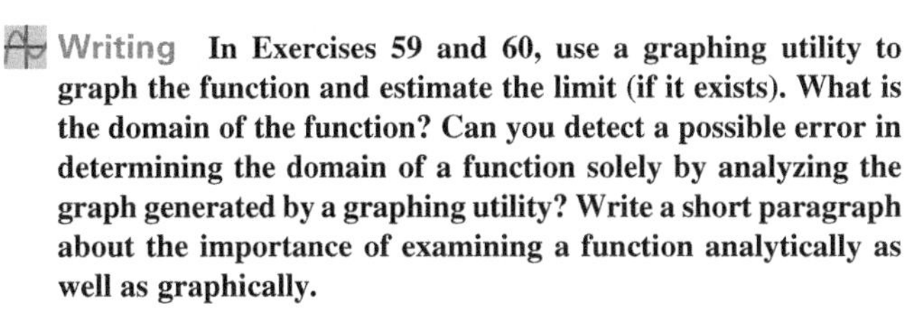

Writing **In Exercises 59 and 60, use a graphing utility to graph the function and estimate the limit (if it exists). What is the domain of the function? Can you detect a possible error in determining the domain of a function solely by analyzing the graph generated by a graphing utility? Write a short paragraph about the importance of examining a function analytically as well as graphically.**

59. $f(x) = \dfrac{\sqrt{x + 5} - 3}{x - 4}$

$\lim_{x\to 4} f(x)$

60. $f(x) = \dfrac{x - 3}{x^2 - 4x + 3}$

$\lim_{x\to 3} f(x)$

61. Modeling Data For a long-distance phone call, a hotel charges \$9.99 for the first minute and \$0.79 for each additional minute or fraction thereof. A formula for the cost is given by

$$C(t) = 9.99 - 0.79[\![1 - t]\!], \quad t > 0$$

where t is the time in minutes.

(*Note:* $[\![x]\!]$ = greatest integer n such that $n \le x$. For example, $[\![3.2]\!] = 3$ and $[\![-1.6]\!] = -2$.)

(a) Evaluate $C(10.75)$. What does $C(10.75)$ represent?

(b) Use a graphing utility to graph the cost function for $0 < t \le 6$. Does the limit of $C(t)$ as t approaches 3 exist? Explain.

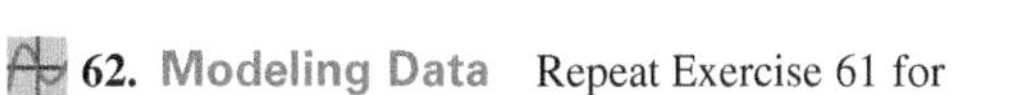

62. Modeling Data Repeat Exercise 61 for

$$C(t) = 5.79 - 0.99[\![1 - t]\!], \quad t > 0.$$

EXPLORING CONCEPTS

63. Finding δ When using the definition of limit to prove that L is the limit of $f(x)$ as x approaches c, you find the largest satisfactory value of δ. Why would any smaller positive value of δ also work?

64. Using the Definition of Limit The definition of limit on page 56 requires that f is a function defined on an open interval containing c, except possibly at c. Why is this requirement necessary?

65. Comparing Functions and Limits If $f(2) = 4$, can you conclude anything about the limit of $f(x)$ as x approaches 2? Explain your reasoning.

66. Comparing Functions and Limits If the limit of $f(x)$ as x approaches 2 is 4, can you conclude anything about $f(2)$? Explain your reasoning.

67. Jewelry A jeweler resizes a ring so that its inner circumference is 6 centimeters.

(a) What is the radius of the ring?

(b) The inner circumference of the ring varies between 5.5 centimeters and 6.5 centimeters. How does the radius vary?

(c) Use the ε-δ definition of limit to describe this situation. Identify ε and δ.

68. Sports

A sporting goods manufacturer designs a golf ball having a volume of 2.48 cubic inches.

(a) What is the radius of the golf ball?

(b) The volume of the golf ball varies between 2.45 cubic inches and 2.51 cubic inches. How does the radius vary?

(c) Use the ε-δ definition of limit to describe this situation. Identify ε and δ.

69. Estimating a Limit Consider the function

$$f(x) = (1 + x)^{1/x}.$$

Estimate

$$\lim_{x\to 0} (1 + x)^{1/x}$$

by evaluating f at x-values near 0. Sketch the graph of f.

The symbol indicates an exercise in which you are instructed to use graphing technology or a symbolic computer algebra system. The solutions of other exercises may also be facilitated by the use of appropriate technology.

70. Estimating a Limit Consider the function

$$f(x) = \frac{|x+1| - |x-1|}{x}.$$

Estimate

$$\lim_{x \to 0} \frac{|x+1| - |x-1|}{x}$$

by evaluating f at x-values near 0. Sketch the graph of f.

71. Graphical Reasoning The statement

$$\lim_{x \to 2} \frac{x^2 - 4}{x - 2} = 4$$

means that for each $\varepsilon > 0$ there corresponds a $\delta > 0$ such that if $0 < |x - 2| < \delta$, then

$$\left| \frac{x^2 - 4}{x - 2} - 4 \right| < \varepsilon.$$

If $\varepsilon = 0.001$, then

$$\left| \frac{x^2 - 4}{x - 2} - 4 \right| < 0.001.$$

Use a graphing utility to graph each side of this inequality. Use the *zoom* feature to find an interval $(2 - \delta, 2 + \delta)$ such that the inequality is true.

72. HOW DO YOU SEE IT? Use the graph of f to identify the values of c for which $\lim_{x \to c} f(x)$ exists.

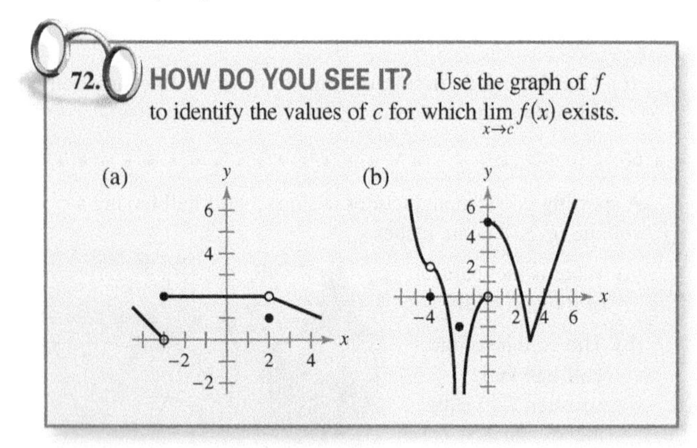

True or False? **In Exercises 73–76, determine whether the statement is true or false. If it is false, explain why or give an example that shows it is false.**

73. If f is undefined at $x = c$, then the limit of $f(x)$ as x approaches c does not exist.

74. If the limit of $f(x)$ as x approaches c is 0, then there must exist a number k such that $f(k) < 0.001$.

75. If $f(c) = L$, then $\lim_{x \to c} f(x) = L$.

76. If $\lim_{x \to c} f(x) = L$, then $f(c) = L$.

Determining a Limit **In Exercises 77 and 78, consider the function $f(x) = \sqrt{x}$.**

77. Is $\lim_{x \to 0.25} \sqrt{x} = 0.5$ a true statement? Explain.

78. Is $\lim_{x \to 0} \sqrt{x} = 0$ a true statement? Explain.

79. Evaluating Limits Use a graphing utility to evaluate

$$\lim_{x \to 0} \frac{\sin nx}{x}$$

for several values of n. What do you notice?

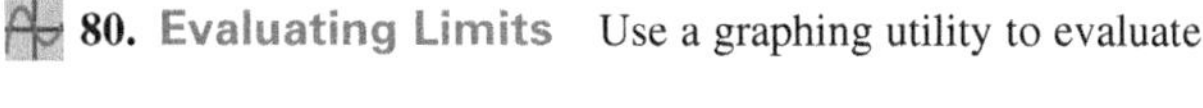

80. Evaluating Limits Use a graphing utility to evaluate

$$\lim_{x \to 0} \frac{\tan nx}{x}$$

for several values of n. What do you notice?

81. Proof Prove that if the limit of $f(x)$ as x approaches c exists, then the limit must be unique. [*Hint:* Let $\lim_{x \to c} f(x) = L_1$ and $\lim_{x \to c} f(x) = L_2$ and prove that $L_1 = L_2$.]

82. Proof Consider the line $f(x) = mx + b$, where $m \neq 0$. Use the ε-δ definition of limit to prove that $\lim_{x \to c} f(x) = mc + b$.

83. Proof Prove that

$$\lim_{x \to c} f(x) = L$$

is equivalent to

$$\lim_{x \to c} [f(x) - L] = 0.$$

84. Proof

(a) Given that

$$\lim_{x \to 0} (3x + 1)(3x - 1)x^2 + 0.01 = 0.01$$

prove that there exists an open interval (a, b) containing 0 such that $(3x + 1)(3x - 1)x^2 + 0.01 > 0$ for all $x \neq 0$ in (a, b).

(b) Given that $\lim_{x \to c} g(x) = L$, where $L > 0$, prove that there exists an open interval (a, b) containing c such that $g(x) > 0$ for all $x \neq c$ in (a, b).

PUTNAM EXAM CHALLENGE

85. Inscribe a rectangle of base b and height h in a circle of radius one, and inscribe an isosceles triangle in a region of the circle cut off by one base of the rectangle (with that side as the base of the triangle). For what value of h do the rectangle and triangle have the same area?

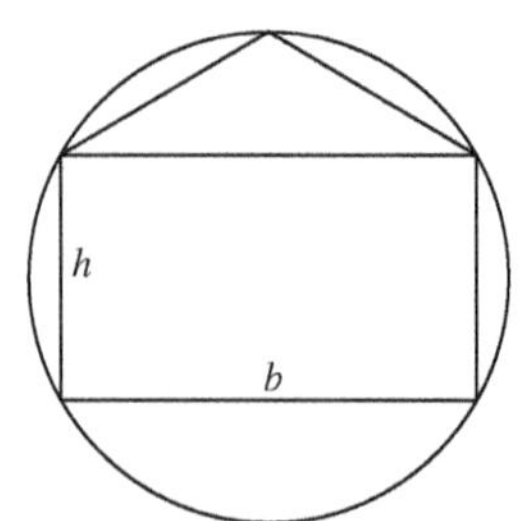

86. A right circular cone has base of radius 1 and height 3. A cube is inscribed in the cone so that one face of the cube is contained in the base of the cone. What is the side-length of the cube?

1.3 Evaluating Limits Analytically

- **Evaluate a limit using properties of limits.**
- **Develop and use a strategy for finding limits.**
- **Evaluate a limit using the dividing out technique.**
- **Evaluate a limit using the rationalizing technique.**
- **Evaluate a limit using the Squeeze Theorem.**

Properties of Limits

In Section 1.2, you learned that the limit of $f(x)$ as x approaches c does not depend on the value of f at $x = c$. It may happen, however, that the limit is precisely $f(c)$. In such cases, you can evaluate the limit by **direct substitution.** That is,

$$\lim_{x \to c} f(x) = f(c). \qquad \text{Substitute } c \text{ for } x.$$

Such *well-behaved* functions are **continuous at c.** You will examine this concept more closely in Section 1.4.

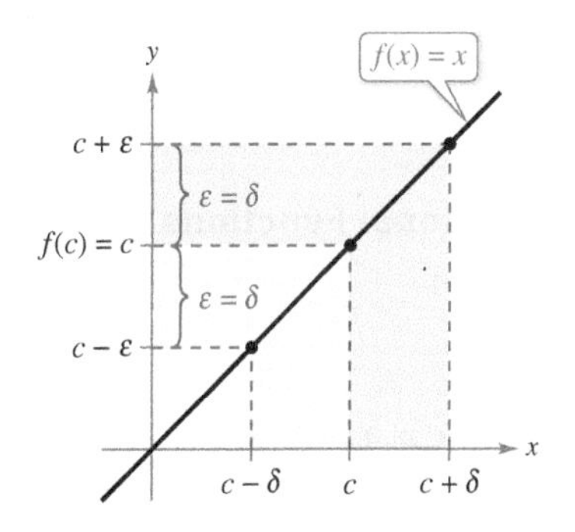

Figure 1.16

THEOREM 1.1 Some Basic Limits

Let b and c be real numbers, and let n be a positive integer.

1. $\lim_{x \to c} b = b$ **2.** $\lim_{x \to c} x = c$ **3.** $\lim_{x \to c} x^n = c^n$

Proof The proofs of Properties 1 and 3 of Theorem 1.1 are left as exercises (see Exercises 107 and 108). To prove Property 2, you need to show that for each $\varepsilon > 0$ there exists a $\delta > 0$ such that $|x - c| < \varepsilon$ whenever $0 < |x - c| < \delta$. To do this, choose $\delta = \varepsilon$. The second inequality then implies the first, as shown in Figure 1.16. ■

EXAMPLE 1 Evaluating Basic Limits

a. $\lim_{x \to 2} 3 = 3$ **b.** $\lim_{x \to -4} x = -4$ **c.** $\lim_{x \to 2} x^2 = 2^2 = 4$ ■

THEOREM 1.2 Properties of Limits

Let b and c be real numbers, let n be a positive integer, and let f and g be functions with the limits

$$\lim_{x \to c} f(x) = L \quad \text{and} \quad \lim_{x \to c} g(x) = K.$$

1. Scalar multiple: $\lim_{x \to c} [b f(x)] = bL$

2. Sum or difference: $\lim_{x \to c} [f(x) \pm g(x)] = L \pm K$

3. Product: $\lim_{x \to c} [f(x)g(x)] = LK$

4. Quotient: $\lim_{x \to c} \dfrac{f(x)}{g(x)} = \dfrac{L}{K}, \quad K \neq 0$

5. Power: $\lim_{x \to c} [f(x)]^n = L^n$

The proof of Property 1 is left as an exercise (see Exercise 109). The proofs of the other properties are given in Appendix A.

The symbol [QR] indicates that a video of this proof is available at *LarsonCalculus.com*.

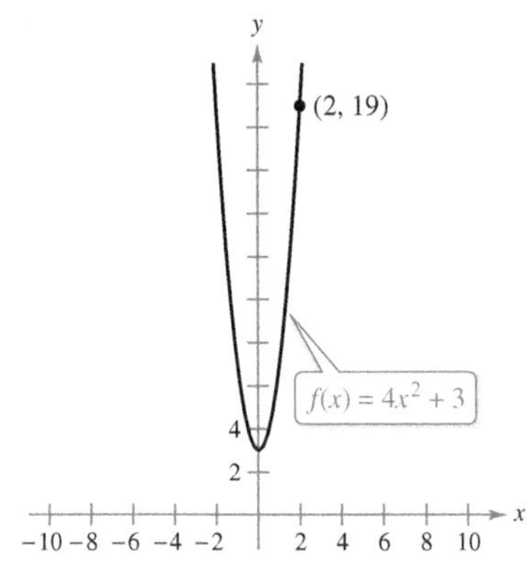

The limit of $f(x)$ as x approaches 2 is 19.
Figure 1.17

EXAMPLE 2 The Limit of a Polynomial

Find the limit: $\lim_{x\to 2} (4x^2 + 3)$.

Solution

$$\lim_{x\to 2} (4x^2 + 3) = \lim_{x\to 2} 4x^2 + \lim_{x\to 2} 3 \qquad \text{Property 2, Theorem 1.2}$$

$$= 4\left(\lim_{x\to 2} x^2\right) + \lim_{x\to 2} 3 \qquad \text{Property 1, Theorem 1.2}$$

$$= 4(2^2) + 3 \qquad \text{Properties 1 and 3, Theorem 1.1}$$

$$= 19 \qquad \text{Simplify.}$$

This limit is reinforced by the graph of $f(x) = 4x^2 + 3$ shown in Figure 1.17. ■

In Example 2, note that the limit (as x approaches 2) of the *polynomial function* $p(x) = 4x^2 + 3$ is simply the value of p at $x = 2$.

$$\lim_{x\to 2} p(x) = p(2) = 4(2^2) + 3 = 19$$

This *direct substitution* property is valid for all polynomial and rational functions with nonzero denominators.

THEOREM 1.3 Limits of Polynomial and Rational Functions

If p is a polynomial function and c is a real number, then

$$\lim_{x\to c} p(x) = p(c).$$

If r is a rational function given by $r(x) = p(x)/q(x)$ and c is a real number such that $q(c) \neq 0$, then

$$\lim_{x\to c} r(x) = r(c) = \frac{p(c)}{q(c)}.$$

EXAMPLE 3 The Limit of a Rational Function

Find the limit: $\lim_{x\to 1} \dfrac{x^2 + x + 2}{x + 1}$.

Solution Because the denominator is not 0 when $x = 1$, you can apply Theorem 1.3 to obtain

$$\lim_{x\to 1} \frac{x^2 + x + 2}{x + 1} = \frac{1^2 + 1 + 2}{1 + 1} = \frac{4}{2} = 2. \qquad \text{See Figure 1.18.}$$ ■

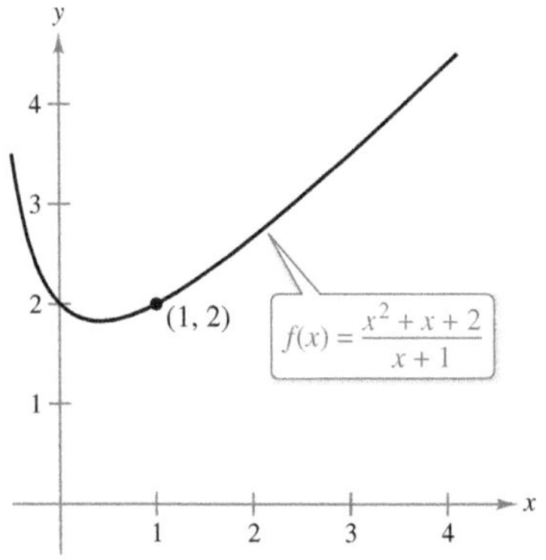

The limit of $f(x)$ as x approaches 1 is 2.
Figure 1.18

Polynomial functions and rational functions are two of the three basic types of algebraic functions. The next theorem deals with the limit of the third type of algebraic function—one that involves a radical.

THEOREM 1.4 The Limit of a Function Involving a Radical

Let n be a positive integer. The limit below is valid for all c when n is odd, and is valid for $c > 0$ when n is even.

$$\lim_{x\to c} \sqrt[n]{x} = \sqrt[n]{c}$$

A proof of this theorem is given in Appendix A.

The next theorem greatly expands your ability to evaluate limits because it shows how to analyze the limit of a composite function.

THEOREM 1.5 The Limit of a Composite Function

If f and g are functions such that $\lim_{x\to c} g(x) = L$ and $\lim_{x\to L} f(x) = f(L)$, then

$$\lim_{x\to c} f(g(x)) = f\left(\lim_{x\to c} g(x)\right) = f(L).$$

A proof of this theorem is given in Appendix A.

EXAMPLE 4 The Limit of a Composite Function

See LarsonCalculus.com for an interactive version of this type of example.

Find the limit.

a. $\lim_{x\to 0} \sqrt{x^2 + 4}$ **b.** $\lim_{x\to 3} \sqrt[3]{2x^2 - 10}$

Solution

a. Because

$$\lim_{x\to 0} (x^2 + 4) = 0^2 + 4 = 4 \quad \text{and} \quad \lim_{x\to 4} \sqrt{x} = \sqrt{4} = 2$$

you can conclude that

$$\lim_{x\to 0} \sqrt{x^2 + 4} = \sqrt{4} = 2.$$

b. Because

$$\lim_{x\to 3} (2x^2 - 10) = 2(3^2) - 10 = 8 \quad \text{and} \quad \lim_{x\to 8} \sqrt[3]{x} = \sqrt[3]{8} = 2$$

you can conclude that

$$\lim_{x\to 3} \sqrt[3]{2x^2 - 10} = \sqrt[3]{8} = 2.$$

You have seen that the limits of many algebraic functions can be evaluated by direct substitution. The six basic trigonometric functions also exhibit this desirable quality, as shown in the next theorem (presented without proof).

THEOREM 1.6 Limits of Trigonometric Functions

Let c be a real number in the domain of the given trigonometric function.

1. $\lim_{x\to c} \sin x = \sin c$ **2.** $\lim_{x\to c} \cos x = \cos c$ **3.** $\lim_{x\to c} \tan x = \tan c$

4. $\lim_{x\to c} \cot x = \cot c$ **5.** $\lim_{x\to c} \sec x = \sec c$ **6.** $\lim_{x\to c} \csc x = \csc c$

EXAMPLE 5 Limits of Trigonometric Functions

a. $\lim_{x\to 0} \tan x = \tan(0) = 0$

b. $\lim_{x\to \pi} (x \cos x) = \left(\lim_{x\to \pi} x\right)\left(\lim_{x\to \pi} \cos x\right) = \pi \cos(\pi) = -\pi$

c. $\lim_{x\to 0} \sin^2 x = \lim_{x\to 0} (\sin x)^2 = 0^2 = 0$

A Strategy for Finding Limits

On the previous three pages, you studied several types of functions whose limits can be evaluated by direct substitution. This knowledge, together with the next theorem, can be used to develop a strategy for finding limits.

THEOREM 1.7 Functions That Agree at All but One Point

Let c be a real number, and let $f(x) = g(x)$ for all $x \neq c$ in an open interval containing c. If the limit of $g(x)$ as x approaches c exists, then the limit of $f(x)$ also exists and

$$\lim_{x \to c} f(x) = \lim_{x \to c} g(x).$$

A proof of this theorem is given in Appendix A.

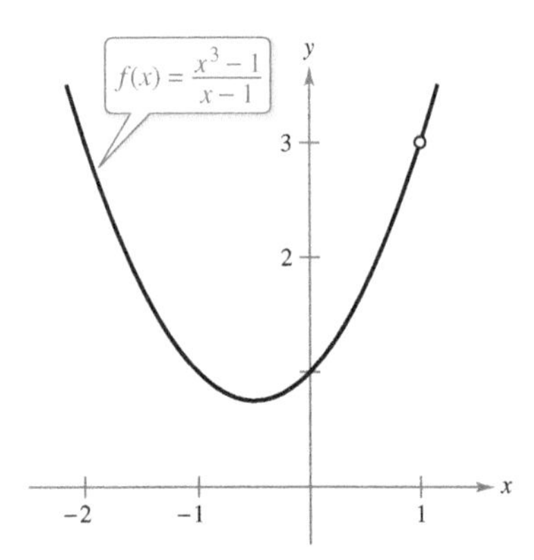

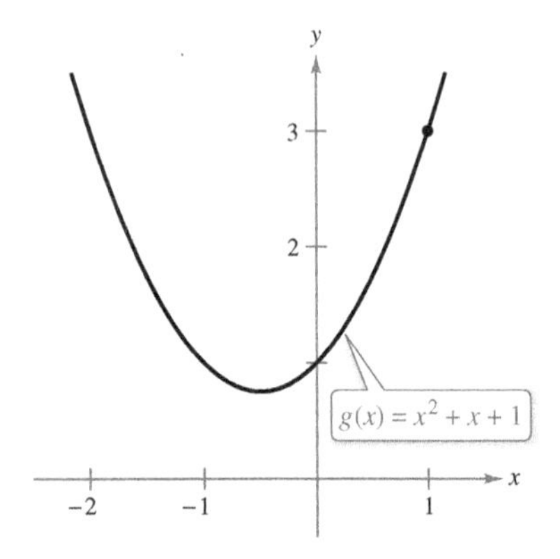

f and g agree at all but one point.
Figure 1.19

EXAMPLE 6 Finding the Limit of a Function

Find the limit.

$$\lim_{x \to 1} \frac{x^3 - 1}{x - 1}$$

Solution Let $f(x) = (x^3 - 1)/(x - 1)$. By factoring and dividing out common factors, you can rewrite f as

$$f(x) = \frac{\cancel{(x-1)}(x^2 + x + 1)}{\cancel{(x-1)}} = x^2 + x + 1 = g(x), \quad x \neq 1.$$

So, for all x-values other than $x = 1$, the functions f and g agree, as shown in Figure 1.19. Because $\lim_{x \to 1} g(x)$ exists, you can apply Theorem 1.7 to conclude that f and g have the same limit at $x = 1$.

$$\lim_{x \to 1} \frac{x^3 - 1}{x - 1} = \lim_{x \to 1} \frac{(x - 1)(x^2 + x + 1)}{x - 1} \qquad \text{Factor.}$$

$$= \lim_{x \to 1} \frac{\cancel{(x-1)}(x^2 + x + 1)}{\cancel{(x-1)}} \qquad \text{Divide out common factor.}$$

$$= \lim_{x \to 1} (x^2 + x + 1) \qquad \text{Apply Theorem 1.7.}$$

$$= 1^2 + 1 + 1 \qquad \text{Use direct substitution.}$$

$$= 3 \qquad \text{Simplify.}$$

REMARK When applying this strategy for finding a limit, remember that some functions do not have a limit (as x approaches c). For instance, the limit below does not exist.

$$\lim_{x \to 1} \frac{x^3 + 1}{x - 1}$$

A Strategy for Finding Limits

1. Learn to recognize which limits can be evaluated by direct substitution. (These limits are listed in Theorems 1.1 through 1.6.)
2. When the limit of $f(x)$ as x approaches c *cannot* be evaluated by direct substitution, try to find a function g that agrees with f for all x other than $x = c$. [Choose g such that the limit of $g(x)$ *can* be evaluated by direct substitution.] Then apply Theorem 1.7 to conclude *analytically* that

   $$\lim_{x \to c} f(x) = \lim_{x \to c} g(x) = g(c).$$

3. Use a *graph* or *table* to reinforce your conclusion.

Dividing Out Technique

Another procedure for finding a limit analytically is the **dividing out technique.** This technique involves dividing out common factors, as shown in Example 7.

EXAMPLE 7 Dividing Out Technique

See LarsonCalculus.com for an interactive version of this type of example.

Find the limit: $\lim_{x\to -3} \dfrac{x^2+x-6}{x+3}$.

Solution Although you are taking the limit of a rational function, you *cannot* apply Theorem 1.3 because the limit of the denominator is 0.

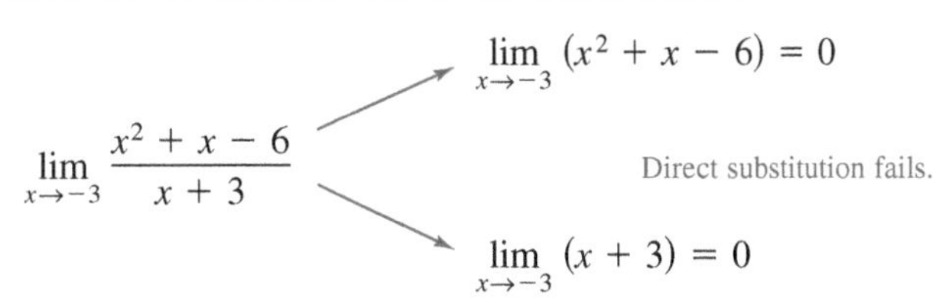

Because the limit of the numerator is also 0, the numerator and denominator have a *common factor* of $(x+3)$. So, for all $x \neq -3$, you can divide out this factor to obtain

$$f(x) = \frac{x^2+x-6}{x+3} = \frac{(x+3)(x-2)}{x+3} = x-2 = g(x), \quad x \neq -3.$$

Using Theorem 1.7, it follows that

$$\lim_{x\to -3} \frac{x^2+x-6}{x+3} = \lim_{x\to -3} (x-2) \qquad \text{Apply Theorem 1.7.}$$
$$= -5. \qquad \text{Use direct substitution.}$$

This result is shown graphically in Figure 1.20. Note that the graph of the function f coincides with the graph of the function $g(x) = x - 2$, except that the graph of f has a hole at the point $(-3, -5)$. ■

REMARK In the solution to Example 7, be sure you see the usefulness of the Factor Theorem of Algebra. This theorem states that if c is a zero of a polynomial function, then $(x - c)$ is a factor of the polynomial. So, when you apply direct substitution to a rational function and obtain

$$r(c) = \frac{p(c)}{q(c)} = \frac{0}{0}$$

you can conclude that $(x - c)$ must be a common factor of both $p(x)$ and $q(x)$.

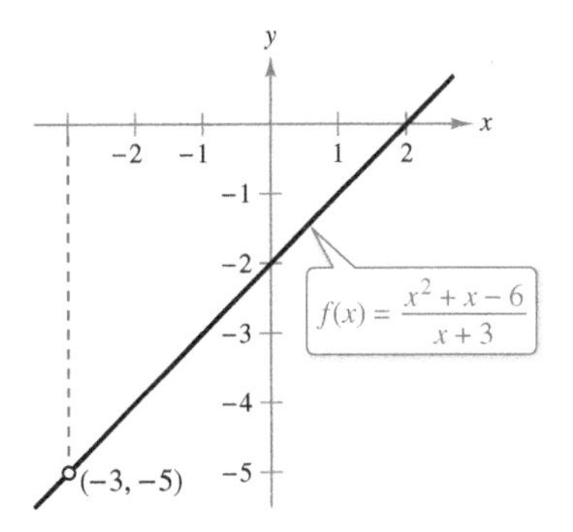

f is undefined when $x = -3$. The limit of $f(x)$ as x approaches -3 is -5.
Figure 1.20

In Example 7, direct substitution produced the meaningless fractional form $0/0$. An expression such as $0/0$ is called an **indeterminate form** because you cannot (from the form alone) determine the limit. When you try to evaluate a limit and encounter this form, remember that you must rewrite the fraction so that the new denominator does not have 0 as its limit. One way to do this is to *divide out common factors*. Another way is to use the *rationalizing technique* shown on the next page.

TECHNOLOGY PITFALL A graphing utility can give misleading information about the graph of a function. For instance, try graphing the function from Example 7

$$f(x) = \frac{x^2+x-6}{x+3}$$

on a graphing utility. On some graphing utilities, the graph may appear to be defined at every real number, as shown in the figure at the right. However, because f is undefined when $x = -3$, you know that the graph of f has a hole at $x = -3$. You can verify this on a graphing utility using the *trace* or *table* feature.

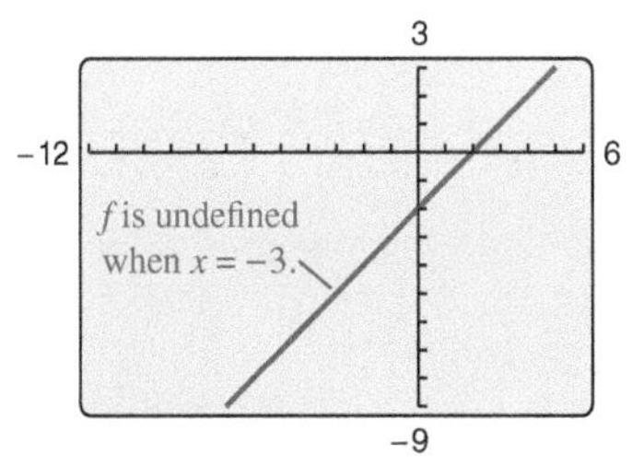

Misleading graph of f

Rationalizing Technique

Another way to find a limit analytically is the **rationalizing technique,** which involves rationalizing either the numerator or denominator of a fractional expression. Recall that rationalizing the numerator (denominator) means multiplying the numerator and denominator by the conjugate of the numerator (denominator). For instance, to rationalize the numerator of

$$\frac{\sqrt{x}+4}{x}$$

multiply the numerator and denominator by the conjugate of $\sqrt{x}+4$, which is

$$\sqrt{x}-4.$$

EXAMPLE 8 Rationalizing Technique

Find the limit: $\displaystyle\lim_{x\to 0}\frac{\sqrt{x+1}-1}{x}$.

Solution By direct substitution, you obtain the indeterminate form 0/0.

$$\lim_{x\to 0}\frac{\sqrt{x+1}-1}{x} \quad \begin{cases} \lim_{x\to 0}\left(\sqrt{x+1}-1\right)=0 \\ \lim_{x\to 0} x = 0 \end{cases}$$

Direct substitution fails.

In this case, you can rewrite the fraction by rationalizing the numerator.

$$\begin{aligned}\frac{\sqrt{x+1}-1}{x} &= \left(\frac{\sqrt{x+1}-1}{x}\right)\left(\frac{\sqrt{x+1}+1}{\sqrt{x+1}+1}\right)\\ &= \frac{(x+1)-1}{x\left(\sqrt{x+1}+1\right)}\\ &= \frac{\cancel{x}}{\cancel{x}\left(\sqrt{x+1}+1\right)}\\ &= \frac{1}{\sqrt{x+1}+1}, \quad x \neq 0\end{aligned}$$

REMARK The rationalizing technique for evaluating limits is based on multiplication by a convenient form of 1. In Example 8, the convenient form is

$$1 = \frac{\sqrt{x+1}+1}{\sqrt{x+1}+1}.$$

Now, using Theorem 1.7, you can evaluate the limit as shown.

$$\begin{aligned}\lim_{x\to 0}\frac{\sqrt{x+1}-1}{x} &= \lim_{x\to 0}\frac{1}{\sqrt{x+1}+1}\\ &= \frac{1}{1+1}\\ &= \frac{1}{2}\end{aligned}$$

A table or a graph can reinforce your conclusion that the limit is $\frac{1}{2}$. (See Figure 1.21.)

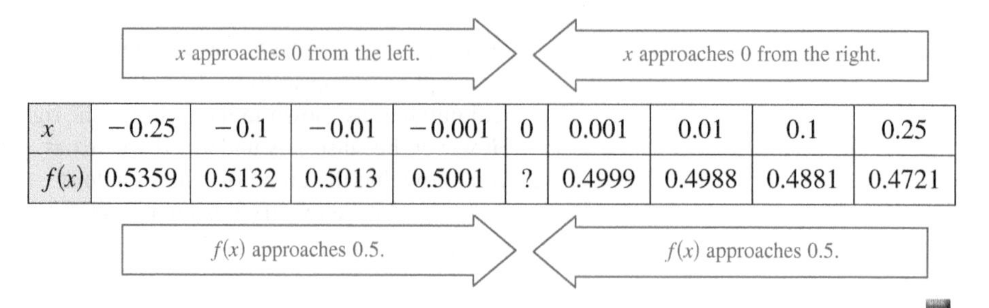

x	-0.25	-0.1	-0.01	-0.001	0	0.001	0.01	0.1	0.25
$f(x)$	0.5359	0.5132	0.5013	0.5001	?	0.4999	0.4988	0.4881	0.4721

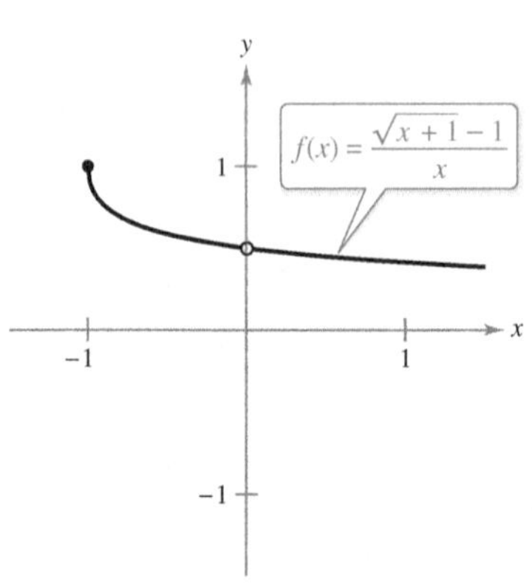

The limit of $f(x)$ as x approaches 0 is $\frac{1}{2}$.
Figure 1.21

The Squeeze Theorem

The next theorem concerns the limit of a function that is squeezed between two other functions, each of which has the same limit at a given x-value, as shown in Figure 1.22.

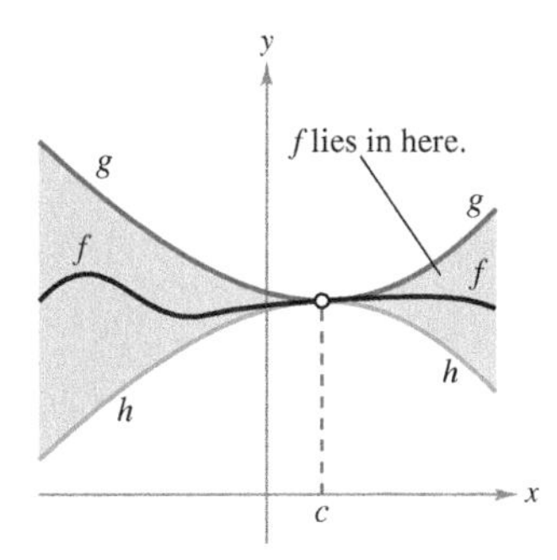

The Squeeze Theorem
Figure 1.22

THEOREM 1.8 The Squeeze Theorem

If $h(x) \le f(x) \le g(x)$ for all x in an open interval containing c, except possibly at c itself, and if

$$\lim_{x\to c} h(x) = L = \lim_{x\to c} g(x)$$

then $\lim_{x\to c} f(x)$ exists and is equal to L.

A proof of this theorem is given in Appendix A.

You can see the usefulness of the Squeeze Theorem (also called the Sandwich Theorem or the Pinching Theorem) in the proof of Theorem 1.9.

THEOREM 1.9 Two Special Trigonometric Limits

1. $\lim_{x\to 0} \dfrac{\sin x}{x} = 1$ **2.** $\lim_{x\to 0} \dfrac{1 - \cos x}{x} = 0$

Proof The proof of the second limit is left as an exercise (see Exercise 121). To avoid the confusion of two different uses of x, the proof of the first limit is presented using the variable θ, where θ is an acute positive angle *measured in radians*. Figure 1.23 shows a circular sector that is squeezed between two triangles.

A circular sector is used to prove Theorem 1.9.
Figure 1.23

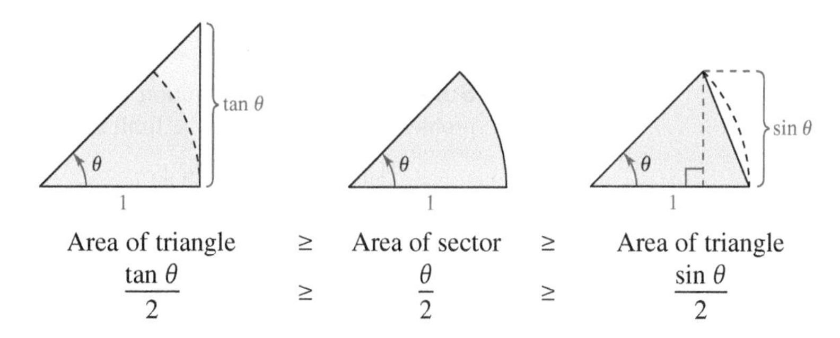

Multiplying each expression by $2/\sin\theta$ produces

$$\frac{1}{\cos\theta} \ge \frac{\theta}{\sin\theta} \ge 1$$

and taking reciprocals and reversing the inequalities yields

$$\cos\theta \le \frac{\sin\theta}{\theta} \le 1.$$

Because $\cos\theta = \cos(-\theta)$ and $(\sin\theta)/\theta = [\sin(-\theta)]/(-\theta)$, you can conclude that this inequality is valid for *all* nonzero θ in the open interval $(-\pi/2, \pi/2)$. Finally, because $\lim_{\theta\to 0} \cos\theta = 1$ and $\lim_{\theta\to 0} 1 = 1$, you can apply the Squeeze Theorem to conclude that

$$\lim_{\theta\to 0} \frac{\sin\theta}{\theta} = 1.$$

EXAMPLE 9 A Limit Involving a Trigonometric Function

Find the limit: $\lim_{x\to 0} \frac{\tan x}{x}$.

Solution Direct substitution yields the indeterminate form 0/0. To solve this problem, you can write tan x as $(\sin x)/(\cos x)$ and obtain

$$\lim_{x\to 0} \frac{\tan x}{x} = \lim_{x\to 0} \left(\frac{\sin x}{x}\right)\left(\frac{1}{\cos x}\right).$$

Now, because

$$\lim_{x\to 0} \frac{\sin x}{x} = 1$$

and

$$\lim_{x\to 0} \frac{1}{\cos x} = 1$$

you can obtain

$$\begin{aligned}\lim_{x\to 0} \frac{\tan x}{x} &= \left(\lim_{x\to 0} \frac{\sin x}{x}\right)\left(\lim_{x\to 0} \frac{1}{\cos x}\right)\\ &= (1)(1)\\ &= 1.\end{aligned}$$

(See Figure 1.24.)

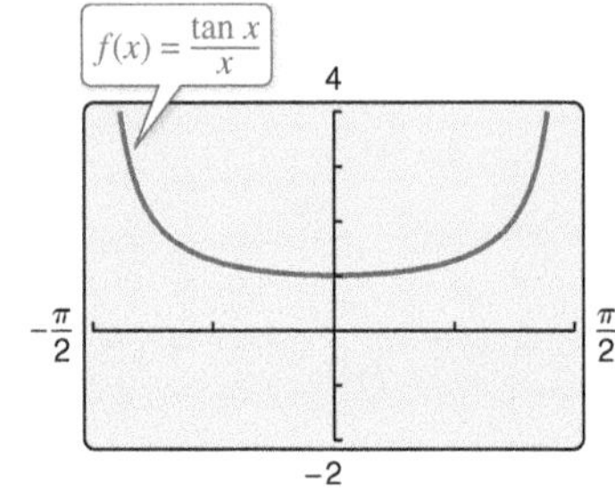

The limit of $f(x)$ as x approaches 0 is 1.
Figure 1.24

REMARK Be sure you understand the mathematical conventions regarding parentheses and trigonometric functions. For instance, in Example 10, sin $4x$ means $\sin(4x)$.

EXAMPLE 10 A Limit Involving a Trigonometric Function

Find the limit: $\lim_{x\to 0} \frac{\sin 4x}{x}$.

Solution Direct substitution yields the indeterminate form 0/0. To solve this problem, you can rewrite the limit as

$$\lim_{x\to 0} \frac{\sin 4x}{x} = 4\left(\lim_{x\to 0} \frac{\sin 4x}{4x}\right). \qquad \text{Multiply and divide by 4.}$$

Now, by letting $y = 4x$ and observing that x approaches 0 if and only if y approaches 0, you can write

$$\begin{aligned}\lim_{x\to 0} \frac{\sin 4x}{x} &= 4\left(\lim_{x\to 0} \frac{\sin 4x}{4x}\right)\\ &= 4\left(\lim_{y\to 0} \frac{\sin y}{y}\right) && \text{Let } y = 4x.\\ &= 4(1) && \text{Apply Theorem 1.9(1).}\\ &= 4.\end{aligned}$$

(See Figure 1.25.)

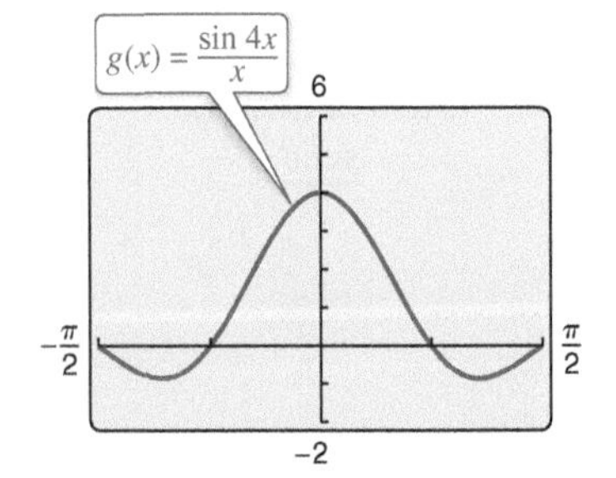

The limit of $g(x)$ as x approaches 0 is 4.
Figure 1.25

TECHNOLOGY Use a graphing utility to confirm the limits in the examples and in the exercise set. For instance, Figures 1.24 and 1.25 show the graphs of

$$f(x) = \frac{\tan x}{x} \quad \text{and} \quad g(x) = \frac{\sin 4x}{x}.$$

Note that the first graph appears to contain the point (0, 1) and the second graph appears to contain the point (0, 4), which lends support to the conclusions obtained in Examples 9 and 10.

1.3 Exercises

See CalcChat.com for tutorial help and worked-out solutions to odd-numbered exercises.

CONCEPT CHECK

1. **Polynomial Function** Describe how to find the limit of a polynomial function $p(x)$ as x approaches c.
2. **Indeterminate Form** What is meant by an indeterminate form?
3. **Squeeze Theorem** In your own words, explain the Squeeze Theorem.
4. **Special Limits** List the two special trigonometric limits.

Finding a Limit **In Exercises 5–22, find the limit.**

5. $\lim_{x\to 2} x^3$

6. $\lim_{x\to -3} x^4$

7. $\lim_{x\to -3} (2x + 5)$

8. $\lim_{x\to 9} (4x - 1)$

9. $\lim_{x\to -3} (x^2 + 3x)$

10. $\lim_{x\to 2} (-x^3 + 1)$

11. $\lim_{x\to -3} (2x^2 + 4x + 1)$

12. $\lim_{x\to 1} (2x^3 - 6x + 5)$

13. $\lim_{x\to 3} \sqrt{x + 8}$

14. $\lim_{x\to 2} \sqrt[3]{12x + 3}$

15. $\lim_{x\to -4} (1 - x)^3$

16. $\lim_{x\to 0} (3x - 2)^4$

17. $\lim_{x\to 2} \dfrac{3}{2x + 1}$

18. $\lim_{x\to -5} \dfrac{5}{x + 3}$

19. $\lim_{x\to 1} \dfrac{x}{x^2 + 4}$

20. $\lim_{x\to 1} \dfrac{3x + 5}{x + 1}$

21. $\lim_{x\to 7} \dfrac{3x}{\sqrt{x + 2}}$

22. $\lim_{x\to 3} \dfrac{\sqrt{x + 6}}{x + 2}$

Finding Limits **In Exercises 23–26, find the limits.**

23. $f(x) = 5 - x,\ g(x) = x^3$

(a) $\lim_{x\to 1} f(x)$ (b) $\lim_{x\to 4} g(x)$ (c) $\lim_{x\to 1} g(f(x))$

24. $f(x) = x + 7,\ g(x) = x^2$

(a) $\lim_{x\to -3} f(x)$ (b) $\lim_{x\to 4} g(x)$ (c) $\lim_{x\to -3} g(f(x))$

25. $f(x) = 4 - x^2,\ g(x) = \sqrt{x + 1}$

(a) $\lim_{x\to 1} f(x)$ (b) $\lim_{x\to 3} g(x)$ (c) $\lim_{x\to 1} g(f(x))$

26. $f(x) = 2x^2 - 3x + 1,\ g(x) = \sqrt[3]{x + 6}$

(a) $\lim_{x\to 4} f(x)$ (b) $\lim_{x\to 21} g(x)$ (c) $\lim_{x\to 4} g(f(x))$

Finding a Limit of a Trigonometric Function **In Exercises 27–36, find the limit of the trigonometric function.**

27. $\lim_{x\to \pi/2} \sin x$

28. $\lim_{x\to \pi} \tan x$

29. $\lim_{x\to 1} \cos \dfrac{\pi x}{3}$

30. $\lim_{x\to 2} \sin \dfrac{\pi x}{12}$

31. $\lim_{x\to 0} \sec 2x$

32. $\lim_{x\to \pi} \cos 3x$

33. $\lim_{x\to 5\pi/6} \sin x$

34. $\lim_{x\to 5\pi/3} \cos x$

35. $\lim_{x\to 3} \tan \dfrac{\pi x}{4}$

36. $\lim_{x\to 7} \sec \dfrac{\pi x}{6}$

Evaluating Limits **In Exercises 37–40, use the information to evaluate the limits.**

37. $\lim_{x\to c} f(x) = \frac{2}{5}$

$\lim_{x\to c} g(x) = 2$

(a) $\lim_{x\to c} [5g(x)]$

(b) $\lim_{x\to c} [f(x) + g(x)]$

(c) $\lim_{x\to c} [f(x)g(x)]$

(d) $\lim_{x\to c} \dfrac{f(x)}{g(x)}$

38. $\lim_{x\to c} f(x) = 2$

$\lim_{x\to c} g(x) = \frac{3}{4}$

(a) $\lim_{x\to c} [4f(x)]$

(b) $\lim_{x\to c} [f(x) + g(x)]$

(c) $\lim_{x\to c} [f(x)g(x)]$

(d) $\lim_{x\to c} \dfrac{f(x)}{g(x)}$

39. $\lim_{x\to c} f(x) = 16$

(a) $\lim_{x\to c} [f(x)]^2$

(b) $\lim_{x\to c} \sqrt{f(x)}$

(c) $\lim_{x\to c} [3f(x)]$

(d) $\lim_{x\to c} [f(x)]^{3/2}$

40. $\lim_{x\to c} f(x) = 27$

(a) $\lim_{x\to c} \sqrt[3]{f(x)}$

(b) $\lim_{x\to c} \dfrac{f(x)}{18}$

(c) $\lim_{x\to c} [f(x)]^2$

(d) $\lim_{x\to c} [f(x)]^{2/3}$

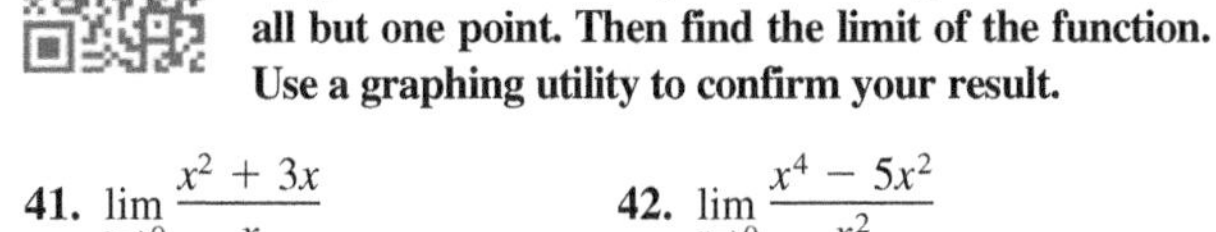

Finding a Limit **In Exercises 41–46, write a simpler function that agrees with the given function at all but one point. Then find the limit of the function. Use a graphing utility to confirm your result.**

41. $\lim_{x\to 0} \dfrac{x^2 + 3x}{x}$

42. $\lim_{x\to 0} \dfrac{x^4 - 5x^2}{x^2}$

43. $\lim_{x\to -1} \dfrac{x^2 - 1}{x + 1}$

44. $\lim_{x\to -2} \dfrac{3x^2 + 5x - 2}{x + 2}$

45. $\lim_{x\to 2} \dfrac{x^3 - 8}{x - 2}$

46. $\lim_{x\to -1} \dfrac{x^3 + 1}{x + 1}$

Finding a Limit **In Exercises 47–62, find the limit.**

47. $\lim_{x\to 0} \dfrac{x}{x^2 - x}$

48. $\lim_{x\to 0} \dfrac{7x^3 - x^2}{x}$

49. $\lim_{x\to 4} \dfrac{x - 4}{x^2 - 16}$

50. $\lim_{x\to 5} \dfrac{5 - x}{x^2 - 25}$

51. $\lim_{x\to -3} \dfrac{x^2 + x - 6}{x^2 - 9}$

52. $\lim_{x\to 2} \dfrac{x^2 + 2x - 8}{x^2 - x - 2}$

53. $\lim_{x\to 4} \dfrac{\sqrt{x + 5} - 3}{x - 4}$

54. $\lim_{x\to 3} \dfrac{\sqrt{x + 1} - 2}{x - 3}$

55. $\lim_{x\to 0} \dfrac{\sqrt{x + 5} - \sqrt{5}}{x}$

56. $\lim_{x\to 0} \dfrac{\sqrt{2 + x} - \sqrt{2}}{x}$

57. $\lim\limits_{x \to 0} \dfrac{[1/(3 + x)] - (1/3)}{x}$

58. $\lim\limits_{x \to 0} \dfrac{[1/(x + 4)] - (1/4)}{x}$

59. $\lim\limits_{\Delta x \to 0} \dfrac{2(x + \Delta x) - 2x}{\Delta x}$

60. $\lim\limits_{\Delta x \to 0} \dfrac{(x + \Delta x)^2 - x^2}{\Delta x}$

61. $\lim\limits_{\Delta x \to 0} \dfrac{(x + \Delta x)^2 - 2(x + \Delta x) + 1 - (x^2 - 2x + 1)}{\Delta x}$

62. $\lim\limits_{\Delta x \to 0} \dfrac{(x + \Delta x)^3 - x^3}{\Delta x}$

Finding a Limit of a Trigonometric Function **In Exercises 63–74, find the limit of the trigonometric function.**

63. $\lim\limits_{x \to 0} \dfrac{\sin x}{5x}$

64. $\lim\limits_{x \to 0} \dfrac{3(1 - \cos x)}{x}$

65. $\lim\limits_{x \to 0} \dfrac{(\sin x)(1 - \cos x)}{x^2}$

66. $\lim\limits_{\theta \to 0} \dfrac{\cos \theta \tan \theta}{\theta}$

67. $\lim\limits_{x \to 0} \dfrac{\sin^2 x}{x}$

68. $\lim\limits_{x \to 0} \dfrac{\tan^2 x}{x}$

69. $\lim\limits_{h \to 0} \dfrac{(1 - \cos h)^2}{h}$

70. $\lim\limits_{\phi \to \pi} \phi \sec \phi$

71. $\lim\limits_{x \to 0} \dfrac{6 - 6 \cos x}{3}$

72. $\lim\limits_{x \to 0} \dfrac{\cos x - \sin x - 1}{2x}$

73. $\lim\limits_{t \to 0} \dfrac{\sin 3t}{2t}$

74. $\lim\limits_{x \to 0} \dfrac{\sin 2x}{\sin 3x}$ $\left[\textit{Hint}: \text{Find } \lim\limits_{x \to 0} \left(\dfrac{2 \sin 2x}{2x}\right)\left(\dfrac{3x}{3 \sin 3x}\right).\right]$

Graphical, Numerical, and Analytic Analysis **In Exercises 75–82, use a graphing utility to graph the function and estimate the limit. Use a table to reinforce your conclusion. Then find the limit by analytic methods.**

75. $\lim\limits_{x \to 0} \dfrac{\sqrt{x + 2} - \sqrt{2}}{x}$

76. $\lim\limits_{x \to 16} \dfrac{4 - \sqrt{x}}{x - 16}$

77. $\lim\limits_{x \to 0} \dfrac{[1/(2 + x)] - (1/2)}{x}$

78. $\lim\limits_{x \to 2} \dfrac{x^5 - 32}{x - 2}$

79. $\lim\limits_{t \to 0} \dfrac{\sin 3t}{t}$

80. $\lim\limits_{x \to 0} \dfrac{\cos x - 1}{2x^2}$

81. $\lim\limits_{x \to 0} \dfrac{\sin x^2}{x}$

82. $\lim\limits_{x \to 0} \dfrac{\sin x}{\sqrt[3]{x}}$

Finding a Limit **In Exercises 83–90, find**

$$\lim_{\Delta x \to 0} \frac{f(x + \Delta x) - f(x)}{\Delta x}.$$

83. $f(x) = 3x - 2$

84. $f(x) = -6x + 3$

85. $f(x) = x^2 - 4x$

86. $f(x) = 3x^2 + 1$

87. $f(x) = 2\sqrt{x}$

88. $f(x) = \sqrt{x} - 5$

89. $f(x) = \dfrac{1}{x + 3}$

90. $f(x) = \dfrac{1}{x^2}$

Using the Squeeze Theorem **In Exercises 91 and 92, use the Squeeze Theorem to find $\lim\limits_{x \to c} f(x)$.**

91. $c = 0$

$4 - x^2 \le f(x) \le 4 + x^2$

92. $c = a$

$b - |x - a| \le f(x) \le b + |x - a|$

Using the Squeeze Theorem **In Exercises 93–96, use a graphing utility to graph the given function and the equations $y = |x|$ and $y = -|x|$ in the same viewing window. Using the graphs to observe the Squeeze Theorem visually, find $\lim\limits_{x \to 0} f(x)$.**

93. $f(x) = |x| \sin x$

94. $f(x) = |x| \cos x$

95. $f(x) = x \sin \dfrac{1}{x}$

96. $f(x) = x \cos \dfrac{1}{x}$

EXPLORING CONCEPTS

97. Functions That Agree at All but One Point

(a) In the context of finding limits, discuss what is meant by two functions that agree at all but one point.

(b) Give an example of two functions that agree at all but one point.

98. Writing Functions Write a function of each specified type that has a limit of 4 as x approaches 8.

(a) linear

(b) polynomial of degree 2

(c) rational

(d) radical

(e) cosine

(f) sine

99. Writing Use a graphing utility to graph

$$f(x) = x, \quad g(x) = \sin x, \quad \text{and} \quad h(x) = \frac{\sin x}{x}$$

in the same viewing window. Compare the magnitudes of $f(x)$ and $g(x)$ when x is close to 0. Use the comparison to write a short paragraph explaining why

$$\lim_{x \to 0} h(x) = 1.$$

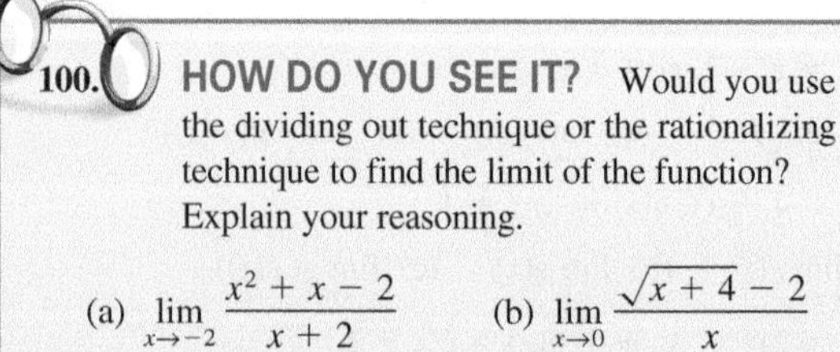

100. HOW DO YOU SEE IT? Would you use the dividing out technique or the rationalizing technique to find the limit of the function? Explain your reasoning.

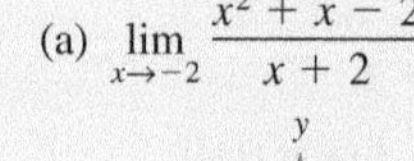

(a) $\lim\limits_{x \to -2} \dfrac{x^2 + x - 2}{x + 2}$

(b) $\lim\limits_{x \to 0} \dfrac{\sqrt{x + 4} - 2}{x}$

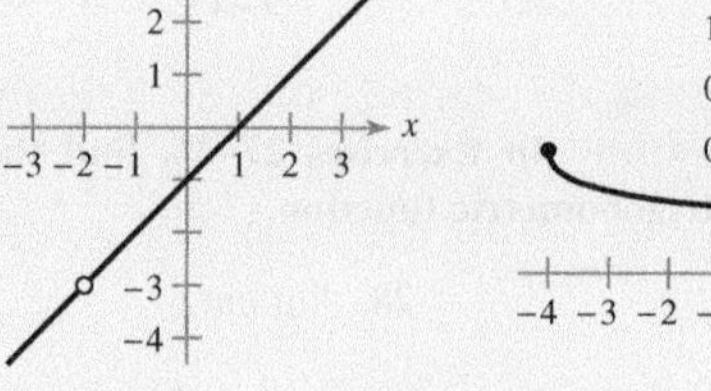

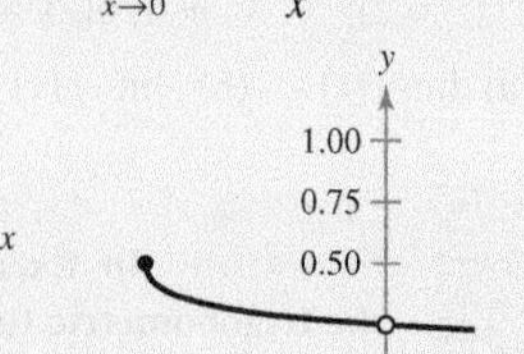

Free-Falling Object

In Exercises 101 and 102, use the position function $s(t) = -16t^2 + 500$, which gives the height (in feet) of an object that has fallen for t seconds from a height of 500 feet. The velocity at time $t = a$ seconds is given by

$$\lim_{t \to a} \frac{s(a) - s(t)}{a - t}.$$

101. A construction worker drops a full paint can from a height of 500 feet. How fast will the paint can be falling after 2 seconds?

102. A construction worker drops a full paint can from a height of 500 feet. When will the paint can hit the ground? At what velocity will the paint can impact the ground?

Free-Falling Object **In Exercises 103 and 104, use the position function $s(t) = -4.9t^2 + 200$, which gives the height (in meters) of an object that has fallen for t seconds from a height of 200 meters. The velocity at time $t = a$ seconds is given by**

$$\lim_{t \to a} \frac{s(a) - s(t)}{a - t}.$$

103. Find the velocity of the object when $t = 3$.

104. At what velocity will the object impact the ground?

105. Finding Functions Find two functions f and g such that $\lim_{x \to 0} f(x)$ and $\lim_{x \to 0} g(x)$ do not exist, but

$$\lim_{x \to 0} [f(x) + g(x)]$$

does exist.

106. Proof Prove that if $\lim_{x \to c} f(x)$ exists and $\lim_{x \to c} [f(x) + g(x)]$ does not exist, then $\lim_{x \to c} g(x)$ does not exist.

107. Proof Prove Property 1 of Theorem 1.1.

108. Proof Prove Property 3 of Theorem 1.1. (You may use Property 3 of Theorem 1.2.)

109. Proof Prove Property 1 of Theorem 1.2.

110. Proof Prove that if $\lim_{x \to c} f(x) = 0$, then $\lim_{x \to c} |f(x)| = 0$.

111. Proof Prove that if $\lim_{x \to c} f(x) = 0$ and $|g(x)| \le M$ for a fixed number M and all $x \ne c$, then $\lim_{x \to c} [f(x)g(x)] = 0$.

112. Proof

(a) Prove that if $\lim_{x \to c} |f(x)| = 0$, then $\lim_{x \to c} f(x) = 0$.

(*Note:* This is the converse of Exercise 110.)

(b) Prove that if $\lim_{x \to c} f(x) = L$, then $\lim_{x \to c} |f(x)| = |L|$.

[*Hint:* Use the inequality $\| f(x)| - |L\| \le |f(x) - L|$.]

Kevin Fleming/Documentary Value/Corbis

113. Think About It Find a function f to show that the converse of Exercise 112(b) is not true. [*Hint:* Find a function f such that $\lim_{x \to c} |f(x)| = |L|$ but $\lim_{x \to c} f(x)$ does not exist.]

114. Think About It When using a graphing utility to generate a table to approximate

$$\lim_{x \to 0} \frac{\sin x}{x}$$

a student concluded that the limit was 0.01745 rather than 1. Determine the probable cause of the error.

True or False? **In Exercises 115–120, determine whether the statement is true or false. If it is false, explain why or give an example that shows it is false.**

115. $\lim_{x \to 0} \frac{|x|}{x} = 1$

116. $\lim_{x \to \pi} \frac{\sin x}{x} = 1$

117. If $f(x) = g(x)$ for all real numbers other than $x = 0$ and $\lim_{x \to 0} f(x) = L$, then $\lim_{x \to 0} g(x) = L$.

118. If $\lim_{x \to c} f(x) = L$, then $f(c) = L$.

119. $\lim_{x \to 2} f(x) = 3$, where $f(x) = \begin{cases} 3, & x \le 2 \\ 0, & x > 2 \end{cases}$

120. If $f(x) < g(x)$ for all $x \ne a$, then $\lim_{x \to a} f(x) < \lim_{x \to a} g(x)$.

121. Proof Prove the second part of Theorem 1.9.

$$\lim_{x \to 0} \frac{1 - \cos x}{x} = 0$$

122. Piecewise Functions Let

$$f(x) = \begin{cases} 0, & \text{if } x \text{ is rational} \\ 1, & \text{if } x \text{ is irrational} \end{cases}$$

and

$$g(x) = \begin{cases} 0, & \text{if } x \text{ is rational} \\ x, & \text{if } x \text{ is irrational} \end{cases}.$$

Find (if possible) $\lim_{x \to 0} f(x)$ and $\lim_{x \to 0} g(x)$.

123. Graphical Reasoning Consider $f(x) = \dfrac{\sec x - 1}{x^2}$.

(a) Find the domain of f.

(b) Use a graphing utility to graph f. Is the domain of f obvious from the graph? If not, explain.

(c) Use the graph of f to approximate $\lim_{x \to 0} f(x)$.

(d) Confirm your answer to part (c) analytically.

124. Approximation

(a) Find $\lim_{x \to 0} \dfrac{1 - \cos x}{x^2}$.

(b) Use your answer to part (a) to derive the approximation $\cos x \approx 1 - \frac{1}{2}x^2$ for x near 0.

(c) Use your answer to part (b) to approximate $\cos(0.1)$.

(d) Use a calculator to approximate $\cos(0.1)$ to four decimal places. Compare the result with part (c).

1.4 Continuity and One-Sided Limits

- **Determine continuity at a point and continuity on an open interval.**
- **Determine one-sided limits and continuity on a closed interval.**
- **Use properties of continuity.**
- **Understand and use the Intermediate Value Theorem.**

Continuity at a Point and on an Open Interval

In mathematics, the term *continuous* has much the same meaning as it has in everyday usage. Informally, to say that a function f is continuous at $x = c$ means that there is no interruption in the graph of f at c. That is, its graph is unbroken at c, and there are no holes, jumps, or gaps. Figure 1.26 identifies three values of x at which the graph of f is *not* continuous. At all other points in the interval (a, b), the graph of f is uninterrupted and **continuous.**

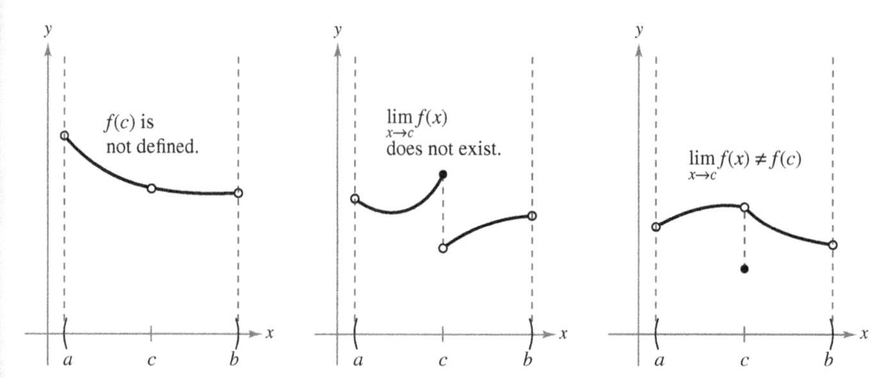

Three conditions exist for which the graph of f is not continuous at $x = c$.
Figure 1.26

Exploration

Informally, you might say that a function is *continuous* on an open interval when its graph can be drawn with a pencil without lifting the pencil from the paper. Use a graphing utility to graph each function on the given interval. From the graphs, which functions would you say are continuous on the interval? Do you think you can trust the results you obtained graphically? Explain your reasoning.

	Function	Interval
a.	$y = x^2 + 1$	$(-3, 3)$
b.	$y = \dfrac{1}{x-2}$	$(-3, 3)$
c.	$y = \dfrac{\sin x}{x}$	$(-\pi, \pi)$
d.	$y = \dfrac{x^2-4}{x+2}$	$(-3, 3)$

In Figure 1.26, it appears that continuity at $x = c$ can be destroyed by any one of three conditions.

1. The function is not defined at $x = c$.
2. The limit of $f(x)$ does not exist at $x = c$.
3. The limit of $f(x)$ exists at $x = c$, but it is not equal to $f(c)$.

If *none* of the three conditions is true, then the function f is called **continuous at *c*,** as indicated in the important definition below.

FOR FURTHER INFORMATION For more information on the concept of continuity, see the article "Leibniz and the Spell of the Continuous" by Hardy Grant in *The College Mathematics Journal*. To view this article, go to *MathArticles.com*.

Definition of Continuity

Continuity at a Point
A function f is **continuous at *c*** when these three conditions are met.

1. $f(c)$ is defined.
2. $\lim_{x \to c} f(x)$ exists.
3. $\lim_{x \to c} f(x) = f(c)$

Continuity on an Open Interval
A function is **continuous on an open interval (a, b)** when the function is continuous at each point in the interval. A function that is continuous on the entire real number line $(-\infty, \infty)$ is **everywhere continuous.**

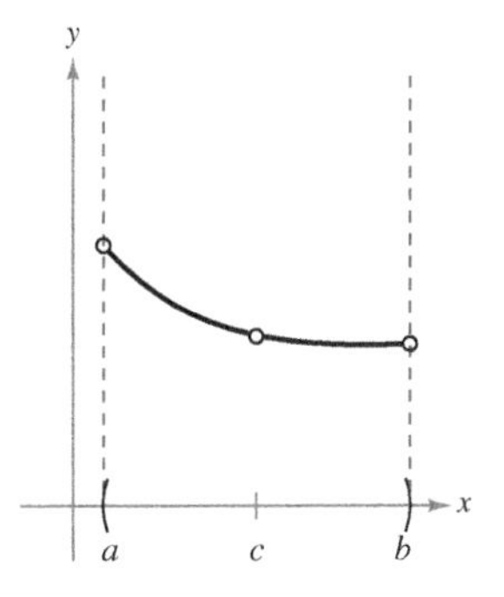

(a) Removable discontinuity

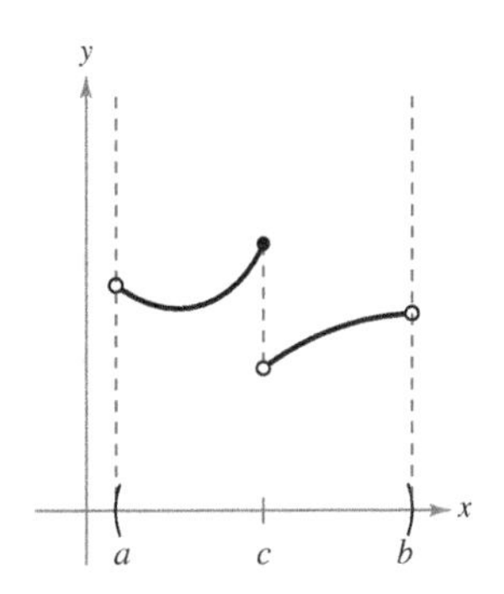

(b) Nonremovable discontinuity

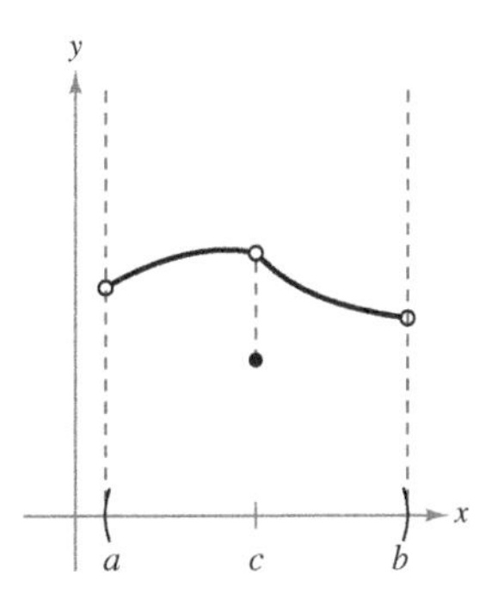

(c) Removable discontinuity
Figure 1.27

Consider an open interval I that contains a real number c. If a function f is defined on I (except possibly at c), and f is not continuous at c, then f is said to have a **discontinuity** at c. Discontinuities fall into two categories: **removable** and **nonremovable**. A discontinuity at c is called removable when f can be made continuous by appropriately defining (or redefining) $f(c)$. For instance, the functions shown in Figures 1.27(a) and (c) have removable discontinuities at c and the function shown in Figure 1.27(b) has a nonremovable discontinuity at c.

EXAMPLE 1 Continuity of a Function

Discuss the continuity of each function.

a. $f(x) = \dfrac{1}{x}$ **b.** $g(x) = \dfrac{x^2 - 1}{x - 1}$ **c.** $h(x) = \begin{cases} x + 1, & x \le 0 \\ x^2 + 1, & x > 0 \end{cases}$ **d.** $y = \sin x$

Solution

a. The domain of f is all nonzero real numbers. From Theorem 1.3, you can conclude that f is continuous at every x-value in its domain. At $x = 0$, f has a nonremovable discontinuity, as shown in Figure 1.28(a). In other words, there is no way to define $f(0)$ so as to make the function continuous at $x = 0$.

b. The domain of g is all real numbers except $x = 1$. From Theorem 1.3, you can conclude that g is continuous at every x-value in its domain. At $x = 1$, the function has a removable discontinuity, as shown in Figure 1.28(b). By defining $g(1)$ as 2, the "redefined" function is continuous for all real numbers.

c. The domain of h is all real numbers. The function h is continuous on $(-\infty, 0)$ and $(0, \infty)$, and because

$$\lim_{x \to 0} h(x) = 1$$

h is continuous on the entire real number line, as shown in Figure 1.28(c).

d. The domain of y is all real numbers. From Theorem 1.6, you can conclude that the function is continuous on its entire domain, $(-\infty, \infty)$, as shown in Figure 1.28(d).

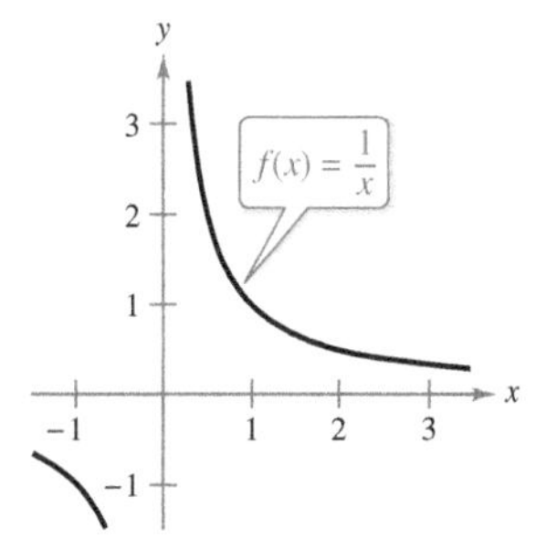

(a) Nonremovable discontinuity at $x = 0$

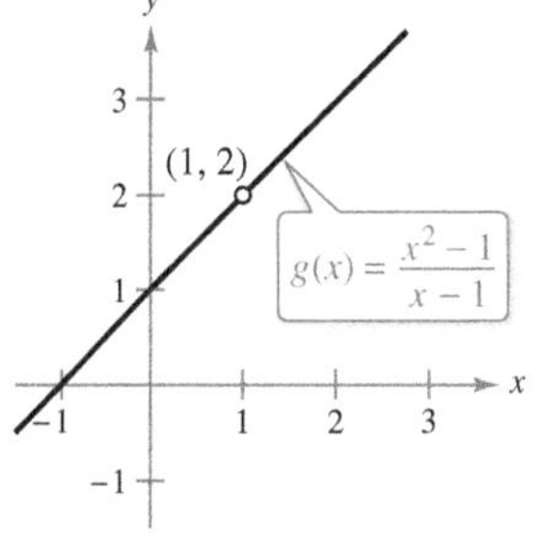

(b) Removable discontinuity at $x = 1$

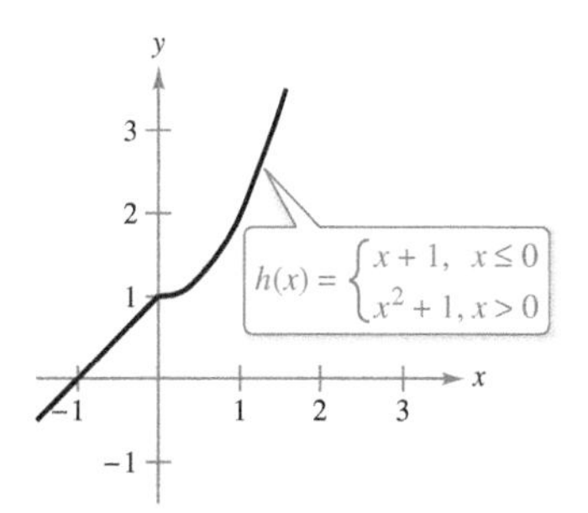

(c) Continuous on entire real number line

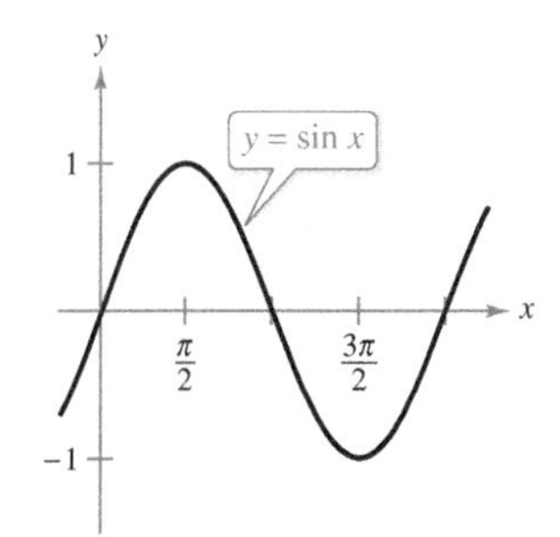

(d) Continuous on entire real number line

Figure 1.28

REMARK Some people may refer to the function in Example 1(a) as "discontinuous," but this terminology can be confusing. Rather than saying that the function is discontinuous, it is more precise to say that the function has a discontinuity at $x = 0$.

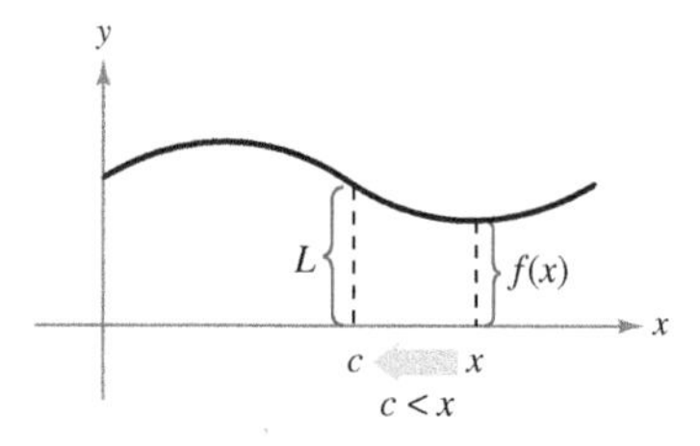

(a) Limit as x approaches c from the right.

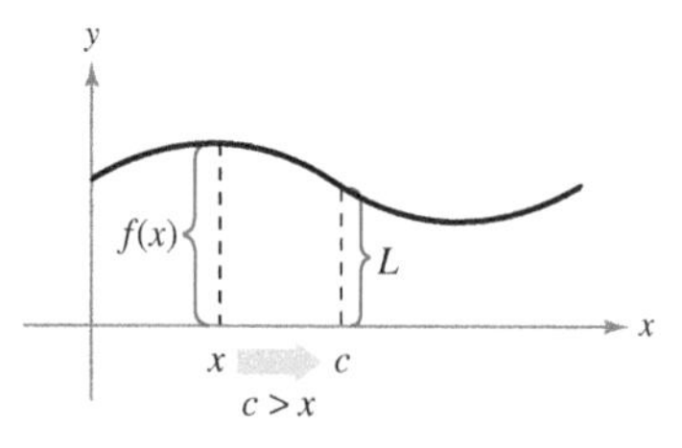

(b) Limit as x approaches c from the left.
Figure 1.29

One-Sided Limits and Continuity on a Closed Interval

To understand continuity on a closed interval, you first need to look at a different type of limit called a **one-sided limit.** For instance, the **limit from the right** (or right-hand limit) means that x approaches c from values greater than c [see Figure 1.29(a)]. This limit is denoted as

$$\lim_{x \to c^+} f(x) = L. \qquad \text{Limit from the right}$$

Similarly, the **limit from the left** (or left-hand limit) means that x approaches c from values less than c [see Figure 1.29(b)]. This limit is denoted as

$$\lim_{x \to c^-} f(x) = L. \qquad \text{Limit from the left}$$

One-sided limits are useful in taking limits of functions involving radicals. For instance, if n is an even integer, then

$$\lim_{x \to 0^+} \sqrt[n]{x} = 0.$$

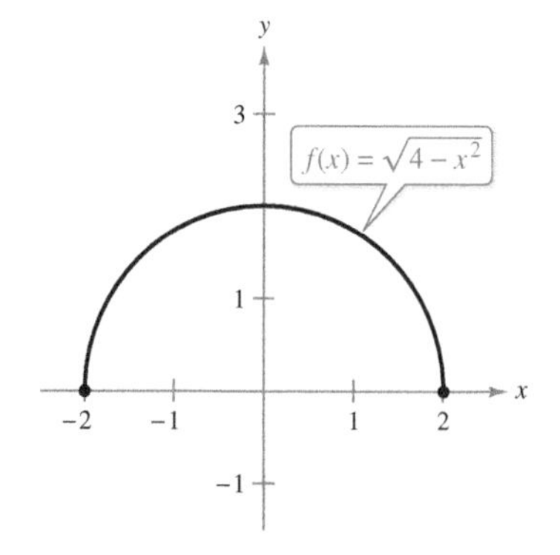

The limit of $f(x)$ as x approaches -2 from the right is 0.
Figure 1.30

EXAMPLE 2 A One-Sided Limit

Find the limit of $f(x) = \sqrt{4 - x^2}$ as x approaches -2 from the right.

Solution As shown in Figure 1.30, the limit as x approaches -2 from the right is

$$\lim_{x \to -2^+} \sqrt{4 - x^2} = 0.$$

One-sided limits can be used to investigate the behavior of **step functions.** One common type of step function is the **greatest integer function** $[\![x]\!]$, defined as

$$[\![x]\!] = \text{greatest integer } n \text{ such that } n \le x. \qquad \text{Greatest integer function}$$

For instance, $[\![2.5]\!] = 2$ and $[\![-2.5]\!] = -3$.

EXAMPLE 3 The Greatest Integer Function

Find the limit of the greatest integer function $f(x) = [\![x]\!]$ as x approaches 0 from the left and from the right.

Solution As shown in Figure 1.31, the limit as x approaches 0 *from the left* is

$$\lim_{x \to 0^-} [\![x]\!] = -1$$

and the limit as x approaches 0 *from the right* is

$$\lim_{x \to 0^+} [\![x]\!] = 0.$$

So, f has a discontinuity at zero because the left- and right-hand limits at zero are different. By similar reasoning, you can see that the greatest integer function has a discontinuity at any integer n.

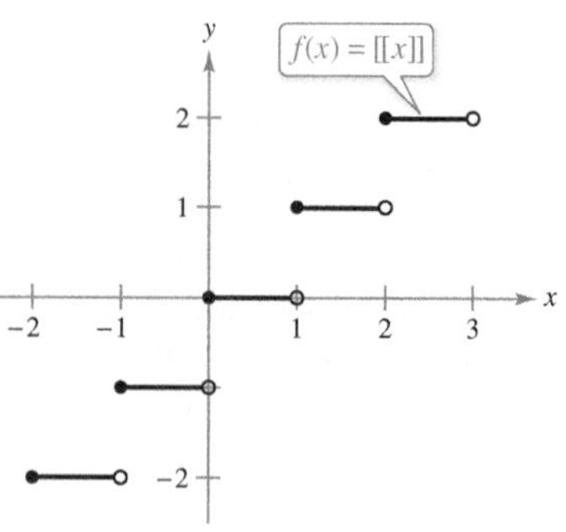

Greatest integer function
Figure 1.31

When the limit from the left is not equal to the limit from the right, the (two-sided) limit *does not exist*. The next theorem makes this more explicit. The proof of this theorem follows directly from the definition of a one-sided limit.

> **THEOREM 1.10 The Existence of a Limit**
>
> Let f be a function, and let c and L be real numbers. The limit of $f(x)$ as x approaches c is L if and only if
>
> $$\lim_{x \to c^-} f(x) = L \quad \text{and} \quad \lim_{x \to c^+} f(x) = L.$$

The concept of a one-sided limit allows you to extend the definition of continuity to closed intervals. Basically, a function is continuous on a closed interval when it is continuous in the interior of the interval and exhibits one-sided continuity at the endpoints. This is stated formally in the next definition.

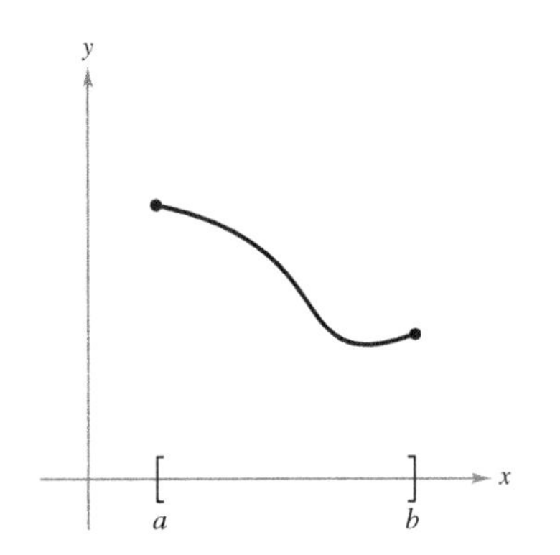

Continuous function on a closed interval
Figure 1.32

> **Definition of Continuity on a Closed Interval**
>
> A function f is **continuous on the closed interval** $[a, b]$ when f is continuous on the open interval (a, b) and
>
> $$\lim_{x \to a^+} f(x) = f(a)$$
>
> and
>
> $$\lim_{x \to b^-} f(x) = f(b).$$
>
> The function f is **continuous from the right** at a and **continuous from the left** at b (see Figure 1.32).

Similar definitions can be made to cover continuity on intervals of the form $(a, b]$ and $[a, b)$ that are neither open nor closed, or on infinite intervals. For example,

$$f(x) = \sqrt{x}$$

is continuous on the infinite interval $[0, \infty)$, and the function

$$g(x) = \sqrt{2 - x}$$

is continuous on the infinite interval $(-\infty, 2]$.

EXAMPLE 4 Continuity on a Closed Interval

Discuss the continuity of

$$f(x) = \sqrt{1 - x^2}.$$

Solution The domain of f is the closed interval $[-1, 1]$. At all points in the open interval $(-1, 1)$, the continuity of f follows from Theorems 1.4 and 1.5. Moreover, because

$$\lim_{x \to -1^+} \sqrt{1 - x^2} = 0 = f(-1) \qquad \text{Continuous from the right}$$

and

$$\lim_{x \to 1^-} \sqrt{1 - x^2} = 0 = f(1) \qquad \text{Continuous from the left}$$

you can conclude that f is continuous on the closed interval $[-1, 1]$, as shown in Figure 1.33.

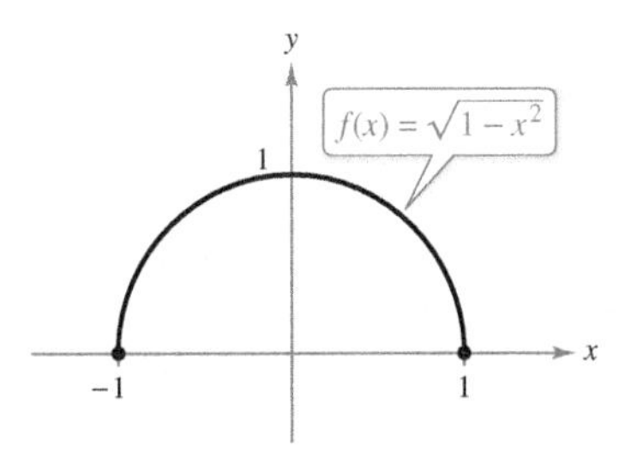

f is continuous on $[-1, 1]$.
Figure 1.33

The next example shows how a one-sided limit can be used to determine the value of absolute zero on the Kelvin scale.

REMARK Charles's Law for gases (assuming constant pressure) can be stated as

$$V = kT$$

where V is volume, k is a constant, and T is temperature.

EXAMPLE 5 Charles's Law and Absolute Zero

On the Kelvin scale, *absolute zero* is the temperature 0 K. Although temperatures very close to 0 K have been produced in laboratories, absolute zero has never been attained. In fact, evidence suggests that absolute zero *cannot* be attained. How did scientists determine that 0 K is the "lower limit" of the temperature of matter? What is absolute zero on the Celsius scale?

Solution The determination of absolute zero stems from the work of the French physicist Jacques Charles (1746–1823). Charles discovered that the volume of gas at a constant pressure increases linearly with the temperature of the gas. The table illustrates this relationship between volume and temperature. To generate the values in the table, one mole of hydrogen is held at a constant pressure of one atmosphere. The volume V is approximated and is measured in liters, and the temperature T is measured in degrees Celsius.

T	-40	-20	0	20	40	60	80
V	19.1482	20.7908	22.4334	24.0760	25.7186	27.3612	29.0038

The points represented by the table are shown in the figure at the right. Moreover, by using the points in the table, you can determine that T and V are related by the linear equation

$$V = 0.08213T + 22.4334.$$

Solving for T, you get an equation for the temperature of the gas.

$$T = \frac{V - 22.4334}{0.08213}$$

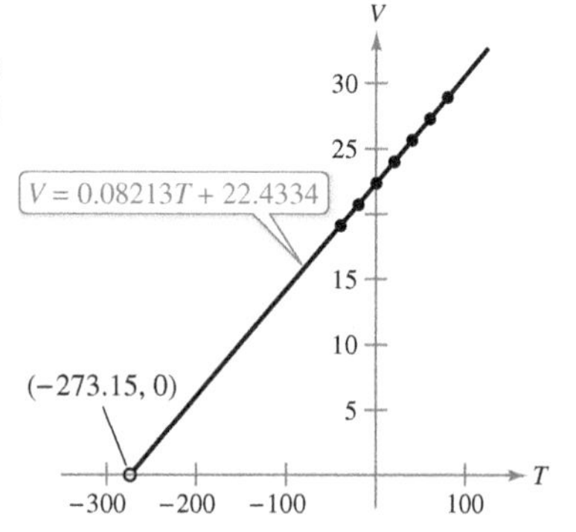

The volume of hydrogen gas depends on its temperature.

By reasoning that the volume of the gas can approach 0 (but can never equal or go below 0), you can determine that the "least possible temperature" is

$$\lim_{V \to 0^+} T = \lim_{V \to 0^+} \frac{V - 22.4334}{0.08213}$$

$$= \frac{0 - 22.4334}{0.08213} \qquad \text{Use direct substitution.}$$

$$\approx -273.15.$$

So, absolute zero on the Kelvin scale (0 K) is approximately $-273.15°$ on the Celsius scale. ■

Liquid helium is used to cool superconducting magnets, such as those used in magnetic resonance imaging (MRI) machines or in the Large Hadron Collider (see above). The magnets are made with materials that only superconduct at temperatures a few degrees above absolute zero. These temperatures are possible with liquid helium because helium becomes a liquid at $-269°$C, or 4.15 K.

The table below shows the temperatures in Example 5 converted to the Fahrenheit scale. Try repeating the solution shown in Example 5 using these temperatures and volumes. Use the result to find the value of absolute zero on the Fahrenheit scale.

T	-40	-4	32	68	104	140	176
V	19.1482	20.7908	22.4334	24.0760	25.7186	27.3612	29.0038

Properties of Continuity

In Section 1.3, you studied several properties of limits. Each of those properties yields a corresponding property pertaining to the continuity of a function. For instance, Theorem 1.11 follows directly from Theorem 1.2.

AUGUSTIN-LOUIS CAUCHY (1789–1857)

The concept of a continuous function was first introduced by Augustin-Louis Cauchy in 1821. The definition given in his text *Cours d'Analyse* stated that indefinite small changes in y were the result of indefinite small changes in x. "$\ldots f(x)$ will be called a *continuous* function if ... the numerical values of the difference $f(x + \alpha) - f(x)$ decrease indefinitely with those of $\alpha \ldots$."

See LarsonCalculus.com to read more of this biography.

THEOREM 1.11 Properties of Continuity

If b is a real number and f and g are continuous at $x = c$, then the functions listed below are also continuous at c.

1. Scalar multiple: bf
2. Sum or difference: $f \pm g$
3. Product: fg
4. Quotient: $\dfrac{f}{g}$, $g(c) \neq 0$

A proof of this theorem is given in Appendix A.

It is important for you to be able to recognize functions that are continuous at every point in their domains. The list below summarizes the functions you have studied so far that are continuous at every point in their domains.

1. Polynomial: $p(x) = a_n x^n + a_{n-1} x^{n-1} + \cdots + a_1 x + a_0$
2. Rational: $r(x) = \dfrac{p(x)}{q(x)}$, $q(x) \neq 0$
3. Radical: $f(x) = \sqrt[n]{x}$
4. Trigonometric: $\sin x$, $\cos x$, $\tan x$, $\cot x$, $\sec x$, $\csc x$

By combining Theorem 1.11 with this list, you can conclude that a wide variety of elementary functions are continuous at every point in their domains.

EXAMPLE 6 Applying Properties of Continuity

See LarsonCalculus.com for an interactive version of this type of example.

By Theorem 1.11, it follows that each of the functions below is continuous at every point in its domain.

$$f(x) = x + \sin x, \quad f(x) = 3 \tan x, \quad f(x) = \frac{x^2 + 1}{\cos x}$$

The next theorem, which is a consequence of Theorem 1.5, allows you to determine the continuity of *composite* functions such as

$$f(x) = \sin 3x, \quad f(x) = \sqrt{x^2 + 1}, \quad \text{and} \quad f(x) = \tan \frac{1}{x}.$$

REMARK One consequence of Theorem 1.12 is that when f and g satisfy the given conditions, you can determine the limit of $f(g(x))$ as x approaches c to be

$$\lim_{x \to c} f(g(x)) = f(g(c)).$$

THEOREM 1.12 Continuity of a Composite Function

If g is continuous at c and f is continuous at $g(c)$, then the composite function given by $(f \circ g)(x) = f(g(x))$ is continuous at c.

Proof By the definition of continuity, $\lim_{x \to c} g(x) = g(c)$ and $\lim_{x \to g(c)} f(x) = f(g(c))$. Apply Theorem 1.5 with $L = g(c)$ to obtain $\lim_{x \to c} f(g(x)) = f\left(\lim_{x \to c} g(x)\right) = f(g(c))$. So, $(f \circ g)(x) = f(g(x))$ is continuous at c. ■

AS400 DB/Bettmann/Corbis

EXAMPLE 7 Testing for Continuity

Describe the interval(s) on which each function is continuous.

a. $f(x) = \tan x$ **b.** $g(x) = \begin{cases} \sin \dfrac{1}{x}, & x \neq 0 \\ 0, & x = 0 \end{cases}$ **c.** $h(x) = \begin{cases} x \sin \dfrac{1}{x}, & x \neq 0 \\ 0, & x = 0 \end{cases}$

Solution

a. The tangent function $f(x) = \tan x$ is undefined at

$$x = \frac{\pi}{2} + n\pi, \quad n \text{ is an integer.}$$

At all other points, f is continuous. So, $f(x) = \tan x$ is continuous on the open intervals

$$\ldots, \left(-\frac{3\pi}{2}, -\frac{\pi}{2}\right), \left(-\frac{\pi}{2}, \frac{\pi}{2}\right), \left(\frac{\pi}{2}, \frac{3\pi}{2}\right), \ldots$$

as shown in Figure 1.34(a).

b. Because $y = 1/x$ is continuous except at $x = 0$ and the sine function is continuous for all real values of x, it follows from Theorem 1.12 that

$$y = \sin \frac{1}{x}$$

is continuous at all real values except $x = 0$. At $x = 0$, the limit of $g(x)$ does not exist (see Example 5, Section 1.2). So, g is continuous on the intervals $(-\infty, 0)$ and $(0, \infty)$, as shown in Figure 1.34(b).

c. This function is similar to the function in part (b) except that the oscillations are damped by the factor x. Using the Squeeze Theorem, you obtain

$$-|x| \leq x \sin \frac{1}{x} \leq |x|, \quad x \neq 0$$

and you can conclude that

$$\lim_{x \to 0} h(x) = 0.$$

So, h is continuous on the entire real number line, as shown in Figure 1.34(c).

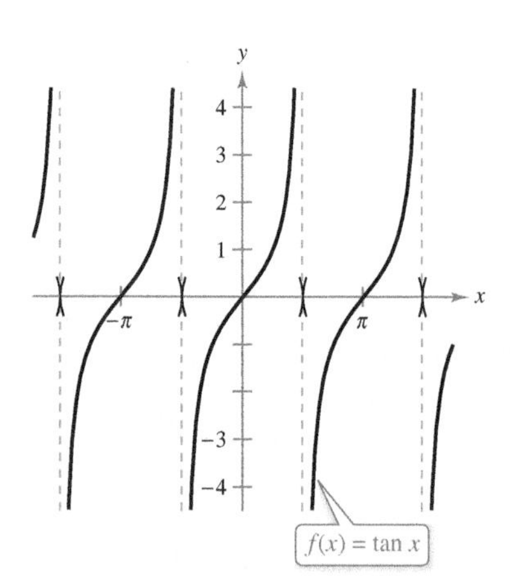

(a) f is continuous on each open interval in its domain.

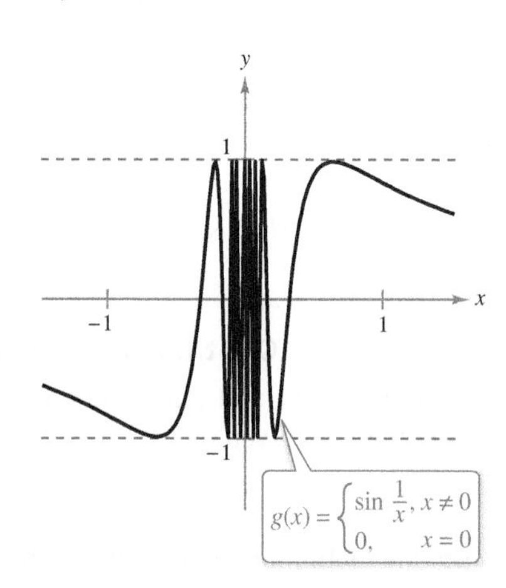

(b) g is continuous on $(-\infty, 0)$ and $(0, \infty)$.

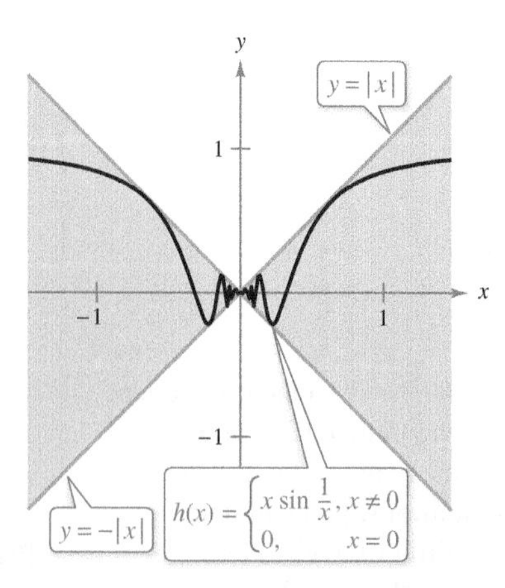

(c) h is continuous on the entire real number line.

Figure 1.34

The Intermediate Value Theorem

Theorem 1.13 is an important theorem concerning the behavior of functions that are continuous on a closed interval.

> **THEOREM 1.13 Intermediate Value Theorem**
>
> If f is continuous on the closed interval $[a, b]$, $f(a) \neq f(b)$, and k is any number between $f(a)$ and $f(b)$, then there is at least one number c in $[a, b]$ such that
>
> $$f(c) = k.$$

REMARK The Intermediate Value Theorem tells you that at least one number c exists, but it does not provide a method for finding c. Such theorems are called **existence theorems.** By referring to a text on advanced calculus, you will find that a proof of this theorem is based on a property of real numbers called *completeness*. The Intermediate Value Theorem states that for a continuous function f, if x takes on all values between a and b, then $f(x)$ must take on all values between $f(a)$ and $f(b)$.

As an example of the application of the Intermediate Value Theorem, consider a person's height. A girl is 5 feet tall on her thirteenth birthday and 5 feet 2 inches tall on her fourteenth birthday. Then, for any height h between 5 feet and 5 feet 2 inches, there must have been a time t when her height was exactly h. This seems reasonable because human growth is continuous and a person's height does not abruptly change from one value to another.

The Intermediate Value Theorem guarantees the existence of *at least one* number c in the closed interval $[a, b]$. There may, of course, be more than one number c such that

$$f(c) = k$$

as shown in Figure 1.35. A function that is not continuous does not necessarily exhibit the intermediate value property. For example, the graph of the function shown in Figure 1.36 jumps over the horizontal line

$$y = k$$

and for this function there is no value of c in $[a, b]$ such that $f(c) = k$.

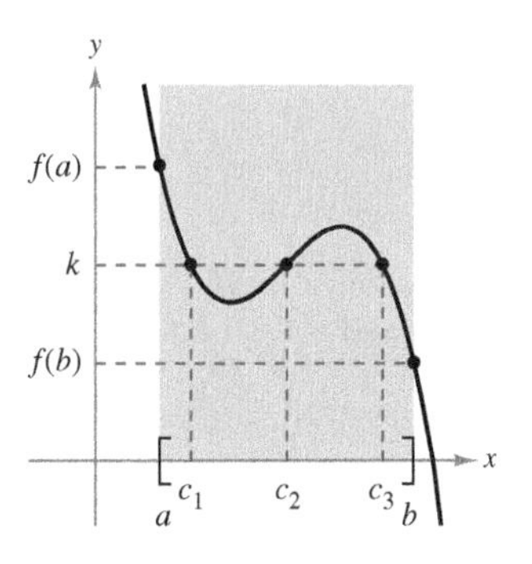

f is continuous on $[a, b]$.
[There exist three c's such that $f(c) = k$.]
Figure 1.35

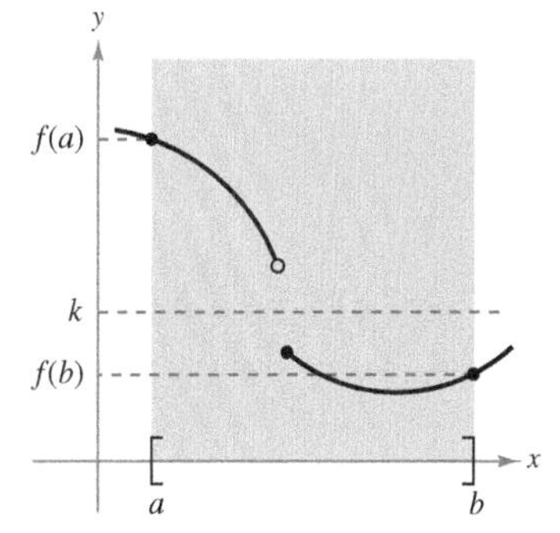

f is not continuous on $[a, b]$.
[There are no c's such that $f(c) = k$.]
Figure 1.36

The Intermediate Value Theorem often can be used to locate the zeros of a function that is continuous on a closed interval. Specifically, if f is continuous on $[a, b]$ and $f(a)$ and $f(b)$ differ in sign, then the Intermediate Value Theorem guarantees the existence of at least one zero of f in the closed interval $[a, b]$.

EXAMPLE 8 An Application of the Intermediate Value Theorem

Use the Intermediate Value Theorem to show that the polynomial function

$$f(x) = x^3 + 2x - 1$$

has a zero in the interval $[0, 1]$.

Solution Note that f is continuous on the closed interval $[0, 1]$. Because

$$f(0) = 0^3 + 2(0) - 1 = -1 \quad \text{and} \quad f(1) = 1^3 + 2(1) - 1 = 2$$

it follows that $f(0) < 0$ and $f(1) > 0$. You can therefore apply the Intermediate Value Theorem to conclude that there must be some c in $[0, 1]$ such that

$$f(c) = 0 \qquad f \text{ has a zero in the closed interval } [0, 1].$$

as shown in Figure 1.37.

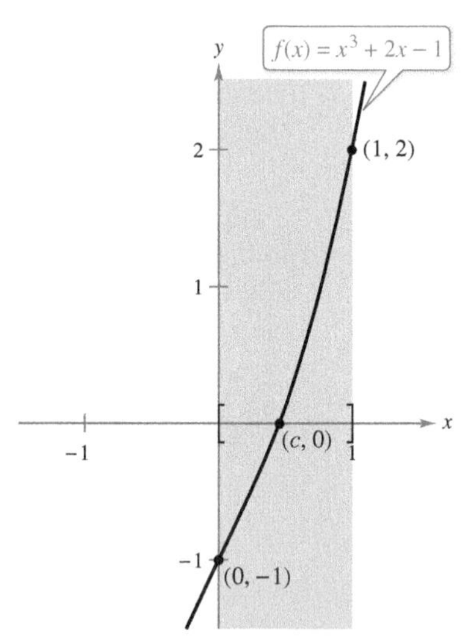

f is continuous on $[0, 1]$ with $f(0) < 0$ and $f(1) > 0$.
Figure 1.37

The **bisection method** for approximating the real zeros of a continuous function is similar to the method used in Example 8. If you know that a zero exists in the closed interval $[a, b]$, then the zero must lie in the interval $[a, (a + b)/2]$ or $[(a + b)/2, b]$. From the sign of $f([a + b]/2)$, you can determine which interval contains the zero. By repeatedly bisecting the interval, you can "close in" on the zero of the function.

TECHNOLOGY You can use the *root* or *zero* feature of a graphing utility to approximate the real zeros of a continuous function. Using this feature, the zero of the function in Example 8, $f(x) = x^3 + 2x - 1$, is approximately 0.453, as shown in the figure.

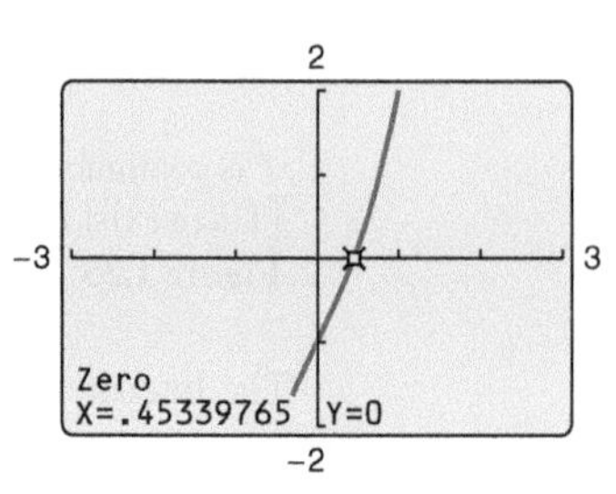

Zero of $f(x) = x^3 + 2x - 1$

1.4 Exercises

See CalcChat.com for tutorial help and worked-out solutions to odd-numbered exercises.

CONCEPT CHECK

1. **Continuity** In your own words, describe what it means for a function to be continuous at a point.
2. **One-Sided Limits** What is the value of c?

$$\lim_{x \to c^+} 2\sqrt{x+1} = 0$$

3. **Existence of a Limit** Determine whether $\lim_{x \to 3} f(x)$ exists. Explain.

$$\lim_{x \to 3^-} f(x) = 1 \quad \text{and} \quad \lim_{x \to 3^+} f(x) = 1$$

4. **Intermediate Value Theorem** In your own words, explain the Intermediate Value Theorem.

Limits and Continuity **In Exercises 5–10, use the graph to determine each limit, and discuss the continuity of the function.**

(a) $\lim_{x \to c^+} f(x)$ (b) $\lim_{x \to c^-} f(x)$ (c) $\lim_{x \to c} f(x)$

5.

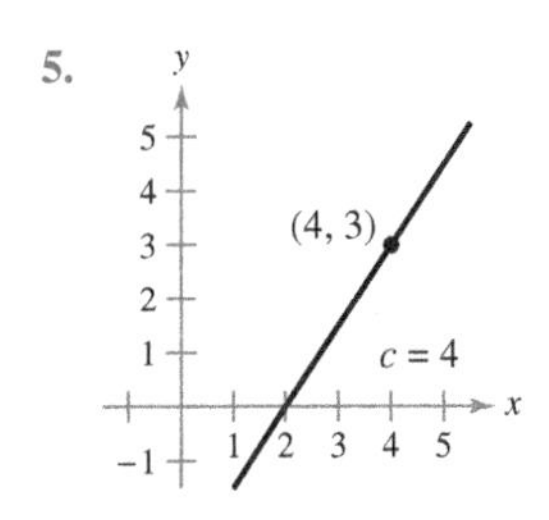

6.

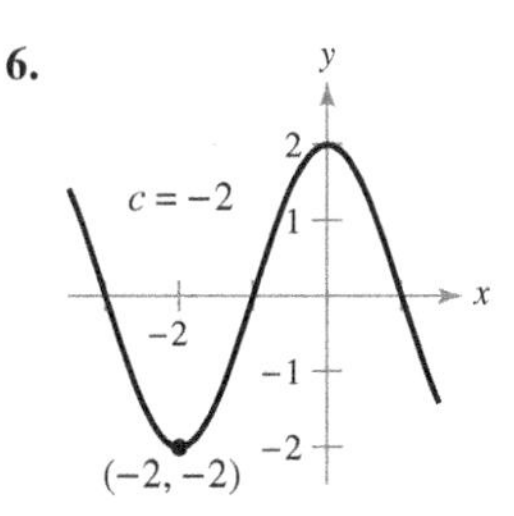

7.

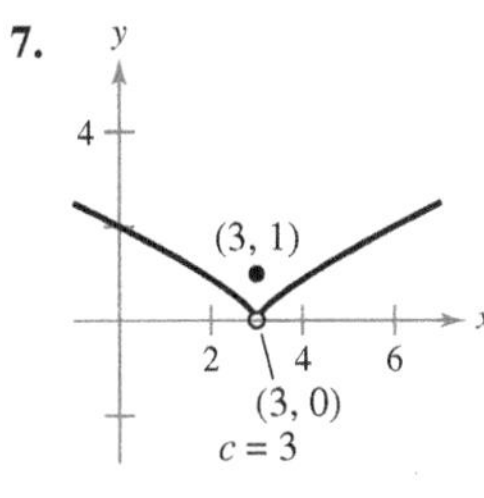

8.

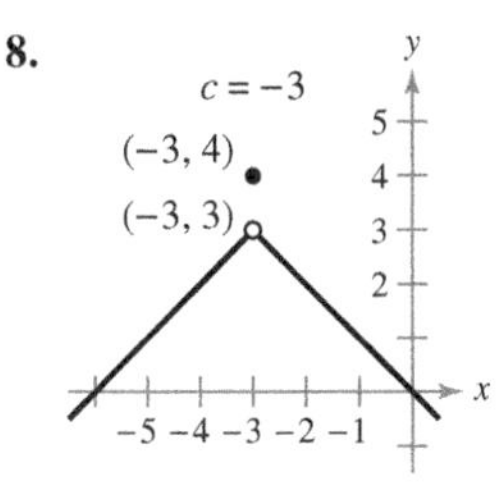

9.

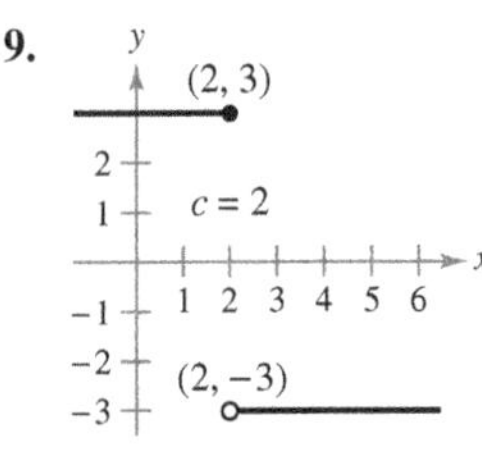

10.

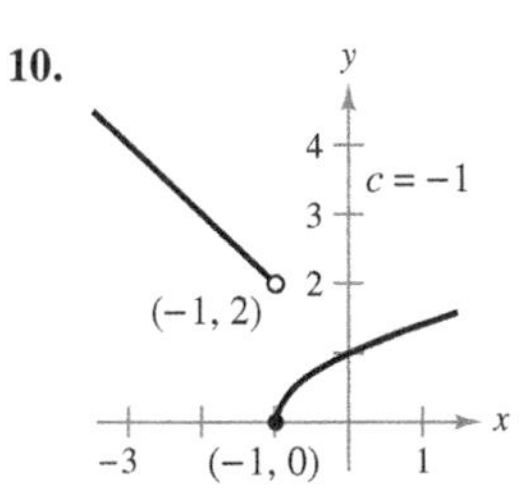

Finding a Limit **In Exercises 11–30, find the limit (if it exists). If it does not exist, explain why.**

11. $\lim_{x \to 8^+} \dfrac{1}{x+8}$

12. $\lim_{x \to 3^+} \dfrac{2}{x+3}$

13. $\lim_{x \to 5^+} \dfrac{x-5}{x^2-25}$

14. $\lim_{x \to 4^+} \dfrac{4-x}{x^2-16}$

15. $\lim_{x \to -3^-} \dfrac{x}{\sqrt{x^2-9}}$

16. $\lim_{x \to 4^-} \dfrac{\sqrt{x}-2}{x-4}$

17. $\lim_{x \to 0^-} \dfrac{|x|}{x}$

18. $\lim_{x \to 10^+} \dfrac{|x-10|}{x-10}$

19. $\lim_{\Delta x \to 0^-} \dfrac{\dfrac{1}{x+\Delta x} - \dfrac{1}{x}}{\Delta x}$

20. $\lim_{\Delta x \to 0^+} \dfrac{(x+\Delta x)^2 + x + \Delta x - (x^2+x)}{\Delta x}$

21. $\lim_{x \to 3^-} f(x)$, where $f(x) = \begin{cases} \dfrac{x+2}{2}, & x < 3 \\ \dfrac{12-2x}{3}, & x > 3 \end{cases}$

22. $\lim_{x \to 3} f(x)$, where $f(x) = \begin{cases} x^2 - 4x + 6, & x < 3 \\ -x^2 + 4x - 2, & x \geq 3 \end{cases}$

23. $\lim_{x \to 1} f(x)$, where $f(x) = \begin{cases} x^3 + 1, & x < 1 \\ x + 1, & x \geq 1 \end{cases}$

24. $\lim_{x \to 1^+} f(x)$, where $f(x) = \begin{cases} x, & x \leq 1 \\ 1 - x, & x > 1 \end{cases}$

25. $\lim_{x \to \pi} \cot x$

26. $\lim_{x \to \pi/2} \sec x$

27. $\lim_{x \to 4^-} (5[\![x]\!] - 7)$

28. $\lim_{x \to 2^+} (2x - [\![x]\!])$

29. $\lim_{x \to -1} \left(\left[\!\left[\dfrac{x}{3} \right]\!\right] + 3 \right)$

30. $\lim_{x \to 1} \left(1 - \left[\!\left[-\dfrac{x}{2} \right]\!\right] \right)$

Continuity of a Function **In Exercises 31–34, discuss the continuity of the function.**

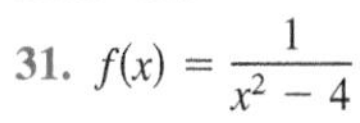

31. $f(x) = \dfrac{1}{x^2-4}$

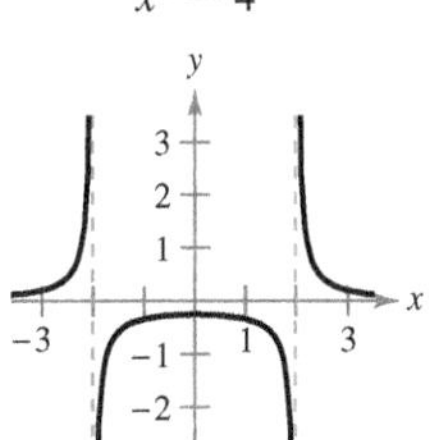

32. $f(x) = \dfrac{x^2-1}{x+1}$

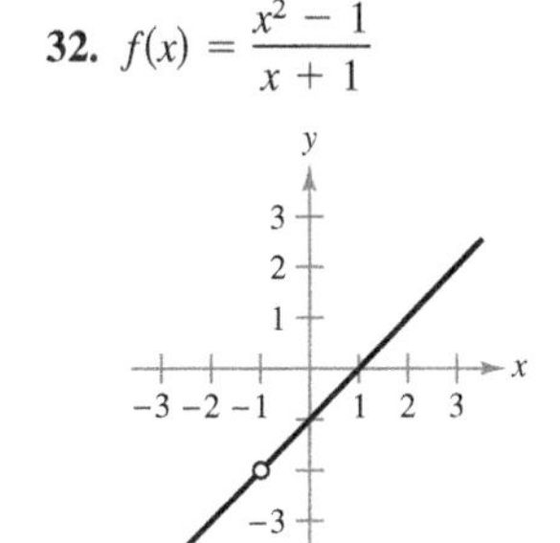

33. $f(x) = \frac{1}{2}[\![x]\!] + x$

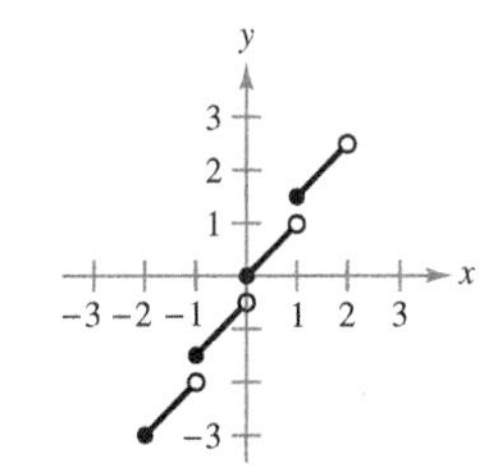

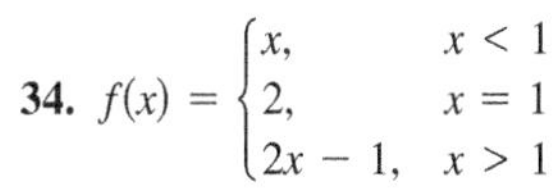

34. $f(x) = \begin{cases} x, & x < 1 \\ 2, & x = 1 \\ 2x - 1, & x > 1 \end{cases}$

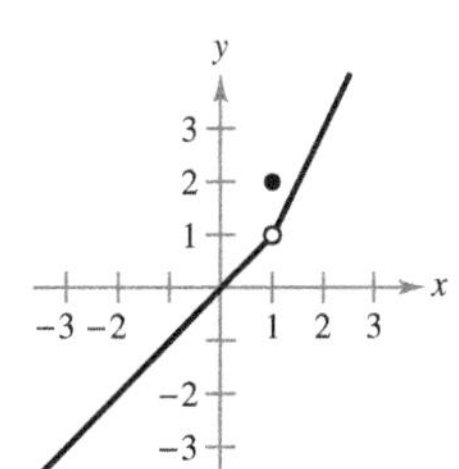

Continuity on a Closed Interval In Exercises 35–38, discuss the continuity of the function on the closed interval.

Function	Interval
35. $g(x) = \sqrt{49 - x^2}$	$[-7, 7]$
36. $f(t) = 3 - \sqrt{9 - t^2}$	$[-3, 3]$
37. $f(x) = \begin{cases} 3 - x, & x \le 0 \\ 3 + \frac{1}{2}x, & x > 0 \end{cases}$	$[-1, 4]$
38. $g(x) = \dfrac{1}{x^2 - 4}$	$[-1, 2]$

Removable and Nonremovable Discontinuities In Exercises 39–58, find the x-values (if any) at which f is not continuous. Which of the discontinuities are removable?

39. $f(x) = \dfrac{6}{x}$

40. $f(x) = \dfrac{4}{x - 6}$

41. $f(x) = \dfrac{1}{4 - x^2}$

42. $f(x) = \dfrac{1}{x^2 + 1}$

43. $f(x) = 3x - \cos x$

44. $f(x) = \sin x - 8x$

45. $f(x) = \dfrac{x}{x^2 - x}$

46. $f(x) = \dfrac{x}{x^2 - 4}$

47. $f(x) = \dfrac{x + 2}{x^2 - 3x - 10}$

48. $f(x) = \dfrac{x + 2}{x^2 - x - 6}$

49. $f(x) = \dfrac{|x + 7|}{x + 7}$

50. $f(x) = \dfrac{2|x - 3|}{x - 3}$

51. $f(x) = \begin{cases} \frac{1}{2}x + 1, & x \le 2 \\ 3 - x, & x > 2 \end{cases}$

52. $f(x) = \begin{cases} -2x, & x \le 2 \\ x^2 - 4x + 1, & x > 2 \end{cases}$

53. $f(x) = \begin{cases} \tan \dfrac{\pi x}{4}, & |x| < 1 \\ x, & |x| \ge 1 \end{cases}$

54. $f(x) = \begin{cases} \csc \dfrac{\pi x}{6}, & |x - 3| \le 2 \\ 2, & |x - 3| > 2 \end{cases}$

55. $f(x) = \csc 2x$

56. $f(x) = \tan \dfrac{\pi x}{2}$

57. $f(x) = [\![x - 8]\!]$

58. $f(x) = 5 - [\![x]\!]$

Making a Function Continuous In Exercises 59–64, find the constant a, or the constants a and b, such that the function is continuous on the entire real number line.

59. $f(x) = \begin{cases} 3x^2, & x \ge 1 \\ ax - 4, & x < 1 \end{cases}$

60. $f(x) = \begin{cases} 3x^3, & x \le 1 \\ ax + 5, & x > 1 \end{cases}$

61. $f(x) = \begin{cases} x^3, & x \le 2 \\ ax^2, & x > 2 \end{cases}$

62. $g(x) = \begin{cases} \dfrac{4 \sin x}{x}, & x < 0 \\ a - 2x, & x \ge 0 \end{cases}$

63. $f(x) = \begin{cases} 2, & x \le -1 \\ ax + b, & -1 < x < 3 \\ -2, & x \ge 3 \end{cases}$

64. $g(x) = \begin{cases} \dfrac{x^2 - a^2}{x - a}, & x \ne a \\ 8, & x = a \end{cases}$

Continuity of a Composite Function In Exercises 65–70, discuss the continuity of the composite function $h(x) = f(g(x))$.

65. $f(x) = x^2$, $g(x) = x - 1$

66. $f(x) = 5x + 1$, $g(x) = x^3$

67. $f(x) = \dfrac{1}{x - 6}$, $g(x) = x^2 + 5$

68. $f(x) = \dfrac{1}{\sqrt{x}}$, $g(x) = x - 1$

69. $f(x) = \tan x$, $g(x) = \dfrac{x}{2}$

70. $f(x) = \sin x$, $g(x) = x^2$

Finding Discontinuities Using Technology In Exercises 71–74, use a graphing utility to graph the function. Use the graph to determine any x-values at which the function is not continuous.

71. $f(x) = [\![x]\!] - x$

72. $h(x) = \dfrac{1}{x^2 + 2x - 15}$

73. $g(x) = \begin{cases} x^2 - 3x, & x > 4 \\ 2x - 5, & x \le 4 \end{cases}$

74. $f(x) = \begin{cases} \dfrac{\cos x - 1}{x}, & x < 0 \\ 5x, & x \ge 0 \end{cases}$

Testing for Continuity In Exercises 75–82, describe the interval(s) on which the function is continuous.

75. $f(x) = \dfrac{x}{x^2 + x + 2}$

76. $f(x) = \dfrac{x + 1}{\sqrt{x}}$

77. $f(x) = 3 - \sqrt{x}$

78. $f(x) = x\sqrt{x + 3}$

79. $f(x) = \sec \dfrac{\pi x}{4}$

80. $f(x) = \cos \dfrac{1}{x}$

81. $f(x) = \begin{cases} \dfrac{x^2 - 1}{x - 1}, & x \ne 1 \\ 2, & x = 1 \end{cases}$

82. $f(x) = \begin{cases} 2x - 4, & x \ne 3 \\ 1, & x = 3 \end{cases}$

Existence of a Zero In Exercises 83–86, explain why the function has at least one zero in the given interval.

Function	Interval
83. $f(x) = \frac{1}{12}x^4 - x^3 + 4$	$[1, 2]$
84. $f(x) = x^3 + 5x - 3$	$[0, 1]$
85. $f(x) = x^2 - 2 - \cos x$	$[0, \pi]$
86. $f(x) = -\dfrac{5}{x} + \tan \dfrac{\pi x}{10}$	$[1, 4]$

Existence of Multiple Zeros In Exercises 87 and 88, explain why the function has at least two zeros in the interval $[1, 5]$.

87. $f(x) = (x - 3)^2 - 2$

88. $f(x) = 2 \cos x$

Using the Intermediate Value Theorem **In Exercises 89–94, use the Intermediate Value Theorem and a graphing utility to approximate the zero of the function in the interval $[0, 1]$. Repeatedly "zoom in" on the graph of the function to approximate the zero accurate to two decimal places. Use the *zero* or *root* feature of the graphing utility to approximate the zero accurate to four decimal places.**

89. $f(x) = x^3 + x - 1$

90. $f(x) = x^4 - x^2 + 3x - 1$

91. $f(x) = \sqrt{x^2 + 17x + 19} - 6$

92. $f(x) = \sqrt{x^4 + 39x + 13} - 4$

93. $g(t) = 2 \cos t - 3t$

94. $h(\theta) = \tan \theta + 3\theta - 4$

Using the Intermediate Value Theorem **In Exercises 95–100, verify that the Intermediate Value Theorem applies to the indicated interval and find the value of c guaranteed by the theorem.**

95. $f(x) = x^2 + x - 1, \quad [0, 5], \quad f(c) = 11$

96. $f(x) = x^2 - 6x + 8, \quad [0, 3], \quad f(c) = 0$

97. $f(x) = \sqrt{x + 7} - 2, \quad [0, 5], \quad f(c) = 1$

98. $f(x) = \sqrt[3]{x} + 8, \quad [-9, -6], \quad f(c) = 6$

99. $f(x) = \dfrac{x - x^3}{x - 4}, \quad [1, 3], \quad f(c) = 3$

100. $f(x) = \dfrac{x^2 + x}{x - 1}, \quad \left[\dfrac{5}{2}, 4\right], \quad f(c) = 6$

EXPLORING CONCEPTS

101. Writing a Function Write a function that is continuous on (a, b) but not continuous on $[a, b]$.

102. Sketching a Graph Sketch the graph of any function f such that

$$\lim_{x\to 3^+} f(x) = 1 \quad \text{and} \quad \lim_{x\to 3^-} f(x) = 0.$$

Is the function continuous at $x = 3$? Explain.

103. Continuity of Combinations of Functions If the functions f and g are continuous for all real x, is $f + g$ always continuous for all real x? Is f/g always continuous for all real x? If either is not continuous, give an example to verify your conclusion.

104. Removable and Nonremovable Discontinuities Describe the difference between a discontinuity that is removable and a discontinuity that is nonremovable. Then give an example of a function that satisfies each description.

(a) A function with a nonremovable discontinuity at $x = 4$

(b) A function with a removable discontinuity at $x = -4$

(c) A function that has both of the characteristics described in parts (a) and (b)

True or False? **In Exercises 105–110, determine whether the statement is true or false. If it is false, explain why or give an example that shows it is false.**

105. If $\lim_{x\to c} f(x) = L$ and $f(c) = L$, then f is continuous at c.

106. If $f(x) = g(x)$ for $x \neq c$ and $f(c) \neq g(c)$, then either f or g is not continuous at c.

107. The Intermediate Value Theorem guarantees that $f(a)$ and $f(b)$ differ in sign when a continuous function f has at least one zero on $[a, b]$.

108. The limit of the greatest integer function as x approaches 0 from the left is -1.

109. A rational function can have infinitely many x-values at which it is not continuous.

110. The function $f(x) = \dfrac{|x - 1|}{x - 1}$ is continuous on $(-\infty, \infty)$.

111. Think About It Describe how the functions

$$f(x) = 3 + [\![x]\!] \quad \text{and} \quad g(x) = 3 - [\![-x]\!]$$

differ.

112. HOW DO YOU SEE IT? Every day you dissolve 28 ounces of chlorine in a swimming pool. The graph shows the amount of chlorine $f(t)$ in the pool after t days. Estimate and interpret $\lim_{t\to 4^-} f(t)$ and $\lim_{t\to 4^+} f(t)$.

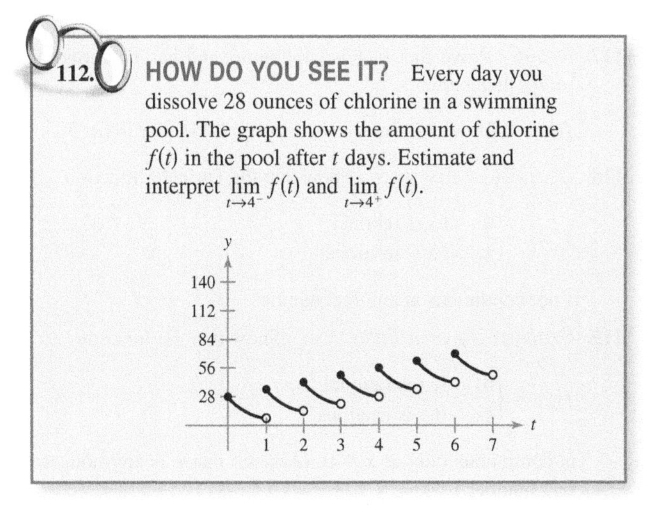

113. Data Plan A cell phone service charges \$10 for the first gigabyte (GB) of data used per month and \$7.50 for each additional gigabyte or fraction thereof. The cost of the data plan is given by

$$C(t) = 10 - 7.5[\![1 - t]\!], \quad t > 0$$

where t is the amount of data used (in GB). Sketch the graph of this function and discuss its continuity.

114. Inventory Management The number of units in inventory in a small company is given by

$$N(t) = 25\left(2\left[\!\!\left[\frac{t + 2}{2}\right]\!\!\right] - t\right)$$

where t is the time in months. Sketch the graph of this function and discuss its continuity. How often must this company replenish its inventory?

115. Déjà Vu At 8:00 A.M. on Saturday, a man begins running up the side of a mountain to his weekend campsite (see figure). On Sunday morning at 8:00 A.M., he runs back down the mountain. It takes him 20 minutes to run up but only 10 minutes to run down. At some point on the way down, he realizes that he passed the same place at exactly the same time on Saturday. Prove that he is correct. [*Hint:* Let $s(t)$ and $r(t)$ be the position functions for the runs up and down, and apply the Intermediate Value Theorem to the function $f(t) = s(t) - r(t)$.]

116. Volume Use the Intermediate Value Theorem to show that for all spheres with radii in the interval $[5, 8]$, there is one with a volume of 1500 cubic centimeters.

117. Proof Prove that if f is continuous and has no zeros on $[a, b]$, then either

$f(x) > 0$ for all x in $[a, b]$ or $f(x) < 0$ for all x in $[a, b]$.

118. Dirichlet Function Show that the Dirichlet function

$$f(x) = \begin{cases} 0, & \text{if } x \text{ is rational} \\ 1, & \text{if } x \text{ is irrational} \end{cases}$$

is not continuous at any real number.

119. Continuity of a Function Show that the function

$$f(x) = \begin{cases} 0, & \text{if } x \text{ is rational} \\ kx, & \text{if } x \text{ is irrational} \end{cases}$$

is continuous only at $x = 0$. (Assume that k is any nonzero real number.)

120. Signum Function The **signum function** is defined by

$$\operatorname{sgn}(x) = \begin{cases} -1, & x < 0 \\ 0, & x = 0. \\ 1, & x > 0 \end{cases}$$

Sketch a graph of $\operatorname{sgn}(x)$ and find the following (if possible).

(a) $\lim_{x\to 0^-} \operatorname{sgn}(x)$ (b) $\lim_{x\to 0^+} \operatorname{sgn}(x)$ (c) $\lim_{x\to 0} \operatorname{sgn}(x)$

121. Modeling Data The table lists the frequency F (in Hertz) of a musical note at various times t (in seconds).

t	0	1	2	3	4	5
F	436	444	434	446	433	444

(a) Plot the data and connect the points with a curve.

(b) Does there appear to be a limiting frequency of the note? Explain.

122. Creating Models A swimmer crosses a pool of width b by swimming in a straight line from $(0, 0)$ to $(2b, b)$. (See figure.)

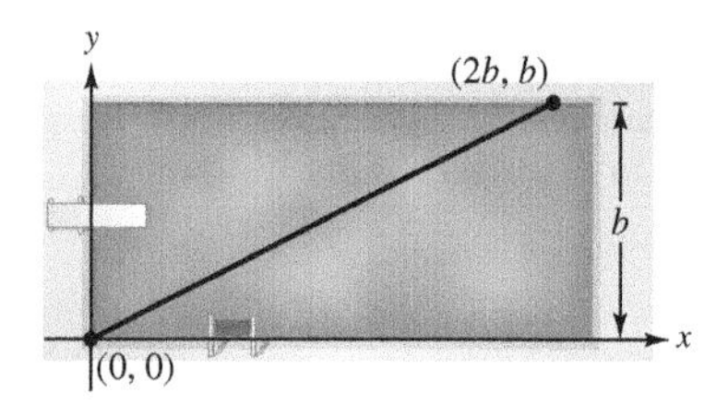

(a) Let f be a function defined as the y-coordinate of the point on the long side of the pool that is nearest the swimmer at any given time during the swimmer's crossing of the pool. Determine the function f and sketch its graph. Is f continuous? Explain.

(b) Let g be the minimum distance between the swimmer and the long sides of the pool. Determine the function g and sketch its graph. Is g continuous? Explain.

123. Making a Function Continuous Find all values of c such that f is continuous on $(-\infty, \infty)$.

$$f(x) = \begin{cases} 1 - x^2, & x \le c \\ x, & x > c \end{cases}$$

124. Proof Prove that for any real number y there exists x in $(-\pi/2, \pi/2)$ such that $\tan x = y$.

125. Making a Function Continuous Let

$$f(x) = \frac{\sqrt{x + c^2} - c}{x}, \quad c > 0.$$

What is the domain of f? How can you define f at $x = 0$ in order for f to be continuous there?

126. Proof Prove that if

$$\lim_{\Delta x\to 0} f(c + \Delta x) = f(c)$$

then f is continuous at c.

127. Continuity of a Function Discuss the continuity of the function $h(x) = x[\![x]\!]$.

128. Proof

(a) Let $f_1(x)$ and $f_2(x)$ be continuous on the closed interval $[a, b]$. If $f_1(a) < f_2(a)$ and $f_1(b) > f_2(b)$, prove that there exists c between a and b such that $f_1(c) = f_2(c)$.

(b) Show that there exists c in $\left[0, \frac{\pi}{2}\right]$ such that $\cos x = x$. Use a graphing utility to approximate c to three decimal places.

PUTNAM EXAM CHALLENGE

129. Prove or disprove: If x and y are real numbers with $y \ge 0$ and $y(y + 1) \le (x + 1)^2$, then $y(y - 1) \le x^2$.

130. Determine all polynomials $P(x)$ such that

$$P(x^2 + 1) = (P(x))^2 + 1 \text{ and } P(0) = 0.$$

These problems were composed by the Committee on the Putnam Prize Competition.

1.5 Infinite Limits

- **Determine infinite limits from the left and from the right.**
- **Find and sketch the vertical asymptotes of the graph of a function.**

Infinite Limits

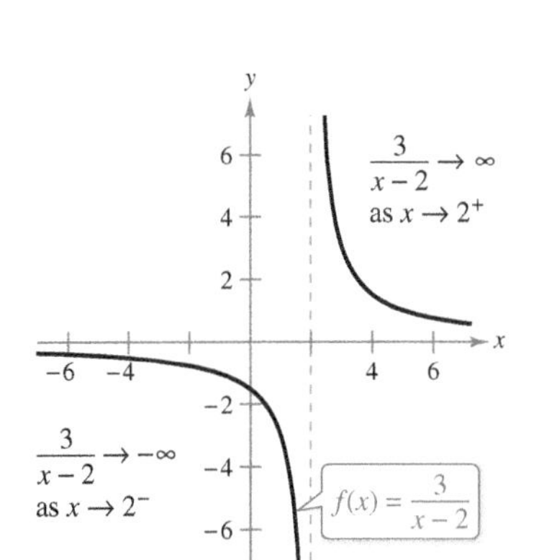

$f(x)$ increases and decreases without bound as x approaches 2.
Figure 1.38

Consider the function $f(x) = 3/(x - 2)$. From Figure 1.38 and the table, you can see that $f(x)$ *decreases without bound* as x approaches 2 from the left, and $f(x)$ *increases without bound* as x approaches 2 from the right.

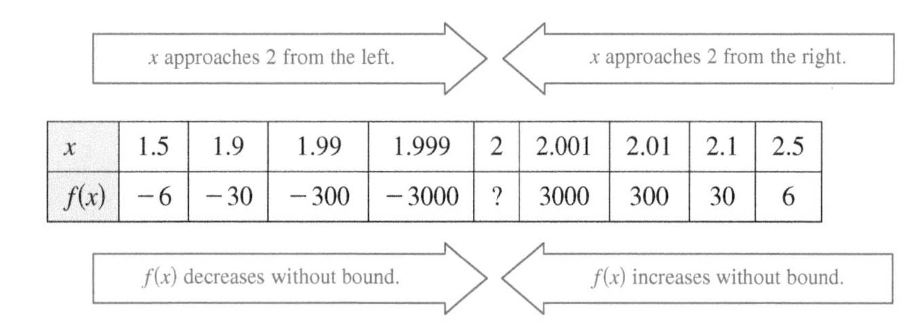

x	1.5	1.9	1.99	1.999	2	2.001	2.01	2.1	2.5
$f(x)$	−6	−30	−300	−3000	?	3000	300	30	6

This behavior is denoted as

$$\lim_{x \to 2^-} \frac{3}{x-2} = -\infty \qquad f(x) \text{ decreases without bound as } x \text{ approaches 2 from the left.}$$

and

$$\lim_{x \to 2^+} \frac{3}{x-2} = \infty. \qquad f(x) \text{ increases without bound as } x \text{ approaches 2 from the right.}$$

The symbols ∞ and $-\infty$ refer to positive infinity and negative infinity, respectively. These symbols do not represent real numbers. They are convenient symbols used to describe unbounded conditions more concisely. A limit in which $f(x)$ increases or decreases without bound as x approaches c is called an **infinite limit.**

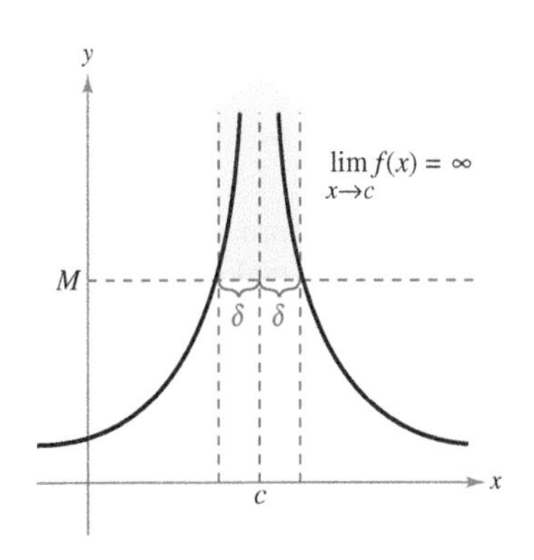

Infinite limits
Figure 1.39

Definition of Infinite Limits

Let f be a function that is defined at every real number in some open interval containing c (except possibly at c itself). The statement

$$\lim_{x \to c} f(x) = \infty$$

means that for each $M > 0$ there exists a $\delta > 0$ such that $f(x) > M$ whenever $0 < |x - c| < \delta$ (see Figure 1.39). Similarly, the statement

$$\lim_{x \to c} f(x) = -\infty$$

means that for each $N < 0$ there exists a $\delta > 0$ such that $f(x) < N$ whenever

$$0 < |x - c| < \delta.$$

To define the **infinite limit from the left,** replace $0 < |x - c| < \delta$ by $c - \delta < x < c$. To define the **infinite limit from the right,** replace $0 < |x - c| < \delta$ by $c < x < c + \delta$.

Be sure you see that the equal sign in the statement $\lim f(x) = \infty$ does not mean that the limit exists! On the contrary, it tells you how the limit **fails to exist** by denoting the unbounded behavior of $f(x)$ as x approaches c.

Exploration

Use a graphing utility to graph each function. For each function, analytically find the single real number c that is not in the domain. Then graphically find the limit (if it exists) of $f(x)$ as x approaches c from the left and from the right.

a. $f(x) = \dfrac{3}{x - 4}$

b. $f(x) = \dfrac{1}{2 - x}$

c. $f(x) = \dfrac{2}{(x - 3)^2}$

d. $f(x) = \dfrac{-3}{(x + 2)^2}$

EXAMPLE 1 Determining Infinite Limits from a Graph

Determine the limit of each function shown in Figure 1.40 as x approaches 1 from the left and from the right.

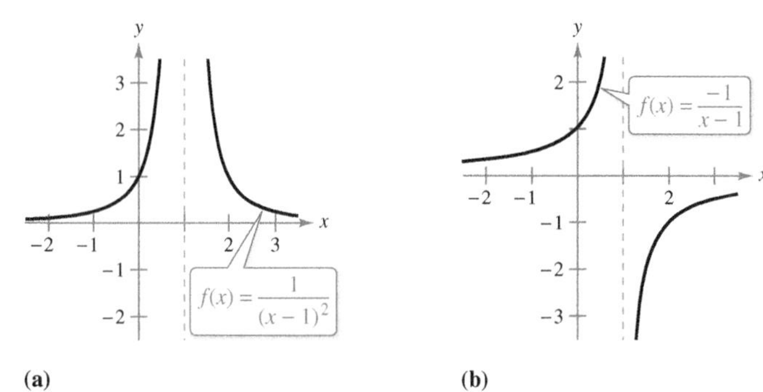

(a) **(b)**

Each graph has an asymptote at $x = 1$.

Figure 1.40

Solution

a. When x approaches 1 from the left or the right, $(x - 1)^2$ is a small positive number. Thus, the quotient $1/(x - 1)^2$ is a large positive number, and $f(x)$ approaches infinity from each side of $x = 1$. So, you can conclude that

$$\lim_{x \to 1} \frac{1}{(x - 1)^2} = \infty. \qquad \text{Limit from each side is infinity.}$$

Figure 1.40(a) confirms this analysis.

b. When x approaches 1 from the left, $x - 1$ is a small negative number. Thus, the quotient $-1/(x - 1)$ is a large positive number, and $f(x)$ approaches infinity from the left of $x = 1$. So, you can conclude that

$$\lim_{x \to 1^-} \frac{-1}{x - 1} = \infty. \qquad \text{Limit from the left side is infinity.}$$

When x approaches 1 from the right, $x - 1$ is a small positive number. Thus, the quotient $-1/(x - 1)$ is a large negative number, and $f(x)$ approaches negative infinity from the right of $x = 1$. So, you can conclude that

$$\lim_{x \to 1^+} \frac{-1}{x - 1} = -\infty. \qquad \text{Limit from the right side is negative infinity.}$$

Figure 1.40(b) confirms this analysis. ■

TECHNOLOGY Remember that you can use a numerical approach to analyze a limit. For instance, you can use a graphing utility to create a table of values to analyze the limit in Example 1(a), as shown in the figure below.

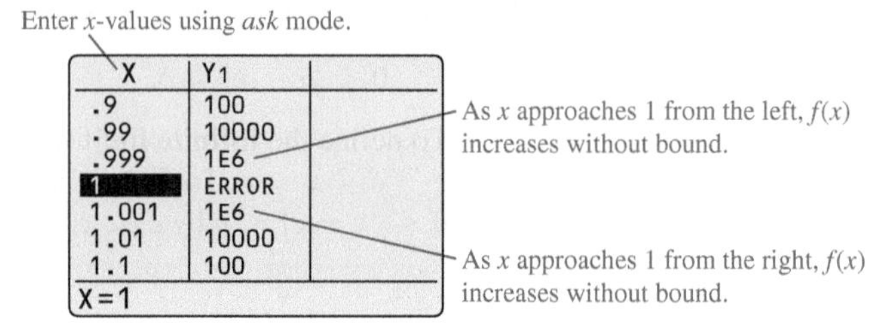

X	Y1
.9	100
.99	10000
.999	1E6
1	ERROR
1.001	1E6
1.01	10000
1.1	100

X=1

Use a graphing utility to make a table of values to analyze the limit in Example 1(b).

Vertical Asymptotes

If it were possible to extend the graphs in Figure 1.40 toward positive and negative infinity, you would see that each graph becomes arbitrarily close to the vertical line $x = 1$. This line is a **vertical asymptote** of the graph of f. (You will study other types of asymptotes in Sections 3.5 and 3.6.)

REMARK If the graph of a function f has a vertical asymptote at $x = c$, then f is *not continuous* at c.

Definition of Vertical Asymptote

If $f(x)$ approaches infinity (or negative infinity) as x approaches c from the right or the left, then the line $x = c$ is a **vertical asymptote** of the graph of f.

In Example 1, note that each of the functions is a *quotient* and that the vertical asymptote occurs at a number at which the denominator is 0 (and the numerator is not 0). The next theorem generalizes this observation.

THEOREM 1.14 Vertical Asymptotes

Let f and g be continuous on an open interval containing c. If $f(c) \neq 0$, $g(c) = 0$, and there exists an open interval containing c such that $g(x) \neq 0$ for all $x \neq c$ in the interval, then the graph of the function

$$h(x) = \frac{f(x)}{g(x)}$$

has a vertical asymptote at $x = c$.

A proof of this theorem is given in Appendix A.

(a)

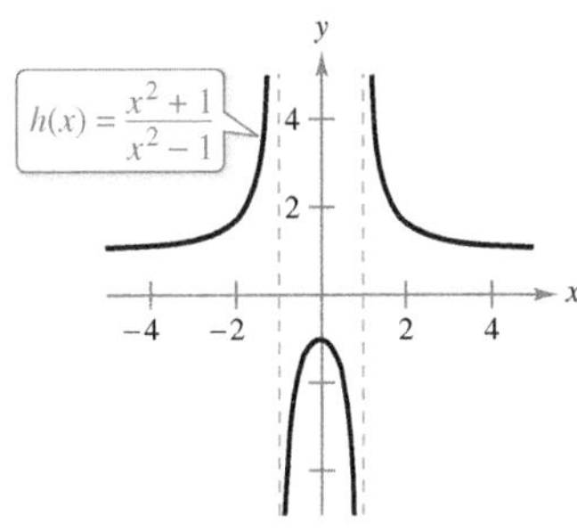

(b)

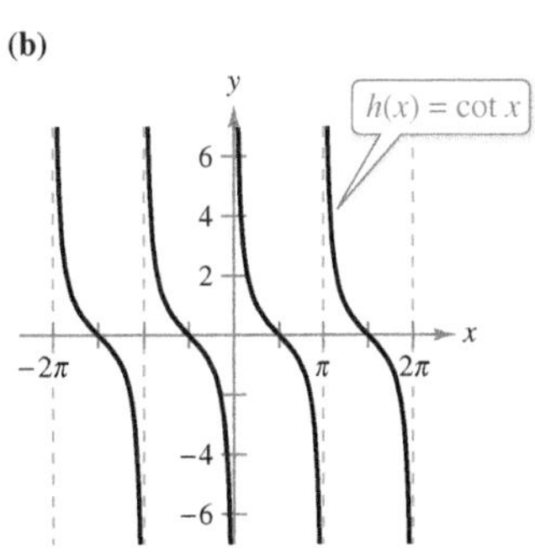

(c)

Functions with vertical asymptotes

Figure 1.41

EXAMPLE 2 Finding Vertical Asymptotes

See LarsonCalculus.com for an interactive version of this type of example.

a. When $x = -1$, the denominator of

$$h(x) = \frac{1}{2(x + 1)}$$

is 0 and the numerator is not 0. So, by Theorem 1.14, you can conclude that $x = -1$ is a vertical asymptote, as shown in Figure 1.41(a).

b. By factoring the denominator as

$$h(x) = \frac{x^2 + 1}{x^2 - 1} = \frac{x^2 + 1}{(x - 1)(x + 1)}$$

you can see that the denominator is 0 at $x = -1$ and $x = 1$. Also, because the numerator is not 0 at these two points, you can apply Theorem 1.14 to conclude that the graph of f has two vertical asymptotes, as shown in Figure 1.41(b).

c. By writing the cotangent function in the form

$$h(x) = \cot x = \frac{\cos x}{\sin x}$$

you can apply Theorem 1.14 to conclude that vertical asymptotes occur at all values of x such that $\sin x = 0$ and $\cos x \neq 0$, as shown in Figure 1.41(c). So, the graph of this function has infinitely many vertical asymptotes. These asymptotes occur at $x = n\pi$, where n is an integer. ■

Theorem 1.14 requires that the value of the numerator at $x = c$ be nonzero. When both the numerator and the denominator are 0 at $x = c$, you obtain the *indeterminate form* $0/0$, and you cannot determine the limit behavior at $x = c$ without further investigation, as illustrated in Example 3.

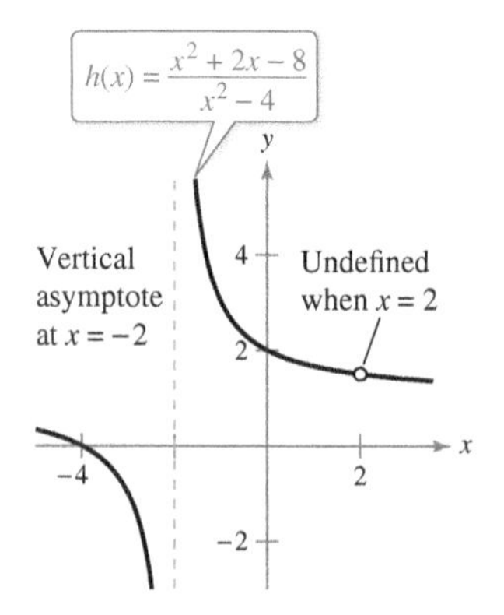

$h(x)$ increases and decreases without bound as x approaches -2.
Figure 1.42

EXAMPLE 3 A Rational Function with Common Factors

Determine all vertical asymptotes of the graph of

$$h(x) = \frac{x^2 + 2x - 8}{x^2 - 4}.$$

Solution Begin by simplifying the expression, as shown.

$$\begin{aligned} h(x) &= \frac{x^2 + 2x - 8}{x^2 - 4} \\ &= \frac{(x + 4)\cancel{(x - 2)}}{(x + 2)\cancel{(x - 2)}} \\ &= \frac{x + 4}{x + 2}, \quad x \neq 2 \end{aligned}$$

At all x-values other than $x = 2$, the graph of h coincides with the graph of $k(x) = (x + 4)/(x + 2)$. So, you can apply Theorem 1.14 to k to conclude that there is a vertical asymptote at $x = -2$, as shown in Figure 1.42. From the graph, you can see that

$$\lim_{x \to -2^-} \frac{x^2 + 2x - 8}{x^2 - 4} = -\infty \quad \text{and} \quad \lim_{x \to -2^+} \frac{x^2 + 2x - 8}{x^2 - 4} = \infty.$$

Note that $x = 2$ is *not* a vertical asymptote.

EXAMPLE 4 Determining Infinite Limits

Find each limit.

$$\lim_{x \to 1^-} \frac{x^2 - 3x}{x - 1} \quad \text{and} \quad \lim_{x \to 1^+} \frac{x^2 - 3x}{x - 1}$$

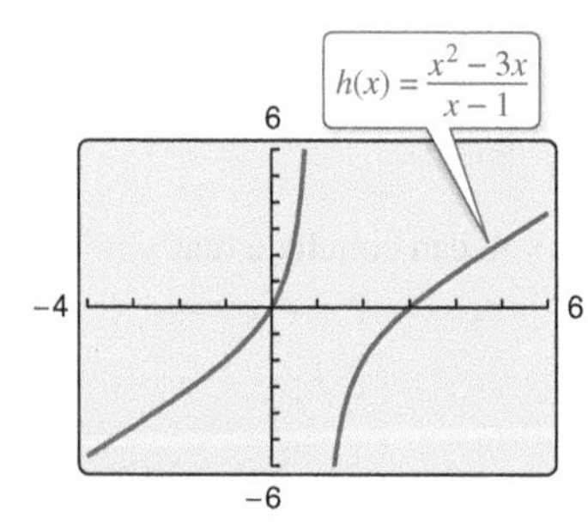

The graph of h has a vertical asymptote at $x = 1$.
Figure 1.43

Solution Because the denominator is 0 when $x = 1$ (and the numerator is not 0), you know that the graph of

$$h(x) = \frac{x^2 - 3x}{x - 1}$$

has a vertical asymptote at $x = 1$. This means that each of the given limits is either ∞ or $-\infty$. You can determine the result by analyzing h at values of x close to 1 or by using a graphing utility. From the graph of h shown in Figure 1.43, you can see that the graph approaches ∞ from the left of $x = 1$ and approaches $-\infty$ from the right of $x = 1$. So, you can conclude that

$$\lim_{x \to 1^-} \frac{x^2 - 3x}{x - 1} = \infty \qquad \text{The limit from the left is infinity.}$$

and

$$\lim_{x \to 1^+} \frac{x^2 - 3x}{x - 1} = -\infty. \qquad \text{The limit from the right is negative infinity.}$$

TECHNOLOGY PITFALL When using a graphing utility, be careful to interpret correctly the graph of a function with a vertical asymptote—some graphing utilities have difficulty drawing this type of graph.

THEOREM 1.15 Properties of Infinite Limits

Let c and L be real numbers, and let f and g be functions such that

$$\lim_{x\to c} f(x) = \infty \quad \text{and} \quad \lim_{x\to c} g(x) = L.$$

1. Sum or difference: $\lim_{x\to c} [f(x) \pm g(x)] = \infty$
2. Product: $\lim_{x\to c} [f(x)g(x)] = \infty, \quad L > 0$

   $\lim_{x\to c} [f(x)g(x)] = -\infty, \quad L < 0$
3. Quotient: $\lim_{x\to c} \dfrac{g(x)}{f(x)} = 0$

Similar properties hold for one-sided limits and for functions for which the limit of $f(x)$ as x approaches c is $-\infty$ [see Example 5(d)].

REMARK Be sure you understand that Property 2 of Theorem 1.15 is not valid when $\lim_{x\to c} g(x) = 0$.

Proof Here is a proof of the sum property. (The proofs of the remaining properties are left as an exercise [see Exercise 70].) To show that the limit of $f(x) + g(x)$ is infinite, choose $M > 0$. You then need to find $\delta > 0$ such that $[f(x) + g(x)] > M$ whenever $0 < |x - c| < \delta$. For simplicity's sake, you can assume L is positive. Let $M_1 = M + 1$. Because the limit of $f(x)$ is infinite, there exists δ_1 such that $f(x) > M_1$ whenever $0 < |x - c| < \delta_1$. Also, because the limit of $g(x)$ is L, there exists δ_2 such that $|g(x) - L| < 1$ whenever $0 < |x - c| < \delta_2$. By letting δ be the smaller of δ_1 and δ_2, you can conclude that $0 < |x - c| < \delta$ implies $f(x) > M + 1$ and $|g(x) - L| < 1$. The second of these two inequalities implies that $g(x) > L - 1$, and adding this to the first inequality, you can write

$$f(x) + g(x) > (M + 1) + (L - 1) = M + L > M.$$

So, you can conclude that

$$\lim_{x\to c} [f(x) + g(x)] = \infty.$$

■

EXAMPLE 5 Determining Limits

a. Because $\lim_{x\to 0} 1 = 1$ and $\lim_{x\to 0} \dfrac{1}{x^2} = \infty$, you can write

$$\lim_{x\to 0} \left(1 + \frac{1}{x^2}\right) = \infty. \qquad \text{Property 1, Theorem 1.15}$$

b. Because $\lim_{x\to 1^-} (x^2 + 1) = 2$ and $\lim_{x\to 1^-} (\cot \pi x) = -\infty$, you can write

$$\lim_{x\to 1^-} \frac{x^2 + 1}{\cot \pi x} = 0. \qquad \text{Property 3, Theorem 1.15}$$

c. Because $\lim_{x\to 0^+} 3 = 3$ and $\lim_{x\to 0^+} \cot x = \infty$, you can write

$$\lim_{x\to 0^+} 3 \cot x = \infty. \qquad \text{Property 2, Theorem 1.15}$$

d. Because $\lim_{x\to 0^-} x^2 = 0$ and $\lim_{x\to 0^-} \dfrac{1}{x} = -\infty$, you can write

$$\lim_{x\to 0^-} \left(x^2 + \frac{1}{x}\right) = -\infty. \qquad \text{Property 1, Theorem 1.15}$$

■

REMARK Note that the solution to Example 5(d) uses Property 1 from Theorem 1.15 for which the limit of $f(x)$ as x approaches c is $-\infty$.

1.5 Exercises

See CalcChat.com for tutorial help and worked-out solutions to odd-numbered exercises.

CONCEPT CHECK

1. **Infinite Limit** In your own words, describe the meaning of an infinite limit. What does ∞ represent?
2. **Vertical Asymptote** In your own words, describe what is meant by a vertical asymptote of a graph.

Determining Infinite Limits from a Graph In Exercises 3–6, determine whether $f(x)$ approaches ∞ or $-\infty$ as x approaches -2 from the left and from the right.

3. $f(x) = 2\left|\dfrac{x}{x^2 - 4}\right|$

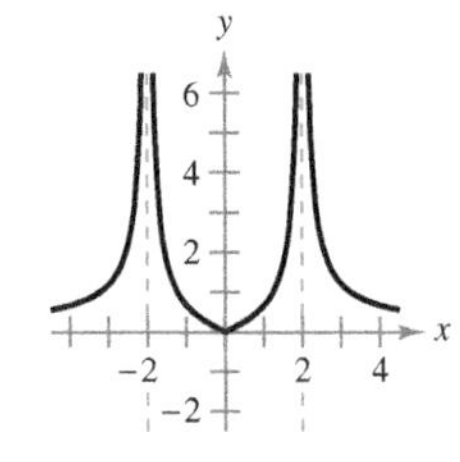

4. $f(x) = \dfrac{1}{x + 2}$

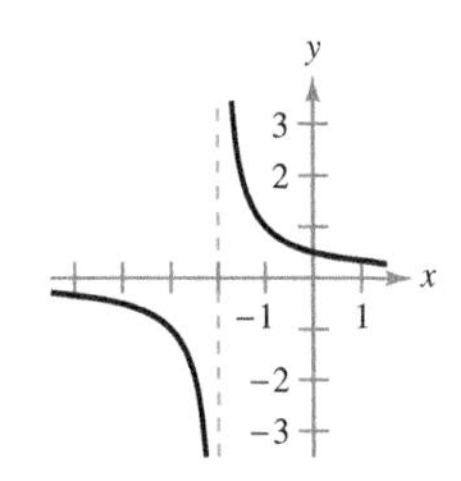

5. $f(x) = \tan \dfrac{\pi x}{4}$

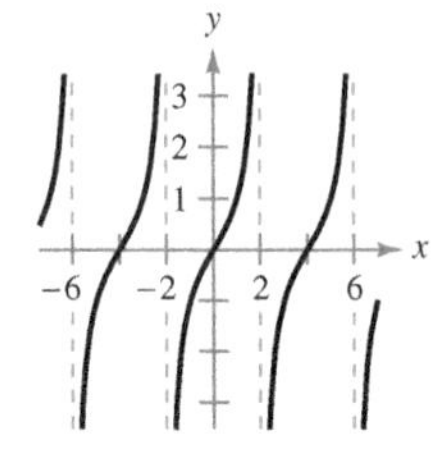

6. $f(x) = \sec \dfrac{\pi x}{4}$

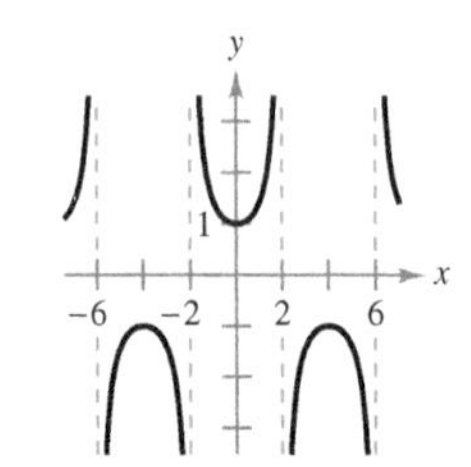

Determining Infinite Limits In Exercises 7–10, determine whether $f(x)$ approaches ∞ or $-\infty$ as x approaches 4 from the left and from the right.

7. $f(x) = \dfrac{1}{x - 4}$
8. $f(x) = \dfrac{-1}{x - 4}$
9. $f(x) = \dfrac{1}{(x - 4)^2}$
10. $f(x) = \dfrac{-1}{(x - 4)^2}$

Numerical and Graphical Analysis In Exercises 11–16, create a table of values for the function and use the result to determine whether $f(x)$ approaches ∞ or $-\infty$ as x approaches -3 from the left and from the right. Use a graphing utility to graph the function to confirm your answer.

11. $f(x) = \dfrac{1}{x^2 - 9}$
12. $f(x) = \dfrac{x}{x^2 - 9}$
13. $f(x) = \dfrac{x^2}{x^2 - 9}$
14. $f(x) = -\dfrac{1}{3 + x}$
15. $f(x) = \cot \dfrac{\pi x}{3}$
16. $f(x) = \tan \dfrac{\pi x}{6}$

Finding Vertical Asymptotes In Exercises 17–32, find the vertical asymptotes (if any) of the graph of the function.

17. $f(x) = \dfrac{1}{x^2}$
18. $f(x) = \dfrac{2}{(x - 3)^3}$
19. $f(x) = \dfrac{x^2}{x^2 - 4}$
20. $f(x) = \dfrac{3x}{x^2 + 9}$
21. $g(t) = \dfrac{t - 1}{t^2 + 1}$
22. $h(s) = \dfrac{3s + 4}{s^2 - 16}$
23. $f(x) = \dfrac{3}{x^2 + x - 2}$
24. $g(x) = \dfrac{x^2 - 5x + 25}{x^3 + 125}$
25. $f(x) = \dfrac{4x^2 + 4x - 24}{x^4 - 2x^3 - 9x^2 + 18x}$
26. $h(x) = \dfrac{x^2 - 9}{x^3 + 3x^2 - x - 3}$
27. $f(x) = \dfrac{x^2 - 2x - 15}{x^3 - 5x^2 + x - 5}$
28. $h(t) = \dfrac{t^2 - 2t}{t^4 - 16}$
29. $f(x) = \csc \pi x$
30. $f(x) = \tan \pi x$
31. $s(t) = \dfrac{t}{\sin t}$
32. $g(\theta) = \dfrac{\tan \theta}{\theta}$

Vertical Asymptote or Removable Discontinuity In Exercises 33–36, determine whether the graph of the function has a vertical asymptote or a removable discontinuity at $x = -1$. Graph the function using a graphing utility to confirm your answer.

33. $f(x) = \dfrac{x^2 - 1}{x + 1}$
34. $f(x) = \dfrac{x^2 - 2x - 8}{x + 1}$
35. $f(x) = \dfrac{\cos(x^2 - 1)}{x + 1}$
36. $f(x) = \dfrac{\sin(x + 1)}{x + 1}$

Finding a One-Sided Limit In Exercises 37–50, find the one-sided limit (if it exists).

37. $\displaystyle\lim_{x \to 2^+} \frac{x}{x - 2}$
38. $\displaystyle\lim_{x \to 2^-} \frac{x^2}{x^2 + 4}$
39. $\displaystyle\lim_{x \to -3^-} \frac{x + 3}{x^2 + x - 6}$
40. $\displaystyle\lim_{x \to (-1/2)^+} \frac{6x^2 + x - 1}{4x^2 - 4x - 3}$
41. $\displaystyle\lim_{x \to 0^-} \left(1 + \frac{1}{x}\right)$
42. $\displaystyle\lim_{x \to 0^+} \left(6 - \frac{1}{x^3}\right)$
43. $\displaystyle\lim_{x \to -4^-} \left(x^2 + \frac{2}{x + 4}\right)$
44. $\displaystyle\lim_{x \to 0^+} \left(x - \frac{1}{x} + 3\right)$
45. $\displaystyle\lim_{x \to 0^+} \left(\sin x + \frac{1}{x}\right)$
46. $\displaystyle\lim_{x \to (\pi/2)^+} \frac{-2}{\cos x}$
47. $\displaystyle\lim_{x \to \pi^+} \frac{\sqrt{x}}{\csc x}$
48. $\displaystyle\lim_{x \to 0^-} \frac{x + 2}{\cot x}$
49. $\displaystyle\lim_{x \to (1/2)^-} x \sec \pi x$
50. $\displaystyle\lim_{x \to (1/2)^+} x^2 \tan \pi x$

Finding a One-Sided Limit Using Technology In Exercises 51 and 52, use a graphing utility to graph the function and determine the one-sided limit.

51. $\lim_{x\to 1^+} \dfrac{x^2+x+1}{x^3-1}$

52. $\lim_{x\to 1^-} \dfrac{x^3-1}{x^2+x+1}$

Determining Limits In Exercises 53 and 54, use the information to determine the limits.

53. $\lim_{x\to c} f(x) = \infty$

$\lim_{x\to c} g(x) = -2$

(a) $\lim_{x\to c} [f(x) + g(x)]$

(b) $\lim_{x\to c} [f(x)g(x)]$

(c) $\lim_{x\to c} \dfrac{g(x)}{f(x)}$

54. $\lim_{x\to c} f(x) = -\infty$

$\lim_{x\to c} g(x) = 3$

(a) $\lim_{x\to c} [f(x) + g(x)]$

(b) $\lim_{x\to c} [f(x)g(x)]$

(c) $\lim_{x\to c} \dfrac{g(x)}{f(x)}$

EXPLORING CONCEPTS

55. Writing a Rational Function Write a rational function with vertical asymptotes at $x = 6$ and $x = -2$, and with a zero at $x = 3$.

56. Rational Function Does the graph of every rational function have a vertical asymptote? Explain.

57. Sketching a Graph Use the graph of the function f (see figure) to sketch the graph of $g(x) = 1/f(x)$ on the interval $[-2, 3]$. To print an enlarged copy of the graph, go to *MathGraphs.com.*

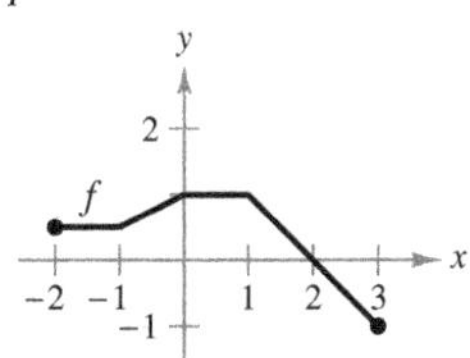

58. Relativity According to the theory of relativity, the mass m of a particle depends on its velocity v. That is,

$$m = \frac{m_0}{\sqrt{1-(v^2/c^2)}},$$ where m_0 is the mass when the particle is at rest and c is the speed of light. Find the limit of the mass as v approaches c from the left.

59. Numerical and Graphical Reasoning Use a graphing utility to complete the table for each function and graph each function to estimate the limit. What is the value of the limit when the power of x in the denominator is greater than 3?

x	1	0.5	0.2	0.1	0.01	0.001	0.0001
$f(x)$							

(a) $\lim_{x\to 0^+} \dfrac{x - \sin x}{x}$

(b) $\lim_{x\to 0^+} \dfrac{x - \sin x}{x^2}$

(c) $\lim_{x\to 0^+} \dfrac{x - \sin x}{x^3}$

(d) $\lim_{x\to 0^+} \dfrac{x - \sin x}{x^4}$

60. HOW DO YOU SEE IT? For a quantity of gas at a constant temperature, the pressure P is inversely proportional to the volume V. What is the limit of P as V approaches 0 from the right? Explain what this means in the context of the problem.

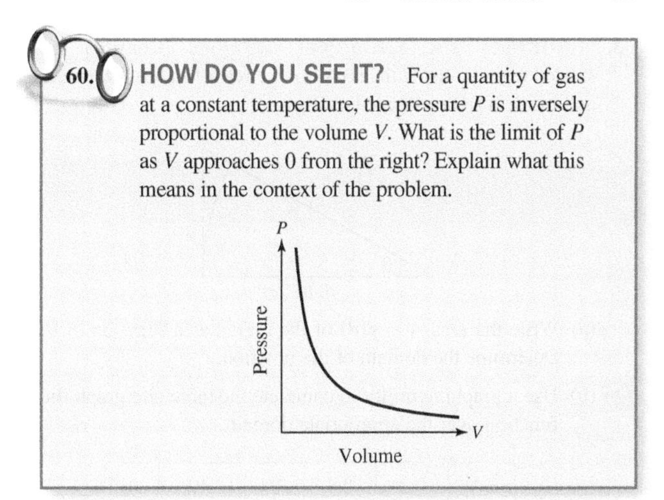

61. Rate of Change A 25-foot ladder is leaning against a house (see figure). If the base of the ladder is pulled away from the house at a rate of 2 feet per second, then the top will move down the wall at a rate of

$$r = \frac{2x}{\sqrt{625 - x^2}} \text{ ft/sec}$$

where x is the distance between the base of the ladder and the house, and r is the rate in feet per second.

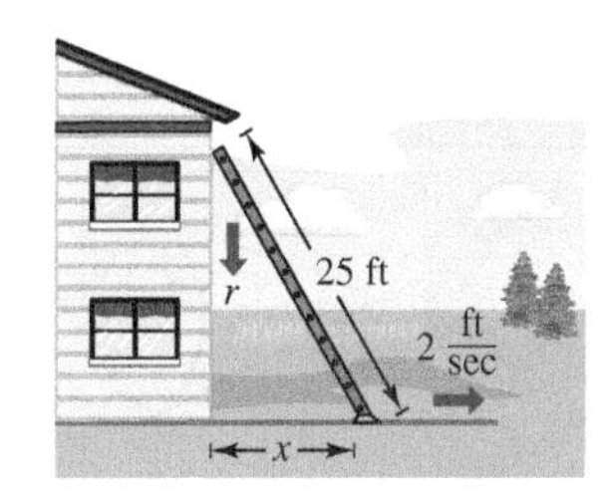

(a) Find the rate r when x is 7 feet.

(b) Find the rate r when x is 15 feet.

(c) Find the limit of r as x approaches 25 from the left.

62. Average Speed

On a trip of d miles to another city, a truck driver's average speed was x miles per hour. On the return trip, the average speed was y miles per hour. The average speed for the round trip was 50 miles per hour.

(a) Verify that

$$y = \frac{25x}{x-25}.$$

What is the domain?

(b) Complete the table.

x	30	40	50	60
y				

Are the values of y different than you expected? Explain.

(c) Find the limit of y as x approaches 25 from the right and interpret its meaning.

iStockphoto.com/WendellandCarolyn

63. Numerical and Graphical Analysis Consider the shaded region outside the sector of a circle of radius 10 meters and inside a right triangle (see figure).

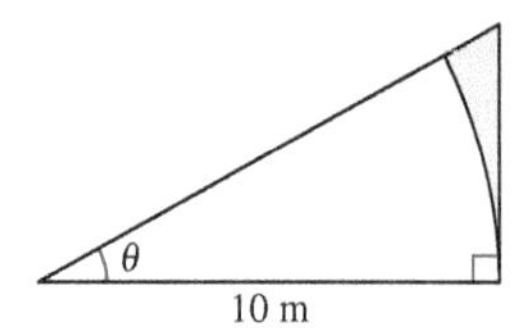

(a) Write the area $A = f(\theta)$ of the region as a function of θ. Determine the domain of the function.

(b) Use a graphing utility to complete the table and graph the function over the appropriate domain.

θ	0.3	0.6	0.9	1.2	1.5
$f(\theta)$					

(c) Find the limit of A as θ approaches $\pi/2$ from the left.

64. Numerical and Graphical Reasoning A crossed belt connects a 20-centimeter pulley (10-cm radius) on an electric motor with a 40-centimeter pulley (20-cm radius) on a saw arbor (see figure). The electric motor runs at 1700 revolutions per minute.

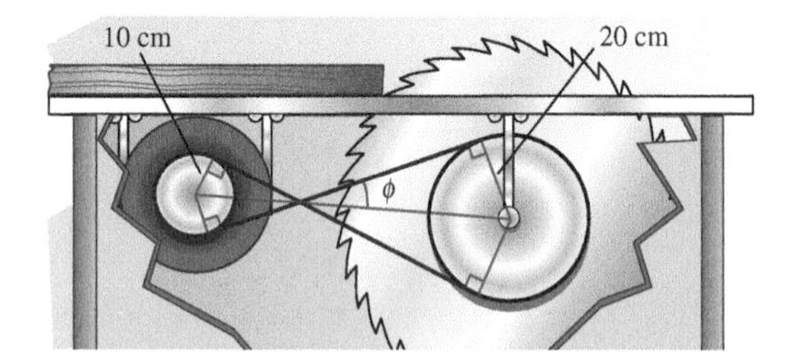

(a) Determine the number of revolutions per minute of the saw.

(b) How does crossing the belt affect the saw in relation to the motor?

(c) Let L be the total length of the belt. Write L as a function of ϕ, where ϕ is measured in radians. What is the domain of the function? (*Hint:* Add the lengths of the straight sections of the belt and the length of the belt around each pulley.)

(d) Use a graphing utility to complete the table.

ϕ	0.3	0.6	0.9	1.2	1.5
L					

(e) Use a graphing utility to graph the function over the appropriate domain.

(f) Find $\lim_{\phi \to (\pi/2)^-} L$.

(g) Use a geometric argument as the basis of a second method of finding the limit in part (f).

(h) Find $\lim_{\phi \to 0^+} L$.

True or False? **In Exercises 65–68, determine whether the statement is true or false. If it is false, explain why or give an example that shows it is false.**

65. The graph of a function cannot cross a vertical asymptote.

66. The graphs of polynomial functions have no vertical asymptotes.

67. The graphs of trigonometric functions have no vertical asymptotes.

68. If f has a vertical asymptote at $x = 0$, then f is undefined at $x = 0$.

69. Finding Functions Find functions f and g such that $\lim_{x \to c} f(x) = \infty$ and $\lim_{x \to c} g(x) = \infty$, but $\lim_{x \to c} [f(x) - g(x)] \neq 0$.

70. Proof Prove the difference, product, and quotient properties in Theorem 1.15.

71. Proof Prove that if $\lim_{x \to c} f(x) = \infty$, then $\lim_{x \to c} \frac{1}{f(x)} = 0$.

72. Proof Prove that if

$$\lim_{x \to c} \frac{1}{f(x)} = 0$$

then $\lim_{x \to c} f(x)$ does not exist.

Infinite Limits **In Exercises 73–76, use the ε–δ definition of infinite limits to prove the statement.**

73. $\lim_{x \to 3^+} \frac{1}{x - 3} = \infty$

74. $\lim_{x \to 5^-} \frac{1}{x - 5} = -\infty$

75. $\lim_{x \to 8^+} \frac{3}{8 - x} = -\infty$

76. $\lim_{x \to 9^-} \frac{6}{9 - x} = \infty$

SECTION PROJECT

Graphs and Limits of Trigonometric Functions

Recall from Theorem 1.9 that the limit of

$$f(x) = \frac{\sin x}{x}$$

as x approaches 0 is 1.

(a) Use a graphing utility to graph the function f on the interval $-\pi \leq x \leq \pi$. Explain how the graph helps confirm this theorem.

(b) Explain how you could use a table of values to confirm the value of this limit numerically.

(c) Graph $g(x) = \sin x$ by hand. Sketch a tangent line at the point $(0, 0)$ and visually estimate the slope of this tangent line.

(d) Let $(x, \sin x)$ be a point on the graph of g near $(0, 0)$, and write a formula for the slope of the secant line joining $(x, \sin x)$ and $(0, 0)$. Evaluate this formula at $x = 0.1$ and $x = 0.01$. Then find the exact slope of the tangent line to g at the point $(0, 0)$.

(e) Sketch the graph of the cosine function $h(x) = \cos x$. What is the slope of the tangent line at the point $(0, 1)$? Use limits to find this slope analytically.

(f) Find the slope of the tangent line to $k(x) = \tan x$ at $(0, 0)$.

Review Exercises

See CalcChat.com for tutorial help and worked-out solutions to odd-numbered exercises.

Precalculus or Calculus **In Exercises 1 and 2, decide whether the problem can be solved using precalculus or whether calculus is required. If the problem can be solved using precalculus, solve it. If the problem seems to require calculus, explain your reasoning and use a graphical or numerical approach to estimate the solution.**

1. Find the distance between the points $(1, 1)$ and $(3, 9)$ along the curve $y = x^2$.

2. Find the distance between the points $(1, 1)$ and $(3, 9)$ along the line $y = 4x - 3$.

Estimating a Limit Numerically **In Exercises 3 and 4, complete the table and use the result to estimate the limit. Use a graphing utility to graph the function to confirm your result.**

3. $\lim\limits_{x \to 3} \dfrac{x - 3}{x^2 - 7x + 12}$

x	2.9	2.99	2.999	3	3.001	3.01	3.1
$f(x)$				?			

4. $\lim\limits_{x \to 0} \dfrac{\sqrt{x + 4} - 2}{x}$

x	−0.1	−0.01	−0.001	0	0.001	0.01	0.1
$f(x)$				?			

Finding a Limit Graphically **In Exercises 5 and 6, use the graph to find the limit (if it exists). If the limit does not exist, explain why.**

5. $h(x) = \left[\!\left[-\dfrac{x}{2} \right]\!\right] + x^2$

6. $g(x) = \dfrac{-2x}{x - 3}$

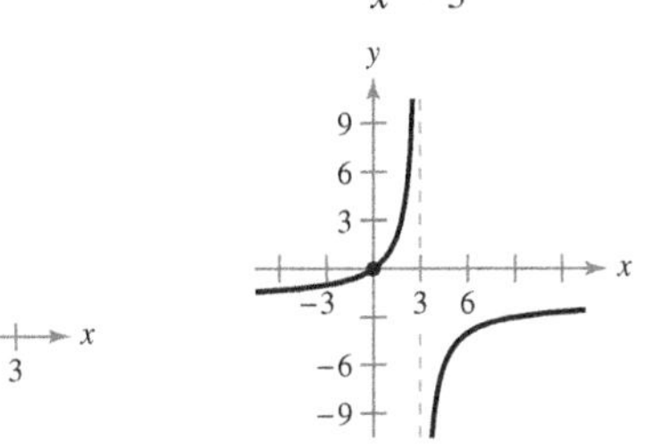

5. (a) $\lim\limits_{x \to 2} h(x)$ (b) $\lim\limits_{x \to 1} h(x)$

6. (a) $\lim\limits_{x \to 3} g(x)$ (b) $\lim\limits_{x \to 0} g(x)$

Using the ε-δ Definition of a Limit **In Exercises 7–10, find the limit L. Then use the ε-δ definition to prove that the limit is L.**

7. $\lim\limits_{x \to 1} (x + 4)$

8. $\lim\limits_{x \to 9} \sqrt{x}$

9. $\lim\limits_{x \to 2} (1 - x^2)$

10. $\lim\limits_{x \to 5} 9$

Finding a Limit **In Exercises 11–28, find the limit.**

11. $\lim\limits_{x \to -6} x^2$

12. $\lim\limits_{x \to 0} (5x - 3)$

13. $\lim\limits_{t \to 4} \sqrt{t + 2}$

14. $\lim\limits_{x \to 2} \sqrt{x^3 + 1}$

15. $\lim\limits_{x \to 27} \left(\sqrt[3]{x} - 1\right)^4$

16. $\lim\limits_{x \to 7} (x - 4)^3$

17. $\lim\limits_{x \to 4} \dfrac{4}{x - 1}$

18. $\lim\limits_{x \to 2} \dfrac{x}{x^2 + 1}$

19. $\lim\limits_{x \to -3} \dfrac{2x^2 + 11x + 15}{x + 3}$

20. $\lim\limits_{t \to 4} \dfrac{t^2 - 16}{t - 4}$

21. $\lim\limits_{x \to 4} \dfrac{\sqrt{x - 3} - 1}{x - 4}$

22. $\lim\limits_{x \to 0} \dfrac{\sqrt{4 + x} - 2}{x}$

23. $\lim\limits_{x \to 0} \dfrac{[1/(x + 1)] - 1}{x}$

24. $\lim\limits_{s \to 0} \dfrac{\left(1/\sqrt{1 + s}\right) - 1}{s}$

25. $\lim\limits_{x \to 0} \dfrac{1 - \cos x}{\sin x}$

26. $\lim\limits_{x \to \pi/4} \dfrac{4x}{\tan x}$

27. $\lim\limits_{\Delta x \to 0} \dfrac{\sin[(\pi/6) + \Delta x] - (1/2)}{\Delta x}$

[*Hint:* $\sin(\theta + \phi) = \sin\theta \cos\phi + \cos\theta \sin\phi$]

28. $\lim\limits_{\Delta x \to 0} \dfrac{\cos(\pi + \Delta x) + 1}{\Delta x}$

[*Hint:* $\cos(\theta + \phi) = \cos\theta \cos\phi - \sin\theta \sin\phi$]

Evaluating a Limit **In Exercises 29–32, evaluate the limit given $\lim\limits_{x \to c} f(x) = -6$ and $\lim\limits_{x \to c} g(x) = \frac{1}{2}$.**

29. $\lim\limits_{x \to c} [f(x)g(x)]$

30. $\lim\limits_{x \to c} \dfrac{f(x)}{g(x)}$

31. $\lim\limits_{x \to c} [f(x) + 2g(x)]$

32. $\lim\limits_{x \to c} [f(x)]^2$

Graphical, Numerical, and Analytic Analysis **In Exercises 33–36, use a graphing utility to graph the function and estimate the limit. Use a table to reinforce your conclusion. Then find the limit by analytic methods.**

33. $\lim\limits_{x \to 0} \dfrac{\sqrt{2x + 9} - 3}{x}$

34. $\lim\limits_{x \to 0} \dfrac{[1/(x + 4)] - (1/4)}{x}$

35. $\lim\limits_{x \to -9} \dfrac{x^3 + 729}{x + 9}$

36. $\lim\limits_{x \to 0} \dfrac{\cos x - 1}{x}$

Free-Falling Object **In Exercises 37 and 38, use the position function $s(t) = -4.9t^2 + 250$, which gives the height (in meters) of an object that has fallen for t seconds from a height of 250 meters. The velocity at time $t = a$ seconds is given by**

$$\lim_{t \to a} \frac{s(a) - s(t)}{a - t}.$$

37. Find the velocity of the object when $t = 4$.

38. When will the object hit the ground? At what velocity will the object impact the ground?

Finding a Limit **In Exercises 39–50, find the limit (if it exists). If it does not exist, explain why.**

39. $\lim\limits_{x \to 3^+} \dfrac{1}{x + 3}$

40. $\lim\limits_{x \to 6^-} \dfrac{x - 6}{x^2 - 36}$

41. $\lim_{x\to 25^+} \frac{\sqrt{x}-5}{x-25}$

42. $\lim_{x\to 3^-} \frac{|x-3|}{x-3}$

43. $\lim_{x\to 2} f(x)$, where $f(x) = \begin{cases} (x-2)^2, & x \le 2 \\ 2-x, & x > 2 \end{cases}$

44. $\lim_{x\to 1^+} g(x)$, where $g(x) = \begin{cases} \sqrt{1-x}, & x \le 1 \\ x+1, & x > 1 \end{cases}$

45. $\lim_{t\to 1} h(t)$, where $h(t) = \begin{cases} t^3+1, & t < 1 \\ \frac{1}{2}(t+1), & t \ge 1 \end{cases}$

46. $\lim_{s\to -2} f(s)$, where $f(s) = \begin{cases} -s^2-4s-2, & s \le -2 \\ s^2+4s+6, & s > -2 \end{cases}$

47. $\lim_{x\to 2^-} (2[\![x]\!]+1)$

48. $\lim_{x\to 4} [\![x-1]\!]$

49. $\lim_{x\to 2^-} \frac{x^2-4}{|x-2|}$

50. $\lim_{x\to 1^+} \sqrt{x(x-1)}$

Continuity on a Closed Interval **In Exercises 51 and 52, discuss the continuity of the function on the closed interval.**

51. $g(x) = \sqrt{8-x^3}$, $[-2, 2]$

52. $h(x) = \frac{3}{5-x}$, $[0, 5]$

Removable and Nonremovable Discontinuities **In Exercises 53–58, find the x-values (if any) at which f is not continuous. Which of the discontinuities are removable?**

53. $f(x) = x^4 - 81x$

54. $f(x) = x^2 - x + 20$

55. $f(x) = \frac{4}{x-5}$

56. $f(x) = \frac{1}{x^2-9}$

57. $f(x) = \frac{x}{x^3-x}$

58. $f(x) = \frac{x+3}{x^2-3x-18}$

59. **Making a Function Continuous** Find the value of c such that the function is continuous on the entire real number line.

$$f(x) = \begin{cases} x+3, & x \le 2 \\ cx+6, & x > 2 \end{cases}$$

60. **Making a Function Continuous** Find the values of b and c such that the function is continuous on the entire real number line.

$$f(x) = \begin{cases} x+1, & 1 < x < 3 \\ x^2+bx+c, & |x-2| \ge 1 \end{cases}$$

Testing for Continuity **In Exercises 61–66, describe the intervals on which the function is continuous.**

61. $f(x) = -3x^2 + 7$

62. $f(x) = \frac{4x^2+7x-2}{x+2}$

63. $f(x) = \sqrt{x} + \cos x$

64. $f(x) = [\![x+3]\!]$

65. $f(x) = \begin{cases} \frac{3x^2-x-2}{x-1}, & x \ne 1 \\ 0, & x = 1 \end{cases}$

66. $f(x) = \begin{cases} 5-x, & x \le 2 \\ 2x-3, & x > 2 \end{cases}$

67. **Using the Intermediate Value Theorem** Use the Intermediate Value Theorem to show that

$$f(x) = 2x^3 - 3$$

has a zero in the interval $[1, 2]$.

68. **Using the Intermediate Value Theorem** Use the Intermediate Value Theorem to show that

$$f(x) = x^2 + x - 2$$

has at least two zeros in the interval $[-3, 3]$.

Using the Intermediate Value Theorem **In Exercises 69 and 70, verify that the Intermediate Value Theorem applies to the indicated interval and find the value of c guaranteed by the theorem.**

69. $f(x) = x^2 + 5x - 4$, $[-1, 2]$, $f(c) = 2$

70. $f(x) = (x-6)^3 + 4$, $[4, 7]$, $f(c) = 3$

Determining Infinite Limits **In Exercises 71 and 72, determine whether $f(x)$ approaches ∞ or $-\infty$ as x approaches 6 from the left and from the right.**

71. $f(x) = \frac{1}{x-6}$

72. $f(x) = \frac{-1}{(x-6)^2}$

Finding Vertical Asymptotes **In Exercises 73–78, find the vertical asymptotes (if any) of the graph of the function.**

73. $f(x) = \frac{3}{x}$

74. $f(x) = \frac{5}{(x-2)^4}$

75. $f(x) = \frac{x^3}{x^2-9}$

76. $h(x) = \frac{6x}{36-x^2}$

77. $f(x) = \sec \frac{\pi x}{2}$

78. $f(x) = \csc \pi x$

Finding a One-Sided Limit **In Exercises 79–88, find the one-sided limit (if it exists).**

79. $\lim_{x\to 1^-} \frac{x^2+2x+1}{x-1}$

80. $\lim_{x\to (1/2)^+} \frac{x}{2x-1}$

81. $\lim_{x\to -1^+} \frac{x+1}{x^3+1}$

82. $\lim_{x\to -1^-} \frac{x+1}{x^4-1}$

83. $\lim_{x\to 0^+} \left(x - \frac{1}{x^3}\right)$

84. $\lim_{x\to 2^-} \frac{1}{\sqrt[3]{x^2-4}}$

85. $\lim_{x\to 0^+} \frac{\sin 4x}{5x}$

86. $\lim_{x\to 0^-} \frac{\sec x^3}{2x}$

87. $\lim_{x\to 0^+} \frac{\csc 2x}{x}$

88. $\lim_{x\to 0^-} \frac{\cos^2 x}{x}$

89. **Environment** A utility company burns coal to generate electricity. The cost C in dollars of removing $p\%$ of the air pollutants in the stack emissions is

$$C = \frac{80{,}000p}{100-p}, \quad 0 \le p < 100.$$

(a) Find the cost of removing 50% of the pollutants.

(b) Find the cost of removing 90% of the pollutants.

(c) Find the limit of C as p approaches 100 from the left and interpret its meaning.

P.S. Problem Solving

See CalcChat.com for tutorial help and worked-out solutions to odd-numbered exercises.

1. Perimeter Let $P(x, y)$ be a point on the parabola $y = x^2$ in the first quadrant. Consider the triangle $\triangle PAO$ formed by P, $A(0, 1)$, and the origin $O(0, 0)$, and the triangle $\triangle PBO$ formed by P, $B(1, 0)$, and the origin (see figure).

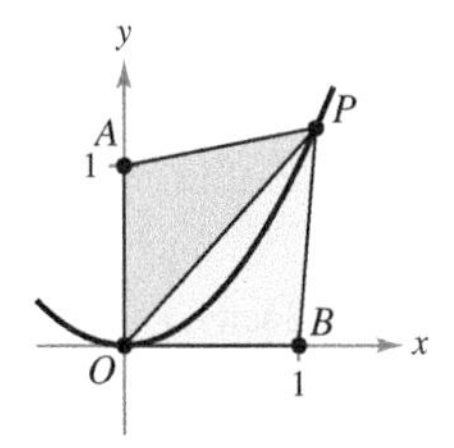

(a) Write the perimeter of each triangle in terms of x.

(b) Let $r(x)$ be the ratio of the perimeters of the two triangles,

$$r(x) = \frac{\text{Perimeter } \triangle PAO}{\text{Perimeter } \triangle PBO}.$$

Complete the table. Calculate $\lim_{x \to 0^+} r(x)$.

x	4	2	1	0.1	0.01
Perimeter $\triangle PAO$					
Perimeter $\triangle PBO$					
$r(x)$					

2. Area Let $P(x, y)$ be a point on the parabola $y = x^2$ in the first quadrant. Consider the triangle $\triangle PAO$ formed by P, $A(0, 1)$, and the origin $O(0, 0)$, and the triangle $\triangle PBO$ formed by P, $B(1, 0)$, and the origin (see figure).

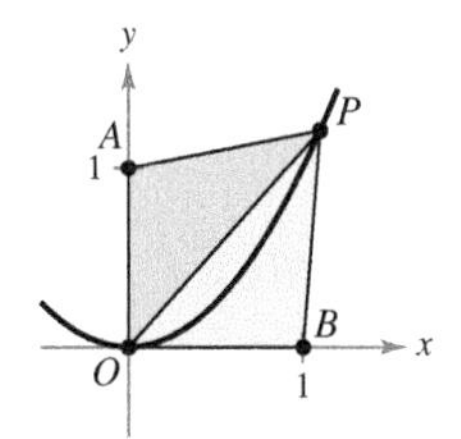

(a) Write the area of each triangle in terms of x.

(b) Let $a(x)$ be the ratio of the areas of the two triangles,

$$a(x) = \frac{\text{Area } \triangle PBO}{\text{Area } \triangle PAO}.$$

Complete the table. Calculate $\lim_{x \to 0^+} a(x)$.

x	4	2	1	0.1	0.01
Area $\triangle PAO$					
Area $\triangle PBO$					
$a(x)$					

3. Area of a Circle

(a) Find the area of a regular hexagon inscribed in a circle of radius 1 (see figure). How close is this area to that of the circle?

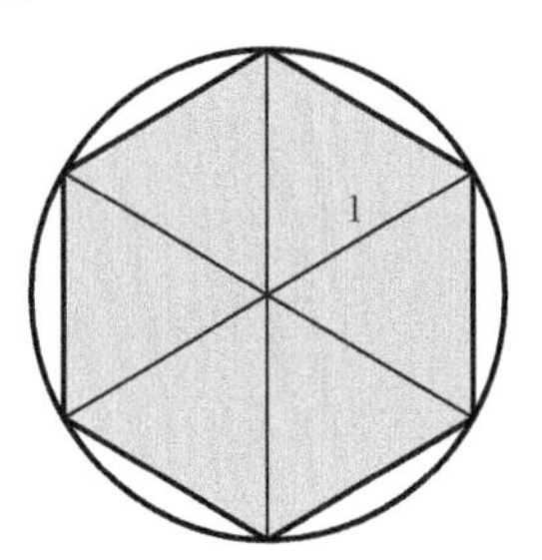

(b) Find the area A_n of an n-sided regular polygon inscribed in a circle of radius 1. Write your answer as a function of n.

(c) Complete the table. What number does A_n approach as n gets larger and larger?

n	6	12	24	48	96
A_n					

4. Tangent Line Let $P(3, 4)$ be a point on the circle $x^2 + y^2 = 25$ (see figure).

(a) What is the slope of the line joining P and $O(0, 0)$?

(b) Find an equation of the tangent line to the circle at P.

(c) Let $Q(x, y)$ be another point on the circle in the first quadrant. Find the slope m_x of the line joining P and Q in terms of x.

(d) Calculate $\lim_{x \to 3} m_x$. How does this number relate to your answer in part (b)?

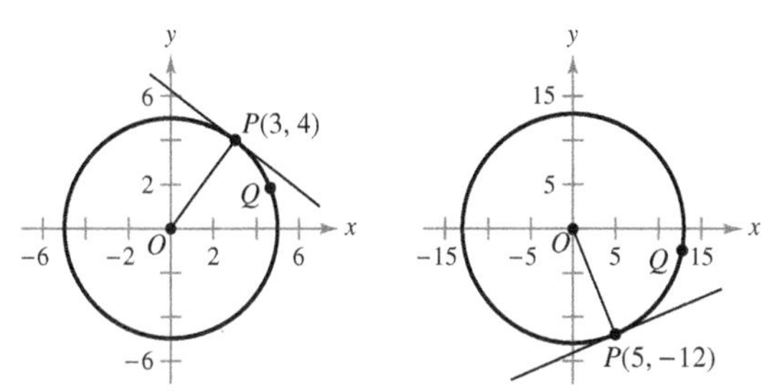

Figure for 4 Figure for 5

5. Tangent Line Let $P(5, -12)$ be a point on the circle $x^2 + y^2 = 169$ (see figure).

(a) What is the slope of the line joining P and $O(0, 0)$?

(b) Find an equation of the tangent line to the circle at P.

(c) Let $Q(x, y)$ be another point on the circle in the fourth quadrant. Find the slope m_x of the line joining P and Q in terms of x.

(d) Calculate $\lim_{x \to 5} m_x$. How does this number relate to your answer in part (b)?

6. Finding Values Find the values of the constants a and b such that

$$\lim_{x\to 0} \frac{\sqrt{a+bx}-\sqrt{3}}{x} = \sqrt{3}.$$

7. Finding Limits Consider the function

$$f(x) = \frac{\sqrt{3+x^{1/3}}-2}{x-1}.$$

(a) Find the domain of f.

(b) Use a graphing utility to graph the function.

(c) Find $\lim_{x\to -27^+} f(x)$.

(d) Find $\lim_{x\to 1} f(x)$.

8. Making a Function Continuous Find all values of the constant a such that f is continuous for all real numbers.

$$f(x) = \begin{cases} \dfrac{ax}{\tan x}, & x \geq 0 \\ a^2 - 2, & x < 0 \end{cases}$$

9. Choosing Graphs Consider the graphs of the four functions g_1, g_2, g_3, and g_4.

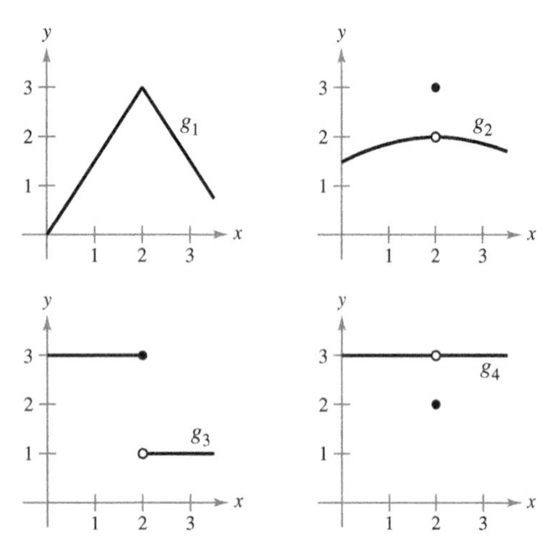

For each given condition of the function f, which of the graphs could be the graph of f?

(a) $\lim_{x\to 2} f(x) = 3$

(b) f is continuous at 2.

(c) $\lim_{x\to 2^-} f(x) = 3$

10. Limits and Continuity Sketch the graph of the function

$$f(x) = \left[\!\left[\frac{1}{x}\right]\!\right].$$

(a) Evaluate $f\left(\frac{1}{4}\right)$, $f(3)$, and $f(1)$.

(b) Evaluate the limits $\lim_{x\to 1^-} f(x)$, $\lim_{x\to 1^+} f(x)$, $\lim_{x\to 0^-} f(x)$, and $\lim_{x\to 0^+} f(x)$.

(c) Discuss the continuity of the function.

11. Limits and Continuity Sketch the graph of the function $f(x) = [\![x]\!] + [\![-x]\!]$.

(a) Evaluate $f(1)$, $f(0)$, $f\left(\frac{1}{2}\right)$, and $f(-2.7)$.

(b) Evaluate the limits $\lim_{x\to 1^-} f(x)$, $\lim_{x\to 1^+} f(x)$, and $\lim_{x\to 1/2} f(x)$.

(c) Discuss the continuity of the function.

12. Escape Velocity To escape Earth's gravitational field, a rocket must be launched with an initial velocity called the **escape velocity.** A rocket launched from the surface of Earth has velocity v (in miles per second) given by

$$v = \sqrt{\frac{2GM}{r} + v_0^2 - \frac{2GM}{R}} \approx \sqrt{\frac{192{,}000}{r} + v_0^2 - 48}$$

where v_0 is the initial velocity, r is the distance from the rocket to the center of Earth, G is the gravitational constant, M is the mass of Earth, and R is the radius of Earth (approximately 4000 miles).

(a) Find the value of v_0 for which you obtain an infinite limit for r as v approaches zero. This value of v_0 is the escape velocity for Earth.

(b) A rocket launched from the surface of the moon has velocity v (in miles per second) given by

$$v = \sqrt{\frac{1920}{r} + v_0^2 - 2.17}.$$

Find the escape velocity for the moon.

(c) A rocket launched from the surface of a planet has velocity v (in miles per second) given by

$$v = \sqrt{\frac{10{,}600}{r} + v_0^2 - 6.99}.$$

Find the escape velocity for this planet. Is the mass of this planet larger or smaller than that of Earth? (Assume that the mean density of this planet is the same as that of Earth.)

13. Pulse Function For positive numbers $a < b$, the **pulse function** is defined as

$$P_{a,b}(x) = H(x-a) - H(x-b) = \begin{cases} 0, & x < a \\ 1, & a \leq x < b \\ 0, & x \geq b \end{cases}$$

where $H(x) = \begin{cases} 1, & x \geq 0 \\ 0, & x < 0 \end{cases}$ is the Heaviside function.

(a) Sketch the graph of the pulse function.

(b) Find the following limits:

(i) $\lim_{x\to a^+} P_{a,b}(x)$ (ii) $\lim_{x\to a^-} P_{a,b}(x)$

(iii) $\lim_{x\to b^+} P_{a,b}(x)$ (iv) $\lim_{x\to b^-} P_{a,b}(x)$

(c) Discuss the continuity of the pulse function.

(d) Why is $U(x) = \dfrac{1}{b-a}P_{a,b}(x)$ called the **unit pulse function?**

14. Proof Let a be a nonzero constant. Prove that if $\lim_{x\to 0} f(x) = L$, then $\lim_{x\to 0} f(ax) = L$. Show by means of an example that a must be nonzero.

2 Differentiation

Rate of Change (Example 2, p. 153)

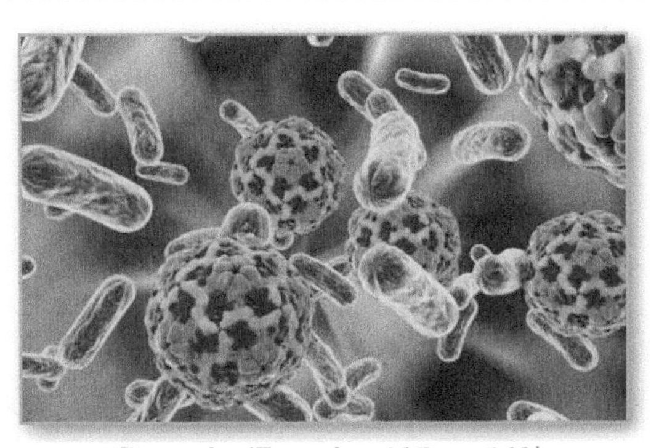

Bacteria (Exercise 107, p. 143)

Acceleration Due to Gravity (Example 10, p. 128)

Velocity of a Falling Object (Example 9, p. 116)

Stopping Distance (Exercise 103, p.121)

2.1 The Derivative and the Tangent Line Problem

- **Find the slope of the tangent line to a curve at a point.**
- **Use the limit definition to find the derivative of a function.**
- **Understand the relationship between differentiability and continuity.**

The Tangent Line Problem

Calculus grew out of four major problems that European mathematicians were working on during the seventeenth century.

1. The tangent line problem (Section 1.1 and this section)
2. The velocity and acceleration problem (Sections 2.2 and 2.3)
3. The minimum and maximum problem (Section 3.1)
4. The area problem (Sections 1.1 and 4.2)

Each problem involves the notion of a limit, and calculus can be introduced with any of the four problems.

A brief introduction to the tangent line problem is given in Section 1.1. Although partial solutions to this problem were given by Pierre de Fermat (1601–1665), René Descartes (1596–1650), Christian Huygens (1629–1695), and Isaac Barrow (1630–1677), credit for the first general solution is usually given to Isaac Newton (1642–1727) and Gottfried Leibniz (1646–1716). Newton's work on this problem stemmed from his interest in optics and light refraction.

ISAAC NEWTON (1642–1727)

In addition to his work in calculus, Newton made revolutionary contributions to physics, including the Law of Universal Gravitation and his three laws of motion.

See LarsonCalculus.com to read more of this biography.

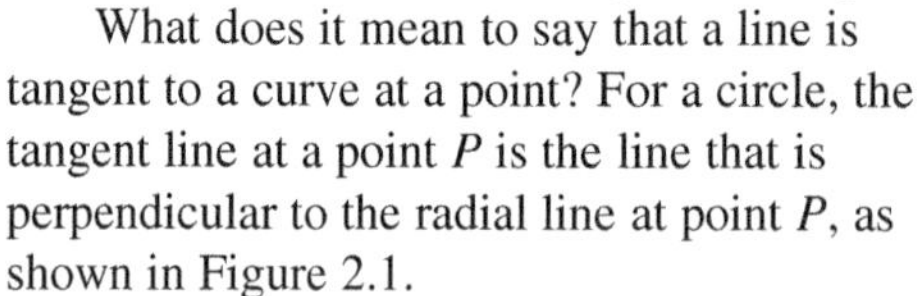

What does it mean to say that a line is tangent to a curve at a point? For a circle, the tangent line at a point P is the line that is perpendicular to the radial line at point P, as shown in Figure 2.1.

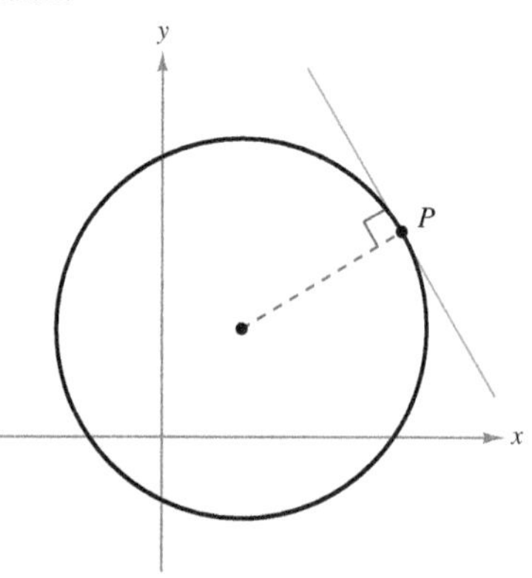

Tangent line to a circle
Figure 2.1

For a general curve, however, the problem is more difficult. For instance, how would you define the tangent lines shown in Figure 2.2? You might say that a line is tangent to a curve at a point P when it touches, but does not cross, the curve at point P. This definition would work for the first curve shown in Figure 2.2 but not for the second. *Or* you might say that a line is tangent to a curve when the line touches or intersects the curve at exactly one point. This definition would work for a circle but not for more general curves, as the third curve in Figure 2.2 shows.

Exploration

Use a graphing utility to graph $f(x) = 2x^3 - 4x^2 + 3x - 5$. On the same screen, graph $y = x - 5$, $y = 2x - 5$, and $y = 3x - 5$. Which of these lines, if any, appears to be tangent to the graph of f at the point $(0, -5)$? Explain your reasoning.

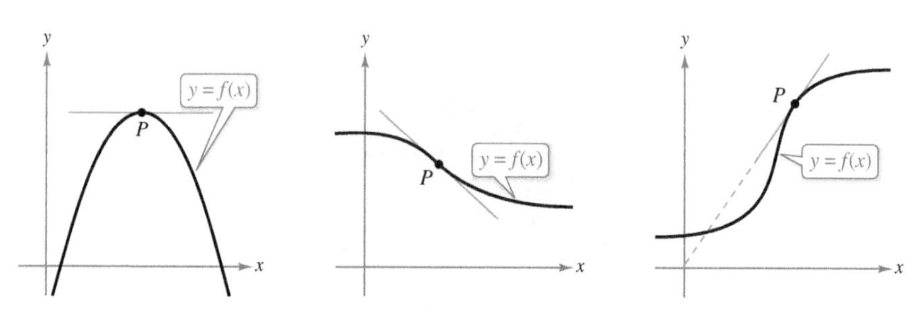

Tangent line to a curve at a point
Figure 2.2

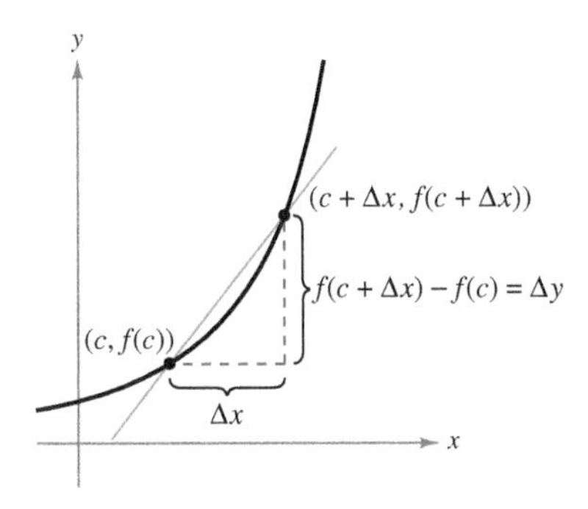

The secant line through $(c, f(c))$ and $(c + \Delta x, f(c + \Delta x))$
Figure 2.3

Essentially, the problem of finding the tangent line at a point P boils down to the problem of finding the *slope* of the tangent line at point P. You can approximate this slope using a **secant line*** through the point of tangency and a second point on the curve, as shown in Figure 2.3. If $(c, f(c))$ is the point of tangency and

$$(c + \Delta x, f(c + \Delta x))$$

is a second point on the graph of f, then the slope of the secant line through the two points is given by substitution into the slope formula

$$m = \frac{y_2 - y_1}{x_2 - x_1}$$

$$m_{\text{sec}} = \frac{f(c + \Delta x) - f(c)}{(c + \Delta x) - c} \qquad \frac{\text{Change in } y}{\text{Change in } x}$$

$$m_{\text{sec}} = \frac{f(c + \Delta x) - f(c)}{\Delta x}. \qquad \text{Slope of secant line}$$

The right-hand side of this equation is a **difference quotient.** The denominator Δx is the **change in x,** and the numerator

$$\Delta y = f(c + \Delta x) - f(c)$$

is the **change in y.**

The beauty of this procedure is that you can obtain more and more accurate approximations of the slope of the tangent line by choosing points closer and closer to the point of tangency, as shown in Figure 2.4.

THE TANGENT LINE PROBLEM

In 1637, mathematician René Descartes stated this about the tangent line problem:

"And I dare say that this is not only the most useful and general problem in geometry that I know, but even that I ever desire to know."

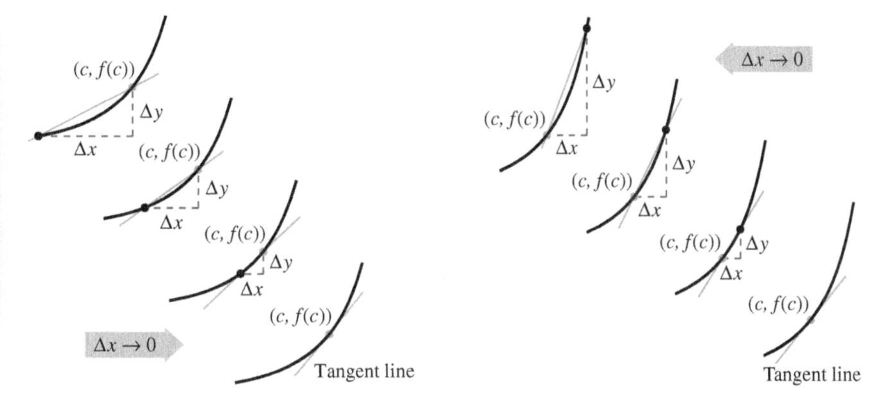

Tangent line approximations
Figure 2.4

Definition of Tangent Line with Slope m

If f is defined on an open interval containing c, and if the limit

$$\lim_{\Delta x \to 0} \frac{\Delta y}{\Delta x} = \lim_{\Delta x \to 0} \frac{f(c + \Delta x) - f(c)}{\Delta x} = m$$

exists, then the line passing through $(c, f(c))$ with slope m is the **tangent line** to the graph of f at the point $(c, f(c))$.

The slope of the tangent line to the graph of f at the point $(c, f(c))$ is also called the **slope of the graph of f at $x = c$.**

* This use of the word *secant* comes from the Latin *secare*, meaning to cut, and is not a reference to the trigonometric function of the same name.

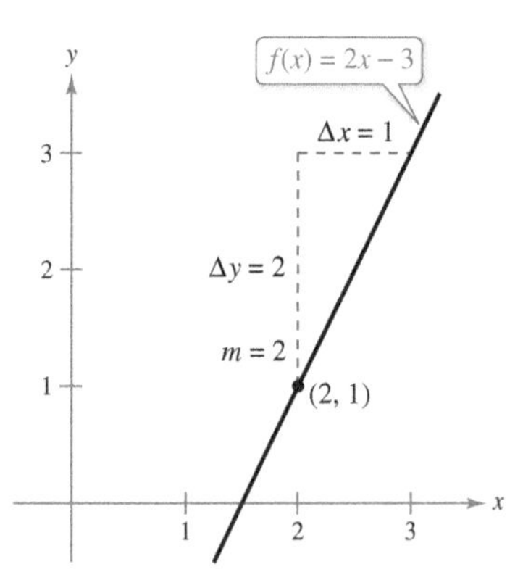

The slope of f at (2, 1) is $m = 2$.
Figure 2.5

EXAMPLE 1 The Slope of the Graph of a Linear Function

To find the slope of the graph of $f(x) = 2x - 3$ when $c = 2$, you can apply the definition of the slope of a tangent line, as shown.

$$\begin{aligned}
\lim_{\Delta x \to 0} \frac{f(2 + \Delta x) - f(2)}{\Delta x} &= \lim_{\Delta x \to 0} \frac{[2(2 + \Delta x) - 3] - [2(2) - 3]}{\Delta x} \\
&= \lim_{\Delta x \to 0} \frac{4 + 2\Delta x - 3 - 4 + 3}{\Delta x} \\
&= \lim_{\Delta x \to 0} \frac{2\cancel{\Delta x}}{\cancel{\Delta x}} \\
&= \lim_{\Delta x \to 0} 2 \\
&= 2
\end{aligned}$$

The slope of f at $(c, f(c)) = (2, 1)$ is $m = 2$, as shown in Figure 2.5. Notice that the limit definition of the slope of f agrees with the definition of the slope of a line as discussed in Section P.2.

The graph of a linear function has the same slope at any point. This is not true of nonlinear functions, as shown in the next example.

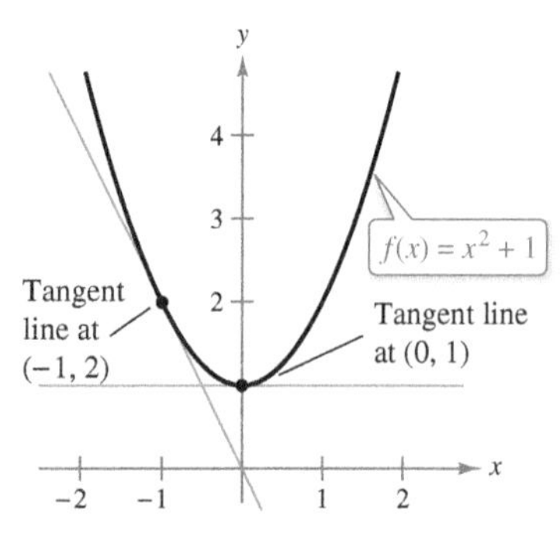

The slope of f at any point $(c, f(c))$ is $m = 2c$.
Figure 2.6

EXAMPLE 2 Tangent Lines to the Graph of a Nonlinear Function

Find the slopes of the tangent lines to the graph of $f(x) = x^2 + 1$ at the points (0, 1) and (−1, 2), as shown in Figure 2.6.

Solution Let $(c, f(c))$ represent an arbitrary point on the graph of f. Then the slope of the tangent line at $(c, f(c))$ can be found as shown below. [Note in the limit process that c is held constant (as Δx approaches 0).]

$$\begin{aligned}
\lim_{\Delta x \to 0} \frac{f(c + \Delta x) - f(c)}{\Delta x} &= \lim_{\Delta x \to 0} \frac{[(c + \Delta x)^2 + 1] - (c^2 + 1)}{\Delta x} \\
&= \lim_{\Delta x \to 0} \frac{c^2 + 2c(\Delta x) + (\Delta x)^2 + 1 - c^2 - 1}{\Delta x} \\
&= \lim_{\Delta x \to 0} \frac{2c(\Delta x) + (\Delta x)^2}{\Delta x} \\
&= \lim_{\Delta x \to 0} (2c + \Delta x) \\
&= 2c
\end{aligned}$$

So, the slope at *any* point $(c, f(c))$ on the graph of f is $m = 2c$. At the point (0, 1), the slope is $m = 2(0) = 0$, and at (−1, 2), the slope is $m = 2(-1) = -2$.

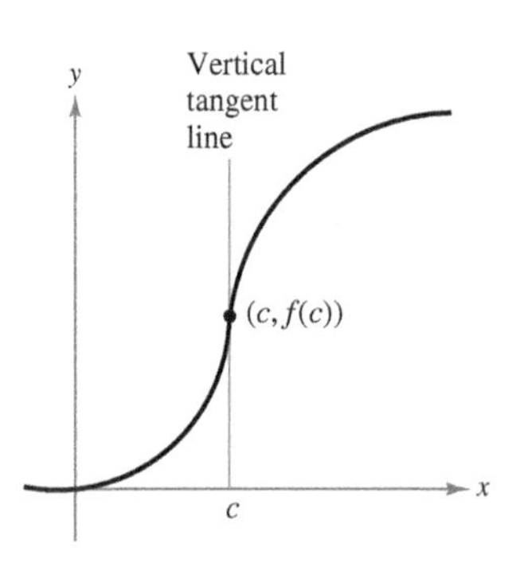

The graph of f has a vertical tangent line at $(c, f(c))$.
Figure 2.7

The definition of a tangent line to a curve does not cover the possibility of a vertical tangent line. For vertical tangent lines, you can use the following definition. If f is continuous at c and

$$\lim_{\Delta x \to 0} \frac{f(c + \Delta x) - f(c)}{\Delta x} = \infty \quad \text{or} \quad \lim_{\Delta x \to 0} \frac{f(c + \Delta x) - f(c)}{\Delta x} = -\infty$$

then the vertical line $x = c$ passing through $(c, f(c))$ is a **vertical tangent line** to the graph of f. For example, the function shown in Figure 2.7 has a vertical tangent line at $(c, f(c))$. When the domain of f is the closed interval $[a, b]$, you can extend the definition of a vertical tangent line to include the endpoints by considering continuity and limits from the right (for $x = a$) and from the left (for $x = b$).

The Derivative of a Function

You have now arrived at a crucial point in the study of calculus. The limit used to define the slope of a tangent line is also used to define one of the two fundamental operations of calculus—**differentiation.**

Definition of the Derivative of a Function

The **derivative** of f at x is

$$f'(x) = \lim_{\Delta x \to 0} \frac{f(x + \Delta x) - f(x)}{\Delta x}$$

provided the limit exists. For all x for which this limit exists, f' is a function of x.

REMARK The notation $f'(x)$ is read as "f prime of x."

Be sure you see that the derivative of a function of x is also a function of x. This "new" function gives the slope of the tangent line to the graph of f at the point $(x, f(x))$, provided that the graph has a tangent line at this point. The derivative can also be used to determine the **instantaneous rate of change** (or simply the **rate of change**) of one variable with respect to another.

The process of finding the derivative of a function is called **differentiation.** A function is **differentiable** at x when its derivative exists at x and is **differentiable on an open interval (a, b)** when it is differentiable at every point in the interval.

FOR FURTHER INFORMATION For more information on the crediting of mathematical discoveries to the first "discoverers," see the article "Mathematical Firsts—Who Done It?" by Richard H. Williams and Roy D. Mazzagatti in *Mathematics Teacher*. To view this article, go to *MathArticles.com*.

In addition to $f'(x)$, other notations are used to denote the derivative of $y = f(x)$. The most common are

$$f'(x), \quad \frac{dy}{dx}, \quad y', \quad \frac{d}{dx}[f(x)], \quad D_x[y]. \qquad \text{Notations for derivatives}$$

The notation dy/dx is read as "the derivative of y *with respect to* x" or simply "*dy*, *dx*." Using limit notation, you can write

$$\frac{dy}{dx} = \lim_{\Delta x \to 0} \frac{\Delta y}{\Delta x} = \lim_{\Delta x \to 0} \frac{f(x + \Delta x) - f(x)}{\Delta x} = f'(x).$$

EXAMPLE 3 Finding the Derivative by the Limit Process

See LarsonCalculus.com for an interactive version of this type of example.

To find the derivative of $f(x) = x^3 + 2x$, use the definition of the derivative as shown.

$$\begin{aligned}
f'(x) &= \lim_{\Delta x \to 0} \frac{f(x + \Delta x) - f(x)}{\Delta x} && \text{Definition of derivative} \\
&= \lim_{\Delta x \to 0} \frac{(x + \Delta x)^3 + 2(x + \Delta x) - (x^3 + 2x)}{\Delta x} \\
&= \lim_{\Delta x \to 0} \frac{x^3 + 3x^2\Delta x + 3x(\Delta x)^2 + (\Delta x)^3 + 2x + 2\Delta x - x^3 - 2x}{\Delta x} \\
&= \lim_{\Delta x \to 0} \frac{3x^2\Delta x + 3x(\Delta x)^2 + (\Delta x)^3 + 2\Delta x}{\Delta x} \\
&= \lim_{\Delta x \to 0} \frac{\cancel{\Delta x}[3x^2 + 3x\Delta x + (\Delta x)^2 + 2]}{\cancel{\Delta x}} \\
&= \lim_{\Delta x \to 0} [3x^2 + 3x\Delta x + (\Delta x)^2 + 2] \\
&= 3x^2 + 2
\end{aligned}$$

REMARK When using the definition to find a derivative of a function, the key is to rewrite the difference quotient so that Δx does not occur as a factor of the denominator.

REMARK Remember that the derivative of a function f is itself a function, which can be used to find the slope of the tangent line at the point $(x, f(x))$ on the graph of f.

EXAMPLE 4 Using the Derivative to Find the Slope at a Point

Find $f'(x)$ for $f(x) = \sqrt{x}$. Then find the slopes of the graph of f at the points $(1, 1)$ and $(4, 2)$. Discuss the behavior of f at $(0, 0)$.

Solution Use the procedure for rationalizing numerators, as discussed in Section 1.3.

$$f'(x) = \lim_{\Delta x \to 0} \frac{f(x + \Delta x) - f(x)}{\Delta x} \qquad \text{Definition of derivative}$$

$$= \lim_{\Delta x \to 0} \frac{\sqrt{x + \Delta x} - \sqrt{x}}{\Delta x}$$

$$= \lim_{\Delta x \to 0} \left(\frac{\sqrt{x + \Delta x} - \sqrt{x}}{\Delta x}\right)\left(\frac{\sqrt{x + \Delta x} + \sqrt{x}}{\sqrt{x + \Delta x} + \sqrt{x}}\right)$$

$$= \lim_{\Delta x \to 0} \frac{(x + \Delta x) - x}{\Delta x\left(\sqrt{x + \Delta x} + \sqrt{x}\right)}$$

$$= \lim_{\Delta x \to 0} \frac{\cancel{\Delta x}}{\cancel{\Delta x}\left(\sqrt{x + \Delta x} + \sqrt{x}\right)}$$

$$= \lim_{\Delta x \to 0} \frac{1}{\sqrt{x + \Delta x} + \sqrt{x}}$$

$$= \frac{1}{2\sqrt{x}}$$

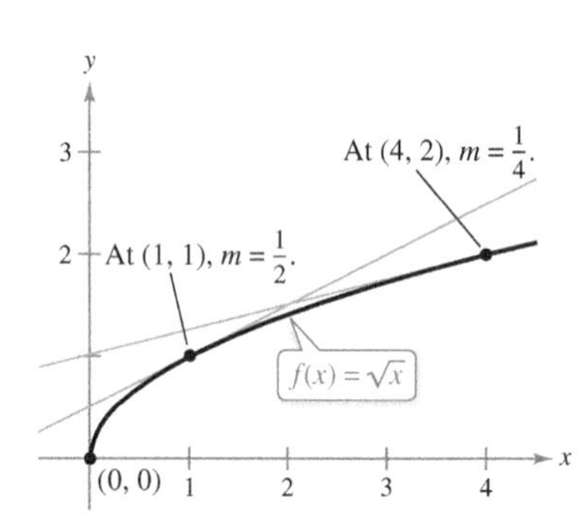

For $x > 0$, the slope of f at $(x, f(x))$ is $m = 1/(2\sqrt{x})$.
Figure 2.8

At the point $(1, 1)$, the slope is $f'(1) = \frac{1}{2}$. At the point $(4, 2)$, the slope is $f'(4) = \frac{1}{4}$. See Figure 2.8. The domain of f' is all $x > 0$, so the slope of f is undefined at $(0, 0)$. Moreover, the graph of f has a vertical tangent line at $(0, 0)$.

REMARK In many applications, it is convenient to use a variable other than x as the independent variable, as shown in Example 5.

EXAMPLE 5 Finding the Derivative of a Function

See LarsonCalculus.com for an interactive version of this type of example.

Find the derivative with respect to t for the function $y = 2/t$.

Solution Considering $y = f(t)$, you obtain

$$\frac{dy}{dt} = \lim_{\Delta t \to 0} \frac{f(t + \Delta t) - f(t)}{\Delta t} \qquad \text{Definition of derivative}$$

$$= \lim_{\Delta t \to 0} \frac{\dfrac{2}{t + \Delta t} - \dfrac{2}{t}}{\Delta t} \qquad f(t + \Delta t) = \frac{2}{t + \Delta t} \text{ and } f(t) = \frac{2}{t}$$

$$= \lim_{\Delta t \to 0} \frac{\dfrac{2t - 2(t + \Delta t)}{t(t + \Delta t)}}{\Delta t} \qquad \text{Combine fractions in numerator.}$$

$$= \lim_{\Delta t \to 0} \frac{-2\cancel{\Delta t}}{\cancel{\Delta t}(t)(t + \Delta t)} \qquad \text{Divide out common factor of } \Delta t.$$

$$= \lim_{\Delta t \to 0} \frac{-2}{t(t + \Delta t)} \qquad \text{Simplify.}$$

$$= -\frac{2}{t^2}. \qquad \text{Evaluate limit as } \Delta t \to 0.$$

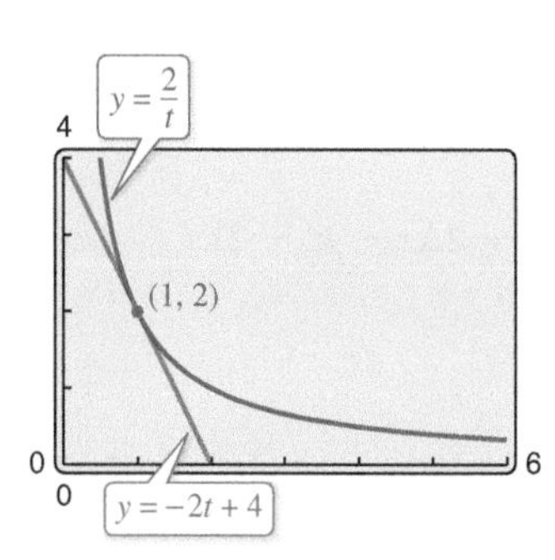

At the point $(1, 2)$, the line $y = -2t + 4$ is tangent to the graph of $y = 2/t$.
Figure 2.9

TECHNOLOGY A graphing utility can be used to reinforce the result given in Example 5. For instance, using the formula $dy/dt = -2/t^2$, you know that the slope of the graph of $y = 2/t$ at the point $(1, 2)$ is $m = -2$. Using the point-slope form, you can find that the equation of the tangent line to the graph at $(1, 2)$ is

$$y - 2 = -2(t - 1) \quad \text{or} \quad y = -2t + 4. \qquad \text{See Figure 2.9.}$$

You can also verify the result using the *tangent* feature of the graphing utility.

Differentiability and Continuity

The alternative limit form of the derivative shown below is useful in investigating the relationship between differentiability and continuity. The derivative of f at c is

$$f'(c) = \lim_{x \to c} \frac{f(x) - f(c)}{x - c}$$ Alternative form of derivative

REMARK A proof of the equivalence of the alternative form of the derivative is given in Appendix A.

provided this limit exists (see Figure 2.10).

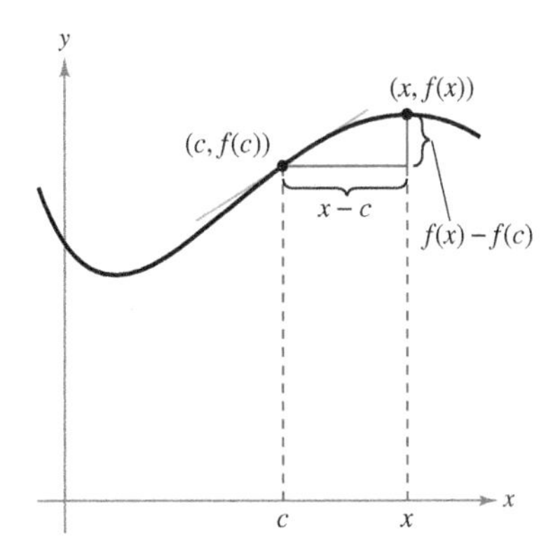

As x approaches c, the secant line approaches the tangent line.
Figure 2.10

Note that the existence of the limit in this alternative form requires that the one-sided limits

$$\lim_{x \to c^-} \frac{f(x) - f(c)}{x - c}$$

and

$$\lim_{x \to c^+} \frac{f(x) - f(c)}{x - c}$$

exist and are equal. These one-sided limits are called the **derivatives from the left and from the right,** respectively. It follows that f is **differentiable on the closed interval** $[a, b]$ when it is differentiable on (a, b) and when the derivative from the right at a and the derivative from the left at b both exist.

When a function is not continuous at $x = c$, it is also not differentiable at $x = c$. For instance, the greatest integer function

$$f(x) = [\![x]\!]$$

is not continuous at $x = 0$, and so it is not differentiable at $x = 0$ (see Figure 2.11). You can verify this by observing that

$$\lim_{x \to 0^-} \frac{f(x) - f(0)}{x - 0} = \lim_{x \to 0^-} \frac{[\![x]\!] - 0}{x} = \infty$$ Derivative from the left

and

$$\lim_{x \to 0^+} \frac{f(x) - f(0)}{x - 0} = \lim_{x \to 0^+} \frac{[\![x]\!] - 0}{x} = 0.$$ Derivative from the right

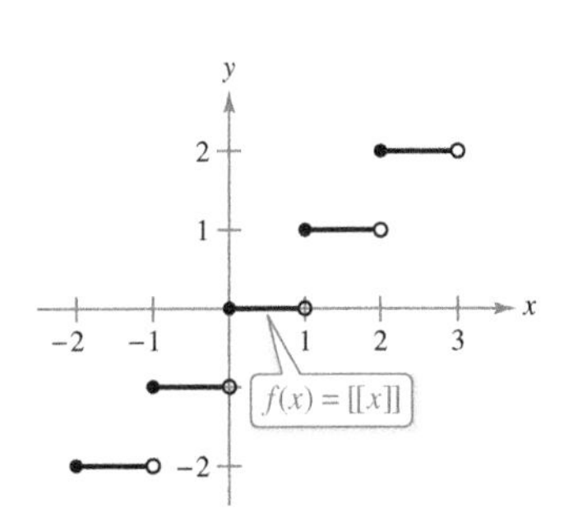

The greatest integer function is not differentiable at $x = 0$ because it is not continuous at $x = 0$.
Figure 2.11

Although it is true that differentiability implies continuity (as shown in Theorem 2.1 on the next page), the converse is not true. That is, it is possible for a function to be continuous at $x = c$ and *not* differentiable at $x = c$. Examples 6 and 7 illustrate this possibility.

EXAMPLE 6 A Graph with a Sharp Turn

See LarsonCalculus.com for an interactive version of this type of example.

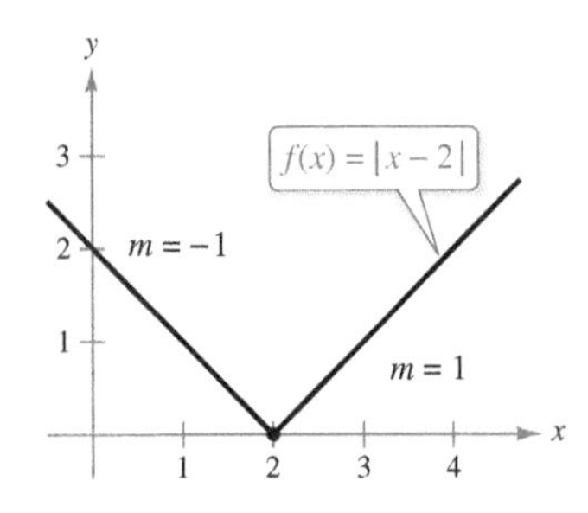

f is not differentiable at $x = 2$ because the derivatives from the left and from the right are not equal.
Figure 2.12

The function $f(x) = |x - 2|$, shown in Figure 2.12, is continuous at $x = 2$. The one-sided limits, however,

$$\lim_{x \to 2^-} \frac{f(x) - f(2)}{x - 2} = \lim_{x \to 2^-} \frac{|x - 2| - 0}{x - 2} = -1 \qquad \text{Derivative from the left}$$

and

$$\lim_{x \to 2^+} \frac{f(x) - f(2)}{x - 2} = \lim_{x \to 2^+} \frac{|x - 2| - 0}{x - 2} = 1 \qquad \text{Derivative from the right}$$

are not equal. So, f is not differentiable at $x = 2$ and the graph of f does not have a tangent line at the point $(2, 0)$.

EXAMPLE 7 A Graph with a Vertical Tangent Line

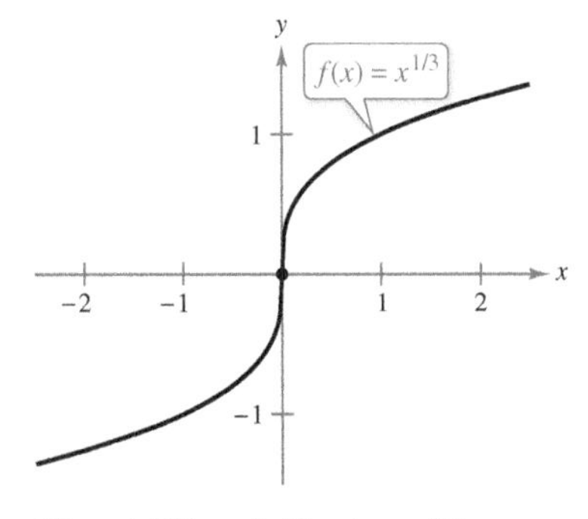

f is not differentiable at $x = 0$ because f has a vertical tangent line at $x = 0$.
Figure 2.13

The function $f(x) = x^{1/3}$ is continuous at $x = 0$, as shown in Figure 2.13. However, because the limit

$$\lim_{x \to 0} \frac{f(x) - f(0)}{x - 0} = \lim_{x \to 0} \frac{x^{1/3} - 0}{x} = \lim_{x \to 0} \frac{1}{x^{2/3}} = \infty$$

is infinite, you can conclude that the tangent line is vertical at $x = 0$. So, f is not differentiable at $x = 0$. ■

From Examples 6 and 7, you can see that a function is not differentiable at a point at which its graph has a sharp turn *or* a vertical tangent line.

THEOREM 2.1 Differentiability Implies Continuity

If f is differentiable at $x = c$, then f is continuous at $x = c$.

TECHNOLOGY Some graphing utilities, such as *Maple*, *Mathematica*, and the *TI-Nspire*, perform symbolic differentiation. Some have a *derivative* feature that performs *numerical differentiation* by finding values of derivatives using the formula

$$f'(x) \approx \frac{f(x + \Delta x) - f(x - \Delta x)}{2\Delta x}$$

where Δx is a small number such as 0.001. Can you see any problems with this definition? For instance, using this definition, what is the value of the derivative of $f(x) = |x|$ when $x = 0$?

Proof You can prove that f is continuous at $x = c$ by showing that $f(x)$ approaches $f(c)$ as $x \to c$. To do this, use the differentiability of f at $x = c$ and consider the following limit.

$$\begin{aligned}\lim_{x \to c} [f(x) - f(c)] &= \lim_{x \to c} \left[(x - c)\left(\frac{f(x) - f(c)}{x - c}\right)\right] \\ &= \left[\lim_{x \to c} (x - c)\right]\left[\lim_{x \to c} \frac{f(x) - f(c)}{x - c}\right] \\ &= (0)[f'(c)] \\ &= 0\end{aligned}$$

Because the difference $f(x) - f(c)$ approaches zero as $x \to c$, you can conclude that $\lim_{x \to c} f(x) = f(c)$. So, f is continuous at $x = c$. ■

The relationship between continuity and differentiability is summarized below.

1. If a function is differentiable at $x = c$, then it is continuous at $x = c$. So, differentiability implies continuity.
2. It is possible for a function to be continuous at $x = c$ and not be differentiable at $x = c$. So, continuity does not imply differentiability (see Examples 6 and 7).

2.1 Exercises

See CalcChat.com for tutorial help and worked-out solutions to odd-numbered exercises.

CONCEPT CHECK

1. **Tangent Line** Describe how to find the slope of the tangent line to the graph of a function at a point.
2. **Notation** List four notation alternatives to $f'(x)$.
3. **Derivative** Describe how to find the derivative of a function using the limit process.
4. **Continuity and Differentiability** Describe the relationship between continuity and differentiability.

Estimating Slope **In Exercises 5 and 6, estimate the slope of the graph at the points (x_1, y_1) and (x_2, y_2).**

5.

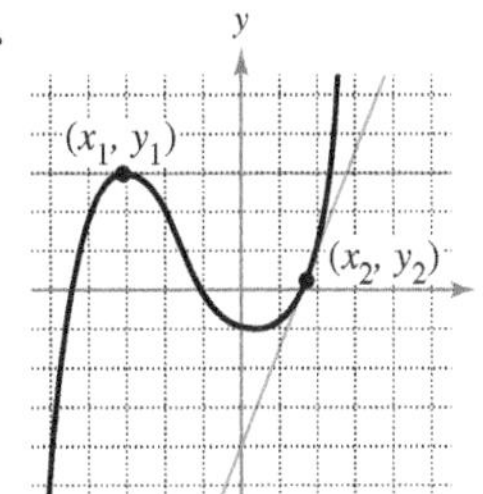

6.

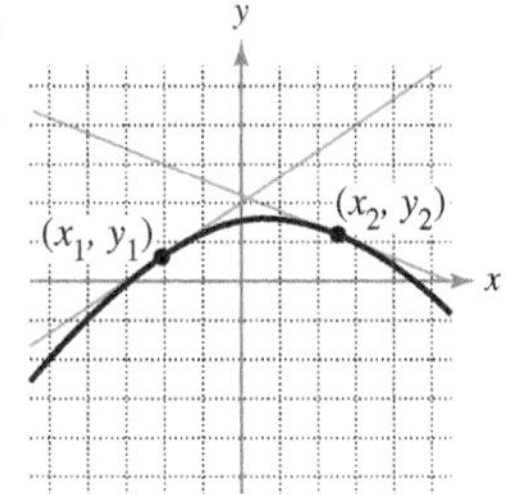

Slopes of Secant Lines **In Exercises 7 and 8, use the graph shown in the figure. To print an enlarged copy of the graph, go to *MathGraphs.com*.**

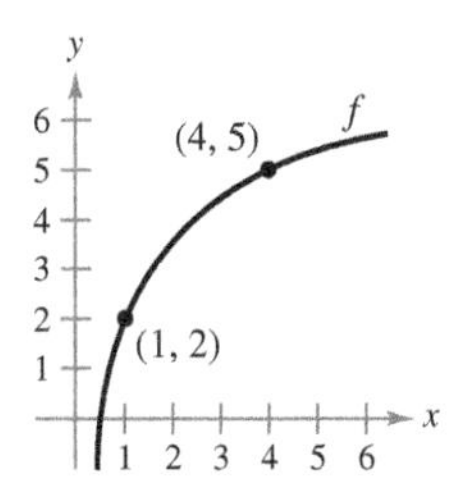

7. Identify or sketch each of the quantities on the figure.
   (a) $f(1)$ and $f(4)$ (b) $f(4) - f(1)$
   (c) $4 - 1$ (d) $y - 2 = \dfrac{f(4) - f(1)}{4 - 1}(x - 1)$

8. Insert the proper inequality symbol (< or >) between the given quantities.
   (a) $\dfrac{f(4) - f(1)}{4 - 1} \quad \square \quad \dfrac{f(4) - f(3)}{4 - 3}$
   (b) $\dfrac{f(4) - f(1)}{4 - 1} \quad \square \quad f'(1)$

Finding the Slope of a Tangent Line **In Exercises 9–14, find the slope of the tangent line to the graph of the function at the given point.**

9. $f(x) = 3 - 5x, \quad (-1, 8)$
10. $g(x) = \frac{3}{2}x + 1, \quad (-2, -2)$
11. $f(x) = 2x^2 - 3, \quad (2, 5)$
12. $f(x) = 5 - x^2, \quad (3, -4)$
13. $f(t) = 3t - t^2, \quad (0, 0)$
14. $h(t) = t^2 + 4t, \quad (1, 5)$

Finding the Derivative by the Limit Process **In Exercises 15–28, find the derivative of the function by the limit process.**

15. $f(x) = 7$
16. $g(x) = -3$
17. $f(x) = -5x$
18. $f(x) = 7x - 3$
19. $h(s) = 3 + \frac{2}{3}s$
20. $f(x) = 5 - \frac{2}{3}x$
21. $f(x) = x^2 + x - 3$
22. $f(x) = x^2 - 5$
23. $f(x) = x^3 - 12x$
24. $g(t) = t^3 + 4t$
25. $f(x) = \dfrac{1}{x - 1}$
26. $f(x) = \dfrac{1}{x^2}$
27. $f(x) = \sqrt{x + 4}$
28. $h(s) = -2\sqrt{s}$

Finding an Equation of a Tangent Line **In Exercises 29–36, (a) find an equation of the tangent line to the graph of f at the given point, (b) use a graphing utility to graph the function and its tangent line at the point, and (c) use the *tangent* feature of a graphing utility to confirm your results.**

29. $f(x) = x^2 + 3, \quad (-1, 4)$
30. $f(x) = x^2 + 2x - 1, \quad (1, 2)$
31. $f(x) = x^3, \quad (2, 8)$
32. $f(x) = x^3 + 1, \quad (-1, 0)$
33. $f(x) = \sqrt{x}, \quad (1, 1)$
34. $f(x) = \sqrt{x - 1}, \quad (5, 2)$
35. $f(x) = x + \dfrac{4}{x}, \quad (-4, -5)$
36. $f(x) = x - \dfrac{1}{x}, \quad (1, 0)$

Finding an Equation of a Tangent Line **In Exercises 37–42, find an equation of the line that is tangent to the graph of f *and* parallel to the given line.**

	Function	Line
37.	$f(x) = -\frac{1}{4}x^2$	$x + y = 0$
38.	$f(x) = 2x^2$	$4x + y + 3 = 0$
39.	$f(x) = x^3$	$3x - y + 1 = 0$
40.	$f(x) = x^3 + 2$	$3x - y - 4 = 0$
41.	$f(x) = \dfrac{1}{\sqrt{x}}$	$x + 2y - 6 = 0$
42.	$f(x) = \dfrac{1}{\sqrt{x - 1}}$	$x + 2y + 7 = 0$

Sketching a Derivative **In Exercises 43–48, sketch the graph of f'. Explain how you found your answer.**

43.

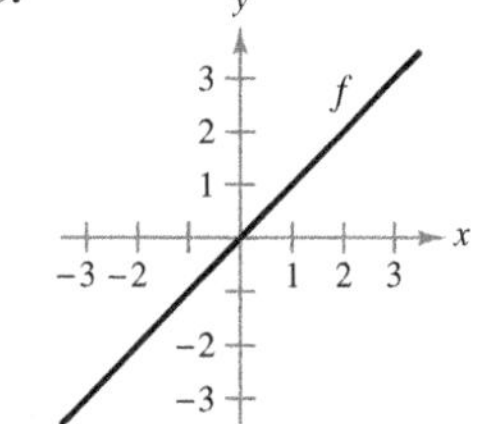

44.

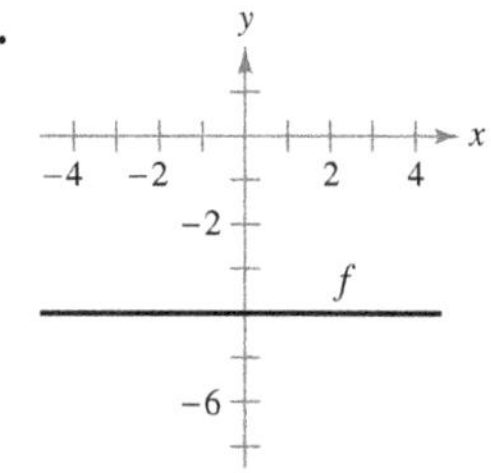

45.

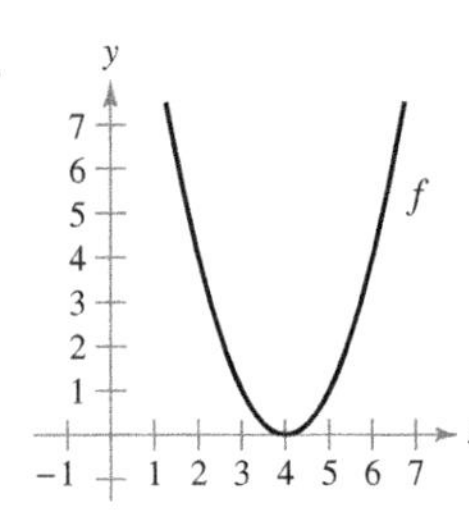

46.

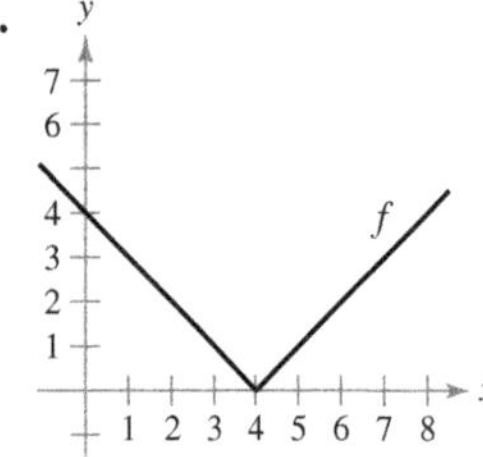

47.

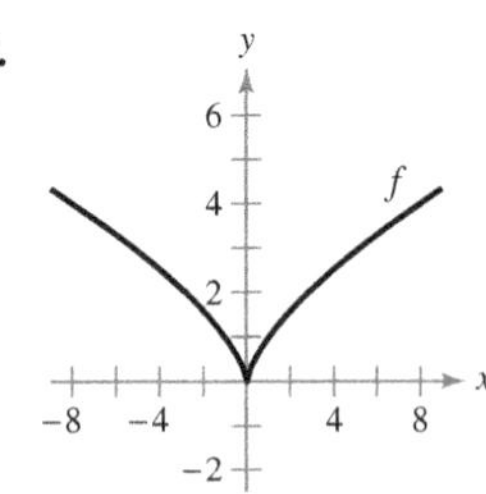

48.

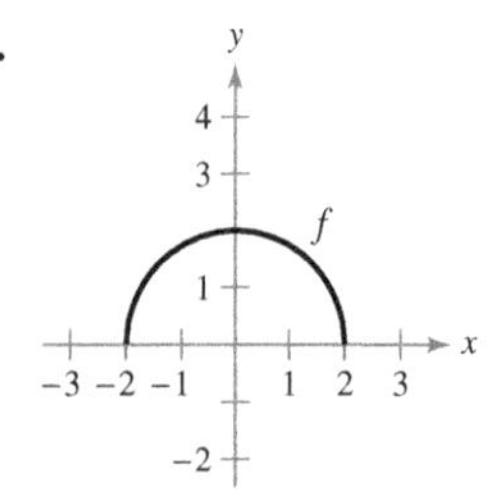

EXPLORING CONCEPTS

49. Sketching a Graph Sketch a graph of a function whose derivative is always negative. Explain how you found the answer.

50. Sketching a Graph Sketch a graph of a function whose derivative is zero at exactly two points. Explain how you found the answer.

51. Domain of the Derivative Do f and f' always have the same domain? Explain.

52. Symmetry of a Graph A function f is symmetric with respect to the origin. Is f' necessarily symmetric with respect to the origin? Explain.

53. Using a Tangent Line The tangent line to the graph of $y = g(x)$ at the point $(4, 5)$ passes through the point $(7, 0)$. Find $g(4)$ and $g'(4)$.

54. Using a Tangent Line The tangent line to the graph of $y = h(x)$ at the point $(-1, 4)$ passes through the point $(3, 6)$. Find $h(-1)$ and $h'(-1)$.

Working Backwards **In Exercises 55–58, the limit represents $f'(c)$ for a function f and a number c. Find f and c.**

55. $\displaystyle\lim_{\Delta x \to 0} \frac{[5 - 3(1 + \Delta x)] - 2}{\Delta x}$

56. $\displaystyle\lim_{\Delta x \to 0} \frac{(-2 + \Delta x)^3 + 8}{\Delta x}$

57. $\displaystyle\lim_{x \to 6} \frac{-x^2 + 36}{x - 6}$

58. $\displaystyle\lim_{x \to 9} \frac{2\sqrt{x} - 6}{x - 9}$

Writing a Function Using Derivatives **In Exercises 59 and 60, identify a function f that has the given characteristics. Then sketch the function.**

59. $f(0) = 2;\ f'(x) = -3$ for $-\infty < x < \infty$

60. $f(0) = 4;\ f'(0) = 0;\ f'(x) < 0$ for $x < 0;\ f'(x) > 0$ for $x > 0$

Finding an Equation of a Tangent Line **In Exercises 61 and 62, find equations of the two tangent lines to the graph of f that pass through the indicated point.**

61. $f(x) = 4x - x^2$

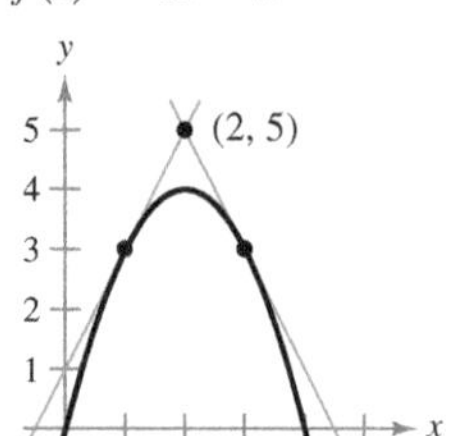

62. $f(x) = x^2$

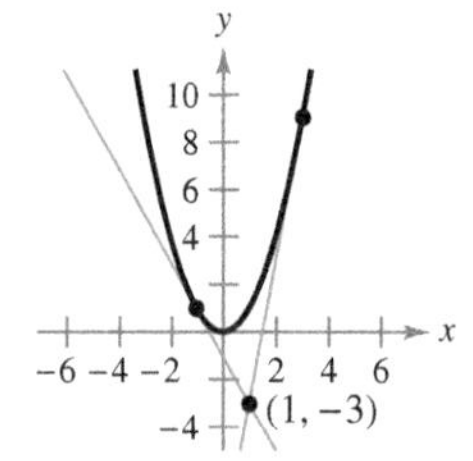

63. Graphical Reasoning Use a graphing utility to graph each function and its tangent lines at $x = -1$, $x = 0$, and $x = 1$. Based on the results, determine whether the slopes of tangent lines to the graph of a function at different values of x are always distinct.

(a) $f(x) = x^2$ (b) $g(x) = x^3$

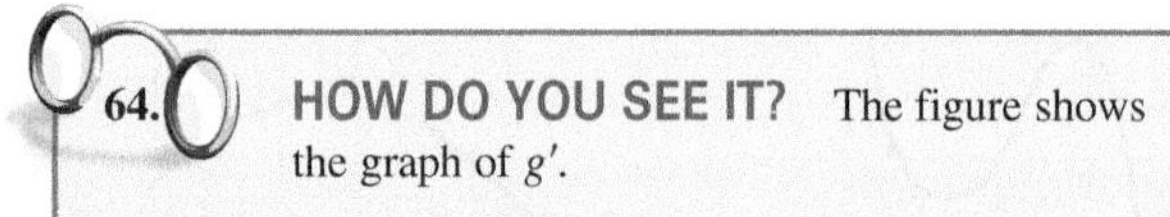

64. HOW DO YOU SEE IT? The figure shows the graph of g'.

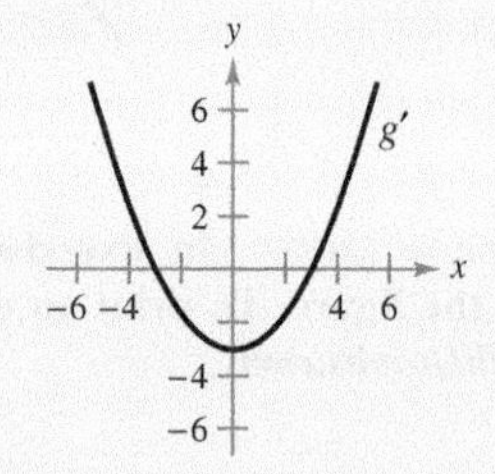

(a) $g'(0) = \square$ (b) $g'(3) = \square$

(c) What can you conclude about the graph of g knowing that $g'(1) = -\frac{8}{3}$?

(d) What can you conclude about the graph of g knowing that $g'(-4) = \frac{7}{3}$?

(e) Is $g(6) - g(4)$ positive or negative? Explain.

(f) Is it possible to find $g(2)$ from the graph? Explain.

65. Graphical Reasoning Consider the function $f(x) = \frac{1}{2}x^2$.

(a) Use a graphing utility to graph the function and estimate the values of $f'(0)$, $f'(\frac{1}{2})$, $f'(1)$, and $f'(2)$.

(b) Use your results from part (a) to determine the values of $f'(-\frac{1}{2})$, $f'(-1)$, and $f'(-2)$.

(c) Sketch a possible graph of f'.

(d) Use the definition of derivative to find $f'(x)$.

66. Graphical Reasoning Consider the function $f(x) = \frac{1}{3}x^3$.

(a) Use a graphing utility to graph the function and estimate the values of $f'(0)$, $f'(\frac{1}{2})$, $f'(1)$, $f'(2)$, and $f'(3)$.

(b) Use your results from part (a) to determine the values of $f'(-\frac{1}{2})$, $f'(-1)$, $f'(-2)$, and $f'(-3)$.

(c) Sketch a possible graph of f'.

(d) Use the definition of derivative to find $f'(x)$.

Approximating a Derivative **In Exercises 67 and 68, evaluate $f(2)$ and $f(2.1)$ and use the results to approximate $f'(2)$.**

67. $f(x) = x(4 - x)$
68. $f(x) = \frac{1}{4}x^3$

Using the Alternative Form of the Derivative **In Exercises 69–76, use the alternative form of the derivative to find the derivative at $x = c$, if it exists.**

69. $f(x) = x^3 + 2x^2 + 1,\ c = -2$
70. $g(x) = x^2 - x,\ c = 1$
71. $g(x) = \sqrt{|x|},\ c = 0$
72. $f(x) = 3/x,\ c = 4$
73. $f(x) = (x - 6)^{2/3},\ c = 6$
74. $g(x) = (x + 3)^{1/3},\ c = -3$
75. $h(x) = |x + 7|,\ c = -7$
76. $f(x) = |x - 6|,\ c = 6$

Determining Differentiability **In Exercises 77–80, describe the x-values at which f is differentiable.**

77. $f(x) = (x + 4)^{2/3}$

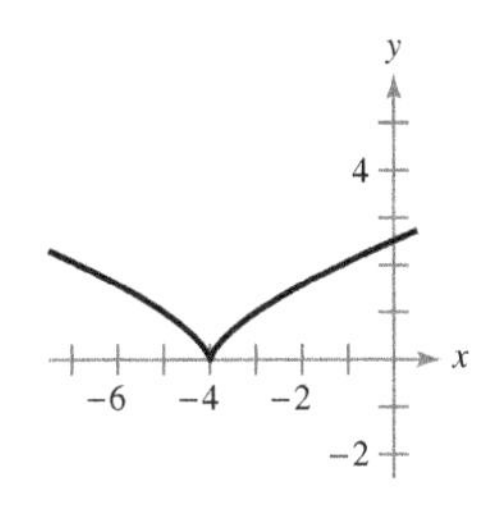

78. $f(x) = \dfrac{x^2}{x^2 - 4}$

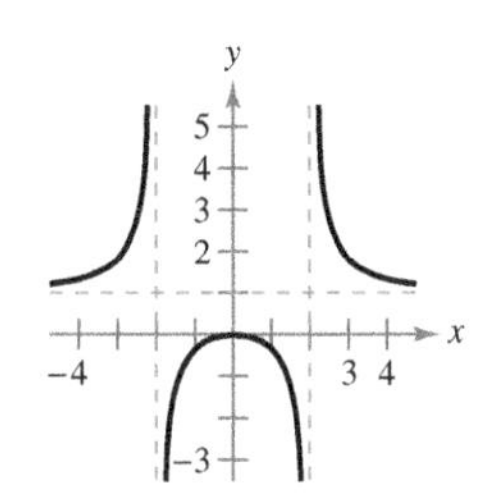

79. $f(x) = \sqrt{x + 1} + 1$

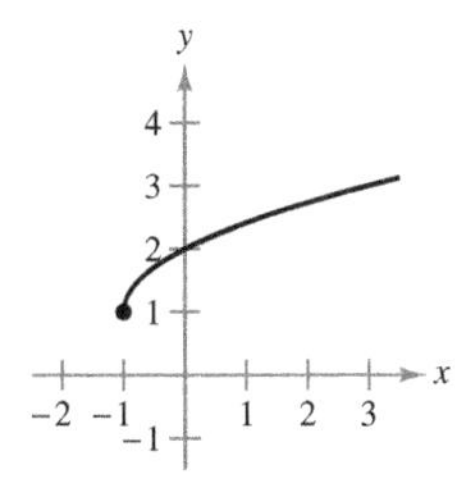

80. $f(x) = \begin{cases} x^2 - 4, & x \le 0 \\ 4 - x^2, & x > 0 \end{cases}$

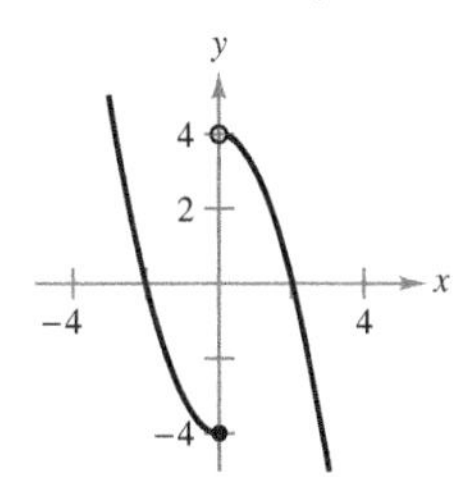

Graphical Analysis **In Exercises 81–84, use a graphing utility to graph the function and find the x-values at which f is differentiable.**

81. $f(x) = |x - 5|$
82. $f(x) = \dfrac{4x}{x - 3}$
83. $f(x) = x^{2/5}$
84. $f(x) = \begin{cases} x^3 - 3x^2 + 3x, & x \le 1 \\ x^2 - 2x, & x > 1 \end{cases}$

Determining Differentiability **In Exercises 85–88, find the derivatives from the left and from the right at $x = 1$ (if they exist). Is the function differentiable at $x = 1$?**

85. $f(x) = |x - 1|$
86. $f(x) = \sqrt{1 - x^2}$
87. $f(x) = \begin{cases} (x - 1)^3, & x \le 1 \\ (x - 1)^2, & x > 1 \end{cases}$
88. $f(x) = (1 - x)^{2/3}$

Determining Differentiability **In Exercises 89 and 90, determine whether the function is differentiable at $x = 2$.**

89. $f(x) = \begin{cases} x^2 + 1, & x \le 2 \\ 4x - 3, & x > 2 \end{cases}$
90. $f(x) = \begin{cases} \frac{1}{2}x + 2, & x < 2 \\ \sqrt{2x}, & x \ge 2 \end{cases}$

91. **Graphical Reasoning** A line with slope m passes through the point $(0, 4)$ and has the equation $y = mx + 4$.
    (a) Write the distance d between the line and the point $(3, 1)$ as a function of m.
    (b) Use a graphing utility to graph the function d in part (a). Based on the graph, is the function differentiable at every value of m? If not, where is it not differentiable?

92. **Conjecture** Consider the functions $f(x) = x^2$ and $g(x) = x^3$.
    (a) Graph f and f' on the same set of axes.
    (b) Graph g and g' on the same set of axes.
    (c) Identify a pattern between f and g and their respective derivatives. Use the pattern to make a conjecture about $h'(x)$ if $h(x) = x^n$, where n is an integer and $n \ge 2$.
    (d) Find $f'(x)$ if $f(x) = x^4$. Compare the result with the conjecture in part (c). Is this a proof of your conjecture? Explain.

True or False? **In Exercises 93–96, determine whether the statement is true or false. If it is false, explain why or give an example that shows it is false.**

93. The slope of the tangent line to the differentiable function f at the point $(2, f(2))$ is

$$\frac{f(2 + \Delta x) - f(2)}{\Delta x}.$$

94. If a function is continuous at a point, then it is differentiable at that point.

95. If a function has derivatives from both the right and the left at a point, then it is differentiable at that point.

96. If a function is differentiable at a point, then it is continuous at that point.

97. **Differentiability and Continuity** Let

$$f(x) = \begin{cases} x \sin \dfrac{1}{x}, & x \ne 0 \\ 0, & x = 0 \end{cases}$$

and

$$g(x) = \begin{cases} x^2 \sin \dfrac{1}{x}, & x \ne 0 \\ 0, & x = 0 \end{cases}.$$

Show that f is continuous, but not differentiable, at $x = 0$. Show that g is differentiable at 0 and find $g'(0)$.

98. **Writing** Use a graphing utility to graph the two functions $f(x) = x^2 + 1$ and $g(x) = |x| + 1$ in the same viewing window. Use the *zoom* and *trace* features to analyze the graphs near the point $(0, 1)$. What do you observe? Which function is differentiable at this point? Write a short paragraph describing the geometric significance of differentiability at a point.

2.2 Basic Differentiation Rules and Rates of Change

- **Find the derivative of a function using the Constant Rule.**
- **Find the derivative of a function using the Power Rule.**
- **Find the derivative of a function using the Constant Multiple Rule.**
- **Find the derivative of a function using the Sum and Difference Rules.**
- **Find the derivatives of the sine function and of the cosine function.**
- **Use derivatives to find rates of change.**

The Constant Rule

In Section 2.1, you used the limit definition to find derivatives. In this and the next two sections, you will be introduced to several "differentiation rules" that allow you to find derivatives without the *direct* use of the limit definition.

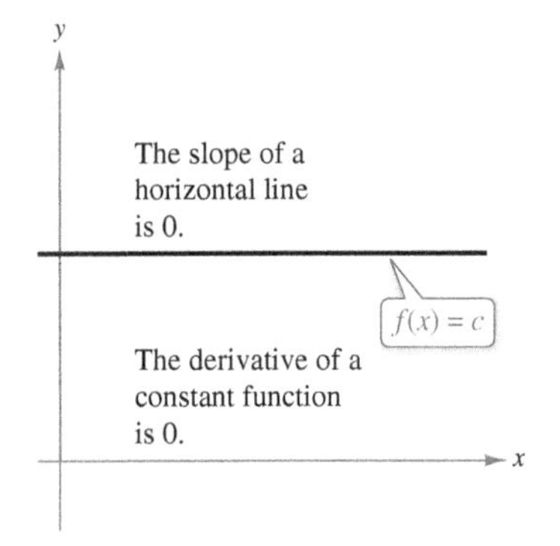

Notice that the Constant Rule is equivalent to saying that the slope of a horizontal line is 0. This demonstrates the relationship between slope and derivative.
Figure 2.14

THEOREM 2.2 The Constant Rule

The derivative of a constant function is 0. That is, if c is a real number, then

$$\frac{d}{dx}[c] = 0.$$ See Figure 2.14.

Proof Let $f(x) = c$. Then, by the limit definition of the derivative,

$$\begin{aligned}\frac{d}{dx}[c] &= f'(x)\\ &= \lim_{\Delta x \to 0} \frac{f(x + \Delta x) - f(x)}{\Delta x}\\ &= \lim_{\Delta x \to 0} \frac{c - c}{\Delta x}\\ &= \lim_{\Delta x \to 0} 0\\ &= 0.\end{aligned}$$

EXAMPLE 1 Using the Constant Rule

	Function	**Derivative**
a.	$y = 7$	$dy/dx = 0$
b.	$f(x) = 0$	$f'(x) = 0$
c.	$s(t) = -3$	$s'(t) = 0$
d.	$y = k\pi^2$, k is constant	$dy/dx = 0$

Exploration

Writing a Conjecture Use the definition of the derivative given in Section 2.1 to find the derivative of each function. What patterns do you see? Use your results to write a conjecture about the derivative of $f(x) = x^n$.

a. $f(x) = x^1$ **b.** $f(x) = x^2$ **c.** $f(x) = x^3$
d. $f(x) = x^4$ **e.** $f(x) = x^{1/2}$ **f.** $f(x) = x^{-1}$

The Power Rule

Before proving the next rule, it is important to review the procedure for expanding a binomial.

$$(x + \Delta x)^2 = x^2 + 2x\Delta x + (\Delta x)^2$$
$$(x + \Delta x)^3 = x^3 + 3x^2\Delta x + 3x(\Delta x)^2 + (\Delta x)^3$$
$$(x + \Delta x)^4 = x^4 + 4x^3\Delta x + 6x^2(\Delta x)^2 + 4x(\Delta x)^3 + (\Delta x)^4$$
$$(x + \Delta x)^5 = x^5 + 5x^4\Delta x + 10x^3(\Delta x)^2 + 10x^2(\Delta x)^3 + 5x(\Delta x)^4 + (\Delta x)^5$$

The general binomial expansion for a positive integer n is

$$(x + \Delta x)^n = x^n + nx^{n-1}(\Delta x) + \underbrace{\frac{n(n-1)x^{n-2}}{2}(\Delta x)^2 + \cdot \cdot \cdot + (\Delta x)^n}_{(\Delta x)^2 \text{ is a factor of these terms.}}.$$

This binomial expansion is used in proving a special case of the Power Rule.

THEOREM 2.3 The Power Rule

If n is a rational number, then the function $f(x) = x^n$ is differentiable and

$$\frac{d}{dx}[x^n] = nx^{n-1}.$$

For f to be differentiable at $x = 0$, n must be a number such that x^{n-1} is defined on an interval containing 0.

REMARK From Example 7 in Section 2.1, you know that the function $f(x) = x^{1/3}$ is defined at $x = 0$ but is not differentiable at $x = 0$. This is because $x^{-2/3}$ is not defined on an interval containing 0.

Proof If n is a positive integer greater than 1, then the binomial expansion produces

$$\frac{d}{dx}[x^n] = \lim_{\Delta x \to 0} \frac{(x + \Delta x)^n - x^n}{\Delta x}$$
$$= \lim_{\Delta x \to 0} \frac{x^n + nx^{n-1}(\Delta x) + \frac{n(n-1)x^{n-2}}{2}(\Delta x)^2 + \cdot \cdot \cdot + (\Delta x)^n - x^n}{\Delta x}$$
$$= \lim_{\Delta x \to 0} \left[nx^{n-1} + \frac{n(n-1)x^{n-2}}{2}(\Delta x) + \cdot \cdot \cdot + (\Delta x)^{n-1} \right]$$
$$= nx^{n-1} + 0 + \cdot \cdot \cdot + 0$$
$$= nx^{n-1}.$$

This proves the case for which n is a positive integer greater than 1. It is left to you to prove the case for $n = 1$. Example 7 in Section 2.3 proves the case for which n is a negative integer. In Exercise 73 in Section 2.5, you are asked to prove the case for which n is rational. (In Section 5.5, the Power Rule will be extended to cover irrational values of n.) ■

When using the Power Rule, the case for which $n = 1$ is best thought of as a separate differentiation rule. That is,

$$\frac{d}{dx}[x] = 1. \qquad \text{Power Rule when } n = 1$$

The slope of the line $y = x$ is 1.
Figure 2.15

This rule is consistent with the fact that the slope of the line $y = x$ is 1, as shown in Figure 2.15.

EXAMPLE 2 Using the Power Rule

Function	Derivative
a. $f(x) = x^3$	$f'(x) = 3x^2$
b. $g(x) = \sqrt[3]{x}$	$g'(x) = \dfrac{d}{dx}[x^{1/3}] = \dfrac{1}{3}x^{-2/3} = \dfrac{1}{3x^{2/3}}$
c. $y = \dfrac{1}{x^2}$	$\dfrac{dy}{dx} = \dfrac{d}{dx}[x^{-2}] = (-2)x^{-3} = -\dfrac{2}{x^3}$

In Example 2(c), note that *before* differentiating, $1/x^2$ was rewritten as x^{-2}. Rewriting is the first step in *many* differentiation problems.

Given: $y = \dfrac{1}{x^2}$ ⇨ Rewrite: $y = x^{-2}$ ⇨ Differentiate: $\dfrac{dy}{dx} = (-2)x^{-3}$ ⇨ Simplify: $\dfrac{dy}{dx} = -\dfrac{2}{x^3}$

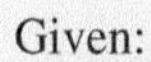

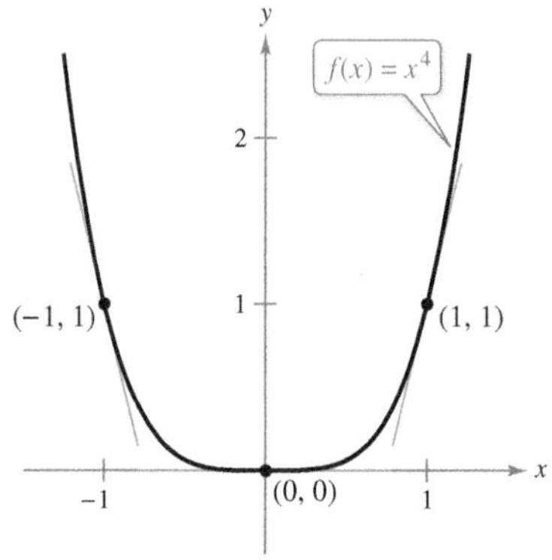

EXAMPLE 3 Finding the Slope of a Graph

See LarsonCalculus.com for an interactive version of this type of example.

Find the slope of the graph of $f(x) = x^4$ for each value of x.

a. $x = -1$ **b.** $x = 0$ **c.** $x = 1$

Solution The slope of a graph at a point is the value of the derivative at that point. The derivative of f is $f'(x) = 4x^3$.

a. When $x = -1$, the slope is $f'(-1) = 4(-1)^3 = -4$. Slope is negative.

b. When $x = 0$, the slope is $f'(0) = 4(0)^3 = 0$. Slope is zero.

c. When $x = 1$, the slope is $f'(1) = 4(1)^3 = 4$. Slope is positive.

See Figure 2.16.

Note that the slope of the graph is negative at the point $(-1, 1)$, the slope is zero at the point $(0, 0)$, and the slope is positive at the point $(1, 1)$.
Figure 2.16

EXAMPLE 4 Finding an Equation of a Tangent Line

See LarsonCalculus.com for an interactive version of this type of example.

Find an equation of the tangent line to the graph of $f(x) = x^2$ when $x = -2$.

Solution To find the *point* on the graph of f, evaluate the original function at $x = -2$.

$$(-2, f(-2)) = (-2, 4) \qquad \text{Point on graph}$$

To find the *slope* of the graph when $x = -2$, evaluate the derivative, $f'(x) = 2x$, at $x = -2$.

$$m = f'(-2) = -4 \qquad \text{Slope of graph at } (-2, 4)$$

Now, using the point-slope form of the equation of a line, you can write

$$y - y_1 = m(x - x_1) \qquad \text{Point-slope form}$$
$$y - 4 = -4[x - (-2)] \qquad \text{Substitute for } y_1, m, \text{ and } x_1.$$
$$y = -4x - 4. \qquad \text{Simplify.}$$

You can check this result using the *tangent* feature of a graphing utility, as shown in Figure 2.17.

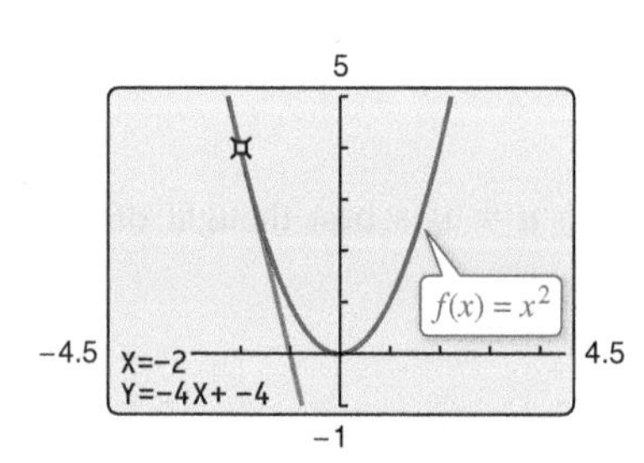

The line $y = -4x - 4$ is tangent to the graph of $f(x) = x^2$ at the point $(-2, 4)$.
Figure 2.17

The Constant Multiple Rule

THEOREM 2.4 The Constant Multiple Rule

If f is a differentiable function and c is a real number, then cf is also differentiable and

$$\frac{d}{dx}[cf(x)] = cf'(x).$$

Proof

$$\begin{aligned}\frac{d}{dx}[cf(x)] &= \lim_{\Delta x \to 0} \frac{cf(x+\Delta x) - cf(x)}{\Delta x} && \text{Definition of derivative}\\ &= \lim_{\Delta x \to 0} c\left[\frac{f(x+\Delta x) - f(x)}{\Delta x}\right] \\ &= c\left[\lim_{\Delta x \to 0} \frac{f(x+\Delta x) - f(x)}{\Delta x}\right] && \text{Apply Theorem 1.2.}\\ &= cf'(x)\end{aligned}$$

Informally, the Constant Multiple Rule states that constants can be factored out of the differentiation process, even when the constants appear in the denominator.

$$\frac{d}{dx}[cf(x)] = c\frac{d}{dx}[f(x)] = cf'(x)$$

$$\frac{d}{dx}\left[\frac{f(x)}{c}\right] = \frac{d}{dx}\left[\left(\frac{1}{c}\right)f(x)\right] = \left(\frac{1}{c}\right)\frac{d}{dx}[f(x)] = \left(\frac{1}{c}\right)f'(x)$$

EXAMPLE 5 Using the Constant Multiple Rule

Function	Derivative
a. $y = 5x^3$	$\frac{dy}{dx} = \frac{d}{dx}[5x^3] = 5\frac{d}{dx}[x^3] = 5(3)x^2 = 15x^2$
b. $y = \frac{2}{x}$	$\frac{dy}{dx} = \frac{d}{dx}[2x^{-1}] = 2\frac{d}{dx}[x^{-1}] = 2(-1)x^{-2} = -\frac{2}{x^2}$
c. $f(t) = \frac{4t^2}{5}$	$f'(t) = \frac{d}{dt}\left[\frac{4}{5}t^2\right] = \frac{4}{5}\frac{d}{dt}[t^2] = \frac{4}{5}(2t) = \frac{8}{5}t$
d. $y = 2\sqrt{x}$	$\frac{dy}{dx} = \frac{d}{dx}[2x^{1/2}] = 2\left(\frac{1}{2}x^{-1/2}\right) = x^{-1/2} = \frac{1}{\sqrt{x}}$
e. $y = \frac{1}{2\sqrt[3]{x^2}}$	$\frac{dy}{dx} = \frac{d}{dx}\left[\frac{1}{2}x^{-2/3}\right] = \frac{1}{2}\left(-\frac{2}{3}\right)x^{-5/3} = -\frac{1}{3x^{5/3}}$
f. $y = -\frac{3x}{2}$	$y' = \frac{d}{dx}\left[-\frac{3}{2}x\right] = -\frac{3}{2}(1) = -\frac{3}{2}$

REMARK Before differentiating functions involving radicals, rewrite the function with rational exponents.

The Constant Multiple Rule and the Power Rule can be combined into one rule. The combination rule is

$$\frac{d}{dx}[cx^n] = cnx^{n-1}.$$

EXAMPLE 6 **Using Parentheses When Differentiating**

	Original Function	Rewrite	Differentiate	Simplify
a.	$y = \frac{5}{2x^3}$	$y = \frac{5}{2}(x^{-3})$	$y' = \frac{5}{2}(-3x^{-4})$	$y' = -\frac{15}{2x^4}$
b.	$y = \frac{5}{(2x)^3}$	$y = \frac{5}{8}(x^{-3})$	$y' = \frac{5}{8}(-3x^{-4})$	$y' = -\frac{15}{8x^4}$
c.	$y = \frac{7}{3x^{-2}}$	$y = \frac{7}{3}(x^2)$	$y' = \frac{7}{3}(2x)$	$y' = \frac{14x}{3}$
d.	$y = \frac{7}{(3x)^{-2}}$	$y = 63(x^2)$	$y' = 63(2x)$	$y' = 126x$

The Sum and Difference Rules

THEOREM 2.5 The Sum and Difference Rules

The sum (or difference) of two differentiable functions f and g is itself differentiable. Moreover, the derivative of $f + g$ (or $f - g$) is the sum (or difference) of the derivatives of f and g.

$$\frac{d}{dx}[f(x) + g(x)] = f'(x) + g'(x) \qquad \text{Sum Rule}$$

$$\frac{d}{dx}[f(x) - g(x)] = f'(x) - g'(x) \qquad \text{Difference Rule}$$

Proof A proof of the Sum Rule follows from Theorem 1.2. (The Difference Rule can be proved in a similar way.)

$$\begin{aligned}
\frac{d}{dx}[f(x) + g(x)] &= \lim_{\Delta x \to 0} \frac{[f(x + \Delta x) + g(x + \Delta x)] - [f(x) + g(x)]}{\Delta x} \\
&= \lim_{\Delta x \to 0} \frac{f(x + \Delta x) + g(x + \Delta x) - f(x) - g(x)}{\Delta x} \\
&= \lim_{\Delta x \to 0} \left[\frac{f(x + \Delta x) - f(x)}{\Delta x} + \frac{g(x + \Delta x) - g(x)}{\Delta x}\right] \\
&= \lim_{\Delta x \to 0} \frac{f(x + \Delta x) - f(x)}{\Delta x} + \lim_{\Delta x \to 0} \frac{g(x + \Delta x) - g(x)}{\Delta x} \\
&= f'(x) + g'(x)
\end{aligned}$$

The Sum and Difference Rules can be extended to any finite number of functions. For instance, if $F(x) = f(x) + g(x) - h(x)$, then $F'(x) = f'(x) + g'(x) - h'(x)$.

REMARK In Example 7(c), note that before differentiating,

$$\frac{3x^2 - x + 1}{x}$$

was rewritten as

$$3x - 1 + \frac{1}{x}.$$

EXAMPLE 7 **Using the Sum and Difference Rules**

	Function	Derivative
a.	$f(x) = x^3 - 4x + 5$	$f'(x) = 3x^2 - 4$
b.	$g(x) = -\frac{x^4}{2} + 3x^3 - 2x$	$g'(x) = -2x^3 + 9x^2 - 2$
c.	$y = \frac{3x^2 - x + 1}{x} = 3x - 1 + \frac{1}{x}$	$y' = 3 - \frac{1}{x^2} = \frac{3x^2 - 1}{x^2}$

FOR FURTHER INFORMATION For the outline of a geometric proof of the derivatives of the sine and cosine functions, see the article "The Spider's Spacewalk Derivation of sin′ and cos′" by Tim Hesterberg in *The College Mathematics Journal*. To view this article, go to *MathArticles.com*.

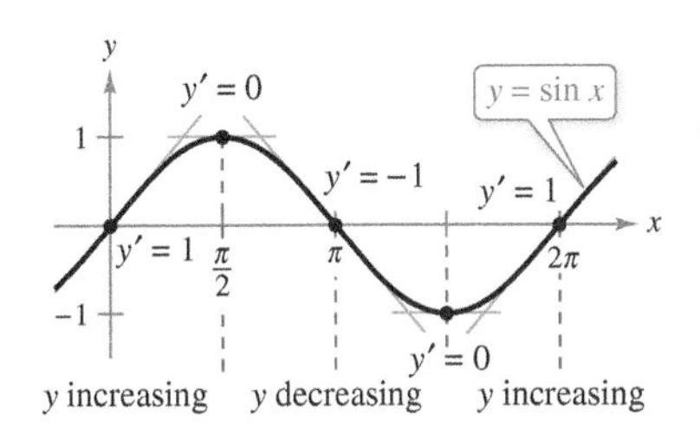

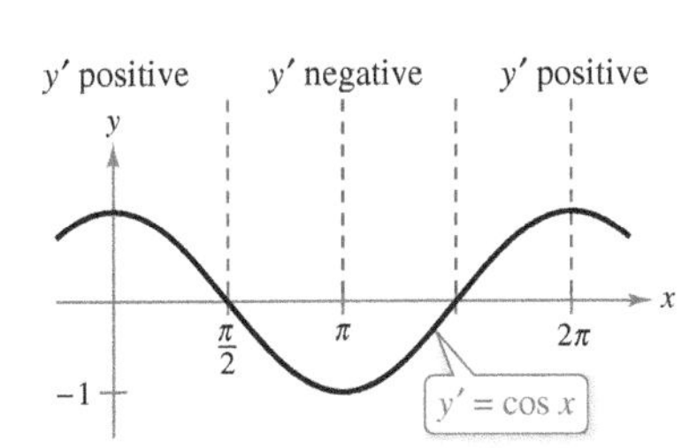

The derivative of the sine function is the cosine function.
Figure 2.18

y = 2 sin x
y = (3/2) sin x
2
−π
π
−2
y = sin x
y = (1/2) sin x

$$\frac{d}{dx}[a \sin x] = a \cos x$$

Figure 2.19

Derivatives of the Sine and Cosine Functions

In Section 1.3, you studied the limits

$$\lim_{\Delta x \to 0} \frac{\sin \Delta x}{\Delta x} = 1 \quad \text{and} \quad \lim_{\Delta x \to 0} \frac{1 - \cos \Delta x}{\Delta x} = 0.$$

These two limits can be used to prove differentiation rules for the sine and cosine functions. (The derivatives of the other four trigonometric functions are discussed in Section 2.3.)

THEOREM 2.6 Derivatives of Sine and Cosine Functions

$$\frac{d}{dx}[\sin x] = \cos x \qquad \frac{d}{dx}[\cos x] = -\sin x$$

Proof Here is a proof of the first rule. (The proof of the second rule is left as an exercise [see Exercise 114].) In the proof, note the use of the trigonometric identity $\sin(x + \Delta x) = \sin x \cos \Delta x + \cos x \sin \Delta x$.

$$\begin{aligned}
\frac{d}{dx}[\sin x] &= \lim_{\Delta x \to 0} \frac{\sin(x + \Delta x) - \sin x}{\Delta x} && \text{Definition of derivative} \\
&= \lim_{\Delta x \to 0} \frac{\sin x \cos \Delta x + \cos x \sin \Delta x - \sin x}{\Delta x} \\
&= \lim_{\Delta x \to 0} \frac{\cos x \sin \Delta x - (\sin x)(1 - \cos \Delta x)}{\Delta x} \\
&= \lim_{\Delta x \to 0} \left[(\cos x)\left(\frac{\sin \Delta x}{\Delta x}\right) - (\sin x)\left(\frac{1 - \cos \Delta x}{\Delta x}\right) \right] \\
&= (\cos x)\left(\lim_{\Delta x \to 0} \frac{\sin \Delta x}{\Delta x}\right) - (\sin x)\left(\lim_{\Delta x \to 0} \frac{1 - \cos \Delta x}{\Delta x}\right) \\
&= (\cos x)(1) - (\sin x)(0) \\
&= \cos x
\end{aligned}$$

This differentiation rule is shown graphically in Figure 2.18. Note that for each x, the *slope* of the sine curve is equal to the value of the cosine. ■

EXAMPLE 8 Derivatives Involving Sines and Cosines

See LarsonCalculus.com for an interactive version of this type of example.

	Function	**Derivative**
a.	$y = 2 \sin x$	$y' = 2 \cos x$
b.	$y = \dfrac{\sin x}{2} = \dfrac{1}{2} \sin x$	$y' = \dfrac{1}{2} \cos x = \dfrac{\cos x}{2}$
c.	$y = x + \cos x$	$y' = 1 - \sin x$
d.	$y = \cos x - \dfrac{\pi}{3} \sin x$	$y' = -\sin x - \dfrac{\pi}{3} \cos x$

■

TECHNOLOGY A graphing utility can provide insight into the interpretation of a derivative. For instance, Figure 2.19 shows the graphs of

$$y = a \sin x$$

for $a = \frac{1}{2}, 1, \frac{3}{2},$ and 2. Estimate the slope of each graph at the point $(0, 0)$. Then verify your estimates analytically by evaluating the derivative of each function when $x = 0$.

Rates of Change

You have seen how the derivative is used to determine slope. The derivative can also be used to determine the rate of change of one variable with respect to another. Applications involving rates of change, sometimes referred to as instantaneous rates of change, occur in a wide variety of fields. A few examples are population growth rates, production rates, water flow rates, velocity, and acceleration.

A common use for rate of change is to describe the motion of an object moving in a straight line. In such problems, it is customary to use either a horizontal or a vertical line with a designated origin to represent the line of motion. On such lines, movement to the right (or upward) is considered to be in the positive direction, and movement to the left (or downward) is considered to be in the negative direction.

The function s that gives the position (relative to the origin) of an object as a function of time t is called a **position function.** If, over a period of time Δt, the object changes its position by the amount

$$\Delta s = s(t + \Delta t) - s(t)$$

then, by the familiar formula

$$\text{Rate} = \frac{\text{distance}}{\text{time}}$$

the **average velocity** is

$$\frac{\text{Change in distance}}{\text{Change in time}} = \frac{\Delta s}{\Delta t}.$$ Average velocity

EXAMPLE 9 Finding Average Velocity of a Falling Object

A billiard ball is dropped from a height of 100 feet. The ball's height s at time t is the position function

$$s = -16t^2 + 100$$ Position function

where s is measured in feet and t is measured in seconds. Find the average velocity over each time interval.

a. $[1, 2]$ **b.** $[1, 1.5]$ **c.** $[1, 1.1]$

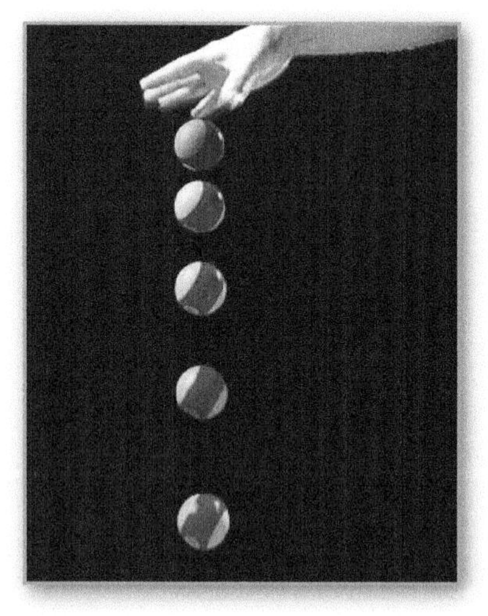

Time-lapse photograph of a free-falling billiard ball

Solution

a. For the interval $[1, 2]$, the object falls from a height of $s(1) = -16(1)^2 + 100 = 84$ feet to a height of $s(2) = -16(2)^2 + 100 = 36$ feet. The average velocity is

$$\frac{\Delta s}{\Delta t} = \frac{36 - 84}{2 - 1} = \frac{-48}{1} = -48 \text{ feet per second.}$$

b. For the interval $[1, 1.5]$, the object falls from a height of 84 feet to a height of $s(1.5) = -16(1.5)^2 + 100 = 64$ feet. The average velocity is

$$\frac{\Delta s}{\Delta t} = \frac{64 - 84}{1.5 - 1} = \frac{-20}{0.5} = -40 \text{ feet per second.}$$

c. For the interval $[1, 1.1]$, the object falls from a height of 84 feet to a height of $s(1.1) = -16(1.1)^2 + 100 = 80.64$ feet. The average velocity is

$$\frac{\Delta s}{\Delta t} = \frac{80.64 - 84}{1.1 - 1} = \frac{-3.36}{0.1} = -33.6 \text{ feet per second.}$$

Note that the average velocities are *negative*, indicating that the object is moving downward.

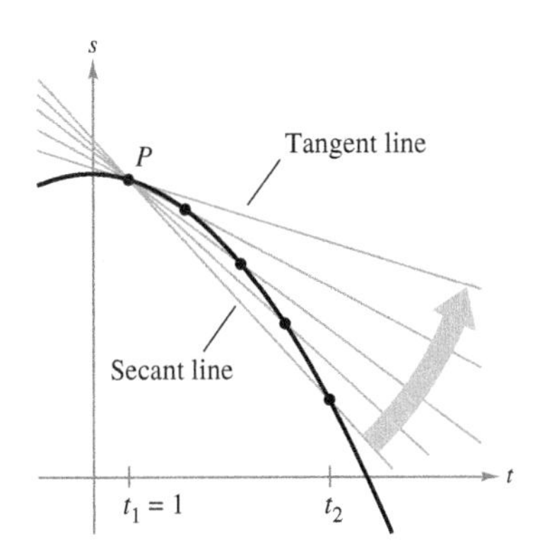

The average velocity between t_1 and t_2 is the slope of the secant line, and the instantaneous velocity at t_1 is the slope of the tangent line.
Figure 2.20

Suppose that in Example 9, you wanted to find the *instantaneous* velocity (or simply the velocity) of the object when $t = 1$. Just as you can approximate the slope of the tangent line by calculating the slope of the secant line, you can approximate the velocity at $t = 1$ by calculating the average velocity over a small interval $[1, 1 + \Delta t]$ (see Figure 2.20). By taking the limit as Δt approaches zero, you obtain the velocity when $t = 1$. Try doing this—you will find that the velocity when $t = 1$ is -32 feet per second.

In general, if $s = s(t)$ is the position function for an object moving along a straight line, then the **velocity** of the object at time t is

$$v(t) = \lim_{\Delta t \to 0} \frac{s(t + \Delta t) - s(t)}{\Delta t} = s'(t).$$ Velocity function

In other words, the velocity function is the derivative of the position function. Velocity can be negative, zero, or positive. The **speed** of an object is the absolute value of its velocity. Speed cannot be negative.

The position of a free-falling object (neglecting air resistance) under the influence of gravity can be represented by the equation

$$s(t) = -\frac{1}{2}gt^2 + v_0t + s_0$$ Position function

where s_0 is the initial height of the object, v_0 is the initial velocity of the object, and g is the acceleration due to gravity. On Earth, the value of g is approximately 32 feet per second per second or 9.8 meters per second per second.

EXAMPLE 10 Using the Derivative to Find Velocity

At time $t = 0$ seconds, a diver jumps from a platform diving board that is 32 feet above the water (see Figure 2.21). The initial velocity of the diver is 16 feet per second. When does the diver hit the water? What is the diver's velocity at impact?

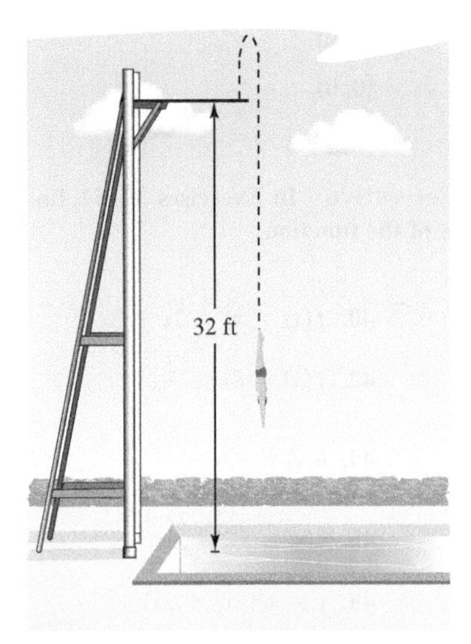

Velocity is positive when an object is rising and is negative when an object is falling. Notice that the diver moves upward for the first half-second because the velocity is positive for $0 < t < \frac{1}{2}$. When the velocity is 0, the diver has reached the maximum height of the dive.
Figure 2.21

Solution

Begin by writing an equation to represent the position of the diver. Using the position function given above with $g = 32$ feet per second per second, $v_0 = 16$ feet per second, and $s_0 = 32$ feet, you can write

$$\begin{aligned} s(t) &= -\frac{1}{2}(32)t^2 + 16t + 32 \\ &= -16t^2 + 16t + 32. \end{aligned}$$ Position function

To find the time t when the diver hits the water, let $s = 0$ and solve for t.

$-16t^2 + 16t + 32 = 0$ Set position function equal to 0.

$-16(t + 1)(t - 2) = 0$ Factor.

$t = -1$ or 2 Solve for t.

Because $t \geq 0$, choose the positive value to conclude that the diver hits the water at $t = 2$ seconds. The velocity at time t is given by the derivative

$s'(t) = -32t + 16.$ Velocity function

So, the velocity at time $t = 2$ is

$s'(2) = -32(2) + 16 = -48$ feet per second.

Notice that the unit for $s'(t)$ is the unit for s (feet) divided by the unit for t (seconds). In general, the unit for $f'(x)$ is the unit for f divided by the unit for x. ■

2.2 Exercises

See CalcChat.com for tutorial help and worked-out solutions to odd-numbered exercises.

CONCEPT CHECK

1. **Constant Rule** What is the derivative of a constant function?
2. **Finding a Derivative** Explain how to find the derivative of the function $f(x) = cx^n$.
3. **Derivatives of Trigonometric Functions** What are the derivatives of the sine and cosine functions?
4. **Average Velocity and Velocity** Describe the difference between average velocity and velocity.

Estimating Slope **In Exercises 5 and 6, use the graph to estimate the slope of the tangent line to $y = x^n$ at the point (1, 1). Verify your answer analytically. To print an enlarged copy of the graph, go to *MathGraphs.com*.**

5. (a) $y = x^{1/2}$

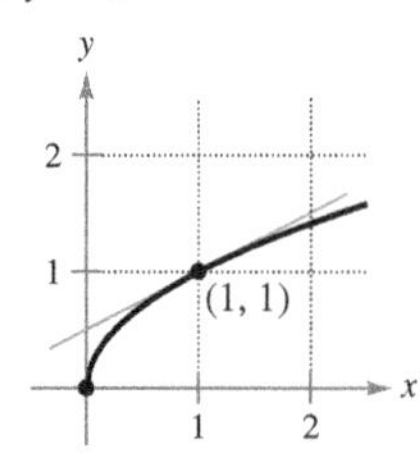

(b) $y = x^3$

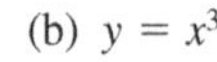

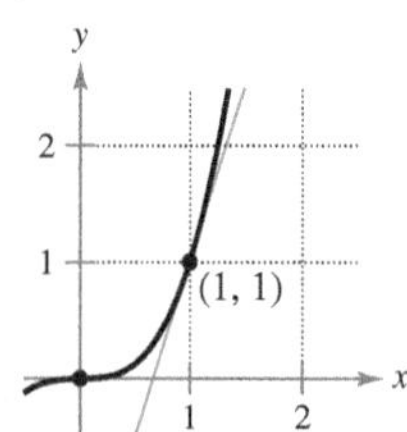

6. (a) $y = x^{-1/2}$

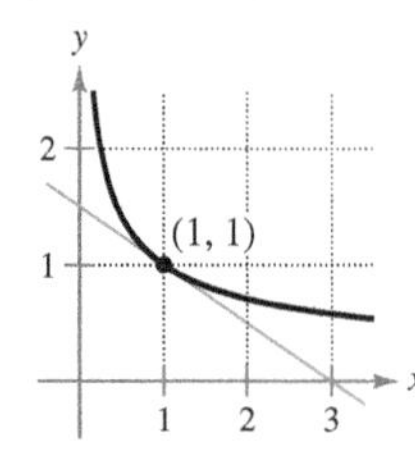

(b) $y = x^{-1}$

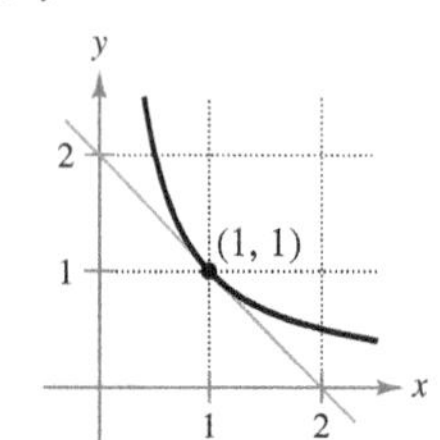

Finding a Derivative **In Exercises 7–26, use the rules of differentiation to find the derivative of the function.**

7. $y = 12$
8. $f(x) = -9$
9. $y = x^7$
10. $y = x^{12}$
11. $y = \dfrac{1}{x^5}$
12. $y = \dfrac{3}{x^7}$
13. $f(x) = \sqrt[9]{x}$
14. $g(x) = \sqrt[4]{x}$
15. $f(x) = x + 11$
16. $g(x) = 6x + 3$
17. $f(t) = -3t^2 + 2t - 4$
18. $y = t^2 - 3t + 1$
19. $g(x) = x^2 + 4x^3$
20. $y = 4x - 3x^3$
21. $s(t) = t^3 + 5t^2 - 3t + 8$
22. $y = 2x^3 + 6x^2 - 1$
23. $y = \dfrac{\pi}{2}\sin\theta$
24. $g(t) = \pi\cos t$
25. $y = x^2 - \frac{1}{2}\cos x$
26. $y = 7x^4 + 2\sin x$

Rewriting a Function Before Differentiating **In Exercises 27–30, complete the table to find the derivative of the function.**

	Original Function	Rewrite	Differentiate	Simplify
27.	$y = \dfrac{2}{7x^4}$			
28.	$y = \dfrac{8}{5x^{-5}}$			
29.	$y = \dfrac{6}{(5x)^3}$			
30.	$y = \dfrac{3}{(2x)^{-2}}$			

Finding the Slope of a Graph **In Exercises 31–38, find the slope of the graph of the function at the given point. Use the *derivative* feature of a graphing utility to confirm your results.**

	Function	Point
31.	$f(x) = \dfrac{8}{x^2}$	$(2, 2)$
32.	$f(t) = 2 - \dfrac{4}{t}$	$(4, 1)$
33.	$f(x) = -\frac{1}{2} + \frac{7}{5}x^3$	$\left(0, -\frac{1}{2}\right)$
34.	$y = 2x^4 - 3$	$(1, -1)$
35.	$y = (4x + 1)^2$	$(0, 1)$
36.	$f(x) = 2(x - 4)^2$	$(2, 8)$
37.	$f(\theta) = 4\sin\theta - \theta$	$(0, 0)$
38.	$g(t) = -2\cos t + 5$	$(\pi, 7)$

Finding a Derivative **In Exercises 39–54, find the derivative of the function.**

39. $f(x) = x^2 + 5 - 3x^{-2}$
40. $f(x) = x^3 - 2x + 3x^{-3}$
41. $g(t) = t^2 - \dfrac{4}{t^3}$
42. $f(x) = 8x + \dfrac{3}{x^2}$
43. $f(x) = \dfrac{x^3 - 3x^2 + 4}{x^2}$
44. $h(x) = \dfrac{4x^3 + 2x + 5}{x}$
45. $g(t) = \dfrac{3t^2 + 4t - 8}{t^{3/2}}$
46. $h(s) = \dfrac{s^5 + 2s + 6}{s^{1/3}}$
47. $y = x(x^2 + 1)$
48. $y = x^2(2x^2 - 3x)$
49. $f(x) = \sqrt{x} - 6\sqrt[3]{x}$
50. $f(t) = t^{2/3} - t^{1/3} + 4$
51. $f(x) = 6\sqrt{x} + 5\cos x$
52. $f(x) = \dfrac{2}{\sqrt[3]{x}} + 3\cos x$
53. $y = \dfrac{1}{(3x)^{-2}} - 5\cos x$
54. $y = \dfrac{3}{(2x)^3} + 2\sin x$

Finding an Equation of a Tangent Line In Exercises 55–58, (a) find an equation of the tangent line to the graph of the function at the given point, (b) use a graphing utility to graph the function and its tangent line at the point, and (c) use the *tangent* feature of a graphing utility to confirm your results.

Function	Point
55. $f(x) = -2x^4 + 5x^2 - 3$	$(1, 0)$
56. $y = x^3 - 3x$	$(2, 2)$
57. $f(x) = \dfrac{2}{\sqrt[4]{x^3}}$	$(1, 2)$
58. $y = (x - 2)(x^2 + 3x)$	$(1, -4)$

Horizontal Tangent Line In Exercises 59–64, determine the point(s) (if any) at which the graph of the function has a horizontal tangent line.

59. $y = x^4 - 2x^2 + 3$

60. $y = x^3 + x$

61. $y = \dfrac{1}{x^2}$

62. $y = x^2 + 9$

63. $y = x + \sin x, \quad 0 \le x < 2\pi$

64. $y = \sqrt{3}x + 2\cos x, \quad 0 \le x < 2\pi$

Finding a Value In Exercises 65–68, find k such that the line is tangent to the graph of the function.

Function	Line
65. $f(x) = k - x^2$	$y = -6x + 1$
66. $f(x) = kx^2$	$y = -2x + 3$
67. $f(x) = \dfrac{k}{x}$	$y = -\dfrac{3}{4}x + 3$
68. $f(x) = k\sqrt{x}$	$y = x + 4$

EXPLORING CONCEPTS

Exploring a Relationship In Exercises 69–72, the relationship between f and g is given. Explain the relationship between f' and g'.

69. $g(x) = f(x) + 6$

70. $g(x) = 2f(x)$

71. $g(x) = -5f(x)$

72. $g(x) = 3f(x) - 1$

A Function and Its Derivative In Exercises 73 and 74, the graphs of a function f and its derivative f' are shown on the same set of coordinate axes. Label the graphs as f or f' and write a short paragraph stating the criteria you used in making your selection. To print an enlarged copy of the graph, go to *MathGraphs.com*.

73.

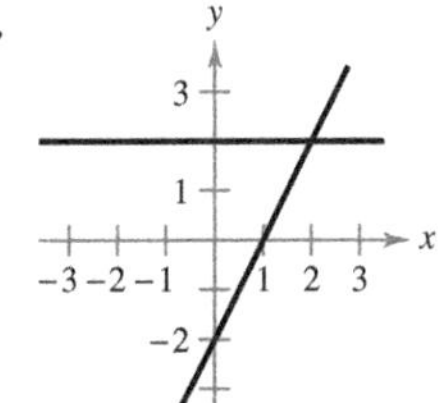

74.

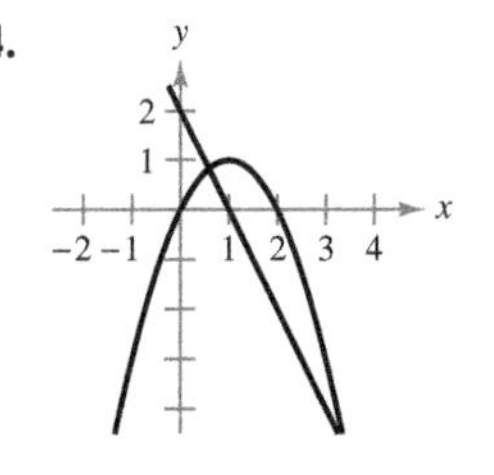

75. **Sketching a Graph** Sketch the graph of a function f such that $f' > 0$ for all x and the rate of change of the function is decreasing.

76. **HOW DO YOU SEE IT?** Use the graph of f to answer each question. To print an enlarged copy of the graph, go to *MathGraphs.com*.

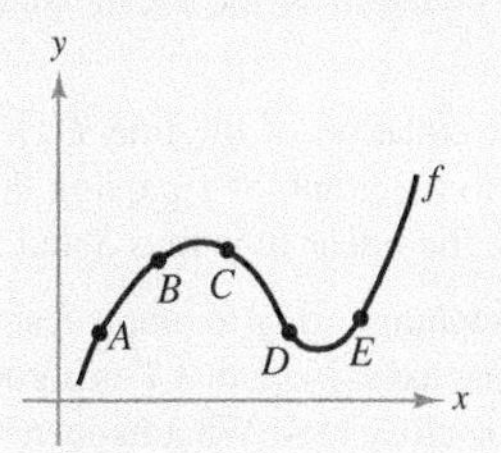

(a) Between which two consecutive points is the average rate of change of the function greatest?

(b) Is the average rate of change of the function between A and B greater than or less than the instantaneous rate of change at B?

(c) Sketch a tangent line to the graph between C and D such that the slope of the tangent line is the same as the average rate of change of the function between C and D.

77. **Finding Equations of Tangent Lines** Sketch the graphs of $y = x^2$ and $y = -x^2 + 6x - 5$, and sketch the two lines that are tangent to both graphs. Find equations of these lines.

78. **Tangent Lines** Show that the graphs of the two equations

$$y = x \quad \text{and} \quad y = \frac{1}{x}$$

have tangent lines that are perpendicular to each other at their point of intersection.

79. **Horizontal Tangent Line** Show that the graph of the function

$$f(x) = 3x + \sin x + 2$$

does not have a horizontal tangent line.

80. **Tangent Line** Show that the graph of the function

$$f(x) = x^5 + 3x^3 + 5x$$

does not have a tangent line with a slope of 3.

Finding an Equation of a Tangent Line In Exercises 81 and 82, find an equation of the tangent line to the graph of the function f through the point (x_0, y_0) not on the graph. To find the point of tangency (x, y) on the graph of f, solve the equation

$$f'(x) = \frac{y_0 - y}{x_0 - x}.$$

81. $f(x) = \sqrt{x}$

$(x_0, y_0) = (-4, 0)$

82. $f(x) = \dfrac{2}{x}$

$(x_0, y_0) = (5, 0)$

83. Linear Approximation Consider the function $f(x) = x^{3/2}$ with the solution point $(4, 8)$.

(a) Use a graphing utility to graph f. Use the *zoom* feature to obtain successive magnifications of the graph in the neighborhood of the point $(4, 8)$. After zooming in a few times, the graph should appear nearly linear. Use the *trace* feature to determine the coordinates of a point near $(4, 8)$. Find an equation of the secant line $S(x)$ through the two points.

(b) Find the equation of the line $T(x) = f'(4)(x - 4) + f(4)$ tangent to the graph of f passing through the given point. Why are the linear functions S and T nearly the same?

(c) Use a graphing utility to graph f and T on the same set of coordinate axes. Note that T is a good approximation of f when x is close to 4. What happens to the accuracy of the approximation as you move farther away from the point of tangency?

(d) Demonstrate the conclusion in part (c) by completing the table.

Δx	-3	-2	-1	-0.5	-0.1	0
$f(4 + \Delta x)$						
$T(4 + \Delta x)$						

Δx	0.1	0.5	1	2	3
$f(4 + \Delta x)$					
$T(4 + \Delta x)$					

84. Linear Approximation Repeat Exercise 83 for the function $f(x) = x^3$, where $T(x)$ is the line tangent to the graph at the point $(1, 1)$. Explain why the accuracy of the linear approximation decreases more rapidly than in Exercise 83.

True or False? **In Exercises 85–90, determine whether the statement is true or false. If it is false, explain why or give an example that shows it is false.**

85. If $f'(x) = g'(x)$, then $f(x) = g(x)$.

86. If $y = x^{a+2} + bx$, then $dy/dx = (a + 2)x^{a+1} + b$.

87. If $y = \pi^2$, then $dy/dx = 2\pi$.

88. If $f(x) = -g(x) + b$, then $f'(x) = -g'(x)$.

89. If $f(x) = 0$, then $f'(x)$ is undefined.

90. If $f(x) = \frac{1}{x^n}$, then $f'(x) = \frac{1}{nx^{n-1}}$.

Finding Rates of Change **In Exercises 91–94, find the average rate of change of the function over the given interval. Compare this average rate of change with the instantaneous rates of change at the endpoints of the interval.**

91. $f(t) = 3t + 5, \quad [1, 2]$

92. $f(t) = t^2 - 7, \quad [3, 3.1]$

93. $f(x) = \frac{-1}{x}, \quad [1, 2]$

94. $f(x) = \sin x, \quad \left[0, \frac{\pi}{6}\right]$

Vertical Motion **In Exercises 95 and 96, use the position function $s(t) = -16t^2 + v_0t + s_0$ for free-falling objects.**

95. A silver dollar is dropped from the top of a building that is 1362 feet tall.

(a) Determine the position and velocity functions for the coin.

(b) Determine the average velocity on the interval $[1, 2]$.

(c) Find the instantaneous velocities when $t = 1$ and $t = 2$.

(d) Find the time required for the coin to reach ground level.

(e) Find the velocity of the coin at impact.

96. A ball is thrown straight down from the top of a 220-foot building with an initial velocity of -22 feet per second. What is its velocity after 3 seconds? What is its velocity after falling 108 feet?

Vertical Motion **In Exercises 97 and 98, use the position function $s(t) = -4.9t^2 + v_0t + s_0$ for free-falling objects.**

97. A projectile is shot upward from the surface of Earth with an initial velocity of 120 meters per second. What is its velocity after 5 seconds? After 10 seconds?

98. A rock is dropped from the edge of a cliff that is 214 meters above water.

(a) Determine the position and velocity functions for the rock.

(b) Determine the average velocity on the interval $[2, 5]$.

(c) Find the instantaneous velocities when $t = 2$ and $t = 5$.

(d) Find the time required for the rock to reach the surface of the water.

(e) Find the velocity of the rock at impact.

99. Think About It The graph of the position function (see figure) represents the distance in miles that a person drives during a 10-minute trip to work. Make a sketch of the corresponding velocity function.

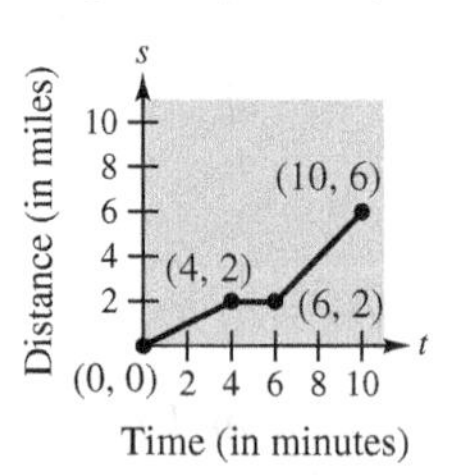

Figure for 99

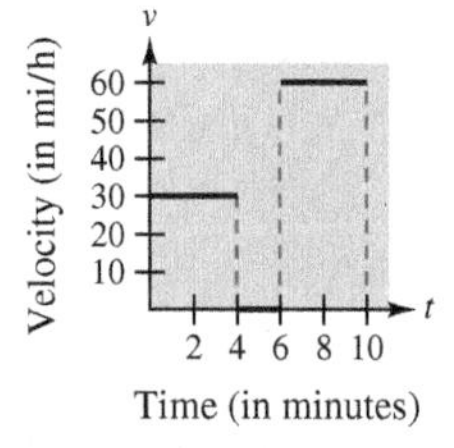

Figure for 100

100. Think About It The graph of the velocity function (see figure) represents the velocity in miles per hour during a 10-minute trip to work. Make a sketch of the corresponding position function.

101. Volume The volume of a cube with sides of length s is given by $V = s^3$. Find the rate of change of the volume with respect to s when $s = 6$ centimeters.

102. Area The area of a square with sides of length s is given by $A = s^2$. Find the rate of change of the area with respect to s when $s = 6$ meters.

103. Modeling Data The stopping distance of an automobile, on dry, level pavement, traveling at a speed v (in kilometers per hour) is the distance R (in meters) the car travels during the reaction time of the driver plus the distance B (in meters) the car travels after the brakes are applied (see figure). The table shows the results of an experiment.

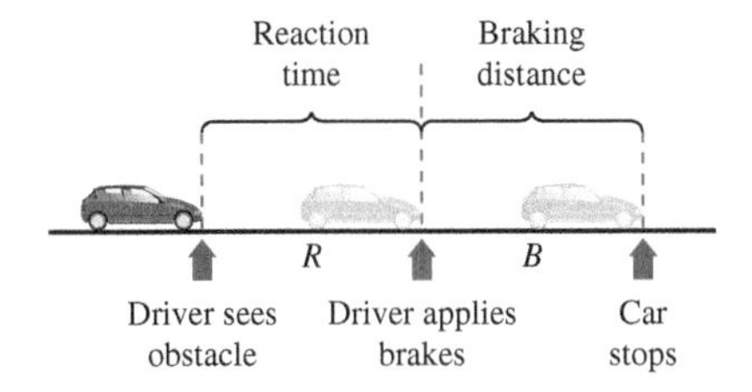

Speed, v	20	40	60	80	100
Reaction Time Distance, R	8.3	16.7	25.0	33.3	41.7
Braking Time Distance, B	2.3	9.0	20.2	35.8	55.9

(a) Use the regression capabilities of a graphing utility to find a linear model for reaction time distance R.

(b) Use the regression capabilities of a graphing utility to find a quadratic model for braking time distance B.

(c) Determine the polynomial giving the total stopping distance T.

(d) Use a graphing utility to graph the functions R, B, and T in the same viewing window.

(e) Find the derivative of T and the rates of change of the total stopping distance for $v = 40$, $v = 80$, and $v = 100$.

(f) Use the results of this exercise to draw conclusions about the total stopping distance as speed increases.

104. Fuel Cost A car is driven 15,000 miles a year and gets x miles per gallon. Assume that the average fuel cost is \$3.48 per gallon. Find the annual cost of fuel C as a function of x and use this function to complete the table.

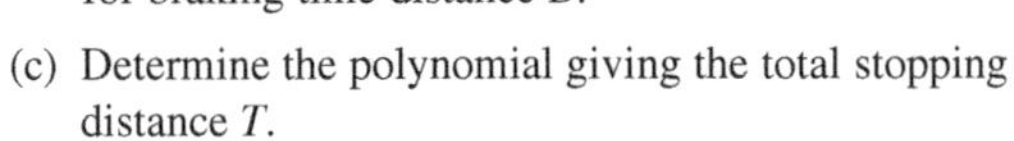

x	10	15	20	25	30	35	40
C							
dC/dx							

Who would benefit more from a one-mile-per-gallon increase in fuel efficiency—the driver of a car that gets 15 miles per gallon or the driver of a car that gets 35 miles per gallon? Explain.

105. Velocity Verify that the average velocity over the time interval $[t_0 - \Delta t, t_0 + \Delta t]$ is the same as the instantaneous velocity at $t = t_0$ for the position function

$$s(t) = -\frac{1}{2}at^2 + c.$$

106. Inventory Management The annual inventory cost C for a manufacturer is

$$C = \frac{1{,}008{,}000}{Q} + 6.3Q$$

where Q is the order size when the inventory is replenished. Find the change in annual cost when Q is increased from 350 to 351 and compare this with the instantaneous rate of change when $Q = 350$.

107. Finding an Equation of a Parabola Find an equation of the parabola $y = ax^2 + bx + c$ that passes through $(0, 1)$ and is tangent to the line $y = x - 1$ at $(1, 0)$.

108. Proof Let (a, b) be an arbitrary point on the graph of $y = 1/x$, $x > 0$. Prove that the area of the triangle formed by the tangent line through (a, b) and the coordinate axes is 2.

109. Tangent Line Find the equation(s) of the tangent line(s) to the graph of the curve $y = x^3 - 9x$ through the point $(1, -9)$ not on the graph.

110. Tangent Line Find the equation(s) of the tangent line(s) to the graph of the parabola $y = x^2$ through the given point not on the graph.

(a) $(0, a)$ (b) $(a, 0)$

Are there any restrictions on the constant a?

Making a Function Differentiable **In Exercises 111 and 112, find a and b such that f is differentiable everywhere.**

111. $f(x) = \begin{cases} ax^3, & x \le 2 \\ x^2 + b, & x > 2 \end{cases}$

112. $f(x) = \begin{cases} \cos x, & x < 0 \\ ax + b, & x \ge 0 \end{cases}$

113. Determining Differentiability Where are the functions $f_1(x) = |\sin x|$ and $f_2(x) = \sin |x|$ differentiable?

114. Proof Prove that $\frac{d}{dx}[\cos x] = -\sin x$.

FOR FURTHER INFORMATION For a geometric interpretation of the derivatives of trigonometric functions, see the article "Sines and Cosines of the Times" by Victor J. Katz in *Math Horizons*. To view this article, go to *MathArticles.com*.

PUTNAM EXAM CHALLENGE

115. Find all differentiable functions $f: \mathbb{R} \to \mathbb{R}$ such that

$$f'(x) = \frac{f(x + n) - f(x)}{n}$$

for all real numbers x and all positive integers n.

2.3 Product and Quotient Rules and Higher-Order Derivatives

- **Find the derivative of a function using the Product Rule.**
- **Find the derivative of a function using the Quotient Rule.**
- **Find the derivative of a trigonometric function.**
- **Find a higher-order derivative of a function.**

The Product Rule

In Section 2.2, you learned that the derivative of the sum of two functions is simply the sum of their derivatives. The rules for the derivatives of the product and quotient of two functions are not as simple.

REMARK A version of the Product Rule that some people prefer is

$$\frac{d}{dx}[f(x)g(x)] = f'(x)g(x) + f(x)g'(x).$$

The advantage of this form is that it generalizes easily to products of three or more factors.

THEOREM 2.7 The Product Rule

The product of two differentiable functions f and g is itself differentiable. Moreover, the derivative of fg is the first function times the derivative of the second, plus the second function times the derivative of the first.

$$\frac{d}{dx}[f(x)g(x)] = f(x)g'(x) + g(x)f'(x)$$

Proof Some mathematical proofs, such as the proof of the Sum Rule, are straightforward. Others involve clever steps that may appear unmotivated to a reader. This proof involves such a step—subtracting and adding the same quantity—which is shown in color.

$$\begin{aligned}
\frac{d}{dx}[f(x)g(x)] &= \lim_{\Delta x \to 0} \frac{f(x+\Delta x)g(x+\Delta x) - f(x)g(x)}{\Delta x} \\
&= \lim_{\Delta x \to 0} \frac{f(x+\Delta x)g(x+\Delta x) - f(x+\Delta x)g(x) + f(x+\Delta x)g(x) - f(x)g(x)}{\Delta x} \\
&= \lim_{\Delta x \to 0} \left[f(x+\Delta x)\frac{g(x+\Delta x) - g(x)}{\Delta x} + g(x)\frac{f(x+\Delta x) - f(x)}{\Delta x}\right] \\
&= \lim_{\Delta x \to 0} \left[f(x+\Delta x)\frac{g(x+\Delta x) - g(x)}{\Delta x}\right] + \lim_{\Delta x \to 0} \left[g(x)\frac{f(x+\Delta x) - f(x)}{\Delta x}\right] \\
&= \lim_{\Delta x \to 0} f(x+\Delta x) \cdot \lim_{\Delta x \to 0} \frac{g(x+\Delta x) - g(x)}{\Delta x} + \lim_{\Delta x \to 0} g(x) \cdot \lim_{\Delta x \to 0} \frac{f(x+\Delta x) - f(x)}{\Delta x} \\
&= f(x)g'(x) + g(x)f'(x)
\end{aligned}$$

Note that $\lim_{\Delta x \to 0} f(x + \Delta x) = f(x)$ because f is given to be differentiable and therefore is continuous. ■

REMARK The proof of the Product Rule for products of more than two factors is left as an exercise (see Exercise 137).

The Product Rule can be extended to cover products involving more than two factors. For example, if f, g, and h are differentiable functions of x, then

$$\frac{d}{dx}[f(x)g(x)h(x)] = f'(x)g(x)h(x) + f(x)g'(x)h(x) + f(x)g(x)h'(x).$$

So, the derivative of $y = x^2 \sin x \cos x$ is

$$\begin{aligned}
\frac{dy}{dx} &= 2x \sin x \cos x + x^2 \cos x \cos x + x^2(\sin x)(-\sin x) \\
&= 2x \sin x \cos x + x^2(\cos^2 x - \sin^2 x).
\end{aligned}$$

THE PRODUCT RULE

When Leibniz originally wrote a formula for the Product Rule, he was motivated by the expression

$(x + dx)(y + dy) - xy$

from which he subtracted $dx\,dy$ (as being negligible) and obtained the differential form $x\,dy + y\,dx$. This derivation resulted in the traditional form of the Product Rule. *(Source: The History of Mathematics by David M. Burton)*

The derivative of a product of two functions is not (in general) given by the product of the derivatives of the two functions. To see this, try comparing the product of the derivatives of

$$f(x) = 3x - 2x^2$$

and

$$g(x) = 5 + 4x$$

with the derivative in Example 1.

EXAMPLE 1 Using the Product Rule

Find the derivative of $h(x) = (3x - 2x^2)(5 + 4x)$.

Solution

$$\begin{aligned} h'(x) &= \underbrace{(3x - 2x^2)}_{\text{First}}\underbrace{\frac{d}{dx}[5 + 4x]}_{\text{Derivative of second}} + \underbrace{(5 + 4x)}_{\text{Second}}\underbrace{\frac{d}{dx}[3x - 2x^2]}_{\text{Derivative of first}} && \text{Apply Product Rule.} \\ &= (3x - 2x^2)(4) + (5 + 4x)(3 - 4x) \\ &= (12x - 8x^2) + (15 - 8x - 16x^2) \\ &= -24x^2 + 4x + 15 \end{aligned}$$

In Example 1, you have the option of finding the derivative with or without the Product Rule. To find the derivative without the Product Rule, you can write

$$\begin{aligned} D_x[(3x - 2x^2)(5 + 4x)] &= D_x[-8x^3 + 2x^2 + 15x] \\ &= -24x^2 + 4x + 15. \end{aligned}$$

In the next example, you must use the Product Rule.

EXAMPLE 2 Using the Product Rule

Find the derivative of $y = 3x^2 \sin x$.

Solution

$$\begin{aligned} \frac{d}{dx}[3x^2 \sin x] &= 3x^2 \frac{d}{dx}[\sin x] + \sin x \frac{d}{dx}[3x^2] && \text{Apply Product Rule.} \\ &= 3x^2 \cos x + (\sin x)(6x) \\ &= 3x^2 \cos x + 6x \sin x \\ &= 3x(x \cos x + 2 \sin x) \end{aligned}$$

REMARK In Example 3, notice that you use the Product Rule when both factors of the product are variable, and you use the Constant Multiple Rule when one of the factors is a constant.

EXAMPLE 3 Using the Product Rule

Find the derivative of $y = 2x \cos x - 2 \sin x$.

Solution

$$\begin{aligned} \frac{dy}{dx} &= \underbrace{(2x)\left(\frac{d}{dx}[\cos x]\right) + (\cos x)\left(\frac{d}{dx}[2x]\right)}_{\text{Product Rule}} - \underbrace{2\frac{d}{dx}[\sin x]}_{\text{Constant Multiple Rule}} \\ &= (2x)(-\sin x) + (\cos x)(2) - 2(\cos x) \\ &= -2x \sin x \end{aligned}$$

The Quotient Rule

THEOREM 2.8 The Quotient Rule

The quotient f/g of two differentiable functions f and g is itself differentiable at all values of x for which $g(x) \neq 0$. Moreover, the derivative of f/g is given by the denominator times the derivative of the numerator minus the numerator times the derivative of the denominator, all divided by the square of the denominator.

$$\frac{d}{dx}\left[\frac{f(x)}{g(x)}\right] = \frac{g(x)f'(x) - f(x)g'(x)}{[g(x)]^2}, \quad g(x) \neq 0$$

REMARK From the Quotient Rule, you can see that the derivative of a quotient is not (in general) the quotient of the derivatives.

Proof As with the proof of Theorem 2.7, the key to this proof is subtracting and adding the same quantity—which is shown in color.

$$\frac{d}{dx}\left[\frac{f(x)}{g(x)}\right] = \lim_{\Delta x \to 0} \frac{\dfrac{f(x+\Delta x)}{g(x+\Delta x)} - \dfrac{f(x)}{g(x)}}{\Delta x} \qquad \text{Definition of derivative}$$

$$= \lim_{\Delta x \to 0} \frac{g(x)f(x+\Delta x) - f(x)g(x+\Delta x)}{\Delta x g(x)g(x+\Delta x)}$$

$$= \lim_{\Delta x \to 0} \frac{g(x)f(x+\Delta x) - f(x)g(x) + f(x)g(x) - f(x)g(x+\Delta x)}{\Delta x g(x)g(x+\Delta x)}$$

$$= \frac{\displaystyle\lim_{\Delta x \to 0} \frac{g(x)[f(x+\Delta x) - f(x)]}{\Delta x} - \lim_{\Delta x \to 0} \frac{f(x)[g(x+\Delta x) - g(x)]}{\Delta x}}{\displaystyle\lim_{\Delta x \to 0} [g(x)g(x+\Delta x)]}$$

$$= \frac{g(x)\left[\displaystyle\lim_{\Delta x \to 0} \frac{f(x+\Delta x) - f(x)}{\Delta x}\right] - f(x)\left[\displaystyle\lim_{\Delta x \to 0} \frac{g(x+\Delta x) - g(x)}{\Delta x}\right]}{\displaystyle\lim_{\Delta x \to 0} [g(x)g(x+\Delta x)]}$$

$$= \frac{g(x)f'(x) - f(x)g'(x)}{[g(x)]^2}$$

Note that $\lim_{\Delta x \to 0} g(x + \Delta x) = g(x)$ because g is given to be differentiable and therefore is continuous. ■

TECHNOLOGY A graphing utility can be used to compare the graph of a function with the graph of its derivative. For instance, in Figure 2.22, the graph of the function in Example 4 appears to have two points that have horizontal tangent lines. What are the values of y' at these two points?

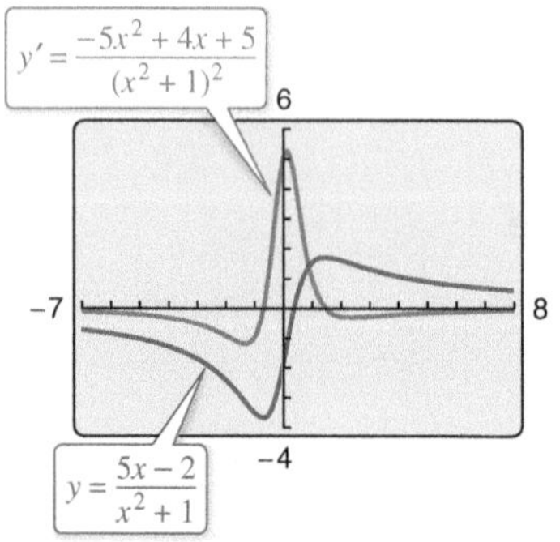

Graphical comparison of a function and its derivative
Figure 2.22

EXAMPLE 4 Using the Quotient Rule

Find the derivative of $y = \dfrac{5x - 2}{x^2 + 1}$.

Solution

$$\frac{d}{dx}\left[\frac{5x - 2}{x^2 + 1}\right] = \frac{(x^2 + 1)\dfrac{d}{dx}[5x - 2] - (5x - 2)\dfrac{d}{dx}[x^2 + 1]}{(x^2 + 1)^2} \qquad \text{Apply Quotient Rule.}$$

$$= \frac{(x^2 + 1)(5) - (5x - 2)(2x)}{(x^2 + 1)^2}$$

$$= \frac{(5x^2 + 5) - (10x^2 - 4x)}{(x^2 + 1)^2}$$

$$= \frac{-5x^2 + 4x + 5}{(x^2 + 1)^2}$$

■

Note the use of parentheses in Example 4. A liberal use of parentheses is recommended for *all* types of differentiation problems. For instance, with the Quotient Rule, it is a good idea to enclose all factors and derivatives in parentheses and to pay special attention to the subtraction required in the numerator.

When differentiation rules were introduced in the preceding section, the need for rewriting *before* differentiating was emphasized. The next example illustrates this point with the Quotient Rule.

EXAMPLE 5 Rewriting Before Differentiating

Find an equation of the tangent line to the graph of $f(x) = \dfrac{3 - (1/x)}{x + 5}$ at $(-1, 1)$.

Solution Begin by rewriting the function.

$$f(x) = \frac{3 - (1/x)}{x + 5} \qquad \text{Write original function.}$$

$$= \frac{x\left(3 - \frac{1}{x}\right)}{x(x + 5)} \qquad \text{Multiply numerator and denominator by } x.$$

$$= \frac{3x - 1}{x^2 + 5x} \qquad \text{Rewrite.}$$

Next, apply the Quotient Rule.

$$f'(x) = \frac{(x^2 + 5x)(3) - (3x - 1)(2x + 5)}{(x^2 + 5x)^2} \qquad \text{Quotient Rule}$$

$$= \frac{(3x^2 + 15x) - (6x^2 + 13x - 5)}{(x^2 + 5x)^2}$$

$$= \frac{-3x^2 + 2x + 5}{(x^2 + 5x)^2} \qquad \text{Simplify.}$$

To find the slope at $(-1, 1)$, evaluate $f'(-1)$.

$$f'(-1) = 0 \qquad \text{Slope of graph at } (-1, 1)$$

Then, using the point-slope form of the equation of a line, you can determine that the equation of the tangent line at $(-1, 1)$ is $y = 1$. See Figure 2.23.

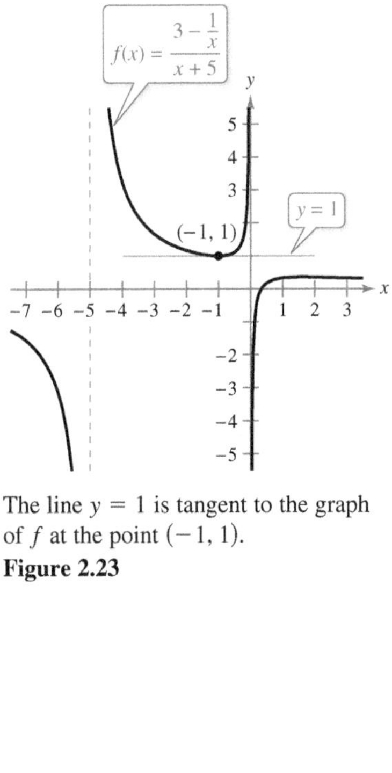

The line $y = 1$ is tangent to the graph of f at the point $(-1, 1)$.
Figure 2.23

Not every quotient needs to be differentiated by the Quotient Rule. For instance, each quotient in the next example can be considered as the product of a constant times a function of x. In such cases, it is more convenient to use the Constant Multiple Rule.

REMARK To see the benefit of using the Constant Multiple Rule for some quotients, try using the Quotient Rule to differentiate the functions in Example 6. You should obtain the same results but with more work.

EXAMPLE 6 Using the Constant Multiple Rule

	Original Function	Rewrite	Differentiate	Simplify
a.	$y = \dfrac{x^2 + 3x}{6}$	$y = \dfrac{1}{6}(x^2 + 3x)$	$y' = \dfrac{1}{6}(2x + 3)$	$y' = \dfrac{2x + 3}{6}$
b.	$y = \dfrac{5x^4}{8}$	$y = \dfrac{5}{8}x^4$	$y' = \dfrac{5}{8}(4x^3)$	$y' = \dfrac{5}{2}x^3$
c.	$y = \dfrac{-3(3x - 2x^2)}{7x}$	$y = -\dfrac{3}{7}(3 - 2x)$	$y' = -\dfrac{3}{7}(-2)$	$y' = \dfrac{6}{7}$
d.	$y = \dfrac{9}{5x^2}$	$y = \dfrac{9}{5}(x^{-2})$	$y' = \dfrac{9}{5}(-2x^{-3})$	$y' = -\dfrac{18}{5x^3}$

In Section 2.2, the Power Rule was proved only for the case in which the exponent n is a positive integer greater than 1. The next example extends the proof to include negative integer exponents.

EXAMPLE 7 **Power Rule: Negative Integer Exponents**

If n is a negative integer, then there exists a positive integer k such that $n = -k$. So, by the Quotient Rule, you can write

$$\begin{aligned}\frac{d}{dx}[x^n] &= \frac{d}{dx}\left[\frac{1}{x^k}\right] \\ &= \frac{x^k(0) - (1)(kx^{k-1})}{(x^k)^2} && \text{Quotient Rule and Power Rule} \\ &= \frac{0 - kx^{k-1}}{x^{2k}} \\ &= -kx^{-k-1} \\ &= nx^{n-1}. && n = -k\end{aligned}$$

So, the Power Rule

$$\frac{d}{dx}[x^n] = nx^{n-1} \qquad \text{Power Rule}$$

is valid for any integer n. In Exercise 73 in Section 2.5, you are asked to prove the case for which n is any rational number. ■

Derivatives of Trigonometric Functions

Knowing the derivatives of the sine and cosine functions, you can use the Quotient Rule to find the derivatives of the four remaining trigonometric functions.

THEOREM 2.9 Derivatives of Trigonometric Functions

$$\frac{d}{dx}[\tan x] = \sec^2 x \qquad \frac{d}{dx}[\cot x] = -\csc^2 x$$

$$\frac{d}{dx}[\sec x] = \sec x \tan x \qquad \frac{d}{dx}[\csc x] = -\csc x \cot x$$

Proof Considering $\tan x = (\sin x)/(\cos x)$ and applying the Quotient Rule, you obtain

$$\begin{aligned}\frac{d}{dx}[\tan x] &= \frac{d}{dx}\left[\frac{\sin x}{\cos x}\right] \\ &= \frac{(\cos x)(\cos x) - (\sin x)(-\sin x)}{\cos^2 x} && \text{Apply Quotient Rule.} \\ &= \frac{\cos^2 x + \sin^2 x}{\cos^2 x} \\ &= \frac{1}{\cos^2 x} \\ &= \sec^2 x.\end{aligned}$$

The proofs of the other three parts of the theorem are left as an exercise (see Exercise 87). ■

REMARK In the proof of Theorem 2.9, note the use of the trigonometric identities

$$\sin^2 x + \cos^2 x = 1$$

and

$$\sec x = \frac{1}{\cos x}.$$

These trigonometric identities and others are listed in Section P.4 and on the formula cards for this text.

EXAMPLE 8 Differentiating Trigonometric Functions

See LarsonCalculus.com for an interactive version of this type of example.

	Function	**Derivative**
a.	$y = x - \tan x$	$\dfrac{dy}{dx} = 1 - \sec^2 x$
b.	$y = x \sec x$	$y' = x(\sec x \tan x) + (\sec x)(1)$ $= (\sec x)(1 + x \tan x)$

REMARK Because of trigonometric identities, the derivative of a trigonometric function can take many forms. This presents a challenge when you are trying to match your answers to those given in the back of the text.

EXAMPLE 9 Different Forms of a Derivative

Differentiate both forms of

$$y = \frac{1 - \cos x}{\sin x} = \csc x - \cot x.$$

Solution

First form: $y = \dfrac{1 - \cos x}{\sin x}$

$$y' = \frac{(\sin x)(\sin x) - (1 - \cos x)(\cos x)}{\sin^2 x}$$

$$= \frac{\sin^2 x - \cos x + \cos^2 x}{\sin^2 x}$$

$$= \frac{1 - \cos x}{\sin^2 x} \qquad \sin^2 x + \cos^2 x = 1$$

Second form: $y = \csc x - \cot x$

$$y' = -\csc x \cot x + \csc^2 x$$

To show that the two derivatives are equal, you can write

$$\frac{1 - \cos x}{\sin^2 x} = \frac{1}{\sin^2 x} - \frac{\cos x}{\sin^2 x}$$

$$= \frac{1}{\sin^2 x} - \left(\frac{1}{\sin x}\right)\left(\frac{\cos x}{\sin x}\right)$$

$$= \csc^2 x - \csc x \cot x.$$

The summary below shows that much of the work in obtaining a simplified form of a derivative occurs *after* differentiating. Note that two characteristics of a simplified form are the absence of negative exponents and the combining of like terms.

	$f'(x)$ After Differentiating	$f'(x)$ After Simplifying
Example 1	$(3x - 2x^2)(4) + (5 + 4x)(3 - 4x)$	$-24x^2 + 4x + 15$
Example 3	$(2x)(-\sin x) + (\cos x)(2) - 2(\cos x)$	$-2x \sin x$
Example 4	$\dfrac{(x^2 + 1)(5) - (5x - 2)(2x)}{(x^2 + 1)^2}$	$\dfrac{-5x^2 + 4x + 5}{(x^2 + 1)^2}$
Example 5	$\dfrac{(x^2 + 5x)(3) - (3x - 1)(2x + 5)}{(x^2 + 5x)^2}$	$\dfrac{-3x^2 + 2x + 5}{(x^2 + 5x)^2}$
Example 9	$\dfrac{(\sin x)(\sin x) - (1 - \cos x)(\cos x)}{\sin^2 x}$	$\dfrac{1 - \cos x}{\sin^2 x}$

Higher-Order Derivatives

Just as you can obtain a velocity function by differentiating a position function, you can obtain an **acceleration** function by differentiating a velocity function. Another way of looking at this is that you can obtain an acceleration function by differentiating a position function *twice*.

$$s(t) \qquad \text{Position function}$$
$$v(t) = s'(t) \qquad \text{Velocity function}$$
$$a(t) = v'(t) = s''(t) \qquad \text{Acceleration function}$$

The function $a(t)$ is the **second derivative** of $s(t)$ and is denoted by $s''(t)$.

The second derivative is an example of a **higher-order derivative.** You can define derivatives of any positive integer order. For instance, the **third derivative** is the derivative of the second derivative. Higher-order derivatives are denoted as shown below.

REMARK The second derivative of a function is the derivative of the first derivative of the function.

First derivative:	y',	$f'(x)$,	$\frac{dy}{dx}$,	$\frac{d}{dx}[f(x)]$,	$D_x[y]$
Second derivative:	y'',	$f''(x)$,	$\frac{d^2y}{dx^2}$,	$\frac{d^2}{dx^2}[f(x)]$,	$D_x^2[y]$
Third derivative:	y''',	$f'''(x)$,	$\frac{d^3y}{dx^3}$,	$\frac{d^3}{dx^3}[f(x)]$,	$D_x^3[y]$
Fourth derivative:	$y^{(4)}$,	$f^{(4)}(x)$,	$\frac{d^4y}{dx^4}$,	$\frac{d^4}{dx^4}[f(x)]$,	$D_x^4[y]$
⋮					
nth derivative:	$y^{(n)}$,	$f^{(n)}(x)$,	$\frac{d^ny}{dx^n}$,	$\frac{d^n}{dx^n}[f(x)]$,	$D_x^n[y]$

EXAMPLE 10 Finding the Acceleration Due to Gravity

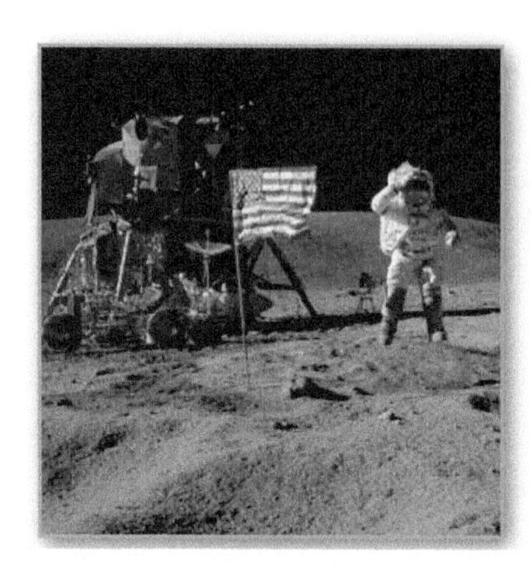

The moon's mass is 7.349×10^{22} kilograms, and Earth's mass is 5.976×10^{24} kilograms. The moon's radius is 1737 kilometers, and Earth's radius is 6378 kilometers. Because the gravitational force on the surface of a planet is directly proportional to its mass and inversely proportional to the square of its radius, the ratio of the gravitational force on Earth to the gravitational force on the moon is

$$\frac{(5.976 \times 10^{24})/6378^2}{(7.349 \times 10^{22})/1737^2} \approx 6.0.$$

Because the moon has no atmosphere, a falling object on the moon encounters no air resistance. In 1971, astronaut David Scott demonstrated that a feather and a hammer fall at the same rate on the moon. The position function for each of these falling objects is

$$s(t) = -0.81t^2 + 2$$

where $s(t)$ is the height in meters and t is the time in seconds, as shown in the figure at the right. What is the ratio of Earth's gravitational force to the moon's?

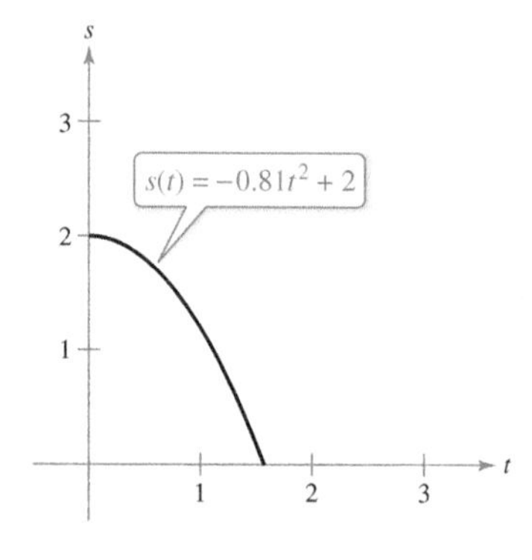

Solution To find the acceleration, differentiate the position function twice.

$$s(t) = -0.81t^2 + 2 \qquad \text{Position function}$$
$$s'(t) = -1.62t \qquad \text{Velocity function}$$
$$s''(t) = -1.62 \qquad \text{Acceleration function}$$

Because $s''(t) = -g$, the acceleration due to gravity on the moon is $g = 1.62$ meters per second per second. The acceleration due to gravity on Earth is 9.8 meters per second per second, so the ratio of Earth's gravitational force to the moon's is

$$\frac{\text{Earth's gravitational force}}{\text{Moon's gravitational force}} = \frac{9.8}{1.62} \approx 6.0.$$

NASA

2.3 Exercises

See CalcChat.com for tutorial help and worked-out solutions to odd-numbered exercises.

CONCEPT CHECK

1. **Product Rule** Describe the Product Rule in your own words.
2. **Quotient Rule** Describe the Quotient Rule in your own words.
3. **Trigonometric Functions** What are the derivatives of $\tan x$, $\cot x$, $\sec x$, and $\csc x$?
4. **Higher-Order Derivative** What is a higher-order derivative?

Using the Product Rule **In Exercises 5–10, use the Product Rule to find the derivative of the function.**

5. $g(x) = (2x - 3)(1 - 5x)$
6. $y = (3x - 4)(x^3 + 5)$
7. $h(t) = \sqrt{t}(1 - t^2)$
8. $g(s) = \sqrt{s}(s^2 + 8)$
9. $f(x) = x^3 \cos x$
10. $g(x) = \sqrt{x} \sin x$

Using the Quotient Rule **In Exercises 11–16, use the Quotient Rule to find the derivative of the function.**

11. $f(x) = \dfrac{x}{x - 5}$
12. $g(t) = \dfrac{3t^2 - 1}{2t + 5}$
13. $h(x) = \dfrac{\sqrt{x}}{x^3 + 1}$
14. $f(x) = \dfrac{x^2}{2\sqrt{x} + 1}$
15. $g(x) = \dfrac{\sin x}{x^2}$
16. $f(t) = \dfrac{\cos t}{t^3}$

Finding and Evaluating a Derivative **In Exercises 17–22, find $f'(x)$ and $f'(c)$.**

Function	Value of c
17. $f(x) = (x^3 + 4x)(3x^2 + 2x - 5)$	$c = 0$
18. $f(x) = (2x^2 - 3x)(9x + 4)$	$c = -1$
19. $f(x) = \dfrac{x^2 - 4}{x - 3}$	$c = 1$
20. $f(x) = \dfrac{x - 4}{x + 4}$	$c = 3$
21. $f(x) = x \cos x$	$c = \dfrac{\pi}{4}$
22. $f(x) = \dfrac{\sin x}{x}$	$c = \dfrac{\pi}{6}$

Using the Constant Multiple Rule **In Exercises 23–28, complete the table to find the derivative of the function without using the Quotient Rule.**

Function	Rewrite	Differentiate	Simplify
23. $y = \dfrac{x^3 + 6x}{3}$			
24. $y = \dfrac{5x^2 - 3}{4}$			
25. $y = \dfrac{6}{7x^2}$			
26. $y = \dfrac{10}{3x^3}$			
27. $y = \dfrac{4x^{3/2}}{x}$			
28. $y = \dfrac{2x}{x^{1/3}}$			

Finding a Derivative **In Exercises 29–40, find the derivative of the algebraic function.**

29. $f(x) = \dfrac{4 - 3x - x^2}{x^2 - 1}$
30. $f(x) = \dfrac{x^2 + 5x + 6}{x^2 - 4}$
31. $f(x) = x\left(1 - \dfrac{4}{x + 3}\right)$
32. $f(x) = x^4\left(1 - \dfrac{2}{x + 1}\right)$
33. $f(x) = \dfrac{3x - 1}{\sqrt{x}}$
34. $f(x) = \sqrt[3]{x}(\sqrt{x} + 3)$
35. $f(x) = \dfrac{2 - \dfrac{1}{x}}{x - 3}$
36. $h(x) = \dfrac{\dfrac{1}{x^2} + 5x}{x + 1}$
37. $g(s) = s^3\left(5 - \dfrac{s}{s + 2}\right)$
38. $g(x) = x^2\left(\dfrac{2}{x} - \dfrac{1}{x + 1}\right)$
39. $f(x) = (2x^3 + 5x)(x - 3)(x + 2)$
40. $f(x) = (x^3 - x)(x^2 + 2)(x^2 + x - 1)$

Finding a Derivative of a Trigonometric Function **In Exercises 41–56, find the derivative of the trigonometric function.**

41. $f(t) = t^2 \sin t$
42. $f(\theta) = (\theta + 1) \cos \theta$
43. $f(t) = \dfrac{\cos t}{t}$
44. $f(x) = \dfrac{\sin x}{x^3}$
45. $f(x) = -x + \tan x$
46. $y = x + \cot x$
47. $g(t) = \sqrt[4]{t} + 6 \csc t$
48. $h(x) = \dfrac{1}{x} - 12 \sec x$
49. $y = \dfrac{3(1 - \sin x)}{2 \cos x}$
50. $y = \dfrac{\sec x}{x}$
51. $y = -\csc x - \sin x$
52. $y = x \sin x + \cos x$
53. $f(x) = x^2 \tan x$
54. $f(x) = \sin x \cos x$
55. $y = 2x \sin x + x^2 \cos x$
56. $h(\theta) = 5\theta \sec \theta + \theta \tan \theta$

Finding a Derivative Using Technology **In Exercises 57 and 58, use a computer algebra system to find the derivative of the function.**

57. $g(x) = \left(\dfrac{x + 1}{x + 2}\right)(2x - 5)$
58. $f(x) = \dfrac{\cos x}{1 - \sin x}$

Finding the Slope of a Graph **In Exercises 59–62, find the slope of the graph of the function at the given point. Use the *derivative* feature of a graphing utility to confirm your results.**

Function	Point
59. $y = \dfrac{1 + \csc x}{1 - \csc x}$	$\left(\dfrac{\pi}{6}, -3\right)$
60. $f(x) = \tan x \cot x$	$(1, 1)$
61. $h(t) = \dfrac{\sec t}{t}$	$\left(\pi, -\dfrac{1}{\pi}\right)$
62. $f(x) = (\sin x)(\sin x + \cos x)$	$\left(\dfrac{\pi}{4}, 1\right)$

Finding an Equation of a Tangent Line **In Exercises 63–68, (a) find an equation of the tangent line to the graph of f at the given point, (b) use a graphing utility to graph the function and its tangent line at the point, and (c) use the *tangent* feature of a graphing utility to confirm your results.**

63. $f(x) = (x^3 + 4x - 1)(x - 2), \quad (1, -4)$

64. $f(x) = (x - 2)(x^2 + 4), \quad (1, -5)$

65. $f(x) = \dfrac{x}{x + 4}, \quad (-5, 5)$

66. $f(x) = \dfrac{x + 3}{x - 3}, \quad (4, 7)$

67. $f(x) = \tan x, \quad \left(\dfrac{\pi}{4}, 1\right)$

68. $f(x) = \sec x, \quad \left(\dfrac{\pi}{3}, 2\right)$

Famous Curves **In Exercises 69–72, find an equation of the tangent line to the graph at the given point. (The graphs in Exercises 69 and 70 are called *Witches of Agnesi*. The graphs in Exercises 71 and 72 are called *serpentines*.)**

69.

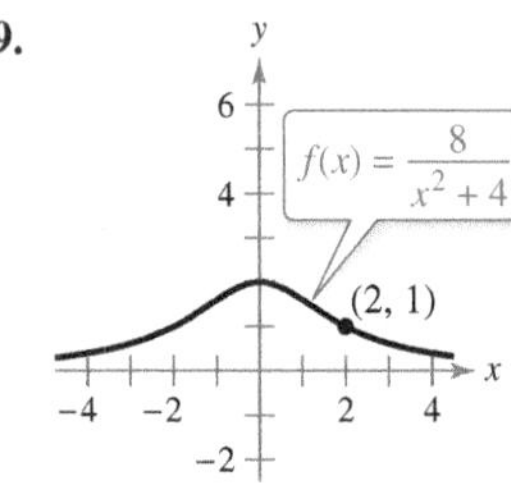

70.

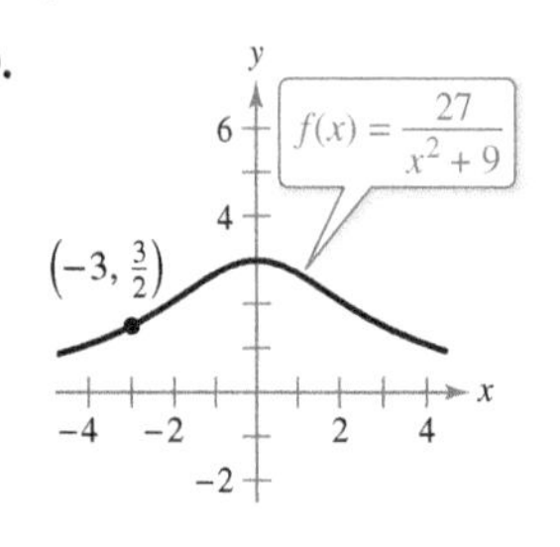

71.

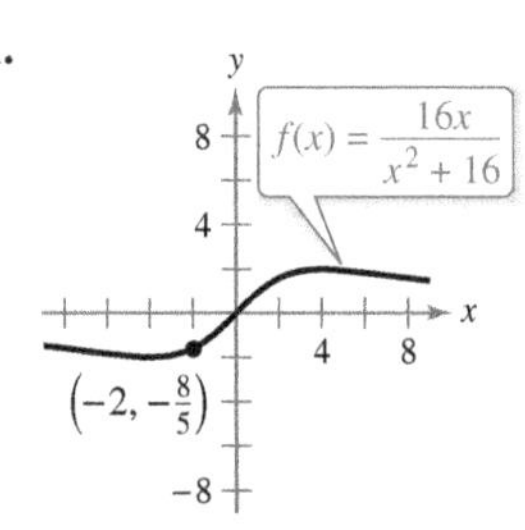

72.

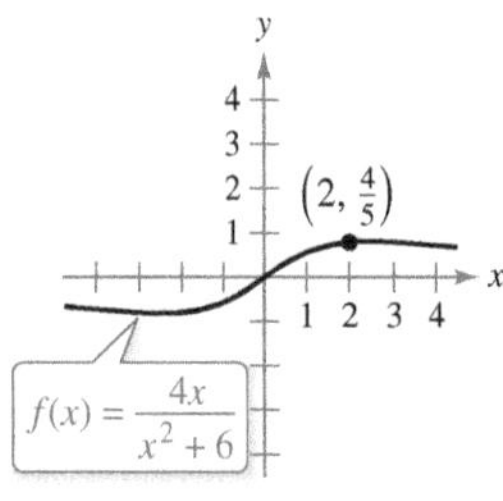

Horizontal Tangent Line **In Exercises 73–76, determine the point(s) at which the graph of the function has a horizontal tangent line.**

73. $f(x) = \dfrac{2x - 1}{x^2}$

74. $f(x) = \dfrac{x^2}{x^2 + 1}$

75. $f(x) = \dfrac{x^2}{x - 1}$

76. $f(x) = \dfrac{x - 4}{x^2 - 7}$

77. **Tangent Lines** Find equations of the tangent lines to the graph of $f(x) = (x + 1)/(x - 1)$ that are parallel to the line $2y + x = 6$. Then graph the function and the tangent lines.

78. **Tangent Lines** Find equations of the tangent lines to the graph of $f(x) = x/(x - 1)$ that pass through the point $(-1, 5)$. Then graph the function and the tangent lines.

Exploring a Relationship **In Exercises 79 and 80, verify that $f'(x) = g'(x)$ and explain the relationship between f and g.**

79. $f(x) = \dfrac{3x}{x + 2}, \quad g(x) = \dfrac{5x + 4}{x + 2}$

80. $f(x) = \dfrac{\sin x - 3x}{x}, \quad g(x) = \dfrac{\sin x + 2x}{x}$

Finding Derivatives **In Exercises 81 and 82, use the graphs of f and g. Let $p(x) = f(x)g(x)$ and $q(x) = f(x)/g(x)$.**

81. (a) Find $p'(1)$.
(b) Find $q'(4)$.

82. (a) Find $p'(4)$.
(b) Find $q'(7)$.

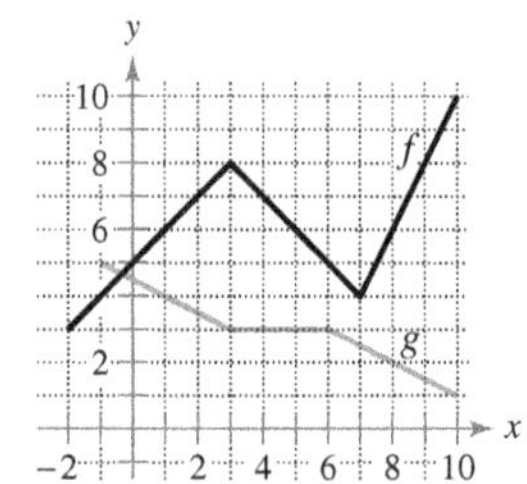

83. **Area** The length of a rectangle is given by $6t + 5$ and its height is $\sqrt{t}$, where t is time in seconds and the dimensions are in centimeters. Find the rate of change of the area with respect to time.

84. **Volume** The radius of a right circular cylinder is given by $\sqrt{t + 2}$ and its height is $\frac{1}{2}\sqrt{t}$, where t is time in seconds and the dimensions are in inches. Find the rate of change of the volume with respect to time.

85. **Inventory Replenishment** The ordering and transportation cost C for the components used in manufacturing a product is

$$C = 100\left(\frac{200}{x^2} + \frac{x}{x + 30}\right), \quad x \geq 1$$

where C is measured in thousands of dollars and x is the order size in hundreds. Find the rate of change of C with respect to x when (a) $x = 10$, (b) $x = 15$, and (c) $x = 20$. What do these rates of change imply about increasing order size?

86. **Population Growth** A population of 500 bacteria is introduced into a culture and grows in number according to the equation

$$P(t) = 500\left(1 + \frac{4t}{50 + t^2}\right)$$

where t is measured in hours. Find the rate at which the population is growing when $t = 2$.

87. Proof Prove each differentiation rule.

(a) $\frac{d}{dx}[\sec x] = \sec x \tan x$

(b) $\frac{d}{dx}[\csc x] = -\csc x \cot x$

(c) $\frac{d}{dx}[\cot x] = -\csc^2 x$

88. Rate of Change Determine whether there exist any values of x in the interval $[0, 2\pi)$ such that the rate of change of $f(x) = \sec x$ and the rate of change of $g(x) = \csc x$ are equal.

89. Modeling Data The table shows the national health care expenditures h (in billions of dollars) in the United States and the population p (in millions) of the United States for the years 2008 through 2013. The year is represented by t, with $t = 8$ corresponding to 2008. *(Source: U.S. Centers for Medicare & Medicaid Services and U.S. Census Bureau)*

Year, t	8	9	10	11	12	13
h	2414	2506	2604	2705	2817	2919
p	304	307	309	311	313	315

(a) Use a graphing utility to find linear models for the health care expenditures $h(t)$ and the population $p(t)$.

(b) Use a graphing utility to graph $h(t)$ and $p(t)$.

(c) Find $A = h(t)/p(t)$, then graph A using a graphing utility. What does this function represent?

(d) Find and interpret $A'(t)$ in the context of the problem.

90. Satellites When satellites observe Earth, they can scan only part of Earth's surface. Some satellites have sensors that can measure the angle θ shown in the figure. Let h represent the satellite's distance from Earth's surface, and let r represent Earth's radius.

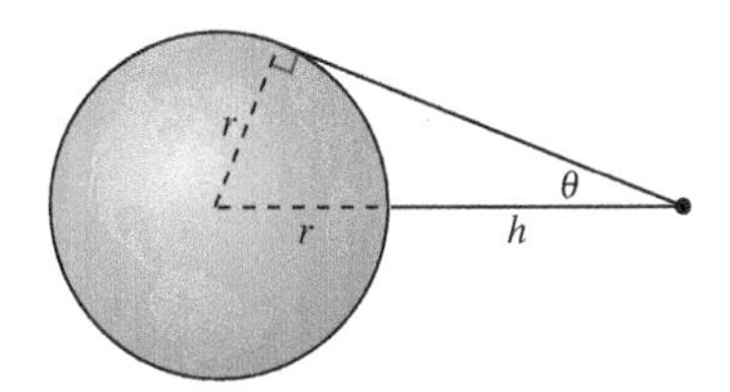

(a) Show that $h = r(\csc \theta - 1)$.

(b) Find the rate at which h is changing with respect to θ when $\theta = 30°$. (Assume $r = 4000$ miles.)

Finding a Second Derivative **In Exercises 91–100, find the second derivative of the function.**

91. $f(x) = x^2 + 7x - 4$

92. $f(x) = 4x^5 - 2x^3 + 5x^2$

93. $f(x) = 4x^{3/2}$

94. $f(x) = x^2 + 3x^{-3}$

95. $f(x) = \frac{x}{x - 1}$

96. $f(x) = \frac{x^2 + 3x}{x - 4}$

97. $f(x) = x \sin x$

98. $f(x) = x \cos x$

99. $f(x) = \csc x$

100. $f(x) = \sec x$

Finding a Higher-Order Derivative **In Exercises 101–104, find the given higher-order derivative.**

101. $f'(x) = x^3 - x^{2/5}, \quad f^{(3)}(x)$

102. $f^{(3)}(x) = \sqrt[5]{x^4}, \quad f^{(4)}(x)$

103. $f''(x) = -\sin x, \quad f^{(8)}(x)$

104. $f^{(4)}(t) = t \cos t, \quad f^{(5)}(t)$

Using Relationships **In Exercises 105–108, use the given information to find $f'(2)$.**

$g(2) = 3$ and $g'(2) = -2$

$h(2) = -1$ and $h'(2) = 4$

105. $f(x) = 2g(x) + h(x)$

106. $f(x) = 4 - h(x)$

107. $f(x) = \frac{g(x)}{h(x)}$

108. $f(x) = g(x)h(x)$

EXPLORING CONCEPTS

109. Higher-Order Derivatives Polynomials of what degree satisfy $f^{(n)} = 0$? Explain your reasoning.

110. Differentiation of Piecewise Functions Describe how you would differentiate a piecewise function. Use your approach to find the first and second derivatives of $f(x) = x|x|$. Explain why $f''(0)$ does not exist.

Identifying Graphs **In Exercises 111 and 112, the graphs of f, f', and f'' are shown on the same set of coordinate axes. Identify each graph. Explain your reasoning. To print an enlarged copy of the graph, go to *MathGraphs.com*.**

111. **112.**

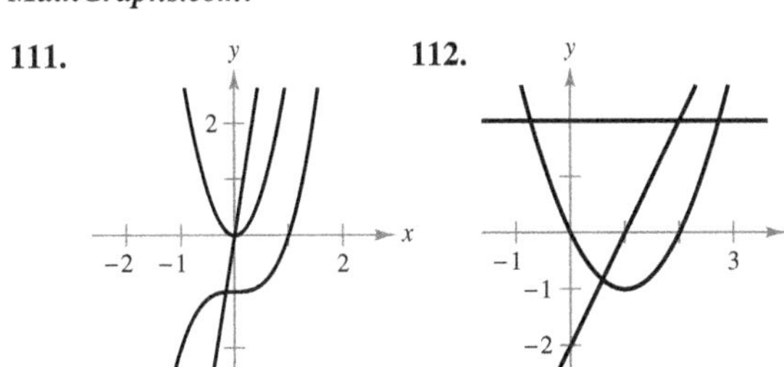

Sketching Graphs **In Exercises 113 and 114, the graph of f is shown. Sketch the graphs of f' and f''. To print an enlarged copy of the graph, go to *MathGraphs.com*.**

113. **114.**

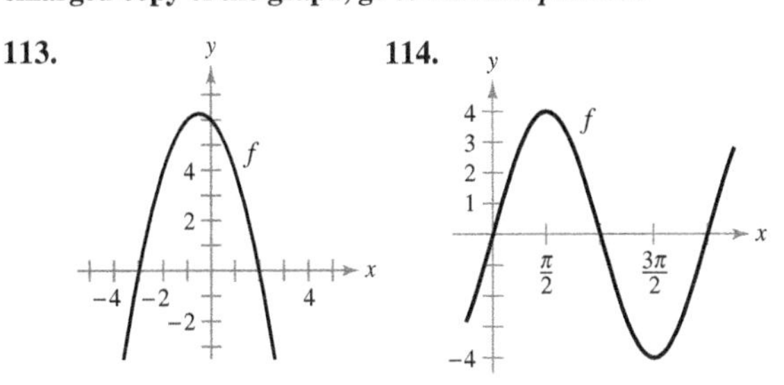

115. Sketching a Graph Sketch the graph of a differentiable function f such that $f(2) = 0$, $f' < 0$ for $-\infty < x < 2$, and $f' > 0$ for $2 < x < \infty$. Explain how you found your answer.

116. Sketching a Graph Sketch the graph of a differentiable function f such that $f > 0$ and $f' < 0$ for all real numbers x. Explain how you found your answer.

117. Acceleration The velocity of an object is

$$v(t) = 36 - t^2, \quad 0 \le t \le 6$$

where v is measured in meters per second and t is the time in seconds. Find the velocity and acceleration of the object when $t = 3$. What can be said about the speed of the object when the velocity and acceleration have opposite signs?

118. Acceleration The velocity of an automobile starting from rest is

$$v(t) = \frac{100t}{2t + 15}$$

where v is measured in feet per second and t is the time in seconds. Find the acceleration at (a) 5 seconds, (b) 10 seconds, and (c) 20 seconds.

119. Stopping Distance A car is traveling at a rate of 66 feet per second (45 miles per hour) when the brakes are applied. The position function for the car is $s(t) = -8.25t^2 + 66t$, where s is measured in feet and t is measured in seconds. Use this function to complete the table and find the average velocity during each time interval.

t	0	1	2	3	4
$s(t)$					
$v(t)$					
$a(t)$					

120. HOW DO YOU SEE IT? The figure shows the graphs of the position, velocity, and acceleration functions of a particle.

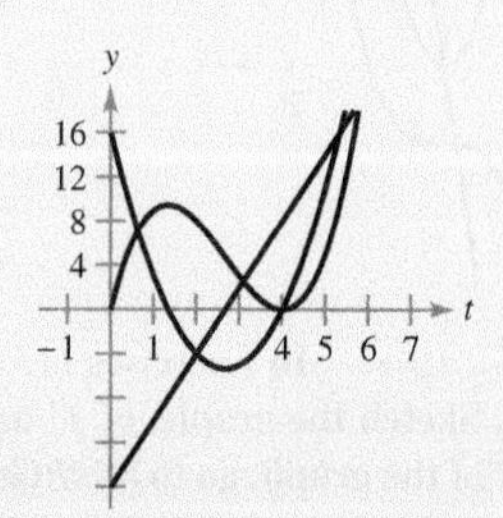

(a) Copy the graphs of the functions shown. Identify each graph. Explain your reasoning. To print an enlarged copy of the graph, go to *MathGraphs.com*.

(b) On your sketch, identify when the particle speeds up and when it slows down. Explain your reasoning.

Finding a Pattern In Exercises 121 and 122, develop a general rule for $f^{(n)}(x)$ given $f(x)$.

121. $f(x) = x^n$

122. $f(x) = \dfrac{1}{x}$

123. Finding a Pattern Consider the function $f(x) = g(x)h(x)$.

(a) Use the Product Rule to generate rules for finding $f''(x)$, $f'''(x)$, and $f^{(4)}(x)$.

(b) Use the results of part (a) to write a general rule for $f^{(n)}(x)$.

124. Finding a Pattern Develop a general rule for the nth derivative of $xf(x)$, where f is a differentiable function of x.

Finding a Pattern In Exercises 125 and 126, find the derivatives of the function f for $n = 1, 2, 3$, and 4. Use the results to write a general rule for $f'(x)$ in terms of n.

125. $f(x) = x^n \sin x$

126. $f(x) = \dfrac{\cos x}{x^n}$

Differential Equations In Exercises 127–130, verify that the function satisfies the differential equation. (A *differential equation* in x and y is an equation that involves x, y, and derivatives of y.)

	Function	Differential Equation
127.	$y = \dfrac{1}{x}, x > 0$	$x^3y'' + 2x^2y' = 0$
128.	$y = 2x^3 - 6x + 10$	$-y''' - xy'' - 2y' = -24x^2$
129.	$y = 2 \sin x + 3$	$y'' + y = 3$
130.	$y = 3 \cos x + \sin x$	$y'' + y = 0$

True or False? In Exercises 131–136, determine whether the statement is true or false. If it is false, explain why or give an example that shows it is false.

131. If $y = f(x)g(x)$, then $\dfrac{dy}{dx} = f'(x)g'(x)$.

132. If $y = (x + 1)(x + 2)(x + 3)(x + 4)$, then $\dfrac{d^5y}{dx^5} = 0$.

133. If $f'(c)$ and $g'(c)$ are zero and $h(x) = f(x)g(x)$, then $h'(c) = 0$.

134. If the position function of an object is linear, then its acceleration is zero.

135. The second derivative represents the rate of change of the first derivative.

136. The function $f(x) = \sin x + c$ satisfies $f^{(n)} = f^{(n+4)}$ for all integers $n \ge 1$.

137. Proof Use the Product Rule twice to prove that if f, g, and h are differentiable functions of x, then

$$\frac{d}{dx}[f(x)g(x)h(x)] = f'(x)g(x)h(x) + f(x)g'(x)h(x) + f(x)g(x)h'(x).$$

138. Think About It Let f and g be functions whose first and second derivatives exist on an interval I. Which of the following formulas is (are) true?

(a) $fg'' - f''g = (fg' - f'g)'$ (b) $fg'' + f''g = (fg)''$

2.4 The Chain Rule

- **Find the derivative of a composite function using the Chain Rule.**
- **Find the derivative of a function using the General Power Rule.**
- **Simplify the derivative of a function using algebra.**
- **Find the derivative of a trigonometric function using the Chain Rule.**

The Chain Rule

This text has yet to discuss one of the most powerful differentiation rules—the **Chain Rule.** This rule deals with composite functions and adds a surprising versatility to the rules discussed in the two previous sections. For example, compare the functions shown below. Those on the left can be differentiated without the Chain Rule, and those on the right are best differentiated with the Chain Rule.

Without the Chain Rule	**With the Chain Rule**
$y = x^2 + 1$	$y = \sqrt{x^2 + 1}$
$y = \sin x$	$y = \sin 6x$
$y = 3x + 2$	$y = (3x + 2)^5$
$y = x + \tan x$	$y = x + \tan x^2$

Basically, the Chain Rule states that if y changes dy/du times as fast as u, and u changes du/dx times as fast as x, then y changes $(dy/du)(du/dx)$ times as fast as x.

EXAMPLE 1 The Derivative of a Composite Function

A set of gears is constructed, as shown in Figure 2.24, such that the second and third gears are on the same axle. As the first axle revolves, it drives the second axle, which in turn drives the third axle. Let y, u, and x represent the numbers of revolutions per minute of the first, second, and third axles, respectively. Find dy/du, du/dx, and dy/dx, and show that

$$\frac{dy}{dx} = \frac{dy}{du} \cdot \frac{du}{dx}.$$

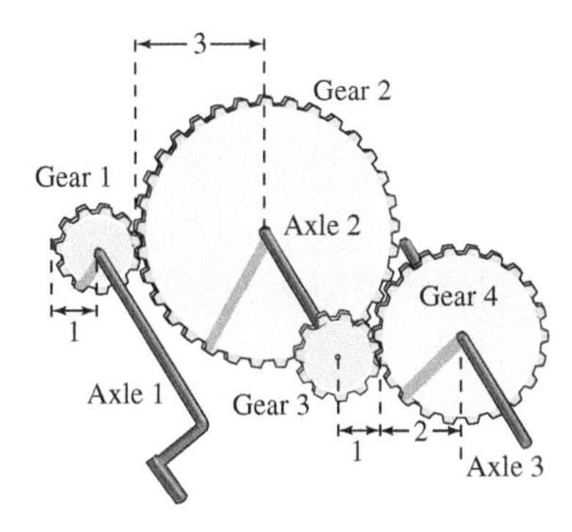

Axle 1: y revolutions per minute
Axle 2: u revolutions per minute
Axle 3: x revolutions per minute
Figure 2.24

Solution Because the circumference of the second gear is three times that of the first, the first axle must make three revolutions to turn the second axle once. Similarly, the second axle must make two revolutions to turn the third axle once, and you can write

$$\frac{dy}{du} = 3 \quad \text{and} \quad \frac{du}{dx} = 2.$$

Combining these two results, you know that the first axle must make six revolutions to turn the third axle once. So, you can write

$$\begin{aligned}\frac{dy}{dx} &= \left[\text{Rate of change of first axle with respect to second axle}\right] \cdot \left[\text{Rate of change of second axle with respect to third axle}\right]\\ &= \frac{dy}{du} \cdot \frac{du}{dx}\\ &= 3 \cdot 2\\ &= 6\\ &= \left[\text{Rate of change of first axle with respect to third axle}\right].\end{aligned}$$

In other words, the rate of change of y with respect to x is the product of the rate of change of y with respect to u and the rate of change of u with respect to x.

Exploration

Using the Chain Rule Each of the following functions can be differentiated using rules that you studied in Sections 2.2 and 2.3. For each function, find the derivative using those rules. Then find the derivative using the Chain Rule. Compare your results. Which method is simpler?

a. $y = \dfrac{2}{3x + 1}$

b. $y = (x + 2)^3$

c. $y = \sin 2x$

Example 1 illustrates a simple case of the Chain Rule. The general rule is stated in the next theorem.

THEOREM 2.10 The Chain Rule

If $y = f(u)$ is a differentiable function of u and $u = g(x)$ is a differentiable function of x, then $y = f(g(x))$ is a differentiable function of x and

$$\frac{dy}{dx} = \frac{dy}{du} \cdot \frac{du}{dx}$$

or, equivalently,

$$\frac{d}{dx}[f(g(x))] = f'(g(x))g'(x).$$

Proof Let $h(x) = f(g(x))$. Then, using the alternative form of the derivative, you need to show that, for $x = c$,

$$h'(c) = f'(g(c))g'(c).$$

An important consideration in this proof is the behavior of g as x approaches c. A problem occurs when there are values of x, other than c, such that

$$g(x) = g(c).$$

Appendix A shows how to use the differentiability of f and g to overcome this problem. For now, assume that $g(x) \neq g(c)$ for values of x other than c. In the proofs of the Product Rule and the Quotient Rule, the same quantity was added and subtracted to obtain the desired form. This proof uses a similar technique—multiplying and dividing by the same (nonzero) quantity. Note that because g is differentiable, it is also continuous, and it follows that $g(x)$ approaches $g(c)$ as x approaches c.

REMARK The alternative limit form of the derivative was given at the end of Section 2.1.

$$h'(c) = \lim_{x \to c} \frac{f(g(x)) - f(g(c))}{x - c} \qquad \text{Alternative form of derivative}$$

$$= \lim_{x \to c} \left[\frac{f(g(x)) - f(g(c))}{x - c} \cdot \frac{g(x) - g(c)}{g(x) - g(c)} \right], \quad g(x) \neq g(c)$$

$$= \lim_{x \to c} \left[\frac{f(g(x)) - f(g(c))}{g(x) - g(c)} \cdot \frac{g(x) - g(c)}{x - c} \right]$$

$$= \left[\lim_{x \to c} \frac{f(g(x)) - f(g(c))}{g(x) - g(c)} \right] \left[\lim_{x \to c} \frac{g(x) - g(c)}{x - c} \right]$$

$$= f'(g(c))g'(c)$$

When applying the Chain Rule, it is helpful to think of the composite function $f \circ g$ as having two parts—an inner part and an outer part.

Outer function

$$y = f(g(x)) = f(u)$$

Inner function

The derivative of $y = f(u)$ is the derivative of the outer function (at the inner function u) *times* the derivative of the inner function.

$$y' = f'(u) \cdot u'$$

EXAMPLE 2 **Decomposition of a Composite Function**

$y = f(g(x))$	$u = g(x)$	$y = f(u)$
a. $y = \dfrac{1}{x+1}$	$u = x + 1$	$y = \dfrac{1}{u}$
b. $y = \sin 2x$	$u = 2x$	$y = \sin u$
c. $y = \sqrt{3x^2 - x + 1}$	$u = 3x^2 - x + 1$	$y = \sqrt{u}$
d. $y = \tan^2 x$	$u = \tan x$	$y = u^2$

EXAMPLE 3 **Using the Chain Rule**

Find dy/dx for

$$y = (x^2 + 1)^3.$$

Solution For this function, you can consider the inside function to be $u = x^2 + 1$ and the outer function to be $y = u^3$. By the Chain Rule, you obtain

$$\frac{dy}{dx} = \underbrace{3(x^2+1)^2}_{\frac{dy}{du}}\underbrace{(2x)}_{\frac{du}{dx}} = 6x(x^2+1)^2.$$

REMARK You could also solve the problem in Example 3 without using the Chain Rule by observing that

$$y = x^6 + 3x^4 + 3x^2 + 1$$

and

$$y' = 6x^5 + 12x^3 + 6x.$$

Verify that this is the same as the derivative in Example 3. Which method would you use to find

$$\frac{d}{dx}[(x^2+1)^{50}]?$$

The General Power Rule

The function in Example 3 is an example of one of the most common types of composite functions, $y = [u(x)]^n$. The rule for differentiating such functions is called the **General Power Rule,** and it is a special case of the Chain Rule.

THEOREM 2.11 The General Power Rule

If $y = [u(x)]^n$, where u is a differentiable function of x and n is a rational number, then

$$\frac{dy}{dx} = n[u(x)]^{n-1}\frac{du}{dx}$$

or, equivalently,

$$\frac{d}{dx}[u^n] = nu^{n-1}u'.$$

Proof Because $y = [u(x)]^n = u^n$, you apply the Chain Rule to obtain

$$\frac{dy}{dx} = \left(\frac{dy}{du}\right)\left(\frac{du}{dx}\right) = \frac{d}{du}[u^n]\frac{du}{dx}.$$

By the (Simple) Power Rule in Section 2.2, you have $D_u[u^n] = nu^{n-1}$, and it follows that

$$\frac{dy}{dx} = nu^{n-1}\frac{du}{dx}.$$

EXAMPLE 4 Applying the General Power Rule

Find the derivative of $f(x) = (3x - 2x^2)^3$.

Solution Let $u = 3x - 2x^2$. Then

$$f(x) = (3x - 2x^2)^3 = u^3$$

and, by the General Power Rule, the derivative is

$$f'(x) = \overbrace{3}^{n}\overbrace{(3x - 2x^2)^2}^{u^{n-1}}\overbrace{\frac{d}{dx}[3x - 2x^2]}^{u'}$$ Apply General Power Rule.

$$= 3(3x - 2x^2)^2(3 - 4x).$$ Differentiate $3x - 2x^2$.

EXAMPLE 5 Differentiating Functions Involving Radicals

Find all points on the graph of

$$f(x) = \sqrt[3]{(x^2 - 1)^2}$$

for which $f'(x) = 0$ and those for which $f'(x)$ does not exist.

Solution Begin by rewriting the function as

$$f(x) = (x^2 - 1)^{2/3}.$$

Then, applying the General Power Rule (with $u = x^2 - 1$) produces

$$f'(x) = \overbrace{\frac{2}{3}}^{n}\overbrace{(x^2 - 1)^{-1/3}}^{u^{n-1}}\overbrace{(2x)}^{u'}$$ Apply General Power Rule.

$$= \frac{4x}{3\sqrt[3]{x^2 - 1}}.$$ Write in radical form.

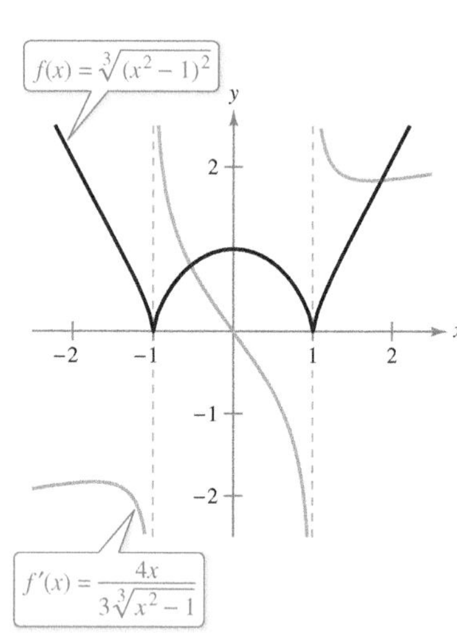

The derivative of f is 0 at $x = 0$ and is undefined at $x = \pm 1$.
Figure 2.25

So, $f'(x) = 0$ when $x = 0$, and $f'(x)$ does not exist when $x = \pm 1$, as shown in Figure 2.25.

EXAMPLE 6 Differentiating Quotients: Constant Numerators

Differentiate the function

$$g(t) = \frac{-7}{(2t - 3)^2}.$$

Solution Begin by rewriting the function as

$$g(t) = -7(2t - 3)^{-2}.$$

Then, applying the General Power Rule (with $u = 2t - 3$) produces

$$g'(t) = \underbrace{(-7)\overbrace{(-2)}^{n}}_{\text{Constant Multiple Rule}}\overbrace{(2t - 3)^{-3}}^{u^{n-1}}\overbrace{(2)}^{u'}$$ Apply General Power Rule.

$$= 28(2t - 3)^{-3}$$ Simplify.

$$= \frac{28}{(2t - 3)^3}.$$ Write with positive exponent.

REMARK Try differentiating the function in Example 6 using the Quotient Rule. You should obtain the same result, but using the Quotient Rule is less efficient than using the General Power Rule.

Simplifying Derivatives

The next three examples demonstrate techniques for simplifying the "raw derivatives" of functions involving products, quotients, and composites.

EXAMPLE 7 Simplifying by Factoring Out the Least Powers

Find the derivative of $f(x) = x^2\sqrt{1 - x^2}$.

Solution

$$
\begin{aligned}
f(x) &= x^2\sqrt{1 - x^2} && \text{Write original function.}\\
&= x^2(1 - x^2)^{1/2} && \text{Rewrite.}\\
f'(x) &= x^2\frac{d}{dx}[(1 - x^2)^{1/2}] + (1 - x^2)^{1/2}\frac{d}{dx}[x^2] && \text{Product Rule}\\
&= x^2\left[\frac{1}{2}(1 - x^2)^{-1/2}(-2x)\right] + (1 - x^2)^{1/2}(2x) && \text{General Power Rule}\\
&= -x^3(1 - x^2)^{-1/2} + 2x(1 - x^2)^{1/2} && \text{Simplify.}\\
&= x(1 - x^2)^{-1/2}[-x^2(1) + 2(1 - x^2)] && \text{Factor.}\\
&= \frac{x(2 - 3x^2)}{\sqrt{1 - x^2}} && \text{Simplify.}
\end{aligned}
$$

EXAMPLE 8 Simplifying the Derivative of a Quotient

> ▷ TECHNOLOGY Symbolic differentiation utilities are capable of differentiating very complicated functions. Often, however, the result is given in unsimplified form. If you have access to such a utility, use it to find the derivatives of the functions given in Examples 7, 8, and 9. Then compare the results with those given in these examples.

$$
\begin{aligned}
f(x) &= \frac{x}{\sqrt[3]{x^2 + 4}} && \text{Original function}\\
&= \frac{x}{(x^2 + 4)^{1/3}} && \text{Rewrite.}\\
f'(x) &= \frac{(x^2 + 4)^{1/3}(1) - x(1/3)(x^2 + 4)^{-2/3}(2x)}{(x^2 + 4)^{2/3}} && \text{Quotient Rule}\\
&= \frac{1}{3}(x^2 + 4)^{-2/3}\left[\frac{3(x^2 + 4) - (2x^2)(1)}{(x^2 + 4)^{2/3}}\right] && \text{Factor.}\\
&= \frac{x^2 + 12}{3(x^2 + 4)^{4/3}} && \text{Simplify.}
\end{aligned}
$$

EXAMPLE 9 Simplifying the Derivative of a Power

▷ See LarsonCalculus.com for an interactive version of this type of example.

$$
\begin{aligned}
y &= \left(\frac{3x - 1}{x^2 + 3}\right)^2 && \text{Original function}\\
y' &= \underset{n}{2}\underbrace{\left(\frac{3x - 1}{x^2 + 3}\right)}_{u^{n-1}}\underbrace{\frac{d}{dx}\left[\frac{3x - 1}{x^2 + 3}\right]}_{u'} && \text{General Power Rule}\\
&= \left[\frac{2(3x - 1)}{x^2 + 3}\right]\left[\frac{(x^2 + 3)(3) - (3x - 1)(2x)}{(x^2 + 3)^2}\right] && \text{Quotient Rule}\\
&= \frac{2(3x - 1)(3x^2 + 9 - 6x^2 + 2x)}{(x^2 + 3)^3} && \text{Multiply.}\\
&= \frac{2(3x - 1)(-3x^2 + 2x + 9)}{(x^2 + 3)^3} && \text{Simplify.}
\end{aligned}
$$

Trigonometric Functions and the Chain Rule

The "Chain Rule versions" of the derivatives of the six trigonometric functions are shown below.

$$\frac{d}{dx}[\sin u] = (\cos u)u' \qquad \frac{d}{dx}[\cos u] = -(\sin u)u'$$

$$\frac{d}{dx}[\tan u] = (\sec^2 u)u' \qquad \frac{d}{dx}[\cot u] = -(\csc^2 u)u'$$

$$\frac{d}{dx}[\sec u] = (\sec u \tan u)u' \qquad \frac{d}{dx}[\csc u] = -(\csc u \cot u)u'$$

EXAMPLE 10 **The Chain Rule and Trigonometric Functions**

a. $y = \sin \overbrace{2x}^{u}$ $\qquad y' = \overbrace{\cos 2x}^{\cos u} \overbrace{\frac{d}{dx}[2x]}^{u'} = (\cos 2x)(2) = 2 \cos 2x$

b. $y = \cos \overbrace{(x-1)}^{u}$ $\qquad y' = \overbrace{-\sin(x-1)}^{-(\sin u)} \overbrace{\frac{d}{dx}[x-1]}^{u'} = -\sin(x-1)$

c. $y = \tan \overbrace{3x}^{u}$ $\qquad y' = \overbrace{\sec^2 3x}^{(\sec^2 u)} \overbrace{\frac{d}{dx}[3x]}^{u'} = (\sec^2 3x)(3) = 3 \sec^2(3x)$ ■

Be sure you understand the mathematical conventions regarding parentheses and trigonometric functions. For instance, in Example 10(a), $\sin 2x$ is written to mean $\sin(2x)$.

EXAMPLE 11 **Parentheses and Trigonometric Functions**

a. $y = \cos 3x^2 = \cos(3x^2)$ $\qquad y' = (-\sin 3x^2)(6x) = -6x \sin 3x^2$

b. $y = (\cos 3)x^2$ $\qquad y' = (\cos 3)(2x) = 2x \cos 3$

c. $y = \cos(3x)^2 = \cos(9x^2)$ $\qquad y' = (-\sin 9x^2)(18x) = -18x \sin 9x^2$

d. $y = \cos^2 x = (\cos x)^2$ $\qquad y' = 2(\cos x)(-\sin x) = -2 \cos x \sin x$

e. $y = \sqrt{\cos x} = (\cos x)^{1/2}$ $\qquad y' = \frac{1}{2}(\cos x)^{-1/2}(-\sin x) = -\frac{\sin x}{2\sqrt{\cos x}}$ ■

To find the derivative of a function of the form $k(x) = f(g(h(x)))$, you need to apply the Chain Rule twice, as shown in Example 12.

EXAMPLE 12 **Repeated Application of the Chain Rule**

$$f(t) = \sin^3 4t \qquad \text{Original function}$$

$$= (\sin 4t)^3 \qquad \text{Rewrite.}$$

$$f'(t) = 3(\sin 4t)^2 \frac{d}{dt}[\sin 4t] \qquad \text{Apply Chain Rule once.}$$

$$= 3(\sin 4t)^2(\cos 4t)\frac{d}{dt}[4t] \qquad \text{Apply Chain Rule a second time.}$$

$$= 3(\sin 4t)^2(\cos 4t)(4)$$

$$= 12 \sin^2 4t \cos 4t \qquad \text{Simplify.}$$

■

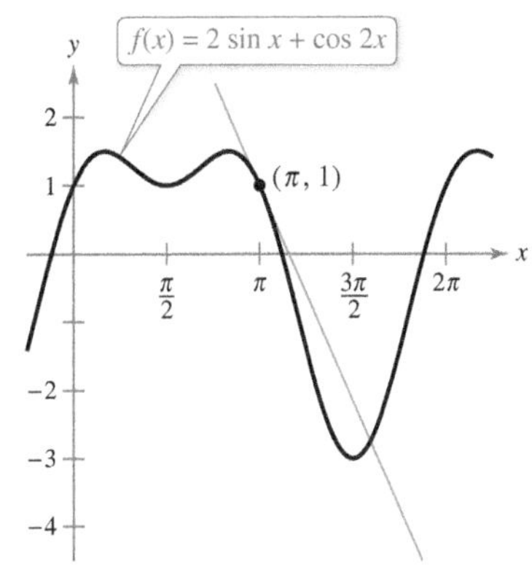

Figure 2.26

EXAMPLE 13 Tangent Line of a Trigonometric Function

Find an equation of the tangent line to the graph of $f(x) = 2\sin x + \cos 2x$ at the point $(\pi, 1)$, as shown in Figure 2.26. Then determine all values of x in the interval $(0, 2\pi)$ at which the graph of f has a horizontal tangent.

Solution Begin by finding $f'(x)$.

$$f(x) = 2\sin x + \cos 2x \qquad \text{Write original function.}$$

$$f'(x) = 2\cos x + (-\sin 2x)(2) \qquad \text{Apply Chain Rule to } \cos 2x.$$

$$= 2\cos x - 2\sin 2x \qquad \text{Simplify.}$$

To find the slope of the tangent line at $(\pi, 1)$, evaluate $f'(\pi)$.

$$f'(\pi) = 2\cos \pi - 2\sin 2\pi \qquad \text{Substitute.}$$

$$= -2 \qquad \text{Slope of tangent line at } (\pi, 1)$$

Now, using the point-slope form of the equation of a line, you can write

$$y - y_1 = m(x - x_1) \qquad \text{Point-slope form}$$

$$y - 1 = -2(x - \pi) \qquad \text{Substitute for } y_1, m, \text{ and } x_1.$$

$$y = -2x + 1 + 2\pi. \qquad \text{Equation of tangent line at } (\pi, 1)$$

You can then determine that $f'(x) = 0$ when $x = \frac{\pi}{6}, \frac{\pi}{2}, \frac{5\pi}{6}$, and $\frac{3\pi}{2}$. So, f has horizontal tangents at $x = \frac{\pi}{6}, \frac{\pi}{2}, \frac{5\pi}{6}$, and $\frac{3\pi}{2}$. ■

This section concludes with a summary of the differentiation rules studied so far. To become skilled at differentiation, you should memorize each rule in words, not symbols. As an aid to memorization, note that the cofunctions (cosine, cotangent, and cosecant) require a negative sign as part of their derivatives.

SUMMARY OF DIFFERENTIATION RULES

General Differentiation Rules

Let c be a real number, let n be a rational number, let u and v be differentiable functions of x, and let f be a differentiable function of u.

Constant Rule:

$$\frac{d}{dx}[c] = 0$$

(Simple) Power Rule:

$$\frac{d}{dx}[x^n] = nx^{n-1}, \quad \frac{d}{dx}[x] = 1$$

Constant Multiple Rule:

$$\frac{d}{dx}[cu] = cu'$$

Sum or Difference Rule:

$$\frac{d}{dx}[u \pm v] = u' \pm v'$$

Product Rule:

$$\frac{d}{dx}[uv] = uv' + vu'$$

Quotient Rule:

$$\frac{d}{dx}\left[\frac{u}{v}\right] = \frac{vu' - uv'}{v^2}$$

Chain Rule:

$$\frac{d}{dx}[f(u)] = f'(u)u'$$

General Power Rule:

$$\frac{d}{dx}[u^n] = nu^{n-1}u'$$

Derivatives of Trigonometric Functions

$$\frac{d}{dx}[\sin x] = \cos x \qquad \frac{d}{dx}[\tan x] = \sec^2 x \qquad \frac{d}{dx}[\sec x] = \sec x \tan x$$

$$\frac{d}{dx}[\cos x] = -\sin x \qquad \frac{d}{dx}[\cot x] = -\csc^2 x \qquad \frac{d}{dx}[\csc x] = -\csc x \cot x$$

2.4 Exercises

See CalcChat.com for tutorial help and worked-out solutions to odd-numbered exercises.

CONCEPT CHECK

1. **Chain Rule** Describe the Chain Rule for the composition of two differentiable functions in your own words.

2. **General Power Rule** What is the difference between the (Simple) Power Rule and the General Power Rule?

Decomposition of a Composite Function In Exercises 3–8, complete the table.

$y = f(g(x))$	$u = g(x)$	$y = f(u)$
3. $y = (6x - 5)^4$		
4. $y = \sqrt[3]{4x + 3}$		
5. $y = \dfrac{1}{3x + 5}$		
6. $y = \dfrac{2}{\sqrt{x^2 + 10}}$		
7. $y = \csc^3 x$		
8. $y = \sin \dfrac{5x}{2}$		

Finding a Derivative In Exercises 9–34, find the derivative of the function.

9. $y = (2x - 7)^3$
10. $y = 5(2 - x^3)^4$
11. $g(x) = 3(4 - 9x)^{5/6}$
12. $f(t) = (9t + 2)^{2/3}$
13. $h(s) = -2\sqrt{5s^2 + 3}$
14. $g(x) = \sqrt{4 - 3x^2}$
15. $y = \sqrt[3]{6x^2 + 1}$
16. $y = 2\sqrt[4]{9 - x^2}$
17. $y = \dfrac{1}{x - 2}$
18. $s(t) = \dfrac{1}{4 - 5t - t^2}$
19. $g(s) = \dfrac{6}{(s^3 - 2)^3}$
20. $y = -\dfrac{3}{(t - 2)^4}$
21. $y = \dfrac{1}{\sqrt{3x + 5}}$
22. $g(t) = \dfrac{1}{\sqrt{t^2 - 2}}$
23. $f(x) = x^2(x - 2)^7$
24. $f(x) = x(2x - 5)^3$
25. $y = x\sqrt{1 - x^2}$
26. $y = x^2\sqrt{16 - x^2}$
27. $y = \dfrac{x}{\sqrt{x^2 + 1}}$
28. $y = \dfrac{x}{\sqrt{x^4 + 4}}$
29. $g(x) = \left(\dfrac{x + 5}{x^2 + 2}\right)^2$
30. $h(t) = \left(\dfrac{t^2}{t^3 + 2}\right)^2$
31. $s(t) = \left(\dfrac{1 + t}{t + 3}\right)^4$
32. $g(x) = \left(\dfrac{3x^2 - 2}{2x + 3}\right)^{-2}$
33. $f(x) = ((x^2 + 3)^5 + x)^2$
34. $g(x) = (2 + (x^2 + 1)^4)^3$

Finding a Derivative of a Trigonometric Function In Exercises 35–54, find the derivative of the trigonometric function.

35. $y = \cos 4x$
36. $y = \sin \pi x$
37. $g(x) = 5 \tan 3x$
38. $h(x) = \sec 6x$
39. $y = \sin(\pi x)^2$
40. $y = \csc(1 - 2x)^2$
41. $h(x) = \sin 2x \cos 2x$
42. $g(\theta) = \sec\left(\frac{1}{2}\theta\right) \tan\left(\frac{1}{2}\theta\right)$
43. $f(x) = \dfrac{\cot x}{\sin x}$
44. $g(v) = \dfrac{\cos v}{\csc v}$
45. $y = 4 \sec^2 x$
46. $g(t) = 5 \cos^2 \pi t$
47. $f(\theta) = \frac{1}{4}\sin^2 2\theta$
48. $h(t) = 2 \cot^2(\pi t + 2)$
49. $f(t) = 3 \sec(\pi t - 1)^2$
50. $y = 5 \cos(\pi x)^2$
51. $y = \sin(3x^2 + \cos x)$
52. $y = \cos(5x + \csc x)$
53. $y = \sin\sqrt{\cot 3\pi x}$
54. $y = \cos\sqrt{\sin(\tan \pi x)}$

Finding a Derivative Using Technology In Exercises 55–60, use a computer algebra system to find the derivative of the function. Then use the utility to graph the function and its derivative on the same set of coordinate axes. Describe the behavior of the function that corresponds to any zeros of the graph of the derivative.

55. $y = \dfrac{\sqrt{x} + 1}{x^2 + 1}$
56. $y = \sqrt{\dfrac{2x}{x + 1}}$
57. $y = \sqrt{\dfrac{x + 1}{x}}$
58. $g(x) = \sqrt{x - 1} + \sqrt{x + 1}$
59. $y = \dfrac{\cos \pi x + 1}{x}$
60. $y = x^2 \tan \dfrac{1}{x}$

Slope of a Tangent Line In Exercises 61 and 62, find the slope of the tangent line to the sine function at the origin. Compare this value with the number of complete cycles in the interval $[0, 2\pi]$.

61.

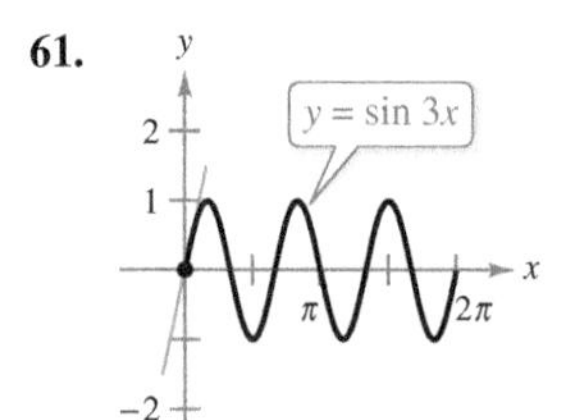

62.

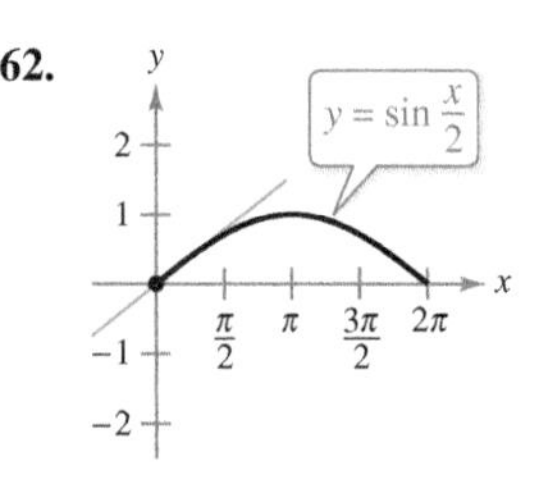

Finding the Slope of a Graph In Exercises 63–70, find the slope of the graph of the function at the given point. Use the *derivative* feature of a graphing utility to confirm your results.

63. $y = \sqrt{x^2 + 8x}, \quad (1, 3)$
64. $y = \sqrt[5]{3x^3 + 4x}, \quad (2, 2)$
65. $f(x) = 5(x^3 - 2)^{-1}, \quad \left(-2, -\frac{1}{2}\right)$
66. $f(x) = \dfrac{1}{(x^2 - 3x)^2}, \quad \left(4, \dfrac{1}{16}\right)$
67. $y = \dfrac{4}{(x + 2)^2}, \quad (0, 1)$
68. $y = \dfrac{4}{(x^2 - 2x)^3}, \quad (1, -4)$

69. $y = 26 - \sec^3 4x, \quad (0, 25)$

70. $y = \dfrac{1}{x} + \sqrt{\cos x}, \quad \left(\dfrac{\pi}{2}, \dfrac{2}{\pi}\right)$

Finding an Equation of a Tangent Line **In Exercises 71–78, (a) find an equation of the tangent line to the graph of the function at the given point, (b) use a graphing utility to graph the function and its tangent line at the point, and (c) use the *tangent* feature of a graphing utility to confirm your results.**

71. $f(x) = \sqrt{2x^2 - 7}, \quad (4, 5)$

72. $f(x) = \frac{1}{3}x\sqrt{x^2 + 5}, \quad (2, 2)$

73. $y = (4x^3 + 3)^2, \quad (-1, 1)$

74. $f(x) = (9 - x^2)^{2/3}, \quad (1, 4)$

75. $f(x) = \sin 8x, \quad (\pi, 0)$

76. $y = \cos 3x, \quad \left(\dfrac{\pi}{4}, -\dfrac{\sqrt{2}}{2}\right)$

77. $f(x) = \tan^2 x, \quad \left(\dfrac{\pi}{4}, 1\right)$

78. $y = 2\tan^3 x, \quad \left(\dfrac{\pi}{4}, 2\right)$

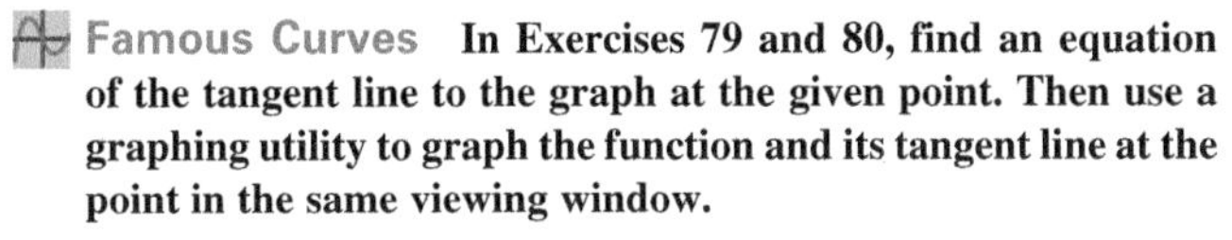

Famous Curves **In Exercises 79 and 80, find an equation of the tangent line to the graph at the given point. Then use a graphing utility to graph the function and its tangent line at the point in the same viewing window.**

79. Semicircle

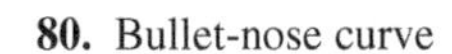

80. Bullet-nose curve

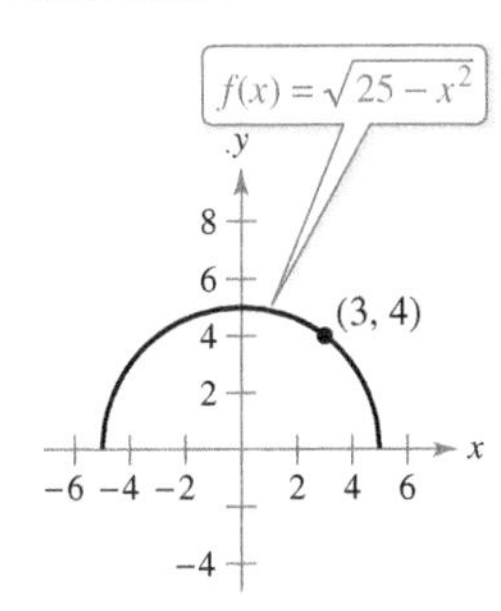

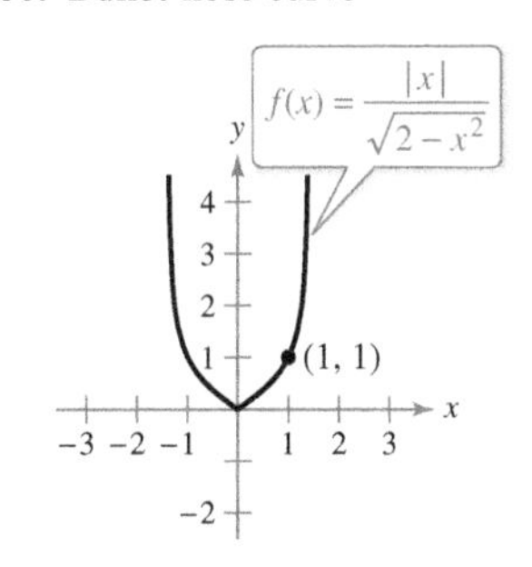

81. Horizontal Tangent Line Determine the point(s) in the interval $(0, 2\pi)$ at which the graph of $f(x) = 2\cos x + \sin 2x$ has a horizontal tangent.

82. Horizontal Tangent Line Determine the point(s) at which the graph of

$$f(x) = \frac{-4x}{\sqrt{2x - 1}}$$

has a horizontal tangent.

Finding a Second Derivative **In Exercises 83–88, find the second derivative of the function.**

83. $f(x) = 5(2 - 7x)^4$

84. $f(x) = 6(x^3 + 4)^3$

85. $f(x) = \dfrac{1}{11x - 6}$

86. $f(x) = \dfrac{8}{(x - 2)^2}$

87. $f(x) = \sin x^2$

88. $f(x) = \sec^2 \pi x$

Evaluating a Second Derivative **In Exercises 89–92, evaluate the second derivative of the function at the given point. Use a computer algebra system to verify your result.**

89. $h(x) = \dfrac{1}{9}(3x + 1)^3, \quad \left(1, \dfrac{64}{9}\right)$

90. $f(x) = \dfrac{1}{\sqrt{x + 4}}, \quad \left(0, \dfrac{1}{2}\right)$

91. $f(x) = \cos x^2, \quad (0, 1)$

92. $g(t) = \tan 2t, \quad \left(\dfrac{\pi}{6}, \sqrt{3}\right)$

EXPLORING CONCEPTS

Identifying Graphs **In Exercises 93 and 94, the graphs of a function f and its derivative f' are shown. Label the graphs as f or f' and write a short paragraph stating the criteria you used in making your selection. To print an enlarged copy of the graph, go to *MathGraphs.com*.**

93.

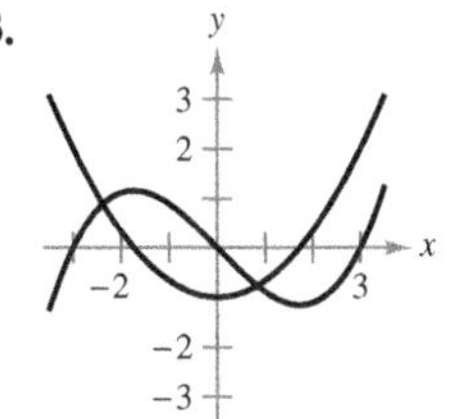

94.

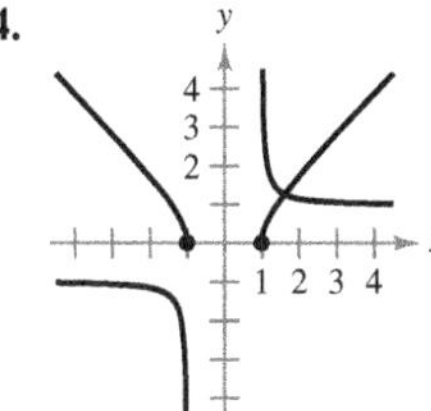

95. Describing Relationships The relationship between f and g is given. Describe the relationship between f' and g'.

(a) $g(x) = f(3x)$ (b) $g(x) = f(x^2)$

96. Comparing Methods Consider the function

$$r(x) = \frac{2x - 5}{(3x + 1)^2}.$$

(a) In general, how do you find the derivative of

$h(x) = \dfrac{f(x)}{g(x)}$ using the Product Rule, where g is a composite function?

(b) Find $r'(x)$ using the Product Rule.

(c) Find $r'(x)$ using the Quotient Rule.

(d) Which method do you prefer? Explain.

97. Think About It The table shows some values of the derivative of an unknown function f. Complete the table by finding the derivative of each transformation of f, if possible.

(a) $g(x) = f(x) - 2$ (b) $h(x) = 2f(x)$

(c) $r(x) = f(-3x)$ (d) $s(x) = f(x + 2)$

x	-2	-1	0	1	2	3
$f'(x)$	4	$\frac{2}{3}$	$-\frac{1}{3}$	-1	-2	-4
$g'(x)$						
$h'(x)$						
$r'(x)$						
$s'(x)$						

98. Using Relationships Given that $g(5) = -3$, $g'(5) = 6$, $h(5) = 3$, and $h'(5) = -2$, find $f'(5)$ for each of the following, if possible. If it is not possible, state what additional information is required.

(a) $f(x) = g(x)h(x)$ (b) $f(x) = g(h(x))$

(c) $f(x) = \dfrac{g(x)}{h(x)}$ (d) $f(x) = [g(x)]^3$

Finding Derivatives **In Exercises 99 and 100, the graphs of f and g are shown. Let $h(x) = f(g(x))$ and $s(x) = g(f(x))$. Find each derivative, if it exists. If the derivative does not exist, explain why.**

99. (a) Find $h'(1)$.

(b) Find $s'(5)$.

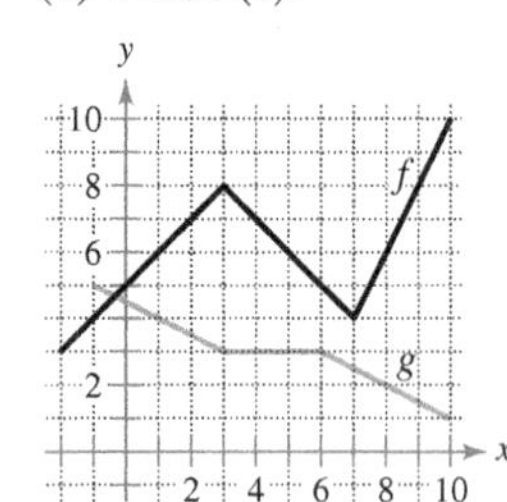

100. (a) Find $h'(3)$.

(b) Find $s'(9)$.

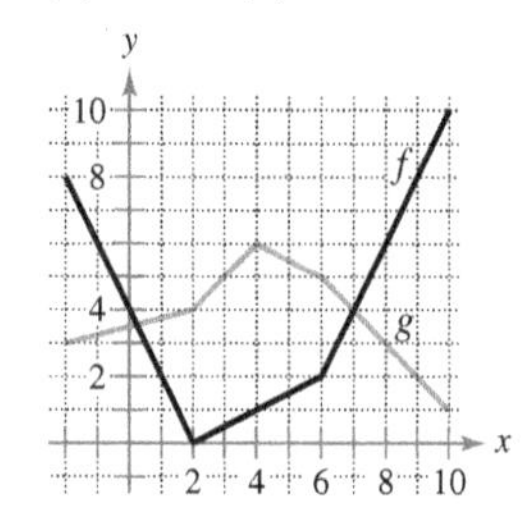

101. Doppler Effect The frequency F of a fire truck siren heard by a stationary observer is

$$F = \frac{132{,}400}{331 \pm v}$$

where $\pm v$ represents the velocity of the accelerating fire truck in meters per second (see figure). Find the rate of change of F with respect to v when

(a) the fire truck is approaching at a velocity of 30 meters per second (use $-v$).

(b) the fire truck is moving away at a velocity of 30 meters per second (use $+v$).

102. Harmonic Motion The displacement from equilibrium of an object in harmonic motion on the end of a spring is

$$y = \frac{1}{3}\cos 12t - \frac{1}{4}\sin 12t$$

where y is measured in feet and t is the time in seconds. Determine the position and velocity of the object when $t = \pi/8$.

103. Pendulum A 15-centimeter pendulum moves according to the equation $\theta = 0.2 \cos 8t$, where θ is the angular displacement from the vertical in radians and t is the time in seconds. Determine the maximum angular displacement and the rate of change of θ when $t = 3$ seconds.

104. Wave Motion A buoy oscillates in simple harmonic motion $y = A \cos \omega t$ as waves move past it. The buoy moves a total of 3.5 feet (vertically) from its low point to its high point. It returns to its high point every 10 seconds.

(a) Write an equation describing the motion of the buoy if it is at its high point at $t = 0$.

(b) Determine the velocity of the buoy as a function of t.

105. Modeling Data The normal daily maximum temperatures T (in degrees Fahrenheit) for Chicago, Illinois, are shown in the table. *(Source: National Oceanic and Atmospheric Administration)*

Month	Jan	Feb	Mar	Apr
Temperature	31.0	35.3	46.6	59.0

Month	May	Jun	Jul	Aug
Temperature	70.0	79.7	84.1	81.9

Month	Sep	Oct	Nov	Dec
Temperature	74.8	62.3	48.2	34.8

(a) Use a graphing utility to plot the data and find a model for the data of the form

$$T(t) = a + b \sin(ct - d)$$

where T is the temperature and t is the time in months, with $t = 1$ corresponding to January.

(b) Use a graphing utility to graph the model. How well does the model fit the data?

(c) Find T' and use a graphing utility to graph T'.

(d) Based on the graph of T', during what times does the temperature change most rapidly? Most slowly? Do your answers agree with your observations of the temperature changes? Explain.

106. HOW DO YOU SEE IT? The cost C (in dollars) of producing x units of a product is $C = 60x + 1350$. For one week, management determined that the number of units produced x at the end of t hours can be modeled by $x = -1.6t^3 + 19t^2 - 0.5t - 1$. The graph shows the cost C in terms of the time t.

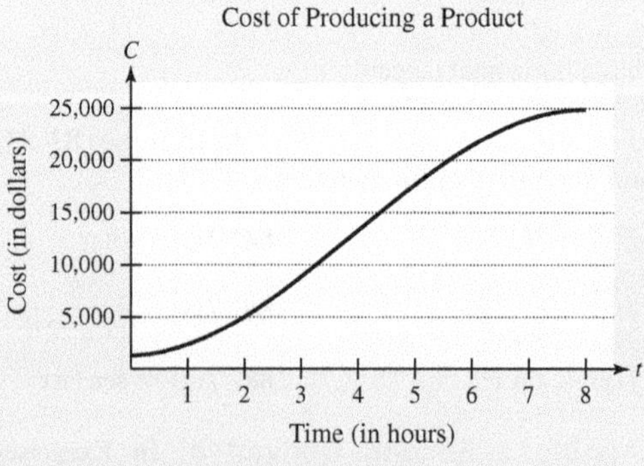

(a) Using the graph, which is greater, the rate of change of the cost after 1 hour or the rate of change of the cost after 4 hours?

(b) Explain why the cost function is not increasing at a constant rate during the eight-hour shift.

107. Biology

The number N of bacteria in a culture after t days is modeled by

$$N = 400\left[1 - \frac{3}{(t^2 + 2)^2}\right].$$

Find the rate of change of N with respect to t when (a) $t = 0$, (b) $t = 1$, (c) $t = 2$, (d) $t = 3$, and (e) $t = 4$. (f) What can you conclude?

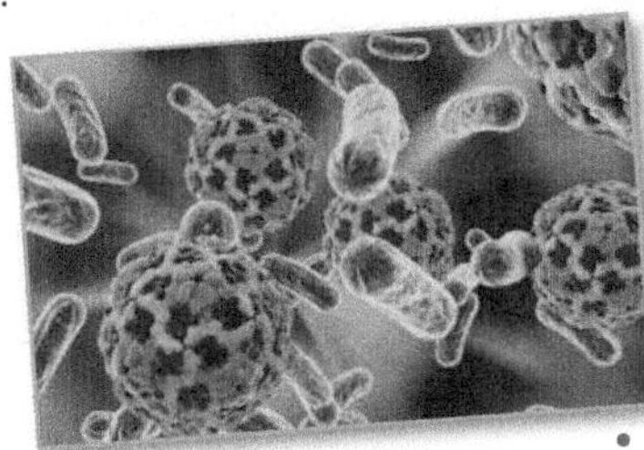

108. Depreciation The value V of a machine t years after it is purchased is inversely proportional to the square root of $t + 1$. The initial value of the machine is \$10,000.

(a) Write V as a function of t.

(b) Find the rate of depreciation when $t = 1$.

(c) Find the rate of depreciation when $t = 3$.

109. Finding a Pattern Consider the function $f(x) = \sin \beta x$, where β is a constant.

(a) Find the first-, second-, third-, and fourth-order derivatives of the function.

(b) Verify that the function and its second derivative satisfy the equation $f''(x) + \beta^2 f(x) = 0$.

(c) Use the results of part (a) to write general rules for the even- and odd-order derivatives $f^{(2k)}(x)$ and $f^{(2k-1)}(x)$.

[*Hint:* $(-1)^k$ is positive if k is even and negative if k is odd.]

110. Conjecture Let f be a differentiable function of period p.

(a) Is the function f' periodic? Verify your answer.

(b) Consider the function $g(x) = f(2x)$. Is the function $g'(x)$ periodic? Verify your answer.

111. Think About It Let $r(x) = f(g(x))$ and $s(x) = g(f(x))$, where f and g are shown in the figure. Find (a) $r'(1)$ and (b) $s'(4)$.

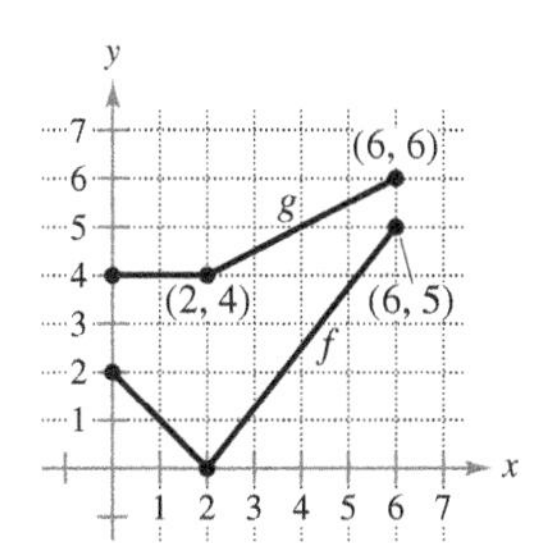

112. Using Trigonometric Functions

(a) Find the derivative of the function $g(x) = \sin^2 x + \cos^2 x$ in two ways.

(b) For $f(x) = \sec^2 x$ and $g(x) = \tan^2 x$, show that

$$f'(x) = g'(x).$$

113. Even and Odd Functions

(a) Show that the derivative of an odd function is even. That is, if $f(-x) = -f(x)$, then $f'(-x) = f'(x)$.

(b) Show that the derivative of an even function is odd. That is, if $f(-x) = f(x)$, then $f'(-x) = -f'(x)$.

114. Proof Let u be a differentiable function of x. Use the fact that $|u| = \sqrt{u^2}$ to prove that

$$\frac{d}{dx}[|u|] = u'\frac{u}{|u|}, \quad u \neq 0.$$

Using Absolute Value In Exercises 115–118, use the result of Exercise 114 to find the derivative of the function.

115. $g(x) = |3x - 5|$

116. $f(x) = |x^2 - 9|$

117. $h(x) = |x| \cos x$

118. $f(x) = |\sin x|$

Linear and Quadratic Approximations The linear and quadratic approximations of a function f at $x = a$ are

$$P_1(x) = f'(a)(x - a) + f(a) \quad \text{and}$$

$$P_2(x) = \tfrac{1}{2}f''(a)(x - a)^2 + f'(a)(x - a) + f(a).$$

In Exercises 119 and 120, (a) find the specified linear and quadratic approximations of f, (b) use a graphing utility to graph f and the approximations, (c) determine whether P_1 or P_2 is the better approximation, and (d) state how the accuracy changes as you move farther from $x = a$.

119. $f(x) = \tan x; \quad a = \frac{\pi}{4}$

120. $f(x) = \sec x; \quad a = \frac{\pi}{6}$

True or False? In Exercises 121–124, determine whether the statement is true or false. If it is false, explain why or give an example that shows it is false.

121. The slope of the function $f(x) = \sin ax$ at the origin is a.

122. The slope of the function $f(x) = \cos bx$ at the origin is $-b$.

123. If y is a differentiable function of u, and u is a differentiable function of x, then y is a differentiable function of x.

124. If y is a differentiable function of u, u is a differentiable function of v, and v is a differentiable function of x, then

$$\frac{dy}{dx} = \frac{dy}{du}\frac{du}{dv}\frac{dv}{dx}.$$

PUTNAM EXAM CHALLENGE

125. Let $f(x) = a_1 \sin x + a_2 \sin 2x + \cdots + a_n \sin nx$, where $a_1, a_2, \ldots, a_n$ are real numbers and where n is a positive integer. Given that $|f(x)| \le |\sin x|$ for all real x, prove that $|a_1 + 2a_2 + \cdots + na_n| \le 1$.

126. Let k be a fixed positive integer. The nth derivative of $\frac{1}{x^k - 1}$ has the form $\frac{P_n(x)}{(x^k - 1)^{n+1}}$ where $P_n(x)$ is a polynomial. Find $P_n(1)$.

2.5 Implicit Differentiation

- **Distinguish between functions written in implicit form and explicit form.**
- **Use implicit differentiation to find the derivative of a function.**

Implicit and Explicit Functions

Up to this point in the text, most functions have been expressed in **explicit form.** For example, in the equation $y = 3x^2 - 5$, the variable y is explicitly written as a function of x. Some functions, however, are only implied by an equation. For instance, the function $y = 1/x$ is defined **implicitly** by the equation

$$xy = 1. \qquad \text{Implicit form}$$

To find dy/dx for this equation, you can write y explicitly as a function of x and then differentiate.

Implicit Form	Explicit Form	Derivative
$xy = 1$	$y = \frac{1}{x} = x^{-1}$	$\frac{dy}{dx} = -x^{-2} = -\frac{1}{x^2}$

This strategy works whenever you can solve for the function explicitly. You cannot, however, use this procedure when you are unable to solve for y as a function of x. For instance, how would you find dy/dx for the equation

$$x^2 - 2y^3 + 4y = 2?$$

For this equation, it is difficult to express y as a function of x explicitly. To find dy/dx, you can use **implicit differentiation.**

To understand how to find dy/dx implicitly, you must realize that the differentiation is taking place *with respect to x*. This means that when you differentiate terms involving x alone, you can differentiate as usual. However, when you differentiate terms involving y, you must apply the Chain Rule, because you are assuming that y is defined implicitly as a differentiable function of x.

EXAMPLE 1 Differentiating with Respect to x

a. $\frac{d}{dx}[x^3] = 3x^2$ — Variables agree: use Simple Power Rule.

Variables agree

b. $\frac{d}{dx}[\overbrace{y^3}^{u^n}] = \overbrace{3y^2}^{nu^{n-1}}\overbrace{\frac{dy}{dx}}^{u'}$ — Variables disagree: use Chain Rule.

Variables disagree

c. $\frac{d}{dx}[x + 3y] = 1 + 3\frac{dy}{dx}$ — Chain Rule: $\frac{d}{dx}[3y] = 3y'$

d.

$$\frac{d}{dx}[xy^2] = x\frac{d}{dx}[y^2] + y^2\frac{d}{dx}[x] \qquad \text{Product Rule}$$

$$= x\left(2y\frac{dy}{dx}\right) + y^2(1) \qquad \text{Chain Rule}$$

$$= 2xy\frac{dy}{dx} + y^2 \qquad \text{Simplify.}$$

Implicit Differentiation

GUIDELINES FOR IMPLICIT DIFFERENTIATION

1. Differentiate both sides of the equation *with respect to x*.
2. Collect all terms involving dy/dx on the left side of the equation and move all other terms to the right side of the equation.
3. Factor dy/dx out of the left side of the equation.
4. Solve for dy/dx.

In Example 2, note that implicit differentiation can produce an expression for dy/dx that contains both x and y.

EXAMPLE 2 Implicit Differentiation

Find dy/dx given that $y^3 + y^2 - 5y - x^2 = -4$.

Solution

1. Differentiate both sides of the equation with respect to x.

$$\frac{d}{dx}[y^3 + y^2 - 5y - x^2] = \frac{d}{dx}[-4]$$

$$\frac{d}{dx}[y^3] + \frac{d}{dx}[y^2] - \frac{d}{dx}[5y] - \frac{d}{dx}[x^2] = \frac{d}{dx}[-4]$$

$$3y^2\frac{dy}{dx} + 2y\frac{dy}{dx} - 5\frac{dy}{dx} - 2x = 0$$

2. Collect the dy/dx terms on the left side of the equation and move all other terms to the right side of the equation.

$$3y^2\frac{dy}{dx} + 2y\frac{dy}{dx} - 5\frac{dy}{dx} = 2x$$

3. Factor dy/dx out of the left side of the equation.

$$\frac{dy}{dx}(3y^2 + 2y - 5) = 2x$$

4. Solve for dy/dx by dividing by $(3y^2 + 2y - 5)$.

$$\frac{dy}{dx} = \frac{2x}{3y^2 + 2y - 5}$$

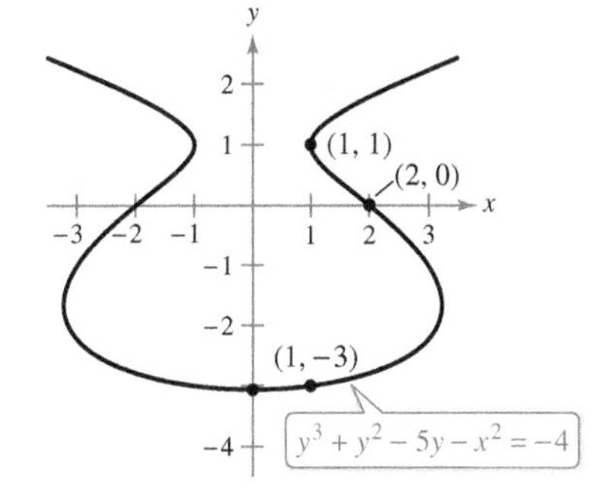

Point on Graph	Slope of Graph
(2, 0)	$-\frac{4}{5}$
(1, −3)	$\frac{1}{8}$
$x = 0$	0
(1, 1)	Undefined

The implicit equation

$$y^3 + y^2 - 5y - x^2 = -4$$

has the derivative

$$\frac{dy}{dx} = \frac{2x}{3y^2 + 2y - 5}.$$

Figure 2.27

To see how you can use an *implicit derivative*, consider the graph shown in Figure 2.27. From the graph, you can see that y is not a function of x. Even so, the derivative found in Example 2 gives a formula for the slope of the tangent line at a point on this graph. The slopes at several points on the graph are shown below the graph.

TECHNOLOGY With most graphing utilities, it is easy to graph an equation that explicitly represents y as a function of x. Graphing other equations, however, can require some ingenuity. For instance, to graph the equation given in Example 2, use a graphing utility, set in *parametric* mode, to graph the parametric representations $x = \sqrt{t^3 + t^2 - 5t + 4}$, $y = t$, and $x = -\sqrt{t^3 + t^2 - 5t + 4}$, $y = t$, for $-5 \le t \le 5$. How does the result compare with the graph shown in Figure 2.27? (You will learn more about this type of representation when you study parametric equations in Section 10.2.)

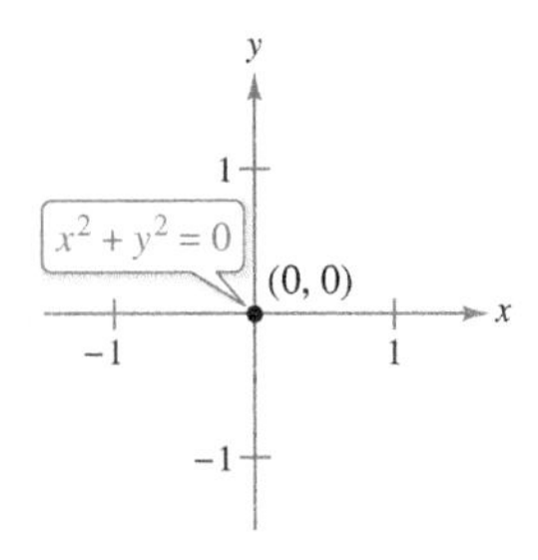

(a)

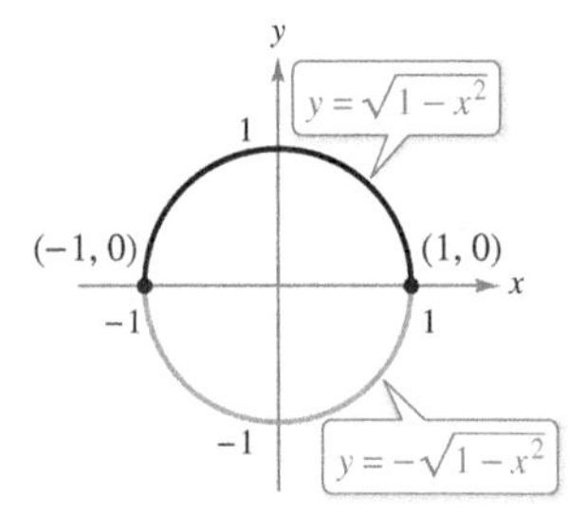

(b)

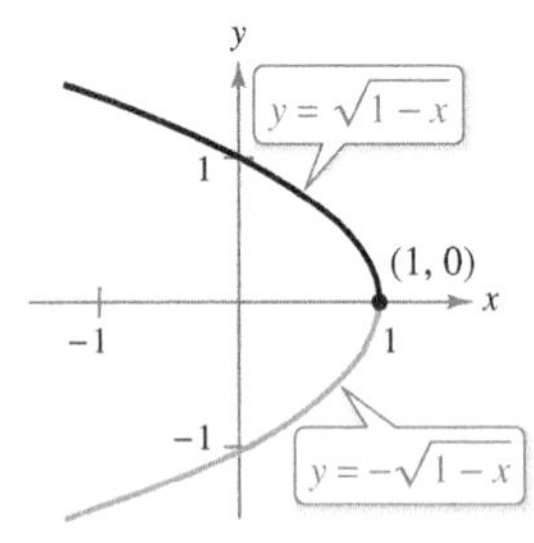

(c)
Some graph segments can be represented by differentiable functions.
Figure 2.28

It is meaningless to solve for dy/dx in an equation that has no solution points. (For example, $x^2 + y^2 = -4$ has no solution points.) If, however, a segment of a graph can be represented by a differentiable function, then dy/dx will have meaning as the slope at each point on the segment. Recall that a function is not differentiable at (a) points with vertical tangents and (b) points at which the function is not continuous.

EXAMPLE 3 Graphs and Differentiable Functions

If possible, represent y as a differentiable function of x.

a. $x^2 + y^2 = 0$ **b.** $x^2 + y^2 = 1$ **c.** $x + y^2 = 1$

Solution

a. The graph of this equation is a single point. So, it does not define y as a differentiable function of x. See Figure 2.28(a).

b. The graph of this equation is the unit circle centered at $(0, 0)$. The upper semicircle is given by the differentiable function

$$y = \sqrt{1 - x^2}, \quad -1 < x < 1$$

and the lower semicircle is given by the differentiable function

$$y = -\sqrt{1 - x^2}, \quad -1 < x < 1.$$

At the points $(-1, 0)$ and $(1, 0)$, the slope of the graph is undefined. See Figure 2.28(b).

c. The upper half of this parabola is given by the differentiable function

$$y = \sqrt{1 - x}, \quad x < 1$$

and the lower half of this parabola is given by the differentiable function

$$y = -\sqrt{1 - x}, \quad x < 1.$$

At the point $(1, 0)$, the slope of the graph is undefined. See Figure 2.28(c).

EXAMPLE 4 Finding the Slope of a Graph Implicitly

See LarsonCalculus.com for an interactive version of this type of example.

Determine the slope of the tangent line to the graph of $x^2 + 4y^2 = 4$ at the point $(\sqrt{2}, -1/\sqrt{2})$. See Figure 2.29.

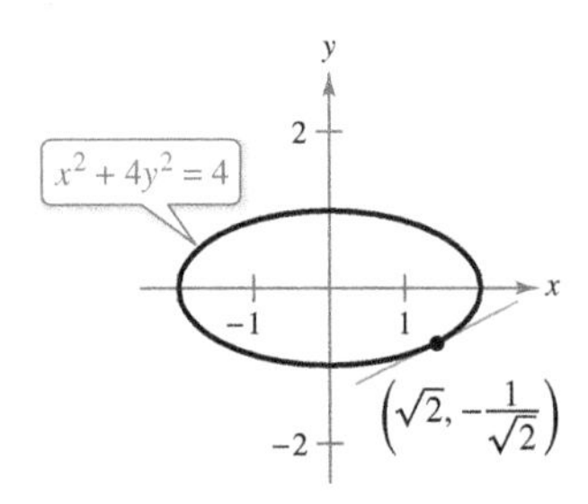

Figure 2.29

Solution

$$x^2 + 4y^2 = 4 \qquad \text{Write original equation.}$$

$$2x + 8y\frac{dy}{dx} = 0 \qquad \text{Differentiate with respect to } x.$$

$$\frac{dy}{dx} = \frac{-2x}{8y} \qquad \text{Solve for } \frac{dy}{dx}.$$

$$= \frac{-x}{4y} \qquad \text{Simplify.}$$

So, at $(\sqrt{2}, -1/\sqrt{2})$, the slope is

$$\frac{dy}{dx} = \frac{-\sqrt{2}}{-4/\sqrt{2}} = \frac{1}{2}. \qquad \text{Evaluate } \frac{dy}{dx} \text{ when } x = \sqrt{2} \text{ and } y = -\frac{1}{\sqrt{2}}.$$

REMARK To see the benefit of implicit differentiation, try doing Example 4 using the explicit function $y = -\frac{1}{2}\sqrt{4 - x^2}$.

EXAMPLE 5 Finding the Slope of a Graph Implicitly

Determine the slope of the graph of

$$3(x^2 + y^2)^2 = 100xy$$

at the point (3, 1).

Solution

$$\frac{d}{dx}[3(x^2 + y^2)^2] = \frac{d}{dx}[100xy]$$

$$3(2)(x^2 + y^2)\left(2x + 2y\frac{dy}{dx}\right) = 100\left[x\frac{dy}{dx} + y(1)\right]$$

$$12y(x^2 + y^2)\frac{dy}{dx} - 100x\frac{dy}{dx} = 100y - 12x(x^2 + y^2)$$

$$[12y(x^2 + y^2) - 100x]\frac{dy}{dx} = 100y - 12x(x^2 + y^2)$$

$$\frac{dy}{dx} = \frac{100y - 12x(x^2 + y^2)}{-100x + 12y(x^2 + y^2)}$$

$$= \frac{25y - 3x(x^2 + y^2)}{-25x + 3y(x^2 + y^2)}$$

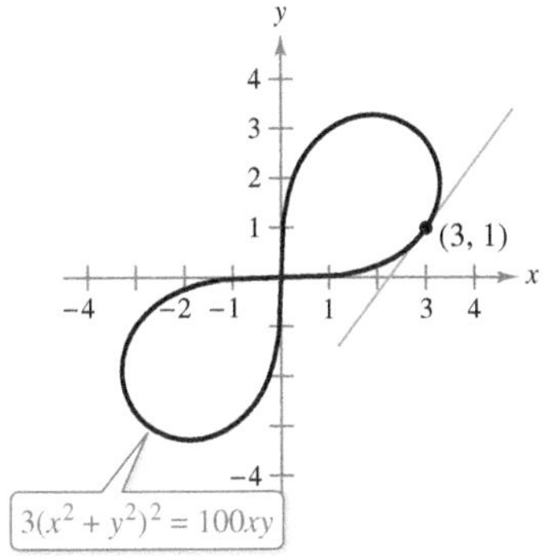

Lemniscate
Figure 2.30

At the point (3, 1), the slope of the graph is

$$\frac{dy}{dx} = \frac{25(1) - 3(3)(3^2 + 1^2)}{-25(3) + 3(1)(3^2 + 1^2)} = \frac{25 - 90}{-75 + 30} = \frac{-65}{-45} = \frac{13}{9}$$

as shown in Figure 2.30. This graph is called a **lemniscate.**

EXAMPLE 6 Determining a Differentiable Function

Find dy/dx implicitly for the equation $\sin y = x$. Then find the largest interval of the form $-a < y < a$ on which y is a differentiable function of x (see Figure 2.31).

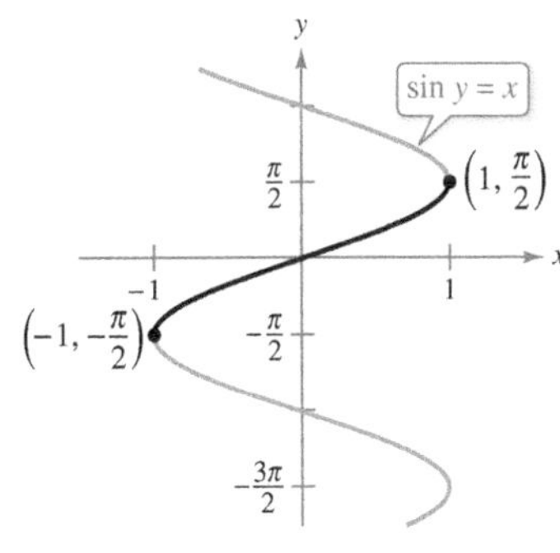

The derivative is $\frac{dy}{dx} = \frac{1}{\sqrt{1 - x^2}}$.

Figure 2.31

Solution

$$\frac{d}{dx}[\sin y] = \frac{d}{dx}[x]$$

$$\cos y\frac{dy}{dx} = 1$$

$$\frac{dy}{dx} = \frac{1}{\cos y}$$

The largest interval about the origin for which y is a differentiable function of x is $-\pi/2 < y < \pi/2$. To see this, note that $\cos y$ is positive for all y in this interval and is 0 at the endpoints. When you restrict y to the interval $-\pi/2 < y < \pi/2$, you should be able to write dy/dx explicitly as a function of x. To do this, you can use

$$\cos y = \sqrt{1 - \sin^2 y}$$

$$= \sqrt{1 - x^2}, \quad -\frac{\pi}{2} < y < \frac{\pi}{2}$$

and conclude that

$$\frac{dy}{dx} = \frac{1}{\sqrt{1 - x^2}}.$$

You will study this example further when inverse trigonometric functions are defined in Section 5.7.

ISAAC BARROW (1630–1677)

The graph in Figure 2.32 is called the **kappa curve** because it resembles the Greek letter kappa, κ. The general solution for the tangent line to this curve was discovered by the English mathematician Isaac Barrow. Newton was Barrow's student, and they corresponded frequently regarding their work in the early development of calculus.

See LarsonCalculus.com to read more of this biography.

With implicit differentiation, the form of the derivative often can be simplified (as in Example 6) by an appropriate use of the *original* equation. A similar technique can be used to find and simplify higher-order derivatives obtained implicitly.

EXAMPLE 7 Finding the Second Derivative Implicitly

Given $x^2 + y^2 = 25$, find $\dfrac{d^2y}{dx^2}$.

Solution Differentiating each term with respect to x produces

$$2x + 2y\frac{dy}{dx} = 0$$

$$2y\frac{dy}{dx} = -2x$$

$$\frac{dy}{dx} = \frac{-2x}{2y}$$

$$= -\frac{x}{y}.$$

Differentiating a second time with respect to x yields

$$\frac{d^2y}{dx^2} = -\frac{(y)(1) - (x)(dy/dx)}{y^2} \qquad \text{Quotient Rule}$$

$$= -\frac{y - (x)(-x/y)}{y^2} \qquad \text{Substitute } -\frac{x}{y} \text{ for } \frac{dy}{dx}.$$

$$= -\frac{y^2 + x^2}{y^3} \qquad \text{Simplify.}$$

$$= -\frac{25}{y^3}. \qquad \text{Substitute 25 for } x^2 + y^2.$$

EXAMPLE 8 Finding a Tangent Line to a Graph

Find the tangent line to the graph of $x^2(x^2 + y^2) = y^2$ at the point $(\sqrt{2}/2, \sqrt{2}/2)$, as shown in Figure 2.32.

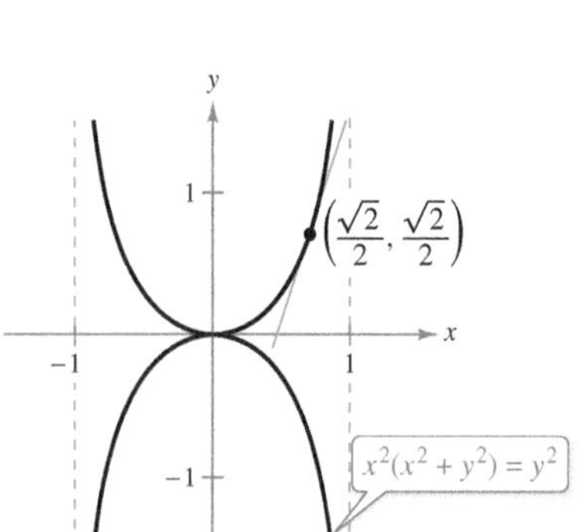

The kappa curve
Figure 2.32

Solution By rewriting and differentiating implicitly, you obtain

$$x^4 + x^2y^2 - y^2 = 0$$

$$4x^3 + x^2\left(2y\frac{dy}{dx}\right) + 2xy^2 - 2y\frac{dy}{dx} = 0$$

$$2y(x^2 - 1)\frac{dy}{dx} = -2x(2x^2 + y^2)$$

$$\frac{dy}{dx} = \frac{x(2x^2 + y^2)}{y(1 - x^2)}.$$

At the point $(\sqrt{2}/2, \sqrt{2}/2)$, the slope is

$$\frac{dy}{dx} = \frac{(\sqrt{2}/2)[2(1/2) + (1/2)]}{(\sqrt{2}/2)[1 - (1/2)]} = \frac{3/2}{1/2} = 3$$

and the equation of the tangent line at this point is

$$y - \frac{\sqrt{2}}{2} = 3\left(x - \frac{\sqrt{2}}{2}\right)$$

$$y = 3x - \sqrt{2}.$$

The Granger Collection, NYC

2.5 Exercises

See CalcChat.com for tutorial help and worked-out solutions to odd-numbered exercises.

CONCEPT CHECK

1. **Explicit and Implicit Functions** Describe the difference between the explicit form of a function and an implicit equation. Give an example of each.
2. **Implicit Differentiation** In your own words, state the guidelines for implicit differentiation.
3. **Implicit Differentiation** Explain when you have to use implicit differentiation to find a derivative.
4. **Chain Rule** How is the Chain Rule applied when finding dy/dx implicitly?

Finding a Derivative **In Exercises 5–20, find dy/dx by implicit differentiation.**

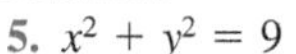

5. $x^2 + y^2 = 9$
6. $x^2 - y^2 = 25$
7. $x^5 + y^5 = 16$
8. $2x^3 + 3y^3 = 64$
9. $x^3 - xy + y^2 = 7$
10. $x^2y + y^2x = -2$
11. $x^3y^3 - y = x$
12. $\sqrt{xy} = x^2y + 1$
13. $x^3 - 3x^2y + 2xy^2 = 12$
14. $x^4y - 8xy + 3xy^2 = 9$
15. $\sin x + 2\cos 2y = 1$
16. $(\sin \pi x + \cos \pi y)^2 = 2$
17. $\csc x = x(1 + \tan y)$
18. $\cot y = x - y$
19. $y = \sin xy$
20. $x = \sec \dfrac{1}{y}$

Finding Derivatives Implicitly and Explicitly **In Exercises 21–24, (a) find two explicit functions by solving the equation for y in terms of x, (b) sketch the graph of the equation and label the parts given by the corresponding explicit functions, (c) differentiate the explicit functions, and (d) find dy/dx implicitly and show that the result is equivalent to that of part (c).**

21. $x^2 + y^2 = 64$
22. $25x^2 + 36y^2 = 300$
23. $16y^2 - x^2 = 16$
24. $x^2 + y^2 - 4x + 6y + 9 = 0$

Finding the Slope of a Graph **In Exercises 25–32, find dy/dx by implicit differentiation. Then find the slope of the graph at the given point.**

25. $xy = 6, \quad (-6, -1)$
26. $3x^3y = 6, \quad (1, 2)$
27. $y^2 = \dfrac{x^2 - 49}{x^2 + 49}, \quad (7, 0)$
28. $4y^3 = \dfrac{x^2 - 36}{x^3 + 36}, \quad (6, 0)$
29. $(x + y)^3 = x^3 + y^3, \quad (-1, 1)$
30. $x^3 + y^3 = 6xy - 1, \quad (2, 3)$
31. $\tan(x + y) = x, \quad (0, 0)$
32. $x \cos y = 1, \quad \left(2, \dfrac{\pi}{3}\right)$

Famous Curves **In Exercises 33–36, find the slope of the tangent line to the graph at the given point.**

33. Witch of Agnesi:
$(x^2 + 4)y = 8$

34. Cissoid:
$(4 - x)y^2 = x^3$

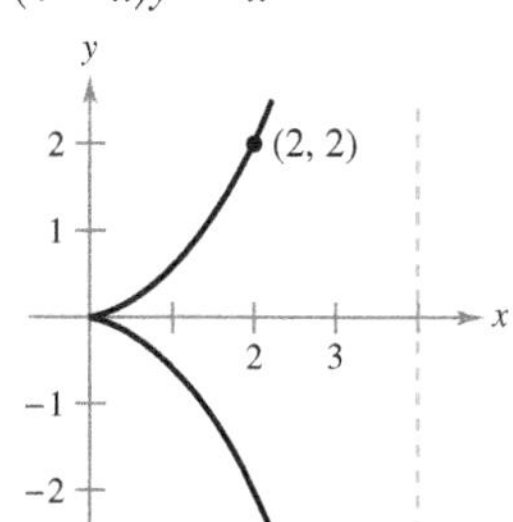

35. Bifolium:
$(x^2 + y^2)^2 = 4x^2y$

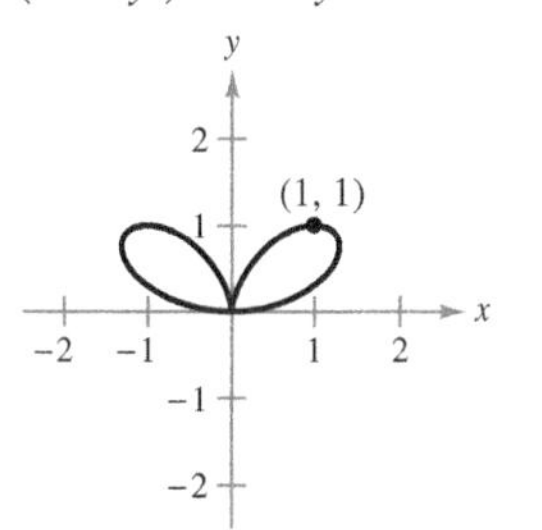

36. Folium of Descartes:
$x^3 + y^3 - 6xy = 0$

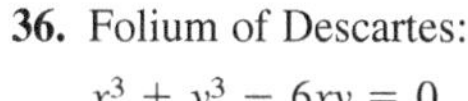

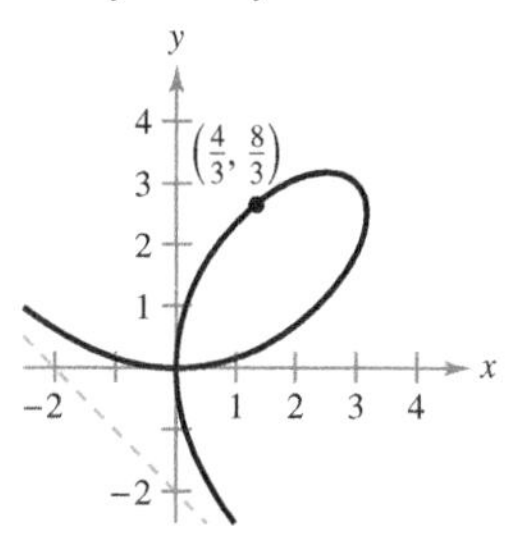

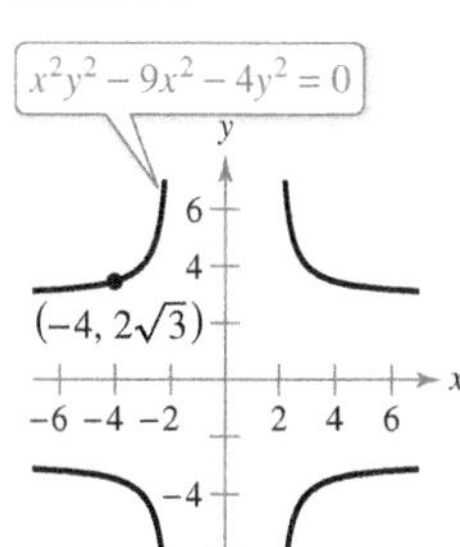

Famous Curves **In Exercises 37–42, find an equation of the tangent line to the graph at the given point. To print an enlarged copy of the graph, go to *MathGraphs.com*.**

37. Parabola

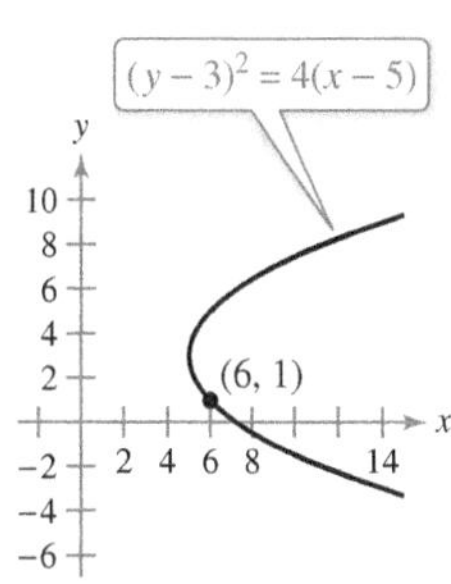

38. Circle

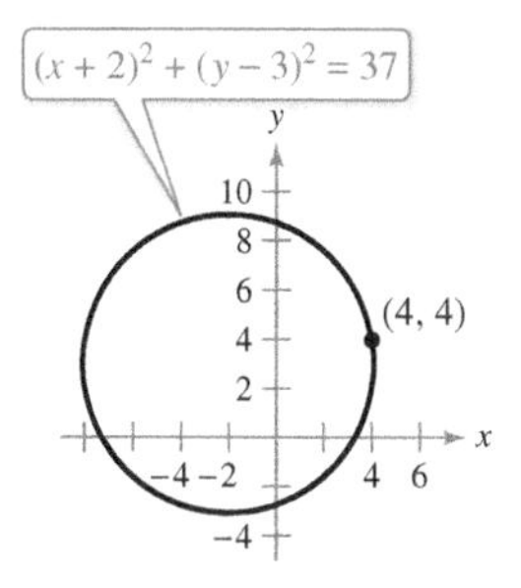

39. Cruciform

40. Astroid

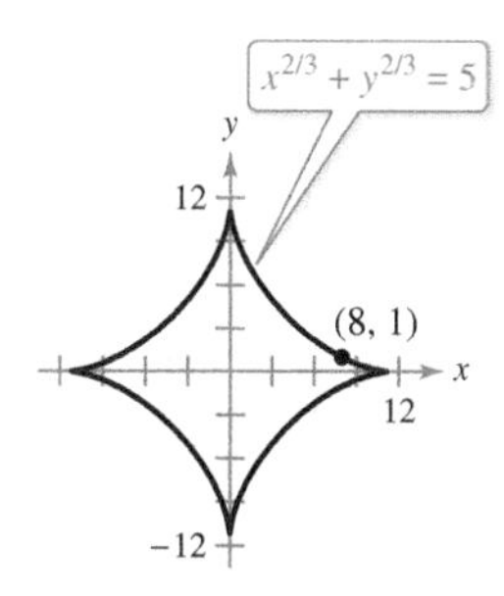

41. Lemniscate

$3(x^2 + y^2)^2 = 100(x^2 - y^2)$

(4, 2)

42. Kappa curve

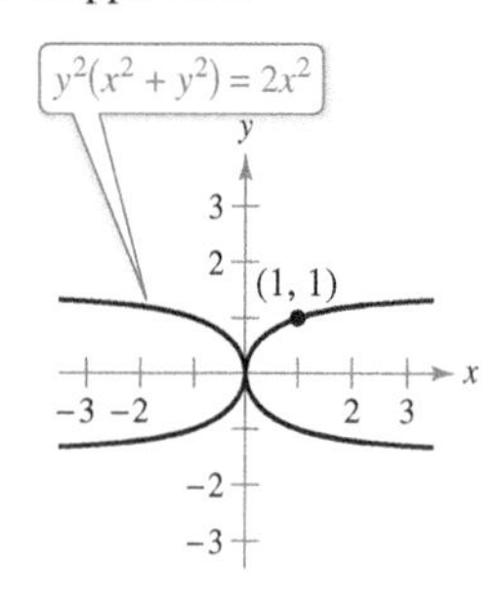

EXPLORING CONCEPTS

43. Implicit and Explicit Forms Write two different equations in implicit form that you can write in explicit form. Then write two different equations in implicit form that you cannot write in explicit form.

44. Think About It Explain why the derivative of $x^2 + y^2 + 2 = 1$ does not mean anything.

45. Ellipse

(a) Use implicit differentiation to find an equation of the tangent line to the ellipse $\dfrac{x^2}{2} + \dfrac{y^2}{8} = 1$ at $(1, 2)$.

(b) Show that the equation of the tangent line to the ellipse $\dfrac{x^2}{a^2} + \dfrac{y^2}{b^2} = 1$ at (x_0, y_0) is $\dfrac{x_0 x}{a^2} + \dfrac{y_0 y}{b^2} = 1$.

46. Hyperbola

(a) Use implicit differentiation to find an equation of the tangent line to the hyperbola $\dfrac{x^2}{6} - \dfrac{y^2}{8} = 1$ at $(3, -2)$.

(b) Show that the equation of the tangent line to the hyperbola $\dfrac{x^2}{a^2} - \dfrac{y^2}{b^2} = 1$ at (x_0, y_0) is $\dfrac{x_0 x}{a^2} - \dfrac{y_0 y}{b^2} = 1$.

Determining a Differentiable Function **In Exercises 47 and 48, find dy/dx implicitly and find the largest interval of the form $-a < y < a$ or $0 < y < a$ such that y is a differentiable function of x. Write dy/dx as a function of x.**

47. $\tan y = x$

48. $\cos y = x$

Finding a Second Derivative **In Exercises 49–54, find d^2y/dx^2 implicitly in terms of x and y.**

49. $x^2 + y^2 = 4$

50. $x^2y - 4x = 5$

51. $x^2y - 2 = 5x + y$

52. $xy - 1 = 2x + y^2$

53. $7xy + \sin x = 2$

54. $3xy - 4\cos x = -6$

Finding an Equation of a Tangent Line **In Exercises 55 and 56, use a graphing utility to graph the equation. Find an equation of the tangent line to the graph at the given point and graph the tangent line in the same viewing window.**

55. $\sqrt{x} + \sqrt{y} = 5$, $(9, 4)$

56. $y^2 = \dfrac{x - 1}{x^2 + 1}$, $\left(2, \dfrac{\sqrt{5}}{5}\right)$

Tangent Lines and Normal Lines **In Exercises 57 and 58, find equations for the tangent line and normal line to the circle at each given point. (The *normal line* at a point is perpendicular to the tangent line at the point.) Use a graphing utility to graph the circle, the tangent lines, and the normal lines.**

57. $x^2 + y^2 = 25$
$(4, 3), (-3, 4)$

58. $x^2 + y^2 = 36$
$(6, 0), (5, \sqrt{11})$

59. Normal Lines Show that the normal line at any point on the circle $x^2 + y^2 = r^2$ passes through the origin.

60. Circles Two circles of radius 4 are tangent to the graph of $y^2 = 4x$ at the point $(1, 2)$. Find equations of these two circles.

Vertical and Horizontal Tangent Lines **In Exercises 61 and 62, find the points at which the graph of the equation has a vertical or horizontal tangent line.**

61. $25x^2 + 16y^2 + 200x - 160y + 400 = 0$

62. $4x^2 + y^2 - 8x + 4y + 4 = 0$

Orthogonal Trajectories **In Exercises 63–66, use a graphing utility to sketch the intersecting graphs of the equations and show that they are orthogonal. [Two graphs are *orthogonal* if at their point(s) of intersection, their tangent lines are perpendicular to each other.]**

63. $2x^2 + y^2 = 6$
$y^2 = 4x$

64. $y^2 = x^3$
$2x^2 + 3y^2 = 5$

65. $x + y = 0$
$x = \sin y$

66. $x^3 = 3(y - 1)$
$x(3y - 29) = 3$

Orthogonal Trajectories **In Exercises 67 and 68, verify that the two families of curves are orthogonal, where C and K are real numbers. Use a graphing utility to graph the two families for two values of C and two values of K.**

67. $xy = C$, $x^2 - y^2 = K$

68. $x^2 + y^2 = C^2$, $y = Kx$

69. Orthogonal Trajectories The figure below shows the topographic map carried by a group of hikers. The hikers are in a wooded area on top of the hill shown on the map, and they decide to follow the path of steepest descent (orthogonal trajectories to the contours on the map). Draw their routes if they start from point A and if they start from point B. Their goal is to reach the road along the top of the map. Which starting point should they use? To print an enlarged copy of the map, go to *MathGraphs.com*.

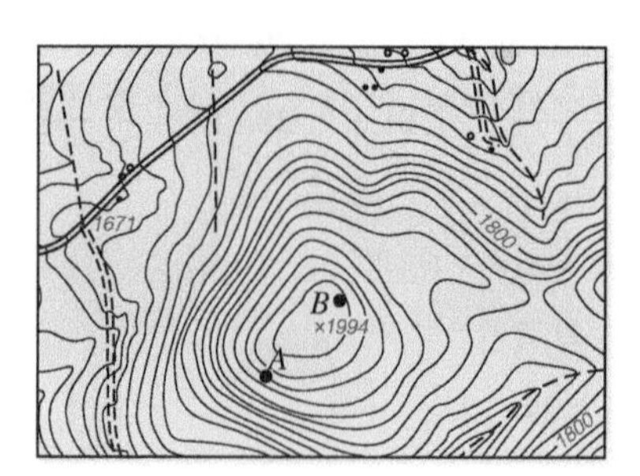

70. HOW DO YOU SEE IT? Use the graph to answer the questions.

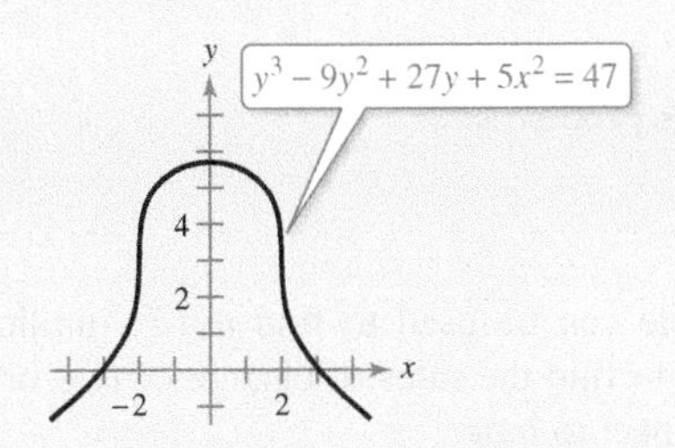

(a) Which is greater, the slope of the tangent line at $x = -3$ or the slope of the tangent line at $x = -1$?

(b) Estimate the point(s) where the graph has a vertical tangent line.

(c) Estimate the point(s) where the graph has a horizontal tangent line.

71. Finding Equations of Tangent Lines Consider the equation $x^4 = 4(4x^2 - y^2)$.

(a) Use a graphing utility to graph the equation.

(b) Find and graph the four tangent lines to the curve for $y = 3$.

(c) Find the exact coordinates of the point of intersection of the two tangent lines in the first quadrant.

72. Tangent Lines and Intercepts Let L be any tangent line to the curve

$$\sqrt{x} + \sqrt{y} = \sqrt{c}.$$

Show that the sum of the x- and y-intercepts of L is c.

73. Proof Prove (Theorem 2.3) that

$$\frac{d}{dx}[x^n] = nx^{n-1}$$

for the case in which n is a rational number. (*Hint:* Write $y = x^{p/q}$ in the form $y^q = x^p$ and differentiate implicitly. Assume that p and q are integers, where $q > 0$.)

74. Slope Find all points on the circle $x^2 + y^2 = 100$ where the slope is $\frac{3}{4}$.

75. Tangent Lines Find equations of both tangent lines to the graph of the ellipse $\frac{x^2}{4} + \frac{y^2}{9} = 1$ that pass through the point $(4, 0)$ not on the graph.

76. Normals to a Parabola The graph shows the normal lines from the point $(2, 0)$ to the graph of the parabola $x = y^2$. How many normal lines are there from the point $(x_0, 0)$ to the graph of the parabola if (a) $x_0 = \frac{1}{4}$, (b) $x_0 = \frac{1}{2}$, and (c) $x_0 = 1$? (d) For what value of x_0 are two of the normal lines perpendicular to each other?

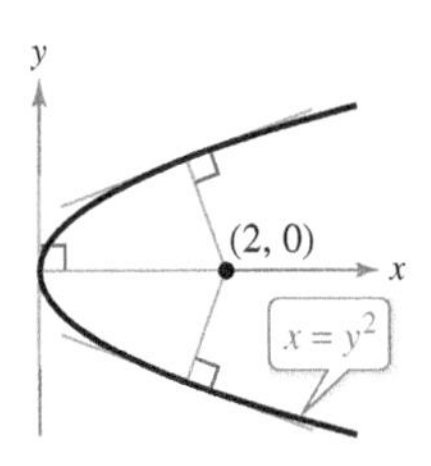

77. Normal Lines (a) Find an equation of the normal line to the ellipse $\frac{x^2}{32} + \frac{y^2}{8} = 1$ at the point $(4, 2)$. (b) Use a graphing utility to graph the ellipse and the normal line. (c) At what other point does the normal line intersect the ellipse?

SECTION PROJECT

Optical Illusions

In each graph below, an optical illusion is created by having lines intersect a family of curves. In each case, the lines appear to be curved. Find the value of dy/dx for the given values.

(a) Circles: $x^2 + y^2 = C^2$

$x = 3, y = 4, C = 5$

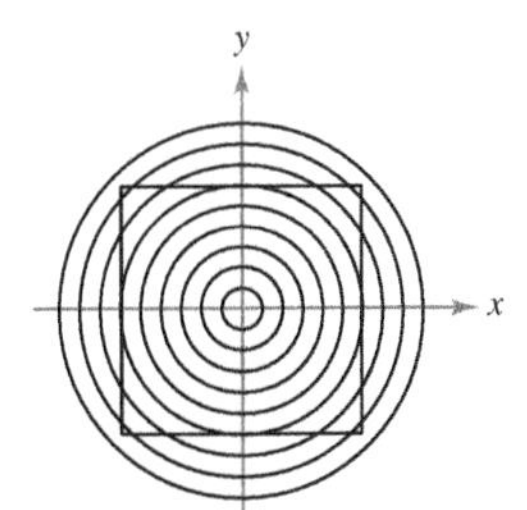

(b) Hyperbolas: $xy = C$

$x = 1, y = 4, C = 4$

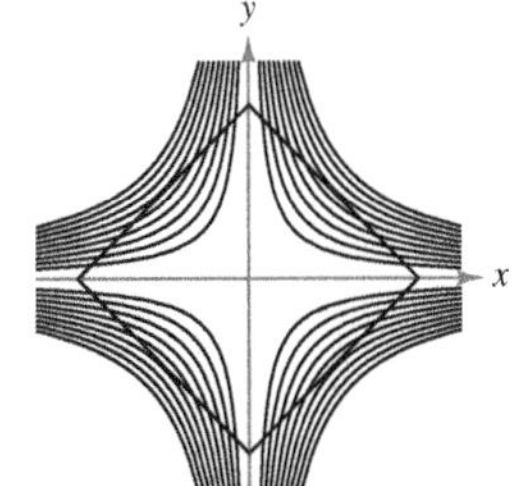

(c) Lines: $ax = by$

$x = \sqrt{3}, y = 3,$

$a = \sqrt{3}, b = 1$

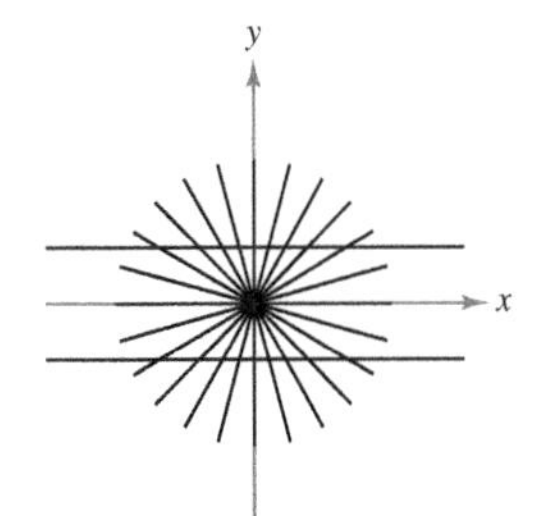

(d) Cosine curves: $y = C \cos x$

$x = \frac{\pi}{3}, y = \frac{1}{3}, C = \frac{2}{3}$

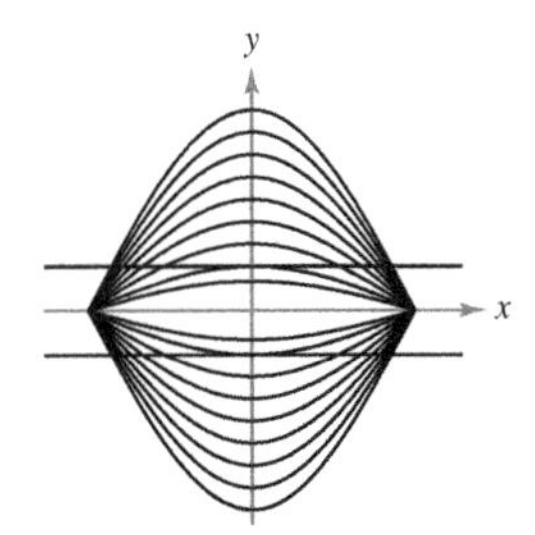

FOR FURTHER INFORMATION For more information on the mathematics of optical illusions, see the article "Descriptive Models for Perception of Optical Illusions" by David A. Smith in *The UMAP Journal.*

2.6 Related Rates

- **Find a related rate.**
- **Use related rates to solve real-life problems.**

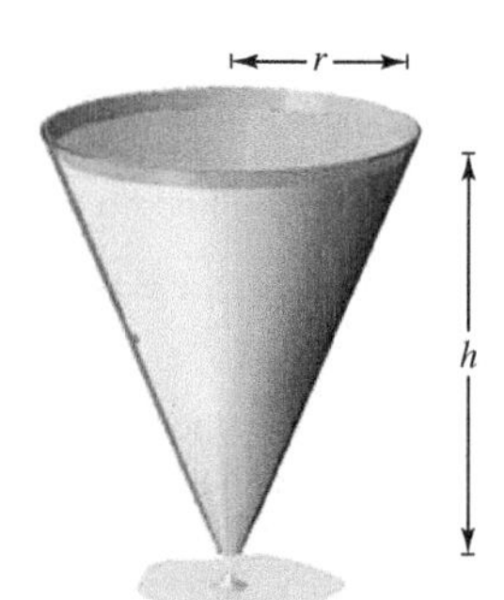

Finding Related Rates

You have seen how the Chain Rule can be used to find dy/dx implicitly. Another important use of the Chain Rule is to find the rates of change of two or more related variables that are changing with respect to *time*.

For example, when water is drained out of a conical tank (see Figure 2.33), the volume V, the radius r, and the height h of the water level are all functions of time t. Knowing that these variables are related by the equation

$$V = \frac{\pi}{3}r^2h \qquad \text{Original equation}$$

you can differentiate implicitly with respect to t to obtain the **related-rate** equation

$$\frac{d}{dt}[V] = \frac{d}{dt}\left[\frac{\pi}{3}r^2h\right]$$

$$\frac{dV}{dt} = \frac{\pi}{3}\left[r^2\frac{dh}{dt} + h\left(2r\frac{dr}{dt}\right)\right] \qquad \text{Differentiate with respect to } t.$$

$$= \frac{\pi}{3}\left(r^2\frac{dh}{dt} + 2rh\frac{dr}{dt}\right).$$

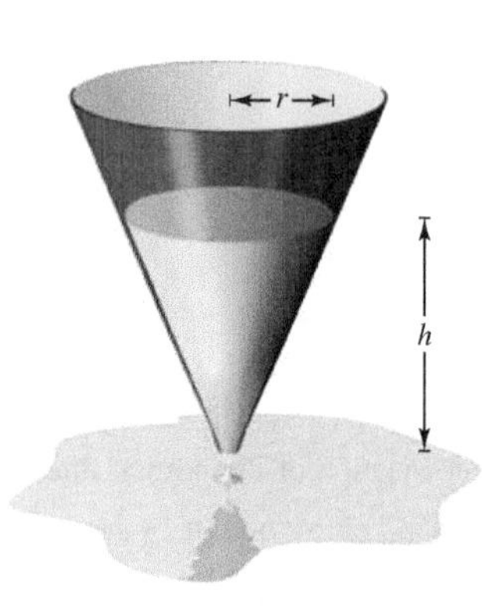

From this equation, you can see that the rate of change of V is related to the rates of change of both h and r.

Exploration

Finding a Related Rate In the conical tank shown in Figure 2.33, the height of the water level is changing at a rate of -0.2 foot per minute and the radius is changing at a rate of -0.1 foot per minute. What is the rate of change of the volume when the radius is $r = 1$ foot and the height is $h = 2$ feet? Does the rate of change of the volume depend on the values of r and h? Explain.

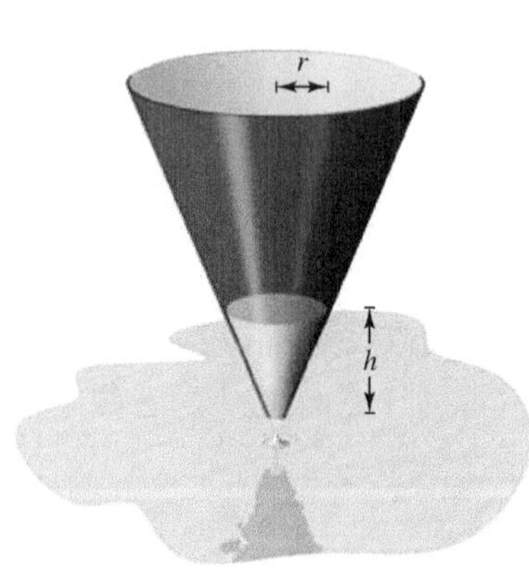

Volume is related to radius and height.
Figure 2.33

EXAMPLE 1 Two Rates That Are Related

The variables x and y are both differentiable functions of t and are related by the equation $y = x^2 + 3$. Find dy/dt when $x = 1$, given that $dx/dt = 2$ when $x = 1$.

Solution Using the Chain Rule, you can differentiate both sides of the equation *with respect to t.*

$$y = x^2 + 3 \qquad \text{Write original equation.}$$

$$\frac{d}{dt}[y] = \frac{d}{dt}[x^2 + 3] \qquad \text{Differentiate with respect to } t.$$

$$\frac{dy}{dt} = 2x\frac{dx}{dt} \qquad \text{Chain Rule}$$

When $x = 1$ and $dx/dt = 2$, you have

$$\frac{dy}{dt} = 2(1)(2) = 4.$$

Problem Solving with Related Rates

In Example 1, you were *given* an equation that related the variables x and y and were asked to find the rate of change of y when $x = 1$.

Equation: $y = x^2 + 3$

Given rate: $\frac{dx}{dt} = 2$ when $x = 1$

Find: $\frac{dy}{dt}$ when $x = 1$

In each of the remaining examples in this section, you must *create* a mathematical model from a verbal description.

EXAMPLE 2 Ripples in a Pond

Total area increases as the outer radius increases.
Figure 2.34

A pebble is dropped into a calm pond, causing ripples in the form of concentric circles, as shown in Figure 2.34. The radius r of the outer ripple is increasing at a constant rate of 1 foot per second. When the radius is 4 feet, at what rate is the total area A of the disturbed water changing?

Solution The variables r and A are related by $A = \pi r^2$. The rate of change of the radius r is $dr/dt = 1$.

Equation: $A = \pi r^2$

Given rate: $\frac{dr}{dt} = 1$ foot per second

Find: $\frac{dA}{dt}$ when $r = 4$ feet

With this information, you can proceed as in Example 1.

$$\frac{d}{dt}[A] = \frac{d}{dt}[\pi r^2] \qquad \text{Differentiate with respect to } t.$$

$$\frac{dA}{dt} = 2\pi r \frac{dr}{dt} \qquad \text{Chain Rule}$$

$$= 2\pi(4)(1) \qquad \text{Substitute 4 for } r \text{ and 1 for } \frac{dr}{dt}.$$

$$= 8\pi \text{ square feet per second} \qquad \text{Simplify.}$$

When the radius is 4 feet, the area is changing at a rate of 8π square feet per second. ■

REMARK When using these guidelines, be sure you perform Step 3 before Step 4. Substituting the known values of the variables before differentiating will produce an inappropriate derivative.

GUIDELINES FOR SOLVING RELATED-RATE PROBLEMS

1. Identify all *given* quantities and quantities *to be determined.* Make a sketch and label the quantities.
2. Write an equation involving the variables whose rates of change either are given or are to be determined.
3. Using the Chain Rule, implicitly differentiate both sides of the equation *with respect to time t.*
4. *After* completing Step 3, substitute into the resulting equation all known values for the variables and their rates of change. Then solve for the required rate of change.

FOR FURTHER INFORMATION To learn more about the history of related-rate problems, see the article "The Lengthening Shadow: The Story of Related Rates" by Bill Austin, Don Barry, and David Berman in *Mathematics Magazine*. To view this article, go to *MathArticles.com*.

The table below lists examples of mathematical models involving rates of change. For instance, the rate of change in the first example is the velocity of a car.

Verbal Statement	Mathematical Model
The velocity of a car after traveling for 1 hour is 50 miles per hour.	$x =$ distance traveled $\frac{dx}{dt} = 50$ mi/h when $t = 1$
Water is being pumped into a swimming pool at a rate of 10 cubic meters per hour.	$V =$ volume of water in pool $\frac{dV}{dt} = 10 \text{ m}^3/\text{h}$
A gear is revolving at a rate of 25 revolutions per minute (1 revolution $= 2\pi$ rad).	$\theta =$ angle of revolution $\frac{d\theta}{dt} = 25(2\pi)$ rad/min
A population of bacteria is increasing at a rate of 2000 per hour.	$x =$ number in population $\frac{dx}{dt} = 2000$ bacteria per hour

EXAMPLE 3 An Inflating Balloon

Air is being pumped into a spherical balloon at a rate of 4.5 cubic feet per minute. Find the rate of change of the radius when the radius is 2 feet.

Solution Let V be the volume of the balloon, and let r be its radius. Because the volume is increasing at a rate of 4.5 cubic feet per minute, you know that at time t the rate of change of the volume is $dV/dt = \frac{9}{2}$. So, the problem can be stated as shown.

Given rate: $\frac{dV}{dt} = \frac{9}{2}$ cubic feet per minute (constant rate)

Find: $\frac{dr}{dt}$ when $r = 2$ feet

To find the rate of change of the radius, you must find an equation that relates the radius r to the volume V.

Equation: $V = \frac{4}{3}\pi r^3$ Volume of a sphere

REMARK The formula for the volume of a sphere and other formulas from geometry are listed on the formula cards for this text.

Differentiating both sides of the equation with respect to t produces

$$\frac{dV}{dt} = 4\pi r^2 \frac{dr}{dt}$$ Differentiate with respect to t.

$$\frac{dr}{dt} = \frac{1}{4\pi r^2}\left(\frac{dV}{dt}\right).$$ Solve for $\frac{dr}{dt}$.

Finally, when $r = 2$, the rate of change of the radius is

$$\frac{dr}{dt} = \frac{1}{4\pi(2)^2}\left(\frac{9}{2}\right) \approx 0.09 \text{ foot per minute.}$$

In Example 3, note that the volume is increasing at a *constant* rate, but the radius is increasing at a *variable* rate. Just because two rates are related does not mean that they are proportional. In this particular case, the radius is growing more and more slowly as t increases. Do you see why?

EXAMPLE 4 The Speed of an Airplane Tracked by Radar

See LarsonCalculus.com for an interactive version of this type of example.

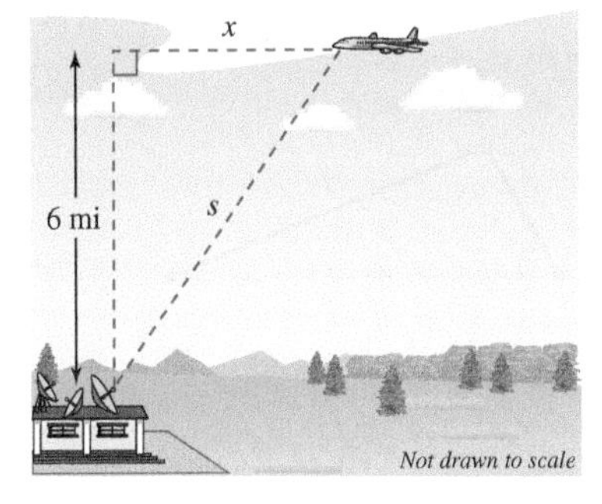

An airplane is flying at an altitude of 6 miles, s miles from the station.
Figure 2.35

An airplane is flying on a flight path that will take it directly over a radar tracking station, as shown in Figure 2.35. The distance s is decreasing at a rate of 400 miles per hour when $s = 10$ miles. What is the speed of the plane?

Solution Let x be the horizontal distance from the station, as shown in Figure 2.35. Notice that when $s = 10$, $x = \sqrt{10^2 - 6^2} = 8$.

Given rate: $ds/dt = -400$ miles per hour when $s = 10$ miles

Find: dx/dt when $s = 10$ miles and $x = 8$ miles

You can find the velocity of the plane as shown.

Equation: $x^2 + 6^2 = s^2$ Pythagorean Theorem

$$2x\frac{dx}{dt} = 2s\frac{ds}{dt}$$ Differentiate with respect to t.

$$\frac{dx}{dt} = \frac{s}{x}\left(\frac{ds}{dt}\right)$$ Solve for $\frac{dx}{dt}$.

$$= \frac{10}{8}(-400)$$ Substitute for s, x, and $\frac{ds}{dt}$.

$$= -500 \text{ miles per hour}$$ Simplify.

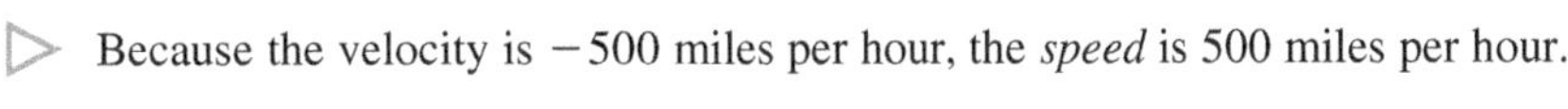

Because the velocity is -500 miles per hour, the *speed* is 500 miles per hour.

REMARK The velocity in Example 4 is negative because x represents a distance that is decreasing.

EXAMPLE 5 A Changing Angle of Elevation

A television camera at ground level is filming the lift-off of a rocket that is rising vertically according to the position equation $s = 50t^2$, where s is measured in feet and t is measured in seconds. The camera is 2000 feet from the launch pad.
Figure 2.36

Find the rate of change in the angle of elevation of the camera shown in Figure 2.36 at 10 seconds after lift-off.

Solution Let θ be the angle of elevation, as shown in Figure 2.36. When $t = 10$, the height s of the rocket is $s = 50t^2 = 50(10)^2 = 5000$ feet.

Given rate: $ds/dt = 100t$ = velocity of rocket (in feet per second)

Find: $d\theta/dt$ when $t = 10$ seconds and $s = 5000$ feet

Using Figure 2.36, you can relate s and θ by the equation $\tan\theta = s/2000$.

Equation: $\tan\theta = \dfrac{s}{2000}$ See Figure 2.36.

$$(\sec^2\theta)\frac{d\theta}{dt} = \frac{1}{2000}\left(\frac{ds}{dt}\right)$$ Differentiate with respect to t.

$$\frac{d\theta}{dt} = \cos^2\theta\,\frac{100t}{2000}$$ Substitute $100t$ for $\frac{ds}{dt}$.

$$= \left(\frac{2000}{\sqrt{s^2 + 2000^2}}\right)^2 \frac{100t}{2000}$$ $\cos\theta = \dfrac{2000}{\sqrt{s^2 + 2000^2}}$

When $t = 10$ and $s = 5000$, you have

$$\frac{d\theta}{dt} = \frac{2000(100)(10)}{5000^2 + 2000^2} = \frac{2}{29} \text{ radian per second.}$$

So, when $t = 10$, θ is changing at a rate of $\frac{2}{29}$ radian per second.

EXAMPLE 6 The Velocity of a Piston

In the engine shown in Figure 2.37, a 7-inch connecting rod is fastened to a crank of radius 3 inches. The crankshaft rotates counterclockwise at a constant rate of 200 revolutions per minute. Find the velocity of the piston when $\theta = \pi/3$.

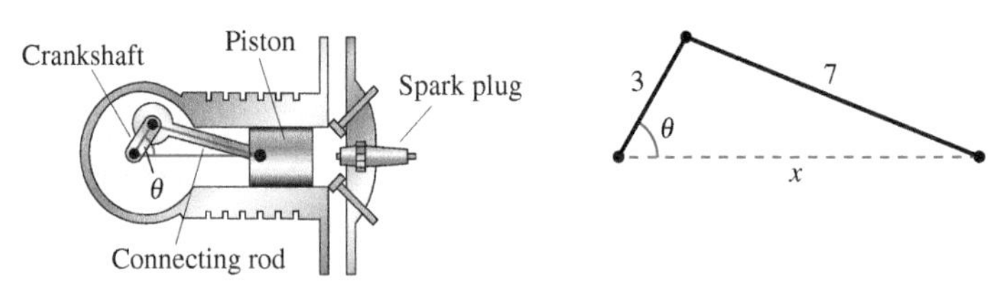

The velocity of a piston is related to the angle of the crankshaft.
Figure 2.37

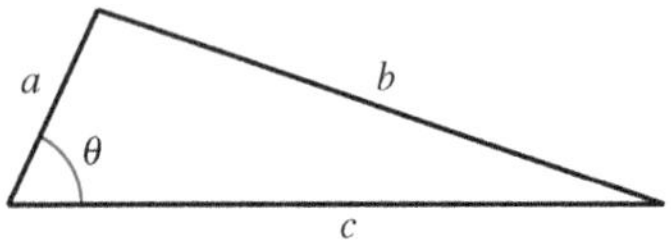

Solution Label the distances as shown in Figure 2.37. Because a complete revolution corresponds to 2π radians, it follows that $d\theta/dt = 200(2\pi) = 400\pi$ radians per minute.

a, b, θ, c

Law of Cosines:
$b^2 = a^2 + c^2 - 2ac\cos\theta$
Figure 2.38

Given rate: $\dfrac{d\theta}{dt} = 400\pi$ radians per minute (constant rate)

Find: $\dfrac{dx}{dt}$ when $\theta = \dfrac{\pi}{3}$

You can use the Law of Cosines (see Figure 2.38) to find an equation that relates x and θ.

Equation:

$$7^2 = 3^2 + x^2 - 2(3)(x)\cos\theta$$

$$0 = 2x\frac{dx}{dt} - 6\left(-x\sin\theta\frac{d\theta}{dt} + \cos\theta\frac{dx}{dt}\right)$$

$$(6\cos\theta - 2x)\frac{dx}{dt} = 6x\sin\theta\frac{d\theta}{dt}$$

$$\frac{dx}{dt} = \frac{6x\sin\theta}{6\cos\theta - 2x}\left(\frac{d\theta}{dt}\right)$$

When $\theta = \pi/3$, you can solve for x as shown.

$$7^2 = 3^2 + x^2 - 2(3)(x)\cos\frac{\pi}{3}$$

$$49 = 9 + x^2 - 6x\left(\frac{1}{2}\right)$$

$$0 = x^2 - 3x - 40$$

$$0 = (x - 8)(x + 5)$$

$$x = 8 \text{ inches}$$ Choose positive solution.

So, when $x = 8$ and $\theta = \pi/3$, the velocity of the piston is

$$\frac{dx}{dt} = \frac{6(8)(\sqrt{3}/2)}{6(1/2) - 16}(400\pi)$$

$$= \frac{9600\pi\sqrt{3}}{-13}$$

$$\approx -4018 \text{ inches per minute.}$$ ■

REMARK The velocity in Example 6 is negative because x represents a distance that is decreasing.

2.6 Exercises

See CalcChat.com for tutorial help and worked-out solutions to odd-numbered exercises.

CONCEPT CHECK

1. **Related-Rate Equation** What is a related-rate equation?

2. **Related Rates** In your own words, state the guidelines for solving related-rate problems.

Using Related Rates **In Exercises 3–6, assume that x and y are both differentiable functions of t and find the required values of dy/dt and dx/dt.**

Equation	Find	Given
3. $y = \sqrt{x}$	(a) $\dfrac{dy}{dt}$ when $x = 4$	$\dfrac{dx}{dt} = 3$
	(b) $\dfrac{dx}{dt}$ when $x = 25$	$\dfrac{dy}{dt} = 2$
4. $y = 3x^2 - 5x$	(a) $\dfrac{dy}{dt}$ when $x = 3$	$\dfrac{dx}{dt} = 2$
	(b) $\dfrac{dx}{dt}$ when $x = 2$	$\dfrac{dy}{dt} = 4$
5. $xy = 4$	(a) $\dfrac{dy}{dt}$ when $x = 8$	$\dfrac{dx}{dt} = 10$
	(b) $\dfrac{dx}{dt}$ when $x = 1$	$\dfrac{dy}{dt} = -6$
6. $x^2 + y^2 = 25$	(a) $\dfrac{dy}{dt}$ when $x = 3, y = 4$	$\dfrac{dx}{dt} = 8$
	(b) $\dfrac{dx}{dt}$ when $x = 4, y = 3$	$\dfrac{dy}{dt} = -2$

Moving Point **In Exercises 7–10, a point is moving along the graph of the given function at the rate dx/dt. Find dy/dt for the given values of x.**

7. $y = 2x^2 + 1$; $\dfrac{dx}{dt} = 2$ centimeters per second

   (a) $x = -1$ (b) $x = 0$ (c) $x = 1$

8. $y = \dfrac{1}{1 + x^2}$; $\dfrac{dx}{dt} = 6$ inches per second

   (a) $x = -2$ (b) $x = 0$ (c) $x = 2$

9. $y = \tan x$; $\dfrac{dx}{dt} = 3$ feet per second

   (a) $x = -\dfrac{\pi}{3}$ (b) $x = -\dfrac{\pi}{4}$ (c) $x = 0$

10. $y = \cos x$; $\dfrac{dx}{dt} = 4$ centimeters per second

   (a) $x = \dfrac{\pi}{6}$ (b) $x = \dfrac{\pi}{4}$ (c) $x = \dfrac{\pi}{3}$

11. **Area** The radius r of a circle is increasing at a rate of 4 centimeters per minute. Find the rates of change of the area when $r = 37$ centimeters.

12. **Area** The length s of each side of an equilateral triangle is increasing at a rate of 13 feet per hour. Find the rate of change of the area when $s = 41$ feet. (*Hint:* The formula for the area of an equilateral triangle is

$$A = \frac{s^2\sqrt{3}}{4}.)$$

13. **Volume** The radius r of a sphere is increasing at a rate of 3 inches per minute.

    (a) Find the rates of change of the volume when $r = 9$ inches and $r = 36$ inches.

    (b) Explain why the rate of change of the volume of the sphere is not constant even though dr/dt is constant.

14. **Radius** A spherical balloon is inflated with gas at the rate of 800 cubic centimeters per minute.

    (a) Find the rates of change of the radius when $r = 30$ centimeters and $r = 85$ centimeters.

    (b) Explain why the rate of change of the radius of the sphere is not constant even though dV/dt is constant.

15. **Volume** All edges of a cube are expanding at a rate of 6 centimeters per second. How fast is the volume changing when each edge is (a) 2 centimeters and (b) 10 centimeters?

16. **Surface Area** All edges of a cube are expanding at a rate of 6 centimeters per second. How fast is the surface area changing when each edge is (a) 2 centimeters and (b) 10 centimeters?

17. **Height** At a sand and gravel plant, sand is falling off a conveyor and onto a conical pile at a rate of 10 cubic feet per minute. The diameter of the base of the cone is approximately three times the altitude. At what rate is the height of the pile changing when the pile is 15 feet high? (*Hint:* The formula for the volume of a cone is $V = \frac{1}{3}\pi r^2 h$.)

18. **Height** The volume of oil in a cylindrical container is increasing at a rate of 150 cubic inches per second. The height of the cylinder is approximately ten times the radius. At what rate is the height of the oil changing when the oil is 35 inches high? (*Hint:* The formula for the volume of a cylinder is $V = \pi r^2 h$.)

19. **Depth** A swimming pool is 12 meters long, 6 meters wide, 1 meter deep at the shallow end, and 3 meters deep at the deep end (see figure). Water is being pumped into the pool at $\frac{1}{4}$ cubic meter per minute, and there is 1 meter of water at the deep end.

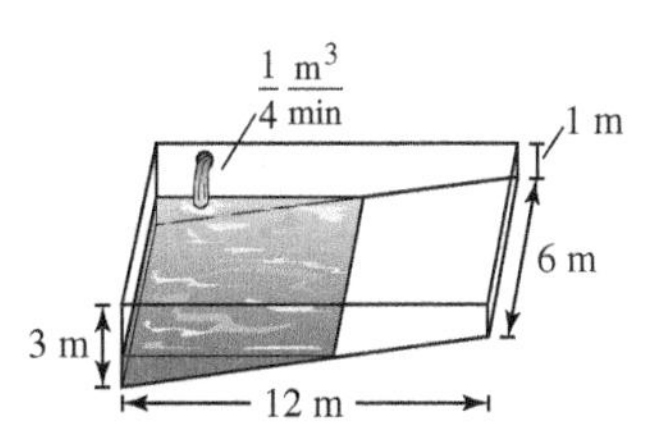

    (a) What percent of the pool is filled?

    (b) At what rate is the water level rising?

20. Depth A trough is 12 feet long and 3 feet across the top (see figure). Its ends are isosceles triangles with altitudes of 3 feet.

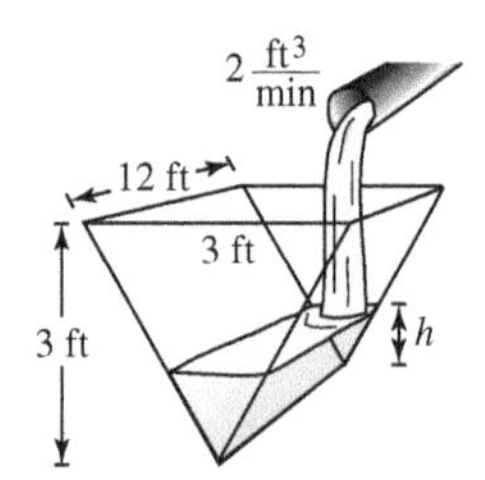

(a) Water is being pumped into the trough at 2 cubic feet per minute. How fast is the water level rising when the depth h is 1 foot?

(b) The water is rising at a rate of $\frac{3}{8}$ inch per minute when $h = 2$ feet. Determine the rate at which water is being pumped into the trough.

21. Moving Ladder A ladder 25 feet long is leaning against the wall of a house (see figure). The base of the ladder is pulled away from the wall at a rate of 2 feet per second.

(a) How fast is the top of the ladder moving down the wall when its base is 7 feet, 15 feet, and 24 feet from the wall?

(b) Consider the triangle formed by the side of the house, the ladder, and the ground. Find the rate at which the area of the triangle is changing when the base of the ladder is 7 feet from the wall.

(c) Find the rate at which the angle between the ladder and the wall of the house is changing when the base of the ladder is 7 feet from the wall.

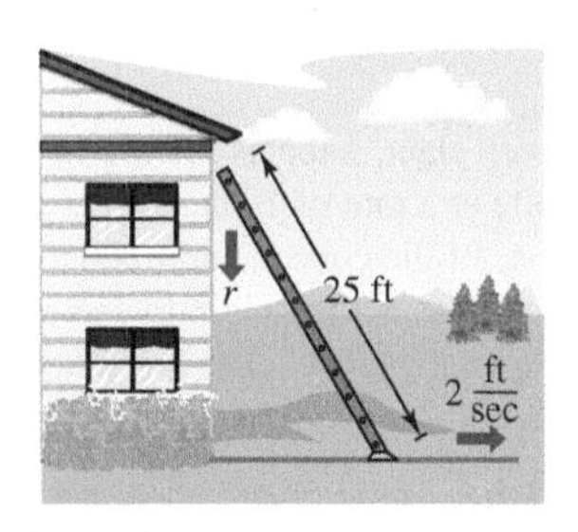

Figure for 21

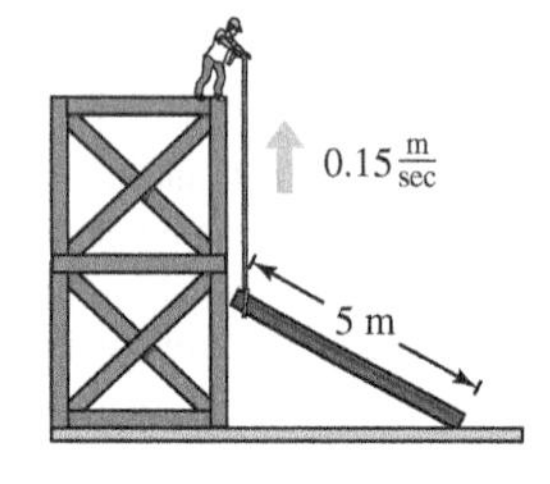

Figure for 22

FOR FURTHER INFORMATION For more information on the mathematics of moving ladders, see the article "The Falling Ladder Paradox" by Paul Scholten and Andrew Simoson in *The College Mathematics Journal*. To view this article, go to *MathArticles.com*.

22. Construction A construction worker pulls a five-meter plank up the side of a building under construction by means of a rope tied to one end of the plank (see figure). Assume the opposite end of the plank follows a path perpendicular to the wall of the building and the worker pulls the rope at a rate of 0.15 meter per second. How fast is the end of the plank sliding along the ground when it is 2.5 meters from the wall of the building?

23. Construction A winch at the top of a 12-meter building pulls a pipe of the same length to a vertical position, as shown in the figure. The winch pulls in rope at a rate of -0.2 meter per second. Find the rate of vertical change and the rate of horizontal change at the end of the pipe when $y = 6$ meters.

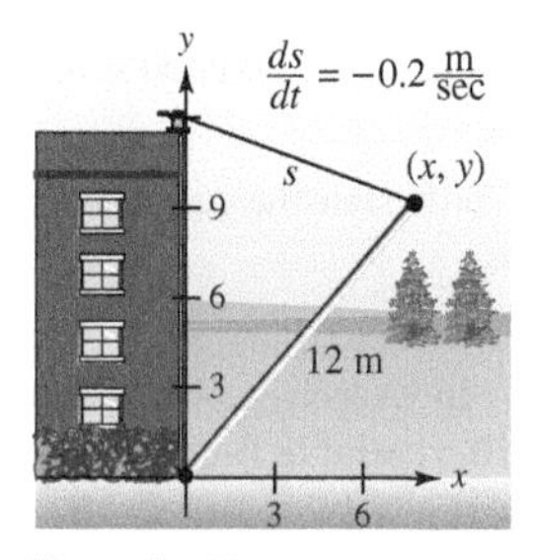

Figure for 23

Figure for 24

24. Boating A boat is pulled into a dock by means of a winch 12 feet above the deck of the boat (see figure).

(a) The winch pulls in rope at a rate of 4 feet per second. Determine the speed of the boat when there is 13 feet of rope out. What happens to the speed of the boat as it gets closer to the dock?

(b) Suppose the boat is moving at a constant rate of 4 feet per second. Determine the speed at which the winch pulls in rope when there is a total of 13 feet of rope out. What happens to the speed at which the winch pulls in rope as the boat gets closer to the dock?

25. Air Traffic Control An air traffic controller spots two planes at the same altitude converging on a point as they fly at right angles to each other (see figure). One plane is 225 miles from the point, moving at 450 miles per hour. The other plane is 300 miles from the point, moving at 600 miles per hour.

(a) At what rate is the distance s between the planes decreasing?

(b) How much time does the air traffic controller have to get one of the planes on a different flight path?

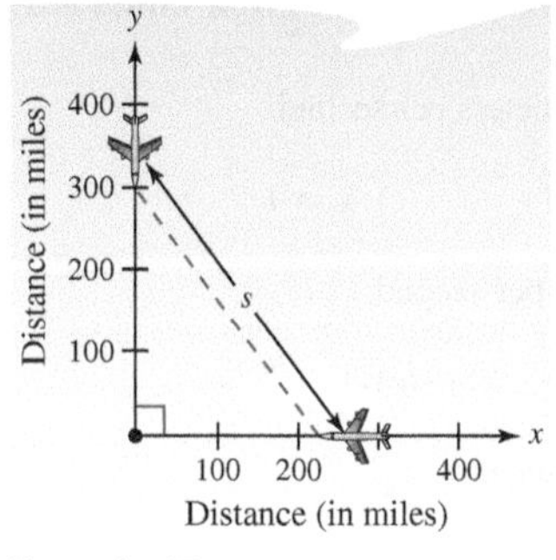

Figure for 25

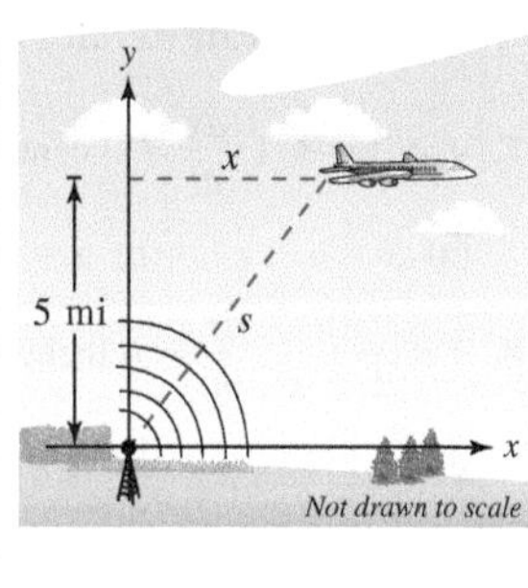

Figure for 26

26. Air Traffic Control An airplane is flying at an altitude of 5 miles and passes directly over a radar antenna (see figure). When the plane is 10 miles away ($s = 10$), the radar detects that the distance s is changing at a rate of 240 miles per hour. What is the speed of the plane?

27. Sports A baseball diamond has the shape of a square with sides 90 feet long (see figure). A player running from second base to third base at a speed of 25 feet per second is 20 feet from third base. At what rate is the player's distance from home plate changing?

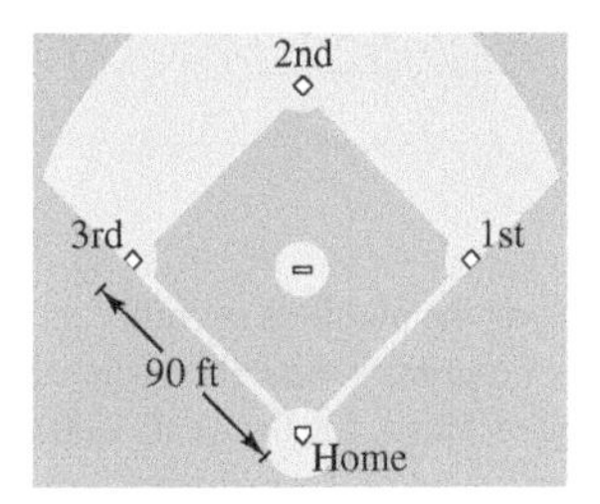

Figure for 27 and 28

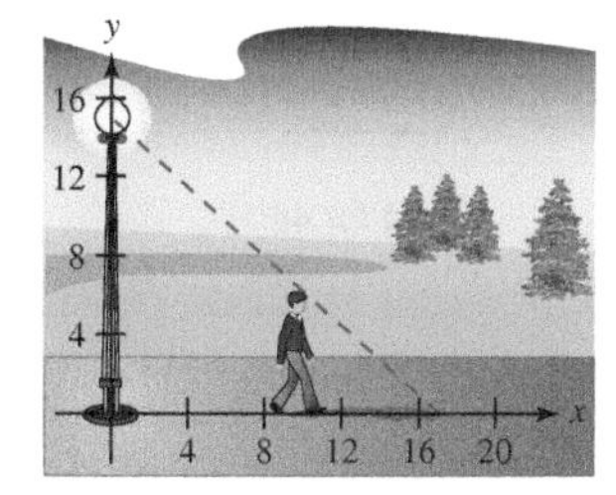

Figure for 29

28. Sports For the baseball diamond in Exercise 27, suppose the player is running from first base to second base at a speed of 25 feet per second. Find the rate at which the distance from home plate is changing when the player is 20 feet from second base.

29. Shadow Length A man 6 feet tall walks at a rate of 5 feet per second away from a light that is 15 feet above the ground (see figure).

(a) When he is 10 feet from the base of the light, at what rate is the tip of his shadow moving?

(b) When he is 10 feet from the base of the light, at what rate is the length of his shadow changing?

30. Shadow Length Repeat Exercise 29 for a man 6 feet tall walking at a rate of 5 feet per second *toward* a light that is 20 feet above the ground (see figure).

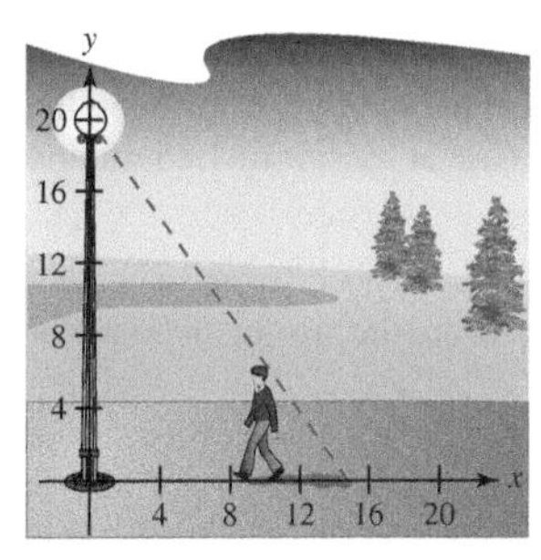

Figure for 30

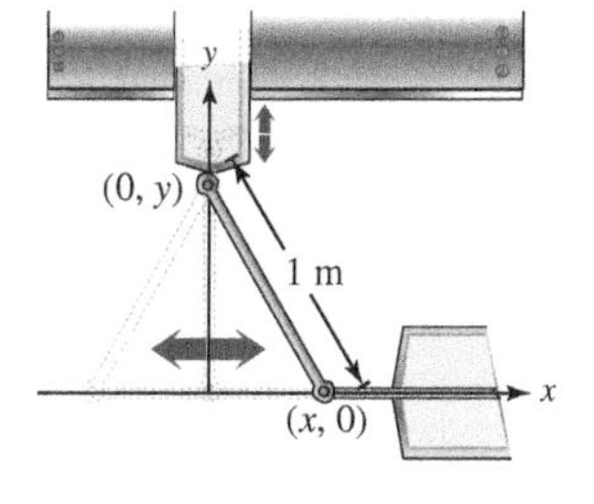

Figure for 31

31. Machine Design The endpoints of a movable rod of length 1 meter have coordinates $(x, 0)$ and $(0, y)$ (see figure). The position of the end on the x-axis is

$$x(t) = \frac{1}{2}\sin\frac{\pi t}{6}$$

where t is the time in seconds.

(a) Find the time of one complete cycle of the rod.

(b) What is the lowest point reached by the end of the rod on the y-axis?

(c) Find the speed of the y-axis endpoint when the x-axis endpoint is $\left(\frac{1}{4}, 0\right)$.

32. Machine Design Repeat Exercise 31 for a position function of $x(t) = \frac{3}{5}\sin \pi t$. Use the point $\left(\frac{3}{10}, 0\right)$ for part (c).

33. Evaporation As a spherical raindrop falls, it reaches a layer of dry air and begins to evaporate at a rate that is proportional to its surface area ($S = 4\pi r^2$). Show that the radius of the raindrop decreases at a constant rate.

34. HOW DO YOU SEE IT? Using the graph of f, (a) determine whether dy/dt is positive or negative given that dx/dt is negative, and (b) determine whether dx/dt is positive or negative given that dy/dt is positive. Explain.

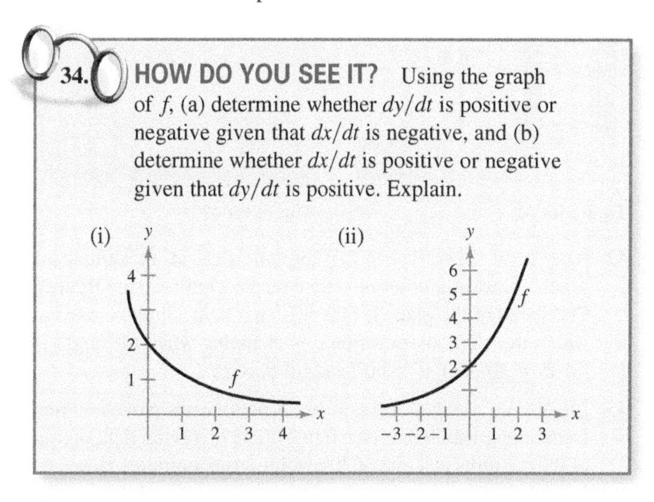

EXPLORING CONCEPTS

35. Think About It Describe the relationship between the rate of change of y and the rate of change of x in each expression. Assume all variables and derivatives are positive.

(a) $\dfrac{dy}{dt} = 3\dfrac{dx}{dt}$ (b) $\dfrac{dy}{dt} = x(L - x)\dfrac{dx}{dt},\quad 0 \le x \le L$

36. Volume Let V be the volume of a cube of side length s that is changing with respect to time. If ds/dt is constant, is dV/dt constant? Explain.

37. Electricity The combined electrical resistance R of two resistors R_1 and R_2, connected in parallel, is given by

$$\frac{1}{R} = \frac{1}{R_1} + \frac{1}{R_2}$$

where R, R_1, and R_2 are measured in ohms. R_1 and R_2 are increasing at rates of 1 and 1.5 ohms per second, respectively. At what rate is R changing when $R_1 = 50$ ohms and $R_2 = 75$ ohms?

38. Electrical Circuit The voltage V in volts of an electrical circuit is $V = IR$, where R is the resistance in ohms and I is the current in amperes. R is increasing at a rate of 2 ohms per second, and V is increasing at a rate of 3 volts per second. At what rate is I changing when $V = 12$ volts and $R = 4$ ohms?

39. Flight Control An airplane is flying in still air with an airspeed of 275 miles per hour. The plane is climbing at an angle of 18°. Find the rate at which the plane is gaining altitude.

40. Angle of Elevation A balloon rises at a rate of 4 meters per second from a point on the ground 50 meters from an observer. Find the rate of change of the angle of elevation of the balloon from the observer when the balloon is 50 meters above the ground.

41. Angle of Elevation A fish is reeled in at a rate of 1 foot per second from a point 10 feet above the water (see figure). At what rate is the angle θ between the line and the water changing when there is a total of 25 feet of line from the end of the rod to the water?

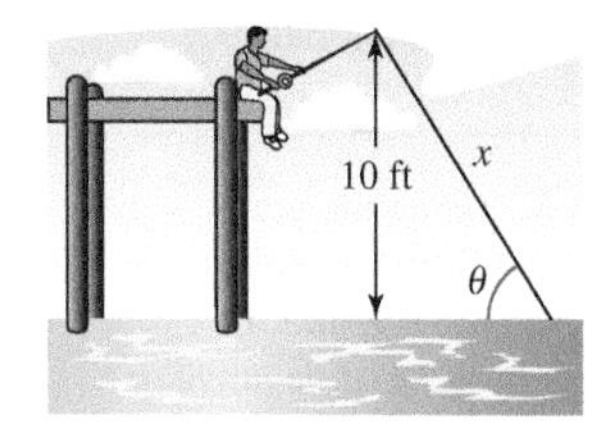

Figure for 41

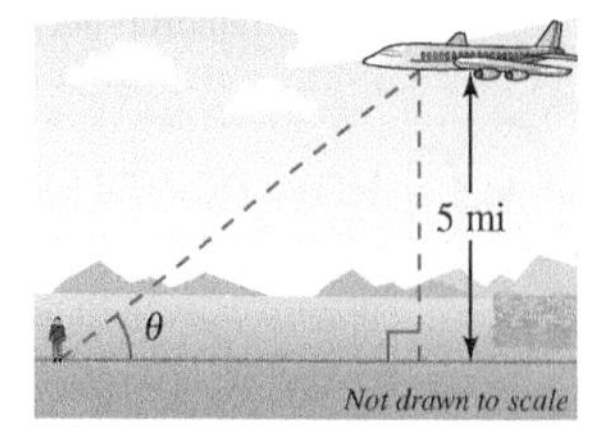

Figure for 42

42. Angle of Elevation An airplane flies at an altitude of 5 miles toward a point directly over an observer (see figure). The speed of the plane is 600 miles per hour. Find the rates at which the angle of elevation θ is changing when the angle is (a) $\theta = 30°$, (b) $\theta = 60°$, and (c) $\theta = 75°$.

43. Linear vs. Angular Speed A patrol car is parked 50 feet from a long warehouse (see figure). The revolving light on top of the car turns at a rate of 30 revolutions per minute. How fast is the light beam moving along the wall when the beam makes angles of (a) $\theta = 30°$, (b) $\theta = 60°$, and (c) $\theta = 70°$ with the perpendicular line from the light to the wall?

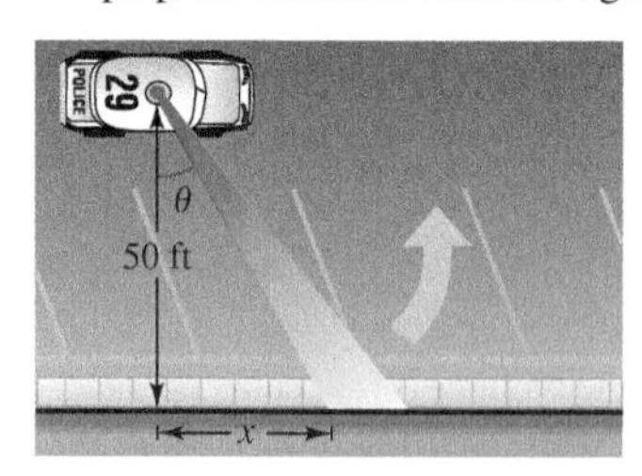

Figure for 43

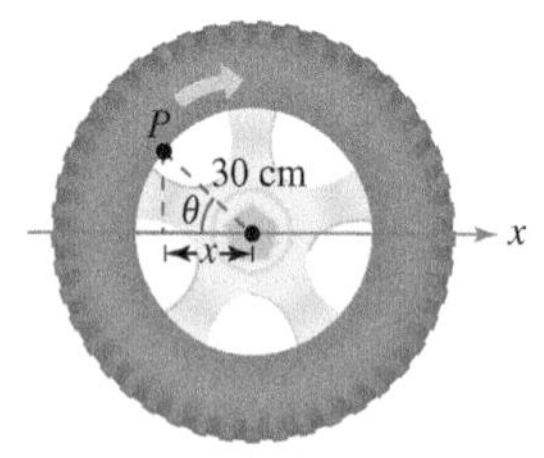

Figure for 44

44. Linear vs. Angular Speed A wheel of radius 30 centimeters revolves at a rate of 10 revolutions per second. A dot is painted at a point P on the rim of the wheel (see figure).

(a) Find dx/dt as a function of θ.

(b) Use a graphing utility to graph the function in part (a).

(c) When is the absolute value of the rate of change of x greatest? When is it least?

(d) Find dx/dt when $\theta = 30°$ and $\theta = 60°$.

45. Area The included angle of the two sides of constant equal length s of an isosceles triangle is θ.

(a) Show that the area of the triangle is given by $A = \frac{1}{2}s^2 \sin \theta$.

(b) The angle θ is increasing at the rate of $\frac{1}{2}$ radian per minute. Find the rates of change of the area when $\theta = \pi/6$ and $\theta = \pi/3$.

46. Security Camera A security camera is centered 50 feet above a 100-foot hallway (see figure). It is easiest to design the camera with a constant angular rate of rotation, but this results in recording the images of the surveillance area at a variable rate. So, it is desirable to design a system with a variable rate of rotation and a constant rate of movement of the scanning beam along the hallway. Find a model for the variable rate of rotation when $|dx/dt| = 2$ feet per second.

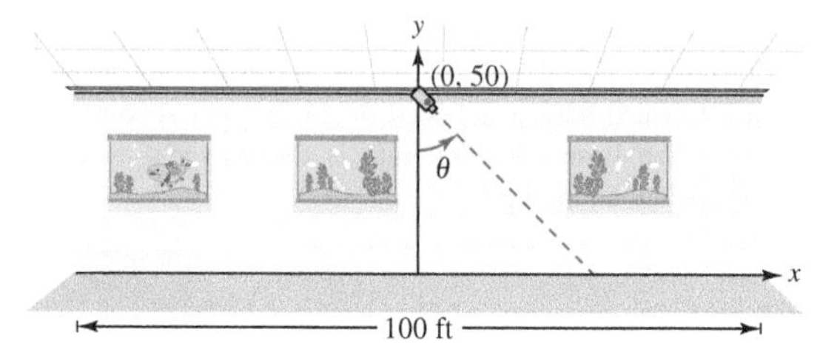

Figure for 46

47. Modeling Data The table shows the numbers (in millions) of participants in the free lunch program f and the reduced price lunch program r in the United States for the years 2007 through 2014. *(Source: U.S. Department of Agriculture)*

Year	2007	2008	2009	2010
f	15.0	15.4	16.3	17.6
r	3.1	3.1	3.2	3.0

Year	2011	2012	2013	2014
f	18.4	18.7	18.9	19.2
r	2.7	2.7	2.6	2.5

(a) Use the regression capabilities of a graphing utility to find a model of the form

$$r(f) = af^3 + bf^2 + cf + d$$

for the data, where t is the time in years, with $t = 7$ corresponding to 2007.

(b) Find dr/dt. Then use the model to estimate dr/dt for $t = 9$ when it is predicted that the number of participants in the free lunch program will increase at the rate of 1.25 million participants per year.

48. Moving Shadow A ball is dropped from a height of 20 meters, 12 meters away from the top of a 20-meter lamppost (see figure). The ball's shadow, caused by the light at the top of the lamppost, is moving along the level ground. How fast is the shadow moving 1 second after the ball is released? *(Submitted by Dennis Gittinger, St. Philips College, San Antonio, TX)*

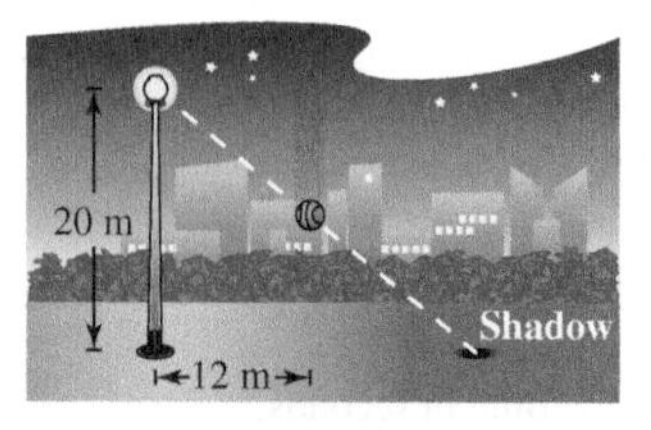

Acceleration In Exercises 49 and 50, find the acceleration of the specified object. (*Hint:* Recall that if a variable is changing at a constant rate, then its acceleration is zero.)

49. Find the acceleration of the top of the ladder described in Exercise 21 when the base of the ladder is 7 feet from the wall.

50. Find the acceleration of the boat in Exercise 24(a) when there is a total of 13 feet of rope out.

Review Exercises

See CalcChat.com for tutorial help and worked-out solutions to odd-numbered exercises.

Finding the Derivative by the Limit Process **In Exercises 1–4, find the derivative of the function by the limit process.**

1. $f(x) = 12$
2. $f(x) = 5x - 4$
3. $f(x) = x^3 - 2x + 1$
4. $f(x) = \dfrac{6}{x}$

Using the Alternative Form of the Derivative **In Exercises 5 and 6, use the alternative form of the derivative to find the derivative at $x = c$, if it exists.**

5. $g(x) = 2x^2 - 3x, \quad c = 2$
6. $f(x) = \dfrac{1}{x + 4}, \quad c = 3$

Determining Differentiability **In Exercises 7 and 8, describe the x-values at which f is differentiable.**

7. $f(x) = (x - 3)^{2/5}$
8. $f(x) = \dfrac{3x}{x + 1}$

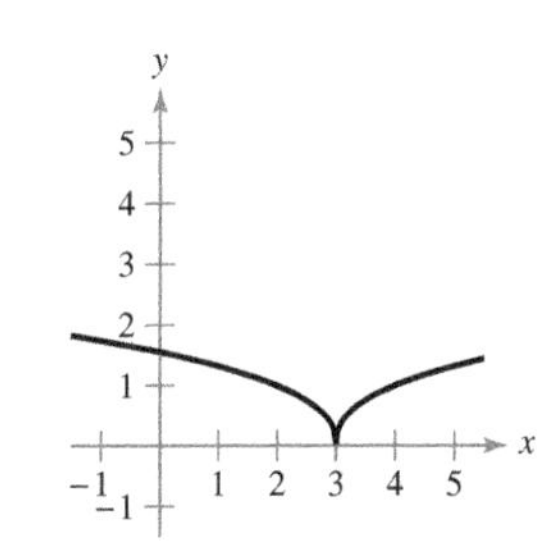

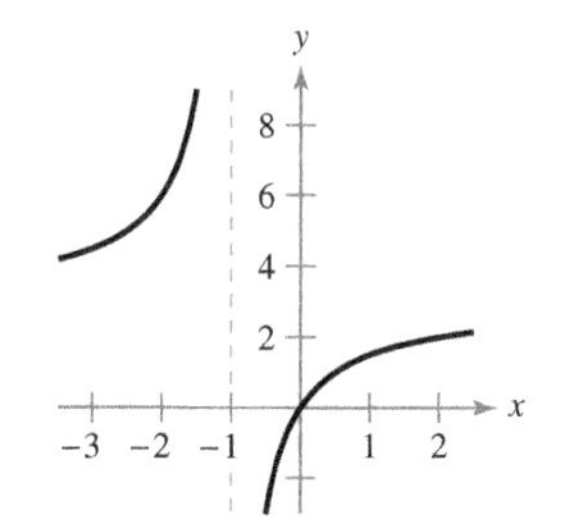

Finding a Derivative **In Exercises 9–20, use the rules of differentiation to find the derivative of the function.**

9. $y = 25$
10. $f(t) = \pi/6$
11. $f(x) = x^3 - 11x^2$
12. $g(s) = 3s^5 - 2s^4$
13. $h(x) = 6\sqrt{x} + 3\sqrt[3]{x}$
14. $f(x) = x^{1/2} - x^{-5/6}$
15. $g(t) = \dfrac{2}{3t^2}$
16. $h(x) = \dfrac{8}{5x^4}$
17. $f(\theta) = 4\theta - 5\sin\theta$
18. $g(\alpha) = 4\cos\alpha + 6$
19. $f(\theta) = 3\cos\theta - \dfrac{\sin\theta}{4}$
20. $g(\alpha) = \dfrac{5\sin\alpha}{3} - 2\alpha$

Finding the Slope of a Graph **In Exercises 21–24, find the slope of the graph of the function at the given point.**

21. $f(x) = \dfrac{27}{x^3}, \quad (3, 1)$
22. $f(x) = 3x^2 - 4x, \quad (1, -1)$
23. $f(x) = 4x^5 + 3x - \sin x, \quad (0, 0)$
24. $f(x) = 5\cos x - 9x, \quad (0, 5)$

25. **Vibrating String** When a guitar string is plucked, it vibrates with a frequency of $F = 200\sqrt{T}$, where F is measured in vibrations per second and the tension T is measured in pounds. Find the rates of change of the frequency when (a) $T = 4$ pounds and (b) $T = 9$ pounds.

26. **Surface Area** The surface area of a cube with sides of length x is given by $S = 6x^2$. Find the rate of change of the surface area with respect to x when $x = 4$ inches.

Vertical Motion **In Exercises 27 and 28, use the position function $s(t) = -16t^2 + v_0t + s_0$ for free-falling objects.**

27. A ball is thrown straight down from the top of a 600-foot building with an initial velocity of -30 feet per second.
    (a) Determine the position and velocity functions for the ball.
    (b) Determine the average velocity on the interval $[1, 3]$.
    (c) Find the instantaneous velocities when $t = 1$ and $t = 3$.
    (d) Find the time required for the ball to reach ground level.
    (e) Find the velocity of the ball at impact.

28. A block is dropped from the top of a 450-foot platform. What is its velocity after 2 seconds? After 5 seconds?

Finding a Derivative **In Exercises 29–40, use the Product Rule or the Quotient Rule to find the derivative of the function.**

29. $f(x) = (5x^2 + 8)(x^2 - 4x - 6)$
30. $g(x) = (2x^3 + 5x)(3x - 4)$
31. $f(x) = (9x - 1)\sin x$
32. $f(t) = 2t^5\cos t$
33. $f(x) = \dfrac{x^2 + x - 1}{x^2 - 1}$
34. $f(x) = \dfrac{2x + 7}{x^2 + 4}$
35. $y = \dfrac{x^4}{\cos x}$
36. $y = \dfrac{\sin x}{x^4}$
37. $y = 3x^2\sec x$
38. $y = -x^2\tan x$
39. $y = x\cos x - \sin x$
40. $g(x) = x^4\cot x + 3x\cos x$

Finding an Equation of a Tangent Line **In Exercises 41–44, find an equation of the tangent line to the graph of f at the given point.**

41. $f(x) = (x + 2)(x^2 + 5), \quad (-1, 6)$
42. $f(x) = (x - 4)(x^2 + 6x - 1), \quad (0, 4)$
43. $f(x) = \dfrac{x + 1}{x - 1}, \quad \left(\dfrac{1}{2}, -3\right)$
44. $f(x) = \dfrac{1 + \cos x}{1 - \cos x}, \quad \left(\dfrac{\pi}{2}, 1\right)$

Finding a Second Derivative **In Exercises 45–52, find the second derivative of the function.**

45. $g(t) = -8t^3 - 5t + 12$
46. $h(x) = 6x^{-2} + 7x^2$
47. $f(x) = 15x^{5/2}$
48. $f(x) = 20\sqrt[5]{x}$
49. $f(\theta) = 3\tan\theta$
50. $h(t) = 10\cos t - 15\sin t$
51. $g(x) = 4\cot x$
52. $h(t) = -12\csc t$

53. Acceleration The velocity of an object is $v(t) = 20 - t^2$, $0 \le t \le 6$, where v is measured in meters per second and t is the time in seconds. Find the velocity and acceleration of the object when $t = 3$.

54. Acceleration The velocity of an automobile starting from rest is

$$v(t) = \frac{90t}{4t + 10}$$

where v is measured in feet per second and t is the time in seconds. Find the acceleration at (a) 1 second, (b) 5 seconds, and (c) 10 seconds.

Finding a Derivative **In Exercises 55–66, find the derivative of the function.**

55. $y = (7x + 3)^4$

56. $y = (x^2 - 6)^3$

57. $y = \dfrac{1}{(x^2 + 5)^3}$

58. $f(x) = \dfrac{1}{(5x + 1)^2}$

59. $y = 5\cos(9x + 1)$

60. $y = -6\sin 3x^4$

61. $y = \dfrac{x}{2} - \dfrac{\sin 2x}{4}$

62. $y = \dfrac{\sec^7 x}{7} - \dfrac{\sec^5 x}{5}$

63. $y = x(6x + 1)^5$

64. $f(s) = (s^2 - 1)^{5/2}(s^3 + 5)$

65. $f(x) = \left(\dfrac{x}{\sqrt{x + 5}}\right)^3$

66. $h(x) = \left(\dfrac{x + 5}{x^2 + 3}\right)^2$

Finding the Slope of a Graph **In Exercises 67–72, find the slope of the graph of the function at the given point.**

67. $f(x) = \sqrt{1 - x^3}, \ (-2, 3)$

68. $f(x) = \sqrt[3]{x^2 - 1}, \ (3, 2)$

69. $f(x) = \dfrac{x + 8}{\sqrt{3x + 1}}, \ (0, 8)$

70. $f(x) = \dfrac{3x + 1}{(4x - 3)^3}, \ (1, 4)$

71. $y = \dfrac{1}{2}\csc 2x, \ \left(\dfrac{\pi}{4}, \dfrac{1}{2}\right)$

72. $y = \csc 3x + \cot 3x, \ \left(\dfrac{\pi}{6}, 1\right)$

Finding a Second Derivative **In Exercises 73–76, find the second derivative of the function.**

73. $y = (8x + 5)^3$

74. $y = \dfrac{1}{5x + 1}$

75. $f(x) = \cot x$

76. $y = x\sin^2 x$

77. Refrigeration The temperature T (in degrees Fahrenheit) of food in a freezer is

$$T = \frac{700}{t^2 + 4t + 10}$$

where t is the time in hours. Find the rate of change of T with respect to t at each of the following times.

(a) $t = 1$ (b) $t = 3$ (c) $t = 5$ (d) $t = 10$

78. Harmonic Motion The displacement from equilibrium of an object in harmonic motion on the end of a spring is

$$y = \frac{1}{4}\cos 8t - \frac{1}{4}\sin 8t$$

where y is measured in feet and t is the time in seconds. Determine the position and velocity of the object when $t = \pi/4$.

Finding a Derivative **In Exercises 79–84, find dy/dx by implicit differentiation.**

79. $x^2 + y^2 = 64$

80. $x^2 + 4xy - y^3 = 6$

81. $x^3y - xy^3 = 4$

82. $\sqrt{xy} = x - 4y$

83. $x\sin y = y\cos x$

84. $\cos(x + y) = x$

Tangent Lines and Normal Lines **In Exercises 85 and 86, find equations for the tangent line and the normal line to the graph of the equation at the given point. (The *normal line* at a point is perpendicular to the tangent line at the point.) Use a graphing utility to graph the equation, the tangent line, and the normal line.**

85. $x^2 + y^2 = 10, \ (3, 1)$

86. $x^2 - y^2 = 20, \ (6, 4)$

87. Rate of Change A point moves along the curve $y = \sqrt{x}$ in such a way that the y-component of the position of the point is increasing at a rate of 2 units per second. At what rate is the x-component changing for each of the following values?

(a) $x = \frac{1}{2}$ (b) $x = 1$ (c) $x = 4$

88. Surface Area All edges of a cube are expanding at a rate of 8 centimeters per second. How fast is the surface area changing when each edge is 6.5 centimeters?

89. Linear vs. Angular Speed A rotating beacon is located 1 kilometer off a straight shoreline (see figure). The beacon rotates at a rate of 3 revolutions per minute. How fast (in kilometers per hour) does the beam of light appear to be moving to a viewer who is $\frac{1}{2}$ kilometer down the shoreline?

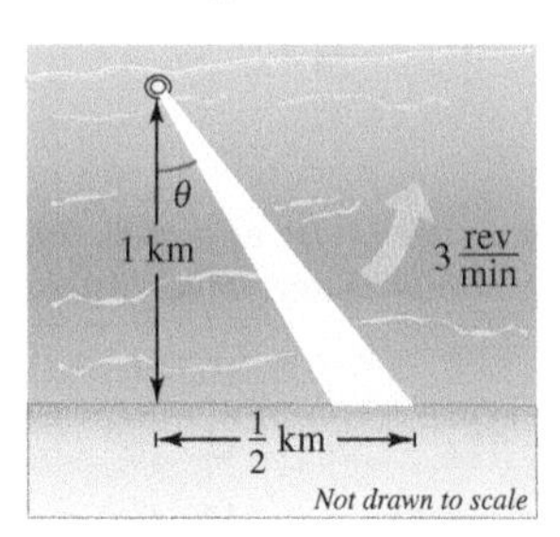

90. Moving Shadow A sandbag is dropped from a balloon at a height of 60 meters when the angle of elevation to the sun is 30° (see figure). The position of the sandbag is

$$s(t) = 60 - 4.9t^2.$$

Find the rate at which the shadow of the sandbag is traveling along the ground when the sandbag is at a height of 35 meters.

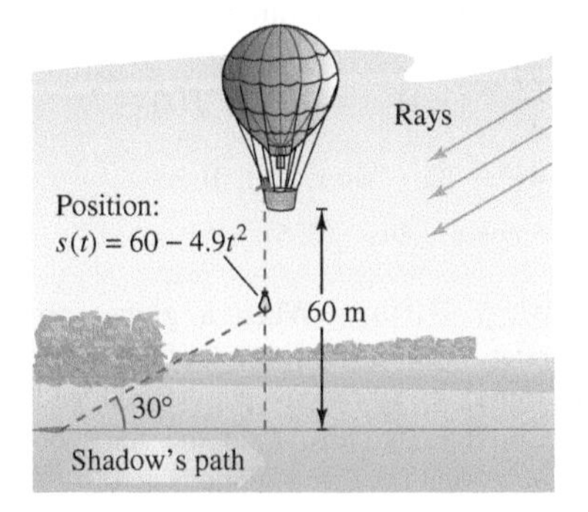

P.S. Problem Solving

See CalcChat.com for tutorial help and worked-out solutions to odd-numbered exercises.

1. Finding Equations of Circles Consider the graph of the parabola $y = x^2$.

(a) Find the radius r of the largest possible circle centered on the y-axis that is tangent to the parabola at the origin, as shown in the figure. This circle is called the **circle of curvature** (see Section 12.5). Find the equation of this circle. Use a graphing utility to graph the circle and parabola in the same viewing window to verify your answer.

(b) Find the center $(0, b)$ of the circle of radius 1 centered on the y-axis that is tangent to the parabola at two points, as shown in the figure. Find the equation of this circle. Use a graphing utility to graph the circle and parabola in the same viewing window to verify your answer.

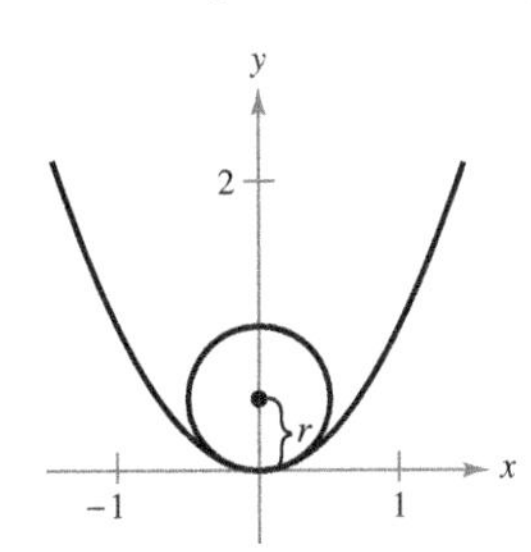

Figure for 1(a)

Figure for 1(b)

2. Finding Equations of Tangent Lines Graph the two parabolas

$$y = x^2 \quad \text{and} \quad y = -x^2 + 2x - 5$$

in the same coordinate plane. Find equations of the two lines that are simultaneously tangent to both parabolas.

3. Finding a Polynomial Find a third-degree polynomial $p(x)$ that is tangent to the line $y = 14x - 13$ at the point $(1, 1)$, and tangent to the line $y = -2x - 5$ at the point $(-1, -3)$.

4. Finding a Function Find a function of the form $f(x) = a + b \cos cx$ that is tangent to the line $y = 1$ at the point $(0, 1)$, and tangent to the line

$$y = x + \frac{3}{2} - \frac{\pi}{4}$$

at the point $\left(\frac{\pi}{4}, \frac{3}{2}\right)$.

5. Tangent Lines and Normal Lines

(a) Find an equation of the tangent line to the parabola $y = x^2$ at the point $(2, 4)$.

(b) Find an equation of the normal line to $y = x^2$ at the point $(2, 4)$. (The *normal line* at a point is perpendicular to the tangent line at the point.) Where does this line intersect the parabola a second time?

(c) Find equations of the tangent line and normal line to $y = x^2$ at the point $(0, 0)$.

(d) Prove that for any point $(a, b) \neq (0, 0)$ on the parabola $y = x^2$, the normal line intersects the graph a second time.

6. Finding Polynomials

(a) Find the polynomial $P_1(x) = a_0 + a_1x$ whose value and slope agree with the value and slope of $f(x) = \cos x$ at the point $x = 0$.

(b) Find the polynomial $P_2(x) = a_0 + a_1x + a_2x^2$ whose value and first two derivatives agree with the value and first two derivatives of $f(x) = \cos x$ at the point $x = 0$. This polynomial is called the second-degree **Taylor polynomial** of $f(x) = \cos x$ at $x = 0$.

(c) Complete the table comparing the values of $f(x) = \cos x$ and $P_2(x)$. What do you observe?

x	-1.0	-0.1	-0.001	0	0.001	0.1	1.0
$\cos x$							
$P_2(x)$							

(d) Find the third-degree Taylor polynomial of $f(x) = \sin x$ at $x = 0$.

7. Famous Curve The graph of the **eight curve**

$$x^4 = a^2(x^2 - y^2), \quad a \neq 0$$

is shown below.

(a) Explain how you could use a graphing utility to graph this curve.

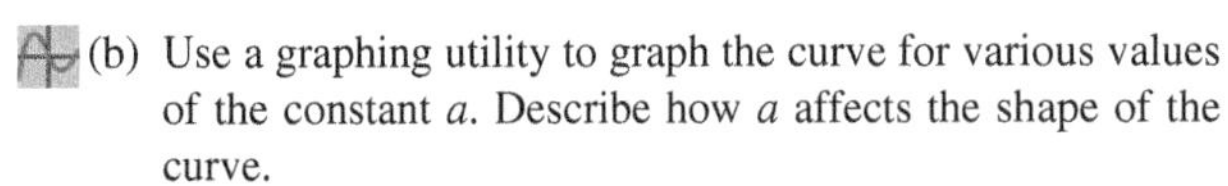

(b) Use a graphing utility to graph the curve for various values of the constant a. Describe how a affects the shape of the curve.

(c) Determine the points on the curve at which the tangent line is horizontal.

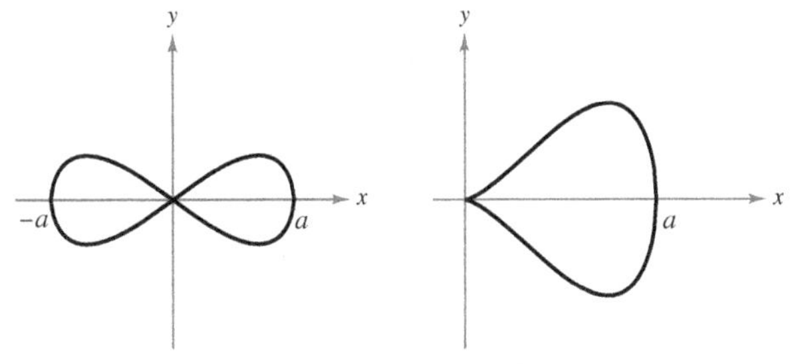

Figure for 7

Figure for 8

8. Famous Curve The graph of the **pear-shaped quartic**

$$b^2y^2 = x^3(a - x), \quad a, b > 0$$

is shown above.

(a) Explain how you could use a graphing utility to graph this curve.

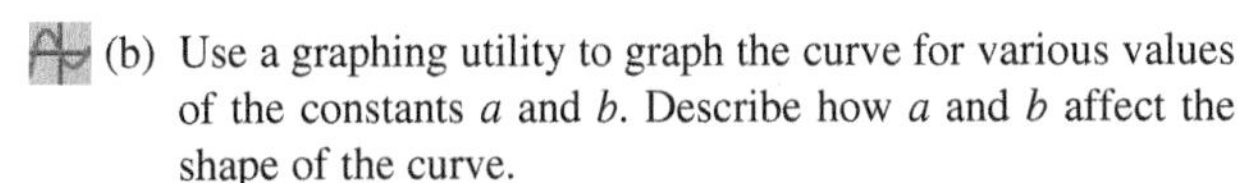

(b) Use a graphing utility to graph the curve for various values of the constants a and b. Describe how a and b affect the shape of the curve.

(c) Determine the points on the curve at which the tangent line is horizontal.

9. Shadow Length A man 6 feet tall walks at a rate of 5 feet per second toward a streetlight that is 30 feet high (see figure). The man's 3-foot-tall child follows at the same speed but 10 feet behind the man. The shadow behind the child is caused by the man at some times and by the child at other times.

(a) Suppose the man is 90 feet from the streetlight. Show that the man's shadow extends beyond the child's shadow.

(b) Suppose the man is 60 feet from the streetlight. Show that the child's shadow extends beyond the man's shadow.

(c) Determine the distance d from the man to the streetlight at which the tips of the two shadows are exactly the same distance from the streetlight.

(d) Determine how fast the tip of the man's shadow is moving as a function of x, the distance between the man and the streetlight. Discuss the continuity of this shadow speed function.

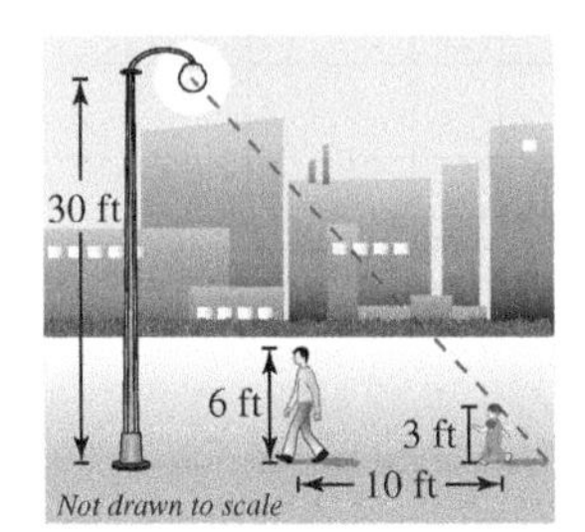

Figure for 9

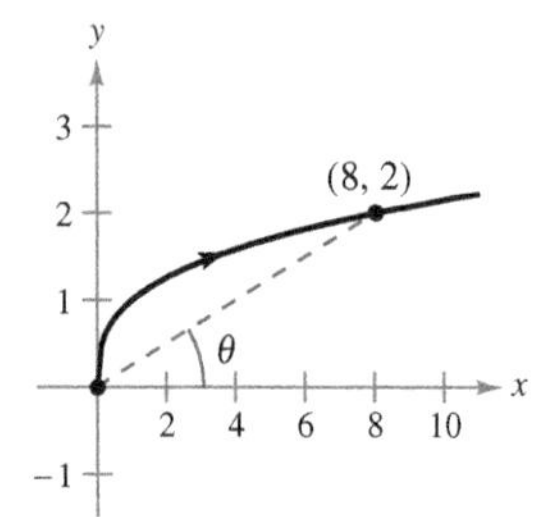

Figure for 10

10. Moving Point A particle is moving along the graph of $y = \sqrt[3]{x}$ (see figure). When $x = 8$, the y-component of the position of the particle is increasing at the rate of 1 centimeter per second.

(a) How fast is the x-component changing at this moment?

(b) How fast is the distance from the origin changing at this moment?

(c) How fast is the angle of inclination θ changing at this moment?

11. Projectile Motion An astronaut standing on the moon throws a rock upward. The height of the rock is

$$s = -\frac{27}{10}t^2 + 27t + 6$$

where s is measured in feet and t is measured in seconds.

(a) Find expressions for the velocity and acceleration of the rock.

(b) Find the time when the rock is at its highest point by finding the time when the velocity is zero. What is the height of the rock at this time?

(c) How does the acceleration of the rock compare with the acceleration due to gravity on Earth?

12. Proof Let E be a function satisfying $E(0) = E'(0) = 1$. Prove that if $E(a + b) = E(a)E(b)$ for all a and b, then E is differentiable and $E'(x) = E(x)$ for all x. Find an example of a function satisfying $E(a + b) = E(a)E(b)$.

13. Proof Let L be a differentiable function for all x. Prove that if $L(a + b) = L(a) + L(b)$ for all a and b, then $L'(x) = L'(0)$ for all x. What does the graph of L look like?

14. Radians and Degrees The fundamental limit

$$\lim_{x \to 0} \frac{\sin x}{x} = 1$$

assumes that x is measured in radians. Suppose you assume that x is measured in degrees instead of radians.

(a) Set your calculator to *degree* mode and complete the table.

z (in degrees)	0.1	0.01	0.0001
$\frac{\sin z}{z}$			

(b) Use the table to estimate

$$\lim_{z \to 0} \frac{\sin z}{z}$$

for z in degrees. What is the exact value of this limit? (*Hint:* $180° = \pi$ radians)

(c) Use the limit definition of the derivative to find

$$\frac{d}{dz}[\sin z]$$

for z in degrees.

(d) Define the new functions

$$S(z) = \sin cz \quad \text{and} \quad C(z) = \cos cz$$

where $c = \pi/180$. Find $S(90)$ and $C(180)$. Use the Chain Rule to calculate

$$\frac{d}{dz}[S(z)].$$

(e) Explain why calculus is made easier by using radians instead of degrees.

15. Acceleration and Jerk If a is the acceleration of an object, then the *jerk* j is defined by $j = a'(t)$.

(a) Use this definition to give a physical interpretation of j.

(b) Find j for the slowing vehicle in Exercise 119 in Section 2.3 and interpret the result.

(c) The figure shows the graphs of the position, velocity, acceleration, and jerk functions of a vehicle. Identify each graph and explain your reasoning.

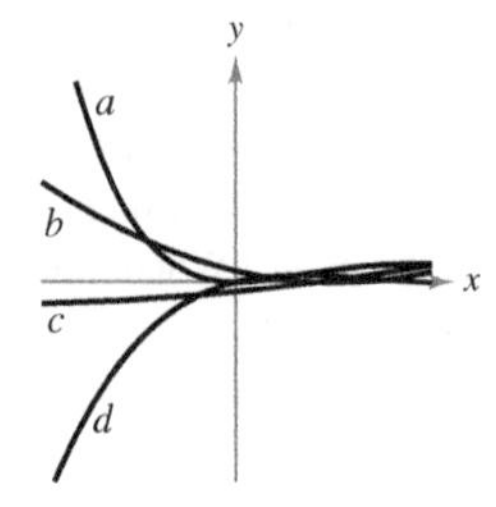

3 Applications of Differentiation

Estimation of Error (Example 3, p. 237)

Offshore Oil Well (Exercise 39, p. 226)

Engine Efficiency (Exercise 51, p. 207)

Path of a Projectile (Example 5, p. 186)

Speed (Exercise 59, p.180)

3.1 Extrema on an Interval

- **Understand the definition of extrema of a function on an interval.**
- **Understand the definition of relative extrema of a function on an open interval.**
- **Find extrema on a closed interval.**

Extrema of a Function

In calculus, much effort is devoted to determining the behavior of a function f on an interval I. Does f have a maximum value on I? Does it have a minimum value? Where is the function increasing? Where is it decreasing? In this chapter, you will learn how derivatives can be used to answer these questions. You will also see why these questions are important in real-life applications.

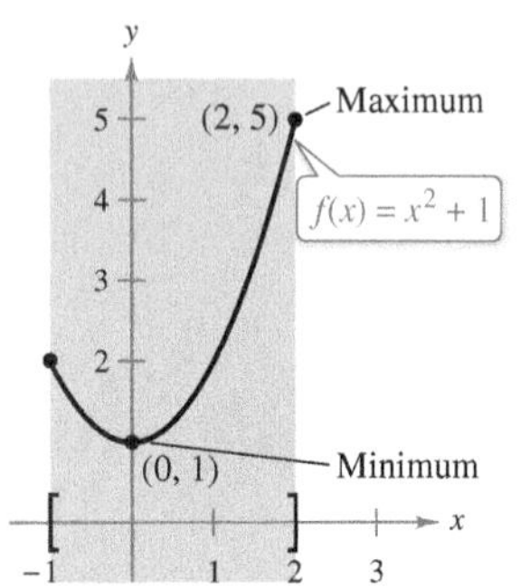

(a) f is continuous, $[-1, 2]$ is closed.

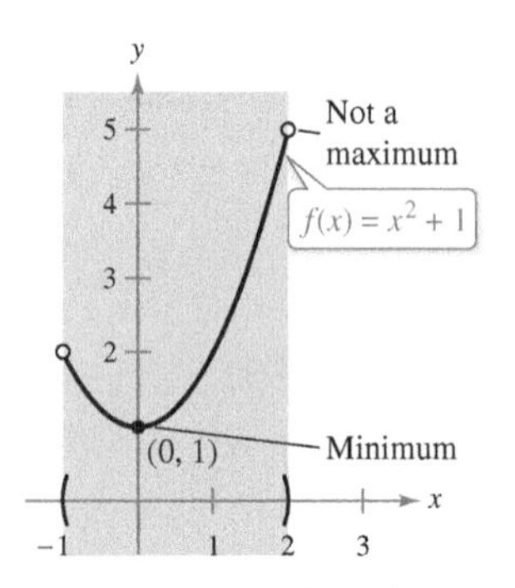

(b) f is continuous, $(-1, 2)$ is open.

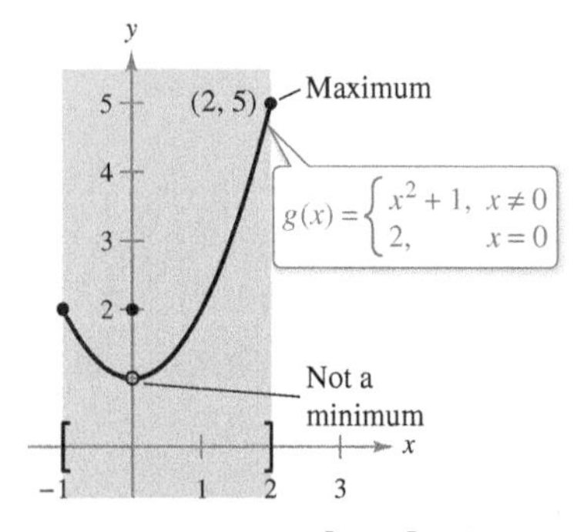

(c) g is not continuous, $[-1, 2]$ is closed.

Figure 3.1

Definition of Extrema

Let f be defined on an interval I containing c.

1. $f(c)$ is the **minimum of f on I** when $f(c) \leq f(x)$ for all x in I.
2. $f(c)$ is the **maximum of f on I** when $f(c) \geq f(x)$ for all x in I.

The minimum and maximum of a function on an interval are the **extreme values,** or **extrema** (the singular form of extrema is extremum), of the function on the interval. The minimum and maximum of a function on an interval are also called the **absolute minimum** and **absolute maximum,** or the **global minimum** and **global maximum,** on the interval. Extrema can occur at interior points or endpoints of an interval (see Figure 3.1). Extrema that occur at the endpoints are called **endpoint extrema.**

A function need not have a minimum or a maximum on an interval. For instance, in Figures 3.1(a) and (b), you can see that the function $f(x) = x^2 + 1$ has both a minimum and a maximum on the closed interval $[-1, 2]$ but does not have a maximum on the open interval $(-1, 2)$. Moreover, in Figure 3.1(c), you can see that continuity (or the lack of it) can affect the existence of an extremum on the interval. This suggests the theorem below. (Although the Extreme Value Theorem is intuitively plausible, a proof of this theorem is not within the scope of this text.)

THEOREM 3.1 The Extreme Value Theorem

If f is continuous on a closed interval $[a, b]$, then f has both a minimum and a maximum on the interval.

Exploration

Finding Minimum and Maximum Values The Extreme Value Theorem (like the Intermediate Value Theorem) is an *existence theorem* because it tells of the existence of minimum and maximum values but does not show how to find these values. Use the *minimum* and *maximum* features of a graphing utility to find the extrema of each function. In each case, do you think the x-values are exact or approximate? Explain your reasoning.

a. $f(x) = x^2 - 4x + 5$ on the closed interval $[-1, 3]$

b. $f(x) = x^3 - 2x^2 - 3x - 2$ on the closed interval $[-1, 3]$

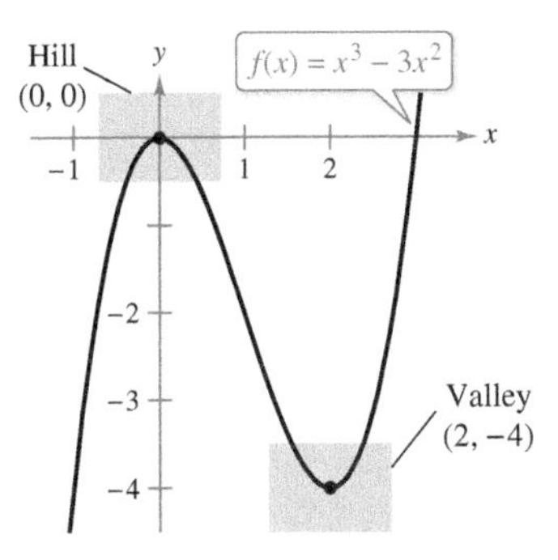

f has a relative maximum at (0, 0) and a relative minimum at (2, −4).
Figure 3.2

Relative Extrema and Critical Numbers

In Figure 3.2, the graph of $f(x) = x^3 - 3x^2$ has a **relative maximum** at the point (0, 0) and a **relative minimum** at the point $(2, -4)$. Informally, for a continuous function, you can think of a relative maximum as occurring on a "hill" on the graph, and a relative minimum as occurring in a "valley" on the graph. Such a hill and valley can occur in two ways. When the hill (or valley) is smooth and rounded, the graph has a horizontal tangent line at the high point (or low point). When the hill (or valley) is sharp and peaked, the graph represents a function that is not differentiable at the high point (or low point).

Definition of Relative Extrema

1. If there is an open interval containing c on which $f(c)$ is a maximum, then $f(c)$ is called a **relative maximum** of f, or you can say that f has a **relative maximum at $(c, f(c))$.**
2. If there is an open interval containing c on which $f(c)$ is a minimum, then $f(c)$ is called a **relative minimum** of f, or you can say that f has a **relative minimum at $(c, f(c))$.**

The plural of relative maximum is relative maxima, and the plural of relative minimum is relative minima. Relative maximum and relative minimum are sometimes called **local maximum** and **local minimum,** respectively.

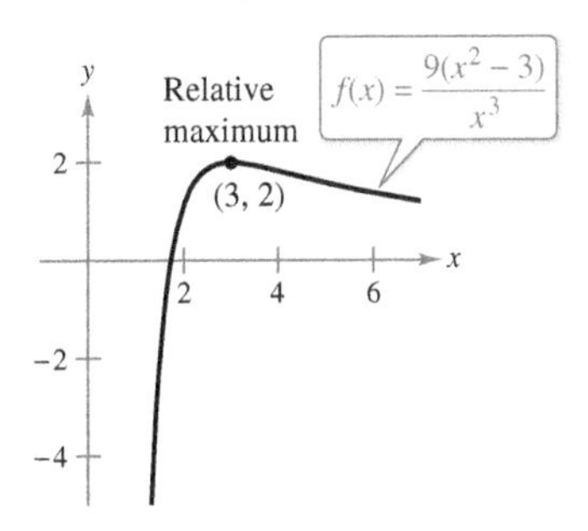

(a) $f'(3) = 0$

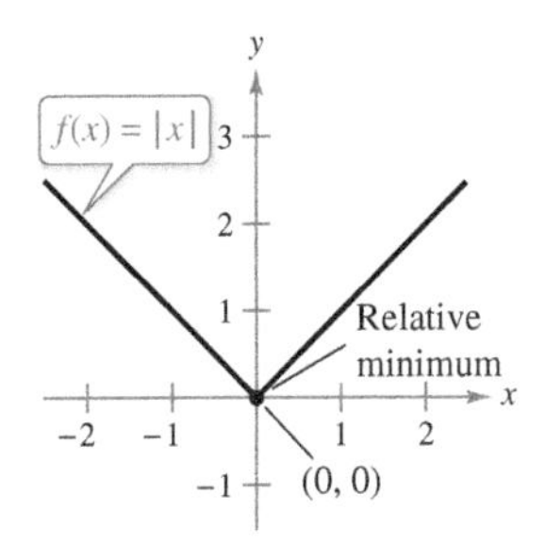

(b) $f'(0)$ does not exist.

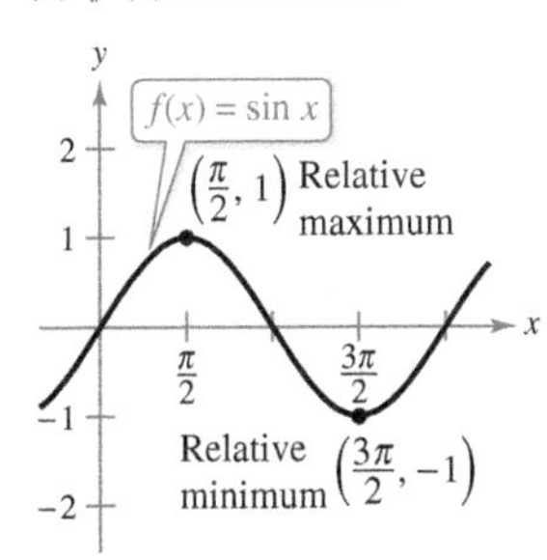

(c) $f'\left(\frac{\pi}{2}\right) = 0$; $f'\left(\frac{3\pi}{2}\right) = 0$
Figure 3.3

Example 1 examines the derivatives of functions at *given* relative extrema. (Much more is said about *finding* the relative extrema of a function in Section 3.3.)

EXAMPLE 1 The Value of the Derivative at Relative Extrema

Find the value of the derivative at each relative extremum shown in Figure 3.3.

Solution

a. The derivative of $f(x) = \dfrac{9(x^2 - 3)}{x^3}$ is

$$f'(x) = \frac{x^3(18x) - (9)(x^2 - 3)(3x^2)}{(x^3)^2} \quad \text{Differentiate using Quotient Rule.}$$

$$= \frac{9(9 - x^2)}{x^4}. \quad \text{Simplify.}$$

At the point (3, 2), the value of the derivative is $f'(3) = 0$. [See Figure 3.3(a).]

b. At $x = 0$, the derivative of $f(x) = |x|$ *does not exist* because the following one-sided limits differ. [See Figure 3.3(b).]

$$\lim_{x \to 0^-} \frac{f(x) - f(0)}{x - 0} = \lim_{x \to 0^-} \frac{|x|}{x} = -1 \quad \text{Limit from the left}$$

$$\lim_{x \to 0^+} \frac{f(x) - f(0)}{x - 0} = \lim_{x \to 0^+} \frac{|x|}{x} = 1 \quad \text{Limit from the right}$$

c. The derivative of $f(x) = \sin x$ is

$$f'(x) = \cos x.$$

At the point $(\pi/2, 1)$, the value of the derivative is $f'(\pi/2) = \cos(\pi/2) = 0$. At the point $(3\pi/2, -1)$, the value of the derivative is $f'(3\pi/2) = \cos(3\pi/2) = 0$. [See Figure 3.3(c).]

Note in Example 1 that at each relative extremum, the derivative either is zero or does not exist. The x-values at these special points are called **critical numbers.** Figure 3.4 illustrates the two types of critical numbers. Notice in the definition that the critical number c has to be in the domain of f, but c does not have to be in the domain of f'.

> **Definition of a Critical Number**
>
> Let f be defined at c. If $f'(c) = 0$ or if f is not differentiable at c, then c is a **critical number** of f.

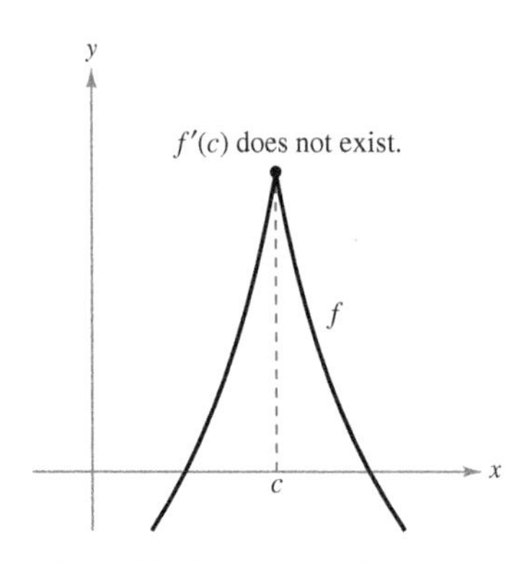

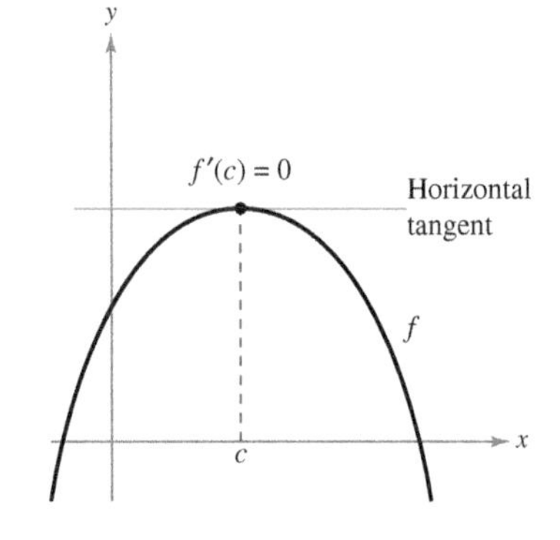

c is a critical number of f.
Figure 3.4

PIERRE DE FERMAT (1601–1665)

For Fermat, who was trained as a lawyer, mathematics was more of a hobby than a profession. Nevertheless, Fermat made many contributions to analytic geometry, number theory, calculus, and probability. In letters to friends, he wrote of many of the fundamental ideas of calculus, long before Newton or Leibniz. For instance, Theorem 3.2 is sometimes attributed to Fermat.

See LarsonCalculus.com to read more of this biography.

> THEOREM 3.2 **Relative Extrema Occur Only at Critical Numbers**
>
> If f has a relative minimum or relative maximum at $x = c$, then c is a critical number of f.
>
> 

Proof

Case 1: If f is *not* differentiable at $x = c$, then, by definition, c is a critical number of f and the theorem is valid.

Case 2: If f is differentiable at $x = c$, then $f'(c)$ must be positive, negative, or 0. Suppose $f'(c)$ is positive. Then

$$f'(c) = \lim_{x \to c} \frac{f(x) - f(c)}{x - c} > 0$$

which implies that there exists an interval (a, b) containing c such that

$$\frac{f(x) - f(c)}{x - c} > 0, \text{ for all } x \neq c \text{ in } (a, b).$$

See Exercise 84(b), Section 1.2.

Because this quotient is positive, the signs of the denominator and numerator must agree. This produces the following inequalities for x-values in the interval (a, b).

Left of c: $x < c$ and $f(x) < f(c)$ ⇨ $f(c)$ is not a relative minimum.

Right of c: $x > c$ and $f(x) > f(c)$ ⇨ $f(c)$ is not a relative maximum.

So, the assumption that $f'(c) > 0$ contradicts the hypothesis that $f(c)$ is a relative extremum. Assuming that $f'(c) < 0$ produces a similar contradiction, you are left with only one possibility—namely, $f'(c) = 0$. So, by definition, c is a critical number of f and the theorem is valid. ■

The Print Collector/Alamy Stock Photo

Finding Extrema on a Closed Interval

Theorem 3.2 states that the relative extrema of a function can occur *only* at the critical numbers of the function. Knowing this, you can use these guidelines to find extrema on a closed interval.

GUIDELINES FOR FINDING EXTREMA ON A CLOSED INTERVAL

To find the extrema of a continuous function f on a closed interval $[a, b]$, use these steps.

1. Find the critical numbers of f in (a, b).
2. Evaluate f at each critical number in (a, b).
3. Evaluate f at each endpoint of $[a, b]$.
4. The least of these values is the minimum. The greatest is the maximum.

The next three examples show how to apply these guidelines. Be sure you see that finding the critical numbers of the function is only part of the procedure. Evaluating the function at the critical numbers *and* the endpoints is the other part.

EXAMPLE 2 Finding Extrema on a Closed Interval

Find the extrema of

$$f(x) = 3x^4 - 4x^3$$

on the interval $[-1, 2]$.

Solution Begin by differentiating the function.

$$f(x) = 3x^4 - 4x^3 \quad \text{Write original function.}$$

$$f'(x) = 12x^3 - 12x^2 \quad \text{Differentiate.}$$

To find the critical numbers of f in the interval $(-1, 2)$, you must find all x-values for which $f'(x) = 0$ and all x-values for which $f'(x)$ does not exist.

$$12x^3 - 12x^2 = 0 \quad \text{Set } f'(x) \text{ equal to 0.}$$

$$12x^2(x - 1) = 0 \quad \text{Factor.}$$

$$x = 0, 1 \quad \text{Critical numbers}$$

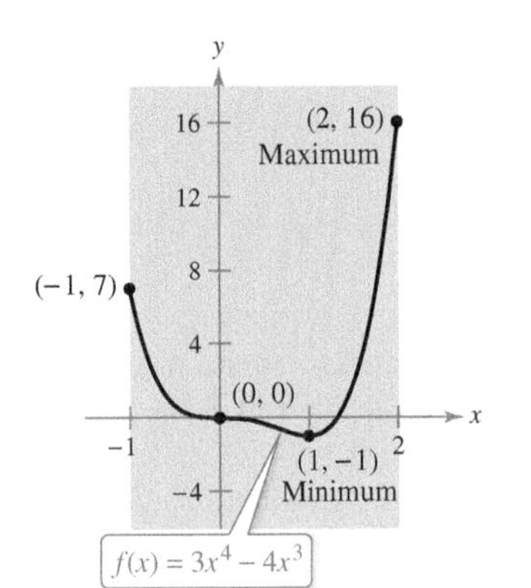

On the closed interval $[-1, 2]$, f has a minimum at $(1, -1)$ and a maximum at $(2, 16)$.
Figure 3.5

Because f' is defined for all x, you can conclude that these are the only critical numbers of f. By evaluating f at these two critical numbers and at the endpoints of $[-1, 2]$, you can determine that the maximum is $f(2) = 16$ and the minimum is $f(1) = -1$, as shown in the table. The graph of f is shown in Figure 3.5.

Left Endpoint	Critical Number	Critical Number	Right Endpoint
$f(-1) = 7$	$f(0) = 0$	$f(1) = -1$ Minimum	$f(2) = 16$ Maximum

In Figure 3.5, note that the critical number $x = 0$ does not yield a relative minimum or a relative maximum. This tells you that the converse of Theorem 3.2 is not true. In other words, *the critical numbers of a function need not produce relative extrema.*

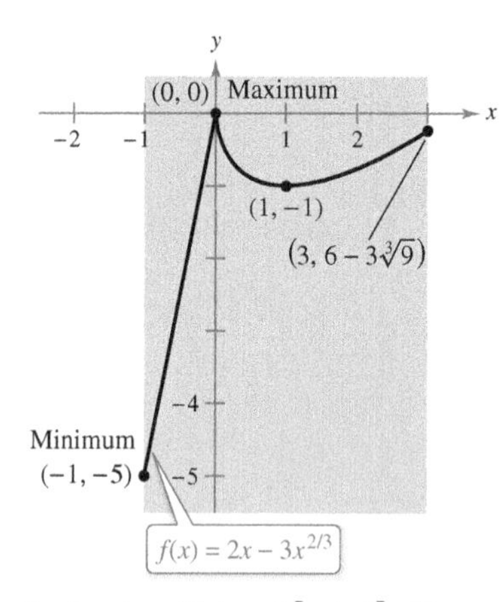

On the closed interval $[-1, 3]$, f has a minimum at $(-1, -5)$ and a maximum at $(0, 0)$.
Figure 3.6

EXAMPLE 3 Finding Extrema on a Closed Interval

Find the extrema of $f(x) = 2x - 3x^{2/3}$ on the interval $[-1, 3]$.

Solution Begin by differentiating the function.

$$f(x) = 2x - 3x^{2/3} \qquad \text{Write original function.}$$

$$f'(x) = 2 - \frac{2}{x^{1/3}} \qquad \text{Differentiate.}$$

$$= 2\left(\frac{x^{1/3} - 1}{x^{1/3}}\right) \qquad \text{Simplify.}$$

From this derivative, you can see that the function has two critical numbers in the interval $(-1, 3)$. The number 1 is a critical number because $f'(1) = 0$, and the number 0 is a critical number because $f'(0)$ does not exist. By evaluating f at these two numbers and at the endpoints of the interval, you can conclude that the minimum is $f(-1) = -5$ and the maximum is $f(0) = 0$, as shown in the table. The graph of f is shown in Figure 3.6.

Left Endpoint	Critical Number	Critical Number	Right Endpoint
$f(-1) = -5$ Minimum	$f(0) = 0$ Maximum	$f(1) = -1$	$f(3) = 6 - 3\sqrt[3]{9} \approx -0.24$

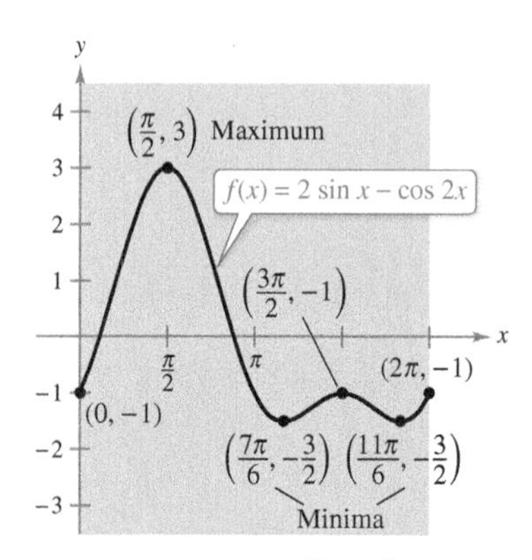

On the closed interval $[0, 2\pi]$, f has two minima at $(7\pi/6, -3/2)$ and $(11\pi/6, -3/2)$ and a maximum at $(\pi/2, 3)$.
Figure 3.7

EXAMPLE 4 Finding Extrema on a Closed Interval

See LarsonCalculus.com for an interactive version of this type of example.

Find the extrema of

$$f(x) = 2 \sin x - \cos 2x$$

on the interval $[0, 2\pi]$.

Solution Begin by differentiating the function.

$$f(x) = 2 \sin x - \cos 2x \qquad \text{Write original function.}$$

$$f'(x) = 2 \cos x + 2 \sin 2x \qquad \text{Differentiate.}$$

$$= 2 \cos x + 4 \cos x \sin x \qquad \sin 2x = 2 \cos x \sin x$$

$$= 2(\cos x)(1 + 2 \sin x) \qquad \text{Factor.}$$

Because f is differentiable for all real x, you can find all critical numbers of f by finding the zeros of its derivative. Considering $2(\cos x)(1 + 2 \sin x) = 0$ in the interval $(0, 2\pi)$, the factor $\cos x$ is zero when $x = \pi/2$ and when $x = 3\pi/2$. The factor $(1 + 2 \sin x)$ is zero when $x = 7\pi/6$ and when $x = 11\pi/6$. By evaluating f at these four critical numbers and at the endpoints of the interval, you can conclude that the maximum is $f(\pi/2) = 3$ and the minimum occurs at *two* points, $f(7\pi/6) = -3/2$ and $f(11\pi/6) = -3/2$, as shown in the table. The graph is shown in Figure 3.7.

Left Endpoint	Critical Number	Critical Number	Critical Number	Critical Number	Right Endpoint
$f(0) = -1$	$f\left(\frac{\pi}{2}\right) = 3$ Maximum	$f\left(\frac{7\pi}{6}\right) = -\frac{3}{2}$ Minimum	$f\left(\frac{3\pi}{2}\right) = -1$	$f\left(\frac{11\pi}{6}\right) = -\frac{3}{2}$ Minimum	$f(2\pi) = -1$

3.1 Exercises

See CalcChat.com for tutorial help and worked-out solutions to odd-numbered exercises.

CONCEPT CHECK

1. **Minimum** What does it mean to say that $f(c)$ is the minimum of f on an interval I?
2. **Extreme Value Theorem** In your own words, describe the Extreme Value Theorem.
3. **Maximum** What is the difference between a relative maximum and an absolute maximum on an interval I?
4. **Critical Numbers** What is a critical number?
5. **Critical Numbers** Explain how to find the critical numbers of a function.
6. **Extrema on a Closed Interval** Explain how to find the extrema of a continuous function on a closed interval $[a, b]$.

The Value of the Derivative at Relative Extrema **In Exercises 7–12, find the value of the derivative (if it exists) at each indicated extremum.**

7. $f(x) = \dfrac{x^2}{x^2 + 4}$

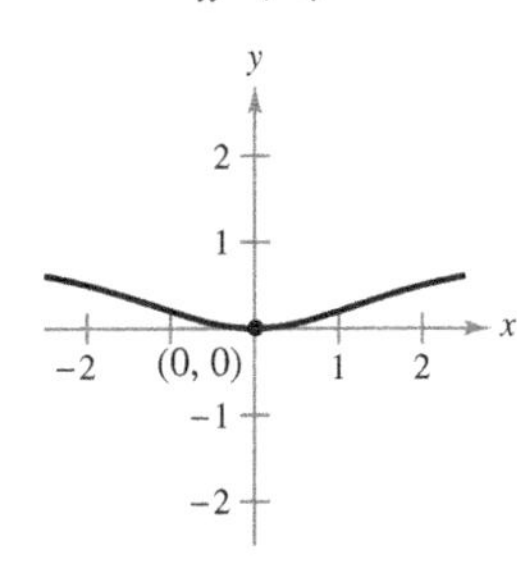

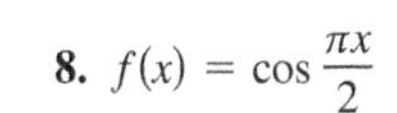

8. $f(x) = \cos \dfrac{\pi x}{2}$

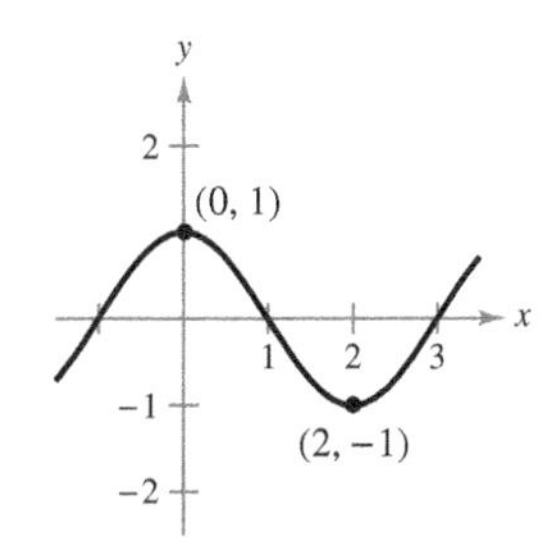

9. $g(x) = x + \dfrac{4}{x^2}$

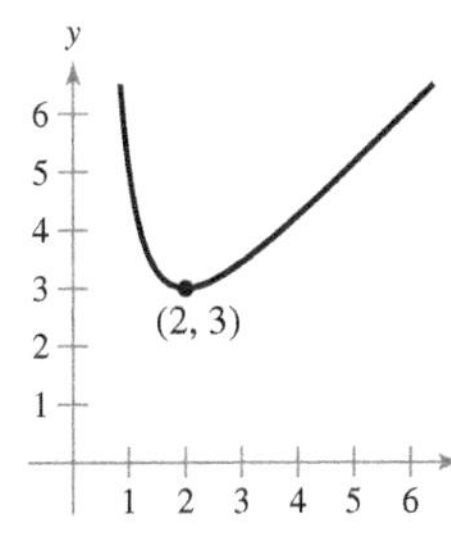

10. $f(x) = -3x\sqrt{x + 1}$

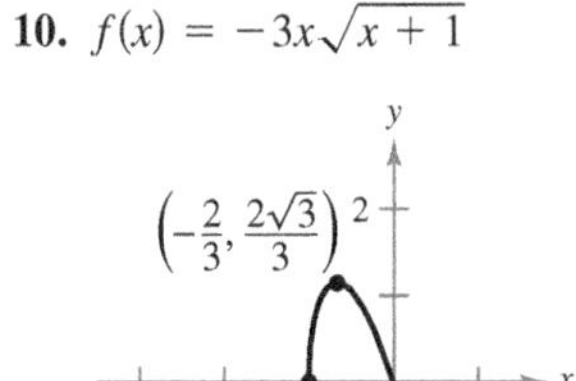

11. $f(x) = (x + 2)^{2/3}$

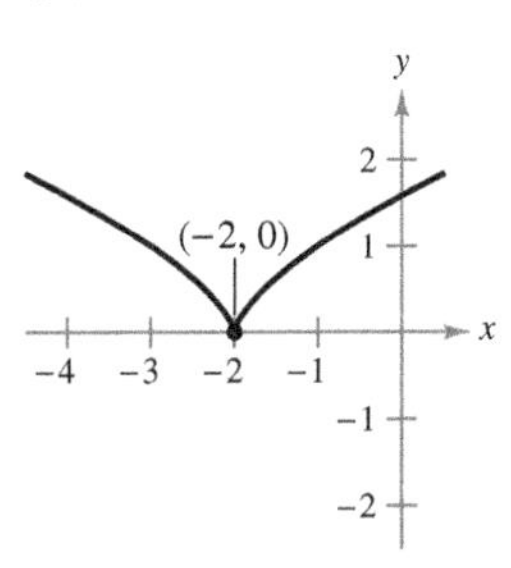

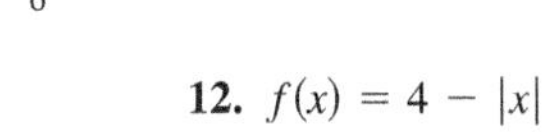

12. $f(x) = 4 - |x|$

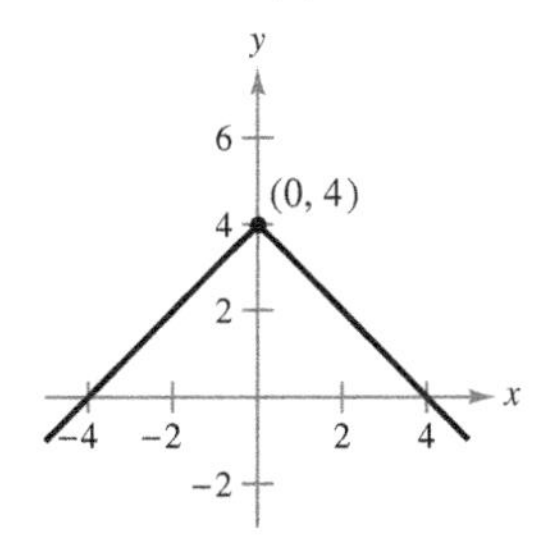

Approximating Critical Numbers **In Exercises 13–16, approximate the critical numbers of the function shown in the graph. Determine whether the function has a relative maximum, a relative minimum, an absolute maximum, an absolute minimum, or none of these at each critical number on the interval shown.**

13.

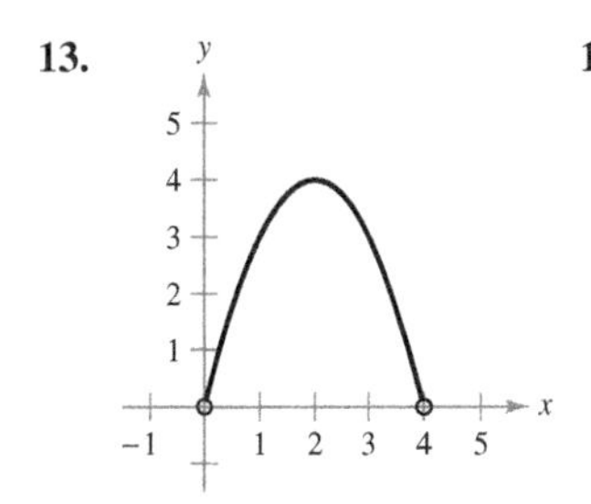

14.

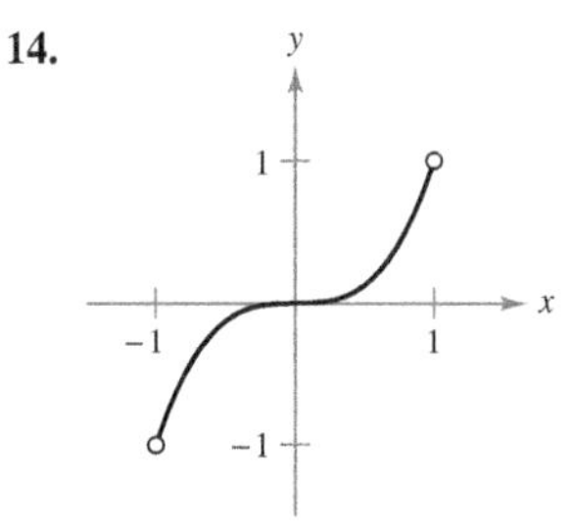

15.

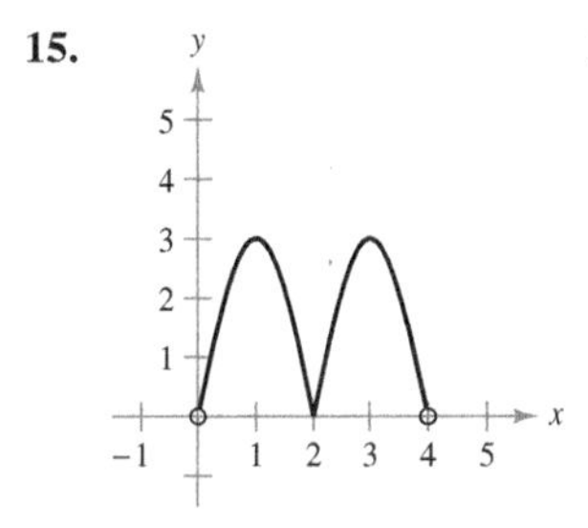

16.

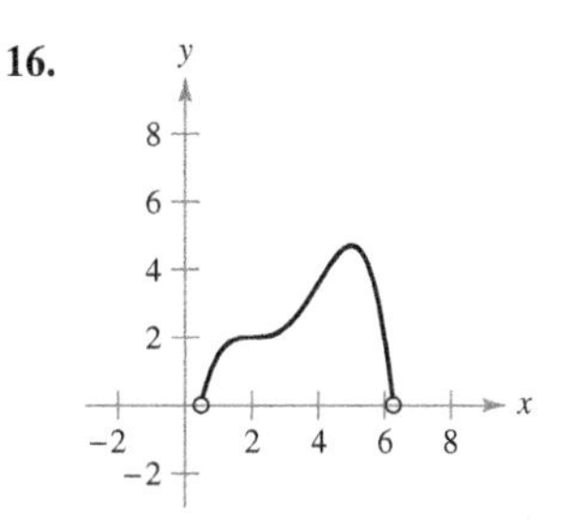

Finding Critical Numbers **In Exercises 17–22, find the critical numbers of the function.**

17. $f(x) = 4x^2 - 6x$
18. $g(x) = x - \sqrt{x}$
19. $g(t) = t\sqrt{4 - t},\ t < 3$
20. $f(x) = \dfrac{4x}{x^2 + 1}$
21. $h(x) = \sin^2 x + \cos x, \quad 0 < x < 2\pi$
22. $f(\theta) = 2 \sec \theta + \tan \theta, \quad 0 < \theta < 2\pi$

Finding Extrema on a Closed Interval **In Exercises 23–40, find the absolute extrema of the function on the closed interval.**

23. $f(x) = 3 - x,\ [-1, 2]$
24. $f(x) = \dfrac{3}{4}x + 2,\ [0, 4]$
25. $h(x) = 5 - 2x^2,\ [-3, 1]$
26. $f(x) = 7x^2 + 1,\ [-1, 2]$
27. $f(x) = x^3 - \dfrac{3}{2}x^2,\ [-1, 2]$
28. $f(x) = 2x^3 - 6x,\ [0, 3]$
29. $y = 3x^{2/3} - 2x,\ [-1, 1]$
30. $g(x) = \sqrt[3]{x},\ [-8, 8]$
31. $g(x) = \dfrac{6x^2}{x - 2},\ [-2, 1]$
32. $h(t) = \dfrac{t}{t + 3},\ [-1, 6]$
33. $y = 3 - |t - 3|,\ [-1, 5]$
34. $g(x) = |x + 4|,\ [-7, 1]$
35. $f(x) = [\![x]\!],\ [-2, 2]$
36. $h(x) = [\![2 - x]\!],\ [-2, 2]$
37. $f(x) = \sin x,\ \left[\dfrac{5\pi}{6}, \dfrac{11\pi}{6}\right]$
38. $g(x) = \sec x,\ \left[-\dfrac{\pi}{6}, \dfrac{\pi}{3}\right]$
39. $y = 3 \cos x,\ [0, 2\pi]$
40. $y = \tan \dfrac{\pi x}{8},\ [0, 2]$

Finding Extrema on an Interval In Exercises 41–44, find the absolute extrema of the function (if any exist) on each interval.

41. $f(x) = 2x - 3$
 (a) $[0, 2]$ (b) $[0, 2)$
 (c) $(0, 2]$ (d) $(0, 2)$

42. $f(x) = 5 - x$
 (a) $[1, 4]$ (b) $[1, 4)$
 (c) $(1, 4]$ (d) $(1, 4)$

43. $f(x) = x^2 - 2x$
 (a) $[-1, 2]$ (b) $(1, 3]$
 (c) $(0, 2)$ (d) $[1, 4)$

44. $f(x) = \sqrt{4 - x^2}$
 (a) $[-2, 2]$ (b) $[-2, 0)$
 (c) $(-2, 2)$ (d) $[1, 2)$

Finding Absolute Extrema Using Technology In Exercises 45–48, use a graphing utility to graph the function and find the absolute extrema of the function on the given interval.

45. $f(x) = \dfrac{3}{x - 1}$, $(1, 4]$

46. $f(x) = \dfrac{2}{2 - x}$, $[0, 2)$

47. $f(x) = \sqrt{x} + \dfrac{\sin x}{3}$, $[0, \pi]$

48. $f(x) = -x + \cos 3\pi x$, $\left[0, \dfrac{\pi}{6}\right]$

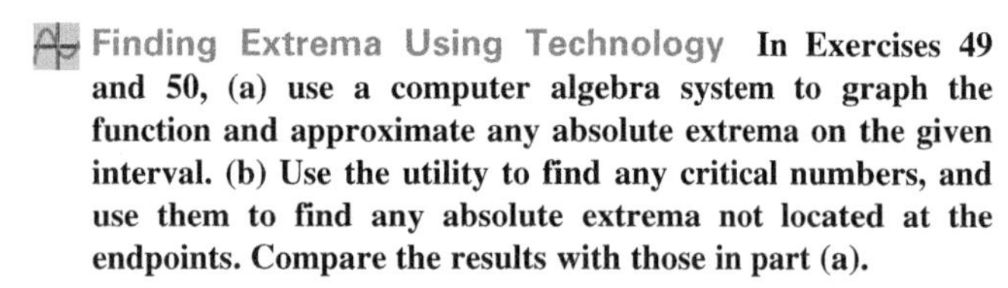

Finding Extrema Using Technology In Exercises 49 and 50, (a) use a computer algebra system to graph the function and approximate any absolute extrema on the given interval. (b) Use the utility to find any critical numbers, and use them to find any absolute extrema not located at the endpoints. Compare the results with those in part (a).

49. $f(x) = 3.2x^5 + 5x^3 - 3.5x$, $[0, 1]$

50. $f(x) = \dfrac{4}{3}x\sqrt{3 - x}$, $[0, 3]$

Finding Maximum Values Using Technology In Exercises 51 and 52, use a computer algebra system to find the maximum value of $|f''(x)|$ on the closed interval. (This value is used in the error estimate for the Trapezoidal Rule, as discussed in Section 8.6.)

51. $f(x) = \sqrt{1 + x^3}$, $[0, 2]$

52. $f(x) = \dfrac{1}{x^2 + 1}$, $\left[\dfrac{1}{2}, 3\right]$

Finding Maximum Values Using Technology In Exercises 53 and 54, use a computer algebra system to find the maximum value of $|f^{(4)}(x)|$ on the closed interval. (This value is used in the error estimate for Simpson's Rule, as discussed in Section 8.6.)

53. $f(x) = (x + 1)^{2/3}$, $[0, 2]$

54. $f(x) = \dfrac{1}{x^2 + 1}$, $[-1, 1]$

55. **Writing** Write a short paragraph explaining why a continuous function on an open interval may not have a maximum or minimum. Illustrate your explanation with a sketch of the graph of such a function.

56. **HOW DO YOU SEE IT?** Determine whether each labeled point is an absolute maximum or minimum, a relative maximum or minimum, or none of these.

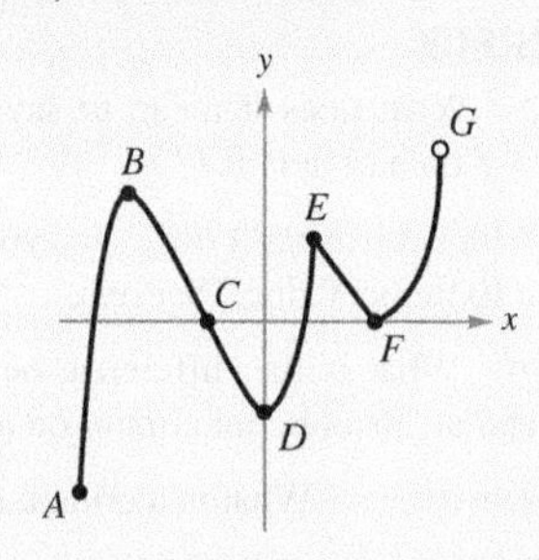

EXPLORING CONCEPTS

Using Graphs In Exercises 57 and 58, determine from the graph whether f has a minimum in the open interval (a, b). Explain your reasoning.

57. (a)

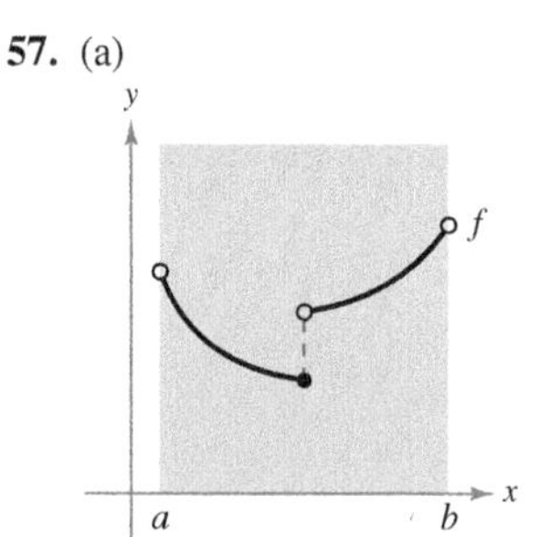

(b)

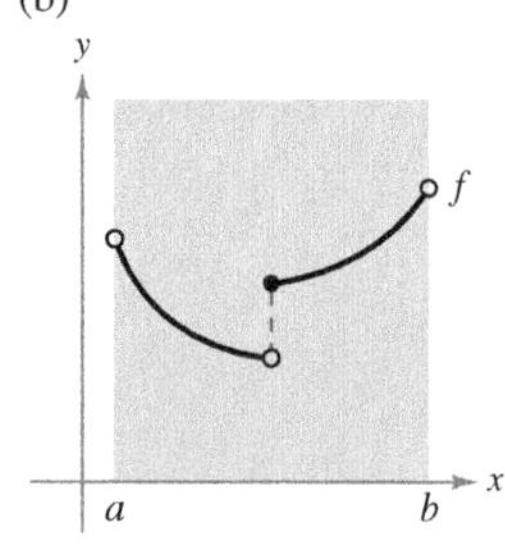

58. (a)

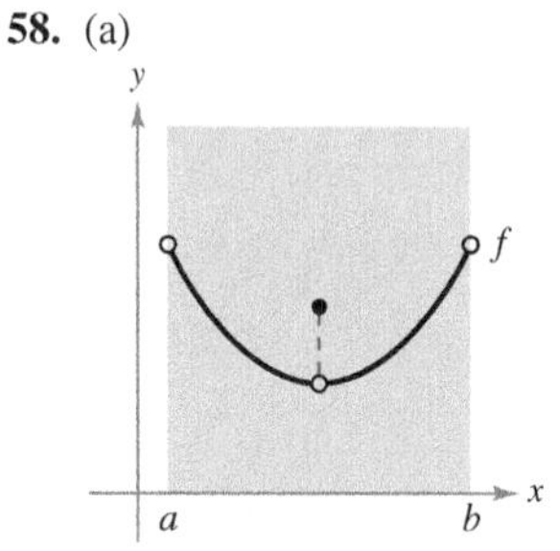

(b)

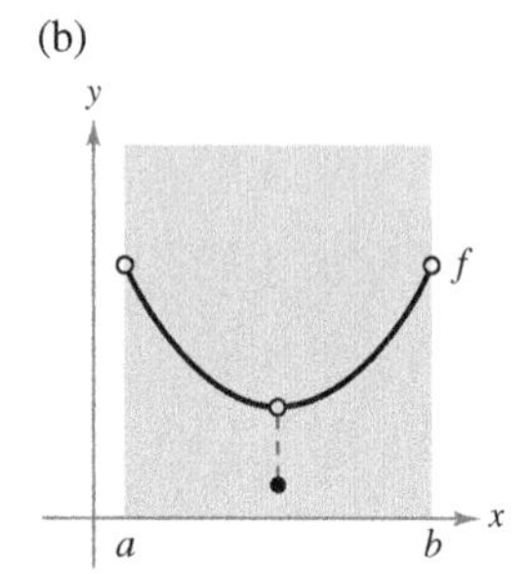

59. **Critical Numbers** Consider the function

$$f(x) = \frac{x - 4}{x + 2}.$$

Is $x = -2$ a critical number of f? Why or why not?

60. **Creating the Graph of a Function** Graph a function on the interval $[-2, 5]$ having the given characteristics.

 Relative minimum at $x = -1$
 Critical number (but no extremum) at $x = 0$
 Absolute maximum at $x = 2$
 Absolute minimum at $x = 5$

61. Power The formula for the power output P of a battery is

$$P = VI - RI^2$$

where V is the electromotive force in volts, R is the resistance in ohms, and I is the current in amperes. Find the current that corresponds to a maximum value of P in a battery for which $V = 12$ volts and $R = 0.5$ ohm. Assume that a 15-ampere fuse bounds the output in the interval $0 \le I \le 15$. Could the power output be increased by replacing the 15-ampere fuse with a 20-ampere fuse? Explain.

62. Lawn Sprinkler A lawn sprinkler is constructed in such a way that $d\theta/dt$ is constant, where θ ranges between 45° and 135° (see figure). The distance the water travels horizontally is

$$x = \frac{v^2 \sin 2\theta}{32}, \quad 45° \le \theta \le 135°$$

where v is the speed of the water. Find dx/dt and explain why this lawn sprinkler does not water evenly. What part of the lawn receives the most water?

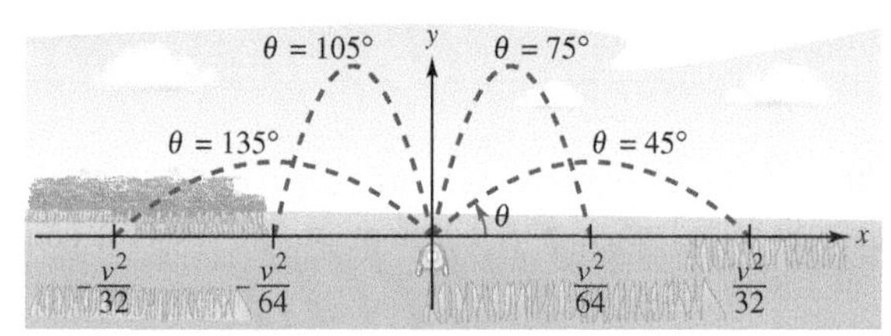

Water sprinkler: $45° \le \theta \le 135°$

FOR FURTHER INFORMATION For more information on the "calculus of lawn sprinklers," see the article "Design of an Oscillating Sprinkler" by Bart Braden in *Mathematics Magazine*. To view this article, go to *MathArticles.com*.

63. Honeycomb The surface area of a cell in a honeycomb is

$$S = 6hs + \frac{3s^2}{2}\left(\frac{\sqrt{3} - \cos \theta}{\sin \theta}\right)$$

where h and s are positive constants and θ is the angle at which the upper faces meet the altitude of the cell (see figure). Find the angle θ $(\pi/6 \le \theta \le \pi/2)$ that minimizes the surface area S.

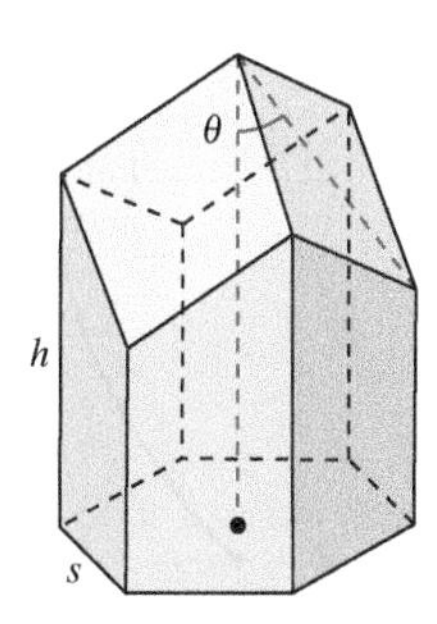

FOR FURTHER INFORMATION For more information on the geometric structure of a honeycomb cell, see the article "The Design of Honeycombs" by Anthony L. Peressini in UMAP Module 502, published by COMAP, Inc., Suite 210, 57 Bedford Street, Lexington, MA.

64. Highway Design In order to build a highway, it is necessary to fill a section of a valley where the grades (slopes) of the sides are 9% and 6% (see figure). The top of the filled region will have the shape of a parabolic arc that is tangent to the two slopes at the points A and B. The horizontal distances from A to the y-axis and from B to the y-axis are both 500 feet.

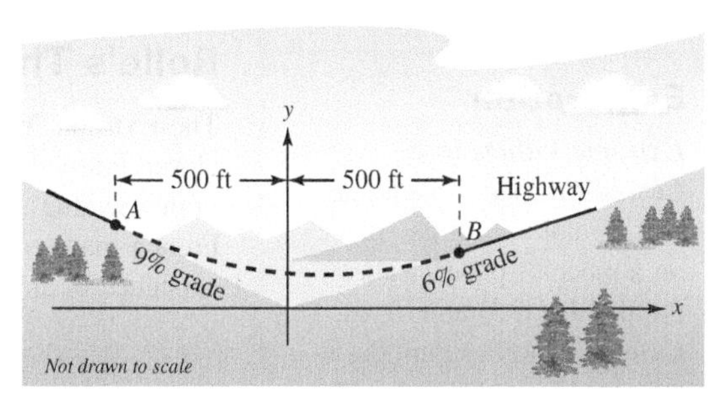

(a) Find the coordinates of A and B.

(b) Find a quadratic function $y = ax^2 + bx + c$ for $-500 \le x \le 500$ that describes the top of the filled region.

(c) Construct a table giving the depths d of the fill for $x = -500, -400, -300, -200, -100, 0, 100, 200, 300, 400$, and 500.

(d) What will be the lowest point on the completed highway? Will it be directly over the point where the two hillsides come together?

True or False? **In Exercises 65–68, determine whether the statement is true or false. If it is false, explain why or give an example that shows it is false.**

65. The maximum of $y = x^2$ on the open interval $(-3, 3)$ is 9.

66. If a function is continuous on a closed interval, then it must have a minimum on the interval.

67. If $x = c$ is a critical number of the function f, then it is also a critical number of the function $g(x) = f(x) + k$, where k is a constant.

68. If $x = c$ is a critical number of the function f, then it is also a critical number of the function $g(x) = f(x - k)$, where k is a constant.

69. Functions Let the function f be differentiable on an interval I containing c. If f has a maximum value at $x = c$, show that $-f$ has a minimum value at $x = c$.

70. Critical Numbers Consider the cubic function $f(x) = ax^3 + bx^2 + cx + d$, where $a \ne 0$. Show that f can have zero, one, or two critical numbers and give an example of each case.

PUTNAM EXAM CHALLENGE

71. Determine all real numbers $a > 0$ for which there exists a nonnegative continuous function $f(x)$ defined on $[0, a]$ with the property that the region $R = \{(x, y); 0 \le x \le a, 0 \le y \le f(x)\}$ has perimeter k units and area k square units for some real number k.

This problem was composed by the Committee on the Putnam Prize Competition.

3.2 Rolle's Theorem and the Mean Value Theorem

- **Understand and use Rolle's Theorem.**
- **Understand and use the Mean Value Theorem.**

Rolle's Theorem

The Extreme Value Theorem (see Section 3.1) states that a continuous function on a closed interval $[a, b]$ must have both a minimum and a maximum on the interval. Both of these values, however, can occur at the endpoints. **Rolle's Theorem,** named after the French mathematician Michel Rolle (1652–1719), gives conditions that guarantee the existence of an extreme value in the *interior* of a closed interval.

Exploration

Extreme Values in a Closed Interval Sketch a rectangular coordinate plane on a piece of paper. Label the points $(1, 3)$ and $(5, 3)$. Using a pencil or pen, draw the graph of a differentiable function f that starts at $(1, 3)$ and ends at $(5, 3)$. Is there at least one point on the graph for which the derivative is zero? Would it be possible to draw the graph so that there is *not* a point for which the derivative is zero? Explain your reasoning.

THEOREM 3.3 Rolle's Theorem

Let f be continuous on the closed interval $[a, b]$ and differentiable on the open interval (a, b). If $f(a) = f(b)$, then there is at least one number c in (a, b) such that $f'(c) = 0$.

Proof Let $f(a) = d = f(b)$.

Case 1: If $f(x) = d$ for all x in $[a, b]$, then f is constant on the interval and, by Theorem 2.2, $f'(x) = 0$ for all x in (a, b).

Case 2: Consider $f(x) > d$ for some x in (a, b). By the Extreme Value Theorem, you know that f has a maximum at some c in the interval. Moreover, because $f(c) > d$, this maximum does not occur at either endpoint. So, f has a maximum in the *open* interval (a, b). This implies that $f(c)$ is a *relative* maximum and, by Theorem 3.2, c is a critical number of f. Finally, because f is differentiable at c, you can conclude that $f'(c) = 0$.

Case 3: When $f(x) < d$ for some x in (a, b), you can use an argument similar to that in Case 2 but involving the minimum instead of the maximum. ■

ROLLE'S THEOREM

French mathematician Michel Rolle first published the theorem that bears his name in 1691. Before this time, however, Rolle was one of the most vocal critics of calculus, stating that it gave erroneous results and was based on unsound reasoning. Later in life, Rolle came to see the usefulness of calculus.

From Rolle's Theorem, you can see that if a function f is continuous on $[a, b]$ and differentiable on (a, b), and if $f(a) = f(b)$, then there must be at least one x-value between a and b at which the graph of f has a horizontal tangent [See Figure 3.8(a)]. When the differentiability requirement is dropped from Rolle's Theorem, f will still have a critical number in (a, b), but it may not yield a horizontal tangent. Such a case is shown in Figure 3.8(b).

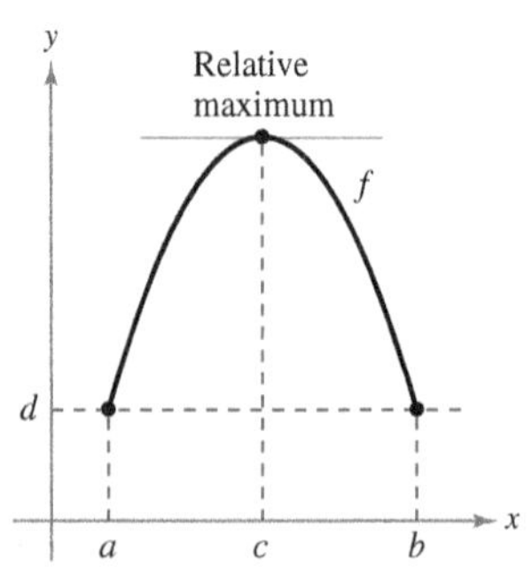

(a) f is continuous on $[a, b]$ and differentiable on (a, b).

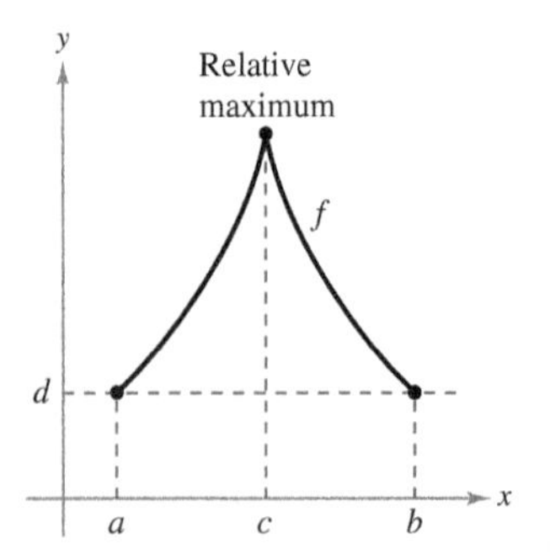

(b) f is continuous on $[a, b]$ but *not* differentiable on (a, b).

Figure 3.8

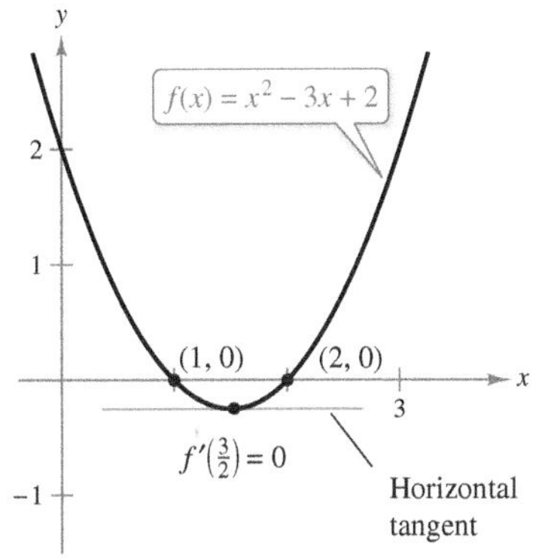

The x-value for which $f'(x) = 0$ is between the two x-intercepts.
Figure 3.9

EXAMPLE 1 Illustrating Rolle's Theorem

Find the two x-intercepts of

$$f(x) = x^2 - 3x + 2$$

and show that $f'(x) = 0$ at some point between the two x-intercepts.

Solution Note that f is differentiable on the entire real number line. Setting $f(x)$ equal to 0 produces

$$x^2 - 3x + 2 = 0 \qquad \text{Set } f(x) \text{ equal to 0.}$$
$$(x - 1)(x - 2) = 0 \qquad \text{Factor.}$$
$$x = 1, 2. \qquad \text{Solve for } x.$$

So, $f(1) = f(2) = 0$, and from Rolle's Theorem you know that there *exists* at least one c in the interval $(1, 2)$ such that $f'(c) = 0$. To *find* such a c, differentiate f to obtain

$$f'(x) = 2x - 3 \qquad \text{Differentiate.}$$

and then determine that $f'(x) = 0$ when $x = \frac{3}{2}$. Note that this x-value lies in the open interval $(1, 2)$, as shown in Figure 3.9. ■

Rolle's Theorem states that when f satisfies the conditions of the theorem, there must be *at least* one point between a and b at which the derivative is 0. There may, of course, be more than one such point, as shown in the next example.

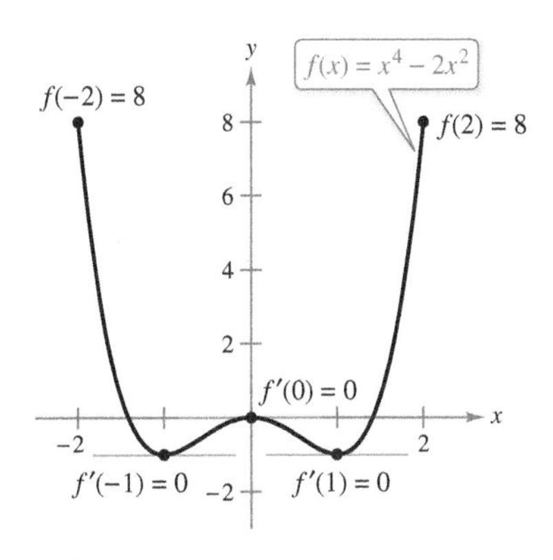

$f'(x) = 0$ for more than one x-value in the interval $(-2, 2)$.
Figure 3.10

EXAMPLE 2 Illustrating Rolle's Theorem

Let $f(x) = x^4 - 2x^2$. Find all values of c in the interval $(-2, 2)$ such that $f'(c) = 0$.

Solution To begin, note that the function satisfies the conditions of Rolle's Theorem. That is, f is continuous on the interval $[-2, 2]$ and differentiable on the interval $(-2, 2)$. Moreover, because $f(-2) = f(2) = 8$, you can conclude that there exists at least one c in $(-2, 2)$ such that $f'(c) = 0$. Because

$$f'(x) = 4x^3 - 4x \qquad \text{Differentiate.}$$

setting the derivative equal to 0 produces

$$4x^3 - 4x = 0 \qquad \text{Set } f'(x) \text{ equal to 0.}$$
$$4x(x - 1)(x + 1) = 0 \qquad \text{Factor.}$$
$$x = 0, 1, -1. \qquad x\text{-values for which } f'(x) = 0$$

So, in the interval $(-2, 2)$, the derivative is zero when $x = -1, 0,$ and 1, as shown in Figure 3.10. ■

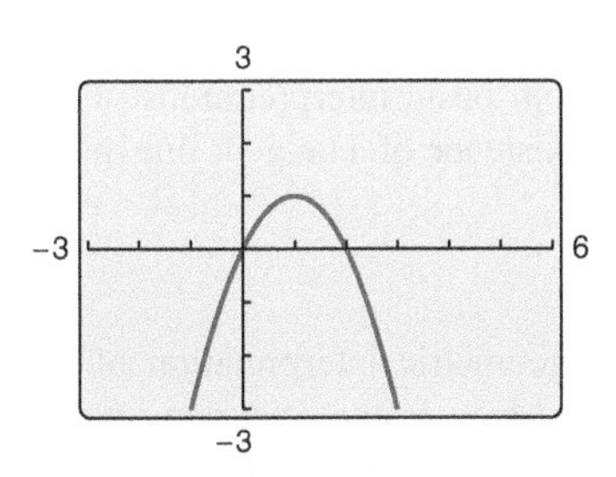

Figure 3.11

▷ **TECHNOLOGY PITFALL** A graphing utility can be used to indicate whether the points on the graphs in Examples 1 and 2 are relative minima or relative maxima of the functions. When using a graphing utility, however, you should keep in mind that it can give misleading pictures of graphs. For example, use a graphing utility to graph

$$f(x) = 1 - (x - 1)^2 - \frac{1}{1000(x - 1)^{1/7} + 1}.$$

With most viewing windows, it appears that the function has a maximum of 1 when $x = 1$, as shown in Figure 3.11. By evaluating the function at $x = 1$, however, you can see that $f(1) = 0$. To determine the behavior of this function near $x = 1$, you need to examine the graph analytically to get the complete picture.

The Mean Value Theorem

Rolle's Theorem can be used to prove another theorem—the **Mean Value Theorem.**

REMARK The "mean" in the Mean Value Theorem refers to the mean (or average) rate of change of f on the interval $[a, b]$.

THEOREM 3.4 The Mean Value Theorem

If f is continuous on the closed interval $[a, b]$ and differentiable on the open interval (a, b), then there exists a number c in (a, b) such that

$$f'(c) = \frac{f(b) - f(a)}{b - a}.$$

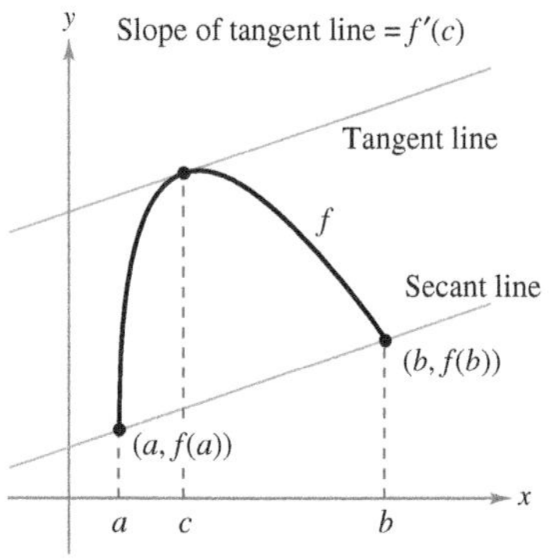

Figure 3.12

Proof Refer to Figure 3.12. The equation of the secant line that passes through the points $(a, f(a))$ and $(b, f(b))$ is

$$y = \left[\frac{f(b) - f(a)}{b - a}\right](x - a) + f(a).$$

Let $g(x)$ be the difference between $f(x)$ and y. Then

$$\begin{aligned} g(x) &= f(x) - y \\ &= f(x) - \left[\frac{f(b) - f(a)}{b - a}\right](x - a) - f(a). \end{aligned}$$

By evaluating g at a and b, you can see that

$$g(a) = 0 = g(b).$$

Because f is continuous on $[a, b]$, it follows that g is also continuous on $[a, b]$. Furthermore, because f is differentiable, g is also differentiable, and you can apply Rolle's Theorem to the function g. So, there exists a number c in (a, b) such that $g'(c) = 0$, which implies that

$$\begin{aligned} g'(c) &= 0 \\ f'(c) - \frac{f(b) - f(a)}{b - a} &= 0. \end{aligned}$$

So, there exists a number c in (a, b) such that

$$f'(c) = \frac{f(b) - f(a)}{b - a}.$$

JOSEPH-LOUIS LAGRANGE (1736–1813)

The Mean Value Theorem was first proved by the famous mathematician Joseph-Louis Lagrange. Born in Italy, Lagrange held a position in the court of Frederick the Great in Berlin for 20 years.

See LarsonCalculus.com to read more of this biography.

Although the Mean Value Theorem can be used directly in problem solving, it is used more often to prove other theorems. In fact, some people consider this to be the most important theorem in calculus—it is closely related to the Fundamental Theorem of Calculus discussed in Section 4.4. For now, you can get an idea of the versatility of the Mean Value Theorem by looking at the results stated in Exercises 77–85 in this section.

The Mean Value Theorem has implications for both basic interpretations of the derivative. Geometrically, the theorem guarantees the existence of a tangent line that is parallel to the secant line through the points

$$(a, f(a)) \quad \text{and} \quad (b, f(b)).$$

as shown in Figure 3.12. Example 3 illustrates this geometric interpretation of the Mean Value Theorem. In terms of rates of change, the Mean Value Theorem implies that there must be a point in the open interval (a, b) at which the instantaneous rate of change is equal to the average rate of change over the interval $[a, b]$. This is illustrated in Example 4.

EXAMPLE 3 Finding a Tangent Line

See LarsonCalculus.com for an interactive version of this type of example.

For $f(x) = 5 - (4/x)$, find all values of c in the open interval $(1, 4)$ such that

$$f'(c) = \frac{f(4) - f(1)}{4 - 1}.$$

Solution The slope of the secant line through $(1, f(1))$ and $(4, f(4))$ is

$$\frac{f(4) - f(1)}{4 - 1} = \frac{4 - 1}{4 - 1} = 1. \qquad \text{Slope of secant line}$$

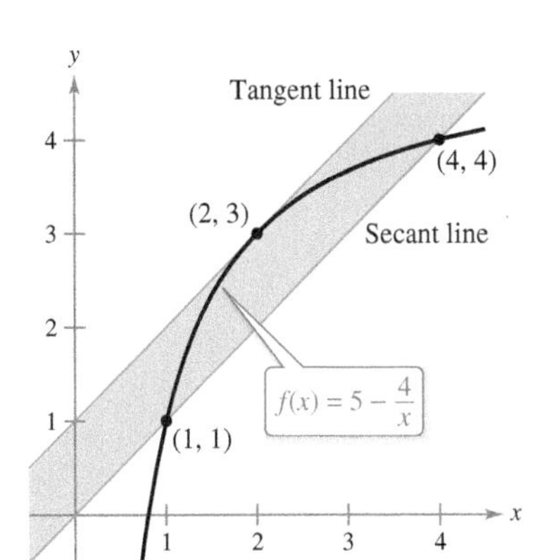

The tangent line at (2, 3) is parallel to the secant line through (1, 1) and (4, 4).
Figure 3.13

Note that the function satisfies the conditions of the Mean Value Theorem. That is, f is continuous on the interval $[1, 4]$ and differentiable on the interval $(1, 4)$. So, there exists at least one number c in $(1, 4)$ such that $f'(c) = 1$. Solving the equation $f'(x) = 1$ yields

$$\frac{4}{x^2} = 1 \qquad \text{Set } f'(x) \text{ equal to 1.}$$

which implies that

$$x = \pm 2.$$

So, in the interval $(1, 4)$, you can conclude that $c = 2$, as shown in Figure 3.13.

EXAMPLE 4 Finding an Instantaneous Rate of Change

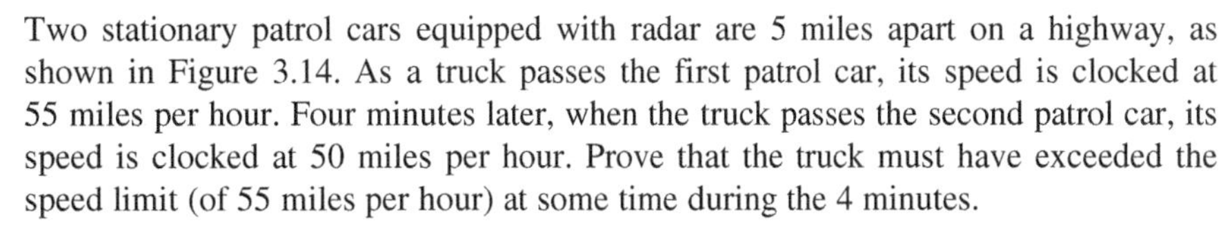

Two stationary patrol cars equipped with radar are 5 miles apart on a highway, as shown in Figure 3.14. As a truck passes the first patrol car, its speed is clocked at 55 miles per hour. Four minutes later, when the truck passes the second patrol car, its speed is clocked at 50 miles per hour. Prove that the truck must have exceeded the speed limit (of 55 miles per hour) at some time during the 4 minutes.

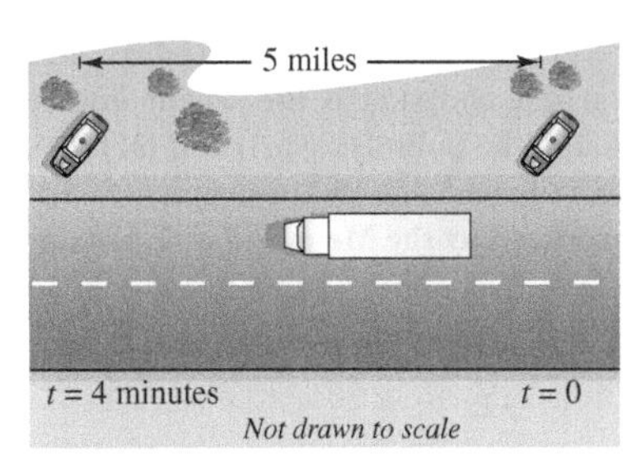

At some time t, the instantaneous velocity is equal to the average velocity over 4 minutes.
Figure 3.14

Solution Let $t = 0$ be the time (in hours) when the truck passes the first patrol car. The time when the truck passes the second patrol car is

$$t = \frac{4}{60} = \frac{1}{15} \text{ hour.}$$

By letting $s(t)$ represent the distance (in miles) traveled by the truck, you have $s(0) = 0$ and $s(\frac{1}{15}) = 5$. So, the average velocity of the truck over the five-mile stretch of highway is

$$\text{Average velocity} = \frac{s(1/15) - s(0)}{(1/15) - 0} = \frac{5}{1/15} = 75 \text{ miles per hour.}$$

Assuming that the position function is differentiable, you can apply the Mean Value Theorem to conclude that the truck must have been traveling at a rate of 75 miles per hour sometime during the 4 minutes. ■

A useful alternative form of the Mean Value Theorem is: If f is continuous on $[a, b]$ and differentiable on (a, b), then there exists a number c in (a, b) such that

$$f(b) = f(a) + (b - a)f'(c). \qquad \text{Alternative form of Mean Value Theorem}$$

When doing the exercises for this section, keep in mind that polynomial functions, rational functions, and trigonometric functions are differentiable at all points in their domains.

3.2 Exercises

See CalcChat.com for tutorial help and worked-out solutions to odd-numbered exercises.

CONCEPT CHECK

1. **Rolle's Theorem** In your own words, describe Rolle's Theorem.
2. **Mean Value Theorem** In your own words, describe the Mean Value Theorem.

Writing In Exercises 3–6, explain why Rolle's Theorem does not apply to the function even though there exist a and b such that $f(a) = f(b)$.

3. $f(x) = \left|\frac{1}{x}\right|, \; [-1, 1]$
4. $f(x) = \cot \frac{x}{2}, \; [\pi, 3\pi]$
5. $f(x) = 1 - |x - 1|, \; [0, 2]$
6. $f(x) = \sqrt{(2 - x^{2/3})^3}, \; [-1, 1]$

Using Rolle's Theorem In Exercises 7–10, find the two x-intercepts of the function f and show that $f'(x) = 0$ at some point between the two x-intercepts.

7. $f(x) = x^2 - x - 2$
8. $f(x) = x^2 + 6x$
9. $f(x) = x\sqrt{x + 4}$
10. $f(x) = -3x\sqrt{x + 1}$

Using Rolle's Theorem In Exercises 11–24, determine whether Rolle's Theorem can be applied to f on the closed interval $[a, b]$. If Rolle's Theorem can be applied, find all values of c in the open interval (a, b) such that $f'(c) = 0$. If Rolle's Theorem cannot be applied, explain why not.

11. $f(x) = -x^2 + 3x, \; [0, 3]$
12. $f(x) = x^2 - 8x + 5, \; [2, 6]$
13. $f(x) = (x - 1)(x - 2)(x - 3), \; [1, 3]$
14. $f(x) = (x - 4)(x + 2)^2, \; [-2, 4]$
15. $f(x) = x^{2/3} - 1, \; [-8, 8]$
16. $f(x) = 3 - |x - 3|, \; [0, 6]$
17. $f(x) = \frac{x^2 - 2x - 3}{x + 2}, \; [-1, 3]$
18. $f(x) = \frac{x^2 - 4}{x - 1}, \; [-2, 2]$
19. $f(x) = \sin x, \; [0, 2\pi]$
20. $f(x) = \cos x, \; [\pi, 3\pi]$
21. $f(x) = \cos \pi x, \; [0, 2]$
22. $f(x) = \sin 3x, \; \left[\frac{\pi}{2}, \frac{7\pi}{6}\right]$
23. $f(x) = \tan x, \; [0, \pi]$
24. $f(x) = \sec x, \; [\pi, 2\pi]$

Using Rolle's Theorem In Exercises 25–28, use a graphing utility to graph the function on the closed interval $[a, b]$. Determine whether Rolle's Theorem can be applied to f on the interval and, if so, find all values of c in the open interval (a, b) such that $f'(c) = 0$.

25. $f(x) = |x| - 1, \; [-1, 1]$
26. $f(x) = x - x^{1/3}, \; [0, 1]$
27. $f(x) = \frac{x}{2} - \sin \frac{\pi x}{6}, \; [-1, 0]$
28. $f(x) = x - \tan \pi x, \; \left[-\frac{1}{4}, \frac{1}{4}\right]$

29. **Vertical Motion** The height of a ball t seconds after it is thrown upward from a height of 6 feet and with an initial velocity of 48 feet per second is

$$f(t) = -16t^2 + 48t + 6.$$

(a) Verify that $f(1) = f(2)$.

(b) According to Rolle's Theorem, what must the velocity be at some time in the interval $(1, 2)$? Find that time.

30. **Reorder Costs** The ordering and transportation cost C for components used in a manufacturing process is approximated by

$$C(x) = 10\left(\frac{1}{x} + \frac{x}{x + 3}\right)$$

where C is measured in thousands of dollars and x is the order size in hundreds.

(a) Verify that $C(3) = C(6)$.

(b) According to Rolle's Theorem, the rate of change of the cost must be 0 for some order size in the interval $(3, 6)$. Find that order size.

Mean Value Theorem In Exercises 31 and 32, copy the graph and sketch the secant line to the graph through the points $(a, f(a))$ and $(b, f(b))$. Then sketch any tangent lines to the graph for each value of c guaranteed by the Mean Value Theorem. To print an enlarged copy of the graph, go to *MathGraphs.com*.

31.

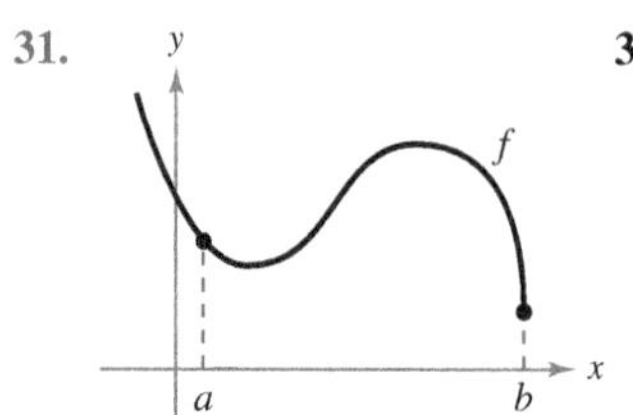

32.

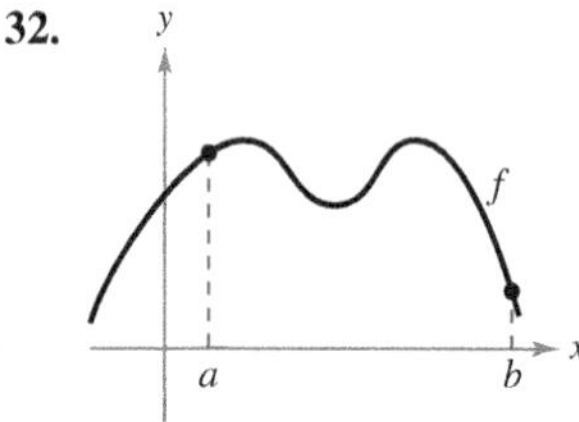

Writing In Exercises 33–36, explain why the Mean Value Theorem does not apply to the function f on the interval $[0, 6]$.

33.

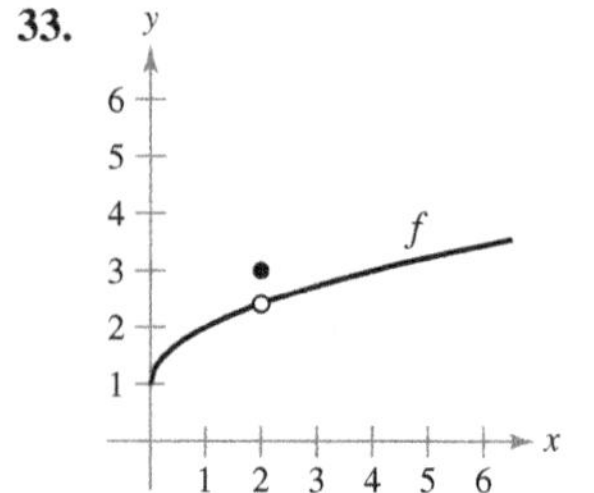

34.

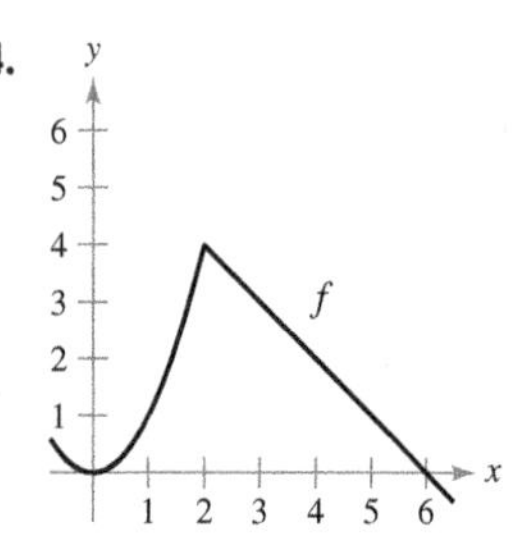

35. $f(x) = \frac{1}{x - 3}$

36. $f(x) = |x - 3|$

37. Mean Value Theorem Consider the graph of the function $f(x) = -x^2 + 5$ (see figure).

(a) Find the equation of the secant line joining the points $(-1, 4)$ and $(2, 1)$.

(b) Use the Mean Value Theorem to determine a point c in the interval $(-1, 2)$ such that the tangent line at c is parallel to the secant line.

(c) Find the equation of the tangent line through c.

(d) Use a graphing utility to graph f, the secant line, and the tangent line.

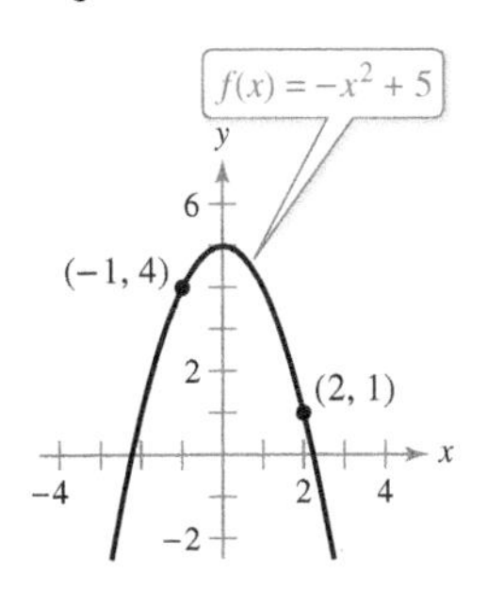

Figure for 37

Figure for 38

38. Mean Value Theorem Consider the graph of the function $f(x) = x^2 - x - 12$ (see figure).

(a) Find the equation of the secant line joining the points $(-2, -6)$ and $(4, 0)$.

(b) Use the Mean Value Theorem to determine a point c in the interval $(-2, 4)$ such that the tangent line at c is parallel to the secant line.

(c) Find the equation of the tangent line through c.

(d) Use a graphing utility to graph f, the secant line, and the tangent line.

Using the Mean Value Theorem In Exercises 39–48, determine whether the Mean Value Theorem can be applied to f on the closed interval $[a, b]$. If the Mean Value Theorem can be applied, find all values of c in the open interval (a, b) such that

$$f'(c) = \frac{f(b) - f(a)}{b - a}.$$

If the Mean Value Theorem cannot be applied, explain why not.

39. $f(x) = 6x^3, \quad [1, 2]$

40. $f(x) = x^6, \quad [-1, 1]$

41. $f(x) = x^3 + 2x + 4, \quad [-1, 0]$

42. $f(x) = x^3 - 3x^2 + 9x + 5, \quad [0, 1]$

43. $f(x) = \frac{x + 2}{x - 1}, \quad [-3, 3]$

44. $f(x) = \frac{x}{x - 5}, \quad [1, 4]$

45. $f(x) = |2x + 1|, \quad [-1, 3]$

46. $f(x) = \sqrt{2 - x}, \quad [-7, 2]$

47. $f(x) = \sin x, \quad [0, \pi]$

48. $f(x) = \cos x + \tan x, \quad [0, \pi]$

Using the Mean Value Theorem In Exercises 49–52, use a graphing utility to (a) graph the function f on the given interval, (b) find and graph the secant line through points on the graph of f at the endpoints of the given interval, and (c) find and graph any tangent lines to the graph of f that are parallel to the secant line.

49. $f(x) = \frac{x}{x + 1}, \quad \left[-\frac{1}{2}, 2\right]$

50. $f(x) = x - 2 \sin x, \quad [-\pi, \pi]$

51. $f(x) = \sqrt{x}, \quad [1, 9]$

52. $f(x) = x^4 - 2x^3 + x^2, \quad [0, 6]$

53. Vertical Motion The height of an object t seconds after it is dropped from a height of 300 meters is

$$s(t) = -4.9t^2 + 300.$$

(a) Find the average velocity of the object during the first 3 seconds.

(b) Use the Mean Value Theorem to verify that at some time during the first 3 seconds of fall, the instantaneous velocity equals the average velocity. Find that time.

54. Sales A company introduces a new product for which the number of units sold S is

$$S(t) = 200\left(5 - \frac{9}{2 + t}\right)$$

where t is the time in months.

(a) Find the average rate of change of S during the first year.

(b) During what month of the first year does $S'(t)$ equal the average rate of change?

EXPLORING CONCEPTS

55. Converse of Rolle's Theorem Let f be continuous on $[a, b]$ and differentiable on (a, b). If there exists c in (a, b) such that $f'(c) = 0$, does it follow that $f(a) = f(b)$? Explain.

56. Rolle's Theorem Let f be continuous on $[a, b]$ and differentiable on (a, b). Also, suppose that $f(a) = f(b)$ and that c is a real number in the interval (a, b) such that $f'(c) = 0$. Find an interval for the function g over which Rolle's Theorem can be applied, and find the corresponding critical number of g, where k is a constant.

(a) $g(x) = f(x) + k$ (b) $g(x) = f(x - k)$

(c) $g(x) = f(kx)$

57. Rolle's Theorem The function

$$f(x) = \begin{cases} 0, & x = 0 \\ 1 - x, & 0 < x \le 1 \end{cases}$$

is differentiable on $(0, 1)$ and satisfies $f(0) = f(1)$. However, its derivative is never zero on $(0, 1)$. Does this contradict Rolle's Theorem? Explain.

58. Mean Value Theorem Can you find a function f such that $f(-2) = -2$, $f(2) = 6$, and $f'(x) < 1$ for all x? Why or why not?

59. Speed A plane begins its takeoff at 2:00 P.M. on a 2500-mile flight. After 5.5 hours, the plane arrives at its destination. Explain why there are at least two times during the flight when the speed of the plane is 400 miles per hour.

60. Temperature When an object is removed from a furnace and placed in an environment with a constant temperature of 90°F, its core temperature is 1500°F. Five hours later, the core temperature is 390°F. Explain why there must exist a time in the interval $(0, 5)$ when the temperature is decreasing at a rate of 222°F per hour.

61. Velocity Two bicyclists begin a race at 8:00 A.M. They both finish the race 2 hours and 15 minutes later. Prove that at some time during the race, the bicyclists are traveling at the same velocity.

62. Acceleration At 9:13 A.M., a sports car is traveling 35 miles per hour. Two minutes later, the car is traveling 85 miles per hour. Prove that at some time during this two-minute interval, the car's acceleration is exactly 1500 miles per hour squared.

63. Think About It Sketch the graph of an arbitrary function f that satisfies the given condition but does not satisfy the conditions of the Mean Value Theorem on the interval $[-5, 5]$.

(a) f is continuous. (b) f is not continuous.

64. HOW DO YOU SEE IT? The figure shows two parts of the graph of a continuous differentiable function f on $[-10, 4]$. The derivative f' is also continuous. To print an enlarged copy of the graph, go to *MathGraphs.com*.

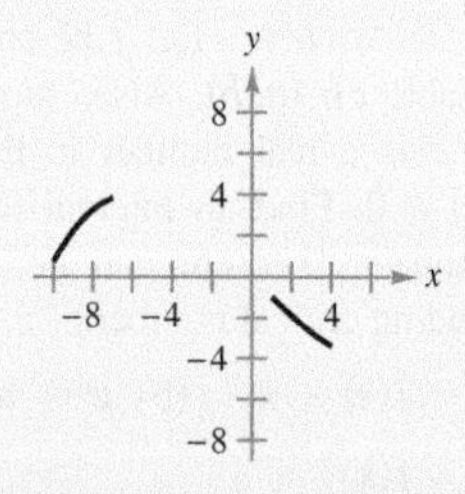

(a) Explain why f must have at least one zero in $[-10, 4]$.

(b) Explain why f' must also have at least one zero in the interval $[-10, 4]$. What are these zeros called?

(c) Make a possible sketch of the function, where f' has one zero on the interval $[-10, 4]$.

Finding a Solution In Exercises 65–68, use the Intermediate Value Theorem and Rolle's Theorem to prove that the equation has exactly one real solution.

65. $x^5 + x^3 + x + 1 = 0$

66. $2x^5 + 7x - 1 = 0$

67. $3x + 1 - \sin x = 0$

68. $2x - 2 - \cos x = 0$

Using a Derivative In Exercises 69–72, find a function f that has the derivative $f'(x)$ and whose graph passes through the given point. Explain your reasoning.

69. $f'(x) = 0, \quad (2, 5)$

70. $f'(x) = 4, \quad (0, 1)$

71. $f'(x) = 2x, \quad (1, 0)$

72. $f'(x) = 6x - 1, \quad (2, 7)$

True or False? In Exercises 73–76, determine whether the statement is true or false. If it is false, explain why or give an example that shows it is false.

73. The Mean Value Theorem can be applied to

$$f(x) = \frac{1}{x}$$

on the interval $[-1, 1]$.

74. If the graph of a function has three x-intercepts, then it must have at least two points at which its tangent line is horizontal.

75. If the graph of a polynomial function has three x-intercepts, then it must have at least two points at which its tangent line is horizontal.

76. The Mean Value Theorem can be applied to $f(x) = \tan x$ on the interval $[0, \pi/4]$.

77. Proof Prove that if $a > 0$ and n is any positive integer, then the polynomial function $p(x) = x^{2n+1} + ax + b$ cannot have two real roots.

78. Proof Prove that if $f'(x) = 0$ for all x in an interval (a, b), then f is constant on (a, b).

79. Proof Let $p(x) = Ax^2 + Bx + C$. Prove that for any interval $[a, b]$, the value c guaranteed by the Mean Value Theorem is the midpoint of the interval.

80. Using Rolle's Theorem

(a) Let $f(x) = x^2$ and $g(x) = -x^3 + x^2 + 3x + 2$. Then $f(-1) = g(-1)$ and $f(2) = g(2)$. Show that there is at least one value c in the interval $(-1, 2)$ where the tangent line to f at $(c, f(c))$ is parallel to the tangent line to g at $(c, g(c))$. Identify c.

(b) Let f and g be differentiable functions on $[a, b]$, where $f(a) = g(a)$ and $f(b) = g(b)$. Show that there is at least one value c in the interval (a, b) where the tangent line to f at $(c, f(c))$ is parallel to the tangent line to g at $(c, g(c))$.

81. Proof Prove that if f is differentiable on $(-\infty, \infty)$ and $f'(x) < 1$ for all real numbers, then f has at most one fixed point. [A *fixed point* of a function f is a real number c such that $f(c) = c$.]

82. Fixed Point Use the result of Exercise 81 to show that $f(x) = \frac{1}{2}\cos x$ has at most one fixed point.

83. Proof Prove that $|\cos a - \cos b| \le |a - b|$ for all a and b.

84. Proof Prove that $|\sin a - \sin b| \le |a - b|$ for all a and b.

85. Using the Mean Value Theorem Let $0 < a < b$. Use the Mean Value Theorem to show that

$$\sqrt{b} - \sqrt{a} < \frac{b - a}{2\sqrt{a}}.$$

3.3 Increasing and Decreasing Functions and the First Derivative Test

- **Determine intervals on which a function is increasing or decreasing.**
- **Apply the First Derivative Test to find relative extrema of a function.**

Increasing and Decreasing Functions

In this section, you will learn how derivatives can be used to *classify* relative extrema as either relative minima or relative maxima. First, it is important to define increasing and decreasing functions.

> **Definitions of Increasing and Decreasing Functions**
>
> A function f is **increasing** on an interval when, for any two numbers x_1 and x_2 in the interval, $x_1 < x_2$ implies $f(x_1) < f(x_2)$.
>
> A function f is **decreasing** on an interval when, for any two numbers x_1 and x_2 in the interval, $x_1 < x_2$ implies $f(x_1) < f(x_2)$.

A function is increasing when, *as x moves to the right*, its graph moves up, and is decreasing when its graph moves down. For example, the function in Figure 3.15 is decreasing on the interval $(-\infty, a)$, is constant on the interval (a, b), and is increasing on the interval (b, ∞). As shown in Theorem 3.5 below, a positive derivative implies that the function is increasing, a negative derivative implies that the function is decreasing, and a zero derivative on an entire interval implies that the function is constant on that interval.

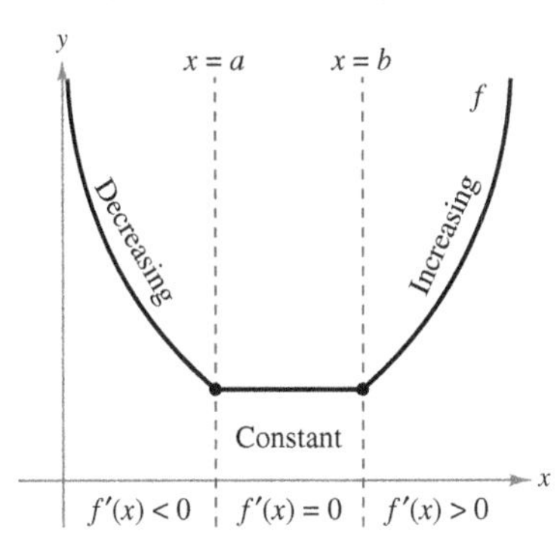

The derivative is related to the slope of a function.
Figure 3.15

> **THEOREM 3.5 Test for Increasing and Decreasing Functions**
>
> Let f be a function that is continuous on the closed interval $[a, b]$ and differentiable on the open interval (a, b).
>
> 1. If $f'(x) > 0$ for all x in (a, b), then f is increasing on $[a, b]$.
> 2. If $f'(x) < 0$ for all x in (a, b), then f is decreasing on $[a, b]$.
> 3. If $f'(x) = 0$ for all x in (a, b), then f is constant on $[a, b]$.
>
> 

REMARK The conclusions in the first two cases of Theorem 3.5 are valid even when $f'(x) = 0$ at a finite number of x-values in (a, b).

Proof To prove the first case, assume that $f'(x) > 0$ for all x in the interval (a, b) and let $x_1 < x_2$ be any two points in the interval. By the Mean Value Theorem, you know that there exists a number c such that $x_1 < c < x_2$, and

$$f'(c) = \frac{f(x_2) - f(x_1)}{x_2 - x_1}.$$

Because $f'(c) > 0$ and $x_2 - x_1 > 0$, you know that $f(x_2) - f(x_1) > 0$, which implies that $f(x_1) < f(x_2)$. So, f is increasing on the interval. The second case has a similar proof (see Exercise 97), and the third case is a consequence of Exercise 78 in Section 3.2. ■

EXAMPLE 1 Intervals on Which f Is Increasing or Decreasing

Find the open intervals on which $f(x) = x^3 - \frac{3}{2}x^2$ is increasing or decreasing.

Solution Note that f is differentiable on the entire real number line and the derivative of f is

$$f(x) = x^3 - \tfrac{3}{2}x^2 \qquad \text{Write original function.}$$

$$f'(x) = 3x^2 - 3x. \qquad \text{Differentiate.}$$

To determine the critical numbers of f, set $f'(x)$ equal to zero.

$$3x^2 - 3x = 0 \qquad \text{Set } f'(x) \text{ equal to 0.}$$

$$3(x)(x - 1) = 0 \qquad \text{Factor.}$$

$$x = 0, 1 \qquad \text{Critical numbers}$$

Because there are no points for which f' does not exist, you can conclude that $x = 0$ and $x = 1$ are the only critical numbers. The table summarizes the testing of the three intervals determined by these two critical numbers.

Figure 3.16

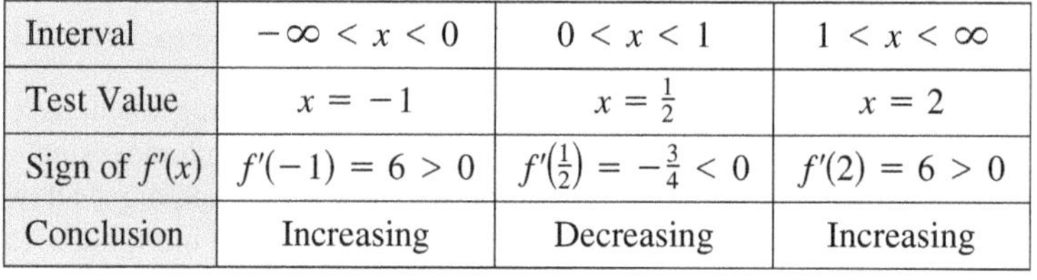

Interval	$-\infty < x < 0$	$0 < x < 1$	$1 < x < \infty$
Test Value	$x = -1$	$x = \frac{1}{2}$	$x = 2$
Sign of $f'(x)$	$f'(-1) = 6 > 0$	$f'(\frac{1}{2}) = -\frac{3}{4} < 0$	$f'(2) = 6 > 0$
Conclusion	Increasing	Decreasing	Increasing

By Theorem 3.5, f is increasing on the intervals $(-\infty, 0)$ and $(1, \infty)$ and decreasing on the interval $(0, 1)$, as shown in Figure 3.16.

Example 1 gives you one instance of how to find intervals on which a function is increasing or decreasing. The guidelines below summarize the steps followed in that example.

GUIDELINES FOR FINDING INTERVALS ON WHICH A FUNCTION IS INCREASING OR DECREASING

Let f be continuous on the interval (a, b). To find the open intervals on which f is increasing or decreasing, use the following steps.

1. Locate the critical numbers of f in (a, b), and use these numbers to determine test intervals.
2. Determine the sign of $f'(x)$ at one test value in each of the intervals.
3. Use Theorem 3.5 to determine whether f is increasing or decreasing on each interval.

These guidelines are also valid when the interval (a, b) is replaced by an interval of the form $(-\infty, b)$, (a, ∞), or $(-\infty, \infty)$.

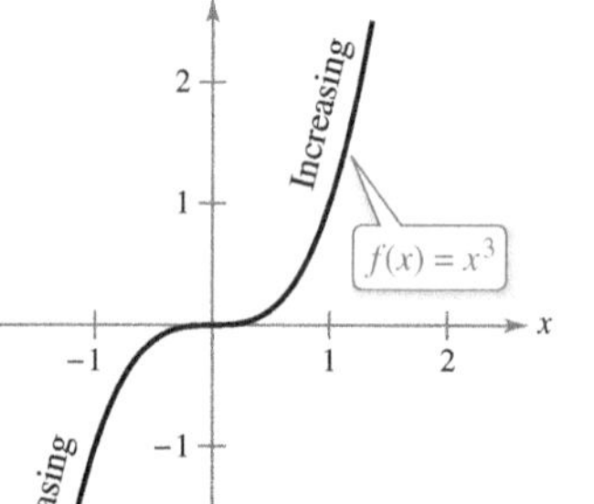

(a) Strictly monotonic function

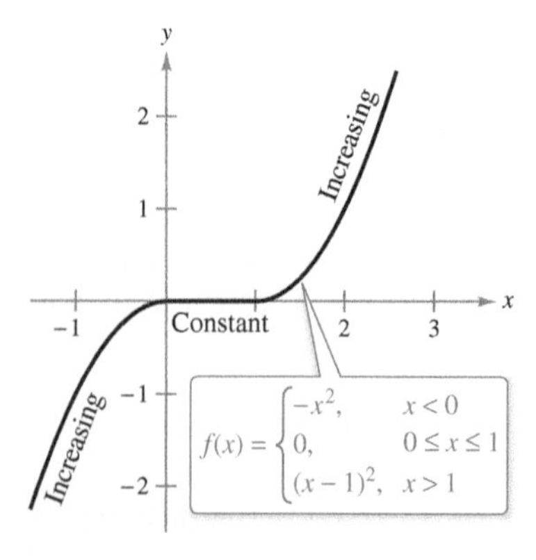

(b) Not strictly monotonic

Figure 3.17

A function is **strictly monotonic** on an interval when it is either increasing on the entire interval or decreasing on the entire interval. For instance, the function $f(x) = x^3$ is strictly monotonic on the entire real number line because it is increasing on the entire real number line, as shown in Figure 3.17(a). The function shown in Figure 3.17(b) is not strictly monotonic on the entire real number line because it is constant on the interval $[0, 1]$.

The First Derivative Test

After you have determined the intervals on which a function is increasing or decreasing, it is not difficult to locate the relative extrema of the function. For instance, in Figure 3.18 (from Example 1), the function

$$f(x) = x^3 - \frac{3}{2}x^2$$

has a relative maximum at the point $(0, 0)$ because f is increasing immediately to the left of $x = 0$ and decreasing immediately to the right of $x = 0$. Similarly, f has a relative minimum at the point $\left(1, -\frac{1}{2}\right)$ because f is decreasing immediately to the left of $x = 1$ and increasing immediately to the right of $x = 1$. The next theorem makes this more explicit.

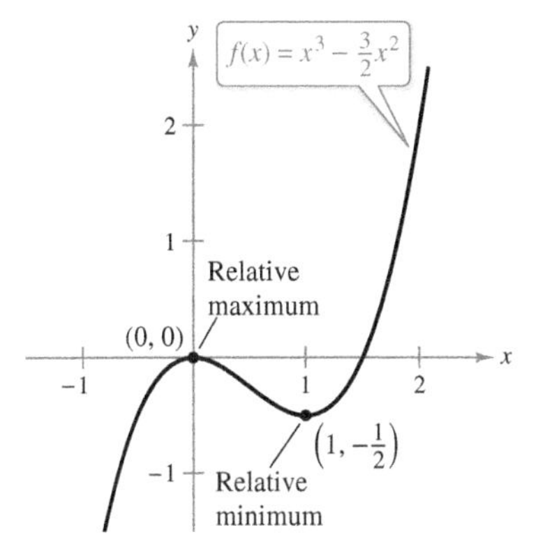

Relative extrema of f
Figure 3.18

THEOREM 3.6 The First Derivative Test

Let c be a critical number of a function f that is continuous on an open interval I containing c. If f is differentiable on the interval, except possibly at c, then $f(c)$ can be classified as follows.

1. If $f'(x)$ changes from negative to positive at c, then f has a *relative minimum* at $(c, f(c))$.
2. If $f'(x)$ changes from positive to negative at c, then f has a *relative maximum* at $(c, f(c))$.
3. If $f'(x)$ is positive on both sides of c or negative on both sides of c, then $f(c)$ is neither a relative minimum nor a relative maximum.

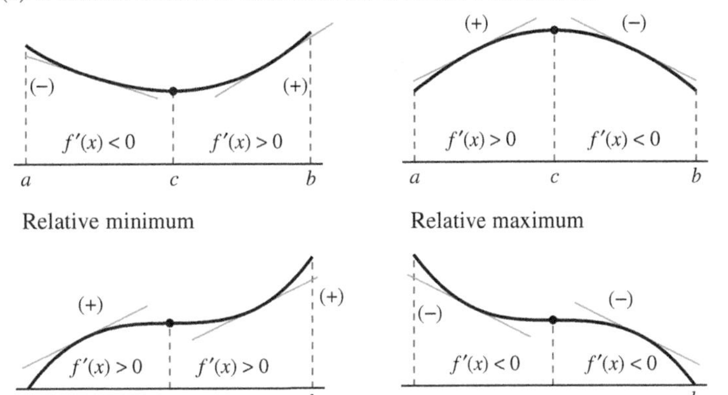

Proof Assume that $f'(x)$ changes from negative to positive at c. Then there exist a and b in I such that

$$f'(x) < 0 \text{ for all } x \text{ in } (a, c) \quad \text{and} \quad f'(x) > 0 \text{ for all } x \text{ in } (c, b).$$

By Theorem 3.5, f is decreasing on $[a, c]$ and increasing on $[c, b]$. So, $f(c)$ is a minimum of f on the open interval (a, b) and, consequently, a relative minimum of f. This proves the first case of the theorem. The second case can be proved in a similar way (see Exercise 98). ∎

EXAMPLE 2 Applying the First Derivative Test

Find the relative extrema of $f(x) = \frac{1}{2}x - \sin x$ in the interval $(0, 2\pi)$.

Solution Note that f is continuous on the interval $(0, 2\pi)$. The derivative of f is $f'(x) = \frac{1}{2} - \cos x$. To determine the critical numbers of f in this interval, set $f'(x)$ equal to 0.

$$\frac{1}{2} - \cos x = 0 \qquad \text{Set } f'(x) \text{ equal to 0.}$$

$$\cos x = \frac{1}{2}$$

$$x = \frac{\pi}{3}, \frac{5\pi}{3} \qquad \text{Critical numbers}$$

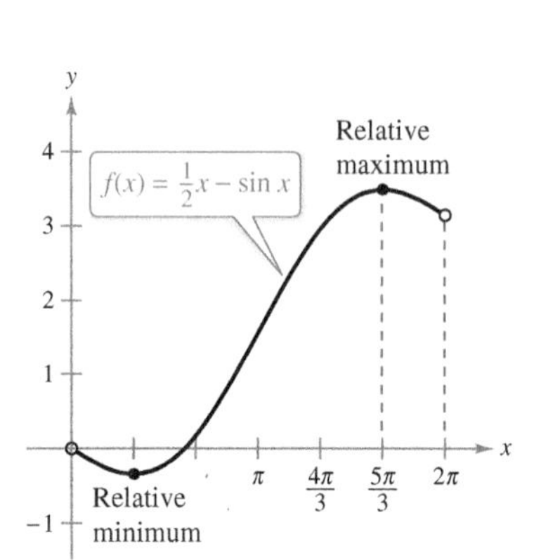

A relative minimum occurs where f changes from decreasing to increasing, and a relative maximum occurs where f changes from increasing to decreasing.
Figure 3.19

Because there are no points for which f' does not exist, you can conclude that $x = \pi/3$ and $x = 5\pi/3$ are the only critical numbers. The table summarizes the testing of the three intervals determined by these two critical numbers. By applying the First Derivative Test, you can conclude that f has a relative minimum at the point where $x = \pi/3$ and a relative maximum at the point where $x = 5\pi/3$, as shown in Figure 3.19.

Interval	$0 < x < \frac{\pi}{3}$	$\frac{\pi}{3} < x < \frac{5\pi}{3}$	$\frac{5\pi}{3} < x < 2\pi$
Test Value	$x = \frac{\pi}{4}$	$x = \pi$	$x = \frac{7\pi}{4}$
Sign of $f'(x)$	$f'\left(\frac{\pi}{4}\right) < 0$	$f'(\pi) > 0$	$f'\left(\frac{7\pi}{4}\right) < 0$
Conclusion	Decreasing	Increasing	Decreasing

EXAMPLE 3 Applying the First Derivative Test

Find the relative extrema of $f(x) = (x^2 - 4)^{2/3}$.

Solution Begin by noting that f is continuous on the entire real number line. The derivative of f

$$f'(x) = \frac{2}{3}(x^2 - 4)^{-1/3}(2x) \qquad \text{General Power Rule}$$

$$= \frac{4x}{3(x^2 - 4)^{1/3}} \qquad \text{Simplify.}$$

is 0 when $x = 0$ and does not exist when $x = \pm 2$. So, the critical numbers are $x = -2$, $x = 0$, and $x = 2$. The table summarizes the testing of the four intervals determined by these three critical numbers. By applying the First Derivative Test, you can conclude that f has a relative minimum at the point $(-2, 0)$, a relative maximum at the point $(0, \sqrt[3]{16})$, and another relative minimum at the point $(2, 0)$, as shown in Figure 3.20.

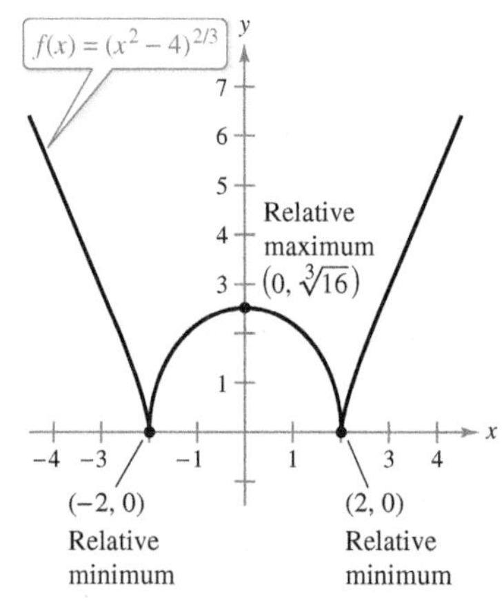

Figure 3.20

Interval	$-\infty < x < -2$	$-2 < x < 0$	$0 < x < 2$	$2 < x < \infty$
Test Value	$x = -3$	$x = -1$	$x = 1$	$x = 3$
Sign of $f'(x)$	$f'(-3) < 0$	$f'(-1) > 0$	$f'(1) < 0$	$f'(3) > 0$
Conclusion	Decreasing	Increasing	Decreasing	Increasing

Note that in Examples 1 and 2, the given functions are differentiable on the entire real number line. For such functions, the only critical numbers are those for which $f'(x) = 0$. Example 3 concerns a function that has two types of critical numbers—those for which $f'(x) = 0$ and those for which f is not differentiable.

When using the First Derivative Test, be sure to consider the domain of the function. For instance, in the next example, the function

$$f(x) = \frac{x^4 + 1}{x^2}$$

is not defined when $x = 0$. This x-value must be used with the critical numbers to determine the test intervals.

EXAMPLE 4 Applying the First Derivative Test

See LarsonCalculus.com for an interactive version of this type of example.

Find the relative extrema of $f(x) = \dfrac{x^4 + 1}{x^2}$.

Solution Note that f is not defined when $x = 0$.

$$f(x) = x^2 + x^{-2} \quad \text{Rewrite original function.}$$

$$f'(x) = 2x - 2x^{-3} \quad \text{Differentiate.}$$

$$= 2x - \frac{2}{x^3} \quad \text{Rewrite with positive exponent.}$$

$$= \frac{2(x^4 - 1)}{x^3} \quad \text{Simplify.}$$

$$= \frac{2(x^2 + 1)(x - 1)(x + 1)}{x^3} \quad \text{Factor.}$$

So, $f'(x)$ is zero at $x = \pm 1$. Moreover, because $x = 0$ is not in the domain of f, you should use this x-value along with the critical numbers to determine the test intervals.

$x = \pm 1$ Critical numbers, $f'(\pm 1) = 0$

$x = 0$ 0 is not in the domain of f.

The table summarizes the testing of the four intervals determined by these three x-values. By applying the First Derivative Test, you can conclude that f has one relative minimum at the point $(-1, 2)$ and another at the point $(1, 2)$, as shown in Figure 3.21.

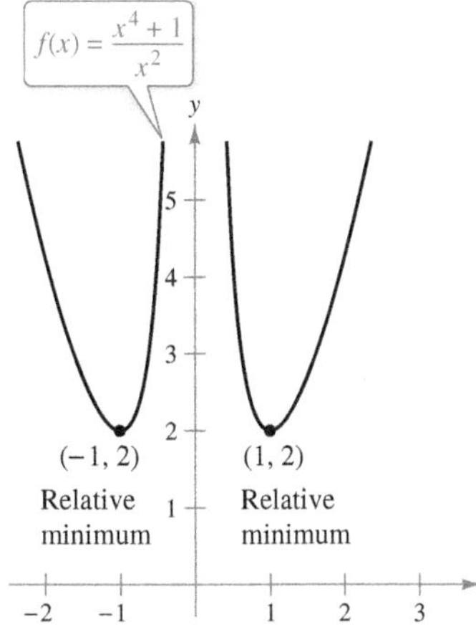

x-values that are not in the domain of f, as well as critical numbers, determine test intervals for f'.
Figure 3.21

Interval	$-\infty < x < -1$	$-1 < x < 0$	$0 < x < 1$	$1 < x < \infty$
Test Value	$x = -2$	$x = -\frac{1}{2}$	$x = \frac{1}{2}$	$x = 2$
Sign of $f'(x)$	$f'(-2) < 0$	$f'(-\frac{1}{2}) > 0$	$f'(\frac{1}{2}) < 0$	$f'(2) > 0$
Conclusion	Decreasing	Increasing	Decreasing	Increasing

TECHNOLOGY The most difficult step in applying the First Derivative Test is finding the values for which the derivative is equal to 0. For instance, the values of x for which the derivative of

$$f(x) = \frac{x^4 + 1}{x^2 + 1}$$

is equal to zero are $x = 0$ and $x = \pm\sqrt{\sqrt{2} - 1}$. If you have access to technology that can perform symbolic differentiation and solve equations, use it to apply the First Derivative Test to this function.

When a projectile is propelled from ground level and air resistance is neglected, the object will travel farthest with an initial angle of 45°. When, however, the projectile is propelled from a point above ground level, the angle that yields a maximum horizontal distance is not 45° (see Example 5).

EXAMPLE 5 The Path of a Projectile

Neglecting air resistance, the path of a projectile that is propelled at an angle θ is

$$y = -\frac{g \sec^2 \theta}{2v_0^2}x^2 + (\tan \theta)x + h, \quad 0 \le \theta \le \frac{\pi}{2}$$

where y is the height, x is the horizontal distance, g is the acceleration due to gravity, v_0 is the initial velocity, and h is the initial height. (This equation is derived in Section 12.3.) Let $g = 32$ feet per second per second, $v_0 = 24$ feet per second, and $h = 9$ feet. What value of θ will produce a maximum horizontal distance?

Solution To find the distance the projectile travels, let $y = 0$, $g = 32$, $v_0 = 24$, and $h = 9$. Then substitute these values in the given equation as shown.

$$-\frac{g \sec^2 \theta}{2v_0^2}x^2 + (\tan \theta)x + h = y$$

$$-\frac{32 \sec^2 \theta}{2(24^2)}x^2 + (\tan \theta)x + 9 = 0$$

$$-\frac{\sec^2 \theta}{36}x^2 + (\tan \theta)x + 9 = 0$$

Next, solve for x using the Quadratic Formula with $a = (-\sec^2 \theta)/36$, $b = \tan \theta$, and $c = 9$.

$$x = \frac{-b \pm \sqrt{b^2 - 4ac}}{2a}$$

$$x = \frac{-\tan \theta \pm \sqrt{(\tan \theta)^2 - 4[(-\sec^2 \theta)/36](9)}}{2[(-\sec^2 \theta)/36]}$$

$$x = \frac{-\tan \theta \pm \sqrt{\tan^2 \theta + \sec^2 \theta}}{(-\sec^2 \theta)/18}$$

$$x = 18(\cos \theta)\left(\sin \theta + \sqrt{\sin^2 \theta + 1}\right), \quad x \ge 0$$

At this point, you need to find the value of θ that produces a maximum value of x. Applying the First Derivative Test by hand would be very tedious. Using technology to solve the equation $dx/d\theta = 0$, however, eliminates most of the messy computations. The result is that the maximum value of x occurs when

$$\theta \approx 0.61548 \text{ radian}, \quad \text{or} \quad 35.3°.$$

This conclusion is reinforced by sketching the path of the projectile for different values of θ, as shown in Figure 3.22. Of the three paths shown, note that the distance traveled is greatest for $\theta = 35°$.

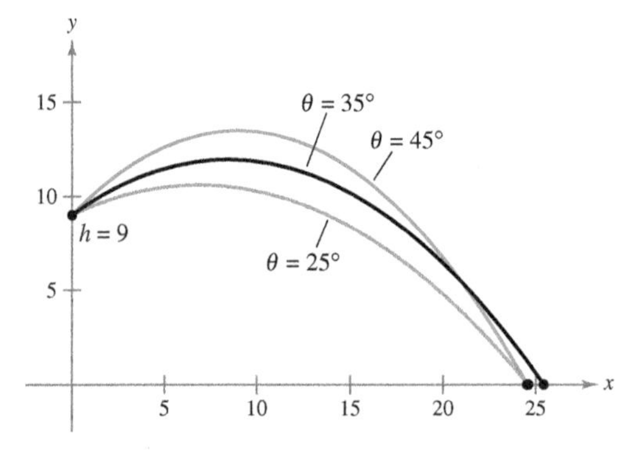

The path of a projectile with initial angle θ
Figure 3.22

3.3 Exercises

See CalcChat.com for tutorial help and worked-out solutions to odd-numbered exercises.

CONCEPT CHECK

1. **Increasing and Decreasing Functions** Describe the Test for Increasing and Decreasing Functions in your own words.
2. **First Derivative Test** Describe the First Derivative Test in your own words.

Using a Graph **In Exercises 3 and 4, use the graph of f to find (a) the largest open interval on which f is increasing and (b) the largest open interval on which f is decreasing.**

3.

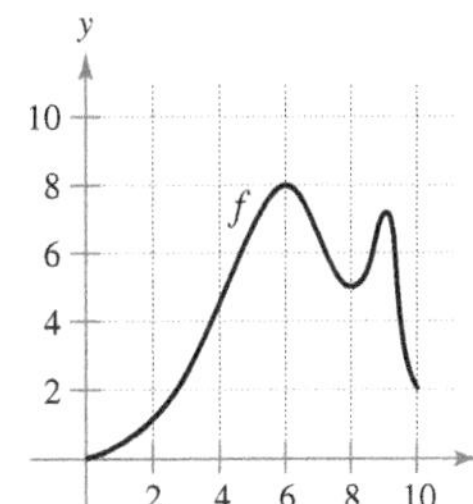

4.

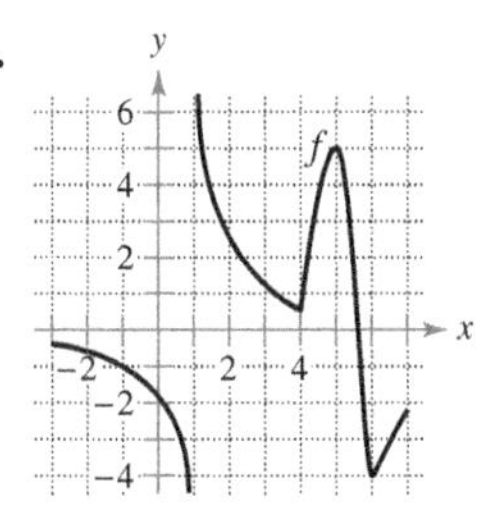

Using a Graph **In Exercises 5–10, use the graph to estimate the open intervals on which the function is increasing or decreasing. Then find the open intervals analytically.**

5. $y = -(x + 1)^2$

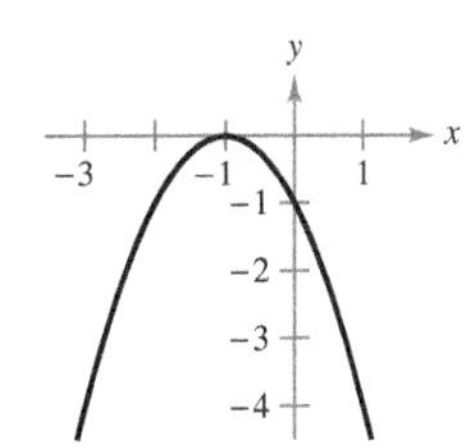

6. $f(x) = x^2 - 6x + 8$

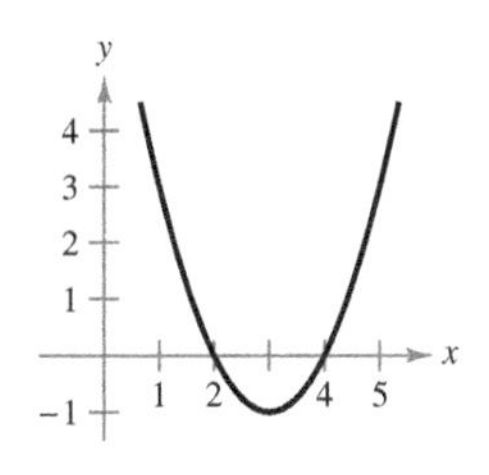

7. $y = \dfrac{x^3}{4} - 3x$

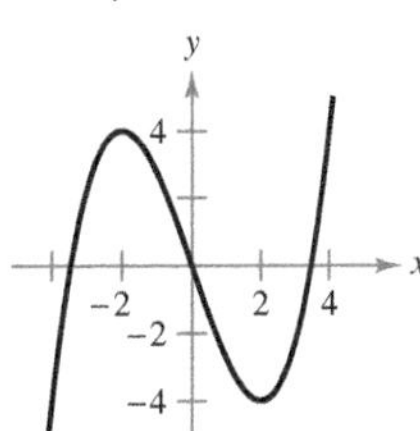

8. $f(x) = x^4 - 2x^2$

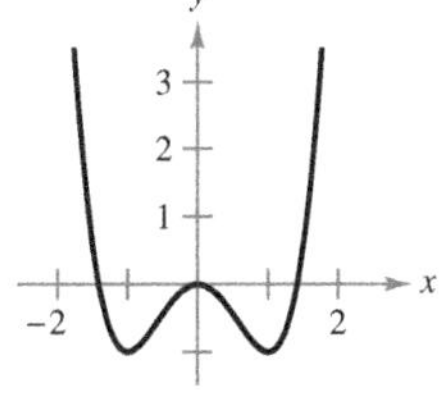

9. $f(x) = \dfrac{1}{(x + 1)^2}$

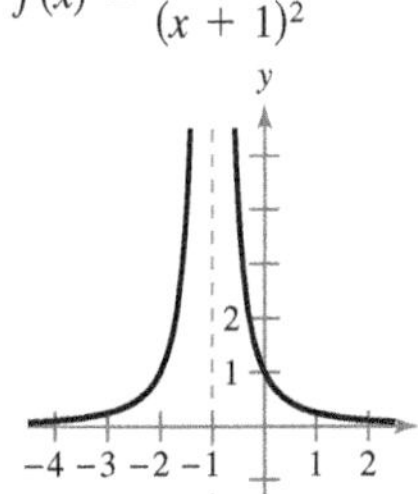

10. $y = \dfrac{x^2}{2x - 1}$

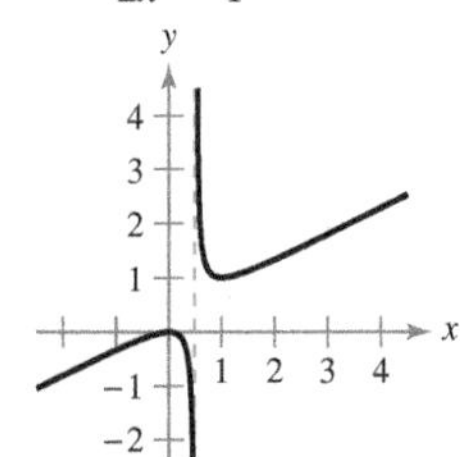

Intervals on Which a Function Is Increasing or Decreasing **In Exercises 11–18, find the open intervals on which the function is increasing or decreasing.**

11. $g(x) = x^2 - 2x - 8$

12. $h(x) = 12x - x^3$

13. $y = x\sqrt{16 - x^2}$

14. $y = x + \dfrac{9}{x}$

15. $f(x) = \sin x - 1, \quad 0 < x < 2\pi$

16. $f(x) = \cos \dfrac{3x}{2}, \quad 0 < x < 2\pi$

17. $y = x - 2 \cos x, \quad 0 < x < 2\pi$

18. $f(x) = \sin^2 x + \sin x, \quad 0 < x < 2\pi$

Applying the First Derivative Test **In Exercises 19–40, (a) find the critical numbers of f, if any, (b) find the open intervals on which the function is increasing or decreasing, (c) apply the First Derivative Test to identify all relative extrema, and (d) use a graphing utility to confirm your results.**

19. $f(x) = x^2 - 8x$

20. $f(x) = x^2 + 6x + 10$

21. $f(x) = -2x^2 + 4x + 3$

22. $f(x) = -3x^2 - 4x - 2$

23. $f(x) = -7x^3 + 21x + 3$

24. $f(x) = x^3 - 6x^2 + 15$

25. $f(x) = (x - 1)^2(x + 3)$

26. $f(x) = (8 - x)(x + 1)^2$

27. $f(x) = \dfrac{x^5 - 5x}{5}$

28. $f(x) = \dfrac{-x^6 + 6x}{10}$

29. $f(x) = x^{1/3} + 1$

30. $f(x) = x^{2/3} - 4$

31. $f(x) = (x + 2)^{2/3}$

32. $f(x) = (x - 3)^{1/3}$

33. $f(x) = 5 - |x - 5|$

34. $f(x) = |x + 3| - 1$

35. $f(x) = 2x + \dfrac{1}{x}$

36. $f(x) = \dfrac{x}{x - 5}$

37. $f(x) = \dfrac{x^2}{x^2 - 9}$

38. $f(x) = \dfrac{x^2 - 2x + 1}{x + 1}$

39. $f(x) = \begin{cases} 4 - x^2, & x \le 0 \\ -2x, & x > 0 \end{cases}$

40. $f(x) = \begin{cases} 2x + 1, & x \le -1 \\ x^2 - 2, & x > -1 \end{cases}$

Applying the First Derivative Test **In Exercises 41–48, consider the function on the interval $(0, 2\pi)$. (a) Find the open intervals on which the function is increasing or decreasing. (b) Apply the First Derivative Test to identify all relative extrema. (c) Use a graphing utility to confirm your results.**

41. $f(x) = x - 2 \sin x$

42. $f(x) = \sin x \cos x + 5$

43. $f(x) = \sin x + \cos x$

44. $f(x) = \dfrac{x}{2} + \cos x$

45. $f(x) = \cos^2(2x)$

46. $f(x) = \sin x - \sqrt{3} \cos x$

47. $f(x) = \sin^2 x + \sin x$

48. $f(x) = \dfrac{\sin x}{1 + \cos^2 x}$

Finding and Analyzing Derivatives Using Technology **In Exercises 49–54, (a) use a computer algebra system to differentiate the function, (b) sketch the graphs of f and f' on the same set of coordinate axes over the given interval, (c) find the critical numbers of f in the open interval, and (d) find the interval(s) on which f' is positive and the interval(s) on which f' is negative. Compare the behavior of f and the sign of f'.**

49. $f(x) = 2x\sqrt{9 - x^2}, \quad [-3, 3]$

50. $f(x) = 10\left(5 - \sqrt{x^2 - 3x + 16}\right), \quad [0, 5]$

51. $f(t) = t^2 \sin t, \quad [0, 2\pi]$

52. $f(x) = \dfrac{x}{2} + \cos\dfrac{x}{2}, \quad [0, 4\pi]$

53. $f(x) = -3 \sin\dfrac{x}{3}, \quad [0, 6\pi]$

54. $f(x) = 2 \sin 3x + 4 \cos 3x, \quad [0, \pi]$

Comparing Functions **In Exercises 55 and 56, use symmetry, extrema, and zeros to sketch the graph of f. How do the functions f and g differ?**

55. $f(x) = \dfrac{x^5 - 4x^3 + 3x}{x^2 - 1}$

$g(x) = x(x^2 - 3)$

56. $f(t) = \cos^2 t - \sin^2 t$

$g(t) = 1 - 2 \sin^2 t$

Think About It **In Exercises 57–62, the graph of f is shown in the figure. Sketch a graph of the derivative of f. To print an enlarged copy of the graph, go to *MathGraphs.com*.**

57.

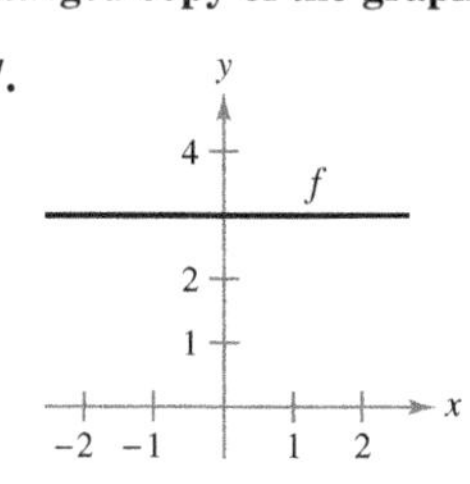

58.

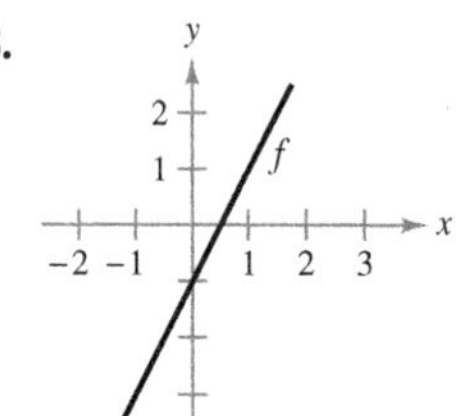

59.

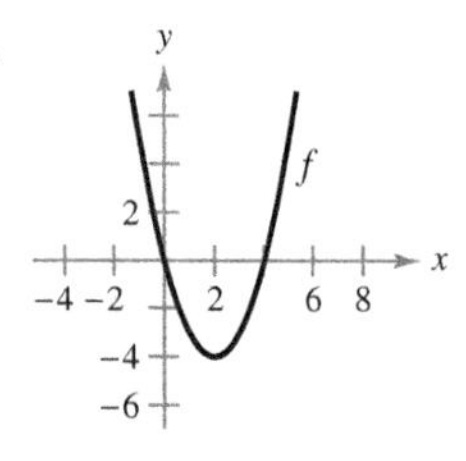

60.

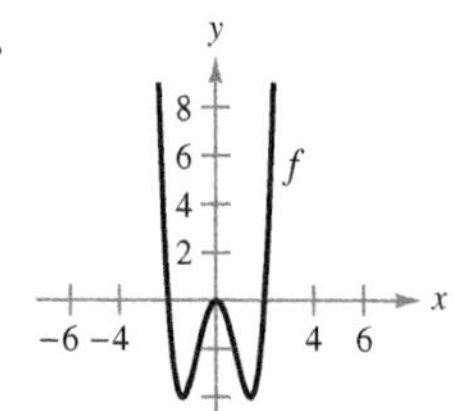

61.

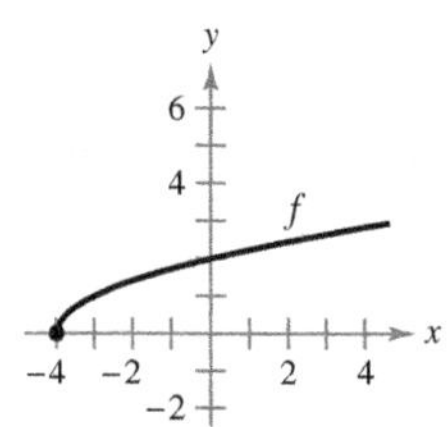

62.

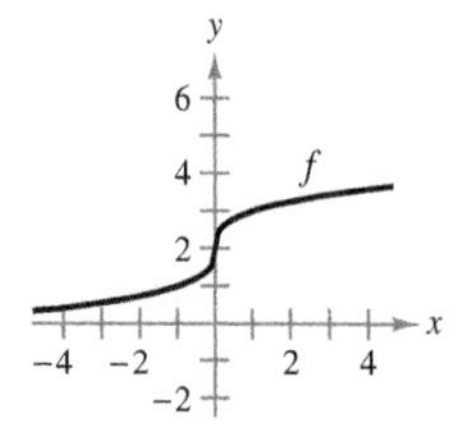

EXPLORING CONCEPTS

Transformations of Functions **In Exercises 63–66, assume that f is differentiable for all x. The signs of f' are as follows.**

$f'(x) > 0$ **on** $(-\infty, -4)$

$f'(x) < 0$ **on** $(-4, 6)$

$f'(x) > 0$ **on** $(6, \infty)$

Supply the appropriate inequality sign for the indicated value of c.

	Function	Sign of $g'(c)$
63.	$g(x) = f(x) + 5$	$g'(0)$ ☐ 0
64.	$g(x) = 3f(x) - 3$	$g'(-5)$ ☐ 0
65.	$g(x) = -f(x)$	$g'(-6)$ ☐ 0
66.	$g(x) = f(x - 10)$	$g'(0)$ ☐ 0

67. Sketching a Graph Sketch the graph of the arbitrary function f such that

$$f'(x)\begin{cases} > 0, & x < 4 \\ \text{undefined}, & x = 4. \\ < 0, & x > 4 \end{cases}$$

68. Increasing Functions Is the sum of two increasing functions always increasing? Explain.

69. Increasing Functions Is the product of two increasing functions always increasing? Explain.

70. HOW DO YOU SEE IT? Use the graph of f' to (a) identify the critical numbers of f, (b) identify the open intervals on which f is increasing or decreasing, and (c) determine whether f has a relative maximum, a relative minimum, or neither at each critical number.

(i)

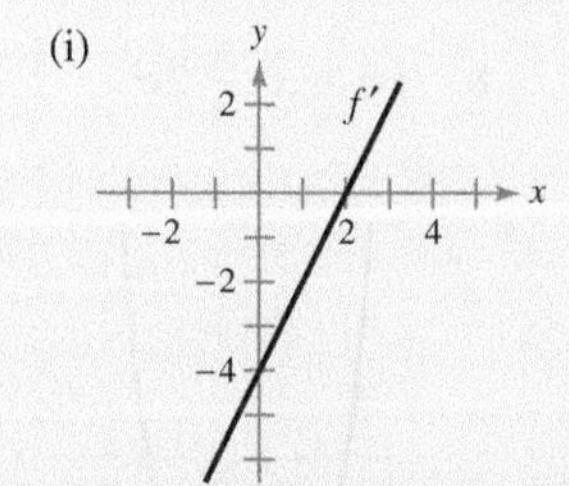

(ii)

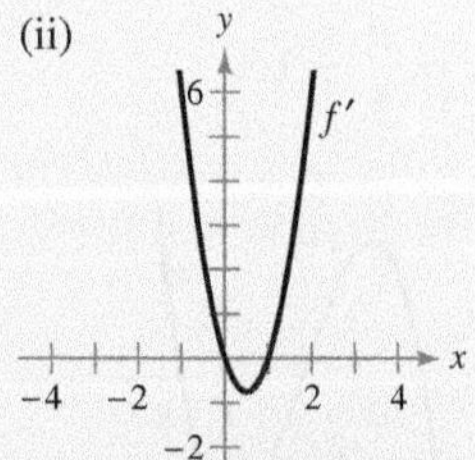

(iii)

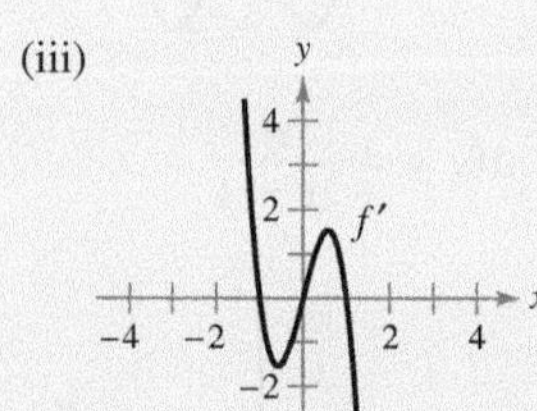

(iv)

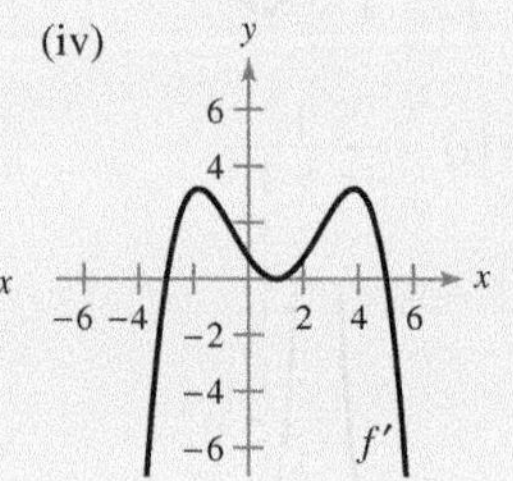

71. Analyzing a Critical Number A differentiable function f has one critical number at $x = 5$. Identify the relative extrema of f at the critical number when $f'(4) = -2.5$ and $f'(6) = 3$.

72. Analyzing a Critical Number A differentiable function f has one critical number at $x = 2$. Identify the relative extrema of f at the critical number when $f'(1) = 2$ and $f'(3) = 6$.

Think About It **In Exercises 73 and 74, the function f is differentiable on the indicated interval. The table shows $f'(x)$ for selected values of x. (a) Sketch the graph of f, (b) approximate the critical numbers, and (c) identify the relative extrema.**

73. f is differentiable on $[-1, 1]$.

x	-1	-0.75	-0.50	-0.25	0
$f'(x)$	-10	-3.2	-0.5	0.8	5.6

x	0.25	0.50	0.75	1
$f'(x)$	3.6	-0.2	-6.7	-20.1

74. f is differentiable on $[0, \pi]$.

x	0	$\pi/6$	$\pi/4$	$\pi/3$	$\pi/2$
$f'(x)$	3.14	-0.23	-2.45	-3.11	0.69

x	$2\pi/3$	$3\pi/4$	$5\pi/6$	π
$f'(x)$	3.00	1.37	-1.14	-2.84

75. Rolling a Ball Bearing A ball bearing is placed on an inclined plane and begins to roll. The angle of elevation of the plane is θ. The distance (in meters) the ball bearing rolls in t seconds is $s(t) = 4.9(\sin \theta)t^2$.

(a) Determine the speed of the ball bearing after t seconds.

(b) Complete the table and use it to determine the value of θ that produces the maximum speed at a particular time.

θ	0	$\pi/4$	$\pi/3$	$\pi/2$	$2\pi/3$	$3\pi/4$	π
$s'(t)$							

76. Modeling Data The end-of-year assets of the Medicare Hospital Insurance Trust Fund (in billions of dollars) for the years 2006 through 2014 are shown.

2006: 305.4 2007: 326.0 2008: 321.3
2009: 304.2 2010: 271.9 2011: 244.2
2012: 220.4 2013: 205.4 2014: 197.3

(Source: U.S. Centers for Medicare and Medicaid Services)

(a) Use the regression capabilities of a graphing utility to find a model of the form $M = at^3 + bt^2 + ct + d$ for the data. Let $t = 6$ represent 2006.

(b) Use a graphing utility to plot the data and graph the model.

(c) Find the maximum value of the model and compare the result with the actual data.

77. Numerical, Graphical, and Analytic Analysis The concentration C of a chemical in the bloodstream t hours after injection into muscle tissue is

$$C(t) = \frac{3t}{27 + t^3}, \quad t \geq 0.$$

(a) Complete the table and use it to approximate the time when the concentration is greatest.

t	0	0.5	1	1.5	2	2.5	3
$C(t)$							

(b) Use a graphing utility to graph the concentration function and use the graph to approximate the time when the concentration is greatest.

(c) Use calculus to determine analytically the time when the concentration is greatest.

78. Numerical, Graphical, and Analytic Analysis Consider the functions $f(x) = x$ and $g(x) = \sin x$ on the interval $(0, \pi)$.

(a) Complete the table and make a conjecture about which is the greater function on the interval $(0, \pi)$.

x	0.5	1	1.5	2	2.5	3
$f(x)$						
$g(x)$						

(b) Use a graphing utility to graph the functions and use the graphs to make a conjecture about which is the greater function on the interval $(0, \pi)$.

(c) Prove that $f(x) > g(x)$ on the interval $(0, \pi)$. [*Hint*: Show that $h'(x) > 0$, where $h = f - g$.]

79. Trachea Contraction Coughing forces the trachea (windpipe) to contract, which affects the velocity v of the air passing through the trachea. The velocity of the air during coughing is

$$v = k(R - r)r^2, \quad 0 \leq r < R$$

where k is a constant, R is the normal radius of the trachea, and r is the radius during coughing. What radius will produce the maximum air velocity?

80. Electrical Resistance The resistance R of a certain type of resistor is

$$R = \sqrt{0.001T^4 - 4T + 100}$$

where R is measured in ohms and the temperature T is measured in degrees Celsius.

(a) Use a computer algebra system to find dR/dT and the critical number of the function. Determine the minimum resistance for this type of resistor.

(b) Use a graphing utility to graph the function R and use the graph to approximate the minimum resistance for this type of resistor.

Motion Along a Line **In Exercises 81–84, the function $s(t)$ describes the motion of a particle along a line. (a) Find the velocity function of the particle at any time $t \geq 0$. (b) Identify the time interval(s) on which the particle is moving in a positive direction. (c) Identify the time interval(s) on which the particle is moving in a negative direction. (d) Identify the time(s) at which the particle changes direction.**

81. $s(t) = 6t - t^2$

82. $s(t) = t^2 - 10t + 29$

83. $s(t) = t^3 - 5t^2 + 4t$

84. $s(t) = t^3 - 20t^2 + 128t - 280$

Motion Along a Line **In Exercises 85 and 86, the graph shows the position of a particle moving along a line. Describe how the position of the particle changes with respect to time.**

85. **86.**

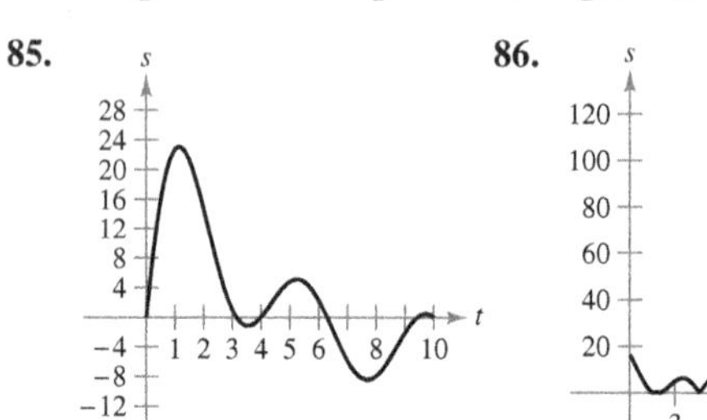

Creating Polynomial Functions **In Exercises 87–90, find a polynomial function**

$$f(x) = a_n x^n + a_{n-1}x^{n-1} + \cdots + a_2x^2 + a_1x + a_0$$

that has only the specified extrema. (a) Determine the minimum degree of the function and give the criteria you used in determining the degree. (b) Using the fact that the coordinates of the extrema are solution points of the function, and that the x-coordinates are critical numbers, determine a system of linear equations whose solution yields the coefficients of the required function. (c) Use a graphing utility to solve the system of equations and determine the function. (d) Use a graphing utility to confirm your result graphically.

87. Relative minimum: $(0, 0)$; Relative maximum: $(2, 2)$

88. Relative minimum: $(0, 0)$; Relative maximum: $(4, 1000)$

89. Relative minima: $(0, 0)$, $(4, 0)$; Relative maximum: $(2, 4)$

90. Relative minimum: $(1, 2)$; Relative maxima: $(-1, 4)$, $(3, 4)$

True or False? **In Exercises 91–96, determine whether the statement is true or false. If it is false, explain why or give an example that shows it is false.**

91. There is no function with an infinite number of critical points.

92. The function $f(x) = x$ has no extrema on any open interval.

93. Every nth-degree polynomial has $(n - 1)$ critical numbers.

94. An nth-degree polynomial has at most $(n - 1)$ critical numbers.

95. There is a relative extremum at each critical number.

96. The relative maxima of the function f are $f(1) = 4$ and $f(3) = 10$. Therefore, f has at least one minimum for some x in the interval $(1, 3)$.

97. Proof Prove the second case of Theorem 3.5.

98. Proof Prove the second case of Theorem 3.6.

99. Proof Use the definitions of increasing and decreasing functions to prove that

$$f(x) = x^3$$

is increasing on $(-\infty, \infty)$.

100. Proof Use the definitions of increasing and decreasing functions to prove that

$$f(x) = \frac{1}{x}$$

is decreasing on $(0, \infty)$.

PUTNAM EXAM CHALLENGE

101. Find the minimum value of

$$|\sin x + \cos x + \tan x + \cot x + \sec x + \csc x|$$

for real numbers x.

SECTION PROJECT

Even Fourth-Degree Polynomials

(a) Graph each of the fourth-degree polynomials below. Then find the critical numbers, the open intervals on which the function is increasing or decreasing, and the relative extrema.

(i) $f(x) = x^4 + 1$

(ii) $f(x) = x^4 + 2x^2 + 1$

(iii) $f(x) = x^4 - 2x^2 + 1$

(b) Consider the fourth-degree polynomial

$$f(x) = x^4 + ax^2 + b.$$

(i) Show that there is one critical number when $a = 0$. Then find the open intervals on which the function is increasing or decreasing.

(ii) Show that there is one critical number when $a > 0$. Then find the open intervals on which the function is increasing or decreasing.

(iii) Show that there are three critical numbers when $a < 0$. Then find the open intervals on which the function is increasing or decreasing.

(iv) Show that there are no real zeros when

$$a^2 < 4b.$$

(v) Determine the possible number of zeros when

$$a^2 \geq 4b.$$

Explain your reasoning.

3.4 Concavity and the Second Derivative Test

- **Determine intervals on which a function is concave upward or concave downward.**
- **Find any points of inflection of the graph of a function.**
- **Apply the Second Derivative Test to find relative extrema of a function.**

Concavity

You have already seen that locating the intervals on which a function f increases or decreases helps to describe its graph. In this section, you will see how locating the intervals on which f' increases or decreases can be used to determine where the graph of f is *curving upward* or *curving downward*.

Definition of Concavity

Let f be differentiable on an open interval I. The graph of f is **concave upward** on I when f' is increasing on the interval and **concave downward** on I when f' is decreasing on the interval.

The following graphical interpretation of concavity is useful. (See Appendix A for a proof of these results.)

1. Let f be differentiable on an open interval I. If the graph of f is concave *upward* on I, then the graph of f lies *above* all of its tangent lines on I.
[See Figure 3.23(a).]
2. Let f be differentiable on an open interval I. If the graph of f is concave *downward* on I, then the graph of f lies *below* all of its tangent lines on I.
[See Figure 3.23(b).]

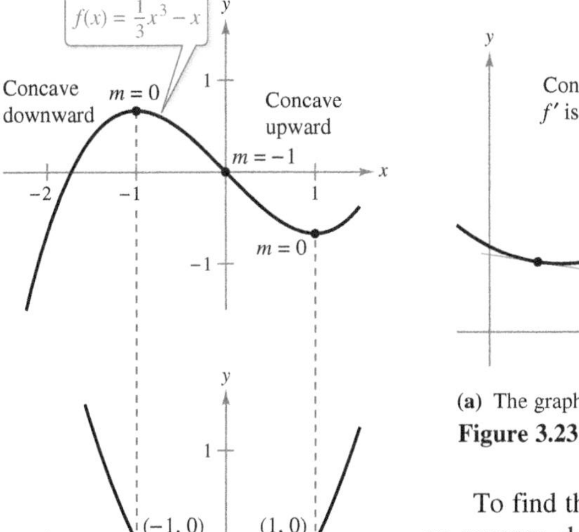

The concavity of f is related to the slope of the derivative.
Figure 3.24

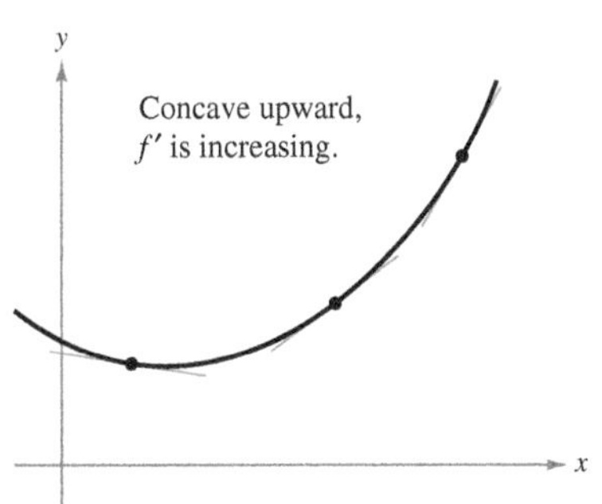

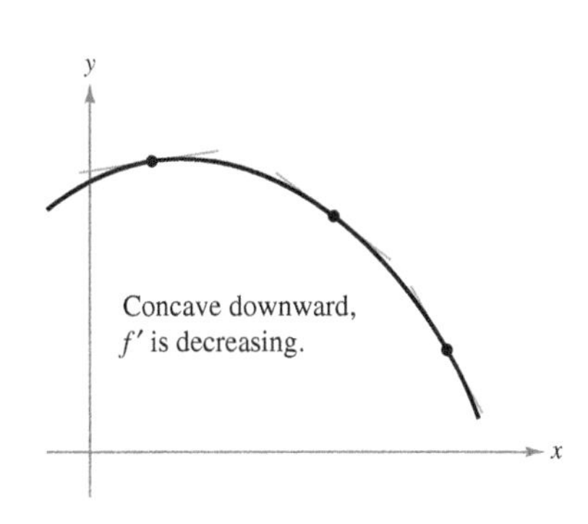

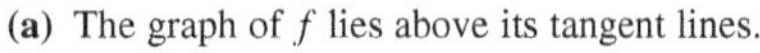

(a) The graph of f lies above its tangent lines. (b) The graph of f lies below its tangent lines.
Figure 3.23

To find the open intervals on which the graph of a function f is concave upward or concave downward, you need to find the intervals on which f' is increasing or decreasing. For instance, the graph of

$$f(x) = \frac{1}{3}x^3 - x$$

is concave downward on the open interval $(-\infty, 0)$ because

$$f'(x) = x^2 - 1$$

is decreasing there. (See Figure 3.24.) Similarly, the graph of f is concave upward on the interval $(0, \infty)$ because f' is increasing on $(0, \infty)$.

The next theorem shows how to use the *second* derivative of a function f to determine intervals on which the graph of f is concave upward or concave downward. A proof of this theorem follows directly from Theorem 3.5 and the definition of concavity.

REMARK A third case of Theorem 3.7 could be that if $f''(x) = 0$ for all x in I, then f is linear. Note, however, that concavity is not defined for a line. In other words, a straight line is neither concave upward nor concave downward.

THEOREM 3.7 Test for Concavity

Let f be a function whose second derivative exists on an open interval I.

1. If $f''(x) > 0$ for all x in I, then the graph of f is concave upward on I.
2. If $f''(x) < 0$ for all x in I, then the graph of f is concave downward on I.

A proof of this theorem is given in Appendix A.

To apply Theorem 3.7, locate the x-values at which $f''(x) = 0$ or $f''(x)$ does not exist. Use these x-values to determine test intervals. Finally, test the sign of $f''(x)$ in each of the test intervals.

EXAMPLE 1 Determining Concavity

Determine the open intervals on which the graph of

$$f(x) = \frac{6}{x^2 + 3}$$

is concave upward or concave downward.

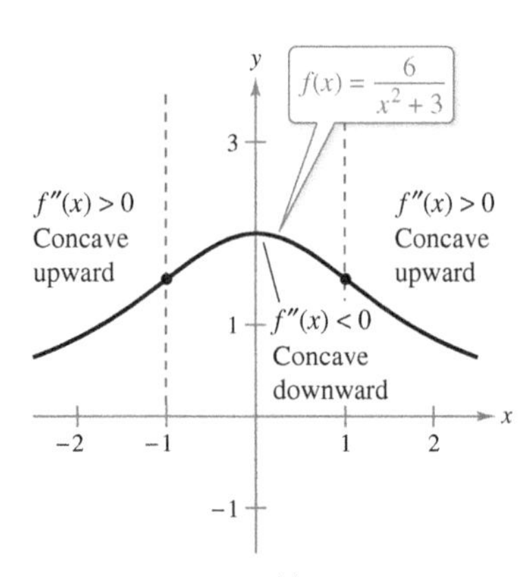

From the sign of $f''(x)$, you can determine the concavity of the graph of f.
Figure 3.25

Solution Begin by observing that f is continuous on the entire real number line. Next, find the second derivative of f.

$$f(x) = 6(x^2 + 3)^{-1} \quad \text{Rewrite original function.}$$

$$f'(x) = (-6)(x^2 + 3)^{-2}(2x) \quad \text{Differentiate.}$$

$$= \frac{-12x}{(x^2 + 3)^2} \quad \text{First derivative}$$

$$f''(x) = \frac{(x^2 + 3)^2(-12) - (-12x)(2)(x^2 + 3)(2x)}{(x^2 + 3)^4} \quad \text{Differentiate.}$$

$$= \frac{36(x^2 - 1)}{(x^2 + 3)^3} \quad \text{Second derivative}$$

Because $f''(x) = 0$ when $x = \pm 1$ and f'' is defined on the entire real number line, you should test f'' in the intervals $(-\infty, -1)$, $(-1, 1)$, and $(1, \infty)$. The results are shown in the table and in Figure 3.25.

Interval	$-\infty < x < -1$	$-1 < x < 1$	$1 < x < \infty$
Test Value	$x = -2$	$x = 0$	$x = 2$
Sign of $f''(x)$	$f''(-2) > 0$	$f''(0) < 0$	$f''(2) > 0$
Conclusion	Concave upward	Concave downward	Concave upward

The function given in Example 1 is continuous on the entire real number line. When there are x-values at which a function is not continuous, these values should be used, along with the points at which $f''(x) = 0$ or $f''(x)$ does not exist, to form the test intervals.

EXAMPLE 2 Determining Concavity

Determine the open intervals on which the graph of

$$f(x) = \frac{x^2 + 1}{x^2 - 4}$$

is concave upward or concave downward.

Solution Differentiating twice produces the following.

$$f(x) = \frac{x^2 + 1}{x^2 - 4} \qquad \text{Write original function.}$$

$$f'(x) = \frac{(x^2 - 4)(2x) - (x^2 + 1)(2x)}{(x^2 - 4)^2} \qquad \text{Differentiate.}$$

$$= \frac{-10x}{(x^2 - 4)^2} \qquad \text{First derivative}$$

$$f''(x) = \frac{(x^2 - 4)^2(-10) - (-10x)(2)(x^2 - 4)(2x)}{(x^2 - 4)^4} \qquad \text{Differentiate.}$$

$$= \frac{10(3x^2 + 4)}{(x^2 - 4)^3} \qquad \text{Second derivative}$$

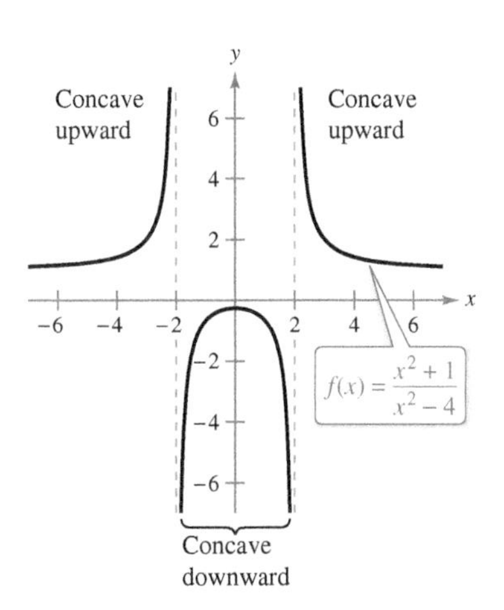

Figure 3.26

There are no points at which $f''(x) = 0$, but at $x = \pm 2$, the function f is not continuous. So, test for concavity in the intervals $(-\infty, -2)$, $(-2, 2)$, and $(2, \infty)$, as shown in the table. The graph of f is shown in Figure 3.26.

Interval	$-\infty < x < -2$	$-2 < x < 2$	$2 < x < \infty$
Test Value	$x = -3$	$x = 0$	$x = 3$
Sign of $f''(x)$	$f''(-3) > 0$	$f''(0) < 0$	$f''(3) > 0$
Conclusion	Concave upward	Concave downward	Concave upward

Points of Inflection

The graph in Figure 3.25 has two points at which the concavity changes. If the tangent line to the graph exists at such a point, then that point is a **point of inflection.** Three types of points of inflection are shown in Figure 3.27.

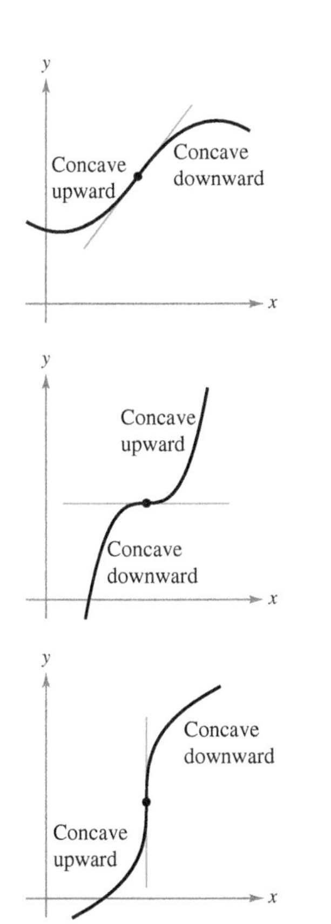

The concavity of f changes at a point of inflection. Note that the graph crosses its tangent line at a point of inflection.
Figure 3.27

Definition of Point of Inflection

Let f be a function that is continuous on an open interval, and let c be a point in the interval. If the graph of f has a tangent line at the point $(c, f(c))$, then this point is a **point of inflection** of the graph of f when the concavity of f changes from upward to downward (or downward to upward) at the point.

The definition of *point of inflection* requires that the tangent line exists at the point of inflection. Some calculus texts do not require this. For instance, after applying the definition above to the function

$$f(x) = \begin{cases} x^3, & x < 0 \\ x^2 + 2x, & x \geq 0 \end{cases}$$

you would conclude that f does *not* have a point of inflection at the origin, even though the concavity of the graph changes from concave downward to concave upward.

To locate *possible* points of inflection, you can determine the values of x for which $f''(x) = 0$ or $f''(x)$ does not exist. This is similar to the procedure for locating relative extrema of f.

THEOREM 3.8 Points of Inflection

If $(c, f(c))$ is a point of inflection of the graph of f, then either $f''(c) = 0$ or $f''(c)$ does not exist.

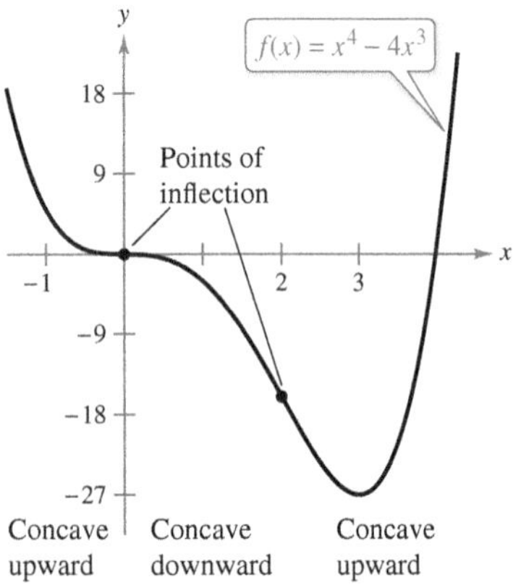

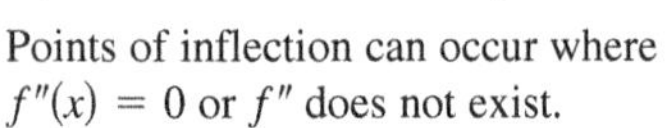

Points of inflection can occur where $f''(x) = 0$ or f'' does not exist.
Figure 3.28

EXAMPLE 3 Finding Points of Inflection

Determine the points of inflection and discuss the concavity of the graph of

$$f(x) = x^4 - 4x^3.$$

Solution Differentiating twice produces the following.

$f(x) = x^4 - 4x^3$ Write original function.

$f'(x) = 4x^3 - 12x^2$ Find first derivative.

$f''(x) = 12x^2 - 24x = 12x(x - 2)$ Find second derivative.

Setting $f''(x) = 0$, you can determine that the possible points of inflection occur at $x = 0$ and $x = 2$. By testing the intervals determined by these x-values, you can conclude that they both yield points of inflection. A summary of this testing is shown in the table, and the graph of f is shown in Figure 3.28.

Interval	$-\infty < x < 0$	$0 < x < 2$	$2 < x < \infty$
Test Value	$x = -1$	$x = 1$	$x = 3$
Sign of $f''(x)$	$f''(-1) > 0$	$f''(1) < 0$	$f''(3) > 0$
Conclusion	Concave upward	Concave downward	Concave upward

Exploration

Consider a general cubic function of the form

$$f(x) = ax^3 + bx^2 + cx + d.$$

You know that the value of d has a bearing on the location of the graph but has no bearing on the value of the first derivative at given values of x. Graphically, this is true because changes in the value of d shift the graph up or down but do not change its basic shape. Use a graphing utility to graph several cubics with different values of c. Then give a graphical explanation of why changes in c do not affect the values of the second derivative.

The converse of Theorem 3.8 is not generally true. That is, it is possible for the second derivative to be 0 at a point that is *not* a point of inflection. For instance, the graph of $f(x) = x^4$ is shown in Figure 3.29. The second derivative is 0 when $x = 0$, but the point $(0, 0)$ is not a point of inflection because the graph of f is concave upward on the intervals $-\infty < x < 0$ and $0 < x < \infty$.

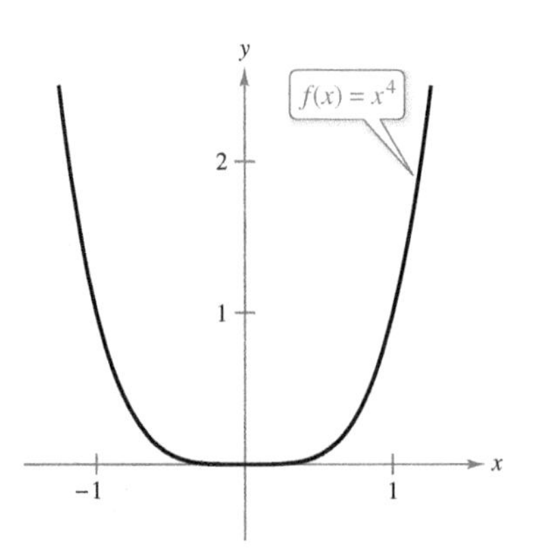

$f''(x) = 0$, but $(0, 0)$ is not a point of inflection.
Figure 3.29

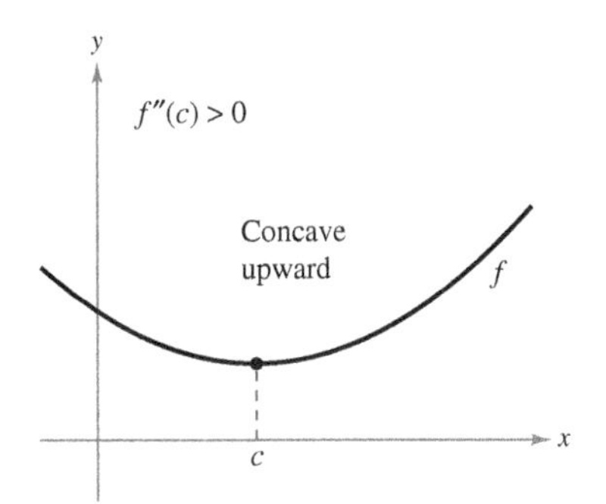

If $f'(c) = 0$ and $f''(c) > 0$, then $f(c)$ is a relative minimum.

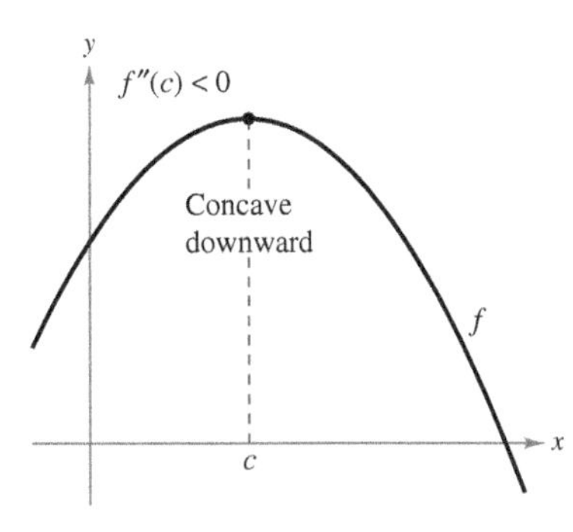

If $f'(c) = 0$ and $f''(c) < 0$, then $f(c)$ is a relative maximum.
Figure 3.30

The Second Derivative Test

In addition to testing for concavity, the second derivative can be used to perform a simple test for relative maxima and minima. The test is based on the fact that if the graph of a function f is concave upward on an open interval containing c, and $f'(c) = 0$, then $f(c)$ must be a relative minimum of f. Similarly, if the graph of a function f is concave downward on an open interval containing c, and $f'(c) = 0$, then $f(c)$ must be a relative maximum of f. (See Figure 3.30.)

THEOREM 3.9 Second Derivative Test

Let f be a function such that $f'(c) = 0$ and the second derivative of f exists on an open interval containing c.

1. If $f''(c) > 0$, then f has a relative minimum at $(c, f(c))$.
2. If $f''(c) < 0$, then f has a relative maximum at $(c, f(c))$.

If $f''(c) = 0$, then the test fails. That is, f may have a relative maximum, a relative minimum, or neither. In such cases, you can use the First Derivative Test.

Proof If $f'(c) = 0$ and $f''(c) > 0$, then there exists an open interval I containing c for which

$$\frac{f'(x) - f'(c)}{x - c} = \frac{f'(x)}{x - c} > 0$$

for all $x \neq c$ in I. If $x < c$, then $x - c < 0$ and $f'(x) < 0$. Also, if $x > c$, then $x - c > 0$ and $f'(x) > 0$. So, $f'(x)$ changes from negative to positive at c, and the First Derivative Test implies that $f(c)$ is a relative minimum. A proof of the second case is left to you. ■

EXAMPLE 4 Using the Second Derivative Test

See LarsonCalculus.com for an interactive version of this type of example.

Find the relative extrema of

$$f(x) = -3x^5 + 5x^3.$$

Solution Begin by finding the first derivative of f.

$$f'(x) = -15x^4 + 15x^2 = 15x^2(1 - x^2)$$

From this derivative, you can see that $x = -1$, 0, and 1 are the only critical numbers of f. By finding the second derivative

$$f''(x) = -60x^3 + 30x = 30x(1 - 2x^2)$$

you can apply the Second Derivative Test as shown below.

Point	$(-1, -2)$	$(0, 0)$	$(1, 2)$
Sign of $f''(x)$	$f''(-1) > 0$	$f''(0) = 0$	$f''(1) < 0$
Conclusion	Relative minimum	Test fails	Relative maximum

Because the Second Derivative Test fails at $(0, 0)$, you can use the First Derivative Test and observe that f increases to the left and right of $x = 0$. So, $(0, 0)$ is neither a relative minimum nor a relative maximum (even though the graph has a horizontal tangent line at this point). The graph of f is shown in Figure 3.31. ■

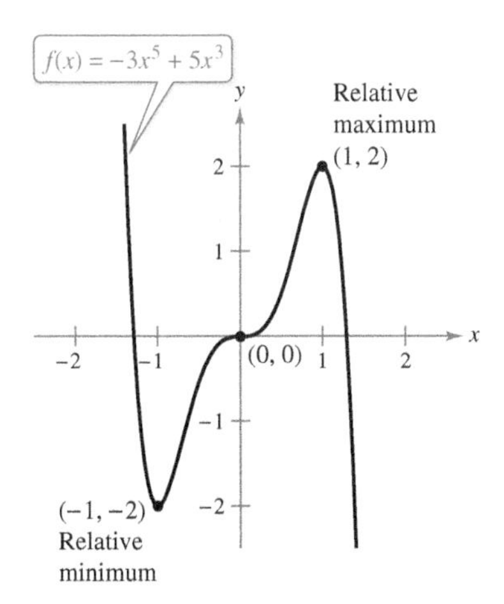

$(0, 0)$ is neither a relative minimum nor a relative maximum.
Figure 3.31

3.4 Exercises

See CalcChat.com for tutorial help and worked-out solutions to odd-numbered exercises.

CONCEPT CHECK

1. **Test for Concavity** Describe the Test for Concavity in your own words.
2. **Second Derivative Test** Describe the Second Derivative Test in your own words.

Using a Graph **In Exercises 3 and 4, the graph of f is shown. State the signs of f' and f'' on the interval (0, 2).**

3. 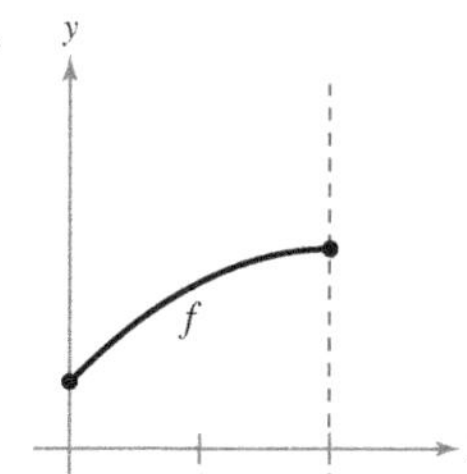

4. 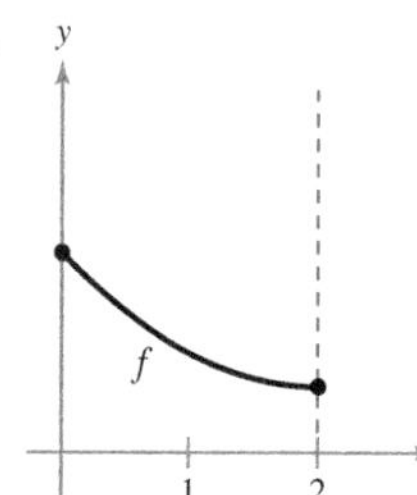

Determining Concavity **In Exercises 5–16, determine the open intervals on which the graph of the function is concave upward or concave downward.**

5. $f(x) = x^2 - 4x + 8$
6. $g(x) = 3x^2 - x^3$
7. $f(x) = x^4 - 3x^3$
8. $h(x) = x^5 - 5x + 2$
9. $f(x) = \dfrac{24}{x^2 + 12}$
10. $f(x) = \dfrac{2x^2}{3x^2 + 1}$
11. $f(x) = \dfrac{x - 2}{6x + 1}$
12. $f(x) = \dfrac{x + 8}{x - 7}$
13. $f(x) = \dfrac{x^2 + 1}{x^2 - 1}$
14. $h(x) = \dfrac{x^2 - 1}{2x - 1}$
15. $y = 2x - \tan x, \quad \left(-\dfrac{\pi}{2}, \dfrac{\pi}{2}\right)$
16. $y = x + \dfrac{2}{\sin x}, \quad (-\pi, \pi)$

Finding Points of Inflection **In Exercises 17–32, find the points of inflection and discuss the concavity of the graph of the function.**

17. $f(x) = x^3 - 9x^2 + 24x - 18$
18. $f(x) = -x^3 + 6x^2 - 5$
19. $f(x) = 2 - 7x^4$
20. $f(x) = 4 - x - 3x^4$
21. $f(x) = x(x - 4)^3$
22. $f(x) = (x - 2)^3(x - 1)$
23. $f(x) = x\sqrt{x + 3}$
24. $f(x) = x\sqrt{9 - x}$
25. $f(x) = \dfrac{6 - x}{\sqrt{x}}$
26. $f(x) = \dfrac{x + 3}{\sqrt{x}}$
27. $f(x) = \sin\dfrac{x}{2}, \quad [0, 4\pi]$
28. $f(x) = 2\csc\dfrac{3x}{2}, \quad (0, 2\pi)$
29. $f(x) = \sec\left(x - \dfrac{\pi}{2}\right), \quad (0, 4\pi)$
30. $f(x) = \sin x + \cos x, \quad [0, 2\pi]$
31. $f(x) = 2\sin x + \sin 2x, \quad [0, 2\pi]$
32. $f(x) = x + 2\cos x, \quad [0, 2\pi]$

Using the Second Derivative Test **In Exercises 33–44, find all relative extrema of the function. Use the Second Derivative Test where applicable.**

33. $f(x) = 6x - x^2$
34. $f(x) = x^2 + 3x - 8$
35. $f(x) = x^3 - 3x^2 + 3$
36. $f(x) = -x^3 + 7x^2 - 15x$
37. $f(x) = x^4 - 4x^3 + 2$
38. $f(x) = -x^4 + 2x^3 + 8x$
39. $f(x) = x^{2/3} - 3$
40. $f(x) = \sqrt{x^2 + 1}$
41. $f(x) = x + \dfrac{4}{x}$
42. $f(x) = \dfrac{9x - 1}{x + 5}$
43. $f(x) = \cos x - x, \quad [0, 4\pi]$
44. $f(x) = 2\sin x + \cos 2x, \quad [0, 2\pi]$

Finding Extrema and Points of Inflection Using Technology **In Exercises 45–48, use a computer algebra system to analyze the function over the given interval. (a) Find the first and second derivatives of the function. (b) Find any relative extrema and points of inflection. (c) Graph f, f', and f'' on the same set of coordinate axes and state the relationship between the behavior of f and the signs of f' and f''.**

45. $f(x) = 0.2x^2(x - 3)^3, \quad [-1, 4]$
46. $f(x) = x^2\sqrt{6 - x^2}, \quad [-\sqrt{6}, \sqrt{6}]$
47. $f(x) = \sin x - \frac{1}{3}\sin 3x + \frac{1}{5}\sin 5x, \quad [0, \pi]$
48. $f(x) = \sqrt{2x}\sin x, \quad [0, 2\pi]$

EXPLORING CONCEPTS

49. **Sketching a Graph** Consider a function f such that f' is increasing. Sketch graphs of f for (a) $f' < 0$ and (b) $f' > 0$.

50. **Think About It** S represents weekly sales of a product. What can be said of S' and S'' for each of the following statements?
    (a) The rate of change of sales is increasing.
    (b) The rate of change of sales is constant.
    (c) Sales are steady.
    (d) Sales are declining but at a slower rate.
    (e) Sales have bottomed out and have started to rise.

Sketching Graphs **In Exercises 51 and 52, the graph of f is shown. Graph f, f', and f'' on the same set of coordinate axes. To print an enlarged copy of the graph, go to *MathGraphs.com*.**

51. 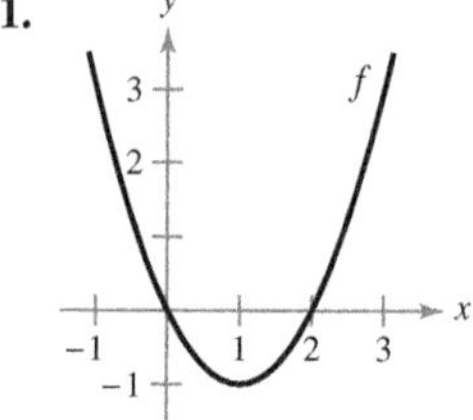

52. 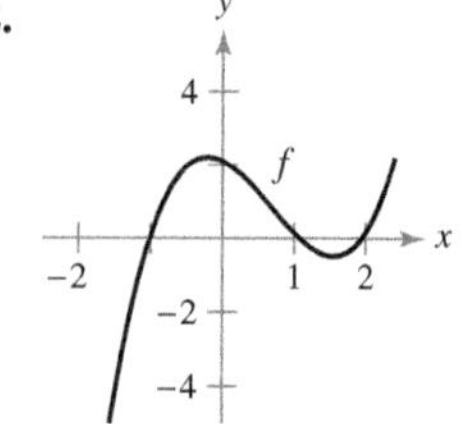

Think About It **In Exercises 53–56, sketch the graph of a function f having the given characteristics.**

53. $f(0) = f(2) = 0$
$f'(x) > 0$ for $x < 1$
$f'(1) = 0$
$f'(x) < 0$ for $x > 1$
$f''(x) < 0$

54. $f(0) = f(2) = 0$
$f'(x) < 0$ for $x < 1$
$f'(1) = 0$
$f'(x) > 0$ for $x > 1$
$f''(x) > 0$

55. $f(2) = f(4) = 0$
$f'(x) < 0$ for $x < 3$
$f'(3)$ does not exist.
$f'(x) > 0$ for $x > 3$
$f''(x) < 0, x \neq 3$

56. $f(1) = f(3) = 0$
$f'(x) > 0$ for $x < 2$
$f'(2)$ does not exist.
$f'(x) < 0$ for $x > 2$
$f''(x) > 0, x \neq 2$

57. **Think About It** The figure shows the graph of f''. Sketch a graph of f. (The answer is not unique.) To print an enlarged copy of the graph, go to *MathGraphs.com*.

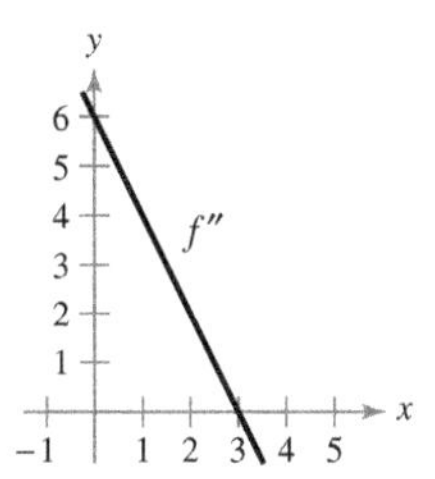

58. **HOW DO YOU SEE IT?** Water is running into the vase shown in the figure at a constant rate.

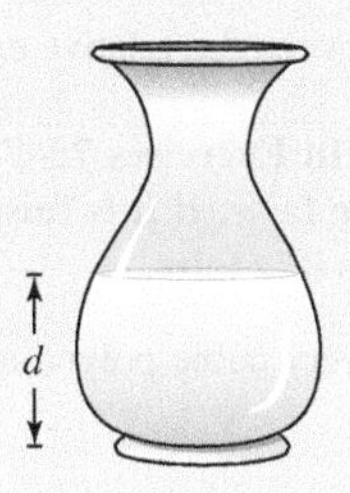

(a) Graph the depth d of water in the vase as a function of time.

(b) Does the function have any extrema? Explain.

(c) Interpret the inflection points of the graph of d.

59. **Conjecture** Consider the function

$$f(x) = (x - 2)^n.$$

(a) Use a graphing utility to graph f for $n = 1, 2, 3,$ and 4. Use the graphs to make a conjecture about the relationship between n and any inflection points of the graph of f.

(b) Verify your conjecture in part (a).

60. **Inflection Point** Consider the function $f(x) = \sqrt[3]{x}$.

(a) Graph the function and identify the inflection point.

(b) Does f'' exist at the inflection point? Explain.

Finding a Cubic Function **In Exercises 61 and 62, find a, b, c, and d such that the cubic function**

$$f(x) = ax^3 + bx^2 + cx + d$$

satisfies the given conditions.

61. Relative maximum: $(3, 3)$
Relative minimum: $(5, 1)$
Inflection point: $(4, 2)$

62. Relative maximum: $(2, 4)$
Relative minimum: $(4, 2)$
Inflection point: $(3, 3)$

63. **Aircraft Glide Path** A small aircraft starts its descent from an altitude of 1 mile, 4 miles west of the runway (see figure).

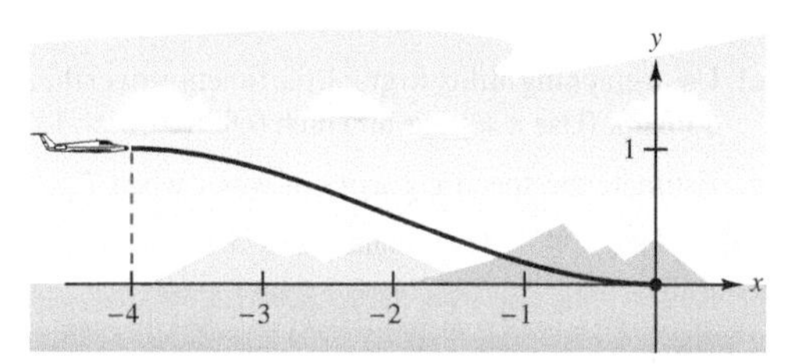

(a) Find the cubic function $f(x) = ax^3 + bx^2 + cx + d$ on the interval $[-4, 0]$ that describes a smooth glide path for the landing.

(b) The function in part (a) models the glide path of the plane. When would the plane be descending at the greatest rate?

FOR FURTHER INFORMATION For more information on this type of modeling, see the article "How Not to Land at Lake Tahoe!" by Richard Barshinger in *The American Mathematical Monthly*. To view this article, go to *MathArticles.com*.

64. **Highway Design** A section of highway connecting two hillsides with grades of 6% and 4% is to be built between two points that are separated by a horizontal distance of 2000 feet (see figure). At the point where the two hillsides come together, there is a 50-foot difference in elevation.

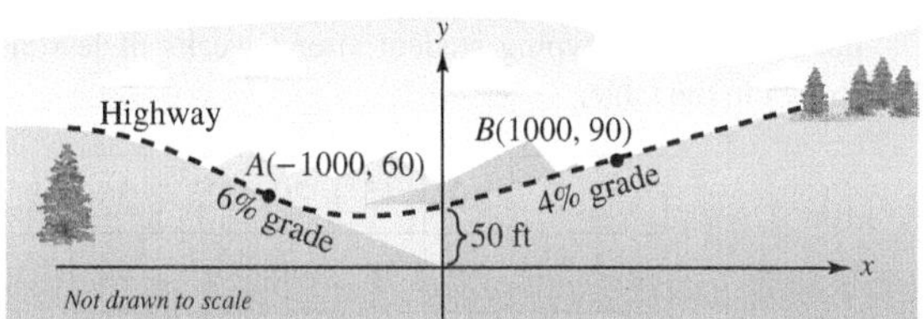

(a) Find the cubic function

$$f(x) = ax^3 + bx^2 + cx + d, \quad -1000 \leq x \leq 1000$$

that describes the section of highway connecting the hillsides. At points A and B, the slope of the model must match the grade of the hillside.

(b) Use a graphing utility to graph the model.

(c) Use a graphing utility to graph the derivative of the model.

(d) Determine the grade at the steepest part of the transitional section of the highway.

65. Average Cost A manufacturer has determined that the total cost C of operating a factory is

$$C = 0.5x^2 + 15x + 5000$$

where x is the number of units produced. At what level of production will the average cost per unit be minimized? (The average cost per unit is C/x.)

66. Specific Gravity A model for the specific gravity of water S is

$$S = \frac{5.755}{10^8}T^3 - \frac{8.521}{10^6}T^2 + \frac{6.540}{10^5}T + 0.99987, \; 0 < T < 25$$

where T is the water temperature in degrees Celsius.

(a) Use the second derivative to determine the concavity of S.

(b) Use a computer algebra system to find the coordinates of the maximum value of the function.

(c) Use a graphing utility to graph the function over the specified domain. (Use a setting in which $0.996 \le S \le 1.001$.)

(d) Estimate the specific gravity of water when $T = 20°$.

67. Sales Growth The annual sales S of a new product are given by

$$S = \frac{5000t^2}{8 + t^2}, \quad 0 \le t \le 3$$

where t is time in years.

(a) Complete the table. Then use it to estimate when the annual sales are increasing at the greatest rate.

t	0.5	1	1.5	2	2.5	3
S						

(b) Use a graphing utility to graph the function S. Then use the graph to estimate when the annual sales are increasing at the greatest rate.

(c) Find the exact time when the annual sales are increasing at the greatest rate.

68. Modeling Data The average typing speeds S (in words per minute) of a typing student after t weeks of lessons are shown in the table.

t	5	10	15	20	25	30
S	28	56	79	90	93	94

A model for the data is

$$S = \frac{100t^2}{65 + t^2}, \quad t > 0.$$

(a) Use a graphing utility to plot the data and graph the model.

(b) Use the second derivative to determine the concavity of S. Compare the result with the graph in part (a).

(c) What is the sign of the first derivative for $t > 0$? By combining this information with the concavity of the model, what inferences can be made about the typing speed as t increases?

Linear and Quadratic Approximations In Exercises 69–72, use a graphing utility to graph the function. Then graph the linear and quadratic approximations

$$P_1(x) = f(a) + f'(a)(x - a)$$

and

$$P_2(x) = f(a) + f'(a)(x - a) + \tfrac{1}{2}f''(a)(x - a)^2$$

in the same viewing window. Compare the values of f, P_1, and P_2 and their first derivatives at $x = a$. How do the approximations change as you move farther away from $x = a$?

Function	Value of a
69. $f(x) = 2(\sin x + \cos x)$	$a = \frac{\pi}{4}$
70. $f(x) = 2(\sin x + \cos x)$	$a = 0$
71. $f(x) = \sqrt{1 - x}$	$a = 0$
72. $f(x) = \frac{\sqrt{x}}{x - 1}$	$a = 2$

73. Determining Concavity Use a graphing utility to graph

$$y = x \sin \frac{1}{x}.$$

Show that the graph is concave downward to the right of

$$x = \frac{1}{\pi}.$$

74. Point of Inflection and Extrema Show that the point of inflection of

$$f(x) = x(x - 6)^2$$

lies midway between the relative extrema of f.

True or False? In Exercises 75–78, determine whether the statement is true or false. If it is false, explain why or give an example that shows it is false.

75. The graph of every cubic polynomial has precisely one point of inflection.

76. The graph of

$$f(x) = \frac{1}{x}$$

is concave downward for $x < 0$ and concave upward for $x > 0$, and thus it has a point of inflection at $x = 0$.

77. If $f'(c) > 0$, then f is concave upward at $x = c$.

78. If $f''(2) = 0$, then the graph of f must have a point of inflection at $x = 2$.

Proof In Exercises 79 and 80, let f and g represent differentiable functions such that $f'' \ne 0$ and $g'' \ne 0$.

79. Show that if f and g are concave upward on the interval (a, b), then $f + g$ is also concave upward on (a, b).

80. Prove that if f and g are positive, increasing, and concave upward on the interval (a, b), then fg is also concave upward on (a, b).

3.5 Limits at Infinity

- **Determine (finite) limits at infinity.**
- **Determine the horizontal asymptotes, if any, of the graph of a function.**
- **Determine infinite limits at infinity.**

Limits at Infinity

This section discusses the "end behavior" of a function on an *infinite* interval. Consider the graph of

$$f(x) = \frac{3x^2}{x^2 + 1}$$

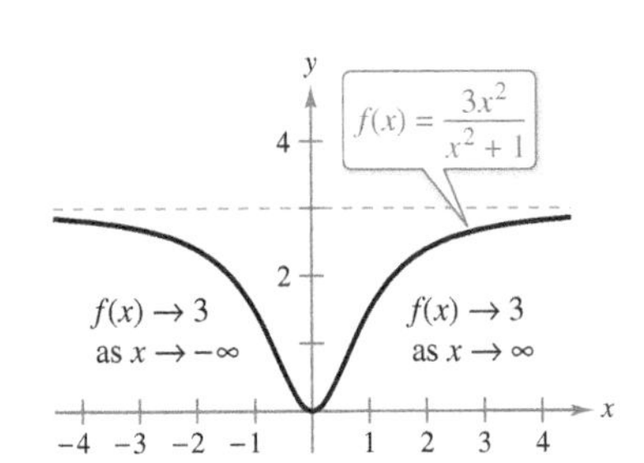

The limit of $f(x)$ as x approaches $-\infty$ or ∞ is 3.
Figure 3.32

as shown in Figure 3.32. Graphically, you can see that $f(x)$ appears to approach 3 as x increases without bound or decreases without bound. You can come to the same conclusions numerically, as shown in the table.

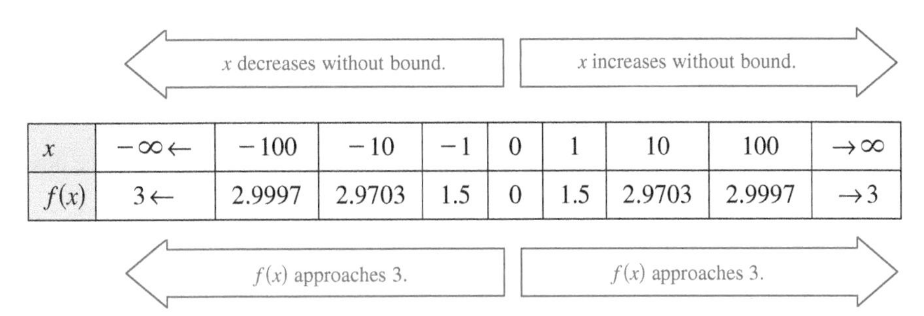

x	$-\infty \leftarrow$	-100	-10	-1	0	1	10	100	$\to \infty$
$f(x)$	$3 \leftarrow$	2.9997	2.9703	1.5	0	1.5	2.9703	2.9997	$\to 3$

The table suggests that $f(x)$ approaches 3 as x increases without bound $(x \to \infty)$. Similarly, $f(x)$ approaches 3 as x decreases without bound $(x \to -\infty)$. These **limits at infinity** are denoted by

$$\lim_{x \to -\infty} f(x) = 3 \qquad \text{Limit at negative infinity}$$

and

$$\lim_{x \to \infty} f(x) = 3. \qquad \text{Limit at positive infinity}$$

REMARK The statement $\lim_{x \to -\infty} f(x) = L$ or $\lim_{x \to \infty} f(x) = L$ means that the limit exists *and* the limit is equal to L.

To say that a statement is true as x increases *without bound* means that for some (large) real number M, the statement is true for *all* x in the interval $\{x: x > M\}$. The next definition uses this concept.

Definition of Limits at Infinity

Let L be a real number.

1. The statement $\lim_{x \to \infty} f(x) = L$ means that for each $\varepsilon > 0$ there exists an $M > 0$ such that $|f(x) - L| < \varepsilon$ whenever $x > M$.
2. The statement $\lim_{x \to -\infty} f(x) = L$ means that for each $\varepsilon > 0$ there exists an $N < 0$ such that $|f(x) - L| < \varepsilon$ whenever $x < N$.

$\lim_{x \to \infty} f(x) = L$

$f(x)$ is within ε units of L as $x \to \infty$.
Figure 3.33

The definition of a limit at infinity is shown in Figure 3.33. In this figure, note that for a given positive number ε, there exists a positive number M such that, for $x > M$, the graph of f will lie between the horizontal lines

$$y = L + \varepsilon \quad \text{and} \quad y = L - \varepsilon.$$

Exploration

Use a graphing utility to graph

$$f(x) = \frac{2x^2 + 4x - 6}{3x^2 + 2x - 16}.$$

Describe all the important features of the graph. Can you find a single viewing window that shows all of these features clearly? Explain your reasoning.

What are the horizontal asymptotes of the graph? How far to the right do you have to move on the graph so that the graph is within 0.001 unit of its horizontal asymptote? Explain your reasoning.

Horizontal Asymptotes

In Figure 3.33, the graph of f approaches the line $y = L$ as x increases without bound. The line $y = L$ is called a **horizontal asymptote** of the graph of f.

Definition of a Horizontal Asymptote

The line $y = L$ is a **horizontal asymptote** of the graph of f when

$$\lim_{x\to-\infty} f(x) = L \quad \text{or} \quad \lim_{x\to\infty} f(x) = L.$$

Note that from this definition, it follows that the graph of a *function* of x can have at most two horizontal asymptotes—one to the right and one to the left.

Limits at infinity have many of the same properties of limits discussed in Section 1.3. For example, if $\lim_{x\to\infty} f(x)$ and $\lim_{x\to\infty} g(x)$ both exist, then

$$\lim_{x\to\infty} [f(x) + g(x)] = \lim_{x\to\infty} f(x) + \lim_{x\to\infty} g(x)$$

and

$$\lim_{x\to\infty} [f(x)g(x)] = \left[\lim_{x\to\infty} f(x)\right]\left[\lim_{x\to\infty} g(x)\right].$$

Similar properties hold for limits at $-\infty$.

When evaluating limits at infinity, the next theorem is helpful.

THEOREM 3.10 Limits at Infinity

If r is a positive rational number and c is any real number, then

$$\lim_{x\to\infty} \frac{c}{x^r} = 0.$$

Furthermore, if x^r is defined when $x < 0$, then

$$\lim_{x\to-\infty} \frac{c}{x^r} = 0.$$

A proof of this theorem is given in Appendix A.

EXAMPLE 1 Finding a Limit at Infinity

Find the limit: $\lim_{x\to\infty} \left(5 - \frac{2}{x^2}\right)$.

Solution Using Theorem 3.10, you can write

$$\lim_{x\to\infty} \left(5 - \frac{2}{x^2}\right) = \lim_{x\to\infty} 5 - \lim_{x\to\infty} \frac{2}{x^2} \qquad \text{Property of limits}$$

$$= 5 - 0$$

$$= 5.$$

So, the line $y = 5$ is a horizontal asymptote to the right. By finding the limit

$$\lim_{x\to-\infty} \left(5 - \frac{2}{x^2}\right) \qquad \text{Limit as } x \to -\infty$$

you can see that $y = 5$ is also a horizontal asymptote to the left. The graph of the function $f(x) = 5 - (2/x^2)$ is shown in Figure 3.34.

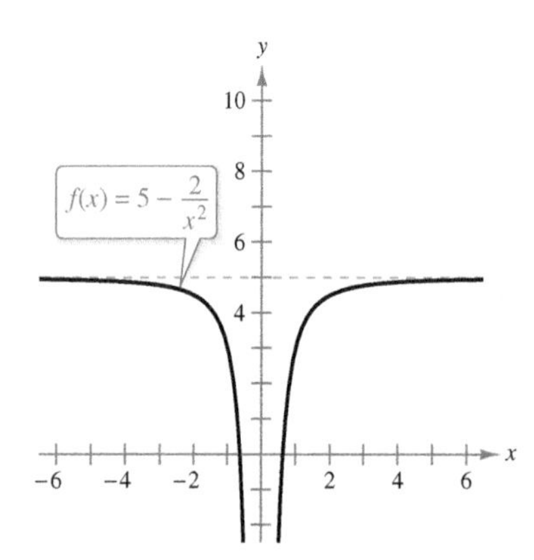

$y = 5$ is a horizontal asymptote.
Figure 3.34

EXAMPLE 2 **Finding a Limit at Infinity**

Find the limit: $\lim_{x\to\infty} \frac{2x-1}{x+1}$.

Solution Note that both the numerator and the denominator approach infinity as x approaches infinity.

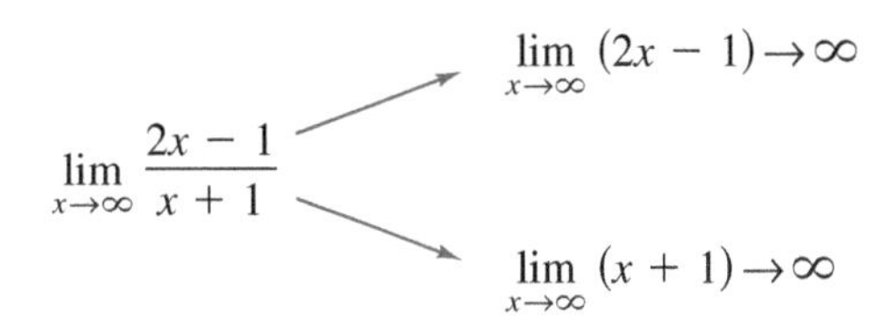

This results in ∞/∞, an **indeterminate form.** To resolve this problem, you can divide both the numerator and the denominator by x. After dividing, the limit may be evaluated as shown.

REMARK When you encounter an indeterminate form such as the one in Example 2, you should divide the numerator and denominator by the highest power of x in the *denominator.*

$$\lim_{x\to\infty} \frac{2x-1}{x+1} = \lim_{x\to\infty} \frac{\frac{2x-1}{x}}{\frac{x+1}{x}} \qquad \text{Divide numerator and denominator by } x.$$

$$= \lim_{x\to\infty} \frac{2-\frac{1}{x}}{1+\frac{1}{x}} \qquad \text{Simplify.}$$

$$= \frac{\lim_{x\to\infty} 2 - \lim_{x\to\infty} \frac{1}{x}}{\lim_{x\to\infty} 1 + \lim_{x\to\infty} \frac{1}{x}} \qquad \text{Take limits of numerator and denominator.}$$

$$= \frac{2-0}{1+0} \qquad \text{Apply Theorem 3.10.}$$

$$= 2$$

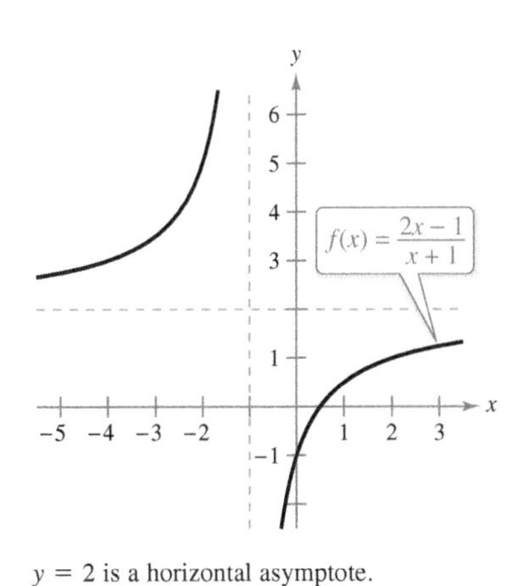

$y = 2$ is a horizontal asymptote.
Figure 3.35

So, the line $y = 2$ is a horizontal asymptote to the right. By taking the limit as $x \to -\infty$, you can see that $y = 2$ is also a horizontal asymptote to the left. The graph of the function is shown in Figure 3.35.

TECHNOLOGY You can test the reasonableness of the limit found in Example 2 by evaluating $f(x)$ for a few large positive values of x. For instance,

$f(100) \approx 1.9703$, $f(1000) \approx 1.9970$, and $f(10{,}000) \approx 1.9997$.

Another way to test the reasonableness of the limit is to use a graphing utility. For instance, in Figure 3.36, the graph of

$$f(x) = \frac{2x-1}{x+1}$$

is shown with the horizontal line $y = 2$. Note that as x increases, the graph of f moves closer and closer to its horizontal asymptote.

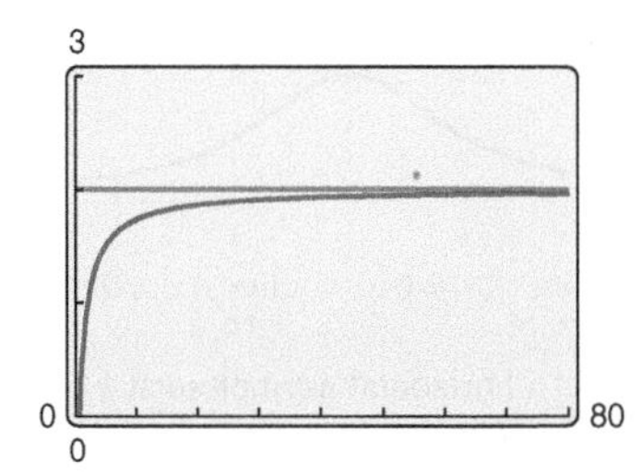

As x increases, the graph of f moves closer and closer to the line $y = 2$.
Figure 3.36

EXAMPLE 3 A Comparison of Three Rational Functions

See LarsonCalculus.com for an interactive version of this type of example.

Find each limit.

a. $\lim_{x\to\infty} \dfrac{2x+5}{3x^2+1}$ **b.** $\lim_{x\to\infty} \dfrac{2x^2+5}{3x^2+1}$ **c.** $\lim_{x\to\infty} \dfrac{2x^3+5}{3x^2+1}$

Solution In each case, attempting to evaluate the limit produces the indeterminate form ∞/∞.

a. Divide both the numerator and the denominator by x^2.

$$\lim_{x\to\infty} \frac{2x+5}{3x^2+1} = \lim_{x\to\infty} \frac{(2/x)+(5/x^2)}{3+(1/x^2)} = \frac{0+0}{3+0} = \frac{0}{3} = 0$$

b. Divide both the numerator and the denominator by x^2.

$$\lim_{x\to\infty} \frac{2x^2+5}{3x^2+1} = \lim_{x\to\infty} \frac{2+(5/x^2)}{3+(1/x^2)} = \frac{2+0}{3+0} = \frac{2}{3}$$

c. Divide both the numerator and the denominator by x^2.

$$\lim_{x\to\infty} \frac{2x^3+5}{3x^2+1} = \lim_{x\to\infty} \frac{2x+(5/x^2)}{3+(1/x^2)} = \frac{\infty}{3}$$

You can conclude that the limit *does not exist* because the numerator increases without bound while the denominator approaches 3.

MARIA GAETANA AGNESI (1718–1799)

Agnesi was one of a handful of women to receive credit for significant contributions to mathematics before the twentieth century. In her early twenties, she wrote the first text that included both differential and integral calculus. By age 30, she was an honorary member of the faculty at the University of Bologna.
See LarsonCalculus.com to read more of this biography.
For more information on the contributions of women to mathematics, see the article "Why Women Succeed in Mathematics" by Mona Fabricant, Sylvia Svitak, and Patricia Clark Kenschaft in *Mathematics Teacher*. To view this article, go to *MathArticles.com.*

Example 3 suggests the guidelines below for finding limits at infinity of rational functions. Use these guidelines to check the results in Example 3.

GUIDELINES FOR FINDING LIMITS AT $\pm\infty$ OF RATIONAL FUNCTIONS

1. If the degree of the numerator is *less than* the degree of the denominator, then the limit of the rational function is 0.
2. If the degree of the numerator is *equal to* the degree of the denominator, then the limit of the rational function is the ratio of the leading coefficients.
3. If the degree of the numerator is *greater than* the degree of the denominator, then the limit of the rational function does not exist.

The guidelines for finding limits at infinity of rational functions seem reasonable when you consider that for large values of x, the highest-power term of the rational function is the most "influential" in determining the limit. For instance,

$$\lim_{x\to\infty} \frac{1}{x^2+1}$$

is 0 because the denominator overpowers the numerator as x increases or decreases without bound, as shown in Figure 3.37.

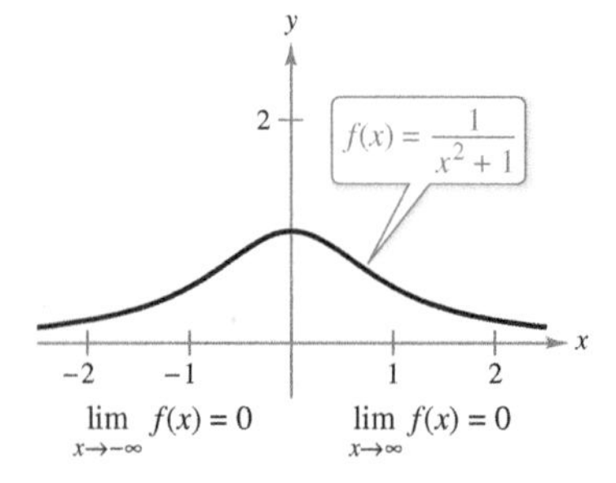

f has a horizontal asymptote at $y = 0$.
Figure 3.37

The function shown in Figure 3.37 is a special case of a type of curve studied by the Italian mathematician Maria Gaetana Agnesi. The general form of this function is

$$f(x) = \frac{8a^3}{x^2+4a^2} \qquad \text{Witch of Agnesi}$$

and, through a mistranslation of the Italian word *vertéré*, the curve has come to be known as the Witch of Agnesi. Agnesi's work with this curve first appeared in a comprehensive text on calculus that was published in 1748.

The Granger Collection, NY

In Figure 3.37, you can see that the function

$$f(x) = \frac{1}{x^2 + 1}$$

approaches the same horizontal asymptote to the right and to the left. This is always true of rational functions. Functions that are not rational, however, may approach different horizontal asymptotes to the right and to the left. This is demonstrated in Example 4.

EXAMPLE 4 A Function with Two Horizontal Asymptotes

Find each limit.

a. $\lim\limits_{x\to\infty} \dfrac{3x - 2}{\sqrt{2x^2 + 1}}$ **b.** $\lim\limits_{x\to-\infty} \dfrac{3x - 2}{\sqrt{2x^2 + 1}}$

Solution

a. For $x > 0$, you can write $x = \sqrt{x^2}$. So, dividing both the numerator and the denominator by x produces

$$\frac{3x - 2}{\sqrt{2x^2 + 1}} = \frac{\frac{3x - 2}{x}}{\frac{\sqrt{2x^2 + 1}}{\sqrt{x^2}}} = \frac{3 - \frac{2}{x}}{\sqrt{\frac{2x^2 + 1}{x^2}}} = \frac{3 - \frac{2}{x}}{\sqrt{2 + \frac{1}{x^2}}}$$

and you can take the limit as follows.

$$\lim_{x\to\infty} \frac{3x - 2}{\sqrt{2x^2 + 1}} = \lim_{x\to\infty} \frac{3 - \frac{2}{x}}{\sqrt{2 + \frac{1}{x^2}}} = \frac{3 - 0}{\sqrt{2 + 0}} = \frac{3}{\sqrt{2}}$$

b. For $x < 0$, you can write $x = -\sqrt{x^2}$. So, dividing both the numerator and the denominator by x produces

$$\frac{3x - 2}{\sqrt{2x^2 + 1}} = \frac{\frac{3x - 2}{x}}{\frac{\sqrt{2x^2 + 1}}{-\sqrt{x^2}}} = \frac{3 - \frac{2}{x}}{-\sqrt{\frac{2x^2 + 1}{x^2}}} = \frac{3 - \frac{2}{x}}{-\sqrt{2 + \frac{1}{x^2}}}$$

and you can take the limit as follows.

$$\lim_{x\to-\infty} \frac{3x - 2}{\sqrt{2x^2 + 1}} = \lim_{x\to-\infty} \frac{3 - \frac{2}{x}}{-\sqrt{2 + \frac{1}{x^2}}} = \frac{3 - 0}{-\sqrt{2 + 0}} = -\frac{3}{\sqrt{2}}$$

The graph of $f(x) = (3x - 2)/\sqrt{2x^2 + 1}$ is shown in Figure 3.38.

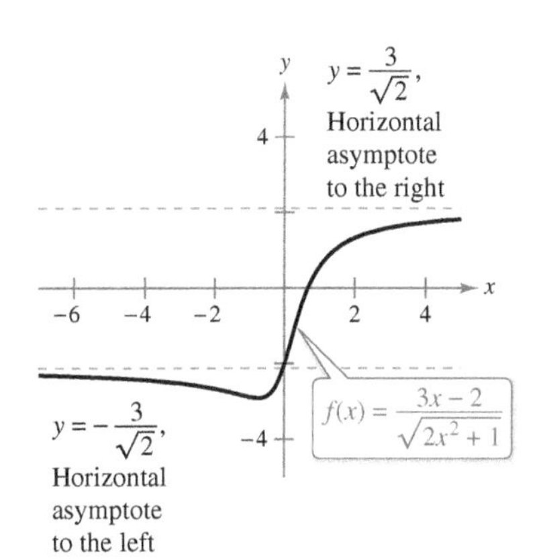

Functions that are not rational may have different right and left horizontal asymptotes.
Figure 3.38

TECHNOLOGY PITFALL If you use a graphing utility to estimate a limit, be sure that you also confirm the estimate analytically—the graphs shown by a graphing utility can be misleading. For instance, Figure 3.39 shows one view of the graph of

$$y = \frac{2x^3 + 1000x^2 + x}{x^3 + 1000x^2 + x + 1000}.$$

From this view, one could be convinced that the graph has $y = 1$ as a horizontal asymptote. An analytical approach shows that the horizontal asymptote is actually $y = 2$. Confirm this by enlarging the viewing window on the graphing utility.

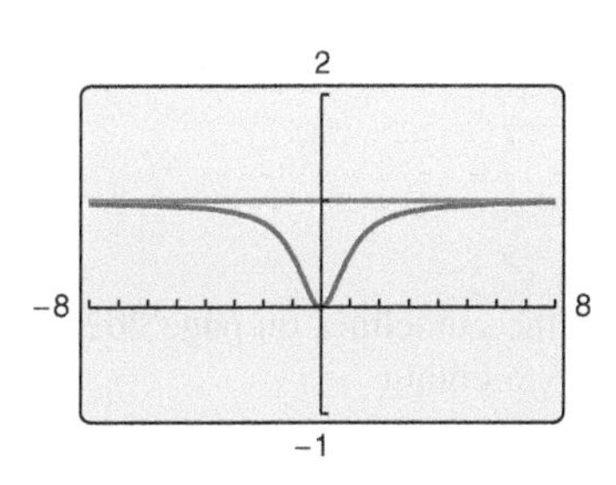

The horizontal asymptote appears to be the line $y = 1$, but it is actually the line $y = 2$.
Figure 3.39

In Section 1.4, Example 7(c), you used the Squeeze Theorem to evaluate a limit involving a trigonometric function. The Squeeze Theorem is also valid for limits at infinity.

EXAMPLE 5 Limits Involving Trigonometric Functions

Find each limit.

a. $\lim_{x\to\infty} \sin x$ **b.** $\lim_{x\to\infty} \frac{\sin x}{x}$

Solution

a. As x approaches infinity, the sine function oscillates between 1 and -1. So, this limit does not exist.

b. Because $-1 \le \sin x \le 1$, it follows that for $x > 0$,

$$-\frac{1}{x} \le \frac{\sin x}{x} \le \frac{1}{x}$$

where

$$\lim_{x\to\infty}\left(-\frac{1}{x}\right) = 0 \quad \text{and} \quad \lim_{x\to\infty}\frac{1}{x} = 0.$$

So, by the Squeeze Theorem, you obtain

$$\lim_{x\to\infty}\frac{\sin x}{x} = 0$$

as shown in Figure 3.40.

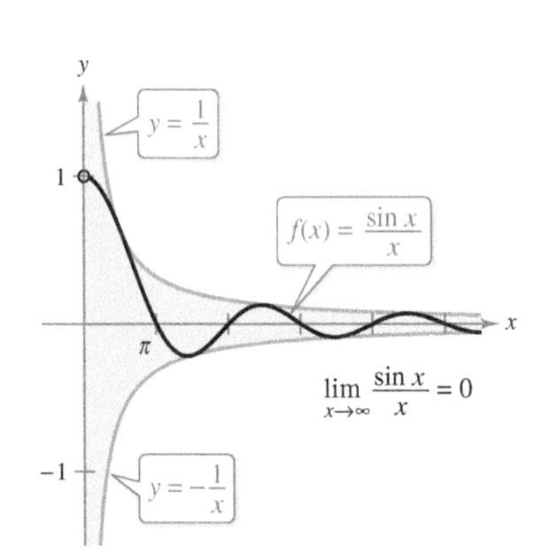

As x increases without bound, $f(x)$ approaches 0.
Figure 3.40

EXAMPLE 6 Oxygen Level in a Pond

Let $f(t)$ measure the level of oxygen in a pond, where $f(t) = 1$ is the normal (unpolluted) level and the time t is measured in weeks. When $t = 0$, organic waste is dumped into the pond, and as the waste material oxidizes, the level of oxygen in the pond is

$$f(t) = \frac{t^2 - t + 1}{t^2 + 1}.$$

What percent of the normal level of oxygen exists in the pond after 1 week? After 2 weeks? After 10 weeks? What is the limit as t approaches infinity?

Solution When $t = 1$, 2, and 10, the levels of oxygen are as shown.

$$f(1) = \frac{1^2 - 1 + 1}{1^2 + 1} = \frac{1}{2} = 50\% \qquad \text{1 week}$$

$$f(2) = \frac{2^2 - 2 + 1}{2^2 + 1} = \frac{3}{5} = 60\% \qquad \text{2 weeks}$$

$$f(10) = \frac{10^2 - 10 + 1}{10^2 + 1} = \frac{91}{101} \approx 90.1\% \qquad \text{10 weeks}$$

To find the limit as t approaches infinity, you can use the guidelines on page 202, or you can divide the numerator and the denominator by t^2 to obtain

$$\lim_{t\to\infty}\frac{t^2 - t + 1}{t^2 + 1} = \lim_{t\to\infty}\frac{1 - (1/t) + (1/t^2)}{1 + (1/t^2)} = \frac{1 - 0 + 0}{1 + 0} = 1 = 100\%.$$

See Figure 3.41.

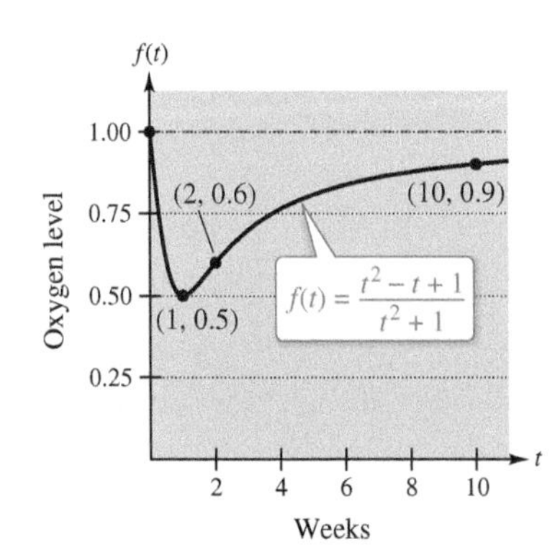

The level of oxygen in a pond approaches the normal level of 1 as t approaches ∞.
Figure 3.41

Infinite Limits at Infinity

Many functions do not approach a finite limit as x increases (or decreases) without bound. For instance, no polynomial function has a finite limit at infinity. The next definition is used to describe the behavior of polynomial and other functions at infinity.

REMARK Determining whether a function has an infinite limit at infinity is useful in analyzing the "end behavior" of its graph. You will see examples of this in Section 3.6 on curve sketching.

> **Definition of Infinite Limits at Infinity**
>
> Let f be a function defined on the interval (a, ∞).
>
> **1.** The statement $\lim_{x\to\infty} f(x) = \infty$ means that for each positive number M, there is a corresponding number $N > 0$ such that $f(x) > M$ whenever $x > N$.
>
> **2.** The statement $\lim_{x\to\infty} f(x) = -\infty$ means that for each negative number M, there is a corresponding number $N > 0$ such that $f(x) < M$ whenever $x > N$.
>
> Similar definitions can be given for the statements
>
> $$\lim_{x\to-\infty} f(x) = \infty \quad \text{and} \quad \lim_{x\to-\infty} f(x) = -\infty.$$

EXAMPLE 7 Finding Infinite Limits at Infinity

Find each limit.

a. $\lim_{x\to\infty} x^3$ **b.** $\lim_{x\to-\infty} x^3$

Solution

a. As x increases without bound, x^3 also increases without bound. So, you can write

$$\lim_{x\to\infty} x^3 = \infty.$$

b. As x decreases without bound, x^3 also decreases without bound. So, you can write

$$\lim_{x\to-\infty} x^3 = -\infty.$$

The graph of $f(x) = x^3$ in Figure 3.42 illustrates these two results. These results agree with the Leading Coefficient Test for polynomial functions as described in Section P.3.

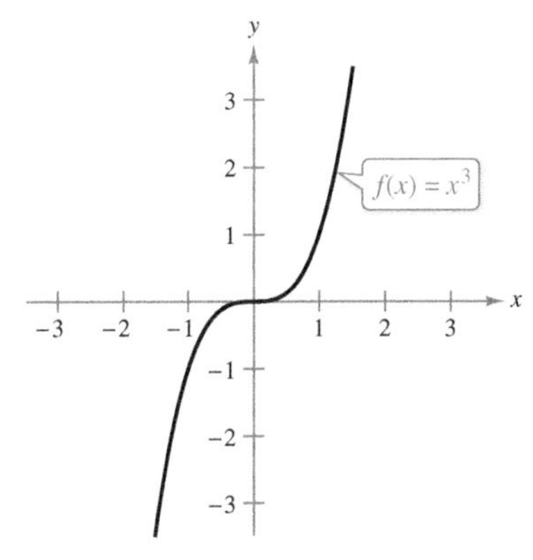

Figure 3.42

EXAMPLE 8 Finding Infinite Limits at Infinity

Find each limit.

a. $\lim_{x\to\infty} \dfrac{2x^2 - 4x}{x + 1}$ **b.** $\lim_{x\to-\infty} \dfrac{2x^2 - 4x}{x + 1}$

Solution One way to evaluate each of these limits is to use long division to rewrite the improper rational function as the sum of a polynomial and a rational function.

a. $$\lim_{x\to\infty} \frac{2x^2 - 4x}{x + 1} = \lim_{x\to\infty} \left(2x - 6 + \frac{6}{x + 1}\right) = \infty$$

b. $$\lim_{x\to-\infty} \frac{2x^2 - 4x}{x + 1} = \lim_{x\to-\infty} \left(2x - 6 + \frac{6}{x + 1}\right) = -\infty$$

The statements above can be interpreted as saying that as x approaches $\pm\infty$, the function $f(x) = (2x^2 - 4x)/(x + 1)$ behaves like the function $g(x) = 2x - 6$. In Section 3.6, you will see that this is graphically described by saying that the line $y = 2x - 6$ is a *slant asymptote* of the graph of f, as shown in Figure 3.43.

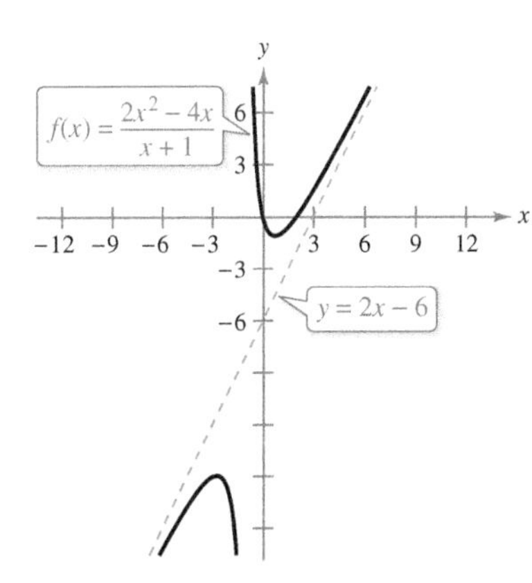

Figure 3.43

3.5 Exercises

See CalcChat.com for tutorial help and worked-out solutions to odd-numbered exercises.

CONCEPT CHECK

1. **Writing** Describe in your own words what each statement means.

   (a) $\lim_{x\to\infty} f(x) = -5$

   (b) $\lim_{x\to-\infty} f(x) = 3$

2. **Horizontal Asymptote** What does it mean for the graph of a function to have a horizontal asymptote?

3. **Horizontal Asymptote** A graph can have a maximum of how many horizontal asymptotes? Explain.

4. **Limits at Infinity** In your own words, summarize the guidelines for finding limits at infinity of rational functions.

Matching **In Exercises 5–10, match the function with its graph using horizontal asymptotes as an aid. [The graphs are labeled (a), (b), (c), (d), (e), and (f).]**

(a)

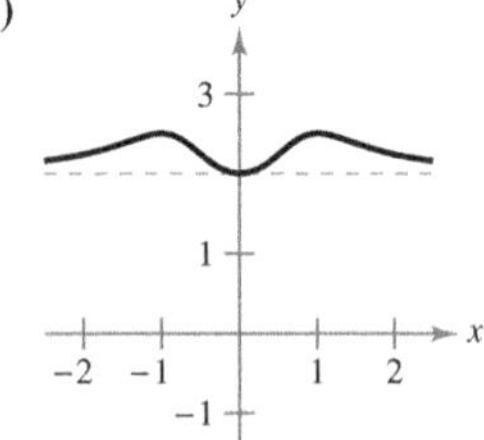

(b)

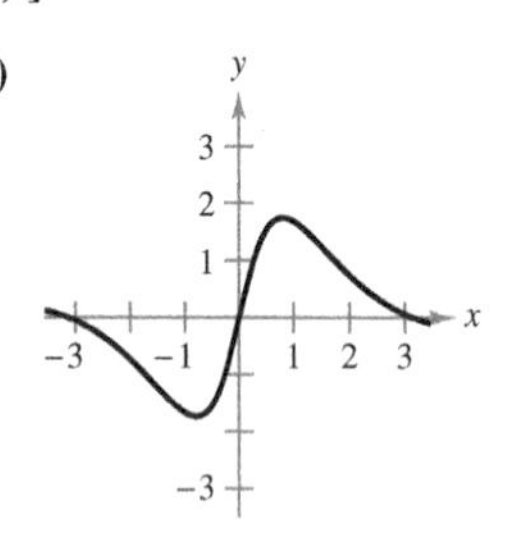

(c)

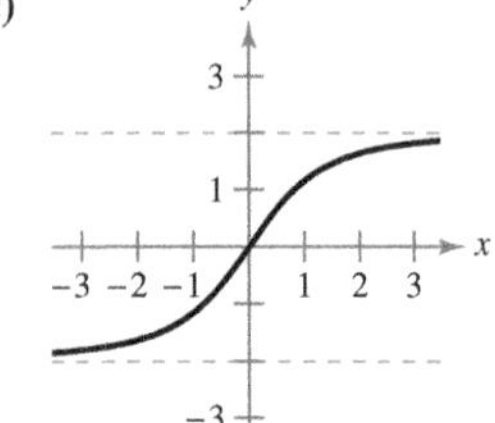

(d)

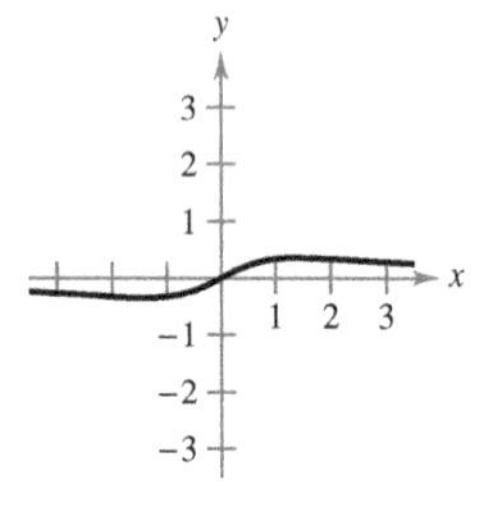

(e)

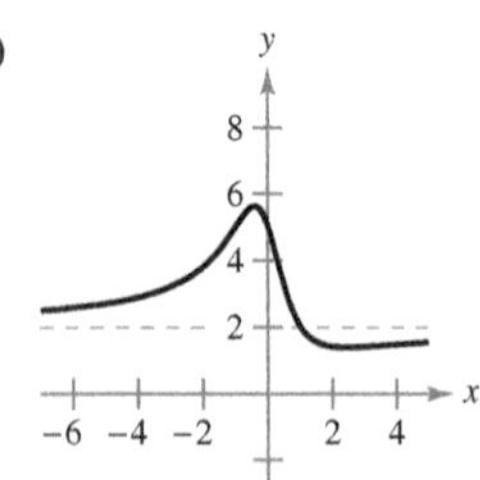

(f)

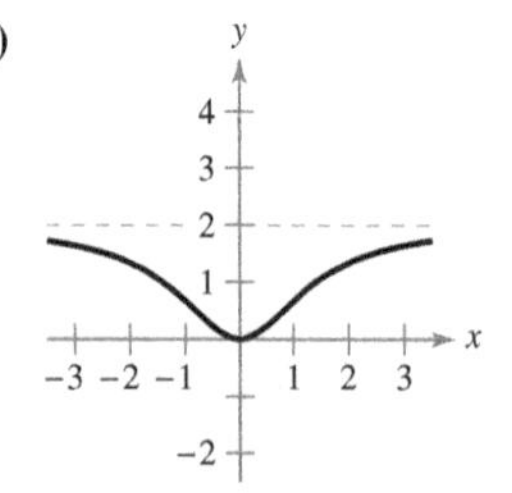

5. $f(x) = \dfrac{2x^2}{x^2+2}$

6. $f(x) = \dfrac{2x}{\sqrt{x^2+2}}$

7. $f(x) = \dfrac{x}{x^2+2}$

8. $f(x) = 2 + \dfrac{x^2}{x^4+1}$

9. $f(x) = \dfrac{4\sin x}{x^2+1}$

10. $f(x) = \dfrac{2x^2-3x+5}{x^2+1}$

Finding Limits at Infinity **In Exercises 11 and 12, find $\lim_{x\to\infty} h(x)$, if it exists.**

11. $f(x) = 5x^3 - 3$

    (a) $h(x) = \dfrac{f(x)}{x^2}$

    (b) $h(x) = \dfrac{f(x)}{x^3}$

    (c) $h(x) = \dfrac{f(x)}{x^4}$

12. $f(x) = -4x^2 + 2x - 5$

    (a) $h(x) = \dfrac{f(x)}{x}$

    (b) $h(x) = \dfrac{f(x)}{x^2}$

    (c) $h(x) = \dfrac{f(x)}{x^3}$

Finding Limits at Infinity **In Exercises 13–16, find each limit, if it exists.**

13. (a) $\lim_{x\to\infty} \dfrac{x^2+2}{x^3-1}$

    (b) $\lim_{x\to\infty} \dfrac{x^2+2}{x^2-1}$

    (c) $\lim_{x\to\infty} \dfrac{x^2+2}{x-1}$

14. (a) $\lim_{x\to\infty} \dfrac{3-2x}{3x^3-1}$

    (b) $\lim_{x\to\infty} \dfrac{3-2x}{3x-1}$

    (c) $\lim_{x\to\infty} \dfrac{3-2x^2}{3x-1}$

15. (a) $\lim_{x\to\infty} \dfrac{5-2x^{3/2}}{3x^2-4}$

    (b) $\lim_{x\to\infty} \dfrac{5-2x^{3/2}}{3x^{3/2}-4}$

    (c) $\lim_{x\to\infty} \dfrac{5-2x^{3/2}}{3x-4}$

16. (a) $\lim_{x\to\infty} \dfrac{5x^{3/2}}{4x^2+1}$

    (b) $\lim_{x\to\infty} \dfrac{5x^{3/2}}{4x^{3/2}+1}$

    (c) $\lim_{x\to\infty} \dfrac{5x^{3/2}}{4\sqrt{x}+1}$

Finding a Limit **In Exercises 17–36, find the limit, if it exists.**

17. $\lim_{x\to\infty} \left(4 + \dfrac{3}{x}\right)$

18. $\lim_{x\to-\infty} \left(\dfrac{5}{x} - \dfrac{x}{3}\right)$

19. $\lim_{x\to\infty} \dfrac{7x+6}{9x-4}$

20. $\lim_{x\to-\infty} \dfrac{4x^2+5}{x^2+3}$

21. $\lim_{x\to-\infty} \dfrac{2x^2+x}{6x^3+2x^2+x}$

22. $\lim_{x\to\infty} \dfrac{5x^3+1}{10x^3-3x^2+7}$

23. $\lim_{x\to-\infty} \dfrac{5x^2}{x+3}$

24. $\lim_{x\to-\infty} \dfrac{x^3-4}{x^2+1}$

25. $\lim_{x\to-\infty} \dfrac{x}{\sqrt{x^2-x}}$

26. $\lim_{x\to-\infty} \dfrac{x}{\sqrt{x^2+1}}$

27. $\lim_{x\to-\infty} \dfrac{2x+1}{\sqrt{x^2-x}}$

28. $\lim_{x\to\infty} \dfrac{5x^2+2}{\sqrt{x^2+3}}$

29. $\lim_{x\to\infty} \dfrac{\sqrt{x^2-1}}{2x-1}$

30. $\lim_{x\to-\infty} \dfrac{\sqrt{x^4-1}}{x^3-1}$

31. $\lim_{x\to\infty} \dfrac{x+1}{(x^2+1)^{1/3}}$

32. $\lim_{x\to-\infty} \dfrac{2x}{(x^6-1)^{1/3}}$

33. $\lim_{x\to\infty} \dfrac{1}{2x+\sin x}$

34. $\lim_{x\to\infty} \cos\dfrac{1}{x}$

35. $\lim_{x\to\infty} \dfrac{\sin 2x}{x}$

36. $\lim_{x\to\infty} \dfrac{x-\cos x}{x}$

Finding Horizontal Asymptotes Using Technology **In Exercises 37–40, use a graphing utility to graph the function and identify any horizontal asymptotes.**

37. $f(x) = \dfrac{|x|}{x+1}$

38. $f(x) = \dfrac{|3x+2|}{x-2}$

39. $f(x) = \dfrac{3x}{\sqrt{x^2+2}}$

40. $f(x) = \dfrac{\sqrt{9x^2-2}}{2x+1}$

Finding a Limit **In Exercises 41 and 42, find the limit. (*Hint:* Let $x = 1/t$ and find the limit as $t \to 0^+$.)**

41. $\lim\limits_{x\to\infty} x \sin \dfrac{1}{x}$

42. $\lim\limits_{x\to\infty} x \tan \dfrac{1}{x}$

Finding a Limit **In Exercises 43–46, find the limit. Use a graphing utility to verify your result. (*Hint:* Treat the expression as a fraction whose denominator is 1, and rationalize the numerator.)**

43. $\lim\limits_{x\to-\infty} \left(x + \sqrt{x^2+3}\right)$

44. $\lim\limits_{x\to\infty} \left(x - \sqrt{x^2+x}\right)$

45. $\lim\limits_{x\to-\infty} \left(3x + \sqrt{9x^2-x}\right)$

46. $\lim\limits_{x\to\infty} \left(4x - \sqrt{16x^2-x}\right)$

Numerical, Graphical, and Analytic Analysis **In Exercises 47–50, use a graphing utility to complete the table and estimate the limit as x approaches infinity. Then use a graphing utility to graph the function and estimate the limit. Finally, find the limit analytically and compare your results with the estimates.**

x	10^0	10^1	10^2	10^3	10^4	10^5	10^6
$f(x)$							

47. $f(x) = x - \sqrt{x(x-1)}$

48. $f(x) = x^2 - x\sqrt{x(x-1)}$

49. $f(x) = x \sin \dfrac{1}{2x}$

50. $f(x) = \dfrac{x+1}{x\sqrt{x}}$

51. **Engine Efficiency** The efficiency (in percent) of an internal combustion engine is

$$\text{Efficiency} = 100\left[1 - \frac{1}{(v_1/v_2)^c}\right]$$

where v_1/v_2 is the ratio of the uncompressed gas to the compressed gas and c is a positive constant dependent on the engine design. Find the limit of the efficiency as the compression ratio approaches infinity.

Straight 8 Photography/Shutterstock.com

52. **Physics** Newton's First Law of Motion and Einstein's Special Theory of Relativity differ concerning the behavior of a particle as its velocity approaches the speed of light c. In the graph, functions N and E represent the velocity v, with respect to time t, of a particle accelerated by a constant force as predicted by Newton and Einstein, respectively. Write limit statements that describe these two theories.

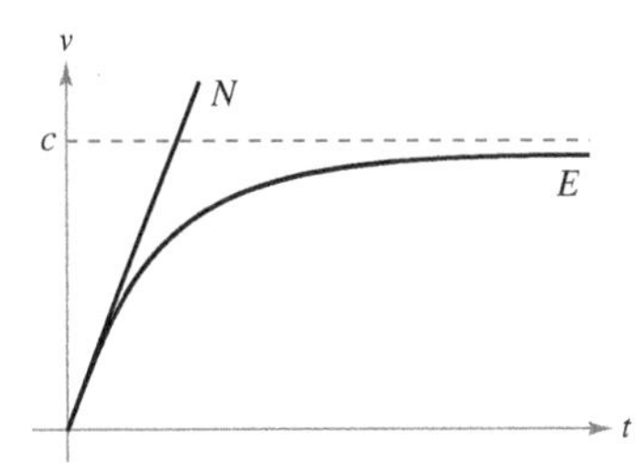

EXPLORING CONCEPTS

53. **Limits** Explain the differences between limits at infinity and infinite limits.

54. **Horizontal Asymptote** Can the graph of a function cross a horizontal asymptote? Explain.

55. **Using Symmetry to Find Limits** If f is a continuous function such that $\lim\limits_{x\to\infty} f(x) = 5$, find, if possible, $\lim\limits_{x\to-\infty} f(x)$ for each specified condition.
   (a) The graph of f is symmetric with respect to the y-axis.
   (b) The graph of f is symmetric with respect to the origin.

56. **HOW DO YOU SEE IT?** The graph shows the temperature T, in degrees Fahrenheit, of molten glass t seconds after it is removed from a kiln.

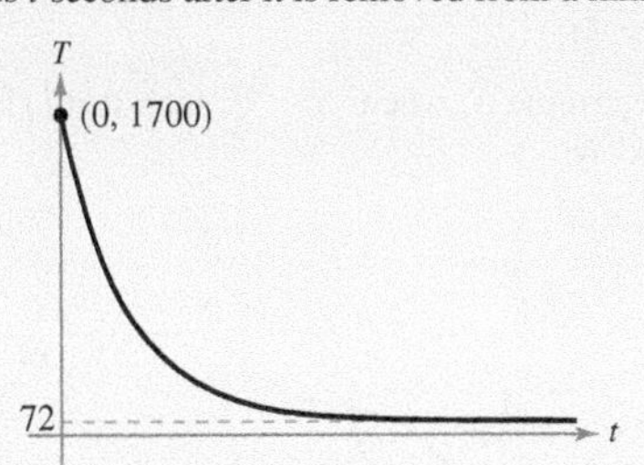

(a) Find $\lim\limits_{t\to0^+} T$. What does this limit represent?
(b) Find $\lim\limits_{t\to\infty} T$. What does this limit represent?

57. **Modeling Data** The average typing speeds S (in words per minute) of a typing student after t weeks of lessons are shown in the table.

t	5	10	15	20	25	30
S	28	56	79	90	93	94

A model for the data is $S = \dfrac{100t^2}{65+t^2}$, $t > 0$.

(a) Use a graphing utility to plot the data and graph the model.
(b) Does there appear to be a limiting typing speed? Explain.

58. Modeling Data A heat probe is attached to the heat exchanger of a heating system. The temperature T (in degrees Celsius) is recorded t seconds after the furnace is started. The results for the first 2 minutes are recorded in the table.

t	0	15	30	45	60
T	25.2	36.9	45.5	51.4	56.0

t	75	90	105	120
T	59.6	62.0	64.0	65.2

(a) Use the regression capabilities of a graphing utility to find a model of the form $T_1 = at^2 + bt + c$ for the data.

(b) Use a graphing utility to graph T_1.

(c) A rational model for the data is

$$T_2 = \frac{1451 + 86t}{58 + t}.$$

Use a graphing utility to graph T_2.

(d) Find $\lim_{t\to\infty} T_2$.

(e) Interpret the result in part (d) in the context of the problem. Is it possible to do this type of analysis using T_1? Explain.

59. Using the Definition of Limits at Infinity The graph of

$$f(x) = \frac{2x^2}{x^2 + 2}$$

is shown (see figure).

(a) Find $L = \lim_{x\to\infty} f(x)$.

(b) Determine x_1 and x_2 in terms of ε.

(c) Determine M, where $M > 0$, such that $|f(x) - L| < \varepsilon$ for $x > M$.

(d) Determine N, where $N < 0$, such that $|f(x) - L| < \varepsilon$ for $x < N$.

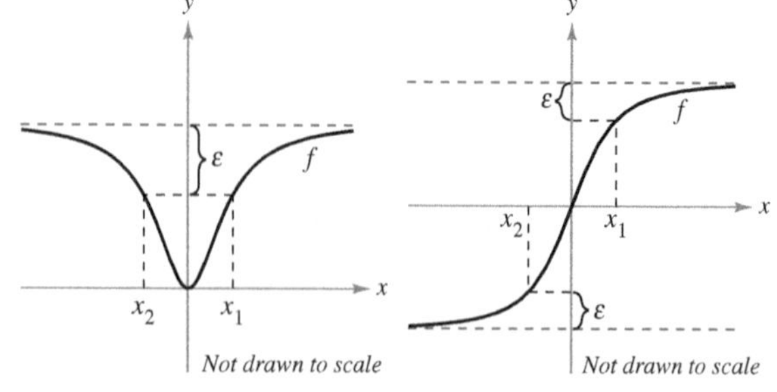

Figure for 59 Figure for 60

60. Using the Definition of Limits at Infinity The graph of $f(x) = \dfrac{6x}{\sqrt{x^2 + 2}}$ is shown (see figure).

(a) Find $L = \lim_{x\to\infty} f(x)$ and $K = \lim_{x\to-\infty} f(x)$.

(b) Determine x_1 and x_2 in terms of ε.

(c) Determine M, where $M > 0$, such that $|f(x) - L| < \varepsilon$ for $x > M$.

(d) Determine N, where $N < 0$, such that $|f(x) - K| < \varepsilon$ for $x < N$.

61. Using the Definition of Limits at Infinity Consider

$$\lim_{x\to\infty} \frac{3x}{\sqrt{x^2 + 3}}.$$

(a) Use the definition of limits at infinity to find the value of M that corresponds to $\varepsilon = 0.5$.

(b) Use the definition of limits at infinity to find the value of M that corresponds to $\varepsilon = 0.1$.

62. Using the Definition of Limits at Infinity Consider

$$\lim_{x\to-\infty} \frac{3x}{\sqrt{x^2 + 3}}.$$

(a) Use the definition of limits at infinity to find the value of N that corresponds to $\varepsilon = 0.5$.

(b) Use the definition of limits at infinity to find the value of N that corresponds to $\varepsilon = 0.1$.

Proof In Exercises 63–66, use the definition of limits at infinity to prove the limit.

63. $\lim_{x\to\infty} \dfrac{1}{x^2} = 0$

64. $\lim_{x\to\infty} \dfrac{2}{\sqrt{x}} = 0$

65. $\lim_{x\to-\infty} \dfrac{1}{x^3} = 0$

66. $\lim_{x\to-\infty} \dfrac{1}{x - 2} = 0$

67. Distance A line with slope m passes through the point $(0, 4)$.

(a) Write the distance d between the line and the point $(3, 1)$ as a function of m. (*Hint:* See Section P.2, Exercise 77.)

(b) Use a graphing utility to graph the equation in part (a).

(c) Find $\lim_{m\to\infty} d(m)$ and $\lim_{m\to-\infty} d(m)$. Interpret the results geometrically.

68. Distance A line with slope m passes through the point $(0, -2)$.

(a) Write the distance d between the line and the point $(4, 2)$ as a function of m. (*Hint:* See Section P.2, Exercise 77.)

(b) Use a graphing utility to graph the equation in part (a).

(c) Find $\lim_{m\to\infty} d(m)$ and $\lim_{m\to-\infty} d(m)$. Interpret the results geometrically.

69. Proof Prove that if

$$p(x) = a_n x^n + \cdots + a_1 x + a_0$$

and

$$q(x) = b_m x^m + \cdots + b_1 x + b_0$$

where $a_n \neq 0$ and $b_m \neq 0$, then

$$\lim_{x\to\infty} \frac{p(x)}{q(x)} = \begin{cases} 0, & n < m \\ \dfrac{a_n}{b_m}, & n = m. \\ \pm\infty, & n > m \end{cases}$$

70. Proof Use the definition of infinite limits at infinity to prove that $\lim_{x\to\infty} x^3 = \infty$.

3.6 A Summary of Curve Sketching

- **Analyze and sketch the graph of a function.**

Analyzing the Graph of a Function

It would be difficult to overstate the importance of using graphs in mathematics. Descartes's introduction of analytic geometry contributed significantly to the rapid advances in calculus that began during the mid-seventeenth century. In the words of Lagrange, "As long as algebra and geometry traveled separate paths their advance was slow and their applications limited. But when these two sciences joined company, they drew from each other fresh vitality and thenceforth marched on at a rapid pace toward perfection."

So far, you have studied several concepts that are useful in analyzing the graph of a function.

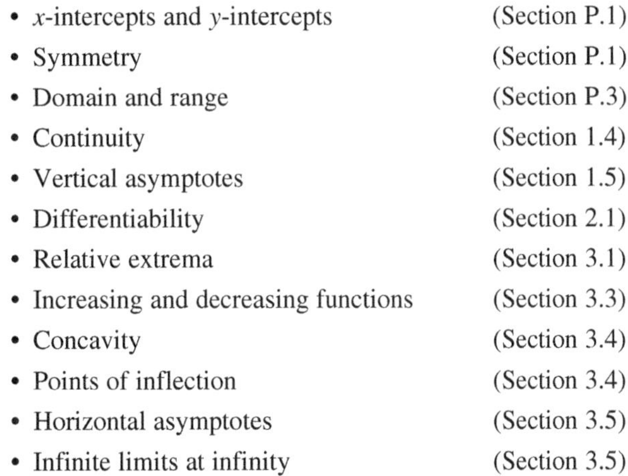
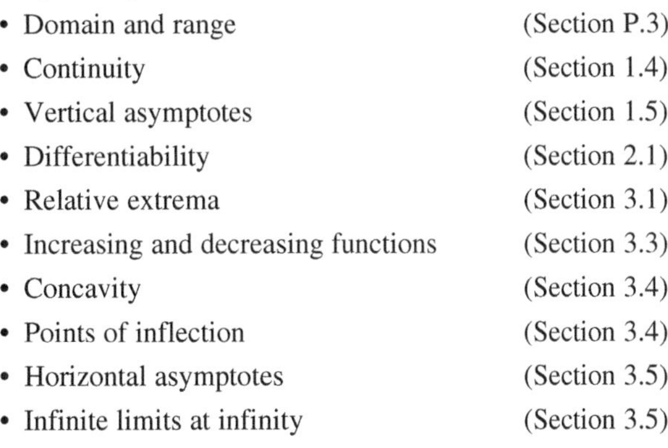

- x-intercepts and y-intercepts (Section P.1)
- Symmetry (Section P.1)
- Domain and range (Section P.3)
- Continuity (Section 1.4)
- Vertical asymptotes (Section 1.5)
- Differentiability (Section 2.1)
- Relative extrema (Section 3.1)
- Increasing and decreasing functions (Section 3.3)
- Concavity (Section 3.4)
- Points of inflection (Section 3.4)
- Horizontal asymptotes (Section 3.5)
- Infinite limits at infinity (Section 3.5)

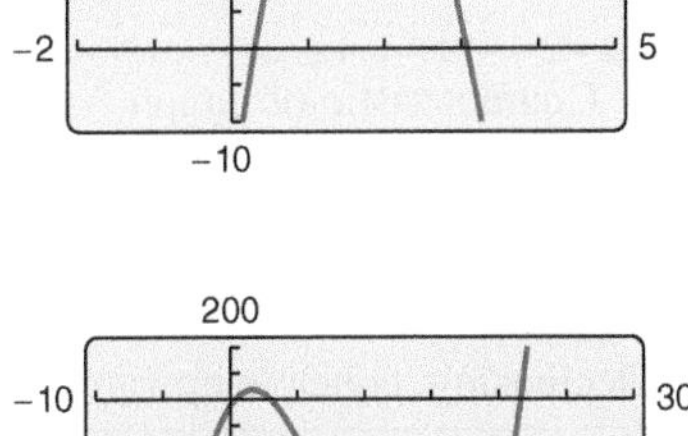

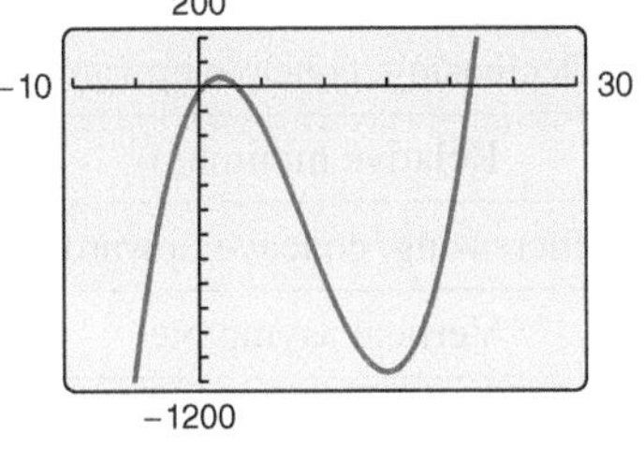

Different viewing windows for the graph of $f(x) = x^3 - 25x^2 + 74x - 20$
Figure 3.44

When you are sketching the graph of a function, either by hand or with a graphing utility, remember that normally you cannot show the *entire* graph. The decision as to which part of the graph you choose to show is often crucial. For instance, which of the viewing windows in Figure 3.44 better represents the graph of

$$f(x) = x^3 - 25x^2 + 74x - 20?$$

By seeing both views, it is clear that the second viewing window gives a more complete representation of the graph. But would a third viewing window reveal other interesting portions of the graph? To answer this, you need to use calculus to interpret the first and second derivatives. To determine a good viewing window for a function, use these guidelines to analyze its graph.

GUIDELINES FOR ANALYZING THE GRAPH OF A FUNCTION

1. Determine the domain and range of the function.
2. Determine the intercepts, asymptotes, and symmetry of the graph.
3. Locate the x-values for which $f'(x)$ and $f''(x)$ either are zero or do not exist. Use the results to determine relative extrema and points of inflection.

REMARK In these guidelines, note the importance of *algebra* (as well as calculus) for solving the equations $f(x) = 0$, $f'(x) = 0$, and $f''(x) = 0$.

EXAMPLE 1 Sketching the Graph of a Rational Function

Analyze and sketch the graph of

$$f(x) = \frac{2(x^2 - 9)}{x^2 - 4}.$$

Solution

Domain:	All real numbers except $x = \pm 2$
Range:	$(-\infty, 2) \cup \left[\frac{9}{2}, \infty\right)$
x-intercepts:	$(-3, 0), (3, 0)$
y-intercept:	$\left(0, \frac{9}{2}\right)$
Vertical asymptotes:	$x = -2, x = 2$
Horizontal asymptote:	$y = 2$
Symmetry:	With respect to y-axis
First derivative:	$f'(x) = \dfrac{20x}{(x^2 - 4)^2}$
Second derivative:	$f''(x) = \dfrac{-20(3x^2 + 4)}{(x^2 - 4)^3}$
Critical number:	$x = 0$
Possible points of inflection:	None
Test intervals:	$(-\infty, -2), (-2, 0), (0, 2), (2, \infty)$

Using calculus, you can be certain that you have determined all characteristics of the graph of f.
Figure 3.45

The table shows how the test intervals are used to determine several characteristics of the graph. The graph of f is shown in Figure 3.45.

	$f(x)$	$f'(x)$	$f''(x)$	Characteristic of Graph
$-\infty < x < -2$		$-$	$-$	Decreasing, concave downward
$x = -2$	Undef.	Undef.	Undef.	Vertical asymptote
$-2 < x < 0$		$-$	$+$	Decreasing, concave upward
$x = 0$	$\frac{9}{2}$	0	$+$	Relative minimum
$0 < x < 2$		$+$	$+$	Increasing, concave upward
$x = 2$	Undef.	Undef.	Undef.	Vertical asymptote
$2 < x < \infty$		$+$	$-$	Increasing, concave downward

FOR FURTHER INFORMATION For more information on the use of technology to graph rational functions, see the article "Graphs of Rational Functions for Computer Assisted Calculus" by Stan Byrd and Terry Walters in *The College Mathematics Journal*. To view this article, go to *MathArticles.com*.

Be sure you understand all of the implications of creating a table such as that shown in Example 1. By using calculus, you can be *sure* that the graph has no relative extrema or points of inflection other than those shown in Figure 3.45.

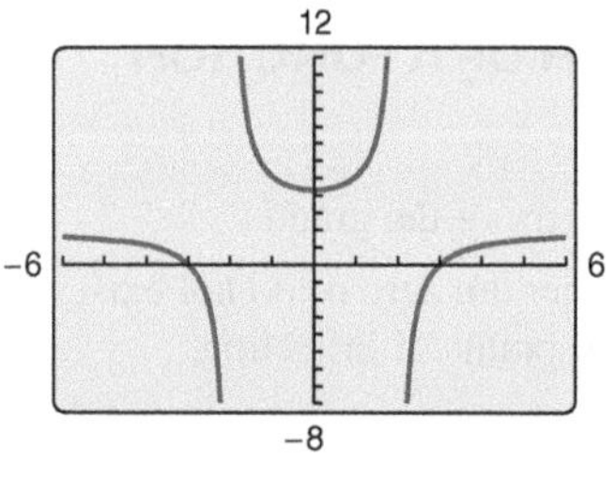

By not using calculus, you may overlook important characteristics of the graph of g.
Figure 3.46

TECHNOLOGY PITFALL Without using the type of analysis outlined in Example 1, it is easy to obtain an incomplete view of the basic characteristics of a graph. For instance, Figure 3.46 shows a view of the graph of

$$g(x) = \frac{2(x^2 - 9)(x - 20)}{(x^2 - 4)(x - 21)}.$$

From this view, it appears that the graph of g is about the same as the graph of f shown in Figure 3.45. The graphs of these two functions, however, differ significantly. Try enlarging the viewing window to see the differences.

EXAMPLE 2 Sketching the Graph of a Rational Function

Analyze and sketch the graph of $f(x) = \dfrac{x^2 - 2x + 4}{x - 2}$.

Solution

Domain:	All real numbers except $x = 2$
Range:	$(-\infty, -2] \cup [6, \infty)$
x-intercepts:	None
y-intercept:	$(0, -2)$
Vertical asymptote:	$x = 2$
Horizontal asymptotes:	None
Symmetry:	None
End behavior:	$\lim\limits_{x\to-\infty} f(x) = -\infty,\ \lim\limits_{x\to\infty} f(x) = \infty$
First derivative:	$f'(x) = \dfrac{x(x-4)}{(x-2)^2}$
Second derivative:	$f''(x) = \dfrac{8}{(x-2)^3}$
Critical numbers:	$x = 0, x = 4$
Possible points of inflection:	None
Test intervals:	$(-\infty, 0), (0, 2), (2, 4), (4, \infty)$

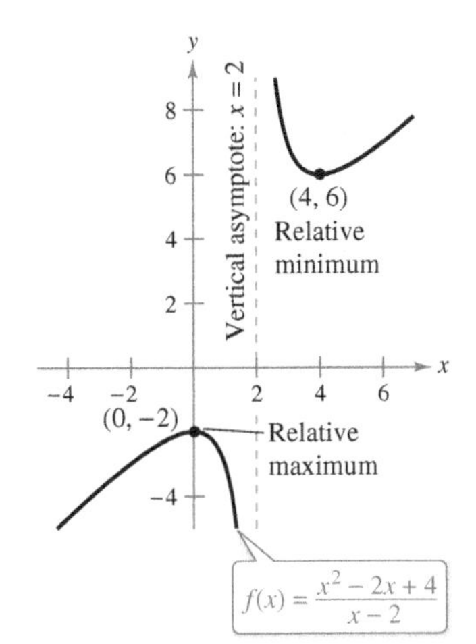

Figure 3.47

The analysis of the graph of f is shown in the table, and the graph is shown in Figure 3.47.

	$f(x)$	$f'(x)$	$f''(x)$	Characteristic of Graph
$-\infty < x < 0$		+	−	Increasing, concave downward
$x = 0$	−2	0	−	Relative maximum
$0 < x < 2$		−	−	Decreasing, concave downward
$x = 2$	Undef.	Undef.	Undef.	Vertical asymptote
$2 < x < 4$		−	+	Decreasing, concave upward
$x = 4$	6	0	+	Relative minimum
$4 < x < \infty$		+	+	Increasing, concave upward

Although the graph of the function in Example 2 has no horizontal asymptote, it does have a slant asymptote. The graph of a rational function (having no common factors and whose denominator is of degree 1 or greater) has a **slant asymptote** when the degree of the numerator exceeds the degree of the denominator by exactly 1. To find the slant asymptote, use long division to rewrite the rational function as the sum of a first-degree polynomial (the slant asymptote) and another rational function.

$$f(x) = \frac{x^2 - 2x + 4}{x - 2} \qquad \text{Write original equation.}$$

$$= x + \frac{4}{x - 2} \qquad \text{Rewrite using long division.}$$

A slant asymptote
Figure 3.48

In Figure 3.48, note that the graph of f approaches the slant asymptote $y = x$ as x approaches $-\infty$ or ∞.

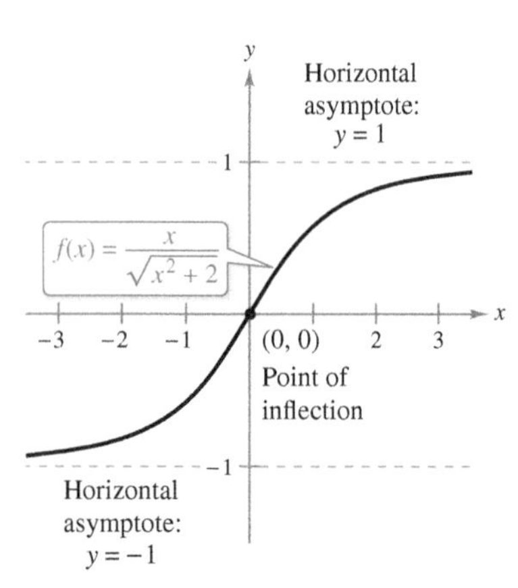

Figure 3.49

EXAMPLE 3 Sketching the Graph of a Radical Function

Analyze and sketch the graph of $f(x) = \dfrac{x}{\sqrt{x^2 + 2}}$.

Solution

$$f'(x) = \frac{2}{(x^2 + 2)^{3/2}} \qquad \text{Find first derivative.}$$

$$f''(x) = -\frac{6x}{(x^2 + 2)^{5/2}} \qquad \text{Find second derivative.}$$

The graph has only one intercept, $(0, 0)$. It has no vertical asymptotes, but it has two horizontal asymptotes: $y = 1$ (to the right) and $y = -1$ (to the left). The function has no critical numbers and one possible point of inflection (at $x = 0$). The domain of the function is all real numbers, and the graph is symmetric with respect to the origin. The analysis of the graph of f is shown in the table, and the graph is shown in Figure 3.49.

	$f(x)$	$f'(x)$	$f''(x)$	Characteristic of Graph
$-\infty < x < 0$		+	+	Increasing, concave upward
$x = 0$	0	+	0	Point of inflection
$0 < x < \infty$		+	−	Increasing, concave downward

EXAMPLE 4 Sketching the Graph of a Radical Function

Analyze and sketch the graph of $f(x) = 2x^{5/3} - 5x^{4/3}$.

Solution

$$f'(x) = \frac{10}{3}x^{1/3}(x^{1/3} - 2) \qquad \text{Find first derivative.}$$

$$f''(x) = \frac{20(x^{1/3} - 1)}{9x^{2/3}} \qquad \text{Find second derivative.}$$

The function has two intercepts: $(0, 0)$ and $\left(\frac{125}{8}, 0\right)$. There are no horizontal or vertical asymptotes. The function has two critical numbers ($x = 0$ and $x = 8$) and two possible points of inflection ($x = 0$ and $x = 1$). The domain is all real numbers. The analysis of the graph of f is shown in the table, and the graph is shown in Figure 3.50.

y Relative maximum (0, 0)

$f(x) = 2x^{5/3} - 5x^{4/3}$

4 8 12 x $\left(\frac{125}{8}, 0\right)$

(1, −3) Point of inflection

−12

−16

(8, −16) Relative minimum

Figure 3.50

	$f(x)$	$f'(x)$	$f''(x)$	Characteristic of Graph
$-\infty < x < 0$		+	−	Increasing, concave downward
$x = 0$	0	0	Undef.	Relative maximum
$0 < x < 1$		−	−	Decreasing, concave downward
$x = 1$	−3	−	0	Point of inflection
$1 < x < 8$		−	+	Decreasing, concave upward
$x = 8$	−16	0	+	Relative minimum
$8 < x < \infty$		+	+	Increasing, concave upward

EXAMPLE 5 Sketching the Graph of a Polynomial Function

See LarsonCalculus.com for an interactive version of this type of example.

Analyze and sketch the graph of

$$f(x) = x^4 - 12x^3 + 48x^2 - 64x.$$

Solution Begin by factoring to obtain

$$f(x) = x^4 - 12x^3 + 48x^2 - 64x$$
$$= x(x-4)^3.$$

Then, using the factored form of $f(x)$, you can perform the following analysis.

Domain: All real numbers
Range: $[-27, \infty)$
x-intercepts: $(0, 0), (4, 0)$
y-intercept: $(0, 0)$
Vertical asymptotes: None
Horizontal asymptotes: None
Symmetry: None
End behavior: $\lim_{x\to-\infty} f(x) = \infty, \lim_{x\to\infty} f(x) = \infty$
First derivative: $f'(x) = 4(x-1)(x-4)^2$
Second derivative: $f''(x) = 12(x-4)(x-2)$
Critical numbers: $x = 1, x = 4$
Possible points of inflection: $x = 2, x = 4$
Test intervals: $(-\infty, 1), (1, 2), (2, 4), (4, \infty)$

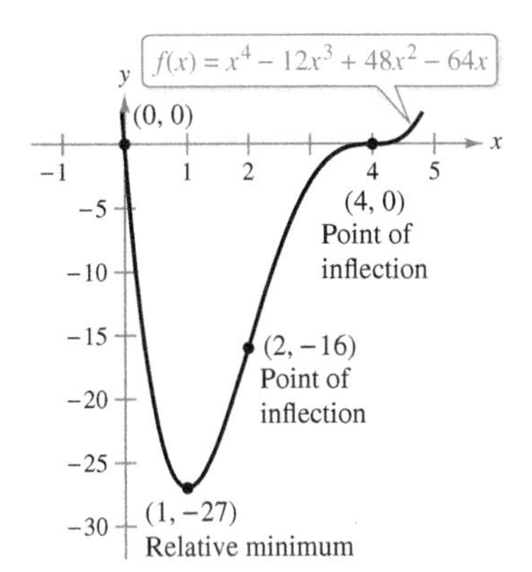

(a)

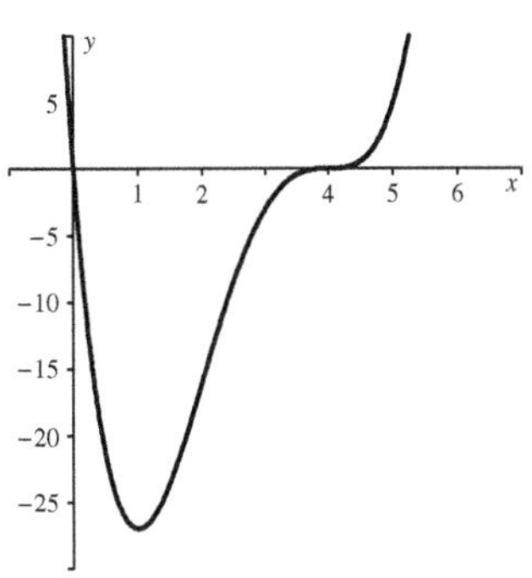

(b)
A polynomial function of even degree must have at least one relative extremum.
Figure 3.51

The analysis of the graph of f is shown in the table, and the graph is shown in Figure 3.51(a). Using a computer algebra system such as *Maple* [see Figure 3.51(b)] can help you verify your analysis.

	$f(x)$	$f'(x)$	$f''(x)$	Characteristic of Graph
$-\infty < x < 1$		$-$	$+$	Decreasing, concave upward
$x = 1$	-27	0	$+$	Relative minimum
$1 < x < 2$		$+$	$+$	Increasing, concave upward
$x = 2$	-16	$+$	0	Point of inflection
$2 < x < 4$		$+$	$-$	Increasing, concave downward
$x = 4$	0	0	0	Point of inflection
$4 < x < \infty$		$+$	$+$	Increasing, concave upward

The fourth-degree polynomial function in Example 5 has one relative minimum and no relative maxima. In general, a polynomial function of degree n can have *at most* $n - 1$ relative extrema and *at most* $n - 2$ points of inflection. Moreover, polynomial functions of even degree must have *at least* one relative extremum.

Remember from the Leading Coefficient Test described in Section P.3 that the "end behavior" of the graph of a polynomial function is determined by its leading coefficient and its degree. For instance, because the polynomial in Example 5 has a positive leading coefficient, the graph rises to the right. Moreover, because the degree is even, the graph also rises to the left.

EXAMPLE 6 Sketching the Graph of a Trigonometric Function

Analyze and sketch the graph of $f(x) = (\cos x)/(1 + \sin x)$.

Solution Because the function has a period of 2π, you can restrict the analysis of the graph to any interval of length 2π. For convenience, choose $[-\pi/2, 3\pi/2]$.

Domain: All real numbers except $x = \dfrac{3 + 4n}{2}\pi$

Range: All real numbers

Period: 2π

x-intercept: $\left(\dfrac{\pi}{2}, 0\right)$

y-intercept: $(0, 1)$

Vertical asymptotes: $x = -\dfrac{\pi}{2},\ x = \dfrac{3\pi}{2}$ See Remark below.

Horizontal asymptotes: None

Symmetry: None

First derivative: $f'(x) = -\dfrac{1}{1 + \sin x}$

Second derivative: $f''(x) = \dfrac{\cos x}{(1 + \sin x)^2}$

Critical numbers: None

Possible points of inflection: $x = \dfrac{\pi}{2}$

Test intervals: $\left(-\dfrac{\pi}{2}, \dfrac{\pi}{2}\right), \left(\dfrac{\pi}{2}, \dfrac{3\pi}{2}\right)$

Vertical asymptote: $x = -\frac{\pi}{2}$
Vertical asymptote: $x = \frac{3\pi}{2}$
$(0, 1)$
$\left(\frac{\pi}{2}, 0\right)$
Point of inflection
$f(x) = \dfrac{\cos x}{1 + \sin x}$

(a)

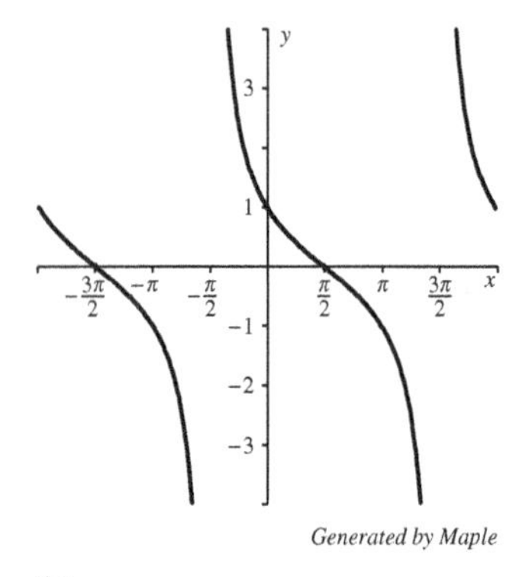

(b)

Figure 3.52

The analysis of the graph of f on the interval $[-\pi/2, 3\pi/2]$ is shown in the table, and the graph is shown in Figure 3.52(a). Compare this with the graph generated by the computer algebra system *Maple* in Figure 3.52(b).

	$f(x)$	$f'(x)$	$f''(x)$	Characteristic of Graph
$x = -\dfrac{\pi}{2}$	Undef.	Undef.	Undef.	Vertical asymptote
$-\dfrac{\pi}{2} < x < \dfrac{\pi}{2}$		$-$	$+$	Decreasing, concave upward
$x = \dfrac{\pi}{2}$	0	$-$	0	Point of inflection
$\dfrac{\pi}{2} < x < \dfrac{3\pi}{2}$		$-$	$-$	Decreasing, concave downward
$x = \dfrac{3\pi}{2}$	Undef.	Undef.	Undef.	Vertical asymptote

REMARK By substituting $-\pi/2$ or $3\pi/2$ into the function, you obtain the indeterminate form $0/0$, which you will study in Section 5.6. To determine that the function has vertical asymptotes at these two values, rewrite f as

$$f(x) = \frac{\cos x}{1 + \sin x} = \frac{(\cos x)(1 - \sin x)}{(1 + \sin x)(1 - \sin x)} = \frac{(\cos x)(1 - \sin x)}{\cos^2 x} = \frac{1 - \sin x}{\cos x}.$$

In this form, it is clear that the graph of f has vertical asymptotes at $x = -\pi/2$ and $3\pi/2$.

3.6 Exercises

See CalcChat.com for tutorial help and worked-out solutions to odd-numbered exercises.

CONCEPT CHECK

1. **Analyzing the Graph of a Function** Name several of the concepts you have learned that are useful for analyzing the graph of a function.
2. **Analyzing a Graph** Explain how to create a table to determine characteristics of a graph. What elements do you include?
3. **Slant Asymptote** Which type of function can have a slant asymptote? How do you determine the equation of a slant asymptote?
4. **Polynomial** What are the maximum numbers of relative extrema and points of inflection that a fifth-degree polynomial can have? Explain.

Matching **In Exercises 5–8, match the graph of the function with the graph of its derivative. [The graphs of the derivatives are labeled (a), (b), (c), and (d).]**

(a)

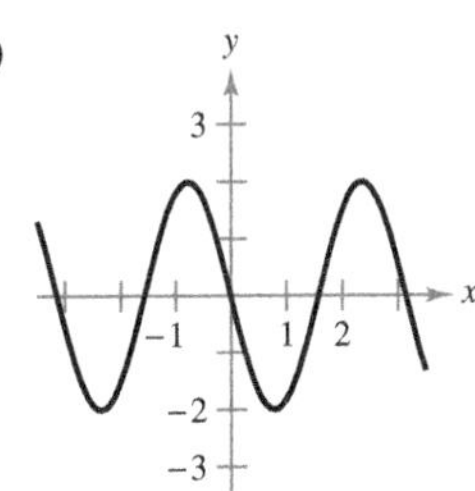

(b)

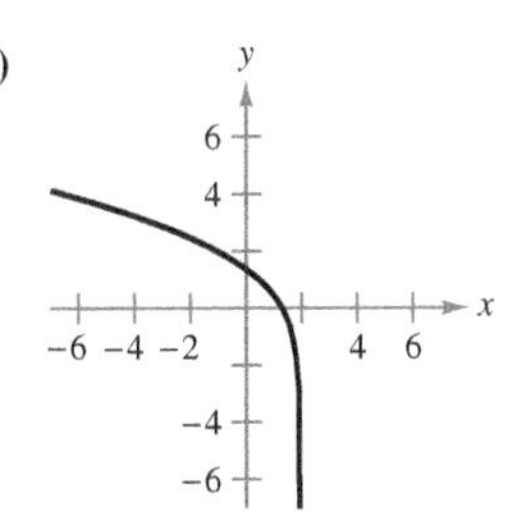

(c)

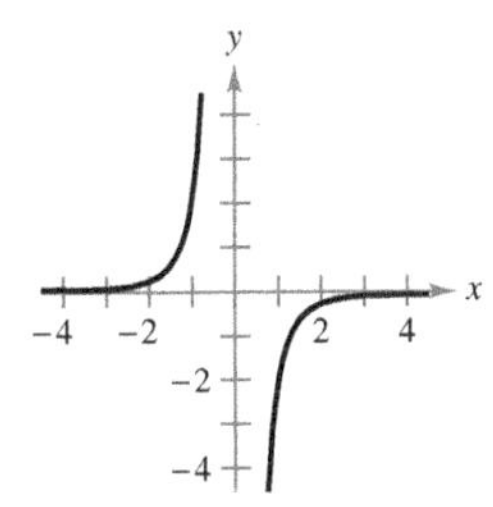

(d)

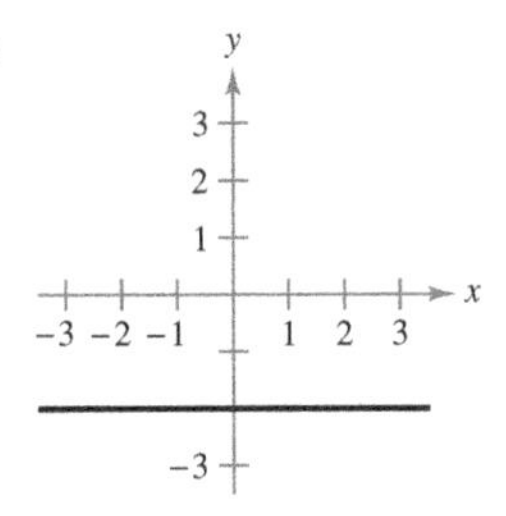

5.

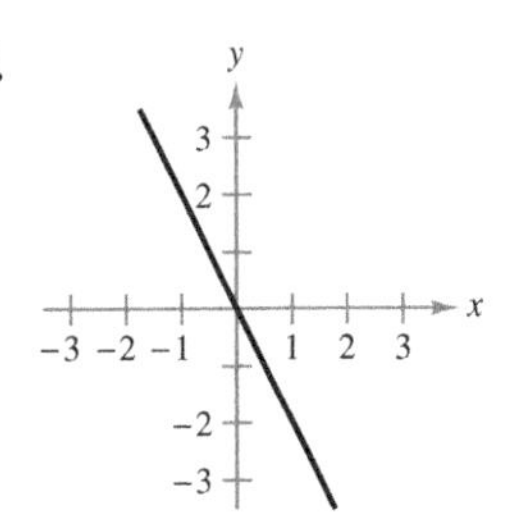

6.

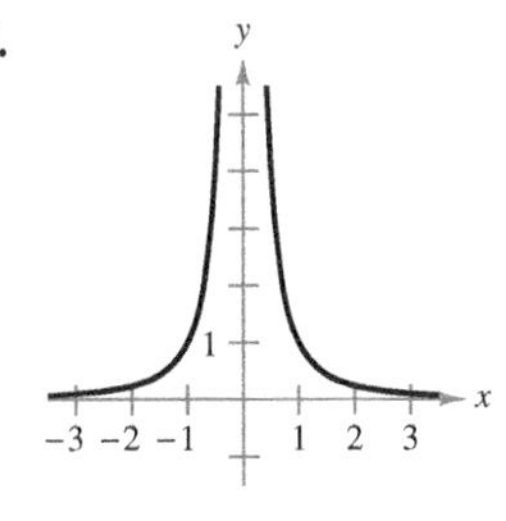

7.

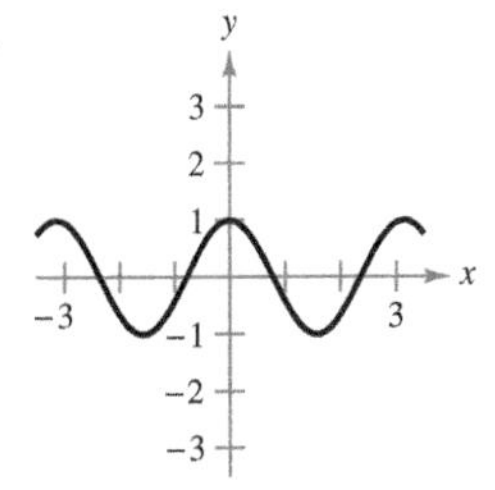

8.

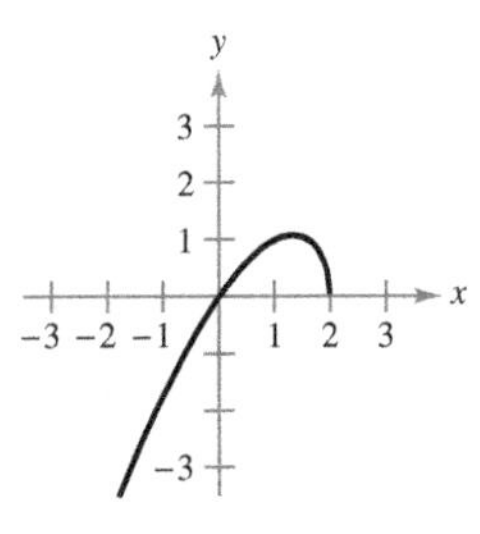

Analyzing the Graph of a Function **In Exercises 9–36, analyze and sketch a graph of the function. Label any intercepts, relative extrema, points of inflection, and asymptotes. Use a graphing utility to verify your results.**

9. $y = \dfrac{1}{x-2} - 3$
10. $y = \dfrac{x}{x^2+1}$
11. $y = \dfrac{x}{1-x}$
12. $y = \dfrac{x-4}{x-3}$
13. $y = \dfrac{x+1}{x^2-4}$
14. $y = \dfrac{2x}{9-x^2}$
15. $y = \dfrac{x^2}{x^2+3}$
16. $y = \dfrac{x^2+1}{x^2-4}$
17. $y = 3 + \dfrac{2}{x}$
18. $f(x) = \dfrac{x-3}{x}$
19. $f(x) = x + \dfrac{32}{x^2}$
20. $y = \dfrac{4}{x^2} + 1$
21. $y = \dfrac{3x}{x^2-1}$
22. $f(x) = \dfrac{x^3}{x^2-9}$
23. $y = \dfrac{x^2-6x+12}{x-4}$
24. $y = \dfrac{-x^2-4x-7}{x+3}$
25. $y = \dfrac{x^3}{\sqrt{x^2-4}}$
26. $y = \dfrac{x}{\sqrt{x^2-4}}$
27. $y = x\sqrt{4-x}$
28. $g(x) = x\sqrt{9-x^2}$
29. $y = 3x^{2/3} - 2x$
30. $y = (x+1)^2 - 3(x+1)^{2/3}$
31. $y = 2 - x - x^3$
32. $y = -\frac{1}{3}(x^3 - 3x + 2)$
33. $y = 3x^4 + 4x^3$
34. $y = -2x^4 + 3x^2$
35. $xy^2 = 9$
36. $x^2y = 9$

Analyzing the Graph of a Function **In Exercises 37–44, analyze and sketch a graph of the function over the given interval. Label any intercepts, relative extrema, points of inflection, and asymptotes. Use a graphing utility to verify your results.**

	Function	Interval
37.	$f(x) = 2x - 4\sin x$	$0 \le x \le 2\pi$
38.	$f(x) = -x + 2\cos x$	$0 \le x \le 2\pi$
39.	$y = \sin x - \frac{1}{18}\sin 3x$	$0 \le x \le 2\pi$
40.	$y = 2(x-2) + \cot x$	$0 < x < \pi$
41.	$y = 2(\csc x + \sec x)$	$0 < x < \dfrac{\pi}{2}$
42.	$y = \sec^2\dfrac{\pi x}{8} - 2\tan\dfrac{\pi x}{8} - 1$	$-3 < x < 3$
43.	$g(x) = x\tan x$	$-\dfrac{3\pi}{2} < x < \dfrac{3\pi}{2}$
44.	$g(x) = x\cot x$	$-2\pi < x < 2\pi$

Analyzing the Graph of a Function Using Technology **In Exercises 45–50, use a computer algebra system to analyze and graph the function. Identify any relative extrema, points of inflection, and asymptotes.**

45. $f(x) = \dfrac{20x}{x^2 + 1} - \dfrac{1}{x}$

46. $f(x) = x + \dfrac{4}{x^2 + 1}$

47. $f(x) = \dfrac{-2x}{\sqrt{x^2 + 7}}$

48. $f(x) = \dfrac{4x}{\sqrt{x^2 + 15}}$

49. $y = \cos x - \frac{1}{4}\cos 2x, \quad 0 \le x \le 2\pi$

50. $y = 2x - \tan x, \quad -\dfrac{\pi}{2} < x < \dfrac{\pi}{2}$

Identifying Graphs **In Exercises 51 and 52, the graphs of f, f', and f'' are shown on the same set of coordinate axes. Identify each graph. Explain your reasoning. To print an enlarged copy of the graph, go to *MathGraphs.com*.**

51.

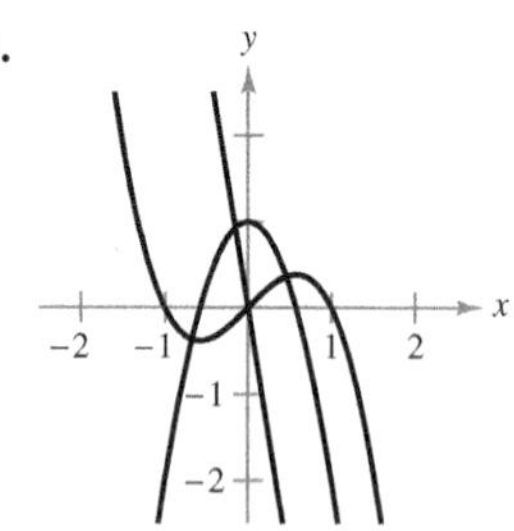

52.

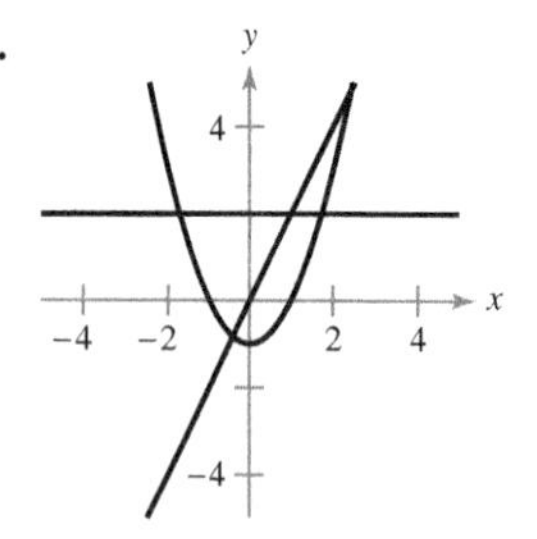

Graphical Reasoning **In Exercises 53–56, use the graph of f' to sketch a graph of f and the graph of f''. To print an enlarged copy of the graph, go to *MathGraphs.com*.**

53.

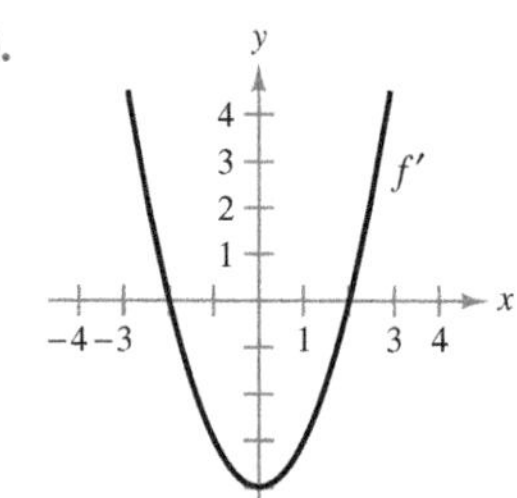

54.

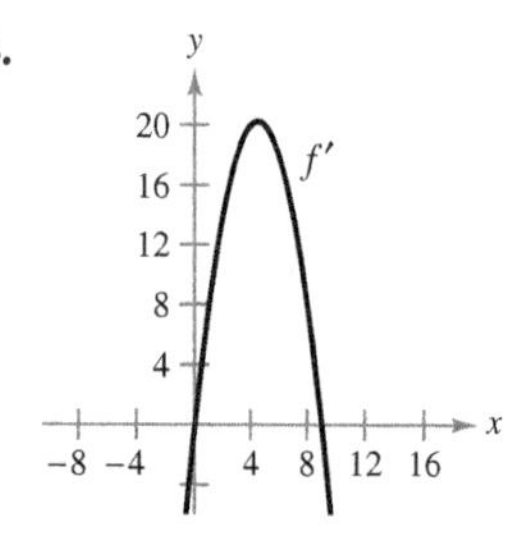

55.

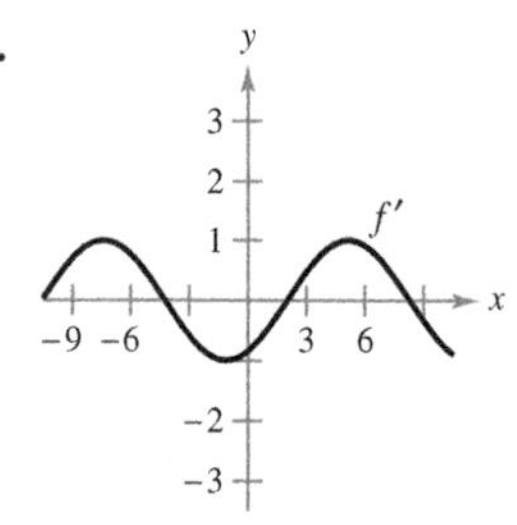

56.

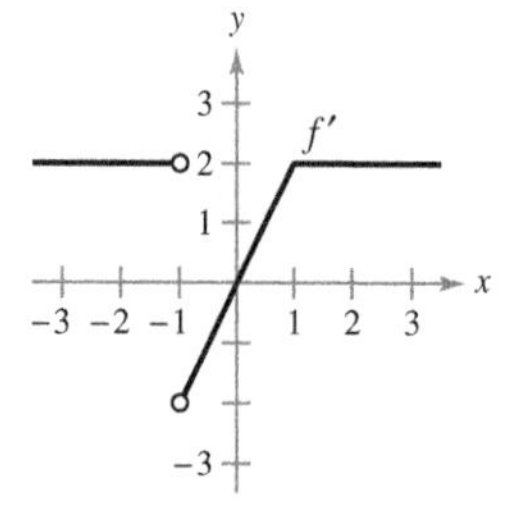

(Submitted by Bill Fox, Moberly Area Community College, Moberly, MO)

57. **Graphical Reasoning** Consider the function

$$f(x) = \frac{\cos^2 \pi x}{\sqrt{x^2 + 1}}, \quad 0 < x < 4.$$

(a) Use a computer algebra system to graph the function and use the graph to approximate the critical numbers visually.

(b) Use a computer algebra system to find f' and approximate the critical numbers. Are the results the same as the visual approximation in part (a)? Explain.

58. **Graphical Reasoning** Consider the function

$$f(x) = \tan(\sin \pi x).$$

(a) Use a graphing utility to graph the function.

(b) Identify any symmetry of the graph.

(c) Is the function periodic? If so, what is the period?

(d) Identify any extrema on $(-1, 1)$.

(e) Use a graphing utility to determine the concavity of the graph on $(0, 1)$.

EXPLORING CONCEPTS

59. **Sketching a Graph** Sketch a graph of a differentiable function f that satisfies the following conditions and has $x = 2$ as its only critical number.

$f'(x) < 0$ for $x < 2$

$f'(x) > 0$ for $x > 2$

$\lim_{x \to -\infty} f(x) = 6$

$\lim_{x \to \infty} f(x) = 6$

60. **Points of Inflection** Is it possible to sketch a graph of a function that satisfies the conditions of Exercise 59 and has *no* points of inflection? Explain.

61. **Using a Derivative** Let $f'(t) < 0$ for all t in the interval $(2, 8)$. Explain why $f(3) > f(5)$.

62. **Using a Derivative** Let $f(0) = 3$ and $2 \le f'(x) \le 4$ for all x in the interval $[-5, 5]$. Determine the greatest and least possible values of $f(2)$.

63. **A Function and Its Derivative** The graph of a function f is shown below. To print an enlarged copy of the graph, go to *MathGraphs.com*.

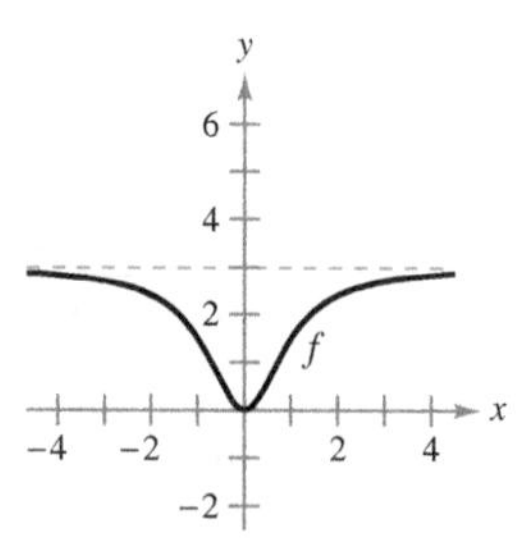

(a) Sketch f'.

(b) Use the graph to estimate $\lim_{x \to \infty} f(x)$ and $\lim_{x \to \infty} f'(x)$.

(c) Explain the answers you gave in part (b).

64. HOW DO YOU SEE IT? The graph of f is shown in the figure.

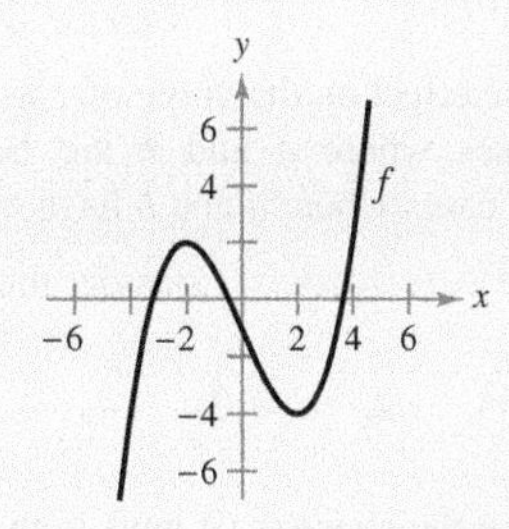

(a) For which values of x is $f'(x)$ zero? Positive? Negative? What do these values mean?

(b) For which values of x is $f''(x)$ zero? Positive? Negative? What do these values mean?

(c) On what open interval is f' an increasing function?

(d) For which value of x is $f'(x)$ minimum? For this value of x, how does the rate of change of f compare with the rates of change of f for other values of x? Explain.

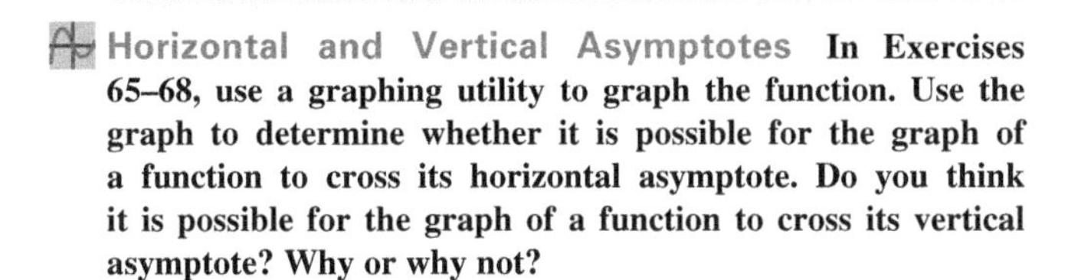

Horizontal and Vertical Asymptotes In Exercises 65–68, use a graphing utility to graph the function. Use the graph to determine whether it is possible for the graph of a function to cross its horizontal asymptote. Do you think it is possible for the graph of a function to cross its vertical asymptote? Why or why not?

65. $f(x) = \dfrac{4(x-1)^2}{x^2-4x+5}$

66. $g(x) = \dfrac{3x^4-5x+3}{x^4+1}$

67. $h(x) = \dfrac{\sin 2x}{x}$

68. $f(x) = \dfrac{\cos 3x}{4x}$

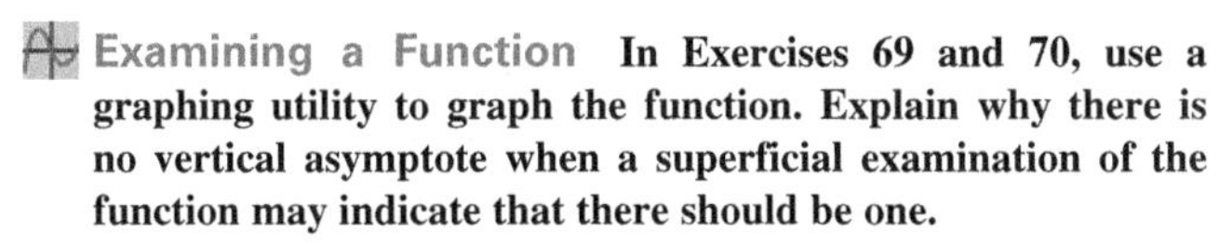

Examining a Function In Exercises 69 and 70, use a graphing utility to graph the function. Explain why there is no vertical asymptote when a superficial examination of the function may indicate that there should be one.

69. $h(x) = \dfrac{6-2x}{3-x}$

70. $g(x) = \dfrac{x^2+x-2}{x-1}$

Slant Asymptote In Exercises 71–76, use a graphing utility to graph the function and determine the slant asymptote of the graph analytically. Zoom out repeatedly and describe how the graph on the display appears to change. Why does this occur?

71. $f(x) = -\dfrac{x^2-3x-1}{x-2}$

72. $g(x) = \dfrac{2x^2-8x-15}{x-5}$

73. $f(x) = \dfrac{2x^3}{x^2+1}$

74. $h(x) = \dfrac{-x^3+x^2+4}{x^2}$

75. $f(x) = \dfrac{x^3-3x^2+2}{x(x-3)}$

76. $f(x) = -\dfrac{x^3-2x^2+2}{2x^2}$

77. Investigation Let $P(x_0, y_0)$ be an arbitrary point on the graph of f that $f'(x_0) \neq 0$, as shown in the figure. Verify each statement.

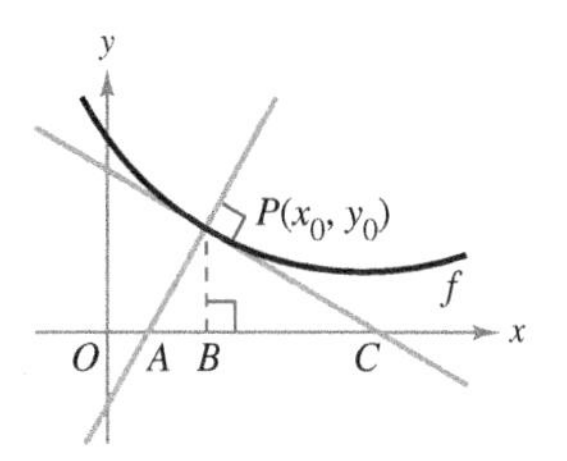

(a) The x-intercept of the tangent line is

$$\left(x_0 - \frac{f(x_0)}{f'(x_0)}, 0\right).$$

(b) The y-intercept of the tangent line is

$$(0, f(x_0) - x_0 f'(x_0)).$$

(c) The x-intercept of the normal line is

$$(x_0 + f(x_0)f'(x_0), 0).$$

(The *normal line* at a point is perpendicular to the tangent line at the point.)

(d) The y-intercept of the normal line is

$$\left(0, y_0 + \frac{x_0}{f'(x_0)}\right).$$

(e) $|BC| = \left|\dfrac{f(x_0)}{f'(x_0)}\right|$

(f) $|PC| = \left|\dfrac{f(x_0)\sqrt{1+[f'(x_0)]^2}}{f'(x_0)}\right|$

(g) $|AB| = |f(x_0)f'(x_0)|$

(h) $|AP| = |f(x_0)|\sqrt{1+[f'(x_0)]^2}$

78. Graphical Reasoning Identify the real numbers x_0, x_1, x_2, x_3, and x_4 in the figure such that each of the following is true.

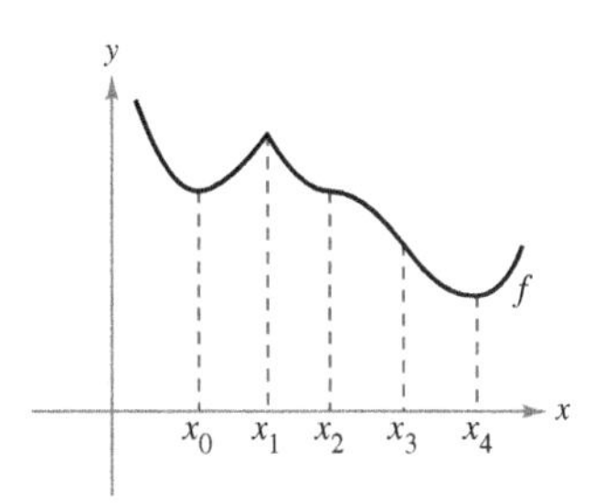

(a) $f'(x) = 0$

(b) $f''(x) = 0$

(c) $f'(x)$ does not exist.

(d) f has a relative maximum.

(e) f has a point of inflection.

Think About It **In Exercises 79–82, create a function whose graph has the given characteristics. (There is more than one correct answer.)**

79. Vertical asymptote: $x = 3$

Horizontal asymptote: $y = 0$

80. Vertical asymptote: $x = -5$

Horizontal asymptote: None

81. Vertical asymptote: $x = 3$

Slant asymptote: $y = 3x + 2$

82. Vertical asymptote: $x = 2$

Slant asymptote: $y = -x$

True or False? **In Exercises 83–86, determine whether the statement is true or false. If it is false, explain why or give an example that shows it is false.**

83. If $f'(x) > 0$ for all real numbers x, then f increases without bound.

84. If $f''(x) < 0$ for all real numbers x, then f decreases without bound.

85. Every rational function has a slant asymptote.

86. Every polynomial function has an absolute maximum and an absolute minimum on $(-\infty, \infty)$.

87. Graphical Reasoning The graph of the first derivative of a function f on the interval $[-7, 5]$ is shown. Use the graph to answer each question.

(a) On what interval(s) is f decreasing?

(b) On what interval(s) is the graph of f concave downward?

(c) At what x-value(s) does f have relative extrema?

(d) At what x-value(s) does the graph of f have a point of inflection?

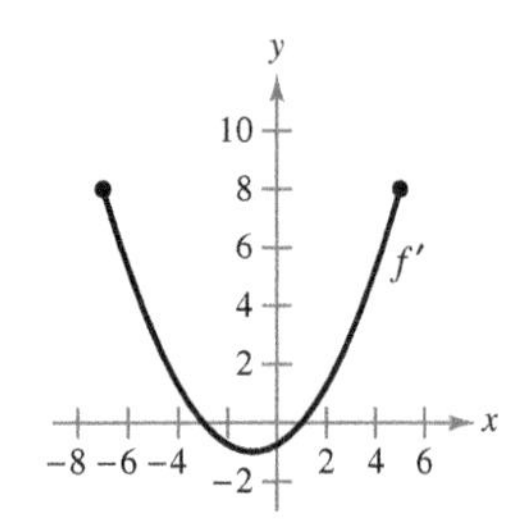

Figure for 87

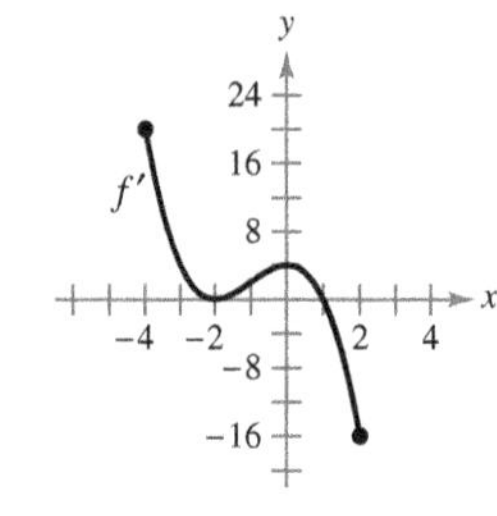

Figure for 88

88. Graphical Reasoning The graph of the first derivative of a function f on the interval $[-4, 2]$ is shown. Use the graph to answer each question.

(a) On what interval(s) is f increasing?

(b) On what interval(s) is the graph of f concave upward?

(c) At what x-value(s) does f have relative extrema?

(d) At what x-value(s) does the graph of f have a point of inflection?

89. Graphical Reasoning Consider the function

$$f(x) = \frac{ax}{(x - b)^2}.$$

Determine the effect on the graph of f as a and b are changed. Consider cases where a and b are both positive or both negative and cases where a and b have opposite signs.

90. Graphical Reasoning Consider the function

$$f(x) = \frac{1}{2}(ax)^2 - ax, \quad a \neq 0.$$

(a) Determine the changes (if any) in the intercepts, extrema, and concavity of the graph of f when a is varied.

(b) In the same viewing window, use a graphing utility to graph the function for four different values of a.

Slant Asymptotes **In Exercises 91 and 92, the graph of the function has two slant asymptotes. Identify each slant asymptote. Then graph the function and its asymptotes.**

91. $y = \sqrt{4 + 16x^2}$ **92.** $y = \sqrt{x^2 + 6x}$

93. Investigation Consider the function

$$f(x) = \frac{2x^n}{x^4 + 1}$$

for nonnegative integer values of n.

(a) Discuss the relationship between the value of n and the symmetry of the graph.

(b) For which values of n will the x-axis be the horizontal asymptote?

(c) For which value of n will $y = 2$ be the horizontal asymptote?

(d) What is the asymptote of the graph when $n = 5$?

(e) Use a graphing utility to graph f for the indicated values of n in the table. Use the graph to determine the number of extrema M and the number of inflection points N of the graph.

n	0	1	2	3	4	5
M						
N						

PUTNAM EXAM CHALLENGE

94. Let $f(x)$ be defined for $a \leq x \leq b$. Assuming appropriate properties of continuity and derivability, prove for $a < x < b$ that

$$\frac{\dfrac{f(x) - f(a)}{x - a} - \dfrac{f(b) - f(a)}{b - a}}{x - b} = \frac{1}{2}f''(\varepsilon),$$

where ε is some number between a and b.

This problem was composed by the Committee on the Putnam Prize Competition.

3.7 Optimization Problems

- **Solve applied minimum and maximum problems.**

Applied Minimum and Maximum Problems

One of the most common applications of calculus involves the determination of minimum and maximum values. Consider how frequently you hear or read terms such as greatest profit, least cost, least time, greatest voltage, optimum size, least size, greatest strength, and greatest distance. Before outlining a general problem-solving strategy for such problems, consider the next example.

EXAMPLE 1 Finding Maximum Volume

A manufacturer wants to design an open box having a square base and a surface area of 108 square inches, as shown in Figure 3.53. What dimensions will produce a box with maximum volume?

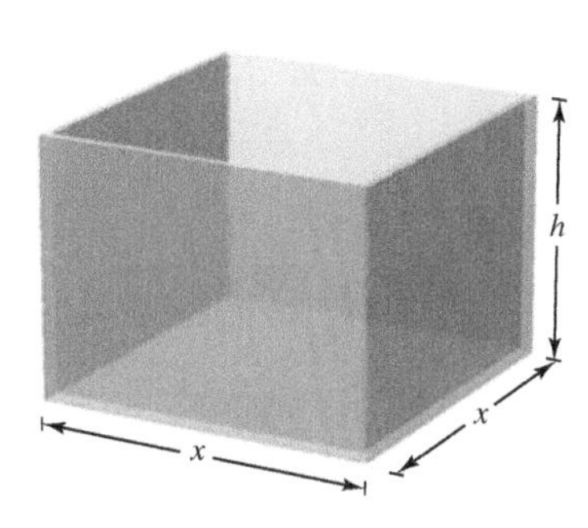

Open box with square base:
$S = x^2 + 4xh = 108$
Figure 3.53

Solution Because the box has a square base, its volume is

$$V = x^2h. \quad \text{Primary equation}$$

This equation is called the **primary equation** because it gives a formula for the quantity to be optimized. The surface area of the box is

$$S = (\text{area of base}) + (\text{area of four sides})$$
$$108 = x^2 + 4xh. \quad \text{Secondary equation}$$

Because V is to be maximized, you want to write V as a function of just one variable. To do this, you can solve the equation $x^2 + 4xh = 108$ for h in terms of x to obtain $h = (108 - x^2)/(4x)$. Substituting into the primary equation produces

$$V = x^2h \quad \text{Function of two variables}$$
$$= x^2\left(\frac{108 - x^2}{4x}\right) \quad \text{Substitute for } h.$$
$$= 27x - \frac{x^3}{4}. \quad \text{Function of one variable}$$

Before finding which x-value will yield a maximum value of V, you should determine the *feasible domain*. That is, what values of x make sense in this problem? You know that $V \geq 0$. You also know that x must be nonnegative and that the area of the base $(A = x^2)$ is at most 108. So, the feasible domain is

$$0 \leq x \leq \sqrt{108}. \quad \text{Feasible domain}$$

To maximize V, find its critical numbers on the interval $(0, \sqrt{108})$.

$$\frac{dV}{dx} = 27 - \frac{3x^2}{4} \quad \text{Differentiate with respect to } x.$$
$$27 - \frac{3x^2}{4} = 0 \quad \text{Set derivative equal to 0.}$$
$$3x^2 = 108 \quad \text{Simplify.}$$
$$x = \pm 6 \quad \text{Critical numbers}$$

So, the critical numbers are $x = \pm 6$. You do not need to consider $x = -6$ because it is outside the domain. Evaluating V at the critical number $x = 6$ and at the endpoints of the domain produces $V(0) = 0$, $V(6) = 108$, and $V(\sqrt{108}) = 0$. So, V is maximum when $x = 6$, and the dimensions of the box are 6 inches by 6 inches by 3 inches. ■

TECHNOLOGY You can verify your answer in Example 1 by using a graphing utility to graph the volume function

$$V = 27x - \frac{x^3}{4}.$$

Use a viewing window in which $0 \leq x \leq \sqrt{108} \approx 10.4$ and $0 \leq y \leq 120$, and use the *maximum* or *trace* feature to determine the value of x that produces a maximum volume.

In Example 1, you should realize that there are infinitely many open boxes having 108 square inches of surface area. To begin solving the problem, you might ask yourself which basic shape would seem to yield a maximum volume. Should the box be tall, squat, or nearly cubical?

You might even try calculating a few volumes, as shown in Figure 3.54, to determine whether you can get a better feeling for what the optimum dimensions should be. Remember that you are not ready to begin solving a problem until you have clearly identified what the problem is.

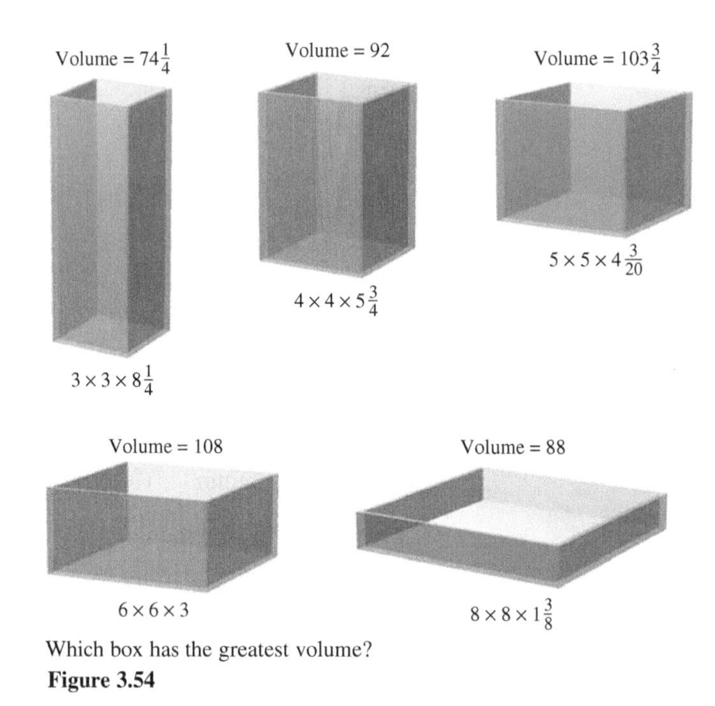

Which box has the greatest volume?
Figure 3.54

Example 1 illustrates the following guidelines for solving applied minimum and maximum problems.

GUIDELINES FOR SOLVING APPLIED MINIMUM AND MAXIMUM PROBLEMS

1. Identify all *given* quantities and all quantities *to be determined.* If possible, make a sketch.
2. Write a **primary equation** for the quantity that is to be maximized or minimized. (A review of several useful formulas from geometry is presented on the formula card inside the back cover.)
3. Reduce the primary equation to one having a *single independent variable.* This may involve the use of **secondary equations** relating the independent variables of the primary equation.
4. Determine the feasible domain of the primary equation. That is, determine the values for which the stated problem makes sense.
5. Determine the desired maximum or minimum value by the calculus techniques discussed in Sections 3.1 through 3.4.

REMARK For Step 5, recall that to determine the maximum or minimum value of a continuous function f on a closed interval, you should compare the values of f at its critical numbers with the values of f at the endpoints of the interval.

EXAMPLE 2 Finding Minimum Distance

See LarsonCalculus.com for an interactive version of this type of example.

Which points on the graph of $y = 4 - x^2$ are closest to the point $(0, 2)$?

Solution Figure 3.55 shows that there are two points at a minimum distance from the point $(0, 2)$. The distance between the point $(0, 2)$ and a point (x, y) on the graph of $y = 4 - x^2$ is

$$d = \sqrt{(x-0)^2 + (y-2)^2}. \qquad \text{Primary equation}$$

Using the secondary equation $y = 4 - x^2$, you can rewrite the primary equation as

$$\begin{aligned} d &= \sqrt{x^2 + (4 - x^2 - 2)^2} \\ &= \sqrt{x^4 - 3x^2 + 4}. \end{aligned}$$

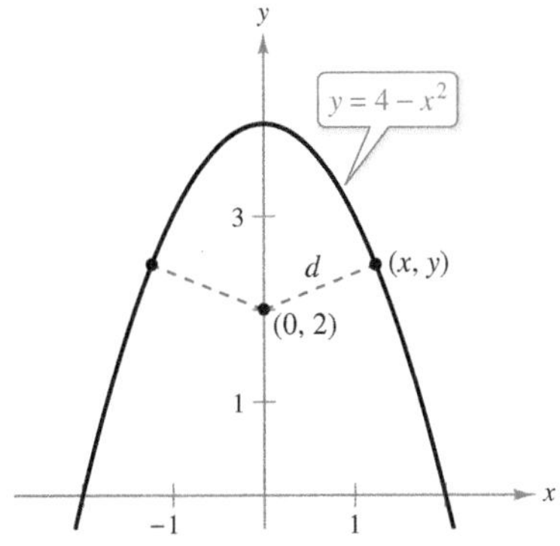

The quantity to be minimized is distance: $d = \sqrt{(x-0)^2 + (y-2)^2}$.
Figure 3.55

Because d is smallest when the expression inside the radical is smallest, you need only find the critical numbers of $f(x) = x^4 - 3x^2 + 4$. Note that the domain of f is the entire real number line. So, there are no endpoints of the domain to consider. Moreover, the derivative of f

$$\begin{aligned} f'(x) &= 4x^3 - 6x \\ &= 2x(2x^2 - 3) \end{aligned}$$

is zero when

$$x = 0, \sqrt{\frac{3}{2}}, -\sqrt{\frac{3}{2}}.$$

Testing these critical numbers using the First Derivative Test verifies that $x = 0$ yields a relative maximum, whereas both $x = \sqrt{3/2}$ and $x = -\sqrt{3/2}$ yield a minimum distance. So, the closest points are $\left(\sqrt{3/2}, 5/2\right)$ and $\left(-\sqrt{3/2}, 5/2\right)$.

EXAMPLE 3 Finding Minimum Area

A rectangular page is to contain 24 square inches of print. The margins at the top and bottom of the page are to be $1\frac{1}{2}$ inches, and the margins on the left and right are to be 1 inch (see Figure 3.56). What should the dimensions of the page be so that the least amount of paper is used?

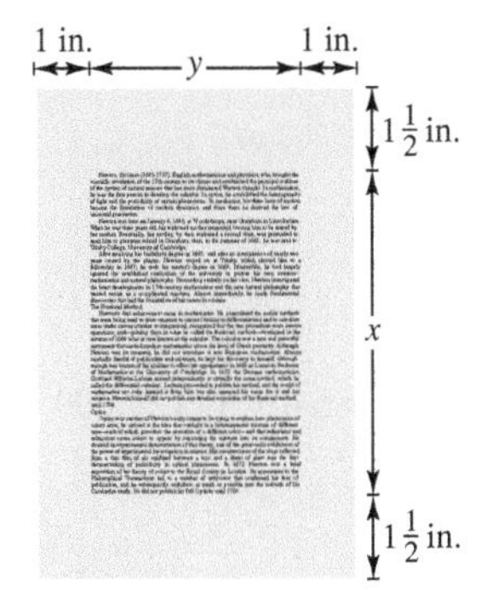

The quantity to be minimized is area: $A = (x + 3)(y + 2)$.
Figure 3.56

Solution Let A be the area to be minimized.

$$A = (x + 3)(y + 2) \qquad \text{Primary equation}$$

The printed area inside the margins is

$$24 = xy. \qquad \text{Secondary equation}$$

Solving this equation for y produces $y = 24/x$. Substituting into the primary equation produces

$$A = (x + 3)\left(\frac{24}{x} + 2\right) = 30 + 2x + \frac{72}{x}. \qquad \text{Function of one variable}$$

Because x must be positive, you are interested only in values of A for $x > 0$. To find the critical numbers, differentiate with respect to x

$$\frac{dA}{dx} = 2 - \frac{72}{x^2}$$

and note that the derivative is zero when $x^2 = 36$, or $x = \pm 6$. So, the critical numbers are $x = \pm 6$. You do not have to consider $x = -6$ because it is outside the domain. The First Derivative Test confirms that A is a minimum when $x = 6$. So, $y = \frac{24}{6} = 4$ and the dimensions of the page should be $x + 3 = 9$ inches by $y + 2 = 6$ inches. ■

EXAMPLE 4 Finding Minimum Length

Two posts, one 12 feet high and the other 28 feet high, stand 30 feet apart. They are to be stayed by two wires, attached to a single stake, running from ground level to the top of each post. Where should the stake be placed to use the least amount of wire?

Solution Let W be the wire length to be minimized. Using Figure 3.57, you can write

$$W = y + z. \qquad \text{Primary equation}$$

In this problem, rather than solving for y in terms of z (or vice versa), you can solve for both y and z in terms of a third variable x, as shown in Figure 3.57. From the Pythagorean Theorem, you obtain

$$x^2 + 12^2 = y^2$$
$$(30 - x)^2 + 28^2 = z^2$$

which implies that

$$y = \sqrt{x^2 + 144}$$
$$z = \sqrt{x^2 - 60x + 1684}.$$

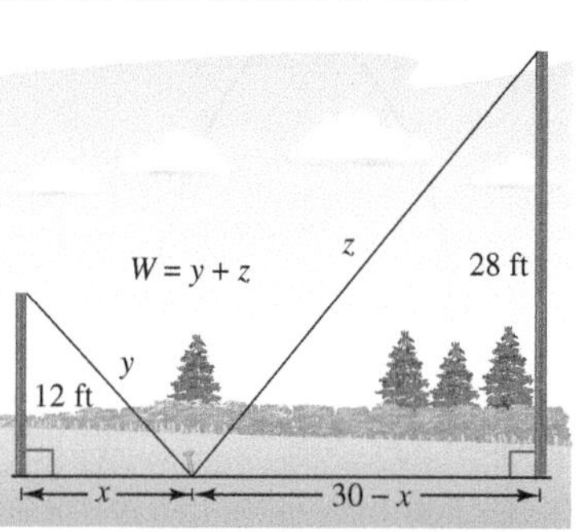

The quantity to be minimized is length. From the diagram, you can see that x varies between 0 and 30.
Figure 3.57

So, you can rewrite the primary equation as

$$\begin{aligned} W &= y + z \\ &= \sqrt{x^2 + 144} + \sqrt{x^2 - 60x + 1684}, \quad 0 \le x \le 30. \end{aligned}$$

Differentiating W with respect to x yields

$$\frac{dW}{dx} = \frac{x}{\sqrt{x^2 + 144}} + \frac{x - 30}{\sqrt{x^2 - 60x + 1684}}.$$

By letting $dW/dx = 0$, you obtain

$$\begin{aligned} \frac{x}{\sqrt{x^2 + 144}} + \frac{x - 30}{\sqrt{x^2 - 60x + 1684}} &= 0 \\ \frac{x}{\sqrt{x^2 + 144}} &= \frac{30 - x}{\sqrt{x^2 - 60x + 1684}} \\ x\sqrt{x^2 - 60x + 1684} &= (30 - x)\sqrt{x^2 + 144} \\ x^2(x^2 - 60x + 1684) &= (30 - x)^2(x^2 + 144) \\ x^4 - 60x^3 + 1684x^2 &= x^4 - 60x^3 + 1044x^2 - 8640x + 129{,}600 \\ 640x^2 + 8640x - 129{,}600 &= 0 \\ 320(x - 9)(2x + 45) &= 0 \\ x &= 9, -22.5. \end{aligned}$$

Because $x = -22.5$ is not in the domain and

$$W(0) \approx 53.04, \quad W(9) = 50, \quad \text{and} \quad W(30) \approx 60.31$$

you can conclude that the wires should be staked at 9 feet from the 12-foot pole. ■

60
Minimum
X=9 Y=50
0 30
45

You can confirm the minimum value of W with a graphing utility.
Figure 3.58

TECHNOLOGY From Example 4, you can see that applied optimization problems can involve a lot of algebra. If you have access to a graphing utility, you can confirm that $x = 9$ yields a minimum value of W by graphing

$$W = \sqrt{x^2 + 144} + \sqrt{x^2 - 60x + 1684}$$

as shown in Figure 3.58.

In each of the first four examples, the extreme value occurred at a critical number. Although this happens often, remember that an extreme value can also occur at an endpoint of an interval, as shown in Example 5.

EXAMPLE 5 An Endpoint Maximum

Four feet of wire is to be used to form a square and a circle. How much of the wire should be used for the square and how much should be used for the circle to enclose the maximum total area?

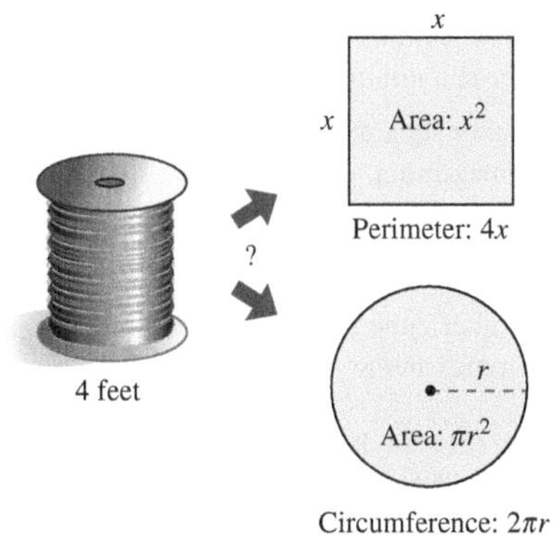

x

x Area: x^2

Perimeter: $4x$

?

r

Area: πr^2

4 feet

Circumference: $2\pi r$

The quantity to be maximized is area: $A = x^2 + \pi r^2$.

Figure 3.59

Solution The total area (see Figure 3.59) is

$$A = (\text{area of square}) + (\text{area of circle})$$

$$A = x^2 + \pi r^2. \qquad \text{Primary equation}$$

Because the total length of wire is 4 feet, you obtain

$$4 = (\text{perimeter of square}) + (\text{circumference of circle})$$

$$4 = 4x + 2\pi r. \qquad \text{Secondary equation}$$

So, $r = 2(1 - x)/\pi$, and by substituting into the primary equation you have

$$\begin{aligned} A &= x^2 + \pi\left[\frac{2(1-x)}{\pi}\right]^2 \\ &= x^2 + \frac{4(1-x)^2}{\pi} \\ &= \frac{1}{\pi}(\pi x^2 + 4 - 8x + 4x^2) \\ &= \frac{1}{\pi}[(\pi + 4)x^2 - 8x + 4]. \end{aligned}$$

The feasible domain is $0 \le x \le 1$, restricted by the square's perimeter. Because

$$\frac{dA}{dx} = \frac{2(\pi + 4)x - 8}{\pi}$$

the only critical number in $(0, 1)$ is $x = 4/(\pi + 4) \approx 0.56$. So, using

$$A(0) \approx 1.27, \quad A(0.56) \approx 0.56, \quad \text{and} \quad A(1) = 1$$

you can conclude that the maximum area occurs when $x = 0$. That is, *all* the wire is used for the circle. ■

Exploration

What would the answer be if Example 5 asked for the dimensions needed to enclose the *minimum* total area?

Before doing the section exercises, review the primary equations developed in Examples 1–5. As applications go, these five examples are fairly simple, and yet the resulting primary equations are quite complicated.

$$V = 27x - \frac{x^3}{4} \qquad \text{Example 1}$$

$$d = \sqrt{x^4 - 3x^2 + 4} \qquad \text{Example 2}$$

$$A = 30 + 2x + \frac{72}{x} \qquad \text{Example 3}$$

$$W = \sqrt{x^2 + 144} + \sqrt{x^2 - 60x + 1684} \qquad \text{Example 4}$$

$$A = \frac{1}{\pi}[(\pi + 4)x^2 - 8x + 4] \qquad \text{Example 5}$$

You must expect that real-life applications often involve equations that are *at least as complicated* as these five. Remember that one of the main goals of this course is to learn to use calculus to analyze equations that initially seem formidable.

3.7 Exercises

See CalcChat.com for tutorial help and worked-out solutions to odd-numbered exercises.

CONCEPT CHECK

1. **Writing** In your own words, describe *primary equation, secondary equation,* and *feasible domain.*
2. **Optimization Problems** In your own words, describe the guidelines for solving applied minimum and maximum problems.

3. **Numerical, Graphical, and Analytic Analysis** Find two positive numbers whose sum is 110 and whose product is a maximum.
   (a) Analytically complete six rows of a table such as the one below. (The first two rows are shown.) Use the table to guess the maximum product.

First Number, x	Second Number	Product, P
10	$110 - 10$	$10(110 - 10) = 1000$
20	$110 - 20$	$20(110 - 20) = 1800$

   (b) Write the product P as a function of x.
   (c) Use calculus to find the critical number of the function in part (b). Then find the two numbers.
   (d) Use a graphing utility to graph the function in part (b) and verify the solution from the graph.

4. **Numerical, Graphical, and Analytic Analysis** An open box of maximum volume is to be made from a square piece of material, 24 inches on a side, by cutting equal squares from the corners and turning up the sides (see figure).

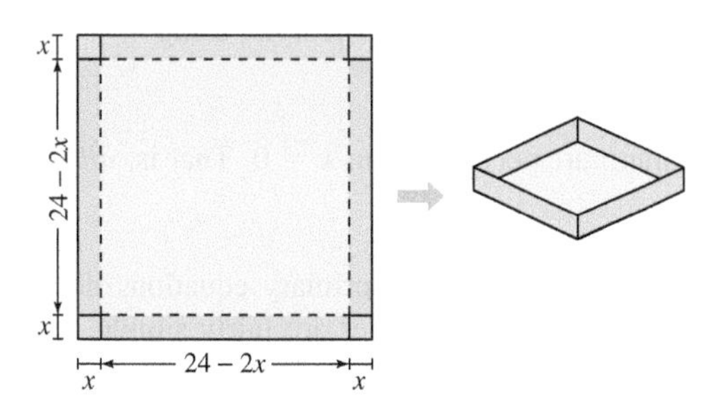

   (a) Analytically complete six rows of a table such as the one below. (The first two rows are shown.) Use the table to guess the maximum volume.

Height, x	Length and Width	Volume, V
1	$24 - 2(1)$	$1[24 - 2(1)]^2 = 484$
2	$24 - 2(2)$	$2[24 - 2(2)]^2 = 800$

   (b) Write the volume V as a function of x.
   (c) Use calculus to find the critical number of the function in part (b). Then find the maximum volume.
   (d) Use a graphing utility to graph the function in part (b) and verify the maximum volume from the graph.

Finding Numbers **In Exercises 5–10, find two positive numbers that satisfy the given requirements.**

5. The sum is S and the product is a maximum.
6. The product is 185 and the sum is a minimum.
7. The product is 147 and the sum of the first number plus three times the second number is a minimum.
8. The sum of the first number squared and the second number is 54 and the product is a maximum.
9. The sum of the first number and twice the second number is 108 and the product is a maximum.
10. The sum of the first number cubed and the second number is 500 and the product is a maximum.

Maximum Area **In Exercises 11 and 12, find the length and width of a rectangle that has the given perimeter and a maximum area.**

11. Perimeter: 80 meters
12. Perimeter: P units

Minimum Perimeter **In Exercises 13 and 14, find the length and width of a rectangle that has the given area and a minimum perimeter.**

13. Area: 49 square feet
14. Area: A square centimeters

Minimum Distance **In Exercises 15 and 16, find the points on the graph of the function that are closest to the given point.**

15. $y = x^2$, $(0, 3)$
16. $y = x^2 - 2$, $(0, -1)$

17. **Minimum Area** A rectangular poster is to contain 648 square inches of print. The margins at the top and bottom of the poster are to be 2 inches, and the margins on the left and right are to be 1 inch. What should the dimensions of the poster be so that the least amount of poster is used?
18. **Minimum Area** A rectangular page is to contain 36 square inches of print. The margins on each side are to be $1\frac{1}{2}$ inches. Find the dimensions of the page such that the least amount of paper is used.
19. **Minimum Length** A farmer plans to fence a rectangular pasture adjacent to a river (see figure). The pasture must contain 405,000 square meters in order to provide enough grass for the herd. No fencing is needed along the river. What dimensions will require the least amount of fencing?

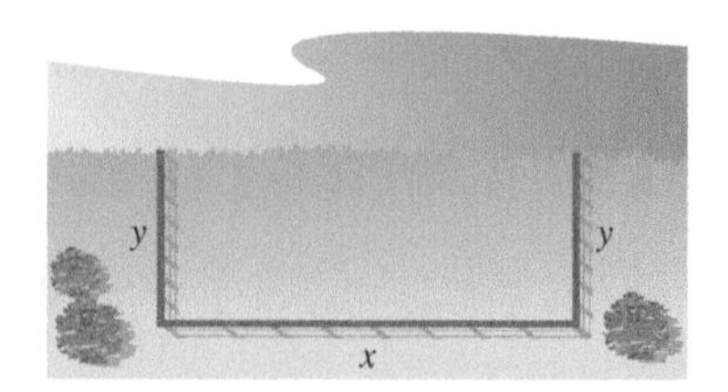

20. Maximum Volume A rectangular solid (with a square base) has a surface area of 337.5 square centimeters. Find the dimensions that will result in a solid with maximum volume.

21. Maximum Area A Norman window is constructed by adjoining a semicircle to the top of an ordinary rectangular window (see figure). Find the dimensions of a Norman window of maximum area when the total perimeter is 16 feet.

22. Maximum Area A rectangle is bounded by the x- and y-axes and the graph of $y = (6 - x)/2$ (see figure). What length and width should the rectangle have so that its area is a maximum?

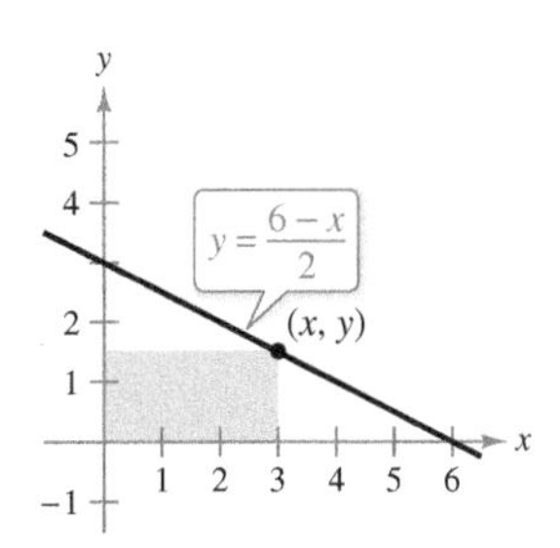

Figure for 22

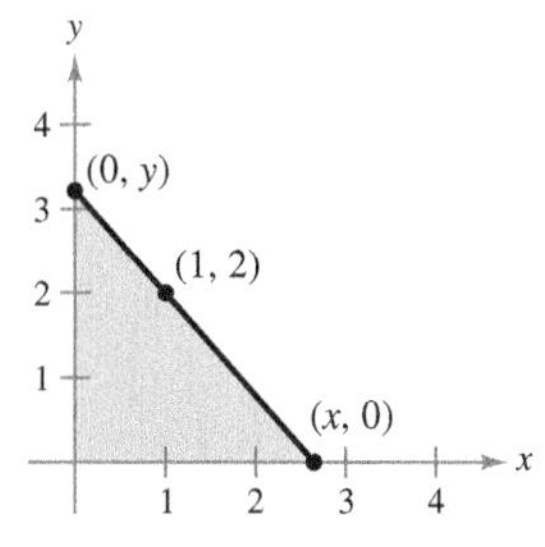

Figure for 23

23. Minimum Length and Minimum Area A right triangle is formed in the first quadrant by the x- and y-axes and a line through the point $(1, 2)$ (see figure).

(a) Write the length L of the hypotenuse as a function of x.

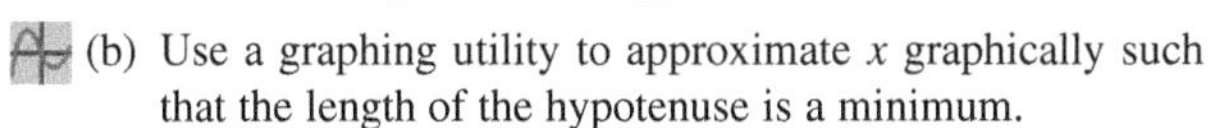
(b) Use a graphing utility to approximate x graphically such that the length of the hypotenuse is a minimum.

(c) Find the vertices of the triangle such that its area is a minimum.

24. Maximum Area Find the area of the largest isosceles triangle that can be inscribed in a circle of radius 6 (see figure).

(a) Solve by writing the area as a function of h.

(b) Solve by writing the area as a function of α.

(c) Identify the type of triangle of maximum area.

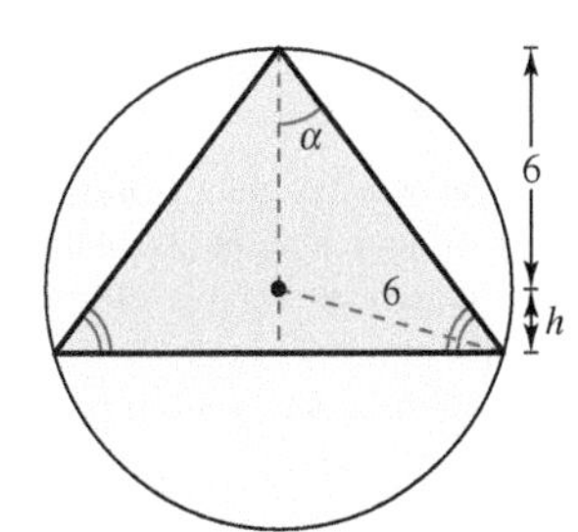

Figure for 24

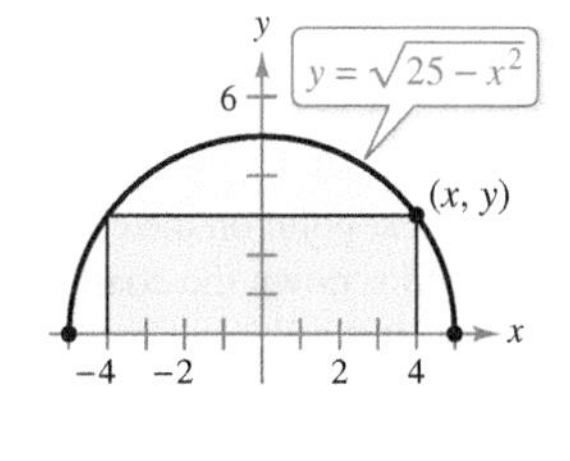

Figure for 25

25. Maximum Area A rectangle is bounded by the x-axis and the semicircle

$$y = \sqrt{25 - x^2}$$

(see figure). What length and width should the rectangle have so that its area is a maximum?

26. Maximum Area Find the dimensions of the largest rectangle that can be inscribed in a semicircle of radius r (see Exercise 25).

27. Numerical, Graphical, and Analytic Analysis An exercise room consists of a rectangle with a semicircle on each end. A 200-meter running track runs around the outside of the room.

(a) Draw a figure to represent the problem. Let x and y represent the length and width of the rectangle, respectively.

(b) Analytically complete six rows of a table such as the one below. (The first two rows are shown.) Use the table to guess the maximum area of the rectangular region.

Length, x	Width, y	Area, xy
10	$\frac{2}{\pi}(100 - 10)$	$(10)\frac{2}{\pi}(100 - 10) \approx 573$
20	$\frac{2}{\pi}(100 - 20)$	$(20)\frac{2}{\pi}(100 - 20) \approx 1019$

(c) Write the area A of the rectangular region as a function of x.

(d) Use calculus to find the critical number of the function in part (c). Then find the maximum area and the dimensions that yield the maximum area.

(e) Use a graphing utility to graph the function in part (c) and verify the maximum area from the graph.

28. Numerical, Graphical, and Analytic Analysis A right circular cylinder is designed to hold 22 cubic inches of a soft drink (approximately 12 fluid ounces).

(a) Analytically complete six rows of a table such as the one below. (The first two rows are shown.)

Radius, r	Height	Surface Area, S
0.2	$\frac{22}{\pi(0.2)^2}$	$2\pi(0.2)\left[0.2 + \frac{22}{\pi(0.2)^2}\right] \approx 220.3$
0.4	$\frac{22}{\pi(0.4)^2}$	$2\pi(0.4)\left[0.4 + \frac{22}{\pi(0.4)^2}\right] \approx 111.0$

(b) Use a graphing utility to generate additional rows of the table. Use the table to estimate the minimum surface area.

(c) Write the surface area S as a function of r.

(d) Use calculus to find the critical number of the function in part (c). Then find the minimum surface area and the dimensions that yield the minimum surface area.

(e) Use a graphing utility to graph the function in part (c) and verify the minimum surface area from the graph.

29. Maximum Volume A rectangular package to be sent by a postal service can have a maximum combined length and girth (perimeter of a cross section) of 108 inches (see figure). Find the dimensions of the package of maximum volume that can be sent. (Assume the cross section is square.)

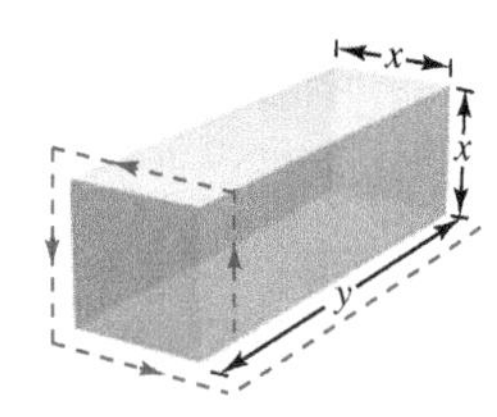

30. Maximum Volume Rework Exercise 29 for a cylindrical package. (The cross section is circular.)

EXPLORING CONCEPTS

31. Surface Area and Volume A shampoo bottle is a right circular cylinder. Because the surface area of the bottle does not change when it is squeezed, is it true that the volume remains the same? Explain.

32. Area and Perimeter The perimeter of a rectangle is 20 feet. Of all possible dimensions, the maximum area is 25 square feet when its length and width are both 5 feet. Are there dimensions that yield a minimum area? Explain.

33. Minimum Surface Area A solid is formed by adjoining two hemispheres to the ends of a right circular cylinder. The total volume of the solid is 14 cubic centimeters. Find the radius of the cylinder that produces the minimum surface area.

34. Minimum Cost An industrial tank of the shape described in Exercise 33 must have a volume of 4000 cubic feet. The hemispherical ends cost twice as much per square foot of surface area as the sides. Find the dimensions that will minimize cost.

35. Minimum Area The sum of the perimeters of an equilateral triangle and a square is 10. Find the dimensions of the triangle and the square that produce a minimum total area.

36. Maximum Area Twenty feet of wire is to be used to form two figures. In each of the following cases, how much wire should be used for each figure so that the total enclosed area is maximum?

(a) Equilateral triangle and square

(b) Square and regular pentagon

(c) Regular pentagon and regular hexagon

(d) Regular hexagon and circle

What can you conclude from this pattern? {*Hint:* The area of a regular polygon with n sides of length x is $A = (n/4)[\cot(\pi/n)]x^2$.}

37. Beam Strength A wooden beam has a rectangular cross section of height h and width w (see figure). The strength S of the beam is directly proportional to the width and the square of the height. What are the dimensions of the strongest beam that can be cut from a round log of diameter 20 inches? (*Hint:* $S = kh^2w$, where k is the proportionality constant.)

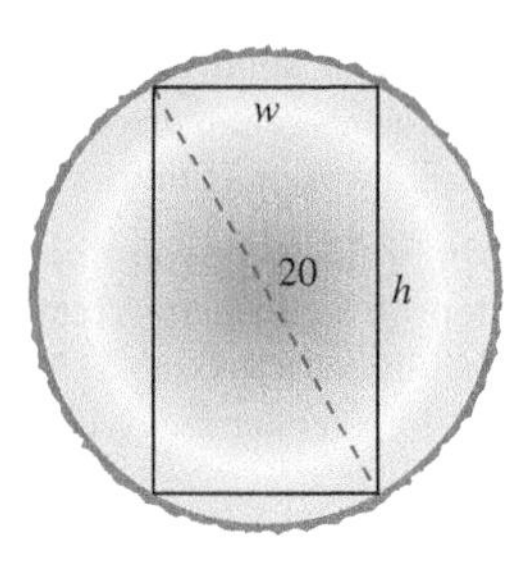

Figure for 37

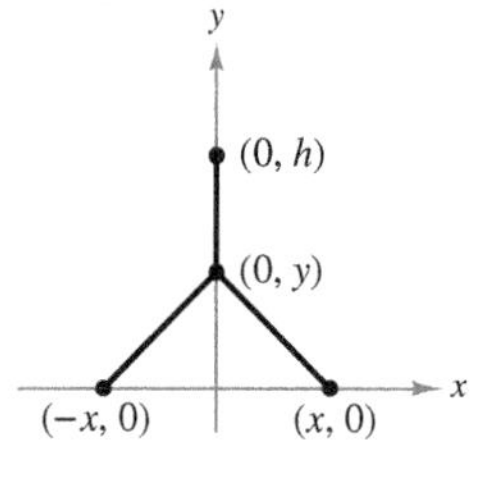

Figure for 38

38. Minimum Length Two factories are located at the coordinates $(-x, 0)$ and $(x, 0)$, and their power supply is at $(0, h)$, as shown in the figure. Find y such that the total length of power line from the power supply to the factories is a minimum.

39. Minimum Cost

An offshore oil well is 2 kilometers off the coast. The refinery is 4 kilometers down the coast. Laying pipe in the ocean is twice as expensive as laying it on land. What path should the pipe follow in order to minimize the cost?

40. Illumination A light source is located over the center of a circular table of diameter 4 feet (see figure). Find the height h of the light source such that the illumination I at the perimeter of the table is maximum when

$$I = \frac{k \sin \alpha}{s^2}$$

where s is the slant height, α is the angle at which the light strikes the table, and k is a constant.

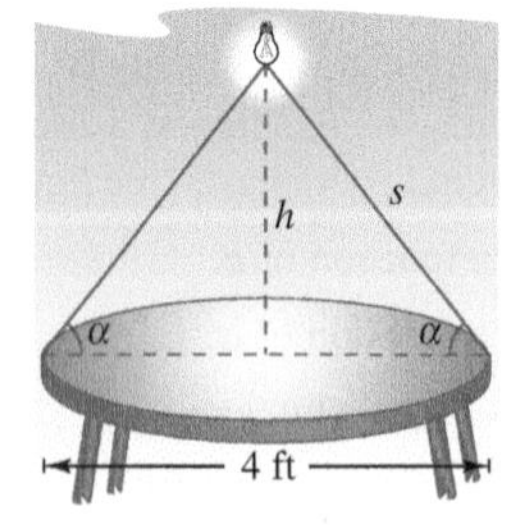

Figure for 40

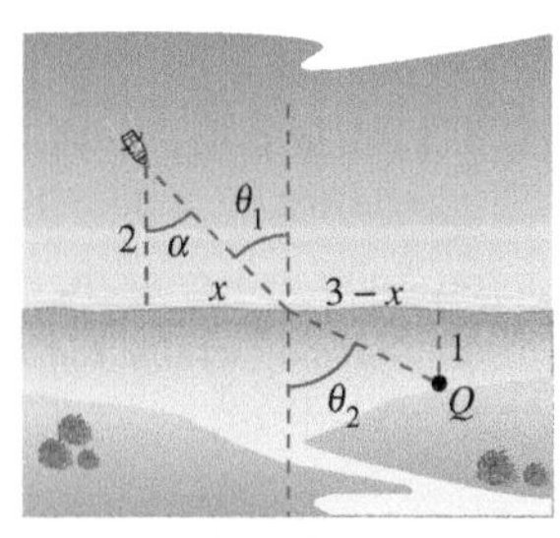

Figure for 41

41. Minimum Time A man is in a boat 2 miles from the nearest point on the coast. He is traveling to a point Q, located 3 miles down the coast and 1 mile inland (see figure). He can row at 2 miles per hour and walk at 4 miles per hour. Toward what point on the coast should he row in order to reach point Q in the least time?

42. **Minimum Time** The conditions are the same as in Exercise 41 except that the man can row at v_1 miles per hour and walk at v_2 miles per hour. If θ_1 and θ_2 are the magnitudes of the angles, show that the man will reach point Q in the least time when

$$\frac{\sin \theta_1}{v_1} = \frac{\sin \theta_2}{v_2}.$$

43. **Minimum Distance** Sketch the graph of

$$f(x) = 2 - 2 \sin x$$

on the interval $[0, \pi/2]$.

(a) Find the distance from the origin to the y-intercept and the distance from the origin to the x-intercept.

(b) Write the distance d from the origin to a point on the graph of f as a function of x.

(c) Use calculus to find the value of x that minimizes the function d on the interval $[0, \pi/2]$. What is the minimum distance? Use a graphing utility to verify your results.

(Submitted by Tim Chapell, Penn Valley Community College, Kansas City, MO)

44. **Minimum Time** When light waves traveling in a transparent medium strike the surface of a second transparent medium, they change direction. This change of direction is called *refraction* and is defined by **Snell's Law of Refraction,**

$$\frac{\sin \theta_1}{v_1} = \frac{\sin \theta_2}{v_2}$$

where θ_1 and θ_2 are the magnitudes of the angles shown in the figure and v_1 and v_2 are the velocities of light in the two media. Show that this problem is equivalent to that in Exercise 42, and that light waves traveling from P to Q follow the path of minimum time.

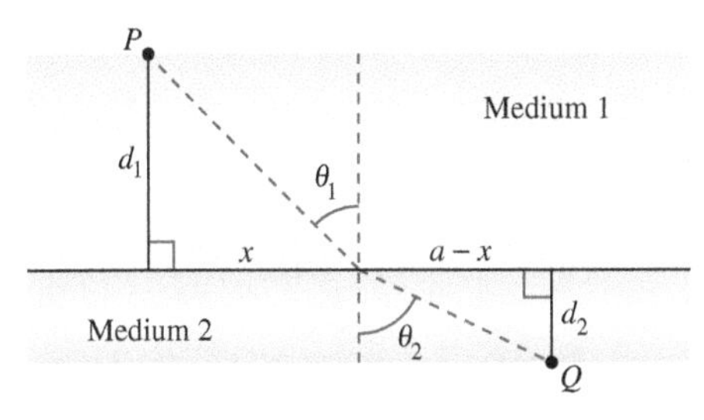

45. **Maximum Volume** A sector with central angle θ is cut from a circle of radius 12 inches (see figure), and the edges of the sector are brought together to form a cone. Find the magnitude of θ such that the volume of the cone is a maximum.

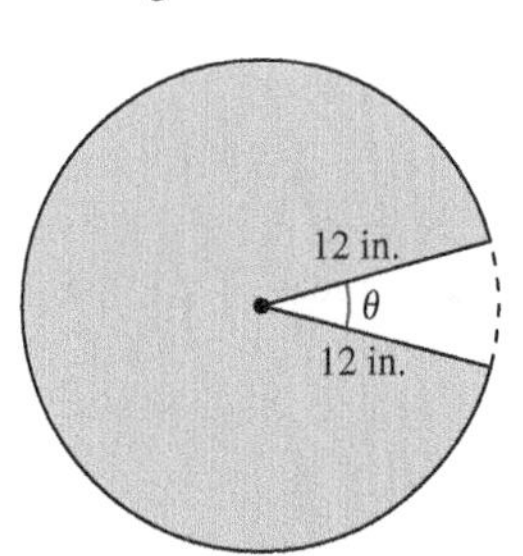

Figure for 45

Figure for 46

46. **Numerical, Graphical, and Analytic Analysis** The cross sections of an irrigation canal are isosceles trapezoids of which three sides are 8 feet long (see figure). Determine the angle of elevation θ of the sides such that the area of the cross sections is a maximum by completing the following.

(a) Analytically complete six rows of a table such as the one below. (The first two rows are shown.)

θ	Base 1	Base 2	Altitude	Area
$10°$	8	$8 + 16 \cos 10°$	$8 \sin 10°$	≈ 22.1
$20°$	8	$8 + 16 \cos 20°$	$8 \sin 20°$	≈ 42.5

(b) Use a graphing utility to generate additional rows of the table. Use the table to estimate the maximum cross-sectional area.

(c) Write the cross-sectional area A as a function of θ.

(d) Use calculus to find the critical number of the function in part (c). Then find the angle that will yield the maximum cross-sectional area. What is the maximum area?

(e) Use a graphing utility to graph the function in part (c) and verify the maximum cross-sectional area.

47. **Maximum Profit** Assume that the amount of money deposited in a bank is proportional to the square of the interest rate the bank pays on this money. Furthermore, the bank can reinvest this money at 8%. Find the interest rate the bank should pay to maximize profit. (Use the simple interest formula.)

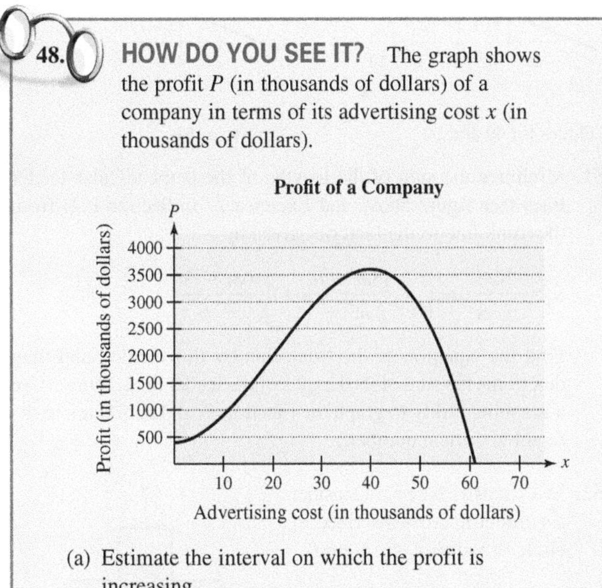

48. **HOW DO YOU SEE IT?** The graph shows the profit P (in thousands of dollars) of a company in terms of its advertising cost x (in thousands of dollars).

(a) Estimate the interval on which the profit is increasing.

(b) Estimate the interval on which the profit is decreasing.

(c) Estimate the amount of money the company should spend on advertising in order to yield a maximum profit.

(d) The *point of diminishing returns* is the point at which the rate of growth of the profit function begins to decline. Estimate the point of diminishing returns.

Minimum Distance In Exercises 49–51, consider a fuel distribution center located at the origin of the rectangular coordinate system (units in miles; see figures). The center supplies three factories with coordinates (4, 1), (5, 6), and (10, 3). A trunk line will run from the distribution center along the line $y = mx$, and feeder lines will run to the three factories. The objective is to find m such that the lengths of the feeder lines are minimized.

49. Minimize the sum of the squares of the lengths of the vertical feeder lines (see figure) given by

$$S_1 = (4m - 1)^2 + (5m - 6)^2 + (10m - 3)^2.$$

Find the equation of the trunk line by this method and then determine the sum of the lengths of the feeder lines.

50. Minimize the sum of the absolute values of the lengths of the vertical feeder lines (see figure) given by

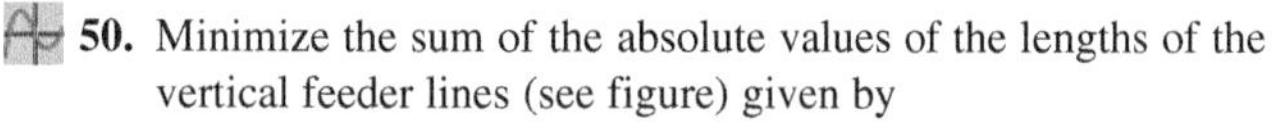

$$S_2 = |4m - 1| + |5m - 6| + |10m - 3|.$$

Find the equation of the trunk line by this method and then determine the sum of the lengths of the feeder lines. (*Hint:* Use a graphing utility to graph the function S_2 and approximate the required critical number.)

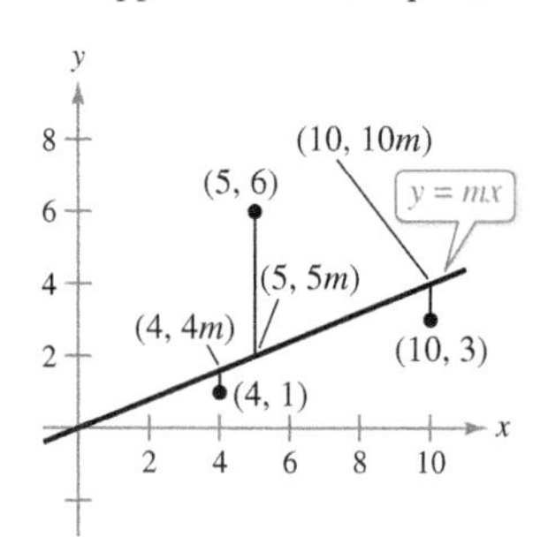

Figure for 49 and 50

Figure for 51

51. Minimize the sum of the lengths of the perpendicular feeder lines (see figure above and Exercise 77 in Section P.2) from the trunk line to the factories given by

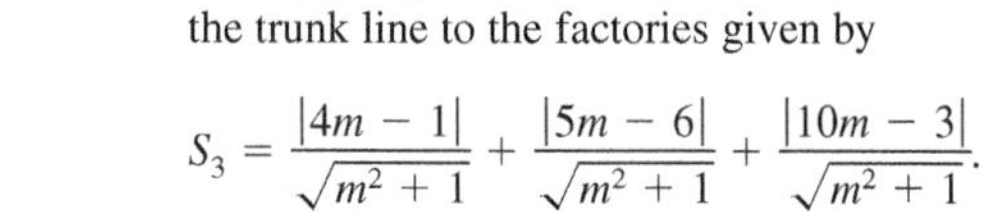

$$S_3 = \frac{|4m - 1|}{\sqrt{m^2 + 1}} + \frac{|5m - 6|}{\sqrt{m^2 + 1}} + \frac{|10m - 3|}{\sqrt{m^2 + 1}}.$$

Find the equation of the trunk line by this method and then determine the sum of the lengths of the feeder lines. (*Hint:* Use a graphing utility to graph the function S_3 and approximate the required critical number.)

52. Maximum Area Consider a symmetric cross inscribed in a circle of radius r (see figure).

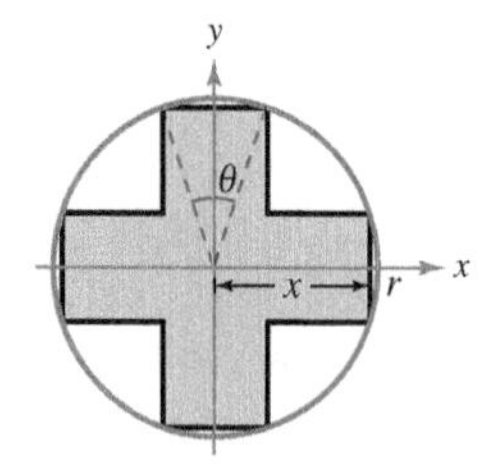

(a) Write the area A of the cross as a function of x and find the value of x that maximizes the area.

(b) Write the area A of the cross as a function of θ and find the value of θ that maximizes the area.

(c) Show that the critical numbers of parts (a) and (b) yield the same maximum area. What is that area?

53. Minimum Distance Find the point on the graph of the equation

$$16x = y^2$$

that is closest to the point (6, 0).

54. Minimum Distance Find the point on the graph of the function

$$x = \sqrt{10y}$$

that is closest to the point (0, 4). (*Hint:* Consider the domain of the function.)

PUTNAM EXAM CHALLENGE

55. Find, with explanation, the maximum value of $f(x) = x^3 - 3x$ on the set of all real numbers x satisfying $x^4 + 36 \leq 13x^2$.

56. Find the minimum value of

$$\frac{(x + 1/x)^6 - (x^6 + 1/x^6) - 2}{(x + 1/x)^3 + (x^3 + 1/x^3)} \quad \text{for } x > 0.$$

SECTION PROJECT

Minimum Time

A woman is at point A on the shore of a circular lake of radius 2 kilometers (see figure). She wants to walk around the lake to point B and then swim to point C in the least amount of time. Point C lies on the diameter through point A. Assume that she can walk at v_1 kilometers per hour and swim at v_2 kilometers per hour, and that $0 \leq \theta \leq \pi$.

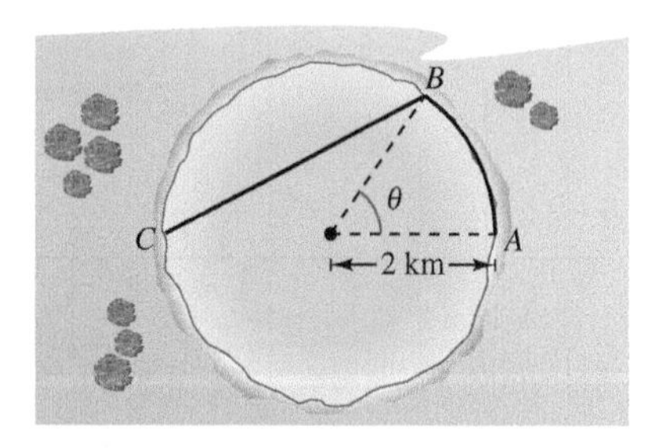

(a) Find the distance walked from point A to point B in terms of θ.

(b) Find the distance swam from point B to point C in terms of θ.

(c) Write the function $f(\theta)$ that represents the total time to move from point A to point C.

(d) Find $f'(\theta)$.

(e) If $v_1 = 5$ and $v_2 = 2$, approximate the critical number(s) of f. Does the critical number(s) correspond to a relative maximum or a relative minimum? Where should point B be located in order to minimize the time for the trip from point A to point C? Explain.

(f) Repeat part (e) for $v_1 = 3$ and $v_2 = 2$.

3.8 Newton's Method

- **Approximate a zero of a function using Newton's Method.**

Newton's Method

In this section, you will study a technique for approximating the real zeros of a function. The technique is called **Newton's Method,** and it uses tangent lines to approximate the graph of the function near its x-intercepts.

To see how Newton's Method works, consider a function f that is continuous on the interval $[a, b]$ and differentiable on the interval (a, b). If $f(a)$ and $f(b)$ differ in sign, then, by the Intermediate Value Theorem, f must have at least one zero in the interval (a, b). To estimate this zero, you choose

$$x = x_1 \qquad \text{First estimate}$$

as shown in Figure 3.60(a). Newton's Method is based on the assumption that the graph of f and the tangent line at $(x_1, f(x_1))$ both cross the x-axis at *about* the same point. Because you can easily calculate the x-intercept for this tangent line, you can use it as a second (and, usually, better) estimate of the zero of f. The tangent line passes through the point $(x_1, f(x_1))$ with a slope of $f'(x_1)$. In point-slope form, the equation of the tangent line is

$$y - f(x_1) = f'(x_1)(x - x_1)$$
$$y = f'(x_1)(x - x_1) + f(x_1).$$

Letting $y = 0$ and solving for x produces

$$x = x_1 - \frac{f(x_1)}{f'(x_1)}.$$

So, from the initial estimate x_1, you obtain a new estimate

$$x_2 = x_1 - \frac{f(x_1)}{f'(x_1)}. \qquad \text{Second estimate [See Figure 3.60(b).]}$$

You can improve on x_2 and calculate yet a third estimate

$$x_3 = x_2 - \frac{f(x_2)}{f'(x_2)}. \qquad \text{Third estimate}$$

Repeated application of this process is called Newton's Method.

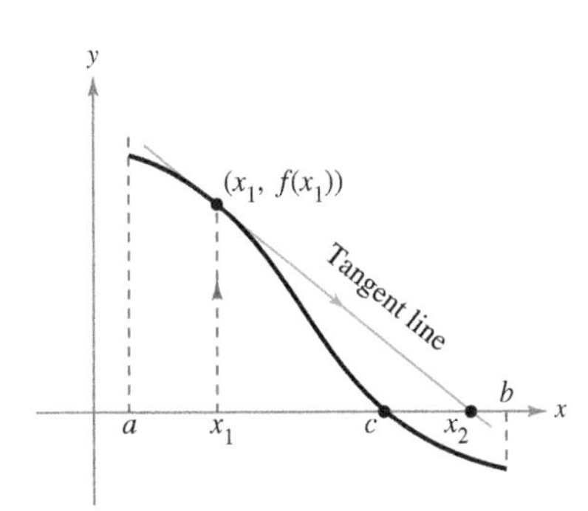

(a)

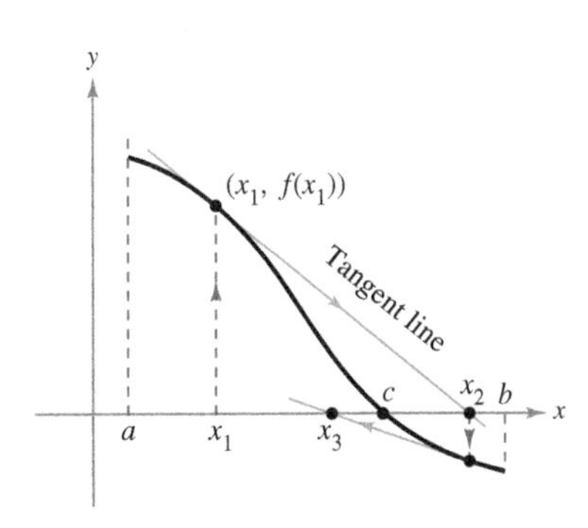

(b)
The x-intercept of the tangent line approximates the zero of f.
Figure 3.60

NEWTON'S METHOD

Isaac Newton first described the method for approximating the real zeros of a function in his text *Method of Fluxions.* Although the book was written in 1671, it was not published until 1736. Meanwhile, in 1690, Joseph Raphson (1648–1715) published a paper describing a method for approximating the real zeros of a function that was very similar to Newton's. For this reason, the method is often referred to as the Newton-Raphson method.

Newton's Method for Approximating the Zeros of a Function

Let $f(c) = 0$, where f is differentiable on an open interval containing c. Then, to approximate c, use these steps.

1. Make an initial estimate x_1 that is close to c. (A graph is helpful.)
2. Determine a new approximation
$$x_{n+1} = x_n - \frac{f(x_n)}{f'(x_n)}.$$
3. When $|x_n - x_{n+1}|$ is within the desired accuracy, let x_{n+1} serve as the final approximation. Otherwise, return to Step 2 and calculate a new approximation.

Each successive application of this procedure is called an **iteration.**

REMARK For many functions, just a few iterations of Newton's Method will produce approximations having very small errors, as shown in Example 1.

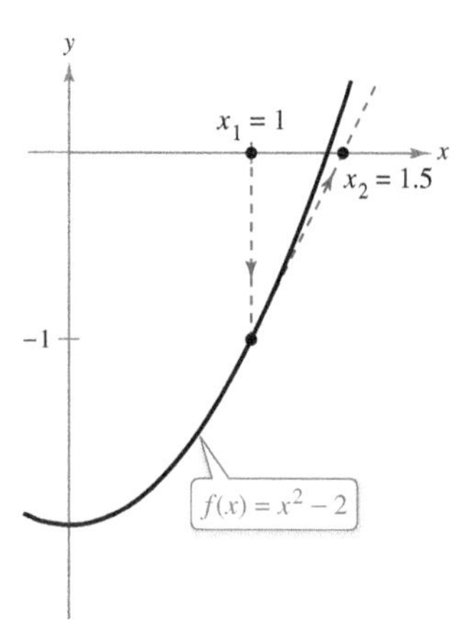

The first iteration of Newton's Method
Figure 3.61

EXAMPLE 1 Using Newton's Method

Calculate three iterations of Newton's Method to approximate a zero of $f(x) = x^2 - 2$. Use $x_1 = 1$ as the initial guess.

Solution Because $f(x) = x^2 - 2$, you have $f'(x) = 2x$, and the iterative formula is

$$x_{n+1} = x_n - \frac{f(x_n)}{f'(x_n)} = x_n - \frac{x_n^2 - 2}{2x_n}.$$

The calculations for three iterations are shown in the table.

n	x_n	$f(x_n)$	$f'(x_n)$	$\frac{f(x_n)}{f'(x_n)}$	$x_n - \frac{f(x_n)}{f'(x_n)}$
1	1.000000	−1.000000	2.000000	−0.500000	1.500000
2	1.500000	0.250000	3.000000	0.083333	1.416667
3	1.416667	0.006945	2.833334	0.002451	1.414216
4	1.414216				

Of course, in this case you know that the two zeros of the function are $\pm\sqrt{2}$. To six decimal places, $\sqrt{2} = 1.414214$. So, after only three iterations of Newton's Method, you have obtained an approximation that is within 0.000002 of an actual root. The first iteration of this process is shown in Figure 3.61.

EXAMPLE 2 Using Newton's Method

See LarsonCalculus.com for an interactive version of this type of example.

Use Newton's Method to approximate the zeros of

$$f(x) = 2x^3 + x^2 - x + 1.$$

Continue the iterations until two successive approximations differ by less than 0.0001.

Solution Begin by sketching a graph of f, as shown in Figure 3.62. From the graph, you can observe that the function has only one zero, which occurs near $x = -1.2$. Next, differentiate f and form the iterative formula

$$x_{n+1} = x_n - \frac{f(x_n)}{f'(x_n)} = x_n - \frac{2x_n^3 + x_n^2 - x_n + 1}{6x_n^2 + 2x_n - 1}.$$

The calculations are shown in the table.

n	x_n	$f(x_n)$	$f'(x_n)$	$\frac{f(x_n)}{f'(x_n)}$	$x_n - \frac{f(x_n)}{f'(x_n)}$
1	−1.20000	0.18400	5.24000	0.03511	−1.23511
2	−1.23511	−0.00771	5.68276	−0.00136	−1.23375
3	−1.23375	0.00001	5.66533	0.00000	−1.23375
4	−1.23375				

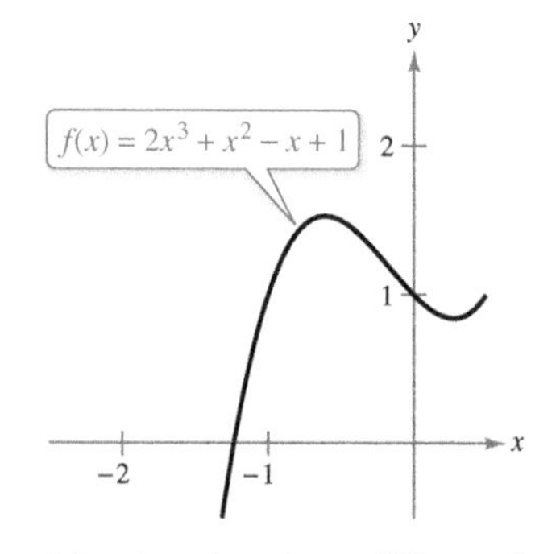

After three iterations of Newton's Method, the zero of f is approximated to the desired accuracy.
Figure 3.62

Because two successive approximations differ by less than the required 0.0001, you can estimate the zero of f to be -1.23375. ■

When, as in Examples 1 and 2, the approximations approach a limit, the sequence of approximations

$$x_1, x_2, x_3, \ldots, x_n, \ldots$$

is said to **converge.** Moreover, when the limit is c, it can be shown that c must be a zero of f.

FOR FURTHER INFORMATION For more on when Newton's Method fails, see the article "No Fooling! Newton's Method Can Be Fooled" by Peter Horton in *Mathematics Magazine*. To view this article, go to *MathArticles.com*.

Newton's Method does not always yield a convergent sequence. One way it can fail to do so is shown in Figure 3.63. Because Newton's Method involves division by $f'(x_n)$, it is clear that the method will fail when the derivative is zero for any x_n in the sequence. When you encounter this problem, you can usually overcome it by choosing a different value for x_1. Another way Newton's Method can fail is shown in the next example.

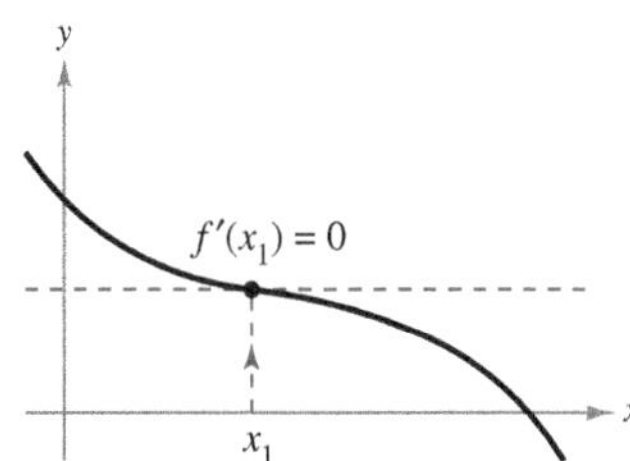

Newton's Method fails to converge when $f'(x_n) = 0$.
Figure 3.63

EXAMPLE 3 An Example in Which Newton's Method Fails

The function $f(x) = x^{1/3}$ is not differentiable at $x = 0$. Show that Newton's Method fails to converge using $x_1 = 0.1$.

Solution Because $f'(x) = \frac{1}{3}x^{-2/3}$, the iterative formula is

$$x_{n+1} = x_n - \frac{f(x_n)}{f'(x_n)} = x_n - \frac{x_n^{1/3}}{\frac{1}{3}x_n^{-2/3}} = x_n - 3x_n = -2x_n.$$

The calculations are shown in the table. This table and Figure 3.64 indicate that x_n continues to increase in magnitude as $n \to \infty$, and so the limit of the sequence does not exist.

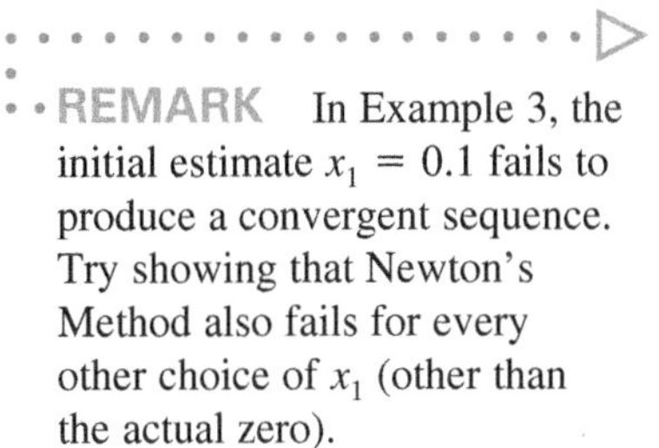

REMARK In Example 3, the initial estimate $x_1 = 0.1$ fails to produce a convergent sequence. Try showing that Newton's Method also fails for every other choice of x_1 (other than the actual zero).

n	x_n	$f(x_n)$	$f'(x_n)$	$\frac{f(x_n)}{f'(x_n)}$	$x_n - \frac{f(x_n)}{f'(x_n)}$
1	0.10000	0.46416	1.54720	0.30000	−0.20000
2	−0.20000	−0.58480	0.97467	−0.60000	0.40000
3	0.40000	0.73681	0.61401	1.20000	−0.80000
4	−0.80000	−0.92832	0.38680	−2.40000	1.60000

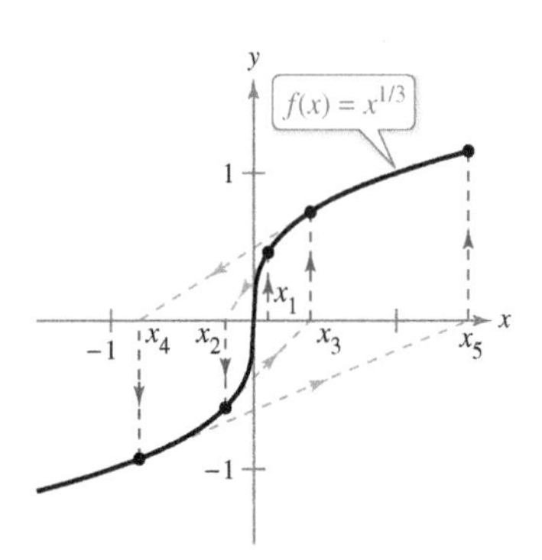

Newton's Method fails to converge for every x-value other than the actual zero of f.
Figure 3.64

It can be shown that a condition sufficient to produce convergence of Newton's Method to a zero of f is that

$$\left|\frac{f(x)f''(x)}{[f'(x)]^2}\right| < 1 \qquad \text{Condition for convergence}$$

on an open interval containing the zero. For instance, in Example 1, this test would yield

$$f(x) = x^2 - 2, \quad f'(x) = 2x, \quad f''(x) = 2,$$

and

$$\left|\frac{f(x)f''(x)}{[f'(x)]^2}\right| = \left|\frac{(x^2-2)(2)}{4x^2}\right| = \left|\frac{1}{2} - \frac{1}{x^2}\right|. \qquad \text{Example 1}$$

On the interval $(1, 3)$, this quantity is less than 1 and therefore the convergence of Newton's Method is guaranteed. On the other hand, in Example 3, you have

$$f(x) = x^{1/3}, \quad f'(x) = \frac{1}{3}x^{-2/3}, \quad f''(x) = -\frac{2}{9}x^{-5/3}$$

and

$$\left|\frac{f(x)f''(x)}{[f'(x)]^2}\right| = \left|\frac{x^{1/3}(-2/9)(x^{-5/3})}{(1/9)(x^{-4/3})}\right| = 2 \qquad \text{Example 3}$$

which is not less than 1 for any value of x, so you cannot conclude that Newton's Method will converge.

You have learned several techniques for finding the zeros of functions. The zeros of some functions, such as

$$f(x) = x^3 - 2x^2 - x + 2$$

can be found by simple algebraic techniques, such as factoring. The zeros of other functions, such as

$$f(x) = x^3 - x + 1$$

cannot be found by *elementary* algebraic methods. This particular function has only one real zero, and by using more advanced algebraic techniques, you can determine the zero to be

$$x = -\sqrt[3]{\frac{3 - \sqrt{23/3}}{6}} - \sqrt[3]{\frac{3 + \sqrt{23/3}}{6}}.$$

Because the *exact* solution is written in terms of square roots and cube roots, it is called a **solution by radicals.**

The determination of radical solutions of a polynomial equation is one of the fundamental problems of algebra. The earliest such result is the Quadratic Formula, which dates back at least to Babylonian times. The general formula for the zeros of a cubic function was developed much later. In the sixteenth century, an Italian mathematician, Jerome Cardan, published a method for finding radical solutions to cubic and quartic equations. Then, for 300 years, the problem of finding a general quintic formula remained open. Finally, in the nineteenth century, the problem was answered independently by two young mathematicians. Niels Henrik Abel, a Norwegian mathematician, and Evariste Galois, a French mathematician, proved that it is not possible to solve a *general* fifth- (or higher-) degree polynomial equation by radicals. Of course, you can solve particular fifth-degree equations, such as $x^5 - 1 = 0$, but Abel and Galois were able to show that no general *radical* solution exists.

NIELS HENRIK ABEL (1802–1829)

EVARISTE GALOIS (1811–1832)

Although the lives of both Abel and Galois were brief, their work in the fields of analysis and abstract algebra was far-reaching.

See LarsonCalculus.com to read a biography about each of these mathematicians.

3.8 Exercises

See CalcChat.com for tutorial help and worked-out solutions to odd-numbered exercises.

CONCEPT CHECK

1. **Newton's Method** In your own words and using a sketch, describe Newton's Method for approximating the zeros of a function.

2. **Failure of Newton's Method** Why does Newton's Method fail when $f'(x_n) = 0$? What does this mean graphically?

Using Newton's Method In Exercises 3–6, calculate two iterations of Newton's Method to approximate a zero of the function using the given initial guess.

3. $f(x) = x^2 - 5, \quad x_1 = 2$

4. $f(x) = x^3 - 3, \quad x_1 = 1.4$

5. $f(x) = \cos x, \quad x_1 = 1.6$

6. $f(x) = \tan x, \quad x_1 = 0.1$

Using Newton's Method In Exercises 7–16, use Newton's Method to approximate the zero(s) of the function. Continue the iterations until two successive approximations differ by less than 0.001. Then find the zero(s) using a graphing utility and compare the results.

7. $f(x) = x^3 + 4$

8. $f(x) = 2 - x^3$

9. $f(x) = x^3 + x - 1$

10. $f(x) = x^5 + x - 1$

11. $f(x) = 5\sqrt{x - 1} - 2x$

12. $f(x) = x - 2\sqrt{x + 1}$

13. $f(x) = x^3 - 3.9x^2 + 4.79x - 1.881$

14. $f(x) = -x^3 + 2.7x^2 + 3.55x - 2.422$

15. $f(x) = 1 - x + \sin x$

16. $f(x) = x^3 - \cos x$

Points of Intersection In Exercises 17–20, apply Newton's Method to approximate the x-value(s) of the indicated point(s) of intersection of the two graphs. Continue the iterations until two successive approximations differ by less than 0.001. [*Hint:* Let $h(x) = f(x) - g(x)$.]

17. $f(x) = 2x + 1$
$g(x) = \sqrt{x + 4}$

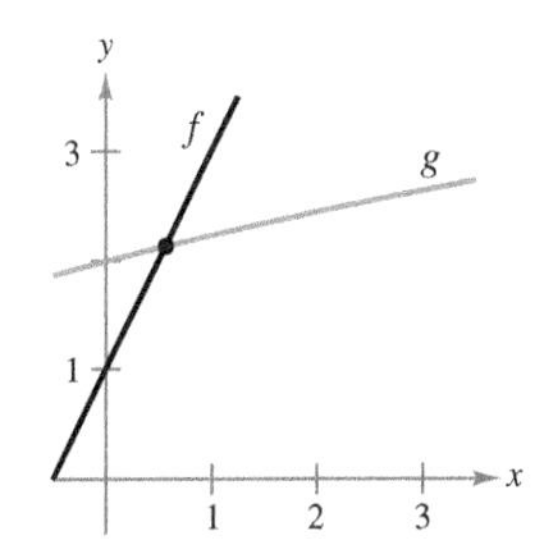

18. $f(x) = 3 - x$
$g(x) = \dfrac{1}{x^2 + 1}$

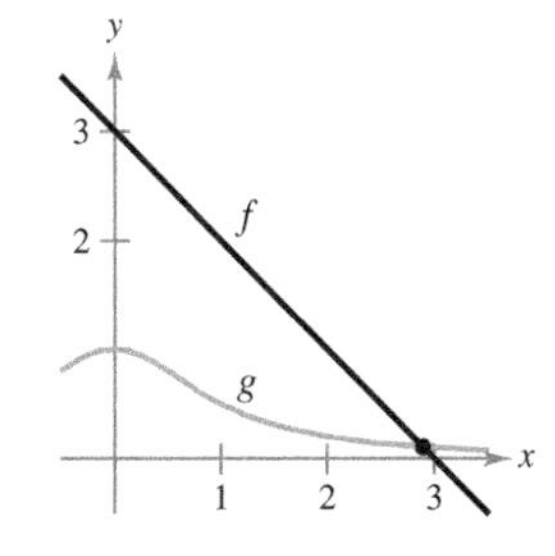

19. $f(x) = x$
$g(x) = \tan x$

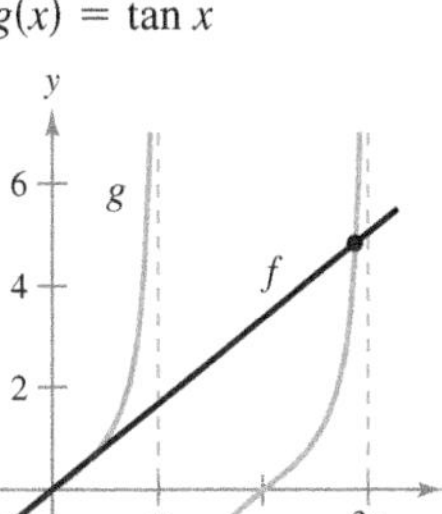

20. $f(x) = x^2$
$g(x) = \cos x$

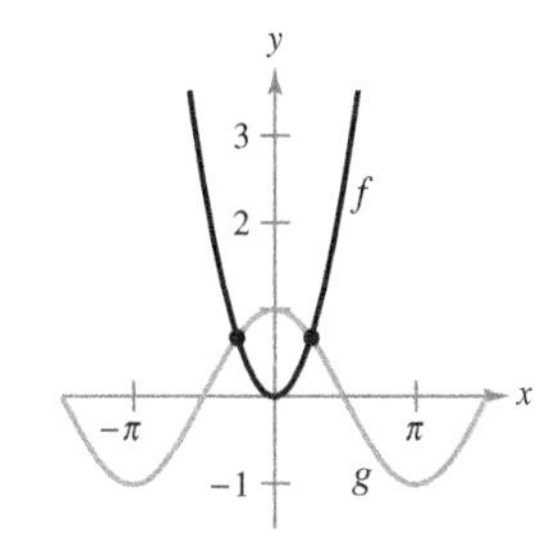

21. **Using Newton's Method** Consider the function $f(x) = x^3 - 3x^2 + 3$.
   (a) Use a graphing utility to graph f.
   (b) Use Newton's Method to approximate a zero with $x_1 = 1$ as the initial guess.
   (c) Repeat part (b) using $x_1 = \frac{1}{4}$ as the initial guess and observe that the result is different.
   (d) To understand why the results in parts (b) and (c) are different, sketch the tangent lines to the graph of f at the points $(1, f(1))$ and $(\frac{1}{4}, f(\frac{1}{4}))$. Describe why it is important to select the initial guess carefully.

22. **Using Newton's Method** Repeat the steps in Exercise 21 for the function $f(x) = \sin x$ with initial guesses of $x_1 = 1.8$ and $x_1 = 3$.

Failure of Newton's Method In Exercises 23 and 24, apply Newton's Method using the given initial guess, and explain why the method fails.

23. $y = 2x^3 - 6x^2 + 6x - 1, \quad x_1 = 1$

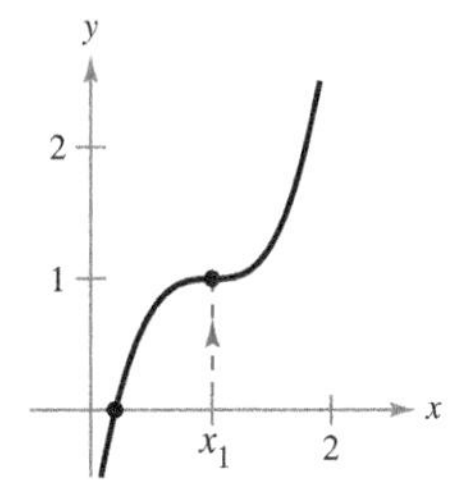

Figure for 23

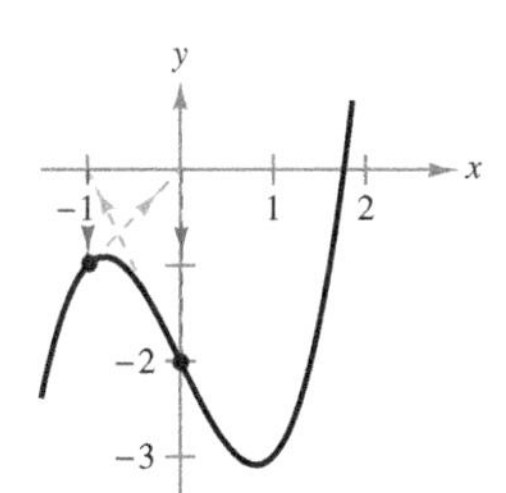

Figure for 24

24. $y = x^3 - 2x - 2, \quad x_1 = 0$

Fixed Point In Exercises 25 and 26, approximate the fixed point of the function to two decimal places. [A *fixed point* of a function f is a real number c such that $f(c) = c$.]

25. $f(x) = \cos x$

26. $f(x) = \cot x, \quad 0 < x < \pi$

EXPLORING CONCEPTS

27. Newton's Method What will be the values of future guesses for x if your initial guess is a zero of f? Explain.

28. Newton's Method Does Newton's Method fail when the initial guess is a relative maximum of f? Explain.

Using Newton's Method Exercises 29–31 present problems similar to exercises from the previous sections of this chapter. In each case, use Newton's Method to approximate the solution.

29. Minimum Distance Find the point on the graph of $f(x) = 4 - x^2$ that is closest to the point $(1, 0)$.

30. Medicine The concentration C of a chemical in the bloodstream t hours after injection into muscle tissue is given by

$$C = \frac{3t^2 + t}{50 + t^3}.$$

When is the concentration the greatest?

31. Minimum Time You are in a boat 2 miles from the nearest point on the coast. You are traveling to a point Q that is 3 miles down the coast and 1 mile inland (see figure). You can row at 3 miles per hour and walk at 4 miles per hour. Toward what point on the coast should you row in order to reach point Q in the least time?

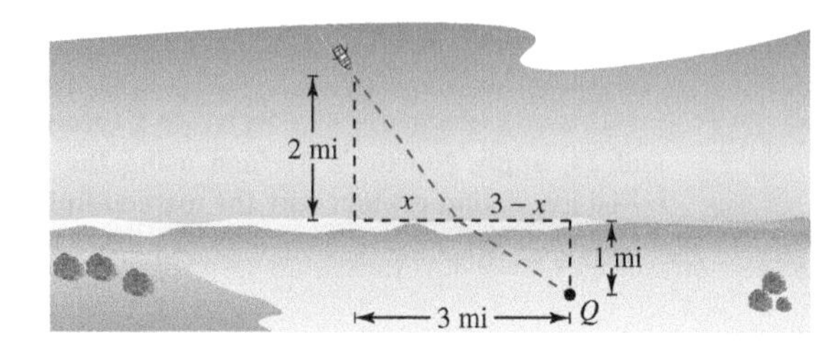

32. HOW DO YOU SEE IT? For what value(s) will Newton's Method fail to converge for the function shown in the graph? Explain your reasoning.

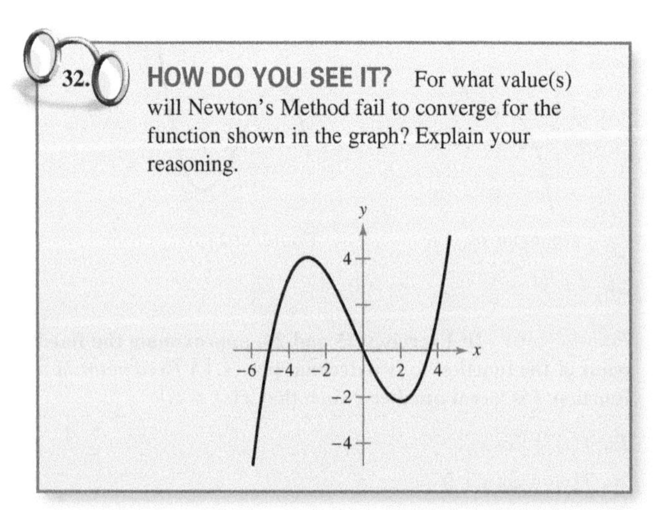

33. Mechanic's Rule The Mechanic's Rule for approximating $\sqrt{a}$, $a > 0$, is

$$x_{n+1} = \frac{1}{2}\left(x_n + \frac{a}{x_n}\right), \quad n = 1, 2, 3, \ldots$$

where x_1 is an approximation of $\sqrt{a}$.

(a) Use Newton's Method and the function $f(x) = x^2 - a$ to derive the Mechanic's Rule.

(b) Use the Mechanic's Rule to approximate $\sqrt{5}$ and $\sqrt{7}$ to three decimal places.

34. Approximating Radicals

(a) Use Newton's Method and the function $f(x) = x^n - a$ to obtain a general rule for approximating $x = \sqrt[n]{a}$.

(b) Use the general rule found in part (a) to approximate $\sqrt[4]{6}$ and $\sqrt[3]{15}$ to three decimal places.

35. Approximating Reciprocals Use Newton's Method to show that the equation $x_{n+1} = x_n(2 - ax_n)$ can be used to approximate $1/a$ when x_1 is an initial guess of the reciprocal of a. Note that this method of approximating reciprocals uses only the operations of multiplication and subtraction. (*Hint:* Consider

$$f(x) = \frac{1}{x} - a.)$$

36. Approximating Reciprocals Use the result of Exercise 35 to approximate (a) $\frac{1}{3}$ and (b) $\frac{1}{11}$ to three decimal places.

True or False? In Exercises 37–40, determine whether the statement is true or false. If it is false, explain why or give an example that shows it is false.

37. The zeros of $f(x) = \dfrac{p(x)}{q(x)}$ coincide with the zeros of $p(x)$.

38. If the coefficients of a polynomial function are all positive, then the polynomial has no positive zeros.

39. If $f(x)$ is a cubic polynomial such that $f'(x)$ is never zero, then any initial guess will force Newton's Method to converge to the zero of f.

40. Newton's Method fails when the initial guess x_1 corresponds to a horizontal tangent line for the graph of f at x_1.

41. Tangent Lines The graph of $f(x) = -\sin x$ has infinitely many tangent lines that pass through the origin. Use Newton's Method to approximate to three decimal places the slope of the tangent line having the greatest slope.

42. Point of Tangency The graph of $f(x) = \cos x$ and a tangent line to f through the origin are shown. Find the coordinates of the point of tangency to three decimal places.

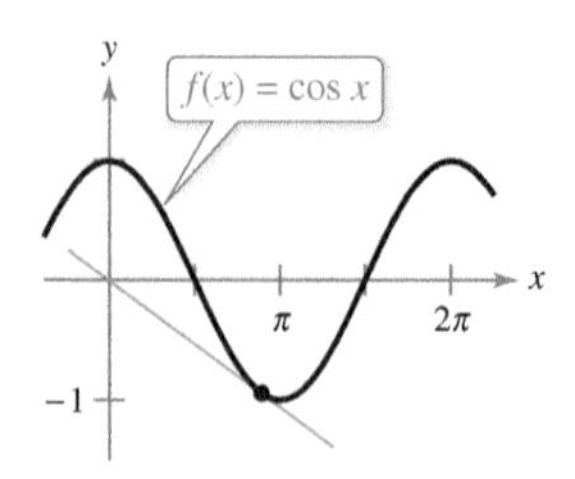

3.9 Differentials

- **Understand the concept of a tangent line approximation.**
- **Compare the value of the differential, dy, with the actual change in y, Δy.**
- **Estimate a propagated error using a differential.**
- **Find the differential of a function using differentiation formulas.**

> **Exploration**
>
> ***Tangent Line Approximation*** Use a graphing utility to graph $f(x) = x^2$. In the same viewing window, graph the tangent line to the graph of f at the point $(1, 1)$. Zoom in twice on the point of tangency. Does your graphing utility distinguish between the two graphs? Use the *trace* feature to compare the two graphs. As the x-values get closer to 1, what can you say about the y-values?

Tangent Line Approximations

Newton's Method (see Section 3.8) is an example of the use of a tangent line to approximate the graph of a function. In this section, you will study other situations in which the graph of a function can be approximated by a straight line.

To begin, consider a function f that is differentiable at c. The equation for the tangent line at the point $(c, f(c))$ is

$$y - f(c) = f'(c)(x - c)$$

$$y = f(c) + f'(c)(x - c)$$

and is called the **tangent line approximation** (or **linear approximation**) **of f at c.** Because c is a constant, y is a linear function of x. Moreover, by restricting the values of x to those sufficiently close to c, the values of y can be used as approximations (to any desired degree of accuracy) of the values of the function f. In other words, as x approaches c, the limit of y is $f(c)$.

EXAMPLE 1 Using a Tangent Line Approximation

See LarsonCalculus.com for an interactive version of this type of example.

Find the tangent line approximation of $f(x) = 1 + \sin x$ at the point $(0, 1)$. Then use a table to compare the y-values of the linear function with those of $f(x)$ on an open interval containing $x = 0$.

Solution The derivative of f is

$$f'(x) = \cos x. \qquad \text{First derivative}$$

So, the equation of the tangent line to the graph of f at the point $(0, 1)$ is

$$y = f(0) + f'(0)(x - 0)$$
$$y = 1 + (1)(x - 0)$$
$$y = 1 + x. \qquad \text{Tangent line approximation}$$

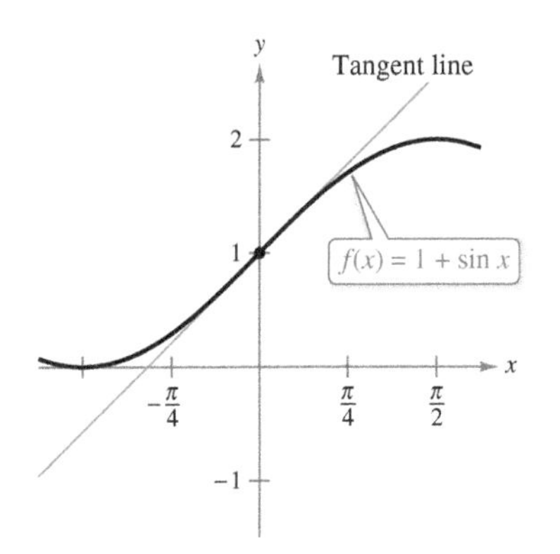

The tangent line approximation of f at the point $(0, 1)$
Figure 3.65

The table compares the values of y given by this linear approximation with the values of $f(x)$ near $x = 0$. Notice that the closer x is to 0, the better the approximation. This conclusion is reinforced by the graph shown in Figure 3.65.

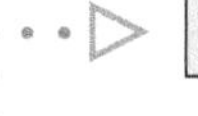

x	-0.5	-0.1	-0.01	0	0.01	0.1	0.5
$f(x) = 1 + \sin x$	0.521	0.9002	0.9900002	1	1.0099998	1.0998	1.479
$y = 1 + x$	0.5	0.9	0.99	1	1.01	1.1	1.5

■

REMARK Be sure you see that this linear approximation of $f(x) = 1 + \sin x$ depends on the point of tangency. At a different point on the graph of f, you would obtain a different tangent line approximation.

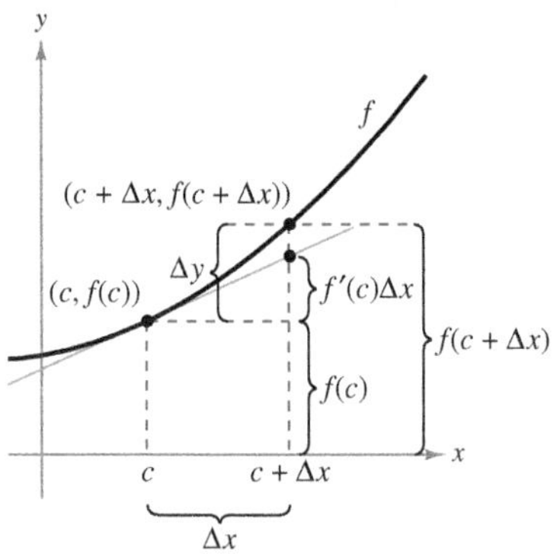

Differentials

When the tangent line to the graph of f at the point $(c, f(c))$

$$y = f(c) + f'(c)(x - c) \qquad \text{Tangent line at } (c, f(c))$$

is used as an approximation of the graph of f, the quantity $x - c$ is called the change in x, and is denoted by Δx, as shown in Figure 3.66. When Δx is small, the change in y (denoted by Δy) can be approximated as shown.

$$\Delta y = f(c + \Delta x) - f(c) \qquad \text{Actual change in } y$$

$$\approx f'(c)\Delta x \qquad \text{Approximate change in } y$$

When Δx is small, $\Delta y = f(c + \Delta x) - f(c)$ is approximated by $f'(c)\Delta x$.
Figure 3.66

For such an approximation, the quantity Δx is traditionally denoted by dx and is called the **differential of *x*.** The expression $f'(x)\,dx$ is denoted by dy and is called the **differential of *y*.**

Definition of Differentials

Let $y = f(x)$ represent a function that is differentiable on an open interval containing x. The **differential of *x*** (denoted by dx) is any nonzero real number. The **differential of *y*** (denoted by dy) is

$$dy = f'(x)\,dx.$$

In many types of applications, the differential of y can be used as an approximation of the change in y. That is,

$$\Delta y \approx dy \quad \text{or} \quad \Delta y \approx f'(x)\,dx.$$

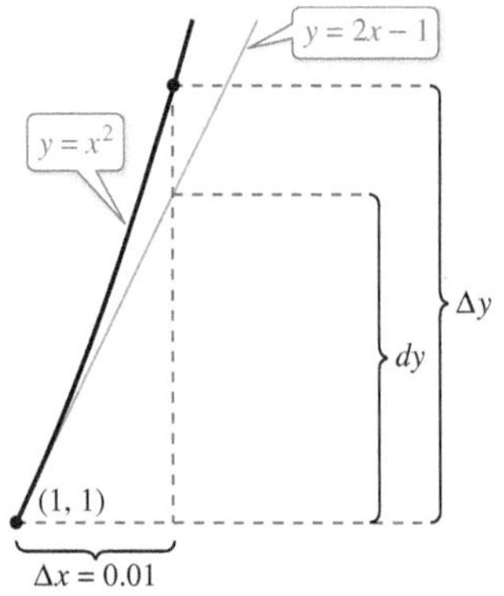

EXAMPLE 2 Comparing Δy and dy

Let $y = x^2$. Find dy when $x = 1$ and $dx = 0.01$. Compare this value with Δy for $x = 1$ and $\Delta x = 0.01$.

Solution Because $y = f(x) = x^2$, you have $f'(x) = 2x$, and the differential dy is

$$dy = f'(x)\,dx = f'(1)(0.01) = 2(0.01) = 0.02. \qquad \text{Differential of } y$$

Now, using $\Delta x = 0.01$, the change in y is

$$\Delta y = f(x + \Delta x) - f(x) = f(1.01) - f(1) = (1.01)^2 - 1^2 = 0.0201.$$

Figure 3.67 shows the geometric comparison of dy and Δy. Try comparing other values of dy and Δy. You will see that the values become closer to each other as dx (or Δx) approaches 0. ■

The change in y, Δy, is approximated by the differential of y, dy.
Figure 3.67

In Example 2, the tangent line to the graph of $f(x) = x^2$ at $x = 1$ is

$$y = 2x - 1. \qquad \text{Tangent line to the graph of } f \text{ at } x = 1.$$

For x-values near 1, this line is close to the graph of f, as shown in Figure 3.67 and in the table.

x	0.5	0.9	0.99	1	1.01	1.1	1.5
$f(x) = x^2$	0.25	0.81	0.9801	1	1.0201	1.21	2.25
$y = 2x - 1$	0	0.8	0.98	1	1.02	1.2	2

Error Propagation

Physicists and engineers tend to make liberal use of the approximation of Δy by dy. One way this occurs in practice is in the estimation of errors propagated by physical measuring devices. For example, if you let x represent the measured value of a variable and let $x + \Delta x$ represent the exact value, then Δx is the *error in measurement.* Finally, if the measured value x is used to compute another value $f(x)$, then the difference between $f(x + \Delta x)$ and $f(x)$ is the **propagated error.**

$$\underbrace{f(x + \overbrace{\Delta x}^{\text{Measurement error}})}_{\text{Exact value}} - \underbrace{f(x)}_{\text{Measured value}} = \overbrace{\Delta y}^{\text{Propagated error}}$$

EXAMPLE 3 Estimation of Error

Ball bearings are used to reduce friction between moving machine parts.

The measured radius of a ball bearing is 0.7 inch, as shown in the figure. The measurement is correct to within 0.01 inch. Estimate the propagated error in the volume V of the ball bearing.

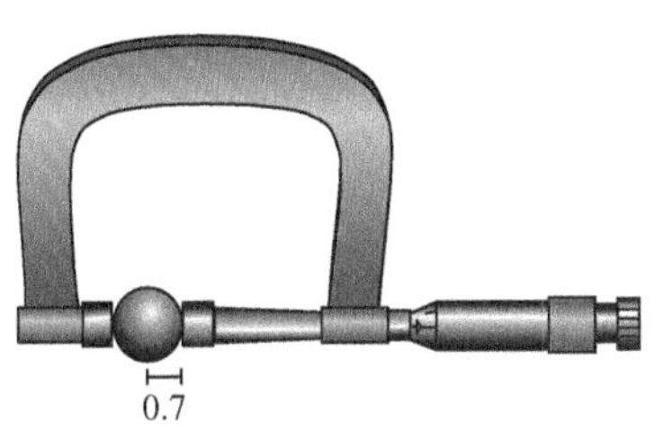

Ball bearing with measured radius that is correct to within 0.01 inch.

Solution The formula for the volume of a sphere is

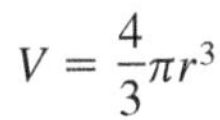

$$V = \frac{4}{3}\pi r^3$$

where r is the radius of the sphere. So, you can write

$$r = 0.7 \qquad \text{Measured radius}$$

and

$$-0.01 \le \Delta r \le 0.01. \qquad \text{Possible error}$$

To approximate the propagated error in the volume, differentiate V to obtain $dV/dr = 4\pi r^2$ and write

$$\begin{aligned} \Delta V &\approx dV && \text{Approximate } \Delta V \text{ by } dV. \\ &= 4\pi r^2\, dr \\ &= 4\pi(0.7)^2(\pm 0.01) && \text{Substitute for } r \text{ and } dr. \\ &\approx \pm 0.06158 \text{ cubic inch.} \end{aligned}$$

So, the volume has a propagated error of about 0.06 cubic inch. ■

Would you say that the propagated error in Example 3 is large or small? The answer is best given in *relative* terms by comparing dV with V. The ratio

$$\begin{aligned} \frac{dV}{V} &= \frac{4\pi r^2\, dr}{\frac{4}{3}\pi r^3} && \text{Ratio of } dV \text{ to } V \\ &= \frac{3\, dr}{r} && \text{Simplify.} \\ &\approx \frac{3(\pm 0.01)}{0.7} && \text{Substitute for } dr \text{ and } r. \\ &\approx \pm 0.0429 \end{aligned}$$

is called the **relative error.** The corresponding **percent error** is approximately 4.29%.

Calculating Differentials

Each of the differentiation rules that you studied in Chapter 2 can be written in **differential form.** For example, let u and v be differentiable functions of x. By the definition of differentials, you have

$$du = u'\,dx$$

and

$$dv = v'\,dx.$$

So, you can write the differential form of the Product Rule as shown below.

$$\begin{aligned} d[uv] &= \frac{d}{dx}[uv]\,dx && \text{Differential of } uv \\ &= [uv' + vu']\,dx && \text{Product Rule} \\ &= uv'\,dx + vu'\,dx \\ &= u\,dv + v\,du \end{aligned}$$

Differential Formulas

Let u and v be differentiable functions of x.

Constant multiple: $d[cu] = c\,du$

Sum or difference: $d[u \pm v] = du \pm dv$

Product: $d[uv] = u\,dv + v\,du$

Quotient: $d\left[\dfrac{u}{v}\right] = \dfrac{v\,du - u\,dv}{v^2}$

EXAMPLE 4 **Finding Differentials**

Function	Derivative	Differential
a. $y = x^2$	$\dfrac{dy}{dx} = 2x$	$dy = 2x\,dx$
b. $y = \sqrt{x}$	$\dfrac{dy}{dx} = \dfrac{1}{2\sqrt{x}}$	$dy = \dfrac{dx}{2\sqrt{x}}$
c. $y = 2\sin x$	$\dfrac{dy}{dx} = 2\cos x$	$dy = 2\cos x\,dx$
d. $y = x\cos x$	$\dfrac{dy}{dx} = -x\sin x + \cos x$	$dy = (-x\sin x + \cos x)\,dx$
e. $y = \dfrac{1}{x}$	$\dfrac{dy}{dx} = -\dfrac{1}{x^2}$	$dy = -\dfrac{dx}{x^2}$

GOTTFRIED WILHELM LEIBNIZ (1646–1716)

Both Leibniz and Newton are credited with creating calculus. It was Leibniz, however, who tried to broaden calculus by developing rules and formal notation. He often spent days choosing an appropriate notation for a new concept.
See LarsonCalculus.com to read more of this biography.

The notation in Example 4 is called the **Leibniz notation** for derivatives and differentials, named after the German mathematician Gottfried Wilhelm Leibniz. The beauty of this notation is that it provides an easy way to remember several important calculus formulas by making it seem as though the formulas were derived from algebraic manipulations of differentials. For instance, in Leibniz notation, the *Chain Rule*

$$\frac{dy}{dx} = \frac{dy}{du}\frac{du}{dx}$$

would appear to be true because the du's divide out. Even though this reasoning is *incorrect*, the notation does help one remember the Chain Rule.

EXAMPLE 5 Finding the Differential of a Composite Function

$$y = f(x) = \sin 3x \qquad \text{Original function}$$
$$f'(x) = 3 \cos 3x \qquad \text{Apply Chain Rule.}$$
$$dy = f'(x)\,dx = 3 \cos 3x\,dx \qquad \text{Differential form}$$

EXAMPLE 6 Finding the Differential of a Composite Function

$$y = f(x) = (x^2 + 1)^{1/2} \qquad \text{Original function}$$
$$f'(x) = \frac{1}{2}(x^2 + 1)^{-1/2}(2x) = \frac{x}{\sqrt{x^2 + 1}} \qquad \text{Apply Chain Rule.}$$
$$dy = f'(x)\,dx = \frac{x}{\sqrt{x^2 + 1}}\,dx \qquad \text{Differential form}$$

Differentials can be used to approximate function values. To do this for the function given by $y = f(x)$, use the formula

$$f(x + \Delta x) \approx f(x) + dy = f(x) + f'(x)\,dx$$

REMARK This formula is equivalent to the tangent line approximation given earlier in this section.

which is derived from the approximation

$$\Delta y = f(x + \Delta x) - f(x) \approx dy.$$

The key to using this formula is to choose a value for x that makes the calculations easier, as shown in Example 7.

EXAMPLE 7 Approximating Function Values

Use differentials to approximate $\sqrt{16.5}$.

Solution Using $f(x) = \sqrt{x}$, you can write

$$f(x + \Delta x) \approx f(x) + f'(x)\,dx = \sqrt{x} + \frac{1}{2\sqrt{x}}\,dx.$$

Now, choosing $x = 16$ and $dx = 0.5$, you obtain the following approximation.

$$f(x + \Delta x) = \sqrt{16.5} \approx \sqrt{16} + \frac{1}{2\sqrt{16}}(0.5) = 4 + \left(\frac{1}{8}\right)\left(\frac{1}{2}\right) = 4.0625$$

So, $\sqrt{16.5} \approx 4.0625$.

The tangent line approximation to $f(x) = \sqrt{x}$ at $x = 16$ is the line $g(x) = \frac{1}{8}x + 2$. For x-values near 16, the graphs of f and g are close together, as shown in Figure 3.68. For instance,

$$f(16.5) = \sqrt{16.5} \approx 4.0620$$

and

$$g(16.5) = \frac{1}{8}(16.5) + 2 = 4.0625.$$

In fact, if you use a graphing utility to zoom in near the point of tangency (16, 4), you will see that the two graphs appear to coincide. Notice also that as you move farther away from the point of tangency, the linear approximation becomes less accurate.

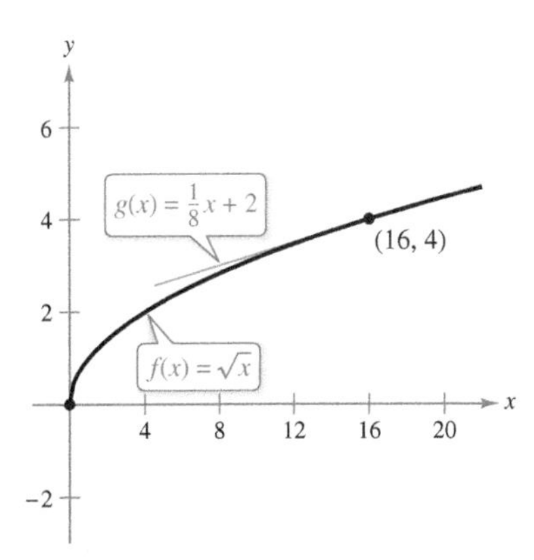

Figure 3.68

3.9 Exercises

See CalcChat.com for tutorial help and worked-out solutions to odd-numbered exercises.

CONCEPT CHECK

1. **Tangent Line Approximations** What is the equation of the tangent line approximation to the graph of a function f at the point $(c, f(c))$?
2. **Differentials** What do the differentials of x and y mean?
3. **Describing Terms** When using differentials, what is meant by the terms *propagated error, relative error,* and *percent error*?
4. **Finding Differentials** Explain how to find a differential of a function.

Using a Tangent Line Approximation **In Exercises 5–10, find the tangent line approximation T to the graph of f at the given point. Then complete the table.**

x	1.9	1.99	2	2.01	2.1
$f(x)$					
$T(x)$					

5. $f(x) = x^2$, $(2, 4)$
6. $f(x) = \dfrac{6}{x^2}$, $\left(2, \dfrac{3}{2}\right)$
7. $f(x) = x^5$, $(2, 32)$
8. $f(x) = \sqrt{x}$, $(2, \sqrt{2})$
9. $f(x) = \sin x$, $(2, \sin 2)$
10. $f(x) = \csc x$, $(2, \csc 2)$

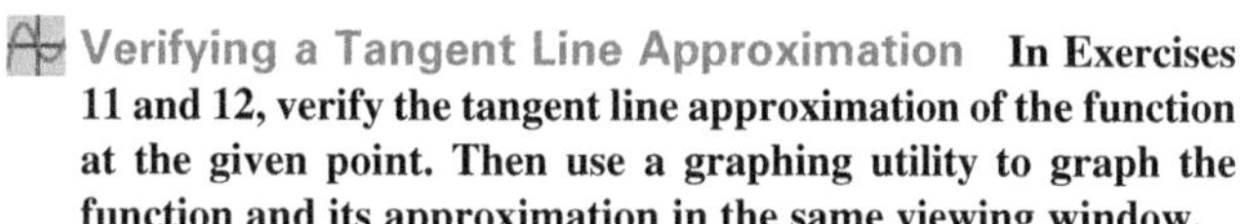

Verifying a Tangent Line Approximation **In Exercises 11 and 12, verify the tangent line approximation of the function at the given point. Then use a graphing utility to graph the function and its approximation in the same viewing window.**

	Function	Approximation	Point
11.	$f(x) = \sqrt{x + 4}$	$y = 2 + \dfrac{x}{4}$	$(0, 2)$
12.	$f(x) = \tan x$	$y = x$	$(0, 0)$

Comparing Δy and dy **In Exercises 13–18, use the information to find and compare Δy and dy.**

	Function	x-Value	Differential of x
13.	$y = 0.5x^3$	$x = 1$	$\Delta x = dx = 0.1$
14.	$y = 6 - 2x^2$	$x = -2$	$\Delta x = dx = 0.1$
15.	$y = x^4 + 1$	$x = -1$	$\Delta x = dx = 0.01$
16.	$y = 2 - x^4$	$x = 2$	$\Delta x = dx = 0.01$
17.	$y = x - 2x^3$	$x = 3$	$\Delta x = dx = 0.001$
18.	$y = 7x^2 - 5x$	$x = -4$	$\Delta x = dx = 0.001$

Finding a Differential **In Exercises 19–28, find the differential dy of the given function.**

19. $y = 3x^2 - 4$
20. $y = 3x^{2/3}$
21. $y = x \tan x$
22. $y = \csc 2x$
23. $y = \dfrac{x + 1}{2x - 1}$
24. $y = \sqrt{x} + \dfrac{1}{\sqrt{x}}$
25. $y = \sqrt{9 - x^2}$
26. $y = x\sqrt{1 - x^2}$
27. $y = 3x - \sin^2 x$
28. $y = \dfrac{\sec^2 x}{x^2 + 1}$

Using Differentials **In Exercises 29 and 30, use differentials and the graph of f to approximate (a) $f(1.9)$ and (b) $f(2.04)$. To print an enlarged copy of the graph, go to *MathGraphs.com*.**

29.

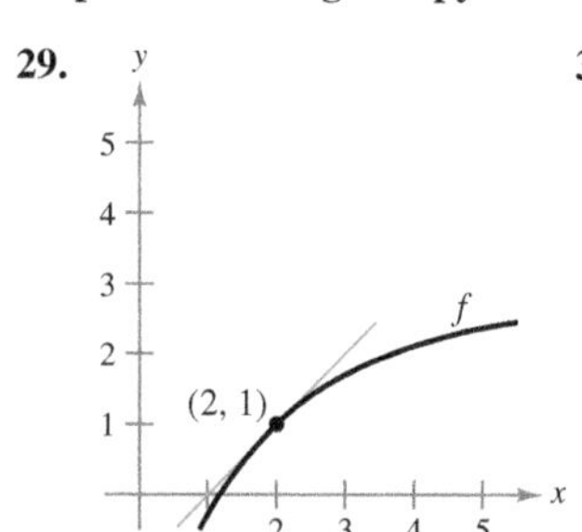

30.

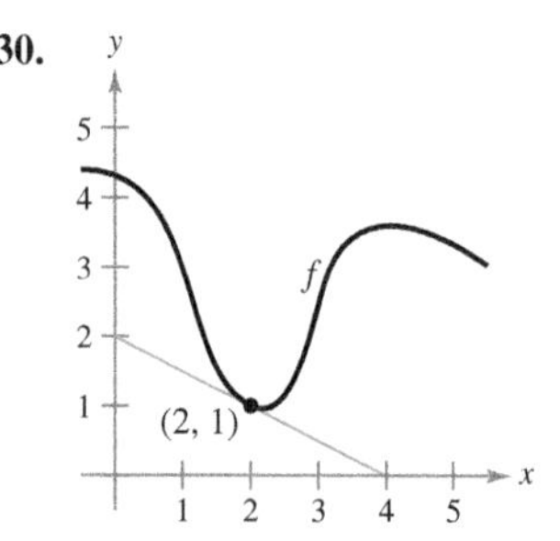

Using Differentials **In Exercises 31 and 32, use differentials and the graph of g' to approximate (a) $g(2.93)$ and (b) $g(3.1)$ given that $g(3) = 8$.**

31.

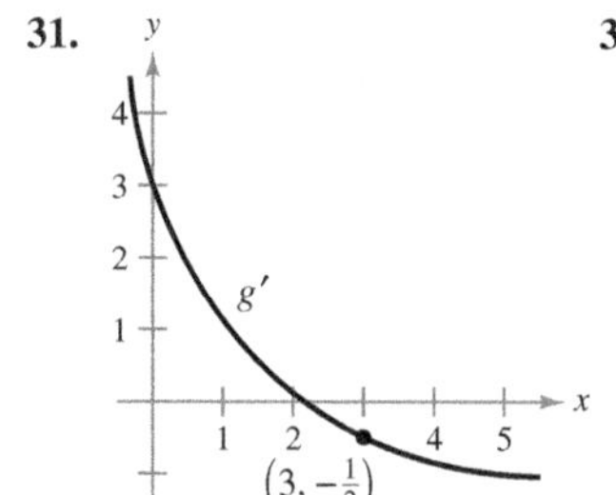

32.

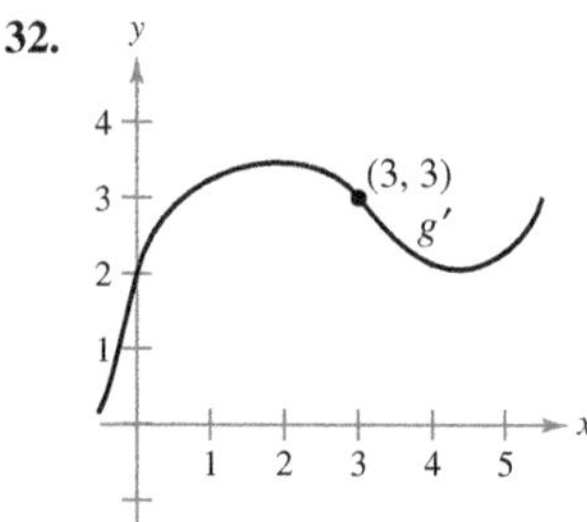

33. **Area** The measurement of the side of a square floor tile is 10 inches, with a possible error of $\frac{1}{32}$ inch.
    (a) Use differentials to approximate the possible propagated error in computing the area of the square.
    (b) Approximate the percent error in computing the area of the square.
34. **Area** The measurements of the base and altitude of a triangle are found to be 36 and 50 centimeters, respectively. The possible error in each measurement is 0.25 centimeter.
    (a) Use differentials to approximate the possible propagated error in computing the area of the triangle.
    (b) Approximate the percent error in computing the area of the triangle.

35. Volume and Surface Area The measurement of the edge of a cube is found to be 15 inches, with a possible error of 0.03 inch.

(a) Use differentials to approximate the possible propagated error in computing the volume of the cube.

(b) Use differentials to approximate the possible propagated error in computing the surface area of the cube.

(c) Approximate the percent errors in parts (a) and (b).

36. Volume and Surface Area The radius of a spherical balloon is measured as 8 inches, with a possible error of 0.02 inch.

(a) Use differentials to approximate the possible propagated error in computing the volume of the sphere.

(b) Use differentials to approximate the possible propagated error in computing the surface area of the sphere.

(c) Approximate the percent errors in parts (a) and (b).

37. Stopping Distance The total stopping distance T of a vehicle is

$$T = 2.5x + 0.5x^2$$

where T is in feet and x is the speed in miles per hour. Approximate the change and percent change in total stopping distance as speed changes from $x = 25$ to $x = 26$ miles per hour.

38. HOW DO YOU SEE IT? The graph shows the profit P (in dollars) from selling x units of an item. Use the graph to determine which is greater, the change in profit when the production level changes from 400 to 401 units or the change in profit when the production level changes from 900 to 901 units. Explain your reasoning.

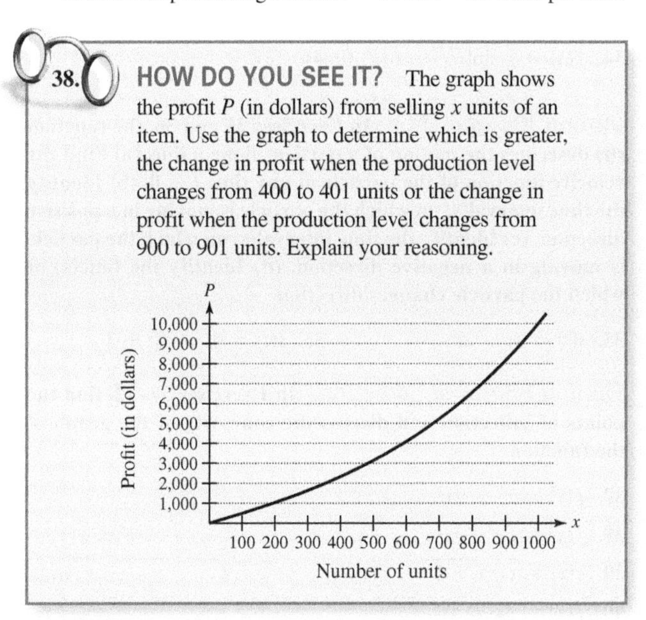

39. Pendulum The period of a pendulum is given by

$$T = 2\pi\sqrt{\frac{L}{g}}$$

where L is the length of the pendulum in feet, g is the acceleration due to gravity, and T is the time in seconds. The pendulum has been subjected to an increase in temperature such that the length has increased by $\frac{1}{2}\%$.

(a) Find the approximate percent change in the period.

(b) Using the result in part (a), find the approximate error in this pendulum clock in 1 day.

40. Ohm's Law A current of I amperes passes through a resistor of R ohms. **Ohm's Law** states that the voltage E applied to the resistor is

$$E = IR.$$

The voltage is constant. Show that the magnitude of the relative error in R caused by a change in I is equal in magnitude to the relative error in I.

41. Projectile Motion The range R of a projectile is

$$R = \frac{v_0^2}{32}(\sin 2\theta)$$

where v_0 is the initial velocity in feet per second and θ is the angle of elevation. Use differentials to approximate the change in the range when $v_0 = 2500$ feet per second and θ is changed from 10° to 11°.

42. Surveying A surveyor standing 50 feet from the base of a large tree measures the angle of elevation to the top of the tree as 71.5°. How accurately must the angle be measured if the percent error in estimating the height of the tree is to be less than 6%?

Approximating Function Values In Exercises 43–46, use differentials to approximate the value of the expression. Compare your answer with that of a calculator.

43. $\sqrt{99.4}$

44. $\sqrt[3]{26}$

45. $\sqrt[4]{624}$

46. $(2.99)^3$

EXPLORING CONCEPTS

47. Comparing Δy and dy Describe the change in accuracy of dy as an approximation for Δy when Δx approaches 0. Use a graph to support your answer.

48. Using Differentials Give a short explanation of why each approximation is valid.

(a) $\sqrt{4.02} \approx 2 + \frac{1}{4}(0.02)$

(b) $\tan 0.05 \approx 0 + 1(0.05)$

True or False? In Exercises 49–53, determine whether the statement is true or false. If it is false, explain why or give an example that shows it is false.

49. If $y = x + c$, then $dy = dx$.

50. If $y = ax + b$, then $\frac{\Delta y}{\Delta x} = \frac{dy}{dx}$.

51. If y is differentiable, then $\lim_{\Delta x \to 0} (\Delta y - dy) = 0$.

52. If $y = f(x)$, f is increasing and differentiable, and $\Delta x > 0$, then $\Delta y \geq dy$.

53. The tangent line approximation at any point for any linear equation is the linear equation itself.

Review Exercises

See CalcChat.com for tutorial help and worked-out solutions to odd-numbered exercises.

Finding Extrema on a Closed Interval **In Exercises 1–8, find the absolute extrema of the function on the closed interval.**

1. $f(x) = x^2 + 5x, \quad [-4, 0]$
2. $f(x) = x^3 + 6x^2, \quad [-6, 1]$
3. $f(x) = \sqrt{x} - 2, \quad [0, 4]$
4. $h(x) = x - 3\sqrt{x}, \quad [0, 9]$
5. $f(x) = \dfrac{4x}{x^2 + 9}, \quad [-4, 4]$
6. $f(x) = \dfrac{x}{\sqrt{x^2 + 1}}, \quad [0, 2]$
7. $g(x) = 2x + 5 \cos x, \quad [0, 2\pi]$
8. $f(x) = \sin 2x, \quad [0, 2\pi]$

Using Rolle's Theorem **In Exercises 9–12, determine whether Rolle's Theorem can be applied to f on the closed interval $[a, b]$. If Rolle's Theorem can be applied, find all values of c in the open interval (a, b) such that $f'(c) = 0$. If Rolle's Theorem cannot be applied, explain why not.**

9. $f(x) = x^3 - 3x - 6, \quad [-1, 2]$
10. $f(x) = (x - 2)(x + 3)^2, \quad [-3, 2]$
11. $f(x) = \dfrac{x^2}{1 - x^2}, \quad [-2, 2]$
12. $f(x) = \sin 2x, \quad [-\pi, \pi]$

Using the Mean Value Theorem **In Exercises 13–18, determine whether the Mean Value Theorem can be applied to f on the closed interval $[a, b]$. If the Mean Value Theorem can be applied, find all values of c in the open interval (a, b) such that**

$$f'(c) = \frac{f(b) - f(a)}{b - a}.$$

If the Mean Value Theorem cannot be applied, explain why not.

13. $f(x) = x^{2/3}, \quad [1, 8]$
14. $f(x) = \dfrac{1}{x}, \quad [1, 4]$
15. $f(x) = |5 - x|, \quad [2, 6]$
16. $f(x) = 2x - 3\sqrt{x}, \quad [-1, 1]$
17. $f(x) = x - \cos x, \quad \left[-\dfrac{\pi}{2}, \dfrac{\pi}{2}\right]$
18. $f(x) = \sqrt{x} - 2x, \quad [0, 4]$

19. **Mean Value Theorem** Can the Mean Value Theorem be applied to the function

$$f(x) = \frac{1}{x^2}$$

on the interval $[-2, 1]$? Explain.

20. **Using the Mean Value Theorem**
    (a) For the function $f(x) = Ax^2 + Bx + C$, determine the value of c guaranteed by the Mean Value Theorem on the interval $[x_1, x_2]$.
    (b) Demonstrate the result of part (a) for $f(x) = 2x^2 - 3x + 1$ on the interval $[0, 4]$.

Intervals on Which a Function Is Increasing or Decreasing **In Exercises 21–26, find the open intervals on which the function is increasing or decreasing.**

21. $f(x) = x^2 + 3x - 12$
22. $h(x) = (x + 2)^{1/3} + 8$
23. $f(x) = (x - 1)^2(2x - 5)$
24. $g(x) = (x + 1)^3$
25. $h(x) = \sqrt{x}(x - 3), \quad x > 0$
26. $f(x) = \sin x + \cos x, \quad 0 < x < 2\pi$

Applying the First Derivative Test **In Exercises 27–34, (a) find the critical numbers of f, if any, (b) find the open intervals on which the function is increasing or decreasing, (c) apply the First Derivative Test to identify all relative extrema, and (d) use a graphing utility to confirm your results.**

27. $f(x) = x^2 - 6x + 5$
28. $f(x) = 4x^3 - 5x$
29. $f(t) = \dfrac{1}{4}t^4 - 8t$
30. $f(x) = \dfrac{x^3 - 8x}{4}$
31. $f(x) = \dfrac{x + 4}{x^2}$
32. $f(x) = \dfrac{x^2 - 3x - 4}{x - 2}$
33. $f(x) = \cos x - \sin x, \quad (0, 2\pi)$
34. $f(x) = \dfrac{3}{2}\sin\left(\dfrac{\pi x}{2} - 1\right), \quad (0, 4)$

Motion Along a Line **In Exercises 35 and 36, the function $s(t)$ describes the motion of a particle along a line. (a) Find the velocity function of the particle at any time $t \geq 0$. (b) Identify the time interval(s) on which the particle is moving in a positive direction. (c) Identify the time interval(s) on which the particle is moving in a negative direction. (d) Identify the time(s) at which the particle changes direction.**

35. $s(t) = 3t - 2t^2$
36. $s(t) = 6t^3 - 8t + 3$

Finding Points of Inflection **In Exercises 37–42, find the points of inflection and discuss the concavity of the graph of the function.**

37. $f(x) = x^3 - 9x^2$
38. $f(x) = 6x^4 - x^2$
39. $g(x) = x\sqrt{x + 5}$
40. $f(x) = 3x - 5x^3$
41. $f(x) = x + \cos x, \quad [0, 2\pi]$
42. $f(x) = \tan \dfrac{x}{4}, \quad (0, 2\pi)$

Using the Second Derivative Test **In Exercises 43–48, find all relative extrema of the function. Use the Second Derivative Test where applicable.**

43. $f(x) = (x + 9)^2$
44. $f(x) = x^4 - 2x^2 + 6$
45. $g(x) = 2x^2(1 - x^2)$
46. $h(t) = t - 4\sqrt{t + 1}$

47. $f(x) = 2x + \dfrac{18}{x}$

48. $h(x) = x - 2\cos x, \quad [0, 4\pi]$

Think About It **In Exercises 49 and 50, sketch the graph of a function f having the given characteristics.**

49. $f(0) = f(6) = 0$
$f'(3) = f'(5) = 0$
$f'(x) > 0$ for $x < 3$
$f'(x) > 0$ for $3 < x < 5$
$f'(x) < 0$ for $x > 5$
$f''(x) < 0$ for $x < 3$ or $x > 4$
$f''(x) > 0$ for $3 < x < 4$

50. $f(0) = 4, f(6) = 0$
$f'(x) < 0$ for $x < 2$ or $x > 4$
$f'(2)$ does not exist.
$f'(4) = 0$
$f'(x) > 0$ for $2 < x < 4$
$f''(x) < 0$ for $x \neq 2$

51. Writing A newspaper headline states that "The rate of growth of the national deficit is decreasing." What does this mean? What does it imply about the graph of the deficit as a function of time?

52. Inventory Cost The cost of inventory C depends on the ordering and storage costs according to the inventory model

$$C = \left(\frac{Q}{x}\right)s + \left(\frac{x}{2}\right)r.$$

Determine the order size that will minimize the cost, assuming that sales occur at a constant rate, Q is the number of units sold per year, r is the cost of storing one unit for one year, s is the cost of placing an order, and x is the number of units per order.

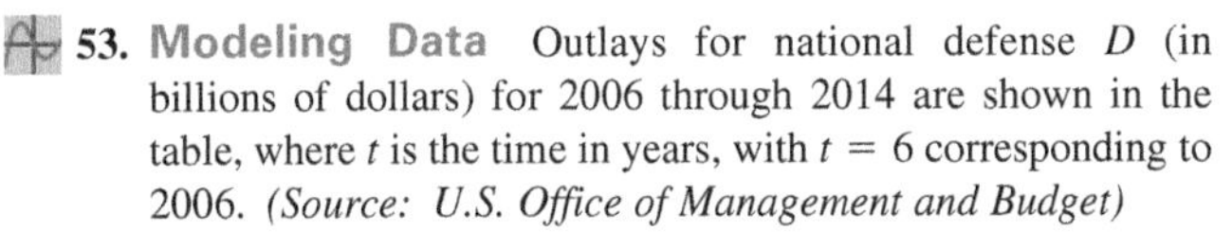

53. Modeling Data Outlays for national defense D (in billions of dollars) for 2006 through 2014 are shown in the table, where t is the time in years, with $t = 6$ corresponding to 2006. *(Source: U.S. Office of Management and Budget)*

t	6	7	8	9	10
D	521.8	551.3	616.1	661.0	693.5

t	11	12	13	14
D	705.6	677.9	633.4	603.5

(a) Use the regression capabilities of a graphing utility to find a model of the form

$$D = at^4 + bt^3 + ct^2 + dt + e$$

for the data.

(b) Use a graphing utility to plot the data and graph the model.

(c) For the years shown in the table, when does the model indicate that the outlay for national defense was at a maximum? When was it at a minimum?

(d) For the years shown in the table, when does the model indicate that the outlay for national defense was increasing at the greatest rate?

54. Modeling Data The manager of a store recorded the annual sales S (in thousands of dollars) of a product over a period of 7 years, as shown in the table, where t is the time in years, with $t = 8$ corresponding to 2008.

t	8	9	10	11	12	13	14
S	8.1	7.3	7.8	9.2	11.3	12.8	12.9

(a) Use the regression capabilities of a graphing utility to find a model of the form

$$S = at^3 + bt^2 + ct + d$$

for the data.

(b) Use a graphing utility to plot the data and graph the model.

(c) Use calculus and the model to find the time t when sales were increasing at the greatest rate.

(d) Do you think the model would be accurate for predicting future sales? Explain.

Finding a Limit **In Exercises 55–64, find the limit, if it exists.**

55. $\displaystyle\lim_{x\to\infty}\left(8 + \frac{1}{x}\right)$

56. $\displaystyle\lim_{x\to-\infty}\frac{1-4x}{x+1}$

57. $\displaystyle\lim_{x\to\infty}\frac{x^2}{1-8x^2}$

58. $\displaystyle\lim_{x\to-\infty}\frac{9x^3+5}{7x^4}$

59. $\displaystyle\lim_{x\to-\infty}\frac{3x^2}{x+5}$

60. $\displaystyle\lim_{x\to-\infty}\frac{\sqrt{x^2+x}}{-2x}$

61. $\displaystyle\lim_{x\to\infty}\frac{5\cos x}{x}$

62. $\displaystyle\lim_{x\to\infty}\frac{x^3}{\sqrt{x^2+2}}$

63. $\displaystyle\lim_{x\to-\infty}\frac{6x}{x+\cos x}$

64. $\displaystyle\lim_{x\to-\infty}\frac{x}{2\sin x}$

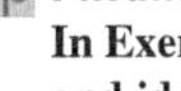

Finding Horizontal Asymptotes Using Technology **In Exercises 65–68, use a graphing utility to graph the function and identify any horizontal asymptotes.**

65. $f(x) = \dfrac{3}{x} + 4$

66. $g(x) = \dfrac{5x^2}{x^2+2}$

67. $f(x) = \dfrac{x}{\sqrt{x^2+6}}$

68. $f(x) = \dfrac{\sqrt{4x^2-1}}{8x+1}$

Analyzing the Graph of a Function **In Exercises 69–78, analyze and sketch a graph of the function. Label any intercepts, relative extrema, points of inflection, and asymptotes. Use a graphing utility to verify your results.**

69. $f(x) = 4x - x^2$

70. $f(x) = x^4 - 2x^2 + 6$

71. $f(x) = x\sqrt{16 - x^2}$

72. $f(x) = (x^2 - 4)^2$

73. $f(x) = x^{1/3}(x+3)^{2/3}$

74. $f(x) = (x-3)(x+2)^3$

75. $f(x) = \dfrac{5-3x}{x-2}$

76. $f(x) = \dfrac{2x}{1+x^2}$

77. $f(x) = x^3 + x + \dfrac{4}{x}$

78. $f(x) = x^2 + \dfrac{1}{x}$

79. Finding Numbers Find two positive numbers such that the sum of twice the first number and three times the second number is 216 and the product is a maximum.

80. Minimum Distance Find the point on the graph of $f(x) = \sqrt{x}$ that is closest to the point $(6, 0)$.

81. Maximum Area A rancher has 400 feet of fencing with which to enclose two adjacent rectangular corrals (see figure). What dimensions should be used so that the enclosed area will be a maximum?

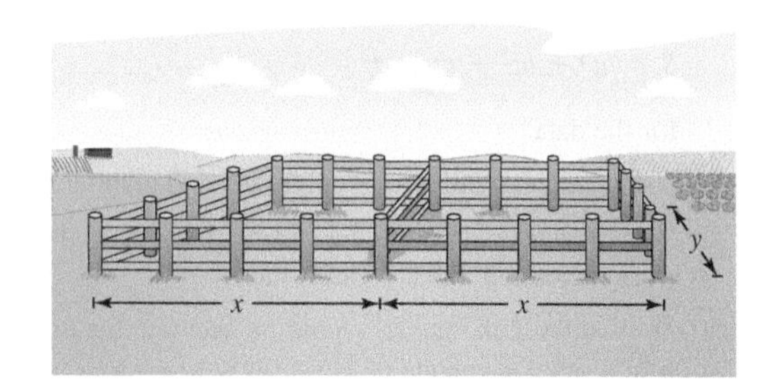

82. Maximum Area Find the dimensions of the rectangle of maximum area, with sides parallel to the coordinate axes, that can be inscribed in the ellipse given by

$$\frac{x^2}{144} + \frac{y^2}{16} = 1.$$

83. Minimum Length A right triangle in the first quadrant has the coordinate axes as sides, and the hypotenuse passes through the point $(1, 8)$. Find the vertices of the triangle such that the length of the hypotenuse is minimum.

84. Minimum Length The wall of a building is to be braced by a beam that must pass over a parallel fence 5 feet high and 4 feet from the building. Find the length of the shortest beam that can be used.

85. Maximum Length Find the length of the longest pipe that can be carried level around a right-angle corner at the intersection of two corridors of widths 4 feet and 6 feet.

86. Maximum Length A hallway of width 6 feet meets a hallway of width 9 feet at right angles. Find the length of the longest pipe that can be carried level around this corner. [*Hint:* If L is the length of the pipe, show that

$$L = 6 \csc \theta + 9 \csc\left(\frac{\pi}{2} - \theta\right)$$

where θ is the angle between the pipe and the wall of the narrower hallway.]

87. Maximum Volume Find the volume of the largest right circular cone that can be inscribed in a sphere of radius r.

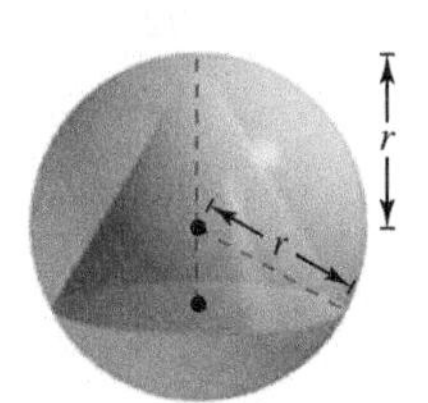

88. Maximum Volume Find the volume of the largest right circular cylinder that can be inscribed in a sphere of radius r.

Using Newton's Method In Exercises 89–92, use Newton's Method to approximate the zero(s) of the function. Continue the iterations until two successive approximations differ by less than 0.001. Then find the zero(s) using a graphing utility and compare the results.

89. $f(x) = x^3 - 3x - 1$

90. $f(x) = x^3 + 2x + 1$

91. $f(x) = x^4 + x^3 - 3x^2 + 2$

92. $f(x) = 3\sqrt{x - 1} - x$

Points of Intersection In Exercises 93 and 94, apply Newton's Method to approximate the x-value(s) of the indicated point(s) of intersection of the two graphs. Continue the iterations until two successive approximations differ by less than 0.001. [*Hint:* Let $h(x) = f(x) - g(x)$.]

93. $f(x) = 1 - x$
$g(x) = x^5 + 2$

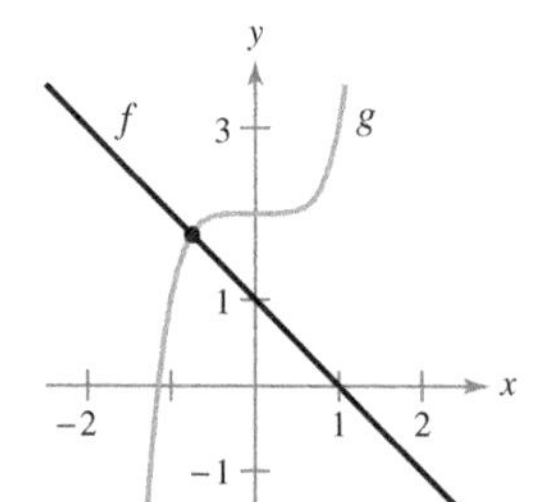

94. $f(x) = \sin x$
$g(x) = x^2 - 2x + 1$

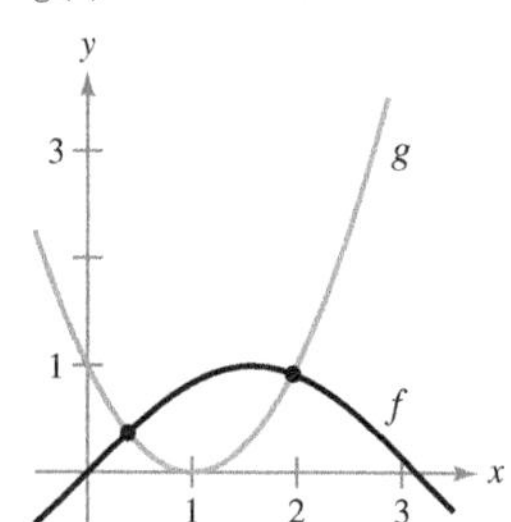

Comparing Δy and dy In Exercises 95 and 96, use the information to find and compare Δy and dy.

	Function	x-Value	Differential of x
95.	$y = 4x^3$	$x = 2$	$\Delta x = dx = 0.1$
96.	$y = x^2 - 5x$	$x = -3$	$\Delta x = dx = 0.01$

Finding a Differential In Exercises 97 and 98, find the differential dy of the given function.

97. $y = x(1 - \cos x)$

98. $y = \sqrt{36 - x^2}$

Approximating Function Values In Exercises 99 and 100, use differentials to approximate the value of the expression. Compare your answer with that of a calculator.

99. $\sqrt{63.9}$

100. $(2.02)^4$

101. Volume and Surface Area The radius of a sphere is measured as 9 centimeters, with a possible error of 0.025 centimeter.

(a) Use differentials to approximate the possible propagated error in computing the volume of the sphere.

(b) Use differentials to approximate the possible propagated error in computing the surface area of the sphere.

(c) Approximate the percent errors in parts (a) and (b).

P.S. Problem Solving

See CalcChat.com for tutorial help and worked-out solutions to odd-numbered exercises.

1. **Relative Extrema** Graph the fourth-degree polynomial

$$p(x) = x^4 + ax^2 + 1$$

for various values of the constant a.

(a) Determine the values of a for which p has exactly one relative minimum.

(b) Determine the values of a for which p has exactly one relative maximum.

(c) Determine the values of a for which p has exactly two relative minima.

(d) Show that the graph of p cannot have exactly two relative extrema.

2. **Relative Extrema**

(a) Graph the fourth-degree polynomial $p(x) = ax^4 - 6x^2$ for $a = -3, -2, -1, 0, 1, 2,$ and 3. For what values of the constant a does p have a relative minimum or relative maximum?

(b) Show that p has a relative maximum for all values of the constant a.

(c) Determine analytically the values of a for which p has a relative minimum.

(d) Let $(x, y) = (x, p(x))$ be a relative extremum of p. Show that (x, y) lies on the graph of $y = -3x^2$. Verify this result graphically by graphing $y = -3x^2$ together with the seven curves from part (a).

3. **Relative Minimum** Let

$$f(x) = \frac{c}{x} + x^2.$$

Determine all values of the constant c such that f has a relative minimum, but no relative maximum.

4. **Points of Inflection**

(a) Let $f(x) = ax^2 + bx + c$, $a \neq 0$, be a quadratic polynomial. How many points of inflection does the graph of f have?

(b) Let $f(x) = ax^3 + bx^2 + cx + d$, $a \neq 0$, be a cubic polynomial. How many points of inflection does the graph of f have?

(c) Suppose the function $y = f(x)$ satisfies the equation

$$\frac{dy}{dx} = ky\left(1 - \frac{y}{L}\right)$$

where k and L are positive constants. Show that the graph of f has a point of inflection at the point where $y = L/2$. (This equation is called the **logistic differential equation.**)

5. **Extended Mean Value Theorem** Prove the **Extended Mean Value Theorem:** If f and f' are continuous on the closed interval $[a, b]$, and if f'' exists in the open interval (a, b), then there exists a number c in (a, b) such that

$$f(b) = f(a) + f'(a)(b - a) + \frac{1}{2}f''(c)(b - a)^2.$$

6. **Illumination** The amount of illumination of a surface is proportional to the intensity of the light source, inversely proportional to the square of the distance from the light source, and proportional to $\sin \theta$, where θ is the angle at which the light strikes the surface. A rectangular room measures 10 feet by 24 feet, with a 10-foot ceiling (see figure). Determine the height at which the light should be placed to allow the corners of the floor to receive as much light as possible.

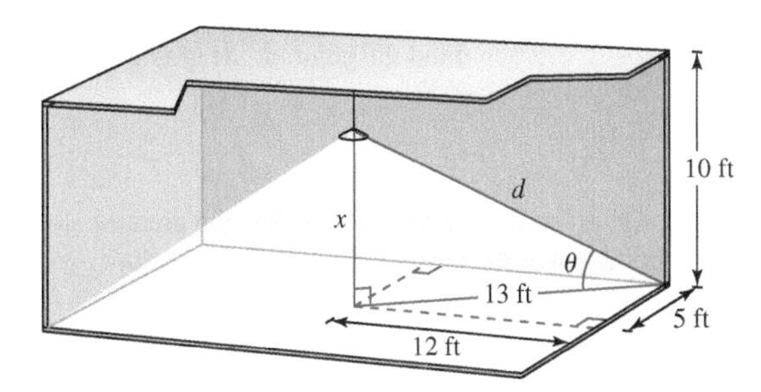

7. **Minimum Distance** Consider a room in the shape of a cube, 4 meters on each side. A bug at point P wants to walk to point Q at the opposite corner, as shown in the figure. Use calculus to determine the shortest path. Explain how you can solve this problem without calculus. (*Hint:* Consider the two walls as one wall.)

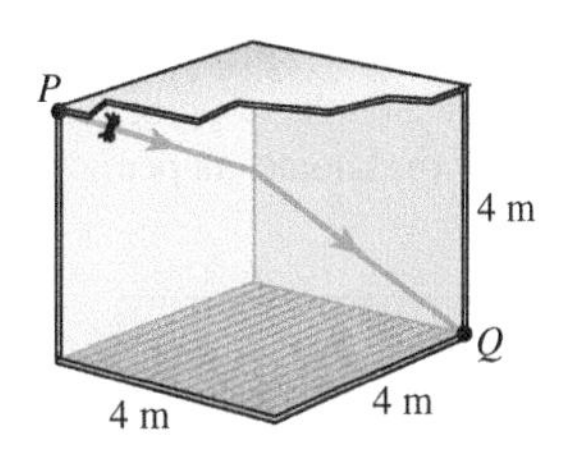

Figure for 7

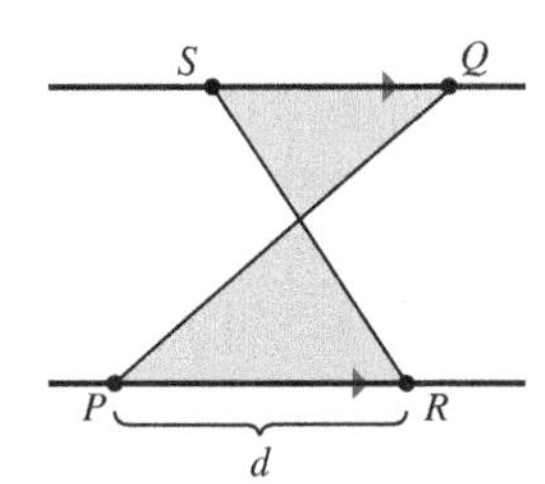

Figure for 8

8. **Areas of Triangles** The line joining P and Q crosses the two parallel lines, as shown in the figure. The point R is d units from P. How far from Q should the point S be positioned so that the sum of the areas of the two shaded triangles is a minimum? So that the sum is a maximum?

9. **Mean Value Theorem** Determine the values a, b, and c such that the function f satisfies the hypotheses of the Mean Value Theorem on the interval $[0, 3]$.

$$f(x) = \begin{cases} 1, & x = 0 \\ ax + b, & 0 < x \le 1 \\ x^2 + 4x + c, & 1 < x \le 3 \end{cases}$$

10. **Mean Value Theorem** Determine the values a, b, c, and d such that the function f satisfies the hypotheses of the Mean Value Theorem on the interval $[-1, 2]$.

$$f(x) = \begin{cases} a, & x = -1 \\ 2, & -1 < x \le 0 \\ bx^2 + c, & 0 < x \le 1 \\ dx + 4, & 1 < x \le 2 \end{cases}$$

11. Proof Let f and g be functions that are continuous on $[a, b]$ and differentiable on (a, b). Prove that if $f(a) = g(a)$ and $g'(x) > f'(x)$ for all x in (a, b), then $g(b) > f(b)$.

12. Proof

(a) Prove that $\lim_{x \to \infty} x^2 = \infty$.

(b) Prove that $\lim_{x \to \infty} \frac{1}{x^2} = 0$.

(c) Let L be a real number. Prove that if $\lim_{x \to \infty} f(x) = L$, then

$$\lim_{y \to 0^+} f\left(\frac{1}{y}\right) = L.$$

13. Tangent Lines Find the point on the graph of

$$y = \frac{1}{1 + x^2}$$

(see figure) where the tangent line has the greatest slope, and the point where the tangent line has the least slope.

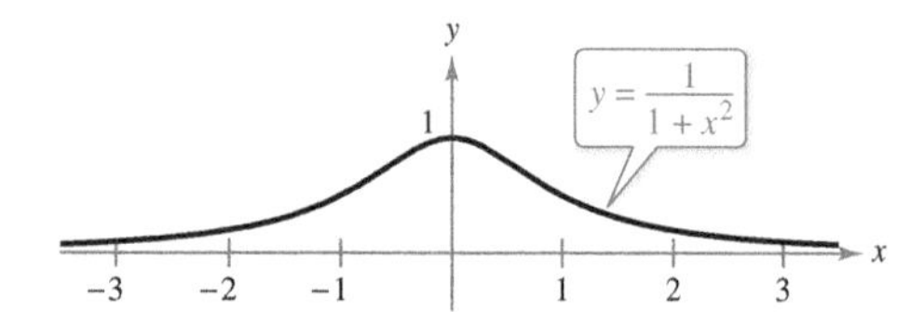

14. Stopping Distance The police department must determine the speed limit on a bridge such that the flow rate of cars is maximum per unit time. The greater the speed limit, the farther apart the cars must be in order to keep a safe stopping distance. Experimental data on the stopping distances d (in meters) for various speeds v (in kilometers per hour) are shown in the table.

v	20	40	60	80	100
d	5.1	13.7	27.2	44.2	66.4

(a) Convert the speeds v in the table to speeds s in meters per second. Use the regression capabilities of a graphing utility to find a model of the form $d(s) = as^2 + bs + c$ for the data.

(b) Consider two consecutive vehicles of average length 5.5 meters, traveling at a safe speed on the bridge. Let T be the difference between the times (in seconds) when the front bumpers of the vehicles pass a given point on the bridge. Verify that this difference in times is given by

$$T = \frac{d(s)}{s} + \frac{5.5}{s}.$$

(c) Use a graphing utility to graph the function T and estimate the speed s that minimizes the time between vehicles.

(d) Use calculus to determine the speed that minimizes T. What is the minimum value of T? Convert the required speed to kilometers per hour.

(e) Find the optimal distance between vehicles for the speed found in part (d).

15. Darboux's Theorem Prove **Darboux's Theorem:** Let f be differentiable on the closed interval $[a, b]$ such that $f'(a) = y_1$ and $f'(b) = y_2$. If d lies between y_1 and y_2, then there exists c in (a, b) such that $f'(c) = d$.

16. Maximum Area The figures show a rectangle, a circle, and a semicircle inscribed in a triangle bounded by the coordinate axes and the first-quadrant portion of the line with intercepts (3, 0) and (0, 4). Find the dimensions of each inscribed figure such that its area is maximum. State whether calculus was helpful in finding the required dimensions. Explain your reasoning.

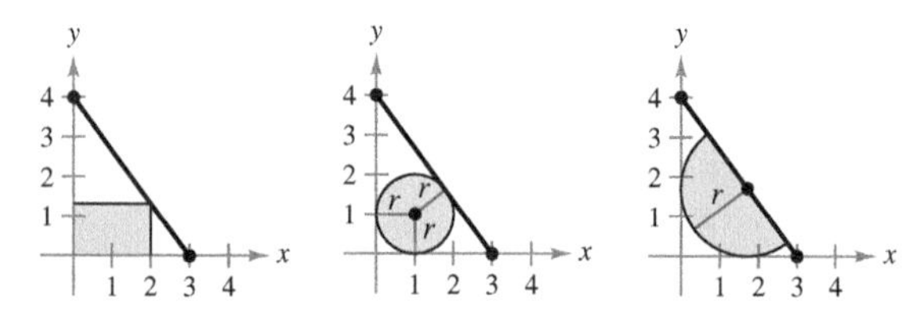

17. Point of Inflection Show that the cubic polynomial $p(x) = ax^3 + bx^2 + cx + d$ has exactly one point of inflection (x_0, y_0), where

$$x_0 = \frac{-b}{3a} \quad \text{and} \quad y_0 = \frac{2b^3}{27a^2} - \frac{bc}{3a} + d.$$

Use these formulas to find the point of inflection of $p(x) = x^3 - 3x^2 + 2$.

18. Minimum Length A legal-sized sheet of paper (8.5 inches by 14 inches) is folded so that corner P touches the opposite 14-inch edge at R (see figure). (*Note:* $PQ = \sqrt{C^2 - x^2}$.)

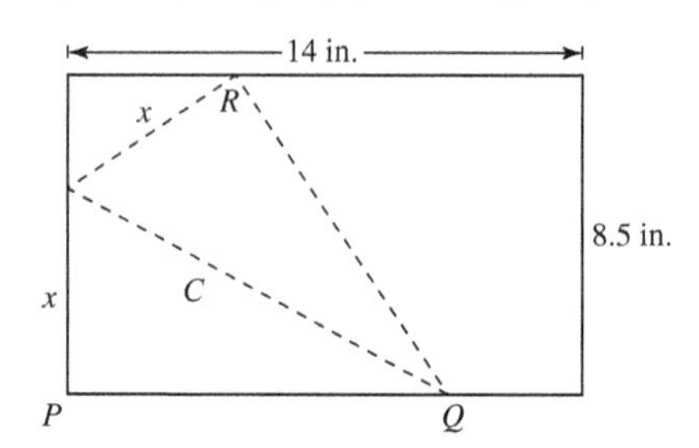

(a) Show that $C^2 = \frac{2x^3}{2x - 8.5}$.

(b) What is the domain of C?

(c) Determine the x-value that minimizes C.

(d) Determine the minimum length C.

19. Quadratic Approximation The polynomial

$$P(x) = c_0 + c_1(x - a) + c_2(x - a)^2$$

is the quadratic approximation of the function f at $(a, f(a))$ when $P(a) = f(a)$, $P'(a) = f'(a)$, and $P''(a) = f''(a)$.

(a) Find the quadratic approximation of

$$f(x) = \frac{x}{x + 1}$$

at (0, 0).

(b) Use a graphing utility to graph $P(x)$ and $f(x)$ in the same viewing window.

4 Integration

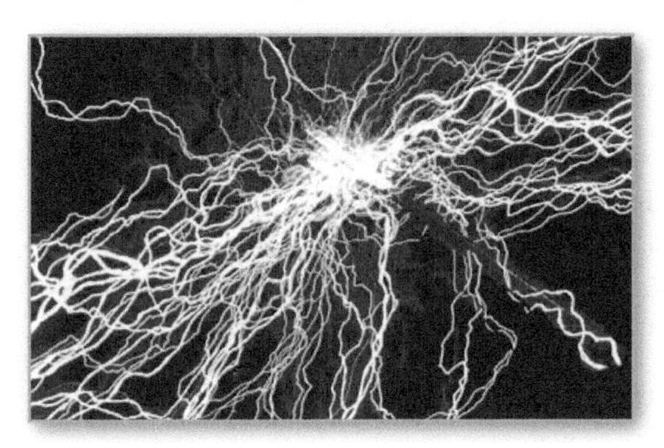

Electricity *(Exercise 86, p. 307)*

Amount of Chemical Flowing into a Tank *(Example 9, p. 290)*

The Speed of Sound *(Example 5, p. 286)*

Seating Capacity *(Exercise 77, p. 269)*

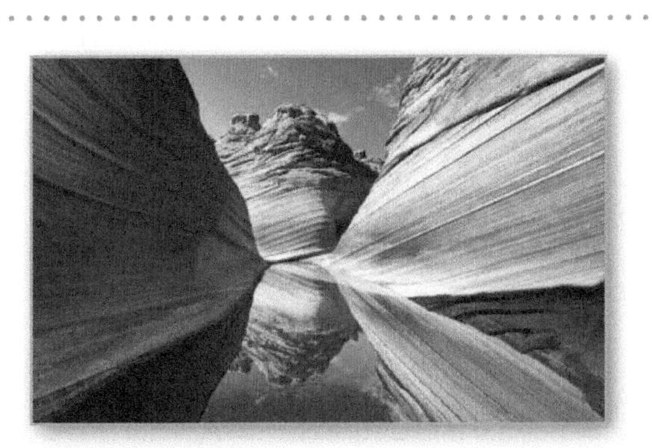

Grand Canyon *(Exercise 62, p. 257)*

4.1 Antiderivatives and Indefinite Integration

- **Write the general solution of a differential equation and use indefinite integral notation for antiderivatives.**
- **Use basic integration rules to find antiderivatives.**
- **Find a particular solution of a differential equation.**

Antiderivatives

To find a function F whose derivative is $f(x) = 3x^2$, you might use your knowledge of derivatives to conclude that

$$F(x) = x^3 \quad \text{because} \quad \frac{d}{dx}[x^3] = 3x^2.$$

The function F is an *antiderivative* of f.

Exploration

Finding Antiderivatives
For each derivative, describe the original function F.

a. $F'(x) = 2x$
b. $F'(x) = x$
c. $F'(x) = x^2$
d. $F'(x) = \frac{1}{x^2}$
e. $F'(x) = \frac{1}{x^3}$
f. $F'(x) = \cos x$

What strategy did you use to find F?

Definition of Antiderivative

A function F is an **antiderivative** of f on an interval I when $F'(x) = f(x)$ for all x in I.

Note that F is called *an* antiderivative of f rather than *the* antiderivative of f. To see why, observe that

$$F_1(x) = x^3, \quad F_2(x) = x^3 - 5, \quad \text{and} \quad F_3(x) = x^3 + 97$$

are all antiderivatives of $f(x) = 3x^2$. In fact, for any constant C, the function $F(x) = x^3 + C$ is an antiderivative of f.

THEOREM 4.1 Representation of Antiderivatives

If F is an antiderivative of f on an interval I, then G is an antiderivative of f on the interval I if and only if G is of the form $G(x) = F(x) + C$ for all x in I, where C is a constant.

Proof The proof of Theorem 4.1 in one direction is straightforward. That is, if $G(x) = F(x) + C$, $F'(x) = f(x)$, and C is a constant, then

$$G'(x) = \frac{d}{dx}[F(x) + C] = F'(x) + 0 = f(x).$$

To prove this theorem in the other direction, assume that G is an antiderivative of f. Define a function H such that

$$H(x) = G(x) - F(x).$$

For any two points a and b $(a < b)$ in the interval, H is continuous on $[a, b]$ and differentiable on (a, b). By the Mean Value Theorem,

$$H'(c) = \frac{H(b) - H(a)}{b - a}$$

for some c in (a, b). However, $H'(c) = 0$, so $H(a) = H(b)$. Because a and b are arbitrary points in the interval, you know that H is a constant function C. So, $G(x) - F(x) = C$ and it follows that $G(x) = F(x) + C$. ■

Using Theorem 4.1, you can represent the entire family of antiderivatives of a function by adding a constant to a *known* antiderivative. For example, knowing that

$$D_x[x^2] = 2x$$

you can represent the family of *all* antiderivatives of $f(x) = 2x$ by

$$G(x) = x^2 + C \qquad \text{Family of all antiderivatives of } f(x) = 2x$$

where C is a constant. The constant C is called the **constant of integration.** The family of functions represented by G is the **general antiderivative** of f, and $G(x) = x^2 + C$ is the **general solution** of the *differential equation*

$$G'(x) = 2x. \qquad \text{Differential equation}$$

A **differential equation** in x and y is an equation that involves x, y, and derivatives of y. For instance,

$$y' = 3x \quad \text{and} \quad y' = x^2 + 1$$

are examples of differential equations.

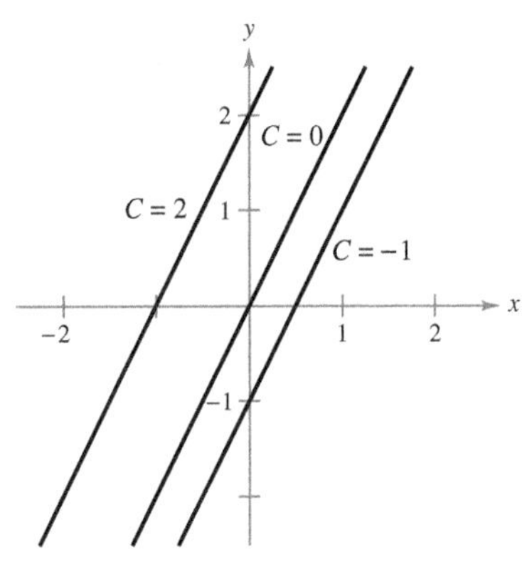

Functions of the form $y = 2x + C$
Figure 4.1

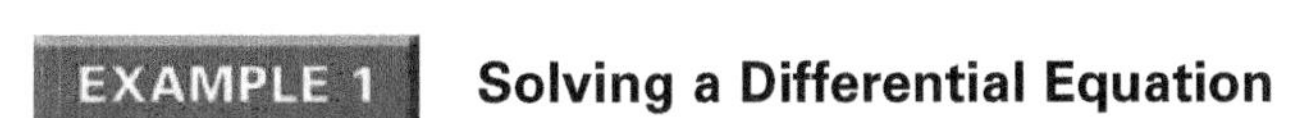

EXAMPLE 1 Solving a Differential Equation

Find the general solution of the differential equation $dy/dx = 2$.

Solution To begin, you need to find a function whose derivative is 2. One such function is

$$y = 2x. \qquad 2x \text{ is } an \text{ antiderivative of } 2.$$

Now, you can use Theorem 4.1 to conclude that the general solution of the differential equation is

$$y = 2x + C. \qquad \text{General solution}$$

The graphs of several functions of the form $y = 2x + C$ are shown in Figure 4.1. ■

When solving a differential equation of the form

$$\frac{dy}{dx} = f(x)$$

it is convenient to write it in the equivalent differential form

$$dy = f(x)\,dx.$$

The operation of finding all solutions of this equation is called **antidifferentiation** (or **indefinite integration**) and is denoted by an integral sign $\int$. The general solution is denoted by

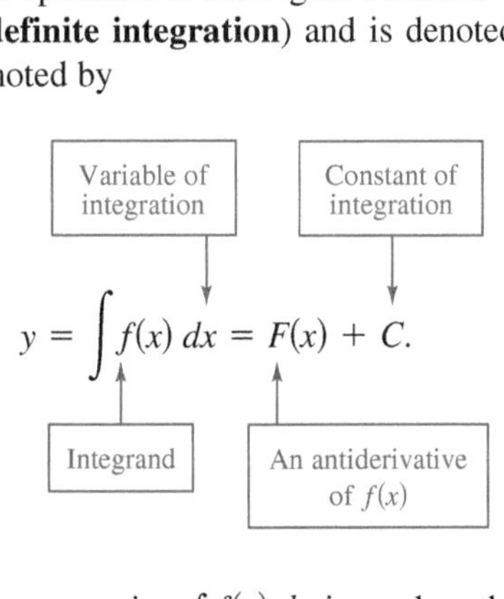

REMARK In this text, the notation $\int f(x)\,dx = F(x) + C$ means that F is an antiderivative of f *on an interval.*

The expression $\int f(x)\,dx$ is read as the *antiderivative of f with respect to x.* So, the differential dx serves to identify x as the variable of integration. The term **indefinite integral** is a synonym for antiderivative.

Basic Integration Rules

The inverse nature of integration and differentiation can be verified by substituting $F'(x)$ for $f(x)$ in the indefinite integration definition to obtain

$$\int F'(x)\,dx = F(x) + C.$$

Integration is the "inverse" of differentiation.

Moreover, if $\int f(x)\,dx = F(x) + C$, then

$$\frac{d}{dx}\left[\int f(x)\,dx\right] = f(x).$$

Differentiation is the "inverse" of integration.

These two equations allow you to obtain integration formulas directly from differentiation formulas, as shown in the following summary.

Basic Integration Rules

Differentiation Formula	Integration Formula	
$\frac{d}{dx}[C] = 0$	$\int 0\,dx = C$	
$\frac{d}{dx}[kx] = k$	$\int k\,dx = kx + C$	
$\frac{d}{dx}[kf(x)] = kf'(x)$	$\int kf(x)\,dx = k\int f(x)\,dx$	
$\frac{d}{dx}[f(x) \pm g(x)] = f'(x) \pm g'(x)$	$\int [f(x) \pm g(x)]\,dx = \int f(x)\,dx \pm \int g(x)\,dx$	
$\frac{d}{dx}[x^n] = nx^{n-1}$	$\int x^n\,dx = \frac{x^{n+1}}{n+1} + C, \quad n \neq -1$	Power Rule
$\frac{d}{dx}[\sin x] = \cos x$	$\int \cos x\,dx = \sin x + C$	
$\frac{d}{dx}[\cos x] = -\sin x$	$\int \sin x\,dx = -\cos x + C$	
$\frac{d}{dx}[\tan x] = \sec^2 x$	$\int \sec^2 x\,dx = \tan x + C$	
$\frac{d}{dx}[\sec x] = \sec x \tan x$	$\int \sec x \tan x\,dx = \sec x + C$	
$\frac{d}{dx}[\cot x] = -\csc^2 x$	$\int \csc^2 x\,dx = -\cot x + C$	
$\frac{d}{dx}[\csc x] = -\csc x \cot x$	$\int \csc x \cot x\,dx = -\csc x + C$	

Note that the Power Rule for Integration has the restriction that $n \neq -1$. The integration formula for

$$\int \frac{1}{x}\,dx$$

must wait until the introduction of the natural logarithmic function in Chapter 5.

REMARK In Example 2, note that the general pattern of integration is similar to that of differentiation.

Original integral

Rewrite

Integrate

Simplify

EXAMPLE 2 Describing Antiderivatives

$$\int 3x\,dx = 3\int x\,dx \qquad \text{Constant Multiple Rule}$$
$$= 3\int x^1\,dx \qquad \text{Rewrite } x \text{ as } x^1.$$
$$= 3\left(\frac{x^2}{2}\right) + C \qquad \text{Power Rule } (n = 1)$$
$$= \frac{3}{2}x^2 + C \qquad \text{Simplify.}$$

The antiderivatives of $3x$ are of the form $\frac{3}{2}x^2 + C$, where C is any constant. ■

When finding indefinite integrals, a strict application of the basic integration rules tends to produce complicated constants of integration. For instance, in Example 2, the solution could have been written as

$$\int 3x\,dx = 3\int x\,dx = 3\left(\frac{x^2}{2} + C\right) = \frac{3}{2}x^2 + 3C.$$

Because C represents *any* constant, it is both cumbersome and unnecessary to write $3C$ as the constant of integration. So, $\frac{3}{2}x^2 + 3C$ is written in the simpler form $\frac{3}{2}x^2 + C$.

TECHNOLOGY Some software programs, such as *Maple* and *Mathematica*, are capable of performing integration symbolically. If you have access to such a symbolic integration utility, try using it to find the indefinite integrals in Example 3.

EXAMPLE 3 Rewriting Before Integrating

See LarsonCalculus.com for an interactive version of this type of example.

	Original Integral	Rewrite	Integrate	Simplify
a.	$\int \frac{1}{x^3}\,dx$	$\int x^{-3}\,dx$	$\frac{x^{-2}}{-2} + C$	$-\frac{1}{2x^2} + C$
b.	$\int \sqrt{x}\,dx$	$\int x^{1/2}\,dx$	$\frac{x^{3/2}}{3/2} + C$	$\frac{2}{3}x^{3/2} + C$
c.	$\int 2\sin x\,dx$	$2\int \sin x\,dx$	$2(-\cos x) + C$	$-2\cos x + C$

REMARK The basic integration rules allow you to integrate any polynomial function.

EXAMPLE 4 Integrating Polynomial Functions

a. $\int dx = \int 1\,dx$ Integrand is understood to be 1.

$= x + C$ Integrate.

b. $\int (x + 2)\,dx = \int x\,dx + \int 2\,dx$

$= \frac{x^2}{2} + C_1 + 2x + C_2$ Integrate.

$= \frac{x^2}{2} + 2x + C$ $C = C_1 + C_2$

The second line in the solution is usually omitted.

c. $\int (3x^4 - 5x^2 + x)\,dx = 3\left(\frac{x^5}{5}\right) - 5\left(\frac{x^3}{3}\right) + \frac{x^2}{2} + C$

$= \frac{3}{5}x^5 - \frac{5}{3}x^3 + \frac{1}{2}x^2 + C$ ■

REMARK Before you begin the exercise set, be sure you realize that one of the most important steps in integration is *rewriting the integrand* in a form that fits one of the basic integration rules.

EXAMPLE 5 Rewriting Before Integrating

$$\int \frac{x+1}{\sqrt{x}}\,dx = \int \left(\frac{x}{\sqrt{x}} + \frac{1}{\sqrt{x}}\right) dx \quad \text{Rewrite as two fractions.}$$

$$= \int (x^{1/2} + x^{-1/2})\,dx \quad \text{Rewrite with fractional exponents.}$$

$$= \frac{x^{3/2}}{3/2} + \frac{x^{1/2}}{1/2} + C \quad \text{Integrate.}$$

$$= \frac{2}{3}x^{3/2} + 2x^{1/2} + C \quad \text{Simplify.}$$

$$= \frac{2}{3}\sqrt{x}(x+3) + C$$

When integrating quotients, do not integrate the numerator and denominator separately. This is no more valid in integration than it is in differentiation. For instance, in Example 5, be sure you understand that

$$\int \frac{x+1}{\sqrt{x}}\,dx = \frac{2}{3}\sqrt{x}(x+3) + C$$

is not the same as

$$\frac{\int (x+1)\,dx}{\int \sqrt{x}\,dx} = \frac{\frac{1}{2}x^2 + x + C_1}{\frac{2}{3}x\sqrt{x} + C_2}.$$

EXAMPLE 6 Rewriting Before Integrating

$$\int \frac{\sin x}{\cos^2 x}\,dx = \int \left(\frac{1}{\cos x}\right)\left(\frac{\sin x}{\cos x}\right) dx \quad \text{Rewrite as a product.}$$

$$= \int \sec x \tan x\,dx \quad \text{Rewrite using trigonometric identities.}$$

$$= \sec x + C \quad \text{Integrate.}$$

EXAMPLE 7 Rewriting Before Integrating

	Original Integral	Rewrite	Integrate	Simplify
a.	$\int \frac{2}{\sqrt{x}}\,dx$	$2\int x^{-1/2}\,dx$	$2\left(\frac{x^{1/2}}{1/2}\right) + C$	$4x^{1/2} + C$
b.	$\int (t^2+1)^2\,dt$	$\int (t^4 + 2t^2 + 1)\,dt$	$\frac{t^5}{5} + 2\left(\frac{t^3}{3}\right) + t + C$	$\frac{1}{5}t^5 + \frac{2}{3}t^3 + t + C$
c.	$\int \frac{x^3+3}{x^2}\,dx$	$\int (x + 3x^{-2})\,dx$	$\frac{x^2}{2} + 3\left(\frac{x^{-1}}{-1}\right) + C$	$\frac{1}{2}x^2 - \frac{3}{x} + C$
d.	$\int \sqrt[3]{x}(x-4)\,dx$	$\int (x^{4/3} - 4x^{1/3})\,dx$	$\frac{x^{7/3}}{7/3} - 4\left(\frac{x^{4/3}}{4/3}\right) + C$	$\frac{3}{7}x^{7/3} - 3x^{4/3} + C$

As you do the exercises, note that you can check your answer to an antidifferentiation problem by differentiating. For instance, in Example 7(a), you can check that $4x^{1/2} + C$ is the correct antiderivative by differentiating the answer to obtain

$$D_x[4x^{1/2} + C] = 4\left(\frac{1}{2}\right)x^{-1/2} = \frac{2}{\sqrt{x}}. \quad \text{Use differentiation to check antiderivative.}$$

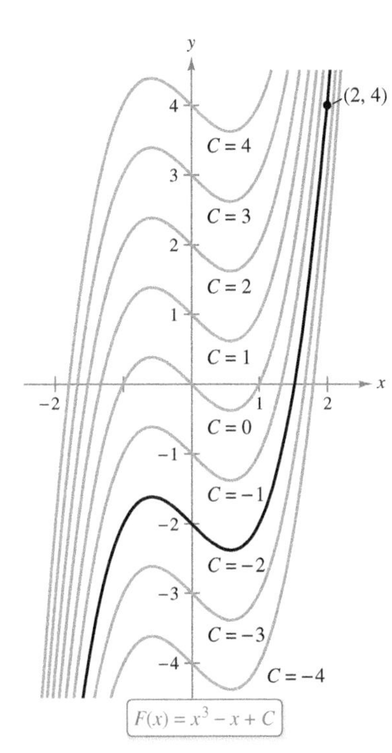

The particular solution that satisfies the initial condition $F(2) = 4$ is $F(x) = x^3 - x - 2$.
Figure 4.2

Initial Conditions and Particular Solutions

You have already seen that the equation $y = \int f(x)\,dx$ has many solutions (each differing from the others by a constant). This means that the graphs of any two antiderivatives of f are vertical translations of each other. For example, Figure 4.2 shows the graphs of several antiderivatives of the form

$$y = \int (3x^2 - 1)\,dx = x^3 - x + C \qquad \text{General solution}$$

for various integer values of C. Each of these antiderivatives is a solution of the differential equation

$$\frac{dy}{dx} = 3x^2 - 1.$$

In many applications of integration, you are given enough information to determine a **particular solution.** To do this, you need only know the value of $y = F(x)$ for one value of x. This information is called an **initial condition.** For example, in Figure 4.2, only one curve passes through the point $(2, 4)$. To find this curve, you can use the general solution

$$F(x) = x^3 - x + C \qquad \text{General solution}$$

and the initial condition

$$F(2) = 4. \qquad \text{Initial condition}$$

By using the initial condition in the general solution, you can determine that

$$F(2) = 8 - 2 + C = 4$$

which implies that $C = -2$. So, you obtain

$$F(x) = x^3 - x - 2. \qquad \text{Particular solution}$$

EXAMPLE 8 Finding a Particular Solution

Find the general solution of

$$F'(x) = \frac{1}{x^2}, \quad x > 0$$

and find the particular solution that satisfies the initial condition $F(1) = 0$.

Solution To find the general solution, integrate to obtain

$$F(x) = \int \frac{1}{x^2}\,dx \qquad F(x) = \int F'(x)\,dx$$

$$= \int x^{-2}\,dx \qquad \text{Rewrite as a power.}$$

$$= \frac{x^{-1}}{-1} + C \qquad \text{Integrate.}$$

$$= -\frac{1}{x} + C, \quad x > 0. \qquad \text{General solution}$$

Using the initial condition $F(1) = 0$, you can solve for C as follows.

$$F(1) = -\frac{1}{1} + C = 0 \implies C = 1$$

So, the particular solution, as shown in Figure 4.3, is

$$F(x) = -\frac{1}{x} + 1, \quad x > 0. \qquad \text{Particular solution}$$

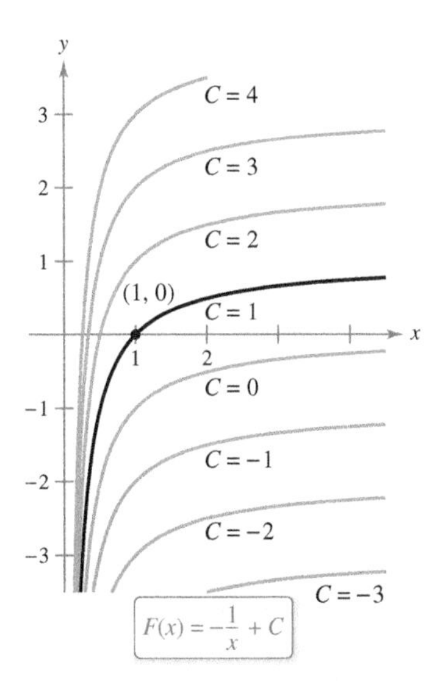

The particular solution that satisfies the initial condition $F(1) = 0$ is $F(x) = -(1/x) + 1, x > 0$.
Figure 4.3

So far in this section, you have been using x as the variable of integration. In applications, it is often convenient to use a different variable. For instance, in the next example, involving *time*, the variable of integration is t.

EXAMPLE 9 Solving a Vertical Motion Problem

A ball is thrown upward with an initial velocity of 64 feet per second from an initial height of 80 feet. [Assume the acceleration is $a(t) = -32$ feet per second per second.]

a. Find the position function giving the height s as a function of the time t.

b. When does the ball hit the ground?

Solution

a. Let $t = 0$ represent the initial time. The two given initial conditions can be written as follows.

$$s(0) = 80 \qquad \text{Initial height is 80 feet.}$$

$$s'(0) = 64 \qquad \text{Initial velocity is 64 feet per second.}$$

Recall that $a(t) = s''(t)$. So, you can write

$$s''(t) = -32$$

$$s'(t) = \int s''(t)\,dt = \int -32\,dt = -32t + C_1.$$

Using the initial velocity, you obtain $s'(0) = 64 = -32(0) + C_1$, which implies that $C_1 = 64$. Next, by integrating $s'(t)$, you obtain

$$s(t) = \int s'(t)\,dt = \int (-32t + 64)\,dt = -16t^2 + 64t + C_2.$$

Using the initial height, you obtain

$$s(0) = 80 = -16(0^2) + 64(0) + C_2$$

which implies that $C_2 = 80$. So, the position function is

$$s(t) = -16t^2 + 64t + 80. \qquad \text{See Figure 4.4.}$$

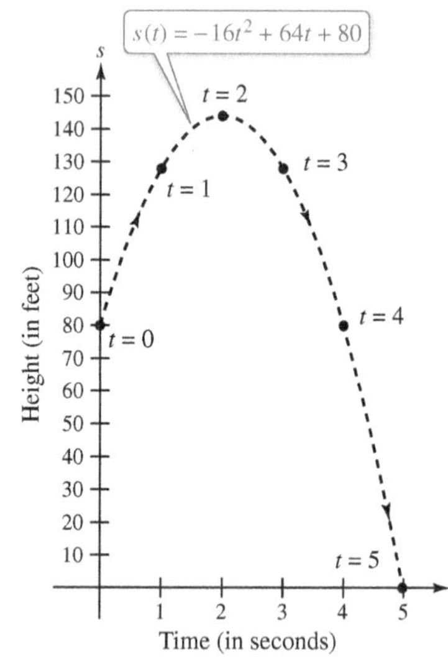

Height of a ball at time t
Figure 4.4

b. Using the position function found in part (a), you can find the time at which the ball hits the ground by solving the equation $s(t) = 0$.

$$-16t^2 + 64t + 80 = 0$$

$$-16(t + 1)(t - 5) = 0$$

$$t = -1, 5$$

Because t must be positive, you can conclude that the ball hits the ground 5 seconds after it was thrown. ■

In Example 9, note that the position function has the form

$$s(t) = -\frac{1}{2}gt^2 + v_0t + s_0$$

where g is the acceleration due to gravity, v_0 is the initial velocity, and s_0 is the initial height, as presented in Section 2.2.

Example 9 shows how to use calculus to analyze vertical motion problems in which the acceleration is determined by a gravitational force. You can use a similar strategy to analyze other linear motion problems (vertical or horizontal) in which the acceleration (or deceleration) is the result of some other force, as you will see in Exercises 65–72.

4.1 Exercises

See CalcChat.com for tutorial help and worked-out solutions to odd-numbered exercises.

CONCEPT CHECK

1. **Antiderivative** What does it mean for a function F to be an antiderivative of a function f on an interval I?
2. **Antiderivatives** Can two different functions both be antiderivatives of the same function? Explain.
3. **Particular Solution** What is a particular solution of a differential equation?
4. **General and Particular Solutions** Describe the difference between the general solution and a particular solution of a differential equation.

Integration and Differentiation **In Exercises 5 and 6, verify the statement by showing that the derivative of the right side equals the integrand on the left side.**

5. $\displaystyle\int\left(-\frac{6}{x^4}\right)dx = \frac{2}{x^3} + C$

6. $\displaystyle\int\left(8x^3 + \frac{1}{2x^2}\right)dx = 2x^4 - \frac{1}{2x} + C$

Solving a Differential Equation **In Exercises 7–10, find the general solution of the differential equation and check the result by differentiation.**

7. $\dfrac{dy}{dt} = 9t^2$

8. $\dfrac{dy}{dt} = 5$

9. $\dfrac{dy}{dx} = x^{3/2}$

10. $\dfrac{dy}{dx} = 2x^{-3}$

Rewriting Before Integrating **In Exercises 11–14, complete the table to find the indefinite integral.**

	Original Integral	Rewrite	Integrate	Simplify
11.	$\int \sqrt[3]{x}\,dx$			
12.	$\int \frac{1}{4x^2}\,dx$			
13.	$\int \frac{1}{x\sqrt{x}}\,dx$			
14.	$\int \frac{1}{(3x)^2}\,dx$			

Finding an Indefinite Integral **In Exercises 15–36, find the indefinite integral and check the result by differentiation.**

15. $\int (x + 7)\,dx$

16. $\int (13 - x)\,dx$

17. $\int (x^5 + 1)\,dx$

18. $\int (9x^8 - 2x - 6)\,dx$

19. $\int (x^{3/2} + 2x + 1)\,dx$

20. $\displaystyle\int\left(\sqrt{x} + \frac{1}{2\sqrt{x}}\right)dx$

21. $\int \sqrt[3]{x^2}\,dx$

22. $\int \left(\sqrt[4]{x^3} + 1\right)dx$

23. $\displaystyle\int \frac{1}{x^5}\,dx$

24. $\displaystyle\int\left(2 - \frac{3}{x^{10}}\right)dx$

25. $\displaystyle\int \frac{x + 6}{\sqrt{x}}\,dx$

26. $\displaystyle\int \frac{x^4 - 3x^2 + 5}{x^4}\,dx$

27. $\int (x + 1)(3x - 2)\,dx$

28. $\int (4t^2 + 3)^2\,dt$

29. $\int (5\cos x + 4\sin x)\,dx$

30. $\int (\sin x - 6\cos x)\,dx$

31. $\int (\csc x \cot x - 2x)\,dx$

32. $\int (\theta^2 + \sec^2\theta)\,d\theta$

33. $\int (\sec^2\theta - \sin\theta)\,d\theta$

34. $\int (\sec y)(\tan y - \sec y)\,dy$

35. $\int (\tan^2 y + 1)\,dy$

36. $\int (4x - \csc^2 x)\,dx$

Finding a Particular Solution **In Exercises 37–44, find the particular solution of the differential equation that satisfies the initial condition(s).**

37. $f'(x) = 6x,\ f(0) = 8$

38. $g'(x) = 4x^2,\ g(-1) = 3$

39. $h'(x) = 7x^6 + 5,\ h(1) = -1$

40. $f'(s) = 10s - 12s^3,\ f(3) = 2$

41. $f''(x) = 2,\ f'(2) = 5,\ f(2) = 10$

42. $f''(x) = 3x^2,\ f'(-1) = -2,\ f(2) = 3$

43. $f''(x) = x^{-3/2},\ f'(4) = 2,\ f(0) = 0$

44. $f''(x) = \sin x,\ f'(0) = 1,\ f(0) = 6$

Slope Field **In Exercises 45 and 46, a differential equation, a point, and a slope field are given. A *slope field* (or *direction field*) consists of line segments with slopes given by the differential equation. These line segments give a visual perspective of the slopes of the solutions of the differential equation. (a) Sketch two approximate solutions of the differential equation on the slope field, one of which passes through the indicated point. (To print an enlarged copy of the graph, go to *MathGraphs.com*.) (b) Use integration and the given point to find the particular solution of the differential equation and use a graphing utility to graph the solution. Compare the result with the sketch in part (a) that passes through the given point.**

45. $\dfrac{dy}{dx} = x^2 - 1,\ (-1, 3)$

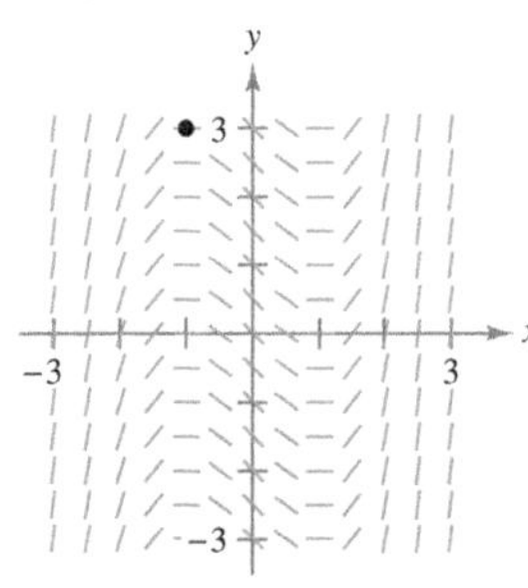

46. $\dfrac{dy}{dx} = -\dfrac{1}{x^2},\ x > 0,\ (1, 3)$

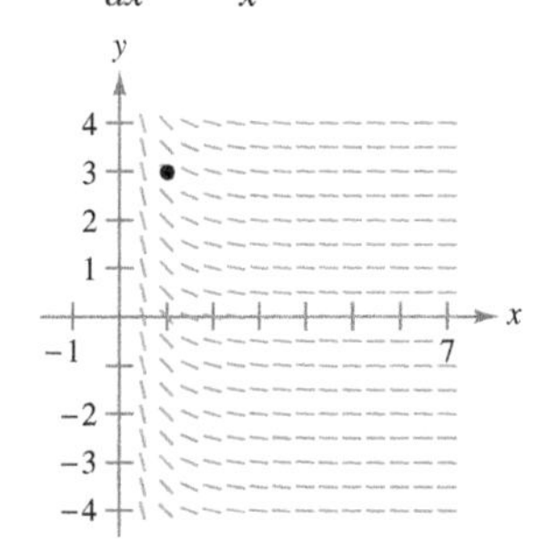

Slope Field **In Exercises 47 and 48, (a) use a graphing utility to graph a slope field for the differential equation, (b) use integration and the given point to find the particular solution of the differential equation, and (c) graph the particular solution and the slope field in the same viewing window.**

47. $\dfrac{dy}{dx} = 2x, (-2, -2)$

48. $\dfrac{dy}{dx} = 2\sqrt{x}, (4, 12)$

EXPLORING CONCEPTS

Sketching a Graph **In Exercises 49 and 50, the graph of the derivative of a function is given. Sketch the graphs of *two* functions that have the given derivative. (There is more than one correct answer.) To print an enlarged copy of the graph, go to *MathGraphs.com*.**

49.

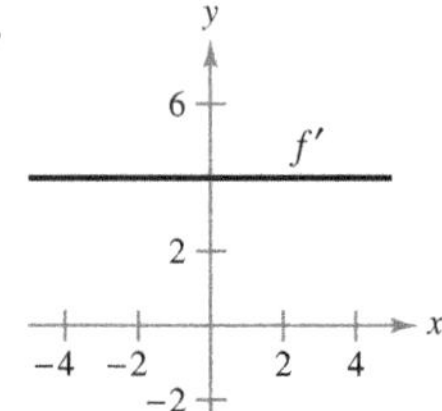

50.

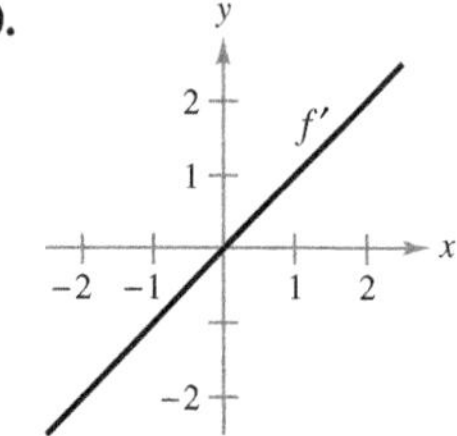

51. Comparing Functions Consider $f(x) = \tan^2 x$ and $g(x) = \sec^2 x$. What do you notice about the derivatives of f and g? What can you conclude about the relationship between f and g?

52. HOW DO YOU SEE IT? Use the graph of f' shown in the figure to answer the following.

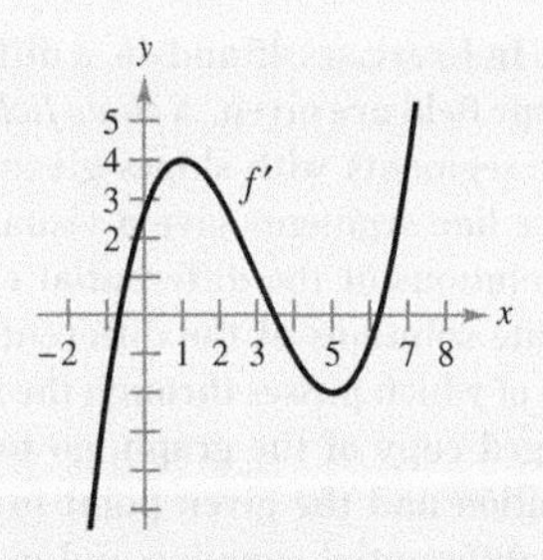

(a) Approximate the slope of f at $x = 4$. Explain.

(b) Is $f(5) - f(4) > 0$? Explain.

(c) Approximate the value of x where f is maximum. Explain.

(d) Approximate any open intervals on which the graph of f is concave upward and any open intervals on which it is concave downward. Approximate the x-coordinates of any points of inflection.

53. Horizontal Tangent Find a function f such that the graph of f has a horizontal tangent at $(2, 0)$ and $f''(x) = 2x$.

54. Sketching Graphs The graphs of f and f' each pass through the origin. Use the graph of f'' shown in the figure to sketch the graphs of f and f'. To print an enlarged copy of the graph, go to *MathGraphs.com*.

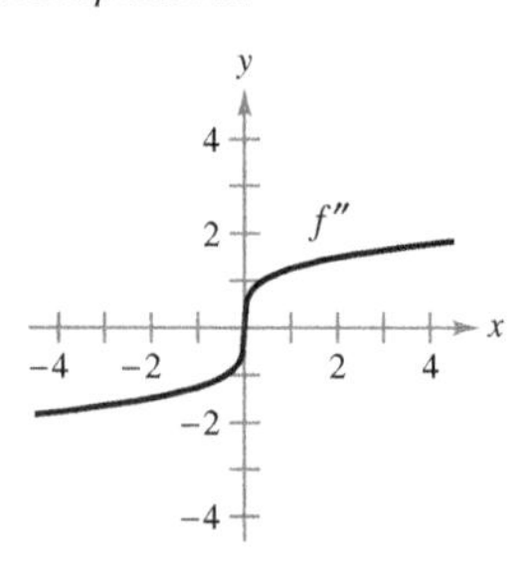

55. Tree Growth An evergreen nursery usually sells a certain type of shrub after 6 years of growth and shaping. The growth rate during those 6 years is approximated by $dh/dt = 1.5t + 5$, where t is the time in years and h is the height in centimeters. The seedlings are 12 centimeters tall when planted ($t = 0$).

(a) Find the height after t years.

(b) How tall are the shrubs when they are sold?

56. Population Growth The rate of growth dP/dt of a population of bacteria is proportional to the square root of t, where P is the population size and t is the time in days ($0 \le t \le 10$). That is,

$$\frac{dP}{dt} = k\sqrt{t}.$$

The initial size of the population is 500. After 1 day, the population has grown to 600. Estimate the population after 7 days.

Vertical Motion **In Exercises 57–59, assume the accleration of the object is $a(t) = -32$ feet per second per second. (Neglect air resistance.)**

57. A ball is thrown vertically upward from a height of 6 feet with an initial velocity of 60 feet per second. How high will the ball go?

58. With what initial velocity must an object be thrown upward (from ground level) to reach the top of the Washington Monument (approximately 550 feet)?

59. A balloon, rising vertically with a velocity of 16 feet per second, releases a sandbag at the instant it is 64 feet above the ground.

(a) How many seconds after its release will the bag strike the ground?

(b) At what velocity will the bag hit the ground?

Vertical Motion **In Exercises 60–62, assume the accleration of the object is $a(t) = -9.8$ meters per second per second. (Neglect air resistance.)**

60. A baseball is thrown upward from a height of 2 meters with an initial velocity of 10 meters per second. Determine its maximum height.

61. With what initial velocity must an object be thrown upward (from a height of 2 meters) to reach a maximum height of 200 meters?

62. Grand Canyon

The Grand Canyon is 1800 meters deep at its deepest point. A rock is dropped from the rim above this point. How long will it take the rock to hit the canyon floor?

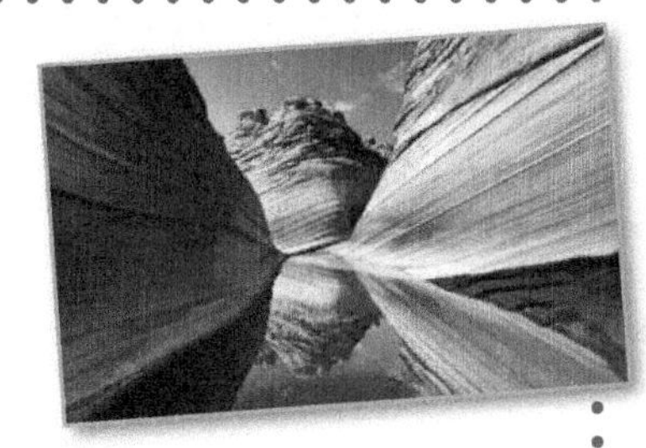

63. Lunar Gravity On the moon, the acceleration of a free-falling object is $a(t) = -1.6$ meters per second per second. A stone is dropped from a cliff on the moon and hits the surface of the moon 20 seconds later. How far did it fall? What was its velocity at impact?

64. Escape Velocity The minimum velocity required for an object to escape Earth's gravitational pull is obtained from the solution of the equation

$$\int v\,dv = -GM\int \frac{1}{y^2}\,dy$$

where v is the velocity of the object projected from Earth, y is the distance from the center of Earth, G is the gravitational constant, and M is the mass of Earth. Show that v and y are related by the equation

$$v^2 = v_0^2 + 2GM\left(\frac{1}{y} - \frac{1}{R}\right)$$

where v_0 is the initial velocity of the object and R is the radius of Earth.

Rectilinear Motion **In Exercises 65–68, consider a particle moving along the x-axis, where $x(t)$ is the position of the particle at time t, $x'(t)$ is its velocity, and $x''(t)$ is its acceleration.**

65. $x(t) = t^3 - 6t^2 + 9t - 2, \quad 0 \le t \le 5$

(a) Find the velocity and acceleration of the particle.

(b) Find the open t-intervals on which the particle is moving to the right.

(c) Find the velocity of the particle when the acceleration is 0.

66. Repeat Exercise 65 for the position function $x(t) = (t - 1)(t - 3)^2, 0 \le t \le 5$.

67. A particle moves along the x-axis at a velocity of $v(t) = 1/\sqrt{t}$, $t > 0$. At time $t = 1$, its position is $x = 4$. Find the acceleration and position functions for the particle.

68. A particle, initially at rest, moves along the x-axis such that its acceleration at time $t > 0$ is given by $a(t) = \cos t$. At time $t = 0$, its position is $x = 3$.

(a) Find the velocity and position functions for the particle.

(b) Find the values of t for which the particle is at rest.

69. Acceleration The maker of an automobile advertises that it takes 13 seconds to accelerate from 25 kilometers per hour to 80 kilometers per hour. Assume the acceleration is constant.

(a) Find the acceleration in meters per second per second.

(b) Find the distance the car travels during the 13 seconds.

70. Deceleration A car traveling at 45 miles per hour is brought to a stop, at constant deceleration, 132 feet from where the brakes are applied.

(a) How far has the car moved when its speed has been reduced to 30 miles per hour?

(b) How far has the car moved when its speed has been reduced to 15 miles per hour?

(c) Draw the real number line from 0 to 132. Plot the points found in parts (a) and (b). What can you conclude?

71. Acceleration At the instant the traffic light turns green, a car that has been waiting at an intersection starts with a constant acceleration of 6 feet per second per second. At the same instant, a truck traveling with a constant velocity of 30 feet per second passes the car.

(a) How far beyond its starting point will the car pass the truck?

(b) How fast will the car be traveling when it passes the truck?

72. Acceleration Assume that a fully loaded plane starting from rest has a constant acceleration while moving down a runway. The plane requires 0.7 mile of runway and a speed of 160 miles per hour in order to lift off. What is the plane's acceleration?

True or False? **In Exercises 73–78, determine whether the statement is true or false. If it is false, explain why or give an example that shows it is false.**

73. The antiderivative of $f(x)$ is unique.

74. Each antiderivative of an nth-degree polynomial function is an $(n + 1)$th-degree polynomial function.

75. If $p(x)$ is a polynomial function, then p has exactly one antiderivative whose graph contains the origin.

76. If $F(x)$ and $G(x)$ are antiderivatives of $f(x)$, then

$$F(x) = G(x) + C.$$

77. If $f'(x) = g(x)$, then $\int g(x)\,dx = f(x) + C$.

78. $\int f(x)g(x)\,dx = \left(\int f(x)\,dx\right)\left(\int g(x)\,dx\right)$

79. Proof Let $s(x)$ and $c(x)$ be two functions satisfying $s'(x) = c(x)$ and $c'(x) = -s(x)$ for all x. If $s(0) = 0$ and $c(0) = 1$, prove that $[s(x)]^2 + [c(x)]^2 = 1$.

80. Think About It Find the general solution of

$$f'(x) = -2x \sin x^2.$$

PUTNAM EXAM CHALLENGE

81. Suppose f and g are non-constant, differentiable, real-valued functions defined on $(-\infty, \infty)$. Furthermore, suppose that for each pair of real numbers x and y,

$$f(x + y) = f(x)f(y) - g(x)g(y) \quad \text{and}$$
$$g(x + y) = f(x)g(y) + g(x)f(y).$$

If $f'(0) = 0$, prove that $(f(x))^2 + (g(x))^2 = 1$ for all x.

4.2 Area

- **Use sigma notation to write and evaluate a sum.**
- **Understand the concept of area.**
- **Approximate the area of a plane region.**
- **Find the area of a plane region using limits.**

Sigma Notation

In the preceding section, you studied antidifferentiation. In this section, you will look further into a problem introduced in Section 1.1—that of finding the area of a region in the plane. At first glance, these two ideas may seem unrelated, but you will discover in Section 4.4 that they are closely related by an extremely important theorem called the Fundamental Theorem of Calculus.

This section begins by introducing a concise notation for sums. This notation is called **sigma notation** because it uses the uppercase Greek letter sigma, written as Σ.

Sigma Notation

The sum of n terms $a_1, a_2, a_3, \ldots, a_n$ is written as

$$\sum_{i=1}^{n} a_i = a_1 + a_2 + a_3 + \cdots + a_n$$

where i is the **index of summation,** a_i is the ***i*th term** of the sum, and the **upper and lower bounds of summation** are n and 1.

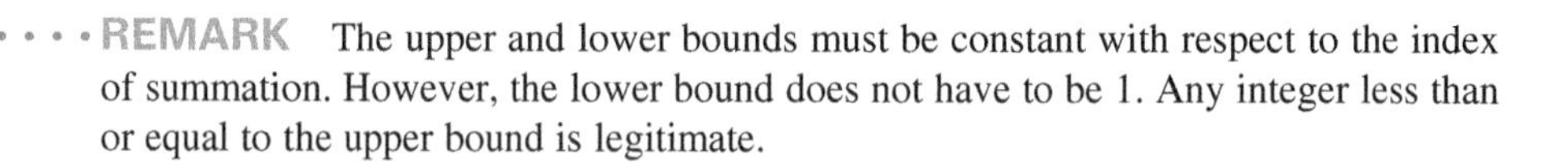

REMARK The upper and lower bounds must be constant with respect to the index of summation. However, the lower bound does not have to be 1. Any integer less than or equal to the upper bound is legitimate.

EXAMPLE 1 Examples of Sigma Notation

a. $\displaystyle\sum_{i=1}^{6} i = 1 + 2 + 3 + 4 + 5 + 6$

b. $\displaystyle\sum_{i=0}^{5} (i + 1) = 1 + 2 + 3 + 4 + 5 + 6$

c. $\displaystyle\sum_{j=3}^{7} j^2 = 3^2 + 4^2 + 5^2 + 6^2 + 7^2$

d. $\displaystyle\sum_{j=1}^{5} \frac{1}{\sqrt{j}} = \frac{1}{\sqrt{1}} + \frac{1}{\sqrt{2}} + \frac{1}{\sqrt{3}} + \frac{1}{\sqrt{4}} + \frac{1}{\sqrt{5}}$

e. $\displaystyle\sum_{k=1}^{n} \frac{1}{n}(k^2 + 1) = \frac{1}{n}(1^2 + 1) + \frac{1}{n}(2^2 + 1) + \cdots + \frac{1}{n}(n^2 + 1)$

f. $\displaystyle\sum_{i=1}^{n} f(x_i)\,\Delta x = f(x_1)\,\Delta x + f(x_2)\,\Delta x + \cdots + f(x_n)\,\Delta x$

From parts (a) and (b), notice that the same sum can be represented in different ways using sigma notation. ■

Although any variable can be used as the index of summation, i, j, and k are often used. Notice in Example 1 that the index of summation does not appear in the terms of the expanded sum.

THE SUM OF THE FIRST 100 INTEGERS

A teacher of Carl Friedrich Gauss (1777–1855) asked him to add all the integers from 1 to 100. When Gauss returned with the correct answer after only a few moments, the teacher could only look at him in astounded silence. This is what Gauss did.

$$\begin{array}{r} 1 + \ \ 2 + \ \ 3 + \cdots + 100 \\ 100 + 99 + 98 + \cdots + \ \ \ 1 \\ \hline 101 + 101 + 101 + \cdots + 101 \end{array}$$

$$\frac{100 \times 101}{2} = 5050$$

This is generalized by Theorem 4.2, Property 2, where

$$\sum_{i=1}^{100} i = \frac{100(101)}{2} = 5050.$$

The properties of summation shown below can be derived using the Associative and Commutative Properties of Addition and the Distributive Property of Addition over Multiplication. (In the first property, k is a constant.)

1. $\displaystyle\sum_{i=1}^{n} ka_i = k\sum_{i=1}^{n} a_i$ **2.** $\displaystyle\sum_{i=1}^{n} (a_i \pm b_i) = \sum_{i=1}^{n} a_i \pm \sum_{i=1}^{n} b_i$

The next theorem lists some useful formulas for sums of powers.

THEOREM 4.2 Summation Formulas

1. $\displaystyle\sum_{i=1}^{n} c = cn$, c is a constant **2.** $\displaystyle\sum_{i=1}^{n} i = \frac{n(n+1)}{2}$

3. $\displaystyle\sum_{i=1}^{n} i^2 = \frac{n(n+1)(2n+1)}{6}$ **4.** $\displaystyle\sum_{i=1}^{n} i^3 = \frac{n^2(n+1)^2}{4}$

A proof of this theorem is given in Appendix A.

EXAMPLE 2 Evaluating a Sum

Evaluate $\displaystyle\sum_{i=1}^{n} \frac{i+1}{n^2}$ for $n = 10$, 100, 1000, and 10,000.

Solution

$$\sum_{i=1}^{n} \frac{i+1}{n^2} = \frac{1}{n^2}\sum_{i=1}^{n} (i+1) \qquad \text{Factor the constant } 1/n^2 \text{ out of sum.}$$

$$= \frac{1}{n^2}\left(\sum_{i=1}^{n} i + \sum_{i=1}^{n} 1\right) \qquad \text{Write as two sums.}$$

$$= \frac{1}{n^2}\left[\frac{n(n+1)}{2} + n\right] \qquad \text{Apply Theorem 4.2.}$$

$$= \frac{1}{n^2}\left[\frac{n^2+3n}{2}\right] \qquad \text{Simplify.}$$

$$= \frac{n+3}{2n} \qquad \text{Simplify.}$$

Now you can evaluate the sum by substituting the appropriate values of n, as shown in the table below.

n	10	100	1000	10,000
$\displaystyle\sum_{i=1}^{n} \frac{i+1}{n^2} = \frac{n+3}{2n}$	0.65000	0.51500	0.50150	0.50015

In the table, note that the sum appears to approach a limit as n increases. Although the discussion of limits at infinity in Section 3.5 applies to a variable x, where x can be any real number, many of the same results hold true for limits involving the variable n, where n is restricted to positive integer values. So, to find the limit of $(n + 3)/2n$ as n approaches infinity, you can write

$$\lim_{n\to\infty} \frac{n+3}{2n} = \lim_{n\to\infty} \left(\frac{n}{2n} + \frac{3}{2n}\right) = \lim_{n\to\infty} \left(\frac{1}{2} + \frac{3}{2n}\right) = \frac{1}{2} + 0 = \frac{1}{2}.$$

FOR FURTHER INFORMATION For a geometric interpretation of summation formulas, see the article "Looking at $\displaystyle\sum_{k=1}^{n} k$ and $\displaystyle\sum_{k=1}^{n} k^2$ Geometrically" by Eric Hegblom in *Mathematics Teacher*. To view this article, go to *MathArticles.com*.

Area

In Euclidean geometry, the simplest type of plane region is a rectangle. Although people often say that the *formula* for the area of a rectangle is

$$A = bh$$

it is actually more proper to say that this is the *definition* of the **area of a rectangle.**

From this definition, you can develop formulas for the areas of many other plane regions. For example, to determine the area of a triangle, you can form a rectangle whose area is twice that of the triangle, as shown in Figure 4.5. Once you know how to find the area of a triangle, you can determine the area of any polygon by subdividing the polygon into triangular regions, as shown in Figure 4.6.

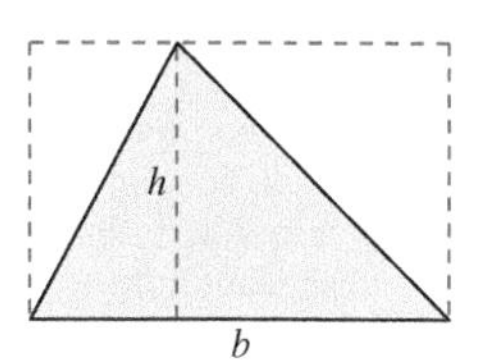

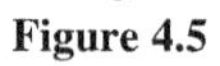

Triangle: $A = \frac{1}{2}bh$
Figure 4.5

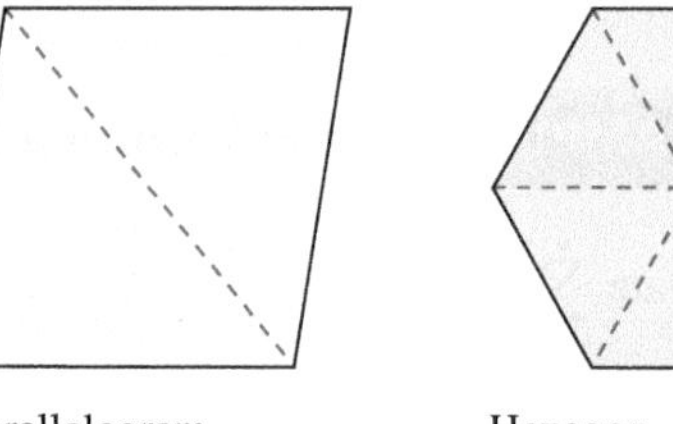

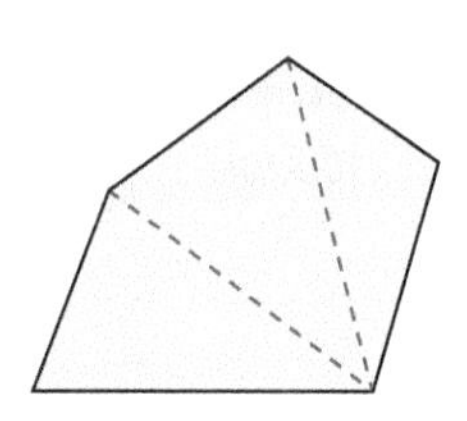

Parallelogram Hexagon Polygon
Figure 4.6

ARCHIMEDES (287–212 B.C.)

Archimedes used the method of exhaustion to derive formulas for the areas of ellipses, parabolic segments, and sectors of a spiral. He is considered to have been the greatest applied mathematician of antiquity.

See LarsonCalculus.com to read more of this biography.

Finding the areas of regions other than polygons is more difficult. The ancient Greeks were able to determine formulas for the areas of some general regions (principally those bounded by conics) by the *exhaustion* method. The clearest description of this method was given by Archimedes. Essentially, the method is a limiting process in which the area is squeezed between two polygons—one inscribed in the region and one circumscribed about the region.

For instance, in Figure 4.7, the area of a circular region is approximated by an n-sided inscribed polygon and an n-sided circumscribed polygon. For each value of n, the area of the inscribed polygon is less than the area of the circle, and the area of the circumscribed polygon is greater than the area of the circle. Moreover, as n increases, the areas of both polygons become better and better approximations of the area of the circle.

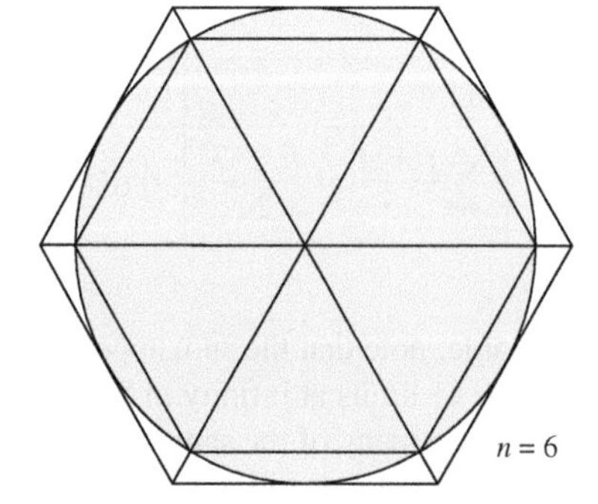

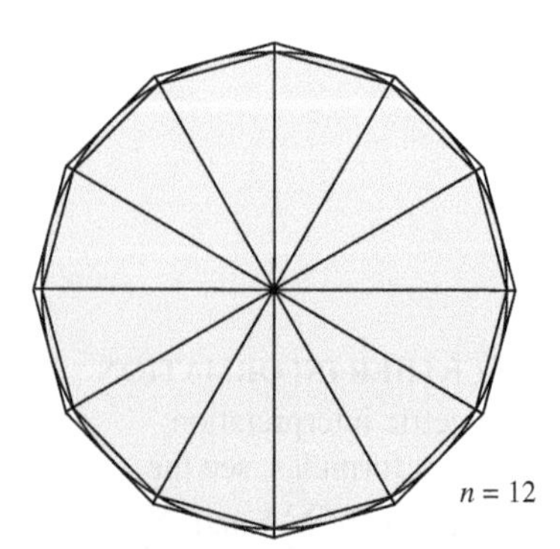

The exhaustion method for finding the area of a circular region
Figure 4.7

FOR FURTHER INFORMATION For an alternative development of the formula for the area of a circle, see the article "Proof Without Words: Area of a Disk is πR^2" by Russell Jay Hendel in *Mathematics Magazine.* To view this article, go to *MathArticles.com.*

A process that is similar to that used by Archimedes to determine the area of a plane region is used in the remaining examples in this section.

The Area of a Plane Region

Recall from Section 1.1 that the origins of calculus are connected to two classic problems: the tangent line problem and the area problem. Example 3 begins the investigation of the area problem.

EXAMPLE 3 Approximating the Area of a Plane Region

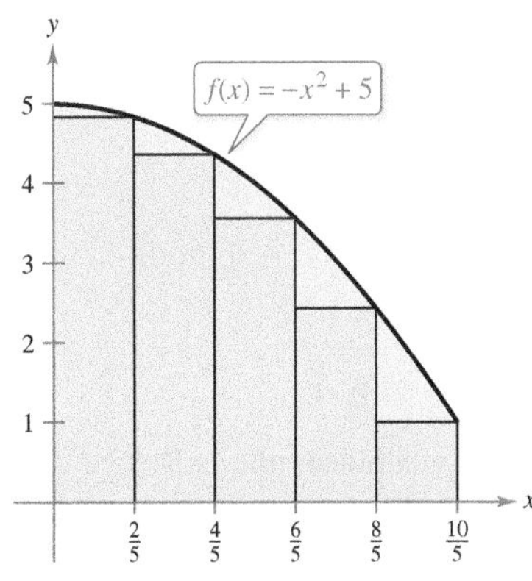

(a) The area of the parabolic region is greater than the area of the rectangles.

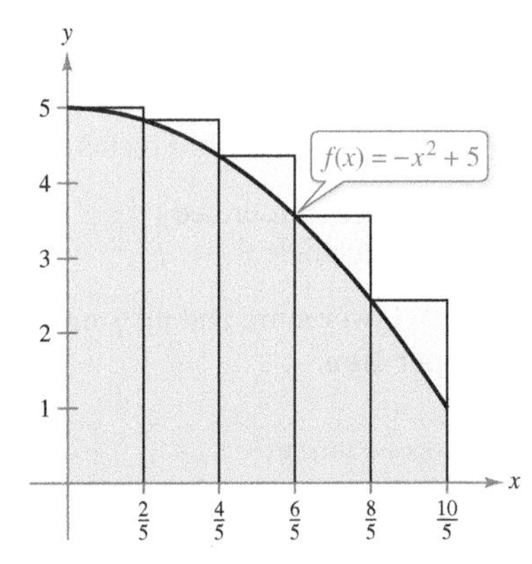

(b) The area of the parabolic region is less than the area of the rectangles.

Figure 4.8

Use the five rectangles in Figures 4.8(a) and (b) to find *two* approximations of the area of the region lying between the graph of

$$f(x) = -x^2 + 5$$

and the x-axis between $x = 0$ and $x = 2$.

Solution

a. The right endpoints of the five intervals are

$$\frac{2}{5}i \qquad \text{Right endpoints}$$

where $i = 1, 2, 3, 4, 5$. The width of each rectangle is $\frac{2}{5}$, and the height of each rectangle can be obtained by evaluating f at the right endpoint of each interval.

$$\left[0, \frac{2}{5}\right], \left[\frac{2}{5}, \frac{4}{5}\right], \left[\frac{4}{5}, \frac{6}{5}\right], \left[\frac{6}{5}, \frac{8}{5}\right], \left[\frac{8}{5}, \frac{10}{5}\right]$$

Evaluate f at the right endpoints of these intervals.

The sum of the areas of the five rectangles is

$$\sum_{i=1}^{5} \overbrace{f\left(\frac{2i}{5}\right)}^{\text{Height}} \overbrace{\left(\frac{2}{5}\right)}^{\text{Width}} = \sum_{i=1}^{5} \left[-\left(\frac{2i}{5}\right)^2 + 5\right]\left(\frac{2}{5}\right) = \frac{162}{25} = 6.48.$$

Because each of the five rectangles lies inside the parabolic region, you can conclude that the area of the parabolic region is greater than 6.48.

b. The left endpoints of the five intervals are

$$\frac{2}{5}(i - 1) \qquad \text{Left endpoints}$$

where $i = 1, 2, 3, 4, 5$. The width of each rectangle is $\frac{2}{5}$, and the height of each rectangle can be obtained by evaluating f at the left endpoint of each interval. So, the sum is

$$\sum_{i=1}^{5} \overbrace{f\left(\frac{2i-2}{5}\right)}^{\text{Height}} \overbrace{\left(\frac{2}{5}\right)}^{\text{Width}} = \sum_{i=1}^{5} \left[-\left(\frac{2i-2}{5}\right)^2 + 5\right]\left(\frac{2}{5}\right) = \frac{202}{25} = 8.08.$$

Because the parabolic region lies within the union of the five rectangular regions, you can conclude that the area of the parabolic region is less than 8.08.

By combining the results in parts (a) and (b), you can conclude that

$$6.48 < (\text{Area of region}) < 8.08.$$

By increasing the number of rectangles used in Example 3, you can obtain closer and closer approximations of the area of the region. For instance, using 25 rectangles of width $\frac{2}{25}$ each, you can conclude that

$$7.1712 < (\text{Area of region}) < 7.4912.$$

Finding Area by the Limit Definition

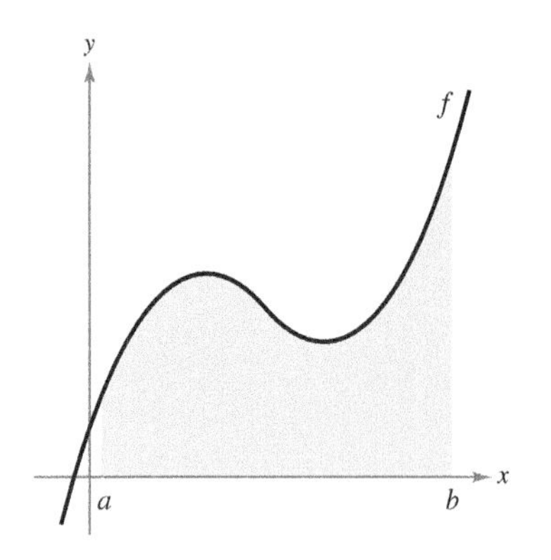

The region under a curve
Figure 4.9

The procedure used in Example 3 can be generalized as follows. Consider a plane region bounded above by the graph of a nonnegative, continuous function

$$y = f(x)$$

as shown in Figure 4.9. The region is bounded below by the x-axis, and the left and right boundaries of the region are the vertical lines $x = a$ and $x = b$.

To approximate the area of the region, begin by subdividing the interval $[a, b]$ into n subintervals, each of width

$$\Delta x = \frac{b - a}{n}$$

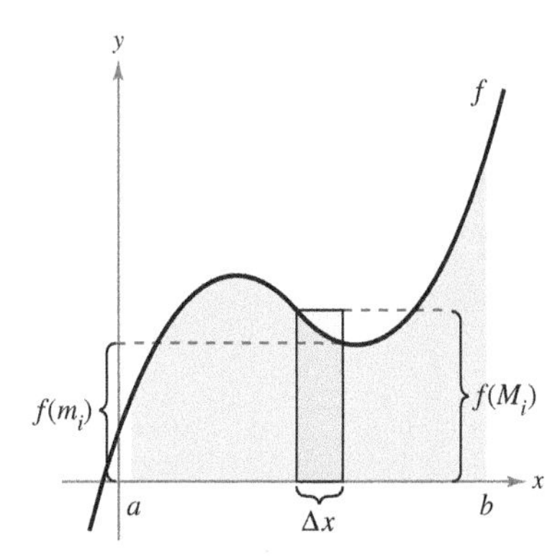

The interval $[a, b]$ is divided into n subintervals of width $\Delta x = \dfrac{b - a}{n}$.
Figure 4.10

as shown in Figure 4.10. The endpoints of the intervals are

$$\underbrace{a + 0(\Delta x)}_{a = x_0} < \underbrace{a + 1(\Delta x)}_{x_1} < \underbrace{a + 2(\Delta x)}_{x_2} < \cdot \cdot \cdot < \underbrace{a + n(\Delta x)}_{x_n = b}.$$

Because f is continuous, the Extreme Value Theorem guarantees the existence of a minimum and a maximum value of $f(x)$ in *each* subinterval.

$f(m_i)$ = Minimum value of $f(x)$ in ith subinterval

$f(M_i)$ = Maximum value of $f(x)$ in ith subinterval

Next, define an **inscribed rectangle** lying *inside* the ith subregion and a **circumscribed rectangle** extending *outside* the ith subregion. The height of the ith inscribed rectangle is $f(m_i)$ and the height of the ith circumscribed rectangle is $f(M_i)$. For *each* i, the area of the inscribed rectangle is less than or equal to the area of the circumscribed rectangle.

$$\left(\begin{array}{c}\text{Area of inscribed} \\ \text{rectangle}\end{array}\right) = f(m_i)\,\Delta x \le f(M_i)\,\Delta x = \left(\begin{array}{c}\text{Area of circumscribed} \\ \text{rectangle}\end{array}\right)$$

The sum of the areas of the inscribed rectangles is called a **lower sum,** and the sum of the areas of the circumscribed rectangles is called an **upper sum.**

$$\text{Lower sum} = s(n) = \sum_{i=1}^{n} f(m_i)\,\Delta x \qquad \text{Area of inscribed rectangles}$$

$$\text{Upper sum} = S(n) = \sum_{i=1}^{n} f(M_i)\,\Delta x \qquad \text{Area of circumscribed rectangles}$$

From Figure 4.11, you can see that the lower sum $s(n)$ is less than or equal to the upper sum $S(n)$. Moreover, the actual area of the region lies between these two sums.

$$s(n) \le (\text{Area of region}) \le S(n)$$

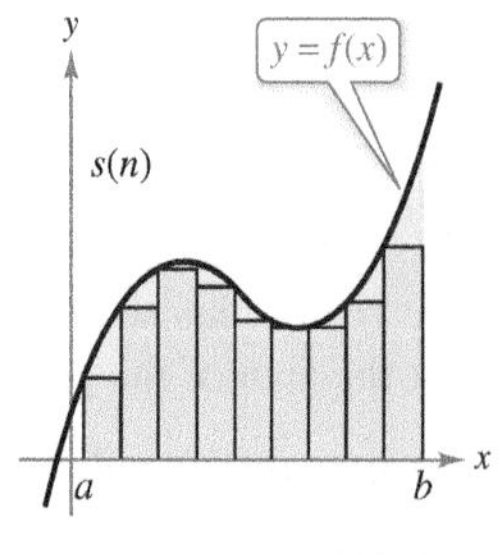

Area of inscribed rectangles is less than area of region.

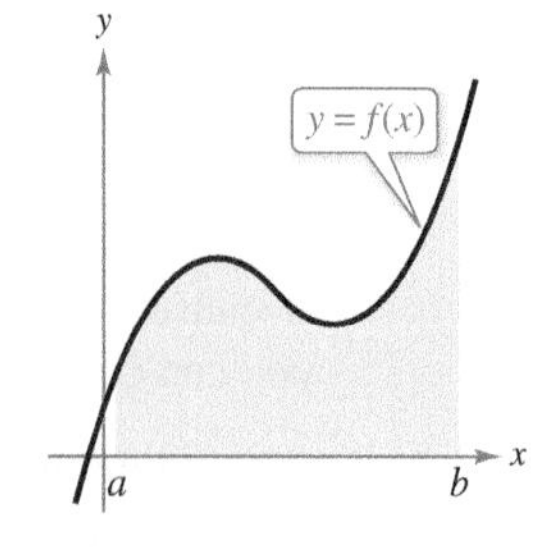

Area of region

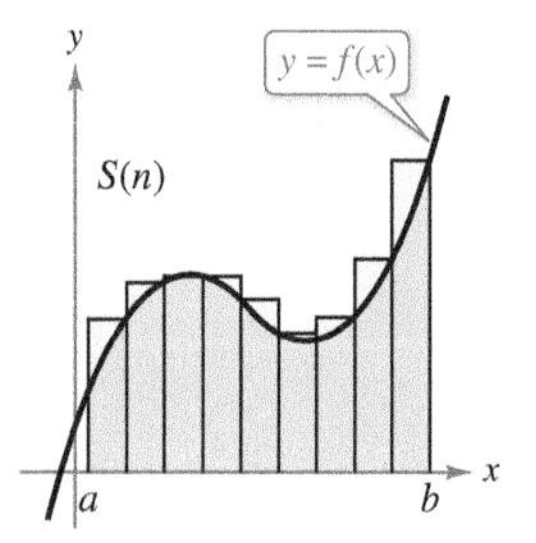

Area of circumscribed rectangles is greater than area of region.

Figure 4.11

EXAMPLE 4 Finding Upper and Lower Sums for a Region

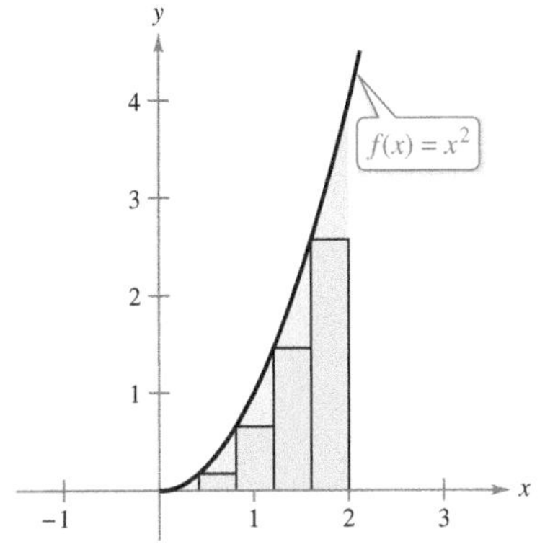

Inscribed rectangles

Circumscribed rectangles

Figure 4.12

Find the upper and lower sums for the region bounded by the graph of $f(x) = x^2$ and the x-axis between $x = 0$ and $x = 2$.

Solution To begin, partition the interval $[0, 2]$ into n subintervals, each of width

$$\Delta x = \frac{b-a}{n} = \frac{2-0}{n} = \frac{2}{n}.$$

Figure 4.12 shows the endpoints of the subintervals and several inscribed and circumscribed rectangles. Because f is increasing on the interval $[0, 2]$, the minimum value on each subinterval occurs at the left endpoint, and the maximum value occurs at the right endpoint.

Left Endpoints

$$m_i = 0 + (i-1)\left(\frac{2}{n}\right) = \frac{2(i-1)}{n}$$

Right Endpoints

$$M_i = 0 + i\left(\frac{2}{n}\right) = \frac{2i}{n}$$

Using the left endpoints, the lower sum is

$$\begin{aligned}
s(n) &= \sum_{i=1}^{n} f(m_i)\,\Delta x \\
&= \sum_{i=1}^{n} f\left[\frac{2(i-1)}{n}\right]\left(\frac{2}{n}\right) \\
&= \sum_{i=1}^{n} \left[\frac{2(i-1)}{n}\right]^2\left(\frac{2}{n}\right) \\
&= \sum_{i=1}^{n} \left(\frac{8}{n^3}\right)(i^2 - 2i + 1) \\
&= \frac{8}{n^3}\left(\sum_{i=1}^{n} i^2 - 2\sum_{i=1}^{n} i + \sum_{i=1}^{n} 1\right) \\
&= \frac{8}{n^3}\left\{\frac{n(n+1)(2n+1)}{6} - 2\left[\frac{n(n+1)}{2}\right] + n\right\} \\
&= \frac{4}{3n^3}(2n^3 - 3n^2 + n) \\
&= \frac{8}{3} - \frac{4}{n} + \frac{4}{3n^2}. \qquad \text{Lower sum}
\end{aligned}$$

Using the right endpoints, the upper sum is

$$\begin{aligned}
S(n) &= \sum_{i=1}^{n} f(M_i)\,\Delta x \\
&= \sum_{i=1}^{n} f\left(\frac{2i}{n}\right)\left(\frac{2}{n}\right) \\
&= \sum_{i=1}^{n} \left(\frac{2i}{n}\right)^2\left(\frac{2}{n}\right) \\
&= \sum_{i=1}^{n} \left(\frac{8}{n^3}\right)i^2 \\
&= \frac{8}{n^3}\left[\frac{n(n+1)(2n+1)}{6}\right] \\
&= \frac{4}{3n^3}(2n^3 + 3n^2 + n) \\
&= \frac{8}{3} + \frac{4}{n} + \frac{4}{3n^2}. \qquad \text{Upper sum}
\end{aligned}$$

> **Exploration**
>
> For the region given in Example 4, evaluate the lower sum
>
> $$s(n) = \frac{8}{3} - \frac{4}{n} + \frac{4}{3n^2}$$
>
> and the upper sum
>
> $$S(n) = \frac{8}{3} + \frac{4}{n} + \frac{4}{3n^2}$$
>
> for $n = 10, 100,$ and 1000. Use your results to determine the area of the region.

Example 4 illustrates some important things about lower and upper sums. First, notice that for any value of n, the lower sum is less than (or equal to) the upper sum.

$$s(n) = \frac{8}{3} - \frac{4}{n} + \frac{4}{3n^2} < \frac{8}{3} + \frac{4}{n} + \frac{4}{3n^2} = S(n)$$

Second, the difference between these two sums lessens as n increases. In fact, when you take the limits as $n \to \infty$, both the lower sum and the upper sum approach $\frac{8}{3}$.

$$\lim_{n\to\infty} s(n) = \lim_{n\to\infty} \left(\frac{8}{3} - \frac{4}{n} + \frac{4}{3n^2}\right) = \frac{8}{3}$$ Lower sum limit

and

$$\lim_{n\to\infty} S(n) = \lim_{n\to\infty} \left(\frac{8}{3} + \frac{4}{n} + \frac{4}{3n^2}\right) = \frac{8}{3}$$ Upper sum limit

The next theorem shows that the equivalence of the limits (as $n \to \infty$) of the upper and lower sums is not mere coincidence. It is true for all functions that are continuous and nonnegative on the closed interval $[a, b]$. The proof of this theorem is best left to a course in advanced calculus.

> **THEOREM 4.3 Limits of the Lower and Upper Sums**
>
> Let f be continuous and nonnegative on the interval $[a, b]$. The limits as $n \to \infty$ of both the lower and upper sums exist and are equal to each other. That is,
>
> $$\begin{aligned}\lim_{n\to\infty} s(n) &= \lim_{n\to\infty} \sum_{i=1}^{n} f(m_i)\,\Delta x \\ &= \lim_{n\to\infty} \sum_{i=1}^{n} f(M_i)\,\Delta x \\ &= \lim_{n\to\infty} S(n)\end{aligned}$$
>
> where $\Delta x = (b - a)/n$ and $f(m_i)$ and $f(M_i)$ are the minimum and maximum values of f on the ith subinterval.

In Theorem 4.3, the same limit is attained for both the minimum value $f(m_i)$ and the maximum value $f(M_i)$. So, it follows from the Squeeze Theorem (Theorem 1.8) that the choice of x in the ith subinterval does not affect the limit. This means that you are free to choose an *arbitrary* x-value in the ith subinterval, as shown in the *definition of the area of a region in the plane.*

> **Definition of the Area of a Region in the Plane**
>
> Let f be continuous and nonnegative on the interval $[a, b]$. (See Figure 4.13.) The area of the region bounded by the graph of f, the x-axis, and the vertical lines $x = a$ and $x = b$ is
>
> $$\text{Area} = \lim_{n\to\infty} \sum_{i=1}^{n} f(c_i)\,\Delta x$$
>
> where $x_{i-1} \le c_i \le x_i$ and
>
> $$\Delta x = \frac{b - a}{n}.$$
>
> 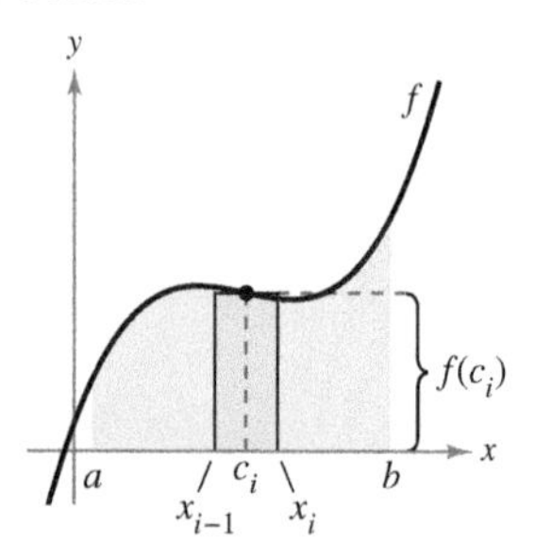
>
>
> The width of the ith subinterval is $\Delta x = x_i - x_{i-1}$.
> **Figure 4.13**

EXAMPLE 5 Finding Area by the Limit Definition

Find the area of the region bounded by the graph of $f(x) = x^3$, the x-axis, and the vertical lines $x = 0$ and $x = 1$, as shown in Figure 4.14.

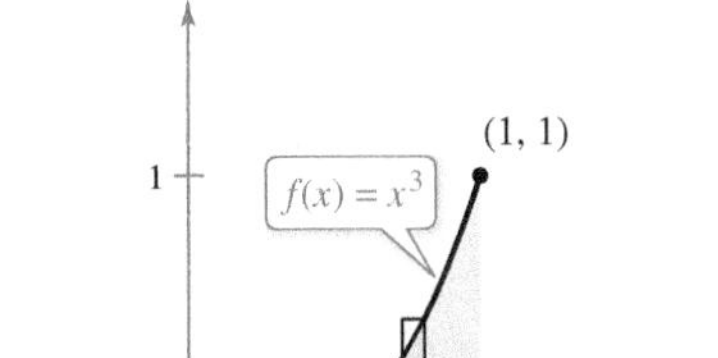

The area of the region bounded by the graph of f, the x-axis, $x = 0$, and $x = 1$ is $\frac{1}{4}$.
Figure 4.14

Solution Begin by noting that f is continuous and nonnegative on the interval $[0, 1]$. Next, partition the interval $[0, 1]$ into n subintervals, each of width $\Delta x = 1/n$. According to the definition of area, you can choose any x-value in the ith subinterval. For this example, the right endpoints $c_i = i/n$ are convenient.

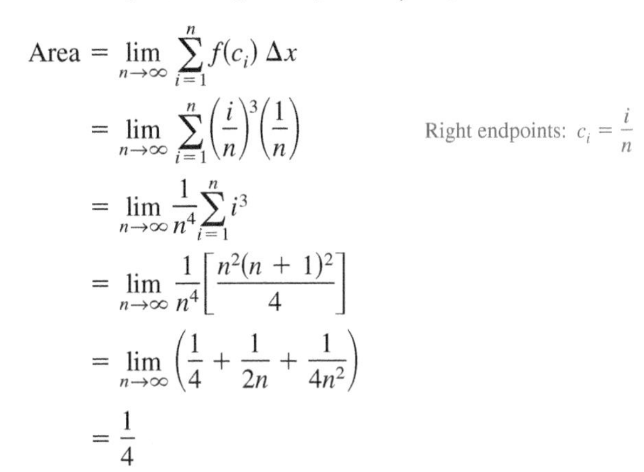

$$\begin{aligned}
\text{Area} &= \lim_{n\to\infty} \sum_{i=1}^{n} f(c_i)\,\Delta x \\
&= \lim_{n\to\infty} \sum_{i=1}^{n} \left(\frac{i}{n}\right)^3\left(\frac{1}{n}\right) && \text{Right endpoints: } c_i = \frac{i}{n} \\
&= \lim_{n\to\infty} \frac{1}{n^4}\sum_{i=1}^{n} i^3 \\
&= \lim_{n\to\infty} \frac{1}{n^4}\left[\frac{n^2(n+1)^2}{4}\right] \\
&= \lim_{n\to\infty} \left(\frac{1}{4} + \frac{1}{2n} + \frac{1}{4n^2}\right) \\
&= \frac{1}{4}
\end{aligned}$$

The area of the region is $\frac{1}{4}$.

EXAMPLE 6 Finding Area by the Limit Definition

See LarsonCalculus.com for an interactive version of this type of example.

Find the area of the region bounded by the graph of $f(x) = 4 - x^2$, the x-axis, and the vertical lines $x = 1$ and $x = 2$, as shown in Figure 4.15.

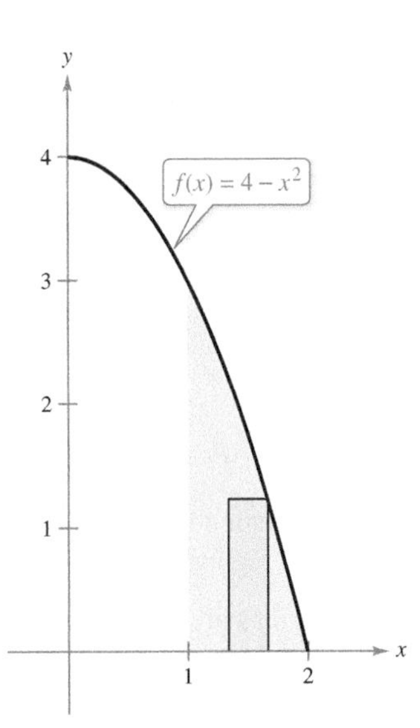

The area of the region bounded by the graph of f, the x-axis, $x = 1$, and $x = 2$ is $\frac{5}{3}$.
Figure 4.15

Solution Note that the function f is continuous and nonnegative on the interval $[1, 2]$. So, begin by partitioning the interval into n subintervals, each of width $\Delta x = 1/n$. Choosing the right endpoint

$$c_i = a + i\Delta x = 1 + \frac{i}{n} \qquad \text{Right endpoints}$$

of each subinterval, you obtain

$$\begin{aligned}
\text{Area} &= \lim_{n\to\infty} \sum_{i=1}^{n} f(c_i)\,\Delta x \\
&= \lim_{n\to\infty} \sum_{i=1}^{n} \left[4 - \left(1 + \frac{i}{n}\right)^2\right]\left(\frac{1}{n}\right) \\
&= \lim_{n\to\infty} \sum_{i=1}^{n} \left(3 - \frac{2i}{n} - \frac{i^2}{n^2}\right)\left(\frac{1}{n}\right) \\
&= \lim_{n\to\infty} \left(\frac{1}{n}\sum_{i=1}^{n} 3 - \frac{2}{n^2}\sum_{i=1}^{n} i - \frac{1}{n^3}\sum_{i=1}^{n} i^2\right) \\
&= \lim_{n\to\infty} \left[3 - \left(1 + \frac{1}{n}\right) - \left(\frac{1}{3} + \frac{1}{2n} + \frac{1}{6n^2}\right)\right] \\
&= 3 - 1 - \frac{1}{3} \\
&= \frac{5}{3}.
\end{aligned}$$

The area of the region is $\frac{5}{3}$.

The next example looks at a region that is bounded by the y-axis (rather than by the x-axis).

EXAMPLE 7 A Region Bounded by the y-axis

Find the area of the region bounded by the graph of $f(y) = y^2$ and the y-axis for $0 \le y \le 1$, as shown in Figure 4.16.

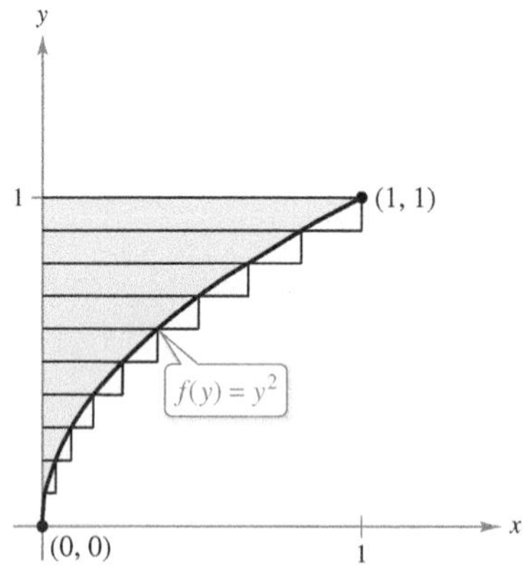

The area of the region bounded by the graph of f and the y-axis for $0 \le y \le 1$ is $\frac{1}{3}$.
Figure 4.16

Solution When f is a continuous, nonnegative function of y, you can still use the same basic procedure shown in Examples 5 and 6. Begin by partitioning the interval $[0, 1]$ into n subintervals, each of width $\Delta y = 1/n$. Then, using the upper endpoints $c_i = i/n$, you obtain

$$\begin{aligned}
\text{Area} &= \lim_{n\to\infty} \sum_{i=1}^{n} f(c_i)\,\Delta y \\
&= \lim_{n\to\infty} \sum_{i=1}^{n} \left(\frac{i}{n}\right)^2 \left(\frac{1}{n}\right) \qquad \text{Upper endpoints: } c_i = \frac{i}{n} \\
&= \lim_{n\to\infty} \frac{1}{n^3} \sum_{i=1}^{n} i^2 \\
&= \lim_{n\to\infty} \frac{1}{n^3}\left[\frac{n(n+1)(2n+1)}{6}\right] \\
&= \lim_{n\to\infty} \left(\frac{1}{3} + \frac{1}{2n} + \frac{1}{6n^2}\right) \\
&= \frac{1}{3}.
\end{aligned}$$

The area of the region is $\frac{1}{3}$.

REMARK You will study other approximation methods in Section 8.6. One of the methods, the Trapezoidal Rule, is similar to the Midpoint Rule.

In Examples 5, 6, and 7, c_i is chosen to be a value that is convenient for calculating the limit. Because each limit gives the exact area for *any* c_i, there is no need to find values that give good approximations when n is small. For an *approximation*, however, you should try to find a value of c_i that gives a good approximation of the area of the ith subregion. In general, a good value to choose is the midpoint of the interval, $c_i = (x_{i-1} + x_i)/2$, and apply the **Midpoint Rule.**

$$\text{Area} \approx \sum_{i=1}^{n} f\left(\frac{x_{i-1} + x_i}{2}\right)\Delta x. \qquad \text{Midpoint Rule}$$

EXAMPLE 8 Approximating Area with the Midpoint Rule

Use the Midpoint Rule with $n = 4$ to approximate the area of the region bounded by the graph of $f(x) = \sin x$ and the x-axis for $0 \le x \le \pi$, as shown in Figure 4.17.

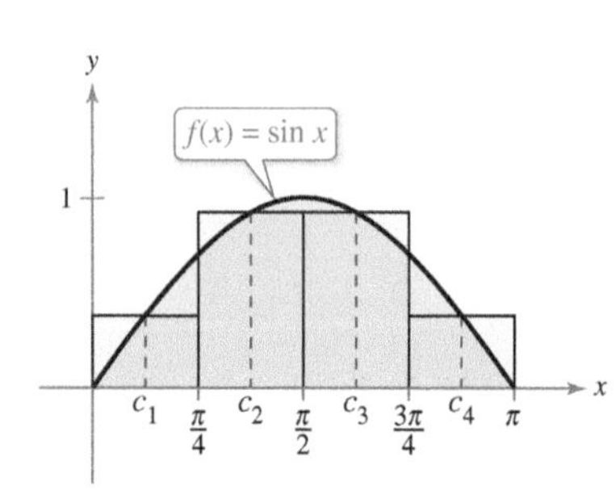

The area of the region bounded by the graph of $f(x) = \sin x$ and the x-axis for $0 \le x \le \pi$ is about 2.052.
Figure 4.17

Solution For $n = 4$, $\Delta x = \pi/4$. The midpoints of the subregions are shown below.

$$c_1 = \frac{0 + (\pi/4)}{2} = \frac{\pi}{8} \qquad c_2 = \frac{(\pi/4) + (\pi/2)}{2} = \frac{3\pi}{8}$$

$$c_3 = \frac{(\pi/2) + (3\pi/4)}{2} = \frac{5\pi}{8} \qquad c_4 = \frac{(3\pi/4) + \pi}{2} = \frac{7\pi}{8}$$

So, the area is approximated by

$$\text{Area} \approx \sum_{i=1}^{n} f(c_i)\,\Delta x = \sum_{i=1}^{4} (\sin c_i)\left(\frac{\pi}{4}\right) = \frac{\pi}{4}\left(\sin\frac{\pi}{8} + \sin\frac{3\pi}{8} + \sin\frac{5\pi}{8} + \sin\frac{7\pi}{8}\right)$$

which is about 2.052.

4.2 Exercises

See CalcChat.com for tutorial help and worked-out solutions to odd-numbered exercises.

CONCEPT CHECK

1. **Sigma Notation** What are the index of summation, the upper bound of summation, and the lower bound of summation for $\sum_{i=3}^{8}(i-4)$?

2. **Sums** What is the value of n?

   (a) $\sum_{i=1}^{n} i = \dfrac{5(5+1)}{2}$ (b) $\sum_{i=1}^{n} i^2 = \dfrac{20(20+1)[2(20)+1]}{6}$

3. **Upper and Lower Sums** In your own words and using appropriate figures, describe the methods of upper sums and lower sums in approximating the area of a region.

4. **Finding Area by the Limit Definition** Explain how to find the area of a plane region using limits.

Finding a Sum In Exercises 5–10, find the sum by adding each term together. Use the summation capabilities of a graphing utility to verify your result.

5. $\sum_{i=1}^{6}(3i+2)$

6. $\sum_{k=3}^{9}(k^2+1)$

7. $\sum_{k=0}^{4}\dfrac{1}{k^2+1}$

8. $\sum_{j=2}^{5}\dfrac{1}{2j}$

9. $\sum_{k=0}^{7} c$

10. $\sum_{i=1}^{4}[(i-1)^2+(i+1)^3]$

Using Sigma Notation In Exercises 11–16, use sigma notation to write the sum.

11. $\dfrac{1}{5(1)}+\dfrac{1}{5(2)}+\dfrac{1}{5(3)}+\cdots+\dfrac{1}{5(11)}$

12. $\dfrac{6}{2+1}+\dfrac{6}{2+2}+\dfrac{6}{2+3}+\cdots+\dfrac{6}{2+11}$

13. $\left[7\left(\dfrac{1}{6}\right)+5\right]+\left[7\left(\dfrac{2}{6}\right)+5\right]+\cdots+\left[7\left(\dfrac{6}{6}\right)+5\right]$

14. $\left[1-\left(\dfrac{1}{4}\right)^2\right]+\left[1-\left(\dfrac{2}{4}\right)^2\right]+\cdots+\left[1-\left(\dfrac{4}{4}\right)^2\right]$

15. $\left[\left(\dfrac{2}{n}\right)^3-\dfrac{2}{n}\right]\left(\dfrac{2}{n}\right)+\cdots+\left[\left(\dfrac{2n}{n}\right)^3-\dfrac{2n}{n}\right]\left(\dfrac{2}{n}\right)$

16. $\left[2\left(1+\dfrac{3}{n}\right)^2\right]\left(\dfrac{3}{n}\right)+\cdots+\left[2\left(1+\dfrac{3n}{n}\right)^2\right]\left(\dfrac{3}{n}\right)$

Evaluating a Sum In Exercises 17–24, use the properties of summation and Theorem 4.2 to evaluate the sum. Use the summation capabilities of a graphing utility to verify your result.

17. $\sum_{i=1}^{12} 7$

18. $\sum_{i=1}^{20} -8$

19. $\sum_{i=1}^{24} 4i$

20. $\sum_{i=1}^{16}(5i-4)$

21. $\sum_{i=1}^{20}(i-1)^2$

22. $\sum_{i=1}^{10}(i^2-1)$

23. $\sum_{i=1}^{7} i(i+3)^2$

24. $\sum_{i=1}^{25}(i^3-2i)$

Evaluating a Sum In Exercises 25–28, use the summation formulas to rewrite the expression without the summation notation. Use the result to find the sums for $n = 10, 100, 1000,$ and $10{,}000$.

25. $\sum_{i=1}^{n}\dfrac{2i+1}{n^2}$

26. $\sum_{j=1}^{n}\dfrac{7j+4}{n^2}$

27. $\sum_{k=1}^{n}\dfrac{6k(k-1)}{n^3}$

28. $\sum_{i=1}^{n}\dfrac{2i^3-3i}{n^4}$

Approximating the Area of a Plane Region In Exercises 29–34, use left and right endpoints and the given number of rectangles to find two approximations of the area of the region between the graph of the function and the x-axis over the given interval.

29. $f(x) = 2x+5,\ [0, 2]$, 4 rectangles

30. $f(x) = 9-x,\ [2, 4]$, 6 rectangles

31. $g(x) = 2x^2-x-1,\ [2, 5]$, 6 rectangles

32. $g(x) = x^2+1,\ [1, 3]$, 8 rectangles

33. $f(x) = \cos x,\ \left[0, \dfrac{\pi}{2}\right]$, 4 rectangles

34. $g(x) = \sin x,\ [0, \pi]$, 6 rectangles

Using Upper and Lower Sums In Exercises 35 and 36, bound the area of the shaded region by approximating the upper and lower sums. Use rectangles of width 1.

35.

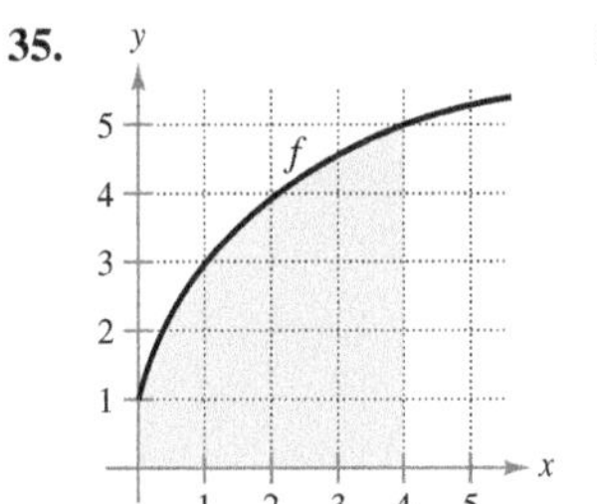

36.

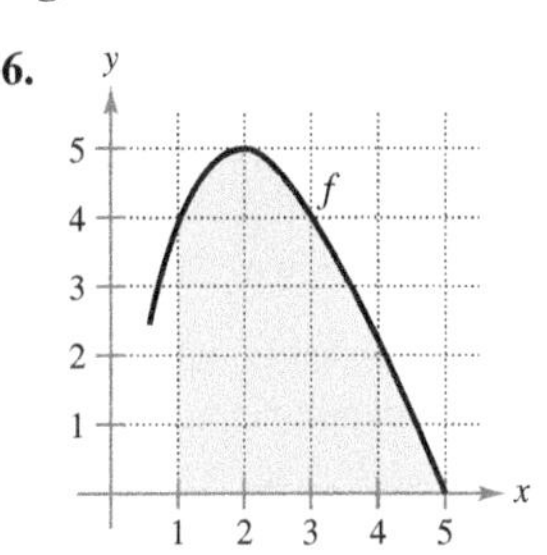

Finding Upper and Lower Sums for a Region In Exercises 37–40, use upper and lower sums to approximate the area of the region using the given number of subintervals (of equal width).

37. $y = \sqrt{x}$

38. $y = \sqrt{x}+2$

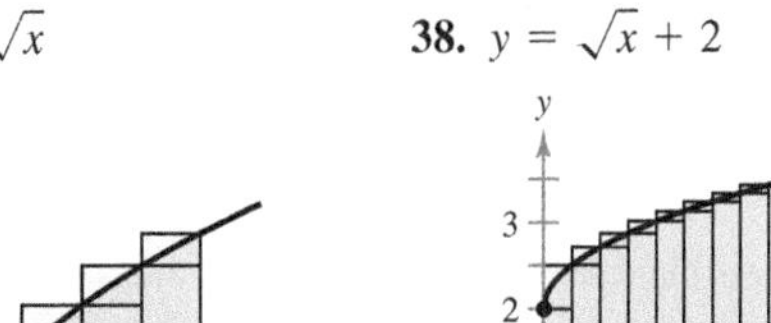

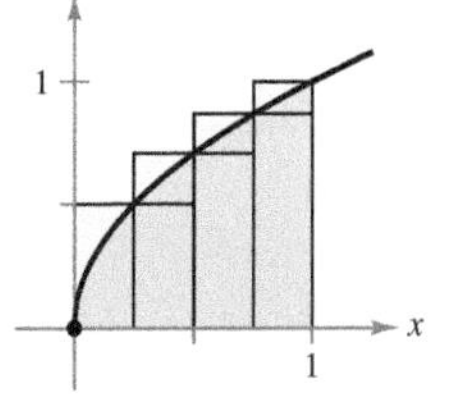

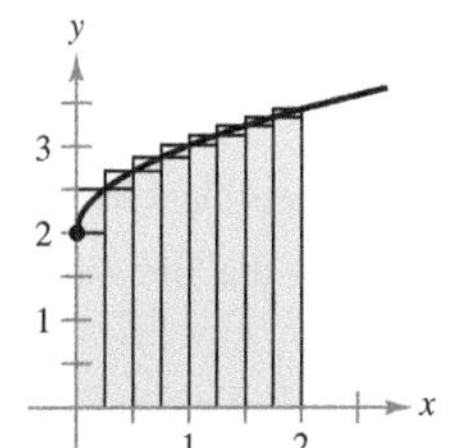

39. $y = \dfrac{1}{x}$

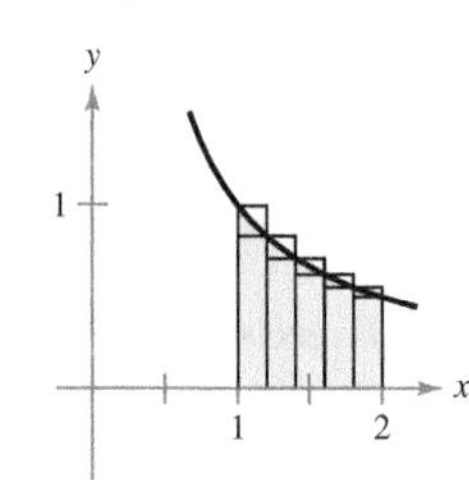

40. $y = \sqrt{1 - x^2}$

Finding Upper and Lower Sums for a Region **In Exercises 41–44, find the upper and lower sums for the region bounded by the graph of the function and the x-axis on the given interval. Leave your answer in terms of n, the number of subintervals.**

Function	Interval
41. $f(x) = 3x$	$[0, 4]$
42. $f(x) = 6 - 2x$	$[1, 2]$
43. $f(x) = 5x^2$	$[0, 1]$
44. $f(x) = 9 - x^2$	$[0, 2]$

45. **Numerical Reasoning** Consider a triangle of area 2 bounded by the graphs of $y = x$, $y = 0$, and $x = 2$.

(a) Sketch the region.

(b) Divide the interval $[0, 2]$ into n subintervals of equal width and show that the endpoints are

$$0 < 1\left(\frac{2}{n}\right) < \cdots < (n-1)\left(\frac{2}{n}\right) < n\left(\frac{2}{n}\right).$$

(c) Show that $s(n) = \sum_{i=1}^{n}\left[(i-1)\left(\frac{2}{n}\right)\right]\left(\frac{2}{n}\right)$.

(d) Show that $S(n) = \sum_{i=1}^{n}\left[i\left(\frac{2}{n}\right)\right]\left(\frac{2}{n}\right)$.

(e) Find $s(n)$ and $S(n)$ for $n = 5, 10, 50$, and 100.

(f) Show that $\lim_{n\to\infty} s(n) = \lim_{n\to\infty} S(n) = 2$.

46. **Numerical Reasoning** Consider a trapezoid of area 4 bounded by the graphs of $y = x$, $y = 0$, $x = 1$, and $x = 3$.

(a) Sketch the region.

(b) Divide the interval $[1, 3]$ into n subintervals of equal width and show that the endpoints are

$$1 < 1 + 1\left(\frac{2}{n}\right) < \cdots < 1 + (n-1)\left(\frac{2}{n}\right) < 1 + n\left(\frac{2}{n}\right).$$

(c) Show that $s(n) = \sum_{i=1}^{n}\left[1 + (i-1)\left(\frac{2}{n}\right)\right]\left(\frac{2}{n}\right)$.

(d) Show that $S(n) = \sum_{i=1}^{n}\left[1 + i\left(\frac{2}{n}\right)\right]\left(\frac{2}{n}\right)$.

(e) Find $s(n)$ and $S(n)$ for $n = 5, 10, 50$, and 100.

(f) Show that $\lim_{n\to\infty} s(n) = \lim_{n\to\infty} S(n) = 4$.

Finding Area by the Limit Definition **In Exercises 47–56, use the limit process to find the area of the region bounded by the graph of the function and the x-axis over the given interval. Sketch the region.**

47. $y = -4x + 5, \quad [0, 1]$

48. $y = 3x - 2, \quad [2, 5]$

49. $y = x^2 + 2, \quad [0, 1]$

50. $y = 5x^2 + 1, \quad [0, 2]$

51. $y = 25 - x^2, \quad [1, 4]$

52. $y = 4 - x^2, \quad [-2, 2]$

53. $y = 27 - x^3, \quad [1, 3]$

54. $y = 2x - x^3, \quad [0, 1]$

55. $y = x^2 - x^3, \quad [-1, 1]$

56. $y = 2x^3 - x^2, \quad [1, 2]$

Finding Area by the Limit Definition **In Exercises 57–62, use the limit process to find the area of the region bounded by the graph of the function and the y-axis over the given y-interval. Sketch the region.**

57. $f(y) = 4y, \quad 0 \le y \le 2$

58. $g(y) = \frac{1}{2}y, \quad 2 \le y \le 4$

59. $f(y) = y^2, \quad 0 \le y \le 5$

60. $y = 3y - y^2, \quad 2 \le y \le 3$

61. $g(y) = 4y^2 - y^3, \quad 1 \le y \le 3$

62. $h(y) = y^3 + 1, \quad 1 \le y \le 2$

Approximating Area with the Midpoint Rule **In Exercises 63–66, use the Midpoint Rule with $n = 4$ to approximate the area of the region bounded by the graph of the function and the x-axis over the given interval.**

63. $f(x) = x^2 + 3, \quad [0, 2]$

64. $f(x) = x^2 + 4x, \quad [0, 4]$

65. $f(x) = \tan x, \quad \left[0, \frac{\pi}{4}\right]$

66. $f(x) = \cos x, \quad \left[0, \frac{\pi}{2}\right]$

EXPLORING CONCEPTS

67. **Approximation** Determine which value best approximates the area of the region bounded by the graph of $f(x) = 4 - x^2$ and the x-axis over the interval $[0, 2]$. Make your selection on the basis of a sketch of the region, not by performing calculations.

(a) -2 (b) 6 (c) 10 (d) 3 (e) 8

68. **Approximation** A function is continuous, nonnegative, concave upward, and decreasing on the interval $[0, a]$. Does using the right endpoints of the subintervals produce an overestimate or an underestimate of the area of the region bounded by the function and the x-axis?

69. **Midpoint Rule** Explain why the Midpoint Rule almost always results in a better area approximation in comparison to the endpoint method.

70. **Midpoint Rule** Does the Midpoint Rule ever give the exact area between a function and the x-axis? Explain.

71. Graphical Reasoning Consider the region bounded by the graphs of $f(x) = 8x/(x + 1)$, $x = 0$, $x = 4$, and $y = 0$, as shown in the figure. To print an enlarged copy of the graph, go to *MathGraphs.com.*

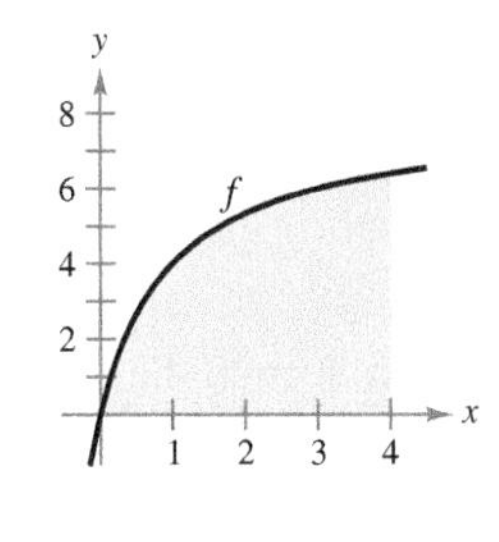

(a) Redraw the figure, and complete and shade the rectangles representing the lower sum when $n = 4$. Find this lower sum.

(b) Redraw the figure, and complete and shade the rectangles representing the upper sum when $n = 4$. Find this upper sum.

(c) Redraw the figure, and complete and shade the rectangles whose heights are determined by the function values at the midpoint of each subinterval when $n = 4$. Find this sum using the Midpoint Rule.

(d) Verify the following formulas for approximating the area of the region using n subintervals of equal width.

Lower sum: $s(n) = \sum_{i=1}^{n} f\left[(i - 1)\frac{4}{n}\right]\left(\frac{4}{n}\right)$

Upper sum: $S(n) = \sum_{i=1}^{n} f\left[(i)\frac{4}{n}\right]\left(\frac{4}{n}\right)$

Midpoint Rule: $M(n) = \sum_{i=1}^{n} f\left[\left(i - \frac{1}{2}\right)\frac{4}{n}\right]\left(\frac{4}{n}\right)$

(e) Use a graphing utility to create a table of values of $s(n)$, $S(n)$, and $M(n)$ for $n = 4, 8, 20, 100$, and 200.

(f) Explain why $s(n)$ increases and $S(n)$ decreases for increasing values of n, as shown in the table in part (e).

72. HOW DO YOU SEE IT? The function shown in the graph below is increasing on the interval $[1, 4]$. The interval will be divided into 12 subintervals.

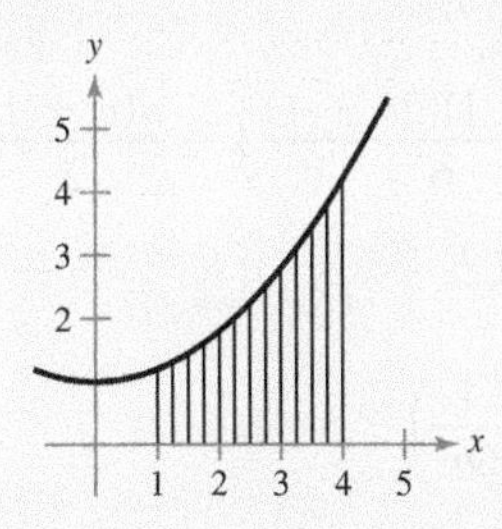

(a) What are the left endpoints of the first and last subintervals?

(b) What are the right endpoints of the first two subintervals?

(c) When using the right endpoints, do the rectangles lie above or below the graph of the function?

(d) What can you conclude about the heights of the rectangles when the function is constant on the given interval?

True or False? **In Exercises 73 and 74, determine whether the statement is true or false. If it is false, explain why or give an example that shows it is false.**

73. The sum of the first n positive integers is $n(n + 1)/2$.

74. If f is continuous and nonnegative on $[a, b]$, then the limits as $n \to \infty$ of its lower sum $s(n)$ and upper sum $S(n)$ both exist and are equal.

75. Writing Use the figure to write a short paragraph explaining why the formula $1 + 2 + \cdots + n = \frac{1}{2}n(n + 1)$ is valid for all positive integers n.

Figure for 75

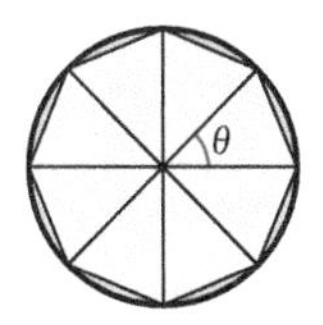

Figure for 76

76. Graphical Reasoning Consider an n-sided regular polygon inscribed in a circle of radius r. Join the vertices of the polygon to the center of the circle, forming n congruent triangles (see figure).

(a) Determine the central angle θ in terms of n.

(b) Show that the area of each triangle is $\frac{1}{2}r^2 \sin \theta$.

(c) Let A_n be the sum of the areas of the n triangles. Find $\lim_{n\to\infty} A_n$.

77. Seating Capacity A teacher places n seats to form the back row of a classroom layout. Each successive row contains two fewer seats than the preceding row. Find a formula for the number of seats used in the layout. (*Hint:* The number of seats in the layout depends on whether n is odd or even.)

78. Proof Prove each formula by mathematical induction. (You may need to review the method of proof by induction from a precalculus text.)

(a) $\sum_{i=1}^{n} 2i = n(n + 1)$ (b) $\sum_{i=1}^{n} i^3 = \frac{n^2(n + 1)^2}{4}$

PUTNAM EXAM CHALLENGE

79. A dart, thrown at random, hits a square target. Assuming that any two parts of the target of equal area are equally likely to be hit, find the probability that the point hit is nearer to the center than to any edge. Write your answer in the form $(a\sqrt{b} + c)/d$, where a, b, c, and d are integers.

4.3 Riemann Sums and Definite Integrals

- **Understand the definition of a Riemann sum.**
- **Evaluate a definite integral using limits and geometric formulas.**
- **Evaluate a definite integral using properties of definite integrals.**

Riemann Sums

In the definition of area given in Section 4.2, the partitions have subintervals of *equal width*. This was done only for computational convenience. The next example shows that it is not necessary to have subintervals of equal width.

EXAMPLE 1 A Partition with Subintervals of Unequal Widths

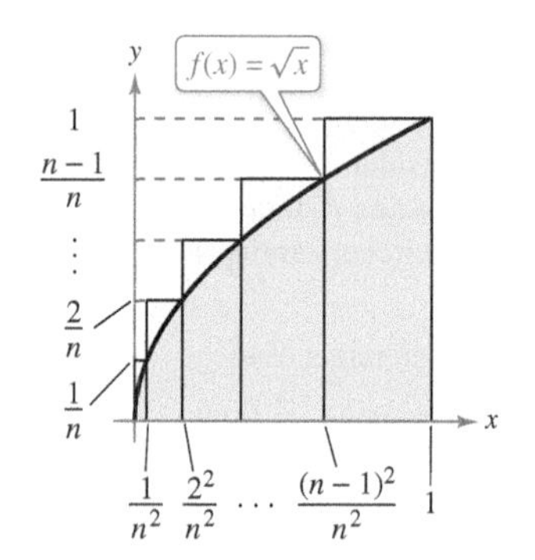

The subintervals do not have equal widths.
Figure 4.18

Consider the region bounded by the graph of $f(x) = \sqrt{x}$ and the x-axis for $0 \le x \le 1$, as shown in Figure 4.18. Evaluate the limit

$$\lim_{n\to\infty} \sum_{i=1}^{n} f(c_i)\,\Delta x_i$$

where c_i is the right endpoint of the partition given by $c_i = i^2/n^2$ and Δx_i is the width of the ith interval.

Solution The width of the ith interval is

$$\begin{aligned}\Delta x_i &= \frac{i^2}{n^2} - \frac{(i-1)^2}{n^2}\\ &= \frac{i^2 - i^2 + 2i - 1}{n^2}\\ &= \frac{2i - 1}{n^2}.\end{aligned}$$

So, the limit is

$$\begin{aligned}\lim_{n\to\infty} \sum_{i=1}^{n} f(c_i)\,\Delta x_i &= \lim_{n\to\infty} \sum_{i=1}^{n} \sqrt{\frac{i^2}{n^2}}\left(\frac{2i-1}{n^2}\right)\\ &= \lim_{n\to\infty} \frac{1}{n^3} \sum_{i=1}^{n} (2i^2 - i)\\ &= \lim_{n\to\infty} \frac{1}{n^3}\left[2\left(\frac{n(n+1)(2n+1)}{6}\right) - \frac{n(n+1)}{2}\right]\\ &= \lim_{n\to\infty} \frac{4n^3 + 3n^2 - n}{6n^3}\\ &= \lim_{n\to\infty} \left(\frac{2}{3} + \frac{1}{2n} - \frac{1}{6n^2}\right)\\ &= \frac{2}{3}.\end{aligned}$$

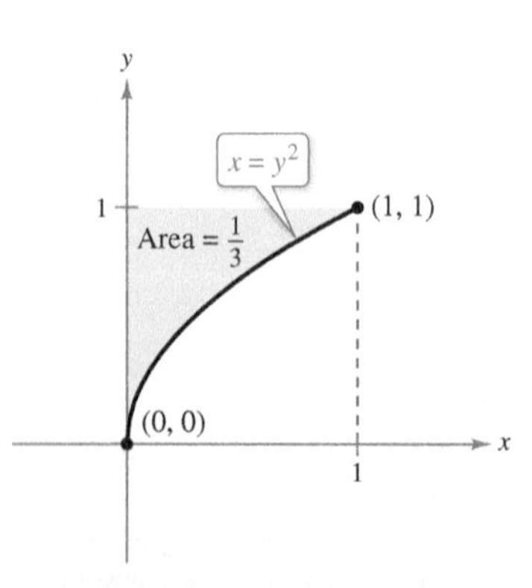

The area of the region bounded by the graph of $x = y^2$ and the y-axis for $0 \le y \le 1$ is $\frac{1}{3}$.
Figure 4.19

From Example 7 in Section 4.2, you know that the region shown in Figure 4.19 has an area of $\frac{1}{3}$. Because the square bounded by $0 \le x \le 1$ and $0 \le y \le 1$ has an area of 1, you can conclude that the area of the region shown in Figure 4.18 has an area of $\frac{2}{3}$. This agrees with the limit found in Example 1, even though that example used a partition having subintervals of unequal widths. The reason this particular partition gave the proper area is that as n increases, the *width of the largest subinterval approaches zero.* This is a key feature of the development of definite integrals.

GEORG FRIEDRICH BERNHARD RIEMANN (1826-1866)

German mathematician Riemann did his most famous work in the areas of non-Euclidean geometry, differential equations, and number theory. It was Riemann's results in physics and mathematics that formed the structure on which Einstein's General Theory of Relativity is based.

See LarsonCalculus.com to read more of this biography.

In Section 4.2, the limit of a sum was used to define the area of a region in the plane. Finding area by this method is only one of *many* applications involving the limit of a sum. A similar approach can be used to determine quantities as diverse as arc lengths, average values, centroids, volumes, work, and surface areas. The next definition is named after Georg Friedrich Bernhard Riemann. Although the definite integral had been defined and used long before Riemann's time, he generalized the concept to cover a broader category of functions.

In the definition of a Riemann sum below, note that the function f has no restrictions other than being defined on the interval $[a, b]$. (In Section 4.2, the function f was assumed to be continuous and nonnegative because you were finding the area under a curve.)

Definition of Riemann Sum

Let f be defined on the closed interval $[a, b]$, and let Δ be a partition of $[a, b]$ given by

$$a = x_0 < x_1 < x_2 < \cdot \cdot \cdot < x_{n-1} < x_n = b$$

where Δx_i is the width of the ith subinterval

$$[x_{i-1}, x_i]. \qquad i\text{th subinterval}$$

If c_i is *any* point in the ith subinterval, then the sum

$$\sum_{i=1}^{n} f(c_i)\,\Delta x_i, \quad x_{i-1} \le c_i \le x_i$$

is called a **Riemann sum** of f for the partition Δ. (The sums in Section 4.2 are examples of Riemann sums, but there are more general Riemann sums than those covered there.)

The width of the largest subinterval of a partition Δ is the **norm** of the partition and is denoted by $\|\Delta\|$. If every subinterval is of equal width, then the partition is **regular** and the norm is denoted by

$$\|\Delta\| = \Delta x = \frac{b - a}{n}. \qquad \text{Regular partition}$$

For a **general partition,** the norm is related to the number of subintervals of $[a, b]$ in the following way.

$$\frac{b - a}{\|\Delta\|} \le n \qquad \text{General partition}$$

So, the number of subintervals in a partition approaches infinity as the norm of the partition approaches 0. That is, $\|\Delta\| \to 0$ implies that $n \to \infty$.

The converse of this statement is not true. For example, let Δ_n be the partition of the interval $[0, 1]$ given by

$$0 < \frac{1}{2^n} < \frac{1}{2^{n-1}} < \cdot \cdot \cdot < \frac{1}{8} < \frac{1}{4} < \frac{1}{2} < 1.$$

As shown in Figure 4.20, for any positive value of n, the norm of the partition Δ_n is $\frac{1}{2}$. So, letting n approach infinity does not force $\|\Delta\|$ to approach 0. In a regular partition, however, the statements

$$\|\Delta\| \to 0 \quad \text{and} \quad n \to \infty$$

are equivalent.

$\|\Delta\| = \frac{1}{2}$

$0 \quad \frac{1}{2^n} \quad \frac{1}{8} \quad \frac{1}{4} \quad \frac{1}{2} \quad 1$

$n \to \infty$ does not imply that $\|\Delta\| \to 0$.
Figure 4.20

Definite Integrals

To define the definite integral, consider the limit

$$\lim_{\|\Delta\|\to 0} \sum_{i=1}^{n} f(c_i)\,\Delta x_i = L.$$

To say that this limit exists means there exists a real number L such that for each $\varepsilon > 0$, there exists a $\delta > 0$ such that for every partition with $\|\Delta\| < \delta$, it follows that

$$\left| L - \sum_{i=1}^{n} f(c_i)\,\Delta x_i \right| < \varepsilon$$

regardless of the choice of c_i in the ith subinterval of each partition Δ.

FOR FURTHER INFORMATION For insight into the history of the definite integral, see the article "The Evolution of Integration" by A. Shenitzer and J. Steprāns in *The American Mathematical Monthly*. To view this article, go to *MathArticles.com*.

Definition of Definite Integral

If f is defined on the closed interval $[a, b]$ and the limit of Riemann sums over partitions Δ

$$\lim_{\|\Delta\|\to 0} \sum_{i=1}^{n} f(c_i)\,\Delta x_i$$

exists (as described above), then f is said to be **integrable** on $[a, b]$ and the limit is denoted by

$$\lim_{\|\Delta\|\to 0} \sum_{i=1}^{n} f(c_i)\,\Delta x_i = \int_a^b f(x)\,dx.$$

The limit is called the **definite integral** of f from a to b. The number a is the **lower limit** of integration, and the number b is the **upper limit** of integration.

REMARK Later in this chapter, you will learn convenient methods for calculating $\int_a^b f(x)\,dx$ for continuous functions. For now, you must use the limit definition.

It is not a coincidence that the notation for definite integrals is similar to that used for indefinite integrals. You will see why in the next section when the Fundamental Theorem of Calculus is introduced. For now, it is important to see that definite integrals and indefinite integrals are different concepts. A definite integral is a *number*, whereas an indefinite integral is a *family of functions.*

Though Riemann sums were defined for functions with very few restrictions, a sufficient condition for a function f to be integrable on $[a, b]$ is that it is continuous on $[a, b]$. A proof of this theorem is beyond the scope of this text.

THEOREM 4.4 Continuity Implies Integrability

If a function f is continuous on the closed interval $[a, b]$, then f is integrable on $[a, b]$. That is, $\int_a^b f(x)\,dx$ exists.

Exploration

The Converse of Theorem 4.4 Is the converse of Theorem 4.4 true? That is, when a function is integrable, does it have to be continuous? Explain your reasoning and give examples.

Describe the relationships among continuity, differentiability, and integrability. Which is the strongest condition? Which is the weakest? Which conditions imply other conditions?

EXAMPLE 2 Evaluating a Definite Integral as a Limit

Evaluate the definite integral $\int_{-2}^{1} 2x\,dx$.

Solution The function $f(x) = 2x$ is integrable on the interval $[-2, 1]$ because it is continuous on $[-2, 1]$. Moreover, the definition of integrability implies that any partition whose norm approaches 0 can be used to determine the limit. For computational convenience, define Δ by subdividing $[-2, 1]$ into n subintervals of equal width

$$\Delta x_i = \Delta x = \frac{b-a}{n} = \frac{3}{n}.$$

Choosing c_i as the right endpoint of each subinterval produces

$$c_i = a + i(\Delta x) = -2 + \frac{3i}{n}.$$

So, the definite integral is

$$\begin{aligned}
\int_{-2}^{1} 2x\,dx &= \lim_{\|\Delta\| \to 0} \sum_{i=1}^{n} f(c_i)\,\Delta x_i \\
&= \lim_{n \to \infty} \sum_{i=1}^{n} f(c_i)\,\Delta x \\
&= \lim_{n \to \infty} \sum_{i=1}^{n} 2\left(-2 + \frac{3i}{n}\right)\left(\frac{3}{n}\right) \\
&= \lim_{n \to \infty} \frac{6}{n} \sum_{i=1}^{n} \left(-2 + \frac{3i}{n}\right) \\
&= \lim_{n \to \infty} \frac{6}{n}\left(-2\sum_{i=1}^{n} 1 + \frac{3}{n}\sum_{i=1}^{n} i\right) \\
&= \lim_{n \to \infty} \frac{6}{n}\left\{-2n + \frac{3}{n}\left[\frac{n(n+1)}{2}\right]\right\} \\
&= \lim_{n \to \infty} \left(-12 + 9 + \frac{9}{n}\right) \\
&= -3.
\end{aligned}$$

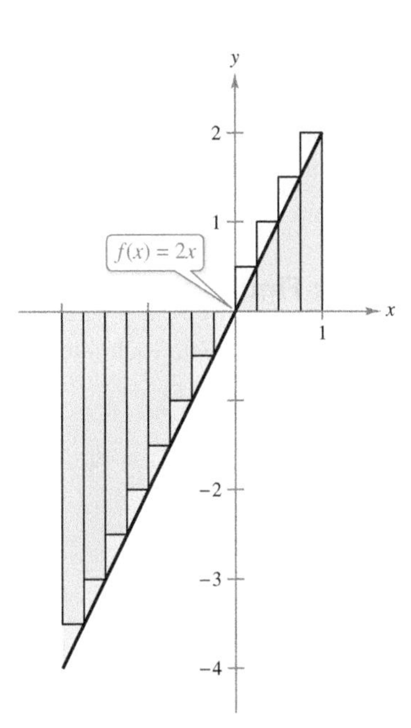

Because the definite integral is negative, it does not represent the area of the region.
Figure 4.21

Because the definite integral in Example 2 is negative, it *does not* represent the area of the region shown in Figure 4.21. Definite integrals can be positive, negative, or zero. For a definite integral to be interpreted as an area (as defined in Section 4.2), the function f must be continuous and nonnegative on $[a, b]$, as stated in the next theorem. The proof of this theorem is straightforward—you simply use the definition of area given in Section 4.2, because it is a Riemann sum.

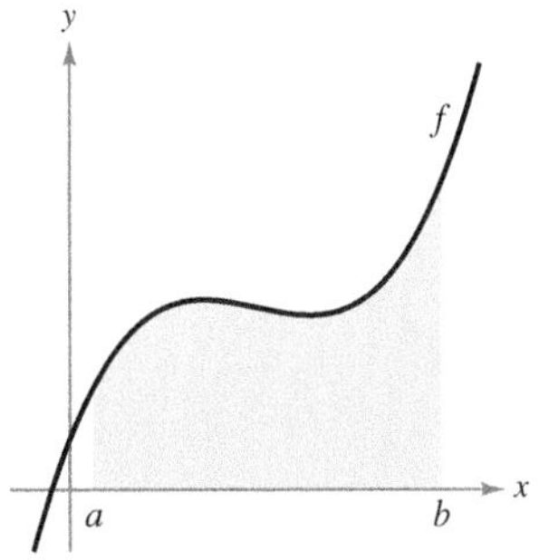

You can use a definite integral to find the area of the region bounded by the graph of f, the x-axis, $x = a$, and $x = b$.
Figure 4.22

THEOREM 4.5 The Definite Integral as the Area of a Region

If f is continuous and nonnegative on the closed interval $[a, b]$, then the area of the region bounded by the graph of f, the x-axis, and the vertical lines $x = a$ and $x = b$ is

$$\text{Area} = \int_a^b f(x)\,dx.$$

(See Figure 4.22.)

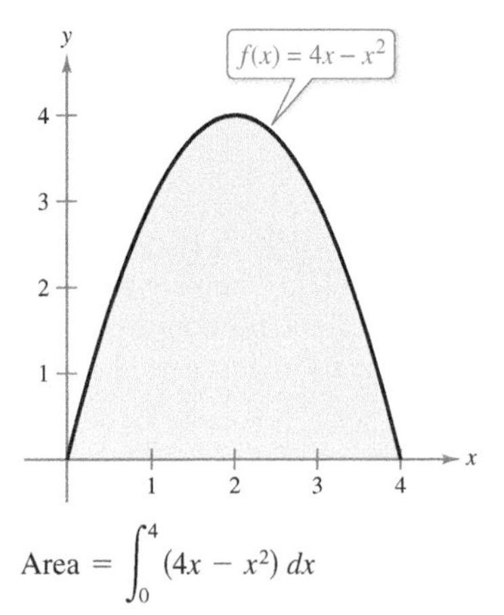

$\text{Area} = \int_0^4 (4x - x^2)\,dx$

Figure 4.23

As an example of Theorem 4.5, consider the region bounded by the graph of

$$f(x) = 4x - x^2$$

and the x-axis, as shown in Figure 4.23. Because f is continuous and nonnegative on the closed interval $[0, 4]$, the area of the region is

$$\text{Area} = \int_0^4 (4x - x^2)\,dx.$$

A straightforward technique for evaluating a definite integral such as this will be discussed in Section 4.4. For now, however, you can evaluate a definite integral in two ways—you can use the limit definition *or* you can check to see whether the definite integral represents the area of a common geometric region, such as a rectangle, triangle, or semicircle.

EXAMPLE 3 Areas of Common Geometric Figures

Sketch the region corresponding to each definite integral. Then evaluate each integral using a geometric formula.

a. $\int_1^3 4\,dx$ **b.** $\int_0^3 (x + 2)\,dx$ **c.** $\int_{-2}^2 \sqrt{4 - x^2}\,dx$

Solution A sketch of each region is shown in Figure 4.24.

a. This region is a rectangle of height 4 and width 2.

$$\int_1^3 4\,dx = (\text{Area of rectangle}) = 4(2) = 8$$

b. This region is a trapezoid with an altitude of 3 and parallel bases of lengths 2 and 5. The formula for the area of a trapezoid is $\frac{1}{2}h(b_1 + b_2)$.

$$\int_0^3 (x + 2)\,dx = (\text{Area of trapezoid}) = \frac{1}{2}(3)(2 + 5) = \frac{21}{2}$$

c. This region is a semicircle of radius 2. The formula for the area of a semicircle is $\frac{1}{2}\pi r^2$.

$$\int_{-2}^2 \sqrt{4 - x^2}\,dx = (\text{Area of semicircle}) = \frac{1}{2}\pi(2^2) = 2\pi$$

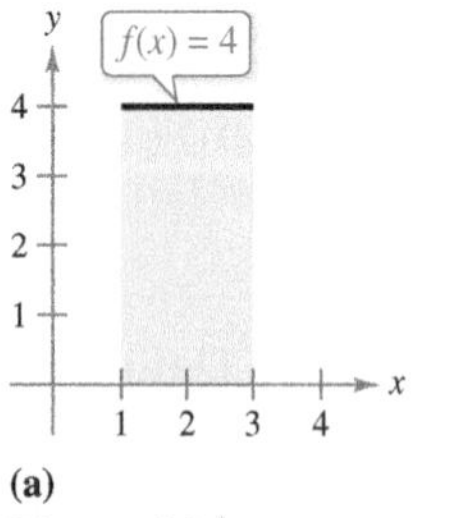

(a)

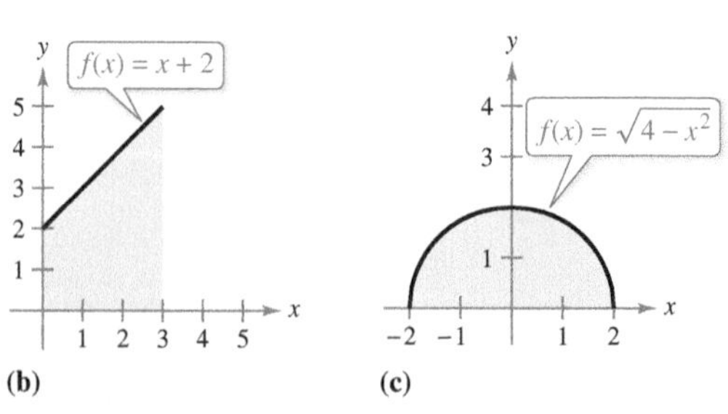

(b) **(c)**

Figure 4.24

The variable of integration in a definite integral is sometimes called a *dummy variable* because it can be replaced by any other variable without changing the value of the integral. For instance, the definite integrals

$$\int_0^3 (x + 2)\,dx \quad \text{and} \quad \int_0^3 (t + 2)\,dt$$

have the same value.

Properties of Definite Integrals

The definition of the definite integral of f on the interval $[a, b]$ specifies that $a < b$. Now, however, it is convenient to extend the definition to cover cases in which $a = b$ or $a > b$. Geometrically, the next two definitions seem reasonable. For instance, it makes sense to define the area of a region of zero width and finite height to be 0.

> **Definitions of Two Special Definite Integrals**
>
> **1.** If f is defined at $x = a$, then $\int_a^a f(x)\,dx = 0$.
>
> **2.** If f is integrable on $[a, b]$, then $\int_b^a f(x)\,dx = -\int_a^b f(x)\,dx$.

EXAMPLE 4 Evaluating Definite Integrals

See LarsonCalculus.com for an interactive version of this type of example.

Evaluate each definite integral.

a. $\int_\pi^\pi \sin x\,dx$ **b.** $\int_3^0 (x + 2)\,dx$

Solution

a. Because the sine function is defined at $x = \pi$, and the upper and lower limits of integration are equal, you can write

$$\int_\pi^\pi \sin x\,dx = 0.$$

b. The integral $\int_3^0 (x + 2)\,dx$ is the same as that given in Example 3(b) except that the upper and lower limits are interchanged. Because the integral in Example 3(b) has a value of $\frac{21}{2}$, you can write

$$\int_3^0 (x + 2)\,dx = -\int_0^3 (x + 2)\,dx = -\frac{21}{2}.$$

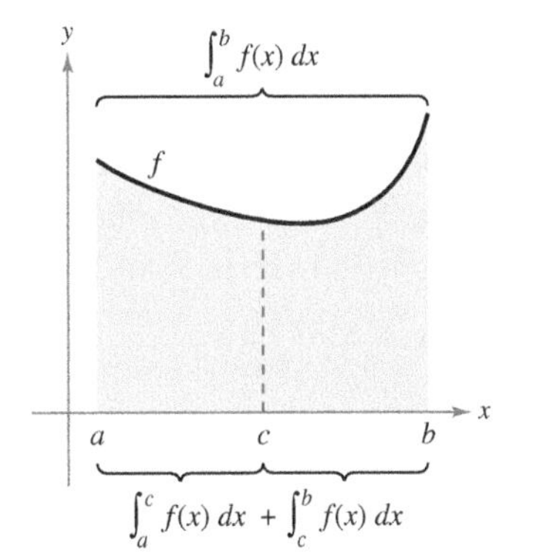

Figure 4.25

In Figure 4.25, the larger region can be divided at $x = c$ into two subregions whose intersection is a line segment. Because the line segment has zero area, it follows that the area of the larger region is equal to the sum of the areas of the two smaller regions.

> **THEOREM 4.6 Additive Interval Property**
>
> If f is integrable on the three closed intervals determined by a, b, and c, then
>
> $$\int_a^b f(x)\,dx = \int_a^c f(x)\,dx + \int_c^b f(x)\,dx.$$ See Figure 4.25.

EXAMPLE 5 Using the Additive Interval Property

$$\int_{-1}^1 |x|\,dx = \int_{-1}^0 -x\,dx + \int_0^1 x\,dx$$ Theorem 4.6

$$= \frac{1}{2} + \frac{1}{2}$$ Area of a triangle

$$= 1$$

Because the definite integral is defined as the limit of a sum, it inherits the properties of summation given at the top of page 259.

THEOREM 4.7 Properties of Definite Integrals

If f and g are integrable on $[a, b]$ and k is a constant, then the functions kf and $f \pm g$ are integrable on $[a, b]$, and

1. $\displaystyle\int_a^b kf(x)\,dx = k\int_a^b f(x)\,dx$

2. $\displaystyle\int_a^b [f(x) \pm g(x)]\,dx = \int_a^b f(x)\,dx \pm \int_a^b g(x)\,dx.$

REMARK Property 2 of Theorem 4.7 can be extended to cover any finite number of functions (see Example 6).

EXAMPLE 6 Evaluation of a Definite Integral

Evaluate $\displaystyle\int_1^3 (-x^2 + 4x - 3)\,dx$ using each of the following values.

$$\int_1^3 x^2\,dx = \frac{26}{3}, \qquad \int_1^3 x\,dx = 4, \qquad \int_1^3 dx = 2$$

Solution

$$\begin{aligned}\int_1^3 (-x^2 + 4x - 3)\,dx &= \int_1^3 (-x^2)\,dx + \int_1^3 4x\,dx + \int_1^3 (-3)\,dx \\ &= -\int_1^3 x^2\,dx + 4\int_1^3 x\,dx - 3\int_1^3 dx \\ &= -\left(\frac{26}{3}\right) + 4(4) - 3(2) \\ &= \frac{4}{3}\end{aligned}$$

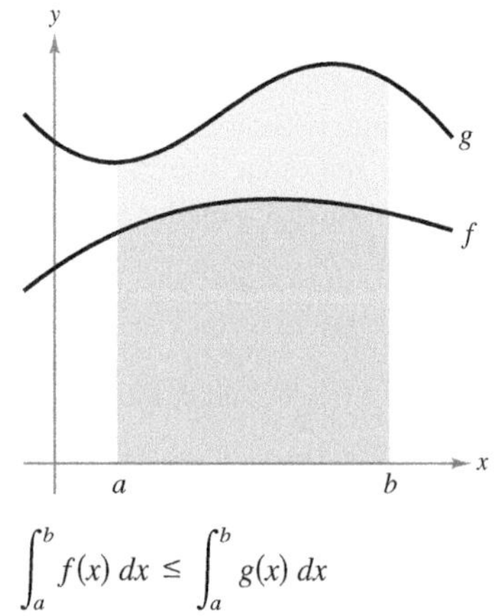

$\displaystyle\int_a^b f(x)\,dx \le \int_a^b g(x)\,dx$

Figure 4.26

If f and g are continuous on the closed interval $[a, b]$ and $0 \le f(x) \le g(x)$ for $a \le x \le b$, then the following properties are true. First, the area of the region bounded by the graph of f and the x-axis (between a and b) must be nonnegative. Second, this area must be less than or equal to the area of the region bounded by the graph of g and the x-axis (between a and b), as shown in Figure 4.26. These two properties are generalized in Theorem 4.8.

THEOREM 4.8 Preservation of Inequality

1. If f is integrable and nonnegative on the closed interval $[a, b]$, then

$$0 \le \int_a^b f(x)\,dx.$$

2. If f and g are integrable on the closed interval $[a, b]$ and $f(x) \le g(x)$ for every x in $[a, b]$, then

$$\int_a^b f(x)\,dx \le \int_a^b g(x)\,dx.$$

A proof of this theorem is given in Appendix A.

4.3 Exercises

See CalcChat.com for tutorial help and worked-out solutions to odd-numbered exercises.

CONCEPT CHECK

1. **Riemann Sum** What does a Riemann sum represent?
2. **Definite Integral** Explain how to find the area of a region using a definite integral in your own words.

Evaluating a Limit In Exercises 3 and 4, use Example 1 as a model to evaluate the limit

$$\lim_{n\to\infty} \sum_{i=1}^{n} f(c_i)\,\Delta x_i$$

over the region bounded by the graphs of the equations.

3. $f(x) = \sqrt{x}, \quad y = 0, \quad x = 0, \quad x = 3 \quad \left(\text{Hint: Let } c_i = \frac{3i^2}{n^2}.\right)$

4. $f(x) = \sqrt[3]{x}, \quad y = 0, \quad x = 0, \quad x = 1 \quad \left(\text{Hint: Let } c_i = \frac{i^3}{n^3}.\right)$

Evaluating a Definite Integral as a Limit In Exercises 5–10, evaluate the definite integral by the limit definition.

5. $\int_2^6 8\,dx$
6. $\int_{-2}^3 x\,dx$
7. $\int_{-1}^1 x^3\,dx$
8. $\int_1^4 4x^2\,dx$
9. $\int_1^2 (x^2 + 1)\,dx$
10. $\int_{-2}^1 (2x^2 + 3)\,dx$

Writing a Limit as a Definite Integral In Exercises 11 and 12, write the limit as a definite integral on the given interval, where c_i is any point in the ith subinterval.

	Limit	Interval
11.	$\lim_{\|\Delta\|\to 0} \sum_{i=1}^{n} (3c_i + 10)\,\Delta x_i$	$[-1, 5]$
12.	$\lim_{\|\Delta\|\to 0} \sum_{i=1}^{n} \sqrt{c_i^2 + 4}\,\Delta x_i$	$[0, 3]$

Writing a Definite Integral In Exercises 13–22, write a definite integral that represents the area of the region. (Do not evaluate the integral.)

13. $f(x) = 5$

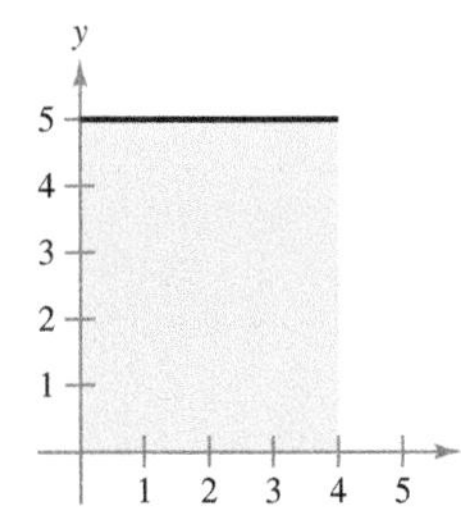

14. $f(x) = 6 - 3x$

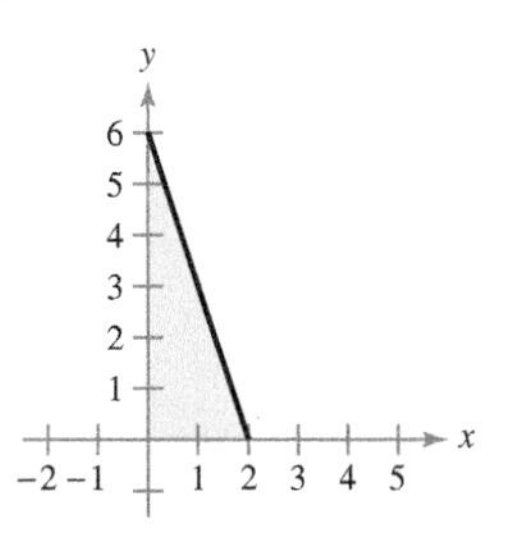

15. $f(x) = 4 - |x|$

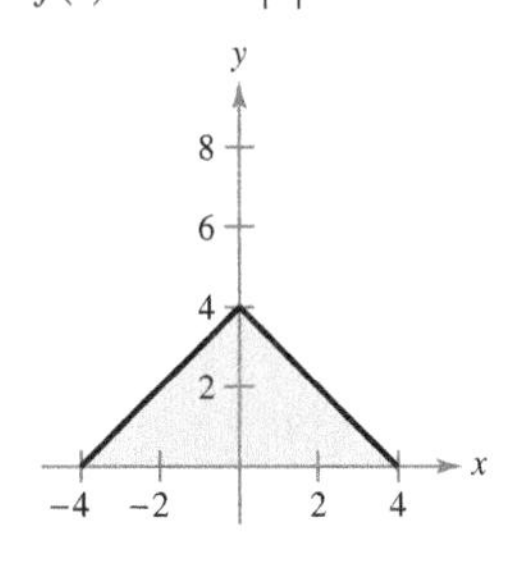

16. $f(x) = x^2$

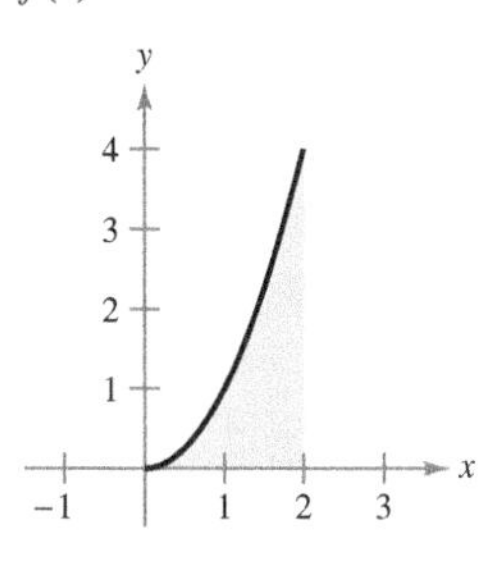

17. $f(x) = 25 - x^2$

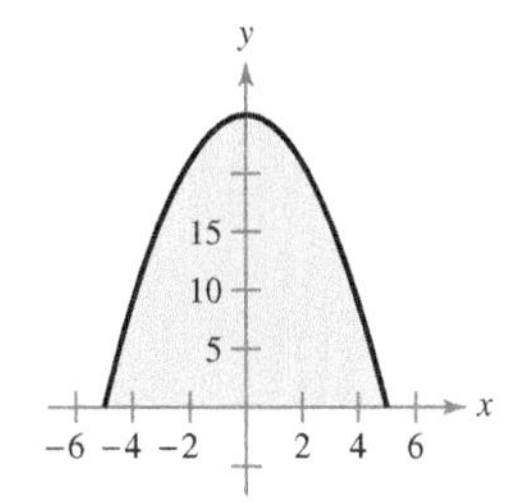

18. $f(x) = \dfrac{4}{x^2 + 2}$

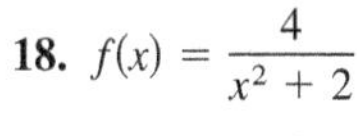

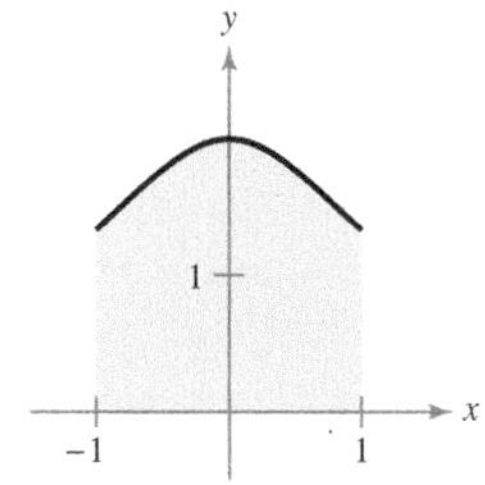

19. $f(x) = \cos x$

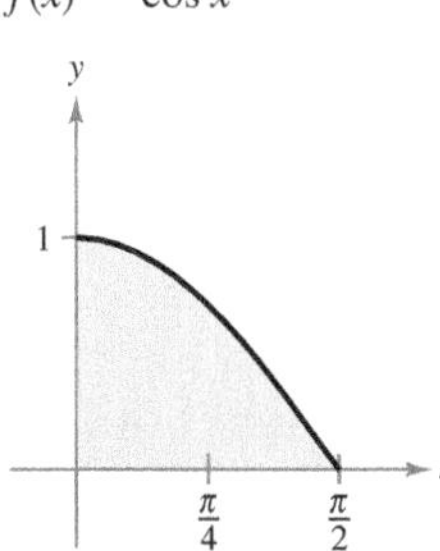

20. $f(x) = \tan x$

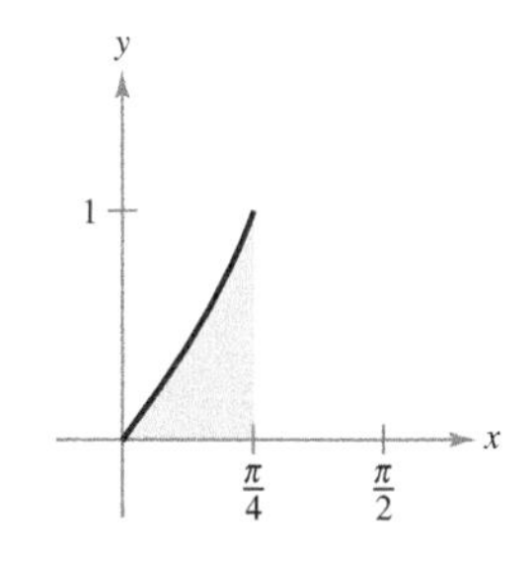

21. $g(y) = y^3$

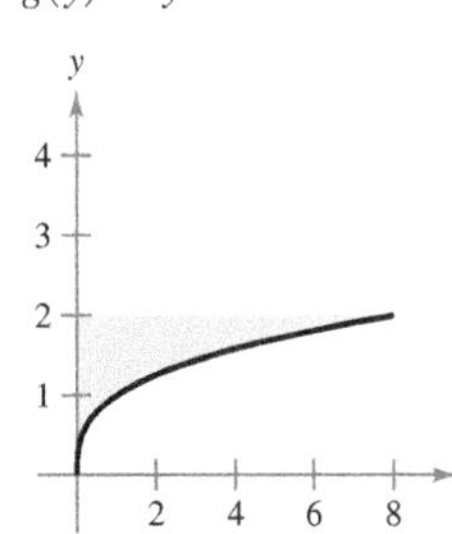

22. $f(y) = (y - 2)^2$

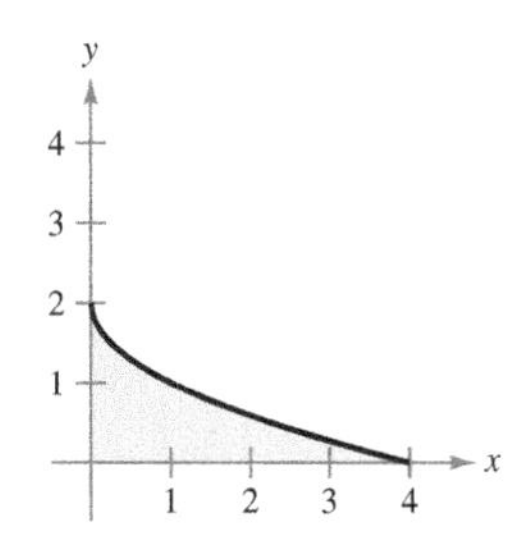

Evaluating a Definite Integral Using a Geometric Formula In Exercises 23–32, sketch the region whose area is given by the definite integral. Then use a geometric formula to evaluate the integral ($a > 0, r > 0$).

23. $\int_0^3 4\,dx$
24. $\int_{-3}^4 9\,dx$
25. $\int_0^4 x\,dx$
26. $\int_0^8 \frac{x}{4}\,dx$

27. $\int_0^2 (3x + 4)\,dx$

28. $\int_0^3 (8 - 2x)\,dx$

29. $\int_{-1}^1 (1 - |x|)\,dx$

30. $\int_{-a}^a (a - |x|)\,dx$

31. $\int_{-7}^7 \sqrt{49 - x^2}\,dx$

32. $\int_{-r}^r \sqrt{r^2 - x^2}\,dx$

Using Properties of Definite Integrals **In Exercises 33–40, evaluate the definite integral using the values below.**

$$\int_2^6 x^3\,dx = 320, \quad \int_2^6 x\,dx = 16, \quad \int_2^6 dx = 4$$

33. $\int_6^2 x^3\,dx$

34. $\int_2^2 x\,dx$

35. $\int_2^6 \frac{1}{4}x^3\,dx$

36. $\int_2^6 -3x\,dx$

37. $\int_2^6 (x - 14)\,dx$

38. $\int_2^6 \left(6x - \frac{1}{8}x^3\right)dx$

39. $\int_2^6 (2x^3 - 10x + 7)\,dx$

40. $\int_2^6 (21 - 5x - x^3)\,dx$

41. Using Properties of Definite Integrals Given

$$\int_0^5 f(x)\,dx = 10 \quad \text{and} \quad \int_5^7 f(x)\,dx = 3$$

evaluate

(a) $\int_0^7 f(x)\,dx.$ (b) $\int_5^0 f(x)\,dx.$

(c) $\int_5^5 f(x)\,dx.$ (d) $\int_0^5 3f(x)\,dx.$

42. Using Properties of Definite Integrals Given

$$\int_0^3 f(x)\,dx = 4 \quad \text{and} \quad \int_3^6 f(x)\,dx = -1$$

evaluate

(a) $\int_0^6 f(x)\,dx.$ (b) $\int_6^3 f(x)\,dx.$

(c) $\int_3^3 f(x)\,dx.$ (d) $\int_3^6 -5f(x)\,dx.$

43. Using Properties of Definite Integrals Given

$$\int_2^6 f(x)\,dx = 10 \quad \text{and} \quad \int_2^6 g(x)\,dx = -2$$

evaluate

(a) $\int_2^6 [f(x) + g(x)]\,dx.$

(b) $\int_2^6 [g(x) - f(x)]\,dx.$

(c) $\int_2^6 2g(x)\,dx.$

(d) $\int_2^6 3f(x)\,dx.$

44. Using Properties of Definite Integrals Given

$$\int_{-1}^1 f(x)\,dx = 0 \quad \text{and} \quad \int_0^1 f(x)\,dx = 5$$

evaluate

(a) $\int_{-1}^0 f(x)\,dx.$ (b) $\int_0^1 f(x)\,dx - \int_{-1}^0 f(x)\,dx.$

(c) $\int_{-1}^1 3f(x)\,dx.$ (d) $\int_0^1 3f(x)\,dx.$

45. Estimating a Definite Integral Use the table of values to find lower and upper estimates of

$$\int_0^{10} f(x)\,dx.$$

Assume that f is a decreasing function.

x	0	2	4	6	8	10
$f(x)$	32	24	12	-4	-20	-36

46. Estimating a Definite Integral Use the table of values to estimate

$$\int_0^6 f(x)\,dx.$$

Use three equal subintervals and the (a) left endpoints, (b) right endpoints, and (c) midpoints. When f is an increasing function, how does each estimate compare with the actual value? Explain your reasoning.

x	0	1	2	3	4	5	6
$f(x)$	-6	0	8	18	30	50	80

47. Think About It The graph of f consists of line segments and a semicircle, as shown in the figure. Evaluate each definite integral by using geometric formulas.

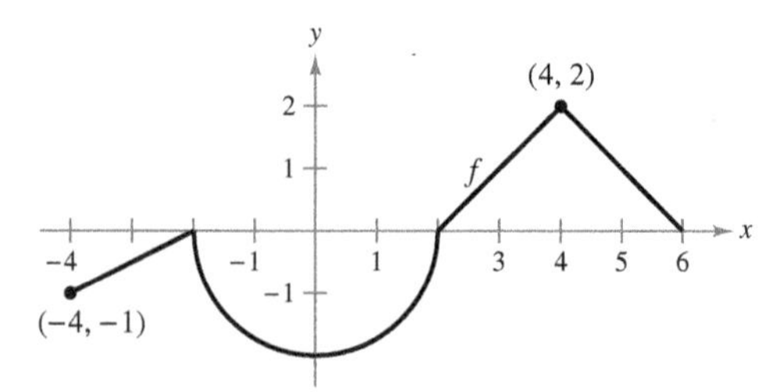

(a) $\int_0^2 f(x)\,dx$ (b) $\int_2^6 f(x)\,dx$

(c) $\int_{-4}^2 f(x)\,dx$

(d) $\int_{-4}^6 f(x)\,dx$

(e) $\int_{-4}^6 |f(x)|\,dx$

(f) $\int_{-4}^6 [f(x) + 2]\,dx$

48. Think About It The graph of f consists of line segments, as shown in the figure. Evaluate each definite integral by using geometric formulas.

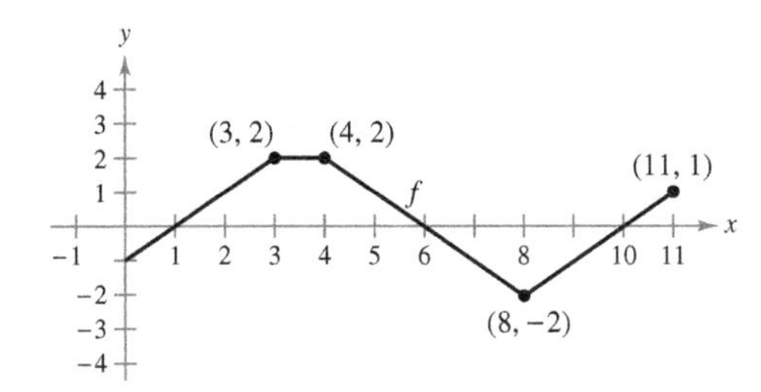

(a) $\int_0^1 -f(x)\,dx$ (b) $\int_3^4 3f(x)\,dx$

(c) $\int_0^7 f(x)\,dx$ (d) $\int_5^{11} f(x)\,dx$

(e) $\int_0^{11} f(x)\,dx$ (f) $\int_4^{10} f(x)\,dx$

49. Think About It Consider a function f that is continuous on the interval $[-5, 5]$ and for which

$$\int_0^5 f(x)\,dx = 4.$$

Evaluate each integral.

(a) $\int_0^5 [f(x) + 2]\,dx$ (b) $\int_{-2}^3 f(x+2)\,dx$

(c) $\int_{-5}^5 f(x)\,dx$, f is even (d) $\int_{-5}^5 f(x)\,dx$, f is odd

50. HOW DO YOU SEE IT? Use the figure to fill in the blank with the symbol <, >, or =. Explain your reasoning.

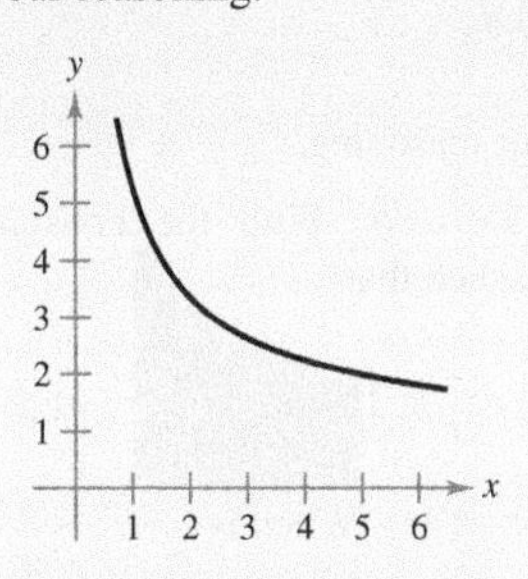

(a) The interval $[1, 5]$ is partitioned into n subintervals of equal width Δx, and x_i is the left endpoint of the ith subinterval.

$$\sum_{i=1}^n f(x_i)\,\Delta x \quad \blacksquare \quad \int_1^5 f(x)\,dx$$

(b) The interval $[1, 5]$ is partitioned into n subintervals of equal width Δx, and x_i is the right endpoint of the ith subinterval.

$$\sum_{i=1}^n f(x_i)\,\Delta x \quad \blacksquare \quad \int_1^5 f(x)\,dx$$

51. Think About It A function f is defined below. Use geometric formulas to find $\int_0^8 f(x)\,dx$.

$$f(x) = \begin{cases} 4, & x < 4 \\ x, & x \ge 4 \end{cases}$$

52. Think About It A function f is defined below. Use geometric formulas to find $\int_0^{12} f(x)\,dx$.

$$f(x) = \begin{cases} 6, & x > 6 \\ -\frac{1}{2}x + 9, & x \le 6 \end{cases}$$

EXPLORING CONCEPTS

Approximation In Exercises 53 and 54, determine which value best approximates the definite integral. Make your selection on the basis of a sketch.

53. $\int_0^4 \sqrt{x}\,dx$

(a) 5 (b) -3 (c) 10 (d) 2 (e) 8

54. $\int_0^{1/2} 4\cos \pi x\,dx$

(a) 4 (b) $\frac{4}{3}$ (c) 16 (d) 2π (e) -6

55. Verifying a Rule Use a graph to explain why

$$\int_a^a f(x)\,dx = 0$$

if f is defined at $x = a$.

56. Verifying a Property Use a graph to explain why

$$\int_a^b kf(x)\,dx = k\int_a^b f(x)\,dx$$

if f is integrable on $[a, b]$ and k is a constant.

57. Using Different Methods Describe two ways to evaluate

$$\int_{-1}^3 (x+2)\,dx.$$

Verify that each method gives the same result.

58. Finding a Function Give an example of a function that is integrable on the interval $[-1, 1]$ but not continuous on $[-1, 1]$.

Finding Values In Exercises 59–62, find possible values of a and b that make the statement true. If possible, use a graph to support your answer. (There may be more than one correct answer.)

59. $\int_{-2}^1 f(x)\,dx + \int_1^5 f(x)\,dx = \int_a^b f(x)\,dx$

60. $\int_{-3}^3 f(x)\,dx + \int_3^6 f(x)\,dx - \int_a^b f(x)\,dx = \int_{-1}^6 f(x)\,dx$

61. $\int_a^b \sin x\,dx < 0$

62. $\int_a^b \cos x\,dx = 0$

True or False? **In Exercises 63–68, determine whether the statement is true or false. If it is false, explain why or give an example that shows it is false.**

63. $\displaystyle\int_a^b [f(x) + g(x)]\,dx = \int_a^b f(x)\,dx + \int_a^b g(x)\,dx$

64. $\displaystyle\int_a^b f(x)g(x)\,dx = \left[\int_a^b f(x)\,dx\right]\left[\int_a^b g(x)\,dx\right]$

65. If the norm of a partition approaches zero, then the number of subintervals approaches infinity.

66. If f is increasing on $[a, b]$, then the minimum value of f on $[a, b]$ is $f(a)$.

67. The value of

$$\int_a^b f(x)\,dx$$

must be positive.

68. The value of

$$\int_2^2 \sin x^2\,dx$$

is 0.

69. Finding a Riemann Sum Find the Riemann sum for $f(x) = x^2 + 3x$ over the interval $[0, 8]$, where

$x_0 = 0,\quad x_1 = 1,\quad x_2 = 3,\quad x_3 = 7,\quad \text{and}\quad x_4 = 8$

and where

$c_1 = 1,\quad c_2 = 2,\quad c_3 = 5,\quad \text{and}\quad c_4 = 8.$

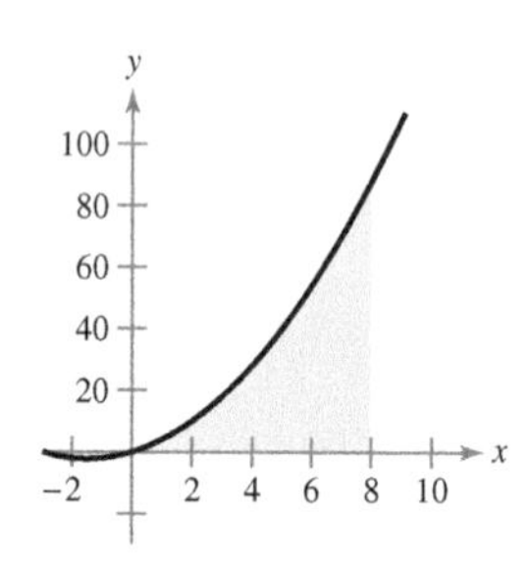

70. Finding a Riemann Sum Find the Riemann sum for $f(x) = \sin x$ over the interval $[0, 2\pi]$, where

$x_0 = 0,\quad x_1 = \frac{\pi}{4},\quad x_2 = \frac{\pi}{3},\quad x_3 = \pi,\quad \text{and}\quad x_4 = 2\pi$

and where

$c_1 = \frac{\pi}{6},\quad c_2 = \frac{\pi}{3},\quad c_3 = \frac{2\pi}{3},\quad \text{and}\quad c_4 = \frac{3\pi}{2}.$

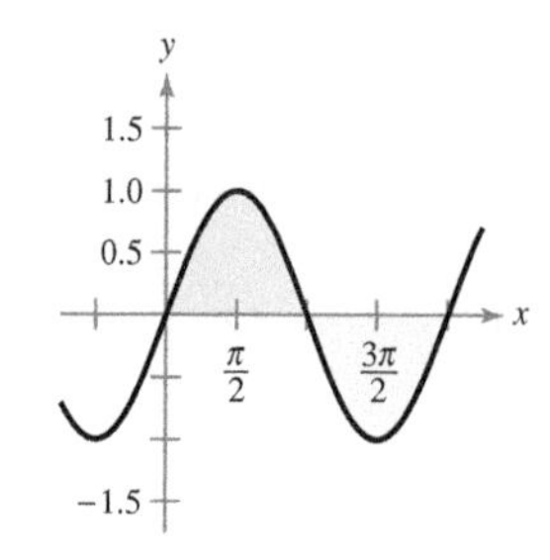

71. Proof Prove that $\displaystyle\int_a^b x\,dx = \frac{b^2 - a^2}{2}.$

72. Proof Prove that $\displaystyle\int_a^b x^2\,dx = \frac{b^3 - a^3}{3}.$

73. Think About It Determine whether the Dirichlet function

$$f(x) = \begin{cases} 1, & x \text{ is rational} \\ 0, & x \text{ is irrational} \end{cases}$$

is integrable on the interval $[0, 1]$. Explain.

74. Finding a Definite Integral The function

$$f(x) = \begin{cases} 0, & x = 0 \\ \frac{1}{x}, & 0 < x \le 1 \end{cases}$$

is defined on $[0, 1]$, as shown in the figure. Show that

$$\int_0^1 f(x)\,dx$$

does not exist. Does this contradict Theorem 4.4? Why or why not?

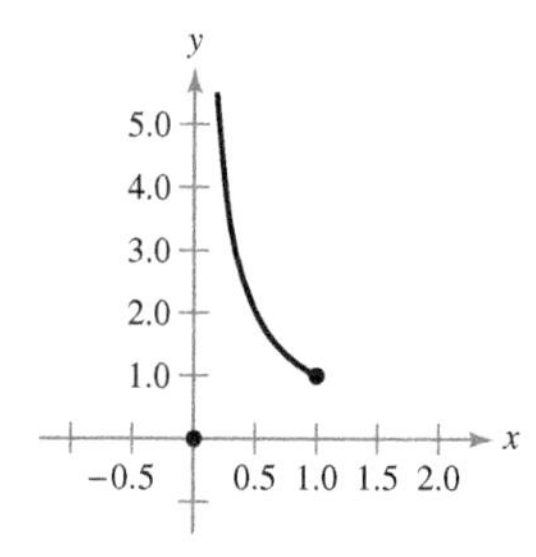

75. Finding Values Find the constants a and b that maximize the value of

$$\int_a^b (1 - x^2)\,dx.$$

Explain your reasoning.

76. Finding Values Find the constants a and b, where $a < 4 < b$, such that

$$\left|\int_a^b (x - 4)\,dx\right| = 16 \text{ and } \int_a^b |x - 4|\,dx = 20.$$

77. Think About It When is

$$\int_a^b f(x)\,dx = \int_a^b |f(x)|\,dx?$$

Explain.

78. Step Function Evaluate, if possible, the integral

$$\int_0^2 [\![x]\!]\,dx.$$

79. Using a Riemann Sum Determine

$$\lim_{n\to\infty} \frac{1}{n^3}(1^2 + 2^2 + 3^2 + \cdots + n^2)$$

by using an appropriate Riemann sum.

4.4 The Fundamental Theorem of Calculus

- **Evaluate a definite integral using the Fundamental Theorem of Calculus.**
- **Understand and use the Mean Value Theorem for Integrals.**
- **Find the average value of a function over a closed interval.**
- **Understand and use the Second Fundamental Theorem of Calculus.**
- **Understand and use the Net Change Theorem.**

The Fundamental Theorem of Calculus

You have now been introduced to the two major branches of calculus: differential calculus (introduced with the tangent line problem) and integral calculus (introduced with the area problem). So far, these two problems might seem unrelated—but there is a very close connection. The connection was discovered independently by Isaac Newton and Gottfried Leibniz and is stated in the **Fundamental Theorem of Calculus.**

Informally, the theorem states that differentiation and (definite) integration are inverse operations, in the same sense that division and multiplication are inverse operations. To see how Newton and Leibniz might have anticipated this relationship, consider the approximations shown in Figure 4.27. The slope of the tangent line was defined using the *quotient* $\Delta y/\Delta x$ (the slope of the secant line). Similarly, the area of a region under a curve was defined using the *product* $\Delta y\Delta x$ (the area of a rectangle). So, at least in the primitive approximation stage, the operations of differentiation and definite integration appear to have an inverse relationship in the same sense that division and multiplication are inverse operations. The Fundamental Theorem of Calculus states that the limit processes (used to define the derivative and definite integral) preserve this inverse relationship.

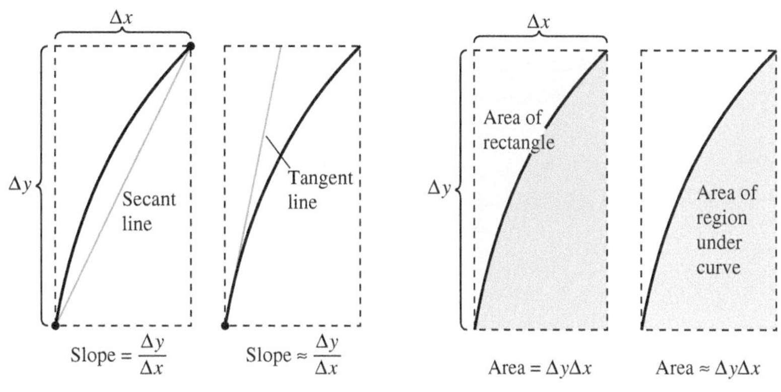

(a) Differentiation (b) Definite integration

Differentiation and definite integration have an "inverse" relationship.

Figure 4.27

> **ANTIDIFFERENTIATION AND DEFINITE INTEGRATION**
>
> Throughout this chapter, you have been using the integral sign to denote an antiderivative (a family of functions) and a definite integral (a number).
>
> Antidifferentiation: $\int f(x)\,dx$ Definite integration: $\int_a^b f(x)\,dx$
>
> The use of the same symbol for both operations makes it appear that they are related. In the early work with calculus, however, it was not known that the two operations were related. The symbol $\int$ was first applied to the definite integral by Leibniz and was derived from the letter S. (Leibniz calculated area as an infinite sum, thus, the letter S.)

THEOREM 4.9 The Fundamental Theorem of Calculus

If a function f is continuous on the closed interval $[a, b]$ and F is an antiderivative of f on the interval $[a, b]$, then

$$\int_a^b f(x)\,dx = F(b) - F(a).$$

Proof The key to the proof is writing the difference $F(b) - F(a)$ in a convenient form. Let Δ be any partition of $[a, b]$.

$$a = x_0 < x_1 < x_2 < \cdot \cdot \cdot < x_{n-1} < x_n = b$$

By pairwise subtraction and addition of like terms, you can write

$$\begin{aligned} F(b) - F(a) &= F(x_n) - F(x_{n-1}) + F(x_{n-1}) - \cdot \cdot \cdot - F(x_1) + F(x_1) - F(x_0) \\ &= \sum_{i=1}^{n} [F(x_i) - F(x_{i-1})]. \end{aligned}$$

By the Mean Value Theorem, you know that there exists a number c_i in the ith subinterval such that

$$F'(c_i) = \frac{F(x_i) - F(x_{i-1})}{x_i - x_{i-1}}.$$

Because $F'(c_i) = f(c_i)$, you can let $\Delta x_i = x_i - x_{i-1}$ and obtain

$$F(b) - F(a) = \sum_{i=1}^{n} f(c_i)\,\Delta x_i.$$

This important equation tells you that by repeatedly applying the Mean Value Theorem, you can always find a collection of c_i's such that the *constant* $F(b) - F(a)$ is a Riemann sum of f on $[a, b]$ for any partition. Theorem 4.4 guarantees that the limit of Riemann sums over the partition with $\|\Delta\| \to 0$ exists. So, taking the limit (as $\|\Delta\| \to 0$) produces

$$F(b) - F(a) = \int_a^b f(x)\,dx.$$

GUIDELINES FOR USING THE FUNDAMENTAL THEOREM OF CALCULUS

1. *Provided you can find* an antiderivative of f, you now have a way to evaluate a definite integral without having to use the limit of a sum.
2. When applying the Fundamental Theorem of Calculus, the notation shown below is convenient.

$$\int_a^b f(x)\,dx = F(x)\Big]_a^b = F(b) - F(a)$$

For instance, to evaluate $\int_1^3 x^3\,dx$, you can write

$$\int_1^3 x^3\,dx = \frac{x^4}{4}\Big]_1^3 = \frac{3^4}{4} - \frac{1^4}{4} = \frac{81}{4} - \frac{1}{4} = 20.$$

3. It is not necessary to include a constant of integration C in the antiderivative.

$$\int_a^b f(x)\,dx = \Big[F(x) + C\Big]_a^b = [F(b) + C] - [F(a) + C] = F(b) - F(a)$$

EXAMPLE 1 Evaluating a Definite Integral

See LarsonCalculus.com for an interactive version of this type of example.

Evaluate each definite integral.

a. $\int_1^2 (x^2 - 3)\,dx$ **b.** $\int_1^4 3\sqrt{x}\,dx$ **c.** $\int_0^{\pi/4} \sec^2 x\,dx$

Solution

a. $\int_1^2 (x^2 - 3)\,dx = \left[\frac{x^3}{3} - 3x\right]_1^2 = \left(\frac{8}{3} - 6\right) - \left(\frac{1}{3} - 3\right) = -\frac{2}{3}$

b. $\int_1^4 3\sqrt{x}\,dx = 3\int_1^4 x^{1/2}\,dx = 3\left[\frac{x^{3/2}}{3/2}\right]_1^4 = 2(4)^{3/2} - 2(1)^{3/2} = 14$

c. $\int_0^{\pi/4} \sec^2 x\,dx = \tan x\Big]_0^{\pi/4} = 1 - 0 = 1$

EXAMPLE 2 A Definite Integral Involving Absolute Value

Evaluate $\int_0^2 |2x - 1|\,dx.$

Solution Using Figure 4.28 and the definition of absolute value, you can rewrite the integrand as shown.

$$|2x - 1| = \begin{cases} -(2x - 1), & x < \frac{1}{2} \\ 2x - 1, & x \geq \frac{1}{2} \end{cases}$$

From this, you can rewrite the integral in two parts.

$$\begin{aligned}\int_0^2 |2x - 1|\,dx &= \int_0^{1/2} -(2x - 1)\,dx + \int_{1/2}^2 (2x - 1)\,dx \\ &= \left[-x^2 + x\right]_0^{1/2} + \left[x^2 - x\right]_{1/2}^2 \\ &= \left(-\frac{1}{4} + \frac{1}{2}\right) - (0 + 0) + (4 - 2) - \left(\frac{1}{4} - \frac{1}{2}\right) \\ &= \frac{5}{2}\end{aligned}$$

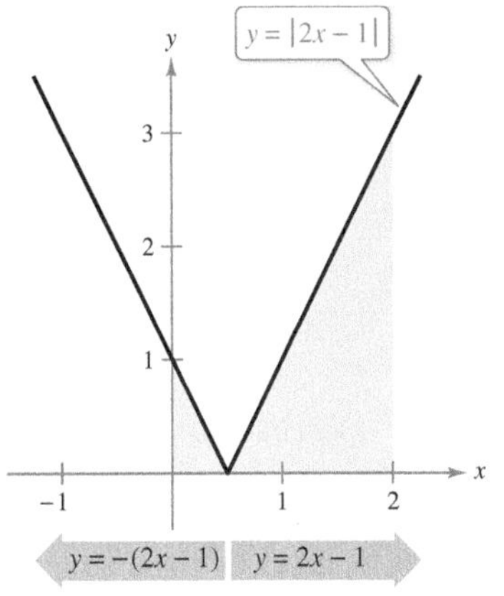

The definite integral of y on $[0, 2]$ is $\frac{5}{2}$.
Figure 4.28

EXAMPLE 3 Using the Fundamental Theorem to Find Area

Find the area of the region bounded by the graph of

$$y = 2x^2 - 3x + 2$$

the x-axis, and the vertical lines $x = 0$ and $x = 2$, as shown in Figure 4.29.

Solution Note that $y > 0$ on the interval $[0, 2]$.

$$\begin{aligned}\text{Area} &= \int_0^2 (2x^2 - 3x + 2)\,dx && \text{Integrate between } x = 0 \text{ and } x = 2. \\ &= \left[\frac{2x^3}{3} - \frac{3x^2}{2} + 2x\right]_0^2 && \text{Find antiderivative.} \\ &= \left(\frac{16}{3} - 6 + 4\right) - (0 - 0 + 0) && \text{Apply Fundamental Theorem.} \\ &= \frac{10}{3} && \text{Simplify.}\end{aligned}$$

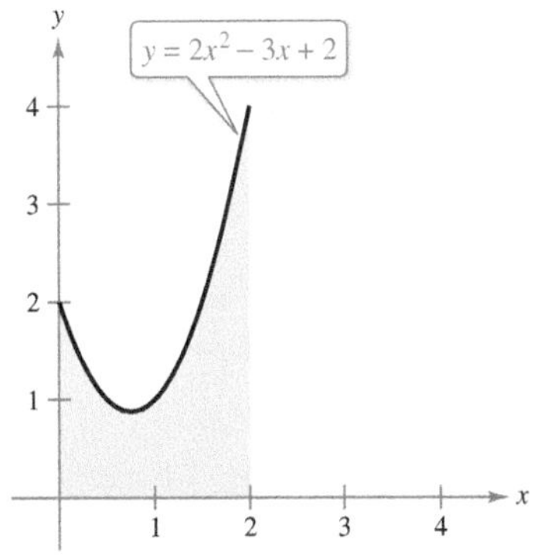

The area of the region bounded by the graph of y, the x-axis, $x = 0$, and $x = 2$ is $\frac{10}{3}$.
Figure 4.29

The Mean Value Theorem for Integrals

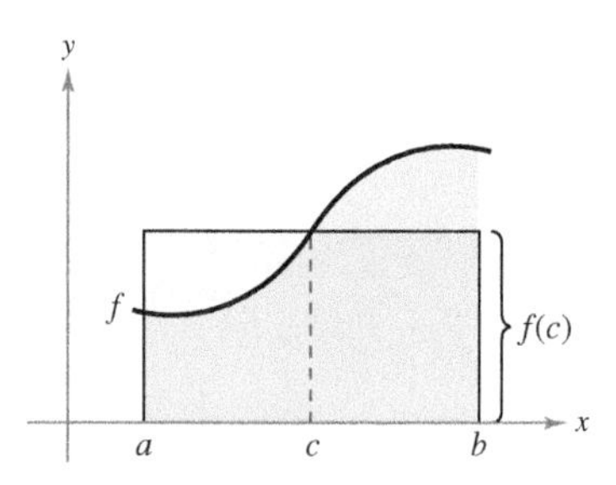

Mean value rectangle:

$f(c)(b-a) = \int_a^b f(x)\,dx$

Figure 4.30

In Section 4.2, you saw that the area of a region under a curve is greater than the area of an inscribed rectangle and less than the area of a circumscribed rectangle. The Mean Value Theorem for Integrals states that somewhere "between" the inscribed and circumscribed rectangles, there is a rectangle whose area is precisely equal to the area of the region under the curve, as shown in Figure 4.30.

THEOREM 4.10 Mean Value Theorem for Integrals

If f is continuous on the closed interval $[a, b]$, then there exists a number c in the closed interval $[a, b]$ such that

$$\int_a^b f(x)\,dx = f(c)(b-a).$$

Proof

Case 1: If f is constant on the interval $[a, b]$, then the theorem is clearly valid because c can be any point in $[a, b]$.

Case 2: If f is not constant on $[a, b]$, then, by the Extreme Value Theorem, you can choose $f(m)$ and $f(M)$ to be the minimum and maximum values of f on $[a, b]$. Because

$$f(m) \le f(x) \le f(M)$$

for all x in $[a, b]$, you can apply Theorem 4.8 to write the following.

$$\int_a^b f(m)\,dx \le \int_a^b f(x)\,dx \le \int_a^b f(M)\,dx \qquad \text{See Figure 4.31.}$$

$$f(m)(b-a) \le \int_a^b f(x)\,dx \le f(M)(b-a) \qquad \text{Apply Fundamental Theorem.}$$

$$f(m) \le \frac{1}{b-a}\int_a^b f(x)\,dx \le f(M) \qquad \text{Divide by } b-a.$$

From the third inequality, you can apply the Intermediate Value Theorem to conclude that there exists some c in $[a, b]$ such that

$$f(c) = \frac{1}{b-a}\int_a^b f(x)\,dx \quad \text{or} \quad f(c)(b-a) = \int_a^b f(x)\,dx.$$

Inscribed rectangle (less than actual area)

$\int_a^b f(m)\,dx = f(m)(b-a)$

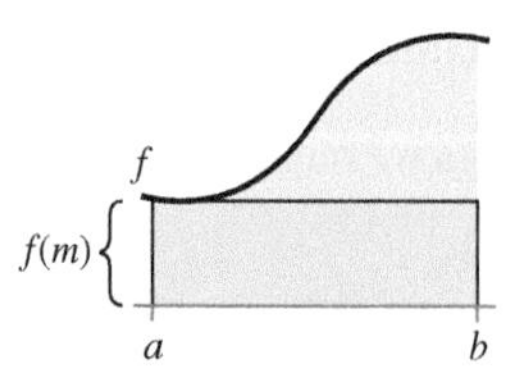

Mean value rectangle (equal to actual area)

$\int_a^b f(x)\,dx$

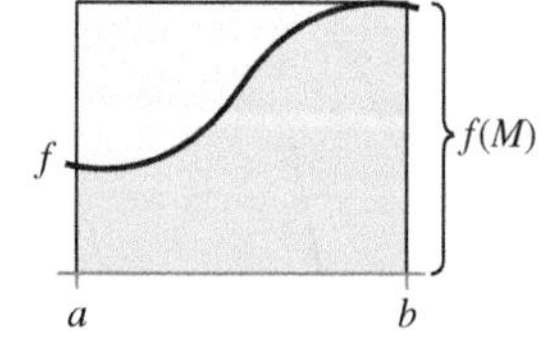

Circumscribed rectangle (greater than actual area)

$\int_a^b f(M)\,dx = f(M)(b-a)$

Figure 4.31

Notice that Theorem 4.10 does not specify how to determine c. It merely guarantees the existence of at least one number c in the interval.

Average Value of a Function

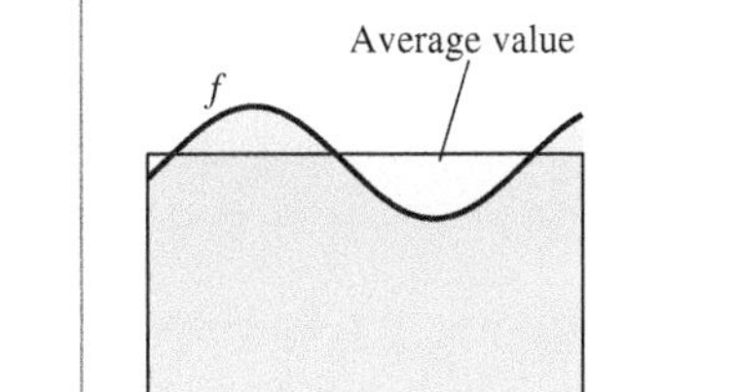

$$\text{Average value} = \frac{1}{b-a}\int_a^b f(x)\,dx$$

Figure 4.32

The value of $f(c)$ given in the Mean Value Theorem for Integrals is called the **average value** of f on the interval $[a, b]$.

> **Definition of the Average Value of a Function on an Interval**
>
> If f is integrable on the closed interval $[a, b]$, then the **average value** of f on the interval is
>
> $$\frac{1}{b-a}\int_a^b f(x)\,dx. \qquad \text{See Figure 4.32.}$$

To see why the average value of f is defined in this way, partition $[a, b]$ into n subintervals of equal width $\Delta x = (b - a)/n$. If c_i is any point in the ith subinterval, then the arithmetic average (or mean) of the function values at the c_i's is

$$a_n = \frac{1}{n}[f(c_1) + f(c_2) + \cdots + f(c_n)]. \qquad \text{Average of } f(c_1), \ldots, f(c_n)$$

By writing the sum using summation notation and then multiplying and dividing by $(b - a)$, you can write the average as

$$\begin{aligned}
a_n &= \frac{1}{n}\sum_{i=1}^{n} f(c_i) && \text{Rewrite using summation notation.}\\
&= \frac{1}{n}\sum_{i=1}^{n} f(c_i)\left(\frac{b-a}{b-a}\right) && \text{Multiply and divide by } (b-a).\\
&= \frac{1}{b-a}\sum_{i=1}^{n} f(c_i)\left(\frac{b-a}{n}\right) && \text{Rewrite.}\\
&= \frac{1}{b-a}\sum_{i=1}^{n} f(c_i)\,\Delta x. && \Delta x = \frac{b-a}{n}
\end{aligned}$$

Finally, taking the limit as $n \to \infty$ produces the average value of f on the interval $[a, b]$, as given in the definition above. In Figure 4.32, notice that the area of the region under the graph of f is equal to the area of the rectangle whose height is the average value.

This development of the average value of a function on an interval is only one of many practical uses of definite integrals to represent summation processes. In Chapter 7, you will study other applications, such as volume, arc length, centers of mass, and work.

EXAMPLE 4 Finding the Average Value of a Function

Find the average value of $f(x) = 3x^2 - 2x$ on the interval $[1, 4]$.

Solution The average value is

$$\begin{aligned}
\frac{1}{b-a}\int_a^b f(x)\,dx &= \frac{1}{4-1}\int_1^4 (3x^2 - 2x)\,dx\\
&= \frac{1}{3}\Big[x^3 - x^2\Big]_1^4\\
&= \frac{1}{3}[64 - 16 - (1 - 1)]\\
&= \frac{48}{3}\\
&= 16. && \text{See Figure 4.33.}
\end{aligned}$$

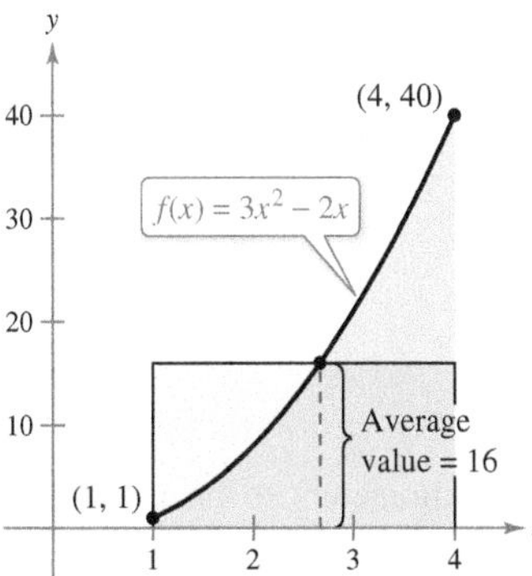

Figure 4.33

The first person to fly at a speed greater than the speed of sound was Charles Yeager. On October 14, 1947, Yeager was clocked at 295.9 meters per second at an altitude of 12.2 kilometers. If Yeager had been flying at an altitude below 11.275 kilometers, this speed would not have "broken the sound barrier." The photo shows an F/A-18F Super Hornet, a supersonic twin-engine strike fighter. A "green Hornet" using a 50/50 mixture of biofuel made from camelina oil became the first U.S. naval tactical aircraft to exceed 1 mach (the speed of sound).

EXAMPLE 5 The Speed of Sound

At different altitudes in Earth's atmosphere, sound travels at different speeds. The speed of sound $s(x)$, in meters per second, can be modeled by

$$s(x) = \begin{cases} -4x + 341, & 0 \le x < 11.5 \\ 295, & 11.5 \le x < 22 \\ \frac{3}{4}x + 278.5, & 22 \le x < 32 \\ \frac{3}{2}x + 254.5, & 32 \le x < 50 \\ -\frac{3}{2}x + 404.5, & 50 \le x \le 80 \end{cases}$$

where x is the altitude in kilometers (see Figure 4.34). What is the average speed of sound over the interval [0, 80]?

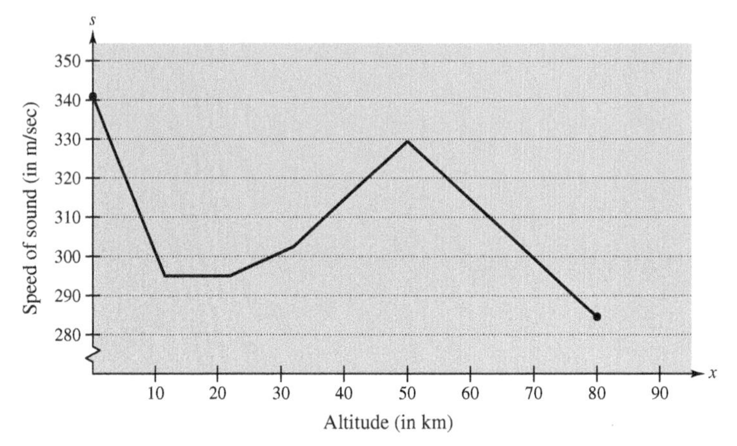

Speed of sound depends on altitude.
Figure 4.34

Solution Begin by integrating $s(x)$ over the interval [0, 80]. To do this, you can break the integral into five parts.

$$\int_0^{11.5} s(x)\,dx = \int_0^{11.5} (-4x + 341)\,dx = \Big[-2x^2 + 341x\Big]_0^{11.5} = 3657$$

$$\int_{11.5}^{22} s(x)\,dx = \int_{11.5}^{22} 295\,dx = \Big[295x\Big]_{11.5}^{22} = 3097.5$$

$$\int_{22}^{32} s(x)\,dx = \int_{22}^{32} \left(\tfrac{3}{4}x + 278.5\right)dx = \Big[\tfrac{3}{8}x^2 + 278.5x\Big]_{22}^{32} = 2987.5$$

$$\int_{32}^{50} s(x)\,dx = \int_{32}^{50} \left(\tfrac{3}{2}x + 254.5\right)dx = \Big[\tfrac{3}{4}x^2 + 254.5x\Big]_{32}^{50} = 5688$$

$$\int_{50}^{80} s(x)\,dx = \int_{50}^{80} \left(-\tfrac{3}{2}x + 404.5\right)dx = \Big[-\tfrac{3}{4}x^2 + 404.5x\Big]_{50}^{80} = 9210$$

By adding the values of the five integrals, you have

$$\int_0^{80} s(x)\,dx = 24{,}640.$$

So, the average speed of sound from an altitude of 0 kilometers to an altitude of 80 kilometers is

$$\text{Average speed} = \frac{1}{80}\int_0^{80} s(x)\,dx = \frac{24{,}640}{80} = 308 \text{ meters per second.}$$

The Second Fundamental Theorem of Calculus

Earlier you saw that the definite integral of f on the interval $[a, b]$ was defined using the constant b as the upper limit of integration and x as the variable of integration. However, a slightly different situation may arise in which the variable x is used in the upper limit of integration. To avoid the confusion of using x in two different ways, t is temporarily used as the variable of integration. (Remember that the definite integral is *not* a function of its variable of integration.)

The Definite Integral as a Number **The Definite Integral as a Function of x**

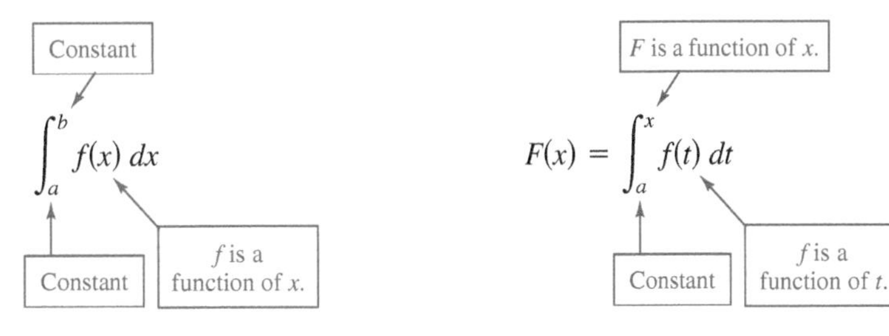

Exploration

Use a graphing utility to graph the function

$$F(x) = \int_0^x \cos t\,dt$$

for $0 \le x \le 2\pi$. Do you recognize this graph? Explain.

EXAMPLE 6 The Definite Integral as a Function

Evaluate the function

$$F(x) = \int_0^x \cos t\,dt$$

at $x = 0, \frac{\pi}{6}, \frac{\pi}{4}, \frac{\pi}{3}$, and $\frac{\pi}{2}$.

Solution You could evaluate five different definite integrals, one for each of the given upper limits. However, it is much simpler to fix x (as a constant) temporarily to obtain

$$\int_0^x \cos t\,dt = \sin t\Big]_0^x$$
$$= \sin x - \sin 0$$
$$= \sin x.$$

Now, using $F(x) = \sin x$, you can obtain the results shown in Figure 4.35.

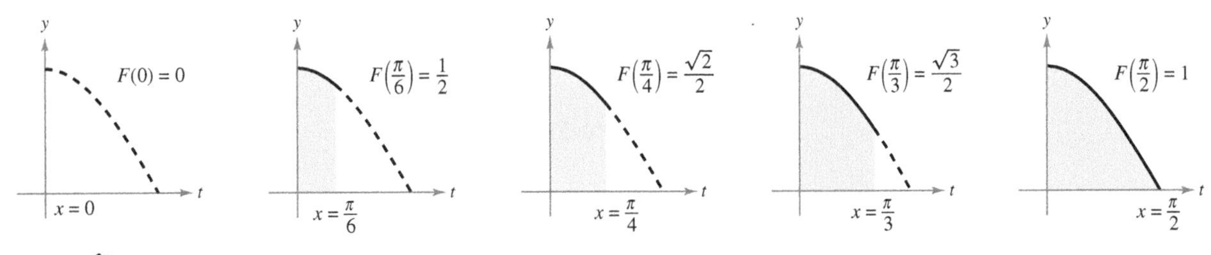

$F(x) = \int_0^x \cos t\,dt$ is the area under the curve $f(t) = \cos t$ from 0 to x.

Figure 4.35

You can think of the function $F(x)$ as *accumulating* the area under the curve $f(t) = \cos t$ from $t = 0$ to $t = x$. For $x = 0$, the area is 0 and $F(0) = 0$. For $x = \pi/2$, $F(\pi/2) = 1$ gives the accumulated area under the cosine curve on the entire interval $[0, \pi/2]$. This interpretation of an integral as an **accumulation function** is used often in applications of integration.

In Example 6, note that the derivative of F is the original integrand (with only the variable changed). That is,

$$\frac{d}{dx}[F(x)] = \frac{d}{dx}[\sin x] = \frac{d}{dx}\left[\int_0^x \cos t\, dt\right] = \cos x.$$

This result is generalized in the next theorem, called the **Second Fundamental Theorem of Calculus.**

THEOREM 4.11 The Second Fundamental Theorem of Calculus

If f is continuous on an open interval I containing a, then, for every x in the interval,

$$\frac{d}{dx}\left[\int_a^x f(t)\, dt\right] = f(x).$$

Proof Begin by defining F as

$$F(x) = \int_a^x f(t)\, dt.$$

Then, by the definition of the derivative, you can write

$$\begin{aligned} F'(x) &= \lim_{\Delta x \to 0} \frac{F(x + \Delta x) - F(x)}{\Delta x} \\ &= \lim_{\Delta x \to 0} \frac{1}{\Delta x}\left[\int_a^{x+\Delta x} f(t)\, dt - \int_a^x f(t)\, dt\right] \\ &= \lim_{\Delta x \to 0} \frac{1}{\Delta x}\left[\int_a^{x+\Delta x} f(t)\, dt + \int_x^a f(t)\, dt\right] \\ &= \lim_{\Delta x \to 0} \frac{1}{\Delta x}\left[\int_x^{x+\Delta x} f(t)\, dt\right]. \end{aligned}$$

From the Mean Value Theorem for Integrals (assuming $\Delta x > 0$), you know there exists a number c in the interval $[x, x + \Delta x]$ such that the integral in the expression above is equal to $f(c)\,\Delta x$. Moreover, because $x \le c \le x + \Delta x$, it follows that $c \to x$ as $\Delta x \to 0$. So, you obtain

$$F'(x) = \lim_{\Delta x \to 0}\left[\frac{1}{\Delta x} f(c)\,\Delta x\right] = \lim_{\Delta x \to 0} f(c) = f(x).$$

A similar argument can be made for $\Delta x < 0$. ∎

Using the area model for definite integrals, the approximation

$$f(x)\,\Delta x \approx \int_x^{x+\Delta x} f(t)\, dt$$

can be viewed as saying that the area of the rectangle of height $f(x)$ and width Δx is approximately equal to the area of the region lying between the graph of f and the x-axis on the interval

$$[x, x + \Delta x]$$

as shown in the figure at the right.

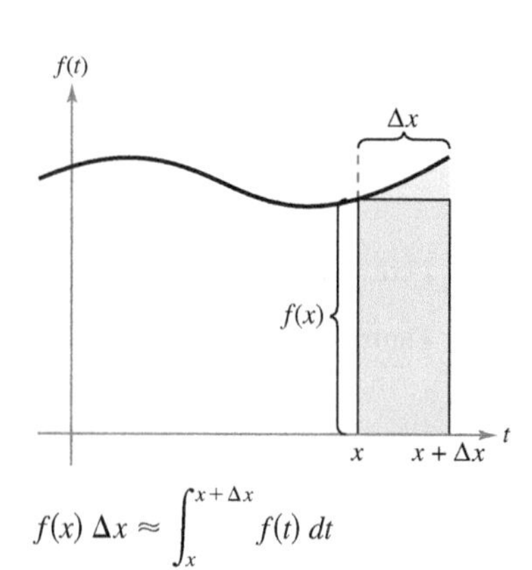

$$f(x)\,\Delta x \approx \int_x^{x+\Delta x} f(t)\, dt$$

Note that the Second Fundamental Theorem of Calculus tells you that when a function is continuous, you can be sure that it has an antiderivative. This antiderivative need not, however, be an elementary function. (Recall the discussion of elementary functions in Section P.3.)

EXAMPLE 7 **The Second Fundamental Theorem of Calculus**

Evaluate $\frac{d}{dx}\left[\int_0^x \sqrt{t^2+1}\,dt\right]$.

Solution Note that $f(t) = \sqrt{t^2+1}$ is continuous on the entire real number line. So, using the Second Fundamental Theorem of Calculus, you can write

$$\frac{d}{dx}\left[\int_0^x \sqrt{t^2+1}\,dt\right] = \sqrt{x^2+1}.$$

The differentiation shown in Example 7 is a straightforward application of the Second Fundamental Theorem of Calculus. The next example shows how this theorem can be combined with the Chain Rule to find the derivative of a function.

EXAMPLE 8 **The Second Fundamental Theorem of Calculus**

Find the derivative of $F(x) = \int_{\pi/2}^{x^3} \cos t\,dt$.

Solution Using $u = x^3$, you can apply the Second Fundamental Theorem of Calculus with the Chain Rule as shown.

$$F'(x) = \frac{dF}{du}\frac{du}{dx} \qquad \text{Chain Rule}$$

$$= \frac{d}{du}[F(x)]\frac{du}{dx} \qquad \text{Definition of } \frac{dF}{du}$$

$$= \frac{d}{du}\left[\int_{\pi/2}^{x^3} \cos t\,dt\right]\frac{du}{dx} \qquad \text{Substitute } \int_{\pi/2}^{x^3} \cos t\,dt \text{ for } F(x).$$

$$= \frac{d}{du}\left[\int_{\pi/2}^{u} \cos t\,dt\right]\frac{du}{dx} \qquad \text{Substitute } u \text{ for } x^3.$$

$$= (\cos u)(3x^2) \qquad \text{Apply Second Fundamental Theorem of Calculus.}$$

$$= (\cos x^3)(3x^2) \qquad \text{Rewrite as function of } x.$$

Because the integrand in Example 8 is easily integrated, you can verify the derivative as follows.

$$F(x) = \int_{\pi/2}^{x^3} \cos t\,dt$$

$$= \sin t\Big]_{\pi/2}^{x^3}$$

$$= \sin x^3 - \sin\frac{\pi}{2}$$

$$= \sin x^3 - 1$$

In this form, you can apply the Chain Rule to verify that the derivative of F is the same as that obtained in Example 8.

$$\frac{d}{dx}[\sin x^3 - 1] = (\cos x^3)(3x^2) \qquad \text{Derivative of } F$$

Net Change Theorem

The Fundamental Theorem of Calculus (Theorem 4.9) states that if f is continuous on the closed interval $[a, b]$ and F is an antiderivative of f on $[a, b]$, then

$$\int_a^b f(x)\,dx = F(b) - F(a).$$

But because $F'(x) = f(x)$, this statement can be rewritten as

$$\int_a^b F'(x)\,dx = F(b) - F(a)$$

where the quantity $F(b) - F(a)$ represents the *net change of* $F(x)$ on the interval $[a, b]$.

THEOREM 4.12 The Net Change Theorem

If $F'(x)$ is the rate of change of a quantity $F(x)$, then the definite integral of $F'(x)$ from a to b gives the total change, or **net change**, of $F(x)$ on the interval $[a, b]$.

$$\int_a^b F'(x)\,dx = F(b) - F(a) \qquad \text{Net change of } F(x)$$

EXAMPLE 9 Using the Net Change Theorem

A chemical flows into a storage tank at a rate of $(180 + 3t)$ liters per minute, where t is the time in minutes and $0 \le t \le 60$. Find the amount of the chemical that flows into the tank during the first 20 minutes.

Solution Let $c(t)$ be the amount of the chemical in the tank at time t. Then $c'(t)$ represents the rate at which the chemical flows into the tank at time t. During the first 20 minutes, the amount that flows into the tank is

$$\begin{aligned}\int_0^{20} c'(t)\,dt &= \int_0^{20} (180 + 3t)\,dt \\ &= \left[180t + \frac{3}{2}t^2\right]_0^{20} \\ &= 3600 + 600 \\ &= 4200.\end{aligned}$$

So, the amount of the chemical that flows into the tank during the first 20 minutes is 4200 liters. ■

Another way to illustrate the Net Change Theorem is to examine the velocity of a particle moving along a straight line, where $s(t)$ is the position at time t. Then its velocity is $v(t) = s'(t)$ and

$$\int_a^b v(t)\,dt = s(b) - s(a).$$

This definite integral represents the net change in position, or **displacement,** of the particle.

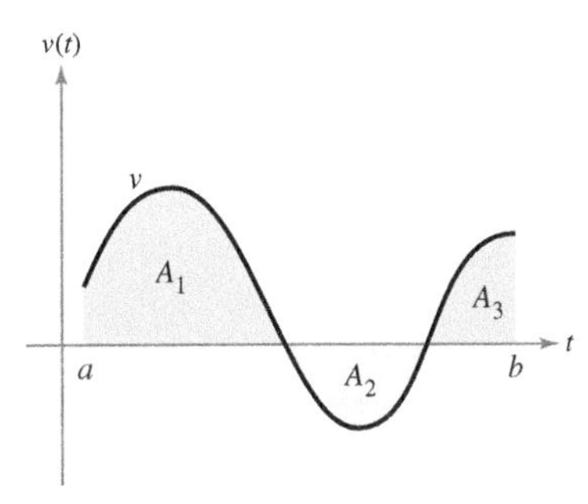

A_1, A_2, and A_3 are the areas of the shaded regions.
Figure 4.36

When calculating the *total* distance traveled by the particle, you must consider the intervals where $v(t) \le 0$ and the intervals where $v(t) \ge 0$. When $v(t) \le 0$, the particle moves to the left, and when $v(t) \ge 0$, the particle moves to the right. To calculate the total distance traveled, integrate the absolute value of velocity $|v(t)|$. So, the **displacement** of the particle on the interval $[a, b]$ is

$$\text{Displacement on } [a, b] = \int_a^b v(t)\, dt = A_1 - A_2 + A_3$$

and the **total distance traveled** by the particle on $[a, b]$ is

$$\text{Total distance traveled on } [a, b] = \int_a^b |v(t)|\, dt = A_1 + A_2 + A_3.$$

(See Figure 4.36.)

EXAMPLE 10 Solving a Particle Motion Problem

The velocity (in feet per second) of a particle moving along a line is

$$v(t) = t^3 - 10t^2 + 29t - 20$$

where t is the time in seconds.

a. What is the displacement of the particle on the time interval $1 \le t \le 5$?

b. What is the total distance traveled by the particle on the time interval $1 \le t \le 5$?

Solution

a. By definition, you know that the displacement is

$$\begin{aligned}\int_1^5 v(t)\, dt &= \int_1^5 (t^3 - 10t^2 + 29t - 20)\, dt\\ &= \left[\frac{t^4}{4} - \frac{10}{3}t^3 + \frac{29}{2}t^2 - 20t\right]_1^5\\ &= \frac{25}{12} - \left(-\frac{103}{12}\right)\\ &= \frac{128}{12}\\ &= \frac{32}{3}.\end{aligned}$$

So, the particle moves $\frac{32}{3}$ feet to the right.

b. To find the total distance traveled, calculate $\int_1^5 |v(t)|\, dt$. Using Figure 4.37 and the fact that $v(t)$ can be factored as $(t - 1)(t - 4)(t - 5)$, you can determine that $v(t) \ge 0$ on $[1, 4]$ and $v(t) \le 0$ on $[4, 5]$. So, the total distance traveled is

$$\begin{aligned}\int_1^5 |v(t)|\, dt &= \int_1^4 v(t)\, dt - \int_4^5 v(t)\, dt\\ &= \int_1^4 (t^3 - 10t^2 + 29t - 20)\, dt - \int_4^5 (t^3 - 10t^2 + 29t - 20)\, dt\\ &= \left[\frac{t^4}{4} - \frac{10}{3}t^3 + \frac{29}{2}t^2 - 20t\right]_1^4 - \left[\frac{t^4}{4} - \frac{10}{3}t^3 + \frac{29}{2}t^2 - 20t\right]_4^5\\ &= \frac{45}{4} - \left(-\frac{7}{12}\right)\\ &= \frac{71}{6} \text{ feet.}\end{aligned}$$

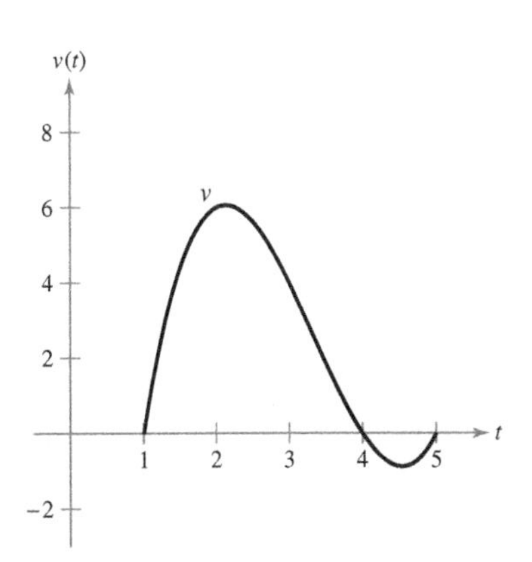

Figure 4.37

4.4 Exercises

See CalcChat.com for tutorial help and worked-out solutions to odd-numbered exercises.

CONCEPT CHECK

1. **Fundamental Theorem of Calculus** Explain how to evaluate a definite integral using the Fundamental Theorem of Calculus.
2. **Mean Value Theorem** Describe the Mean Value Theorem for Integrals in your own words.
3. **Average Value of a Function** Describe the average value of a function on an interval in your own words.
4. **Accumulation Function** Why is

$$F(x) = \int_0^x f(t)\,dt$$

considered an accumulation function?

Graphical Reasoning **In Exercises 5–8, use a graphing utility to graph the integrand. Use the graph to determine whether the definite integral is positive, negative, or zero.**

5. $\displaystyle\int_0^{\pi} \frac{4}{x^2+1}\,dx$
6. $\displaystyle\int_0^{\pi} \cos x\,dx$
7. $\displaystyle\int_{-2}^{2} x\sqrt{x^2+1}\,dx$
8. $\displaystyle\int_{-2}^{2} x\sqrt{2-x}\,dx$

Evaluating a Definite Integral **In Exercises 9–36, evaluate the definite integral. Use a graphing utility to verify your result.**

9. $\displaystyle\int_{-1}^{0} (2x-1)\,dx$
10. $\displaystyle\int_{-1}^{2} (7-3t)\,dt$
11. $\displaystyle\int_{-1}^{1} (t^2-5)\,dt$
12. $\displaystyle\int_{1}^{2} (6x^2-3x)\,dx$
13. $\displaystyle\int_{0}^{1} (2t-1)^2\,dt$
14. $\displaystyle\int_{1}^{4} (8x^3-x)\,dx$
15. $\displaystyle\int_{1}^{2} \left(\frac{3}{x^2}-1\right)dx$
16. $\displaystyle\int_{-2}^{-1} \left(u-\frac{1}{u^2}\right)du$
17. $\displaystyle\int_{1}^{4} \frac{u-2}{\sqrt{u}}\,du$
18. $\displaystyle\int_{-8}^{8} x^{1/3}\,dx$
19. $\displaystyle\int_{-1}^{1} (\sqrt[3]{t}-2)\,dt$
20. $\displaystyle\int_{1}^{8} \sqrt{\frac{2}{x}}\,dx$
21. $\displaystyle\int_{0}^{1} \frac{x-\sqrt{x}}{3}\,dx$
22. $\displaystyle\int_{0}^{2} (6-t)\sqrt{t}\,dt$
23. $\displaystyle\int_{-1}^{0} (t^{1/3}-t^{2/3})\,dt$
24. $\displaystyle\int_{-8}^{-1} \frac{x-x^2}{2\sqrt[3]{x}}\,dx$
25. $\displaystyle\int_{0}^{5} |2x-5|\,dx$
26. $\displaystyle\int_{1}^{4} (3-|x-3|)\,dx$
27. $\displaystyle\int_{0}^{4} |x^2-9|\,dx$
28. $\displaystyle\int_{0}^{4} |x^2-4x+3|\,dx$
29. $\displaystyle\int_{0}^{\pi} (\sin x-7)\,dx$
30. $\displaystyle\int_{0}^{\pi} (2+\cos x)\,dx$
31. $\displaystyle\int_{0}^{\pi/4} \frac{1-\sin^2\theta}{\cos^2\theta}\,d\theta$
32. $\displaystyle\int_{0}^{\pi/4} \frac{\sec^2\theta}{\tan^2\theta+1}\,d\theta$
33. $\displaystyle\int_{-\pi/6}^{\pi/6} \sec^2 x\,dx$
34. $\displaystyle\int_{\pi/4}^{\pi/2} (2-\csc^2 x)\,dx$
35. $\displaystyle\int_{-\pi/3}^{\pi/3} 4\sec\theta\tan\theta\,d\theta$
36. $\displaystyle\int_{-\pi/2}^{\pi/2} (2t+\cos t)\,dt$

Finding the Area of a Region **In Exercises 37–40, find the area of the given region.**

37. $y = x - x^2$

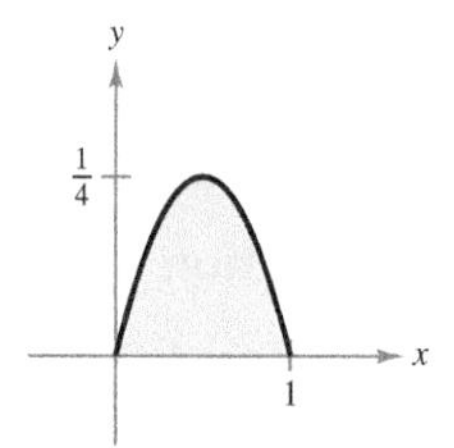

38. $y = \dfrac{1}{x^2}$

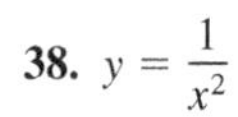

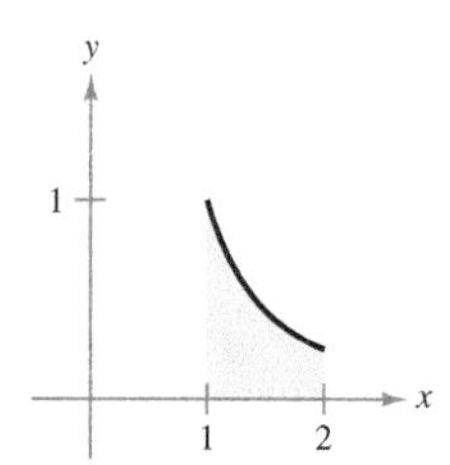

39. $y = \cos x$

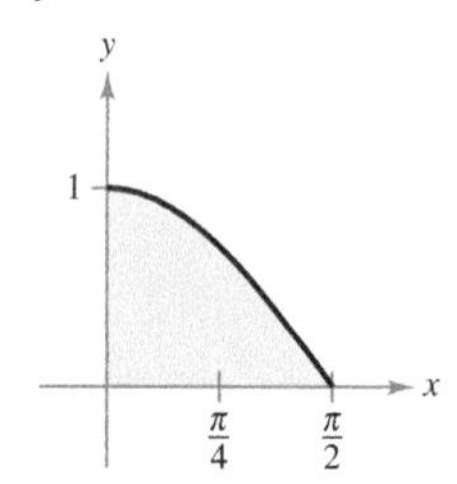

40. $y = x + \sin x$

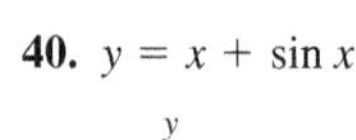

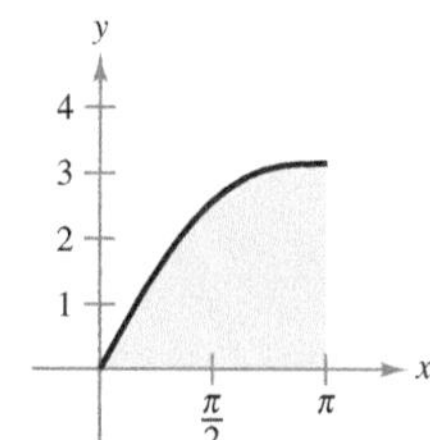

Finding the Area of a Region **In Exercises 41–46, find the area of the region bounded by the graphs of the equations.**

41. $y = 5x^2 + 2,\quad x = 0,\quad x = 2,\quad y = 0$
42. $y = x^3 + 6x,\quad x = 2,\quad y = 0$
43. $y = 1 + \sqrt[3]{x},\quad x = 0,\quad x = 8,\quad y = 0$
44. $y = 2\sqrt{x} - x,\quad y = 0$
45. $y = -x^2 + 4x,\quad y = 0$
46. $y = 1 - x^4,\quad y = 0$

Using the Mean Value Theorem for Integrals **In Exercises 47–52, find the value(s) of c guaranteed by the Mean Value Theorem for Integrals for the function over the given interval.**

47. $f(x) = x^3,\quad [0, 3]$
48. $f(x) = \sqrt{x},\quad [4, 9]$
49. $y = \dfrac{x^2}{4},\quad [0, 6]$
50. $f(x) = \dfrac{9}{x^3},\quad [1, 3]$
51. $f(x) = 2\sec^2 x,\quad \left[-\dfrac{\pi}{4}, \dfrac{\pi}{4}\right]$
52. $f(x) = \cos x,\quad \left[-\dfrac{\pi}{3}, \dfrac{\pi}{3}\right]$

Finding the Average Value of a Function **In Exercises 53–58, find the average value of the function over the given interval and all values of x in the interval for which the function equals its average value.**

53. $f(x) = 4 - x^2, \quad [-2, 2]$

54. $f(x) = \dfrac{4(x^2 + 1)}{x^2}, \quad [1, 3]$

55. $f(x) = x^4 + 7, \quad [0, 2]$

56. $f(x) = 4x^3 - 3x^2, \quad [0, 1]$

57. $f(x) = \sin x, \quad [0, \pi]$

58. $f(x) = \cos x, \quad \left[0, \dfrac{\pi}{2}\right]$

59. **Force** The force F (in newtons) of a hydraulic cylinder in a press is proportional to the square of $\sec x$, where x is the distance (in meters) that the cylinder is extended in its cycle. The domain of F is $[0, \pi/3]$, and $F(0) = 500$.

(a) Find F as a function of x.

(b) Find the average force exerted by the press over the interval $[0, \pi/3]$.

60. **Respiratory Cycle** The volume V, in liters, of air in the lungs during a five-second respiratory cycle is approximated by the model $V = 0.1729t + 0.1522t^2 - 0.0374t^3$, where t is the time in seconds. Approximate the average volume of air in the lungs during one cycle.

61. **Buffon's Needle Experiment** A horizontal plane is ruled with parallel lines 2 inches apart. A two-inch needle is tossed randomly onto the plane. The probability that the needle will touch a line is

$$P = \frac{2}{\pi}\int_0^{\pi/2} \sin\theta \, d\theta$$

where θ is the acute angle between the needle and any one of the parallel lines. Find this probability.

62. **HOW DO YOU SEE IT?** The graph of f is shown in the figure. The shaded region A has an area of 1.5, and $\int_0^6 f(x)\,dx = 3.5$. Use this information to fill in the blanks.

(a) $\displaystyle\int_0^2 f(x)\,dx = \_\_\_$

(b) $\displaystyle\int_2^6 f(x)\,dx = \_\_\_$

(c) $\displaystyle\int_0^6 |f(x)|\,dx = \_\_\_$

(d) $\displaystyle\int_0^2 -2f(x)\,dx = \_\_\_$

(e) $\displaystyle\int_0^6 [2 + f(x)]\,dx = \_\_\_$

(f) The average value of f over the interval $[0, 6]$ is ___.

Evaluating a Definite Integral **In Exercises 63 and 64, find F as a function of x and evaluate it at $x = 2$, $x = 5$, and $x = 8$.**

63. $F(x) = \displaystyle\int_1^x \frac{20}{v^2}\,dv$

64. $F(x) = \displaystyle\int_2^x (t^3 + 2t - 2)\,dt$

Evaluating a Definite Integral **In Exercises 65 and 66, find F as a function of x and evaluate it at $x = 0$, $x = \pi/4$, and $x = \pi/2$.**

65. $F(x) = \displaystyle\int_0^x \cos\theta\,d\theta$

66. $F(x) = \displaystyle\int_{-\pi}^x \sin\theta\,d\theta$

67. **Analyzing a Function** Let

$$g(x) = \int_0^x f(t)\,dt$$

where f is the function whose graph is shown in the figure.

(a) Estimate $g(0)$, $g(2)$, $g(4)$, $g(6)$, and $g(8)$.

(b) Find the largest open interval on which g is increasing. Find the largest open interval on which g is decreasing.

(c) Identify any extrema of g.

(d) Sketch a rough graph of g.

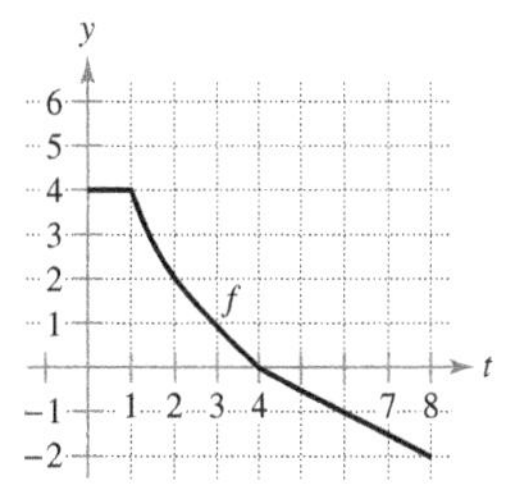

Figure for 67

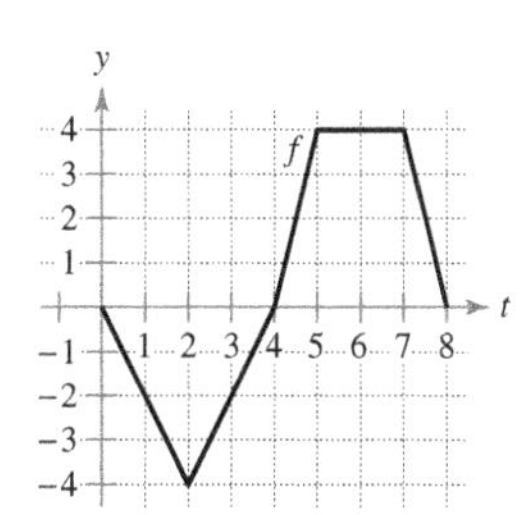

Figure for 68

68. **Analyzing a Function** Let

$$g(x) = \int_0^x f(t)\,dt$$

where f is the function whose graph is shown in the figure.

(a) Estimate $g(0)$, $g(2)$, $g(4)$, $g(6)$, and $g(8)$.

(b) Find the largest open interval on which g is increasing. Find the largest open interval on which g is decreasing.

(c) Identify any extrema of g.

(d) Sketch a rough graph of g.

Finding and Checking an Integral **In Exercises 69–74, (a) integrate to find F as a function of x, and (b) demonstrate the Second Fundamental Theorem of Calculus by differentiating the result in part (a).**

69. $F(x) = \displaystyle\int_0^x (t + 2)\,dt$

70. $F(x) = \displaystyle\int_0^x t(t^2 + 1)\,dt$

71. $F(x) = \displaystyle\int_8^x \sqrt[3]{t}\,dt$

72. $F(x) = \displaystyle\int_4^x t^{3/2}\,dt$

73. $F(x) = \displaystyle\int_{\pi/4}^x \sec^2 t\,dt$

74. $F(x) = \displaystyle\int_{\pi/3}^x \sec t \tan t\,dt$

Using the Second Fundamental Theorem of Calculus In Exercises 75–80, use the Second Fundamental Theorem of Calculus to find $F'(x)$.

75. $F(x) = \int_{-2}^{x} (t^2 - 2t)\, dt$

76. $F(x) = \int_{1}^{x} \frac{t^2}{t^2 + 1}\, dt$

77. $F(x) = \int_{-1}^{x} \sqrt{t^4 + 1}\, dt$

78. $F(x) = \int_{1}^{x} \sqrt[4]{t}\, dt$

79. $F(x) = \int_{1}^{x} \sqrt{t} \csc t\, dt$

80. $F(x) = \int_{0}^{x} \sec^3 t\, dt$

Finding a Derivative In Exercises 81–86, find $F'(x)$.

81. $F(x) = \int_{x}^{x+2} (4t + 1)\, dt$

82. $F(x) = \int_{-x}^{x} t^3\, dt$

83. $F(x) = \int_{0}^{\sin x} \sqrt{t}\, dt$

84. $F(x) = \int_{2}^{x^2} \frac{1}{t^3}\, dt$

85. $F(x) = \int_{0}^{x^3} \sin t^2\, dt$

86. $F(x) = \int_{0}^{2x} \cos t^4\, dt$

87. **Graphical Analysis** Sketch an approximate graph of g on the interval $0 \le x \le 4$, where

$$g(x) = \int_{0}^{x} f(t)\, dt.$$

Identify the x-coordinate of an extremum of g. To print an enlarged copy of the graph, go to *MathGraphs.com*.

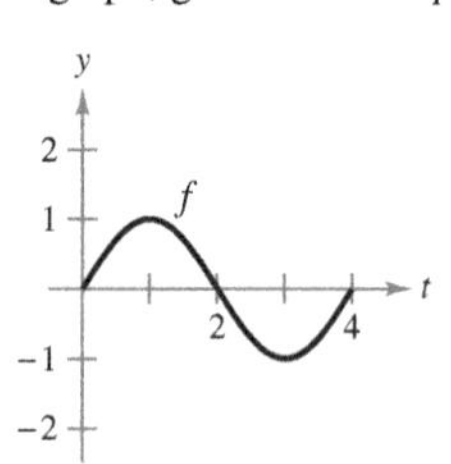

88. **Area** The area A between the graph of the function

$$g(t) = 4 - \frac{4}{t^2}$$

and the t-axis over the interval $[1, x]$ is

$$A(x) = \int_{1}^{x} \left(4 - \frac{4}{t^2}\right) dt.$$

(a) Find the horizontal asymptote of the graph of g.

(b) Integrate to find A as a function of x. Does the graph of A have a horizontal asymptote? Explain.

89. **Water Flow** Water flows from a storage tank at a rate of $(500 - 5t)$ liters per minute. Find the amount of water that flows out of the tank during the first 18 minutes.

90. **Oil Leak** At 1:00 P.M., oil begins leaking from a tank at a rate of $(4 + 0.75t)$ gallons per hour.

(a) How much oil is lost from 1:00 P.M. to 4:00 P.M.?

(b) How much oil is lost from 4:00 P.M. to 7:00 P.M.?

(c) Compare your answers to parts (a) and (b). What do you notice?

91. **Velocity** The graph shows the velocity, in feet per second, of a car accelerating from rest. Use the graph to estimate the distance the car travels in 8 seconds.

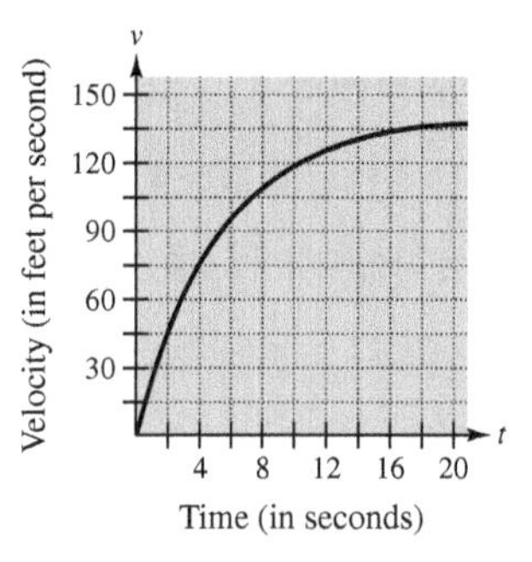

92. **Velocity** The graph shows the velocity, in feet per second, of a decelerating car after the driver applies the brakes. Use the graph to estimate how far the car travels before it comes to a stop.

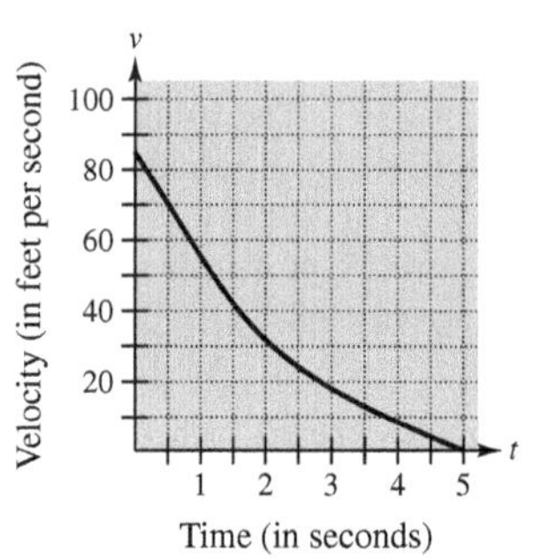

Particle Motion In Exercises 93–98, the velocity function, in feet per second, is given for a particle moving along a straight line, where t is the time in seconds. Find (a) the displacement and (b) the total distance that the particle travels over the given interval.

93. $v(t) = 5t - 7, \quad 0 \le t \le 3$

94. $v(t) = t^2 - t - 12, \quad 1 \le t \le 5$

95. $v(t) = t^3 - 10t^2 + 27t - 18, \quad 1 \le t \le 7$

96. $v(t) = t^3 - 8t^2 + 15t, \quad 0 \le t \le 5$

97. $v(t) = \frac{1}{\sqrt{t}}, \quad 1 \le t \le 4$

98. $v(t) = \cos t, \quad 0 \le t \le 3\pi$

EXPLORING CONCEPTS

99. **Particle Motion** Describe a situation where the displacement and the total distance traveled for a particle are equal.

100. **Rate of Growth** Let $r'(t)$ represent the rate of growth of a dog, in pounds per year. What does $r(t)$ represent? What does $\int_2^6 r'(t)\, dt$ represent about the dog?

101. **Fundamental Theorem of Calculus** Explain why the Fundamental Theorem of Calculus cannot be used to integrate

$$f(x) = \frac{1}{x - c}$$

on any interval containing c.

102. Modeling Data An experimental vehicle is tested on a straight track. It starts from rest, and its velocity v (in meters per second) is recorded every 10 seconds for 1 minute (see table).

t	0	10	20	30	40	50	60
v	0	5	21	40	62	78	83

(a) Use a graphing utility to find a model of the form $v = at^3 + bt^2 + ct + d$ for the data.

(b) Use a graphing utility to plot the data and graph the model.

(c) Approximate the distance traveled by the vehicle during the test.

103. Particle Motion A particle is moving along the x-axis. The position of the particle at time t is given by

$$x(t) = t^3 - 6t^2 + 9t - 2, \quad 0 \le t \le 5.$$

Find the total distance the particle travels in 5 units of time.

104. Particle Motion Repeat Exercise 103 for the position function given by

$$x(t) = (t-1)(t-3)^2, \quad 0 \le t \le 5.$$

Error Analysis In Exercises 105–108, describe why the statement is incorrect.

105. $\displaystyle\int_{-1}^{1} x^{-2}\,dx = \left[-x^{-1}\right]_{-1}^{1} = (-1) - 1 = -2$ X

106. $\displaystyle\int_{-2}^{1} \frac{2}{x^3}\,dx = \left[-\frac{1}{x^2}\right]_{-2}^{1} = -\frac{3}{4}$ X

107. $\displaystyle\int_{\pi/4}^{3\pi/4} \sec^2 x\,dx = \left[\tan x\right]_{\pi/4}^{3\pi/4} = -2$ X

108. $\displaystyle\int_{\pi/2}^{3\pi/2} \csc x \cot x\,dx = \left[-\csc x\right]_{\pi/2}^{3\pi/2} = 2$ X

True or False? In Exercises 109 and 110, determine whether the statement is true or false. If it is false, explain why or give an example that shows it is false.

109. If $F'(x) = G'(x)$ on the interval $[a, b]$, then

$$F(b) - F(a) = G(b) - G(a).$$

110. If $F(b) - F(a) = G(b) - G(a)$, then $F'(x) = G'(x)$ on the interval $[a, b]$.

111. Analyzing a Function Show that the function

$$f(x) = \int_0^{1/x} \frac{1}{t^2+1}\,dt + \int_0^{x} \frac{1}{t^2+1}\,dt$$

is constant for $x > 0$.

112. Finding a Function Find the function $f(x)$ and all values of c such that

$$\int_c^x f(t)\,dt = x^2 + x - 2.$$

113. Finding Values Let

$$G(x) = \int_0^x \left[s\int_0^s f(t)\,dt\right] ds$$

where f is continuous for all real t. Find (a) $G(0)$, (b) $G'(0)$, (c) $G''(x)$, and (d) $G''(0)$.

114. Proof Prove that

$$\frac{d}{dx}\left[\int_{u(x)}^{v(x)} f(t)\,dt\right] = f(v(x))v'(x) - f(u(x))u'(x).$$

PUTNAM EXAM CHALLENGE

115. For each continuous function $f\colon [0, 1] \to \mathbb{R}$, let

$$I(f) = \int_0^1 x^2 f(x)\,dx$$

and

$$J(x) = \int_0^1 x(f(x))^2\,dx.$$

Find the maximum value of $I(f) - J(f)$ over all such functions f.

This problem was composed by the Committee on the Putnam Prize Competition.

SECTION PROJECT

Demonstrating the Fundamental Theorem

Use a graphing utility to graph the function

$$y_1 = \sin^2 t$$

on the interval $0 \le t \le \pi$. Let F be the following function of x.

$$F(x) = \int_0^x \sin^2 t\,dt$$

(a) Complete the table. Explain why the values of F are increasing.

x	0	$\frac{\pi}{6}$	$\frac{\pi}{3}$	$\frac{\pi}{2}$	$\frac{2\pi}{3}$	$\frac{5\pi}{6}$	π
$F(x)$							

(b) Use the integration capabilities of a graphing utility to graph F.

(c) Use the differentiation capabilities of a graphing utility to graph F'. How is this graph related to the graph in part (b)?

(d) Verify that the derivative of

$$y = \frac{1}{2}t - \frac{1}{4}\sin 2t$$

is $\sin^2 t$. Graph y and write a short paragraph about how this graph is related to those in parts (b) and (c).

4.5 Integration by Substitution

- **Use pattern recognition to find an indefinite integral.**
- **Use a change of variables to find an indefinite integral.**
- **Use the General Power Rule for Integration to find an indefinite integral.**
- **Use a change of variables to evaluate a definite integral.**
- **Evaluate a definite integral involving an even or odd function.**

Pattern Recognition

In this section, you will study techniques for integrating composite functions. The discussion is split into two parts—*pattern recognition* and *change of variables.* Both techniques involve a ***u*-substitution.** With pattern recognition, you perform the substitution mentally, and with change of variables, you write the substitution steps.

The role of substitution in integration is comparable to the role of the Chain Rule in differentiation. Recall that for the differentiable functions

$$y = F(u) \quad \text{and} \quad u = g(x)$$

the Chain Rule states that

$$\frac{d}{dx}[F(g(x))] = F'(g(x))g'(x).$$

From the definition of an antiderivative, it follows that

$$\int F'(g(x))g'(x)\,dx = F(g(x)) + C.$$

These results are summarized in the next theorem.

REMARK The statement of Theorem 4.13 does not tell how to distinguish between $f(g(x))$ and $g'(x)$ in the integrand. As you become more experienced at integration, your skill in doing this will increase. Of course, part of the key is familiarity with derivatives.

THEOREM 4.13 Antidifferentiation of a Composite Function

Let g be a function whose range is an interval I, and let f be a function that is continuous on I. If g is differentiable on its domain and F is an antiderivative of f on I, then

$$\int f(g(x))g'(x)\,dx = F(g(x)) + C.$$

Letting $u = g(x)$ gives $du = g'(x)\,dx$ and

$$\int f(u)\,du = F(u) + C.$$

Examples 1 and 2 show how to apply Theorem 4.13 *directly*, by recognizing the presence of $f(g(x))$ and $g'(x)$. Note that the composite function in the integrand has an *outside function* f and an *inside function* g. Moreover, the derivative $g'(x)$ is present as a factor of the integrand.

Outside function

$$\int f(g(x))g'(x)\,dx = F(g(x)) + C$$

Inside function

Derivative of inside function

EXAMPLE 1 Recognizing the $f(g(x))g'(x)$ Pattern

Find $\int (x^2 + 1)^2(2x)\, dx$.

Solution Letting $g(x) = x^2 + 1$, you obtain

$$g'(x) = 2x$$

and

$$f(g(x)) = f(x^2 + 1) = (x^2 + 1)^2.$$

From this, you can recognize that the integrand follows the $f(g(x))g'(x)$ pattern. Using the Power Rule for Integration and Theorem 4.13, you can write

$$\int \overbrace{(x^2 + 1)^2}^{f(g(x))}\overbrace{(2x)}^{g'(x)}\, dx = \frac{1}{3}(x^2 + 1)^3 + C.$$

Try using the Chain Rule to check that the derivative of $\frac{1}{3}(x^2 + 1)^3 + C$ is the integrand of the original integral.

EXAMPLE 2 Recognizing the $f(g(x))g'(x)$ Pattern

Find $\int 5 \cos 5x\, dx$.

Solution Letting $g(x) = 5x$, you obtain

$$g'(x) = 5$$

and

$$f(g(x)) = f(5x) = \cos 5x.$$

From this, you can recognize that the integrand follows the $f(g(x))g'(x)$ pattern. Using the Cosine Rule for Integration and Theorem 4.13, you can write

$$\int \overbrace{(\cos 5x)}^{f(g(x))}\overbrace{(5)}^{g'(x)}\, dx = \sin 5x + C.$$

You can check this by differentiating $\sin 5x + C$ to obtain the original integrand.

TECHNOLOGY Try using a computer algebra system, such as *Maple, Mathematica,* or the *TI-Nspire*, to find the integrals given in Examples 1 and 2. Do you obtain the same antiderivatives that are listed in the examples?

Exploration

Recognizing Patterns The integrand in each of the integrals labeled (a)–(c) fits the pattern $f(g(x))g'(x)$. Identify the pattern and use the result to find the integral.

a. $\int 2x(x^2 + 1)^4\, dx$ **b.** $\int 3x^2\sqrt{x^3 + 1}\, dx$ **c.** $\int (\sec^2 x)(\tan x + 3)\, dx$

The integrals labeled (d)–(f) are similar to (a)–(c). Show how you can multiply and divide by a constant to find these integrals.

d. $\int x(x^2 + 1)^4\, dx$ **e.** $\int x^2\sqrt{x^3 + 1}\, dx$ **f.** $\int (2 \sec^2 x)(\tan x + 3)\, dx$

The integrands in Examples 1 and 2 fit the $f(g(x))g'(x)$ pattern exactly—you only had to recognize the pattern. You can extend this technique considerably with the Constant Multiple Rule

$$\int kf(x)\,dx = k\int f(x)\,dx.$$

Many integrands contain the essential part (the variable part) of $g'(x)$ but are missing a constant multiple. In such cases, you can multiply and divide by the necessary constant multiple, as shown in Example 3.

EXAMPLE 3 Multiplying and Dividing by a Constant

Find the indefinite integral.

$$\int x(x^2 + 1)^2\,dx$$

Solution This is similar to the integral given in Example 1, except that the integrand is missing a factor of 2. Recognizing that $2x$ is the derivative of $x^2 + 1$, you can let

$$g(x) = x^2 + 1$$

and supply the $2x$ as shown.

$$\int x(x^2 + 1)^2\,dx = \int (x^2 + 1)^2\left(\frac{1}{2}\right)(2x)\,dx \qquad \text{Multiply and divide by 2.}$$

$$= \frac{1}{2}\int \underbrace{(x^2 + 1)^2}_{f(g(x))}\underbrace{(2x)}_{g'(x)}\,dx \qquad \text{Constant Multiple Rule}$$

$$= \frac{1}{2}\left[\frac{(x^2 + 1)^3}{3}\right] + C \qquad \text{Integrate.}$$

$$= \frac{1}{6}(x^2 + 1)^3 + C \qquad \text{Simplify.}$$

In practice, most people would not write as many steps as are shown in Example 3. For instance, you could evaluate the integral by simply writing

$$\int x(x^2 + 1)^2\,dx = \frac{1}{2}\int (x^2 + 1)^2\,(2x)\,dx$$

$$= \frac{1}{2}\left[\frac{(x^2 + 1)^3}{3}\right] + C$$

$$= \frac{1}{6}(x^2 + 1)^3 + C.$$

Be sure you see that the *Constant* Multiple Rule applies only to *constants*. You cannot multiply and divide by a variable and then move the variable outside the integral sign. For instance,

$$\int (x^2 + 1)^2\,dx \neq \frac{1}{2x}\int (x^2 + 1)^2\,(2x)\,dx.$$

After all, if it were legitimate to move variable quantities outside the integral sign, you could move the entire integrand out and simplify the whole process. But the result would be incorrect.

Change of Variables for Indefinite Integrals

With a formal **change of variables,** you completely rewrite the integral in terms of u and du (or any other convenient variable). Although this procedure can involve more written steps than the pattern recognition illustrated in Examples 1 through 3, it is useful for complicated integrands. The change of variables technique uses the Leibniz notation for the differential. That is, if $u = g(x)$, then $du = g'(x)\,dx$, and the integral in Theorem 4.13 takes the form

$$\int f(g(x))g'(x)\,dx = \int f(u)\,du = F(u) + C.$$

EXAMPLE 4 Change of Variables

Find $\int \sqrt{2x - 1}\,dx$.

Solution First, let u be the inner function, $u = 2x - 1$. Then calculate the differential du to be $du = 2\,dx$. Now, using $\sqrt{2x - 1} = \sqrt{u}$ and $dx = du/2$, substitute to obtain

$$\int \sqrt{2x - 1}\,dx = \int \sqrt{u}\left(\frac{du}{2}\right) \qquad \text{Integral in terms of } u$$

$$= \frac{1}{2}\int u^{1/2}\,du \qquad \text{Constant Multiple Rule}$$

$$= \frac{1}{2}\left(\frac{u^{3/2}}{3/2}\right) + C \qquad \text{Antiderivative in terms of } u$$

$$= \frac{1}{3}u^{3/2} + C \qquad \text{Simplify.}$$

$$= \frac{1}{3}(2x - 1)^{3/2} + C. \qquad \text{Antiderivative in terms of } x$$

REMARK Because integration is usually more difficult than differentiation, you should always check your answer to an integration problem by differentiating. For instance, in Example 4, you should differentiate $\frac{1}{3}(2x - 1)^{3/2} + C$ to verify that you obtain the original integrand.

EXAMPLE 5 Change of Variables

See LarsonCalculus.com for an interactive version of this type of example.

Find $\int x\sqrt{2x - 1}\,dx$.

Solution As in the previous example, let $u = 2x - 1$ and obtain $dx = du/2$. Because the integrand contains a factor of x, you must also solve for x in terms of u, as shown.

$$u = 2x - 1 \implies x = \frac{u + 1}{2} \qquad \text{Solve for } x \text{ in terms of } u.$$

Now, using substitution, you obtain

$$\int x\sqrt{2x - 1}\,dx = \int \left(\frac{u + 1}{2}\right)u^{1/2}\left(\frac{du}{2}\right)$$

$$= \frac{1}{4}\int (u^{3/2} + u^{1/2})\,du$$

$$= \frac{1}{4}\left(\frac{u^{5/2}}{5/2} + \frac{u^{3/2}}{3/2}\right) + C$$

$$= \frac{1}{10}(2x - 1)^{5/2} + \frac{1}{6}(2x - 1)^{3/2} + C.$$

To complete the change of variables in Example 5, you solved for x in terms of u. Sometimes this is very difficult. Fortunately, it is not always necessary, as shown in the next example.

EXAMPLE 6 Change of Variables

Find $\int \sin^2 3x \cos 3x\, dx$.

Solution Because $\sin^2 3x = (\sin 3x)^2$, you can let $u = \sin 3x$. Then

$$du = (\cos 3x)(3)\, dx.$$

Now, because $\cos 3x\, dx$ is part of the original integral, you can write

$$\frac{du}{3} = \cos 3x\, dx.$$

Substituting u and $du/3$ in the original integral yields

$$\begin{aligned}\int \sin^2 3x \cos 3x\, dx &= \int u^2 \frac{du}{3}\\ &= \frac{1}{3}\int u^2\, du\\ &= \frac{1}{3}\left(\frac{u^3}{3}\right) + C\\ &= \frac{1}{9}\sin^3 3x + C.\end{aligned}$$

You can check this by differentiating.

$$\begin{aligned}\frac{d}{dx}\left[\frac{1}{9}\sin^3 3x + C\right] &= \left(\frac{1}{9}\right)(3)(\sin 3x)^2(\cos 3x)(3)\\ &= \sin^2 3x \cos 3x\end{aligned}$$

Because differentiation produces the original integrand, you know that you have obtained the correct antiderivative. ■

REMARK When making a change of variables, be sure that your answer is written using the same variables as in the original integrand. For instance, in Example 6, you should not leave your answer as

$$\frac{1}{9}u^3 + C$$

but rather, you should replace u by $\sin 3x$.

The steps used for integration by substitution are summarized in the following guidelines.

GUIDELINES FOR MAKING A CHANGE OF VARIABLES

1. Choose a substitution $u = g(x)$. Usually, it is best to choose the *inner* part of a composite function, such as a quantity raised to a power.
2. Compute $du = g'(x)\, dx$.
3. Rewrite the integral in terms of the variable u.
4. Find the resulting integral in terms of u.
5. Replace u by $g(x)$ to obtain an antiderivative in terms of x.
6. Check your answer by differentiating.

So far, you have seen two techniques for applying substitution, and you will see more techniques in the remainder of this section. Each technique differs slightly from the others. You should remember, however, that the goal is the same with each technique—*you are trying to find an antiderivative of the integrand.*

The General Power Rule for Integration

One of the most common u-substitutions involves quantities in the integrand that are raised to a power. Because of the importance of this type of substitution, it is given a special name—the **General Power Rule for Integration.** A proof of this rule follows directly from the (simple) Power Rule for Integration, together with Theorem 4.13.

THEOREM 4.14 The General Power Rule for Integration

If g is a differentiable function of x, then

$$\int [g(x)]^n g'(x)\, dx = \frac{[g(x)]^{n+1}}{n+1} + C, \quad n \neq -1.$$

Equivalently, if $u = g(x)$, then

$$\int u^n\, du = \frac{u^{n+1}}{n+1} + C, \quad n \neq -1.$$

EXAMPLE 7 Substitution and the General Power Rule

a. $$\int 3(3x-1)^4\, dx = \int \overbrace{(3x-1)^4}^{u^4}\overbrace{(3)\, dx}^{du} = \overbrace{\frac{(3x-1)^5}{5}}^{u^5/5} + C$$

b. $$\int (2x+1)(x^2+x)\, dx = \int \overbrace{(x^2+x)^1}^{u^1}\overbrace{(2x+1)\, dx}^{du} = \overbrace{\frac{(x^2+x)^2}{2}}^{u^2/2} + C$$

c. $$\int 3x^2\sqrt{x^3-2}\, dx = \int \overbrace{(x^3-2)^{1/2}}^{u^{1/2}}\overbrace{(3x^2)\, dx}^{du} = \overbrace{\frac{(x^3-2)^{3/2}}{3/2}}^{u^{3/2}/(3/2)} + C = \frac{2}{3}(x^3-2)^{3/2} + C$$

d. $$\int \frac{-4x}{(1-2x^2)^2}\, dx = \int \overbrace{(1-2x^2)^{-2}}^{u^{-2}}\overbrace{(-4x)\, dx}^{du} = \overbrace{\frac{(1-2x^2)^{-1}}{-1}}^{u^{-1}/(-1)} + C = -\frac{1}{1-2x^2} + C$$

e. $$\int \cos^2 x \sin x\, dx = -\int \overbrace{(\cos x)^2}^{u^2}\overbrace{(-\sin x)\, dx}^{du} = -\overbrace{\frac{(\cos x)^3}{3}}^{u^3/3} + C$$

Some integrals whose integrands involve quantities raised to powers cannot be found by the General Power Rule. Consider the two integrals

$$\int x(x^2+1)^2\, dx \quad \text{and} \quad \int (x^2+1)^2\, dx.$$

The substitution

$$u = x^2 + 1$$

works in the first integral but not in the second. In the second, the substitution fails because the integrand lacks the factor x needed for du. Fortunately, *for this particular integral*, you can expand the integrand as

$$(x^2+1)^2 = x^4 + 2x^2 + 1$$

and use the (simple) Power Rule to integrate each term.

Change of Variables for Definite Integrals

When using u-substitution with a definite integral, it is often convenient to determine the limits of integration for the variable u rather than to convert the antiderivative back to the variable x and evaluate at the original limits. This change of variables is stated explicitly in the next theorem. The proof follows from Theorem 4.13 combined with the Fundamental Theorem of Calculus.

THEOREM 4.15 Change of Variables for Definite Integrals

If the function $u = g(x)$ has a continuous derivative on the closed interval $[a, b]$ and f is continuous on the range of g, then

$$\int_a^b f(g(x))g'(x)\,dx = \int_{g(a)}^{g(b)} f(u)\,du.$$

EXAMPLE 8 Change of Variables

Evaluate $\int_0^1 x(x^2 + 1)^3\,dx$.

Solution To evaluate this integral, let $u = x^2 + 1$. Then, you obtain

$$du = 2x\,dx.$$

Before substituting, determine the new upper and lower limits of integration.

Lower Limit	**Upper Limit**
When $x = 0$, $u = 0^2 + 1 = 1$.	When $x = 1$, $u = 1^2 + 1 = 2$.

Now, you can substitute to obtain

$$\int_0^1 x(x^2 + 1)^3\,dx = \frac{1}{2}\int_0^1 (x^2 + 1)^3(2x)\,dx \qquad \text{Integration limits for } x$$

$$= \frac{1}{2}\int_1^2 u^3\,du \qquad \text{Integration limits for } u$$

$$= \frac{1}{2}\left[\frac{u^4}{4}\right]_1^2$$

$$= \frac{1}{2}\left(4 - \frac{1}{4}\right)$$

$$= \frac{15}{8}.$$

Notice that you obtain the same result when you rewrite the antiderivative $\frac{1}{2}(u^4/4)$ in terms of the variable x and evaluate the definite integral at the original limits of integration, as shown below.

$$\frac{1}{2}\left[\frac{u^4}{4}\right]_1^2 = \frac{1}{2}\left[\frac{(x^2 + 1)^4}{4}\right]_0^1$$

$$= \frac{1}{2}\left(4 - \frac{1}{4}\right)$$

$$= \frac{15}{8}$$

■

EXAMPLE 9 Change of Variables

Evaluate the definite integral.

$$\int_1^5 \frac{x}{\sqrt{2x-1}}\,dx$$

Solution To evaluate this integral, let $u = \sqrt{2x-1}$. Then, you obtain

$$u^2 = 2x - 1$$
$$u^2 + 1 = 2x$$
$$\frac{u^2+1}{2} = x$$
$$u\,du = dx. \qquad \text{Differentiate each side.}$$

Before substituting, determine the new upper and lower limits of integration.

Lower Limit

When $x = 1$, $u = \sqrt{2-1} = 1$.

Upper Limit

When $x = 5$, $u = \sqrt{10-1} = 3$.

Now, substitute to obtain

$$\begin{aligned}\int_1^5 \frac{x}{\sqrt{2x-1}}\,dx &= \int_1^3 \frac{1}{u}\left(\frac{u^2+1}{2}\right)u\,du\\ &= \frac{1}{2}\int_1^3 (u^2+1)\,du\\ &= \frac{1}{2}\left[\frac{u^3}{3} + u\right]_1^3\\ &= \frac{1}{2}\left(9 + 3 - \frac{1}{3} - 1\right)\\ &= \frac{16}{3}.\end{aligned}$$

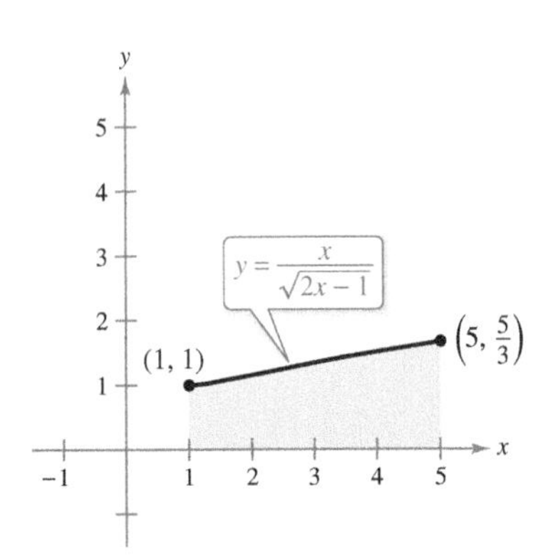

The region before substitution has an area of $\frac{16}{3}$.
Figure 4.38

Geometrically, you can interpret the equation

$$\int_1^5 \frac{x}{\sqrt{2x-1}}\,dx = \int_1^3 \frac{u^2+1}{2}\,du$$

to mean that the two *different* regions shown in Figures 4.38 and 4.39 have the *same* area.

When evaluating definite integrals by substitution, it is possible for the upper limit of integration of the u-variable form to be smaller than the lower limit. When this happens, do not rearrange the limits. Simply evaluate as usual. For example, after substituting $u = \sqrt{1-x}$ in the integral

$$\int_0^1 x^2(1-x)^{1/2}\,dx$$

you obtain $u = \sqrt{1-0} = 1$ when $x = 0$, and $u = \sqrt{1-1} = 0$ when $x = 1$. So, the correct u-variable form of this integral is

$$-2\int_1^0 (1-u^2)^2u^2\,du.$$

Expanding the integrand, you can evaluate this integral as shown.

$$-2\int_1^0 (u^2 - 2u^4 + u^6)\,du = -2\left[\frac{u^3}{3} - \frac{2u^5}{5} + \frac{u^7}{7}\right]_1^0 = -2\left(-\frac{1}{3} + \frac{2}{5} - \frac{1}{7}\right) = \frac{16}{105}$$

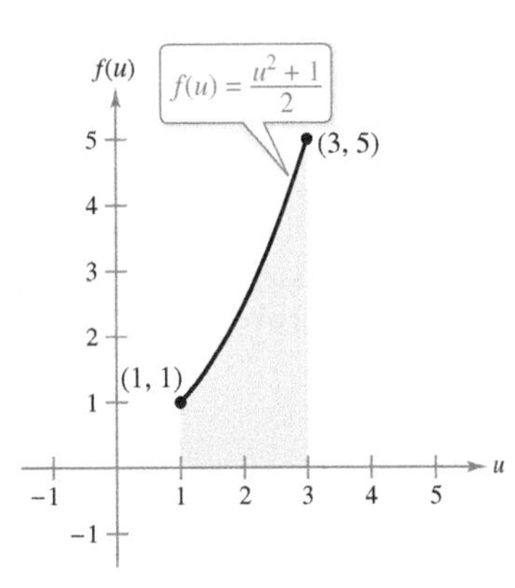

The region after substitution has an area of $\frac{16}{3}$.
Figure 4.39

Integration of Even and Odd Functions

Even with a change of variables, integration can be difficult. Occasionally, you can simplify the evaluation of a definite integral over an interval that is symmetric about the y-axis or about the origin by recognizing the integrand to be an even or odd function (see Figure 4.40).

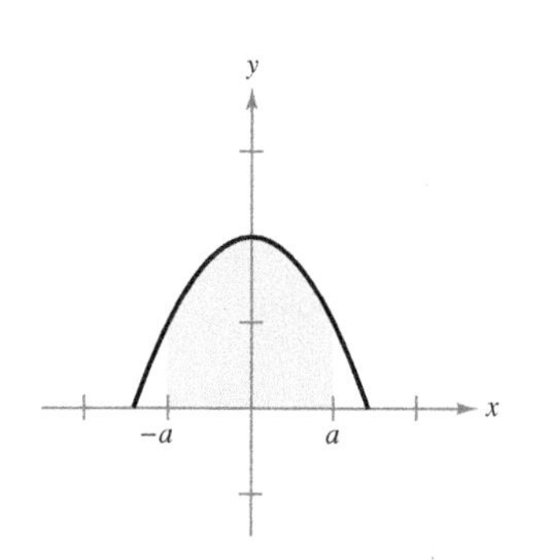

Even function

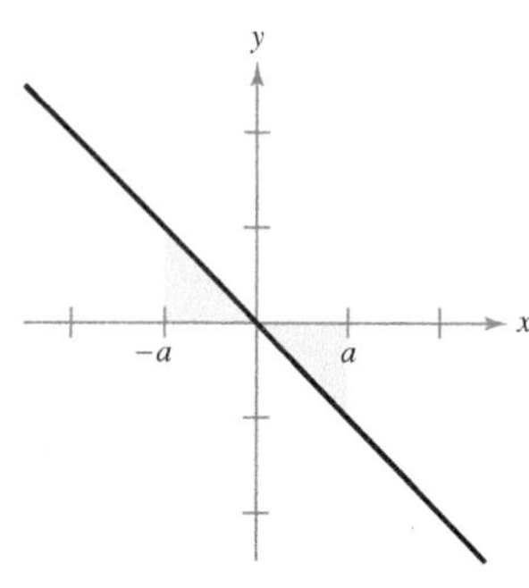

Odd function

Figure 4.40

THEOREM 4.16 Integration of Even and Odd Functions

Let f be integrable on the closed interval $[-a, a]$.

1. If f is an *even* function, then $\int_{-a}^{a} f(x)\,dx = 2\int_{0}^{a} f(x)\,dx.$

2. If f is an *odd* function, then $\int_{-a}^{a} f(x)\,dx = 0.$

Proof Here is the proof of the first property. (The proof of the second property is left to you [see Exercise 101].) Because f is even, you know that

$$f(x) = f(-x).$$

Using Theorem 4.13 with the substitution $u = -x$ produces

$$\int_{-a}^{0} f(x)\,dx = \int_{a}^{0} f(-u)(-du) = -\int_{a}^{0} f(u)\,du = \int_{0}^{a} f(u)\,du = \int_{0}^{a} f(x)\,dx.$$

Finally, using Theorem 4.6, you obtain

$$\begin{aligned}\int_{-a}^{a} f(x)\,dx &= \int_{-a}^{0} f(x)\,dx + \int_{0}^{a} f(x)\,dx\\ &= \int_{0}^{a} f(x)\,dx + \int_{0}^{a} f(x)\,dx\\ &= 2\int_{0}^{a} f(x)\,dx.\end{aligned}$$

EXAMPLE 10 Integration of an Odd Function

Evaluate the definite integral.

$$\int_{-\pi/2}^{\pi/2} (\sin^3 x \cos x + \sin x \cos x)\,dx$$

Solution Letting $f(x) = \sin^3 x \cos x + \sin x \cos x$ produces

$$\begin{aligned}f(-x) &= \sin^3(-x)\cos(-x) + \sin(-x)\cos(-x)\\ &= -\sin^3 x \cos x - \sin x \cos x\\ &= -f(x).\end{aligned}$$

So, f is an odd function, and because f is symmetric about the origin over $[-\pi/2, \pi/2]$, you can apply Theorem 4.16 to conclude that

$$\int_{-\pi/2}^{\pi/2} (\sin^3 x \cos x + \sin x \cos x)\,dx = 0.$$

From Figure 4.41, you can see that the two regions on either side of the y-axis have the same area. However, because one lies below the x-axis and one lies above it, integration produces a cancellation effect. (More will be said about areas below the x-axis in Section 7.1.)

$f(x) = \sin^3 x \cos x + \sin x \cos x$

Because f is an odd function, $\int_{-\pi/2}^{\pi/2} f(x)\,dx = 0.$

Figure 4.41

4.5 Exercises

See CalcChat.com for tutorial help and worked-out solutions to odd-numbered exercises.

CONCEPT CHECK

1. **Constant Multiple Rule** Explain how to use the Constant Multiple Rule when finding an indefinite integral.
2. **Change of Variables** In your own words, summarize the guidelines for making a change of variables when finding an indefinite integral.
3. **The General Power Rule for Integration** Describe the General Power Rule for Integration in your own words.
4. **Analyzing the Integrand** Without integrating, explain why

$$\int_{-2}^{2} x(x^2+1)^2\,dx = 0.$$

Recognizing Patterns **In Exercises 5–8, complete the table by identifying u and du for the integral.**

$\int f(g(x))g'(x)\,dx$	$u = g(x)$	$du = g'(x)\,dx$
5. $\int (5x^2+1)^2(10x)\,dx$		
6. $\int x^2\sqrt{x^3+1}\,dx$		
7. $\int \tan^2 x \sec^2 x\,dx$		
8. $\int \frac{\cos x}{\sin^2 x}\,dx$		

Finding an Indefinite Integral **In Exercises 9–30, find the indefinite integral and check the result by differentiation.**

9. $\int (1+6x)^4(6)\,dx$
10. $\int (x^2-9)^3(2x)\,dx$
11. $\int \sqrt{25-x^2}\,(-2x)\,dx$
12. $\int \sqrt[3]{3-4x^2}(-8x)\,dx$
13. $\int x^3(x^4+3)^2\,dx$
14. $\int x^2(6-x^3)^5\,dx$
15. $\int x^2(2x^3-1)^4\,dx$
16. $\int x(5x^2+4)^3\,dx$
17. $\int t\sqrt{t^2+2}\,dt$
18. $\int t^3\sqrt{2t^4+3}\,dt$
19. $\int 5x\sqrt[3]{1-x^2}\,dx$
20. $\int 6u^6\sqrt{u^7+8}\,du$
21. $\int \frac{7x}{(1-x^2)^3}\,dx$
22. $\int \frac{x^3}{(1+x^4)^2}\,dx$
23. $\int \frac{x^2}{(1+x^3)^2}\,dx$
24. $\int \frac{6x^2}{(4x^3-9)^3}\,dx$
25. $\int \frac{x}{\sqrt{1-x^2}}\,dx$
26. $\int \frac{x^3}{\sqrt{1+x^4}}\,dx$
27. $\int \left(1+\frac{1}{t}\right)^3\left(\frac{1}{t^2}\right)dt$
28. $\int \left(8-\frac{1}{t^4}\right)^2\left(\frac{1}{t^5}\right)dt$
29. $\int \frac{1}{\sqrt{2x}}\,dx$
30. $\int \frac{x}{\sqrt[3]{5x^2}}\,dx$

Differential Equation **In Exercises 31–34, find the general solution of the differential equation.**

31. $\frac{dy}{dx} = 4x + \frac{4x}{\sqrt{16-x^2}}$
32. $\frac{dy}{dx} = \frac{10x^2}{\sqrt{1+x^3}}$
33. $\frac{dy}{dx} = \frac{x+1}{(x^2+2x-3)^2}$
34. $\frac{dy}{dx} = \frac{18-6x^2}{\sqrt{x^3-9x+7}}$

Slope Field **In Exercises 35 and 36, a differential equation, a point, and a slope field are given. A *slope field* (or *direction field*) consists of line segments with slopes given by the differential equation. These line segments give a visual perspective of the slopes of the solutions of the differential equation. (a) Sketch two approximate solutions of the differential equation on the slope field, one of which passes through the given point. (To print an enlarged copy of the graph, go to *MathGraphs.com*.) (b) Use integration and the given point to find the particular solution of the differential equation and use a graphing utility to graph the solution. Compare the result with the sketch in part (a) that passes through the given point.**

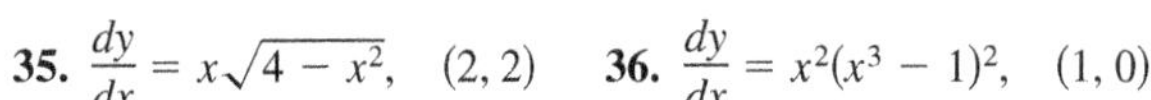

35. $\frac{dy}{dx} = x\sqrt{4-x^2}$, $(2, 2)$
36. $\frac{dy}{dx} = x^2(x^3-1)^2$, $(1, 0)$

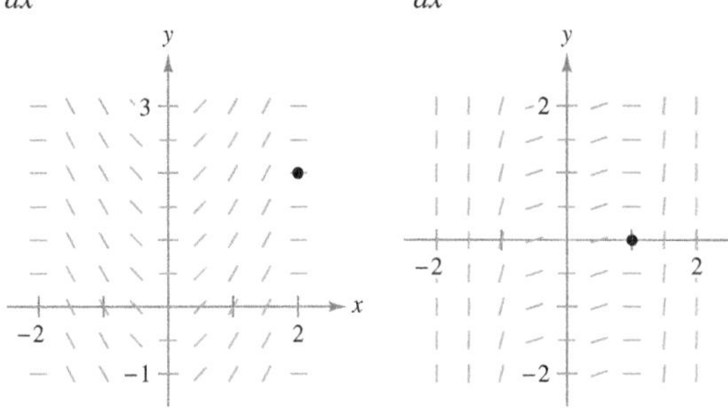

Differential Equation **In Exercises 37 and 38, the graph of a function f is shown. Use the differential equation and the given point to find an equation of the function.**

37. $\frac{dy}{dx} = 18x^2(2x^3+1)^2$
38. $\frac{dy}{dx} = \frac{-48}{(3x+5)^3}$

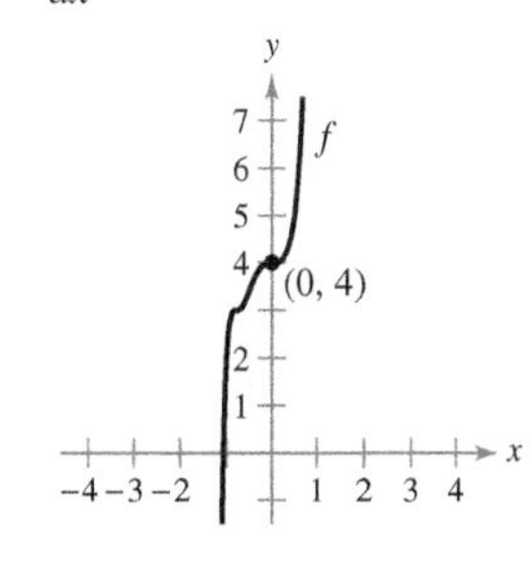

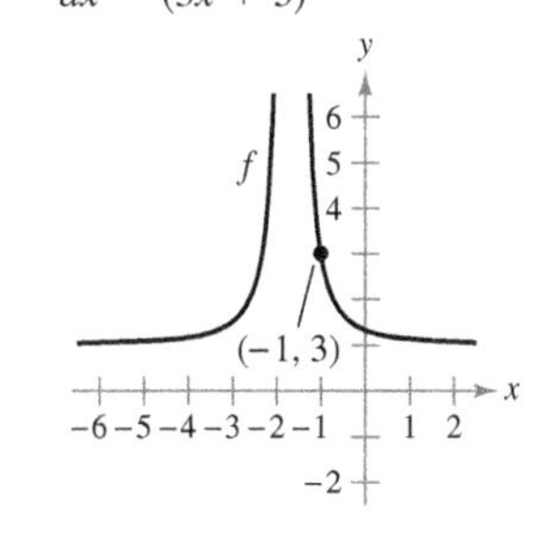

Finding an Indefinite Integral **In Exercises 39–48, find the indefinite integral.**

39. $\int \pi \sin \pi x \, dx$

40. $\int \sin 4x \, dx$

41. $\int \cos 6x \, dx$

42. $\int \csc^2\left(\frac{x}{2}\right) dx$

43. $\int \frac{1}{\theta^2} \cos \frac{1}{\theta} \, d\theta$

44. $\int x \sin x^2 \, dx$

45. $\int \sin 2x \cos 2x \, dx$

46. $\int \sqrt[3]{\tan x} \sec^2 x \, dx$

47. $\int \frac{\csc^2 x}{\cot^3 x} \, dx$

48. $\int \frac{\sin x}{\cos^3 x} \, dx$

Finding an Equation **In Exercises 49–52, find an equation for the function f that has the given derivative and whose graph passes through the given point.**

	Derivative	Point
49.	$f'(x) = -\sin \frac{x}{2}$	$(0, 6)$
50.	$f'(x) = \sec^2 2x$	$\left(\frac{\pi}{2}, 2\right)$
51.	$f'(x) = 2x(4x^2 - 10)^2$	$(2, 10)$
52.	$f'(x) = -2x\sqrt{8 - x^2}$	$(2, 7)$

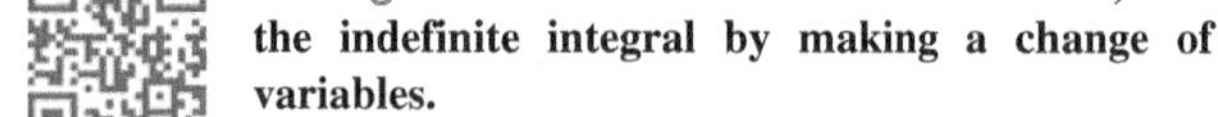

Change of Variables **In Exercises 53–60, find the indefinite integral by making a change of variables.**

53. $\int x\sqrt{x + 6} \, dx$

54. $\int x\sqrt{3x - 4} \, dx$

55. $\int x^2\sqrt{1 - x} \, dx$

56. $\int (x + 1)\sqrt{2 - x} \, dx$

57. $\int \frac{x^2 - 1}{\sqrt{2x - 1}} \, dx$

58. $\int \frac{2x + 1}{\sqrt{x + 4}} \, dx$

59. $\int \cos^3 2x \sin 2x \, dx$

60. $\int \sec^5 7x \tan 7x \, dx$

Evaluating a Definite Integral **In Exercises 61–68, evaluate the definite integral. Use a graphing utility to verify your result.**

61. $\int_{-1}^{1} x(x^2 + 1)^3 \, dx$

62. $\int_{0}^{1} x^3(2x^4 + 1)^2 \, dx$

63. $\int_{1}^{2} 2x^2\sqrt{x^3 + 1} \, dx$

64. $\int_{-1}^{0} x\sqrt{1 - x^2} \, dx$

65. $\int_{0}^{4} \frac{1}{\sqrt{2x + 1}} \, dx$

66. $\int_{0}^{2} \frac{x}{\sqrt{1 + 2x^2}} \, dx$

67. $\int_{1}^{9} \frac{1}{\sqrt{x}(1 + \sqrt{x})^2} \, dx$

68. $\int_{4}^{5} \frac{x}{\sqrt{2x - 6}} \, dx$

Finding the Area of a Region **In Exercises 69–72, find the area of the region. Use a graphing utility to verify your result.**

69. $\int_{0}^{7} x\sqrt[3]{x + 1} \, dx$

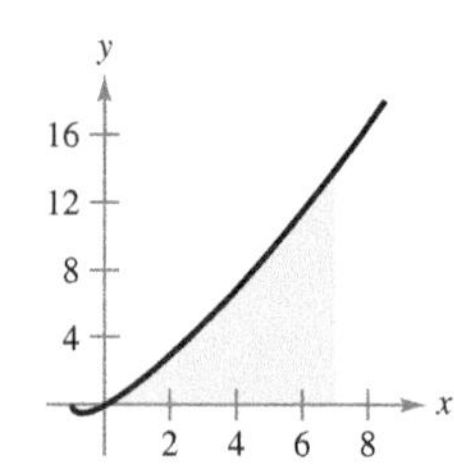

70. $\int_{-2}^{6} x^2\sqrt[3]{x + 2} \, dx$

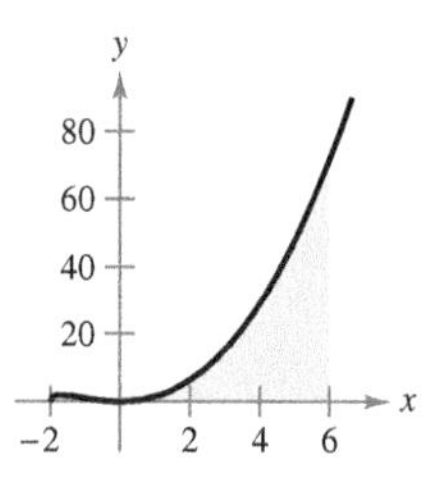

71. $\int_{\pi/2}^{2\pi/3} \sec^2\left(\frac{x}{2}\right) dx$

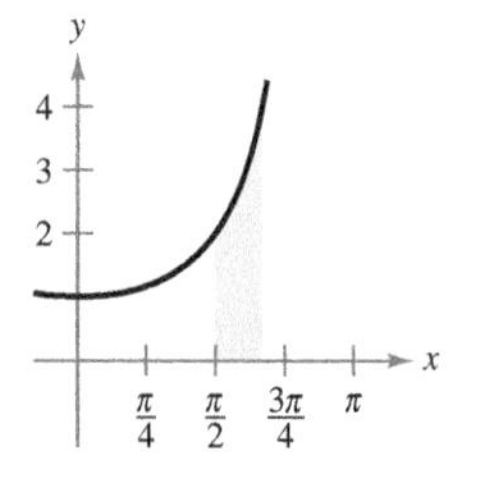

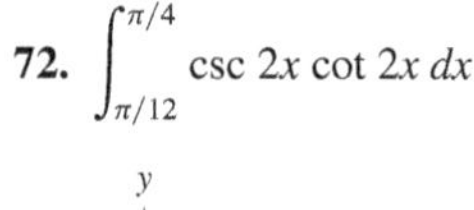

72. $\int_{\pi/12}^{\pi/4} \csc 2x \cot 2x \, dx$

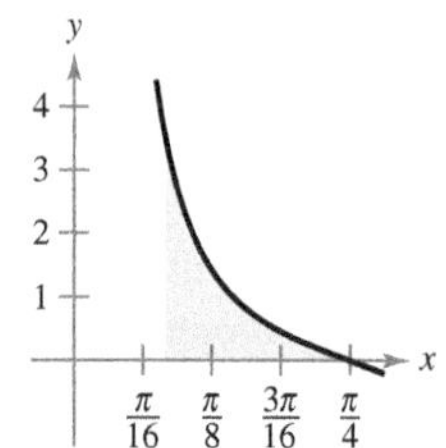

Even and Odd Functions **In Exercises 73–76, evaluate the integral using the properties of even and odd functions as an aid.**

73. $\int_{-2}^{2} x^2(x^2 + 1) \, dx$

74. $\int_{-2}^{2} x(x^2 + 1)^3 \, dx$

75. $\int_{-\pi/2}^{\pi/2} \sin x \cos x \, dx$

76. $\int_{-\pi/2}^{\pi/2} \sin^2 x \cos x \, dx$

77. **Using an Even Function** Use $\int_0^6 x^2 \, dx = 72$ to evaluate each definite integral without using the Fundamental Theorem of Calculus.

(a) $\int_{-6}^{6} x^2 \, dx$ (b) $\int_{-6}^{0} x^2 \, dx$

(c) $\int_{0}^{6} -2x^2 \, dx$ (d) $\int_{-6}^{6} 3x^2 \, dx$

78. **Using Symmetry** Use the symmetry of the graphs of the sine and cosine functions as an aid in evaluating each definite integral.

(a) $\int_{-\pi/4}^{\pi/4} \sin x \, dx$ (b) $\int_{-\pi/4}^{\pi/4} \cos x \, dx$

(c) $\int_{-\pi/2}^{\pi/2} \cos x \, dx$ (d) $\int_{-\pi/2}^{\pi/2} \sin x \cos x \, dx$

Even and Odd Functions **In Exercises 79 and 80, write the integral as the sum of the integral of an odd function and the integral of an even function. Use this simplification to evaluate the integral.**

79. $\int_{-3}^{3} (x^3 + 4x^2 - 3x - 6) \, dx$

80. $\int_{-\pi/2}^{\pi/2} (\sin 4x + \cos 4x) \, dx$

EXPLORING CONCEPTS

81. Choosing an Integral You are asked to find one of the integrals. Which one would you choose? Explain.

(a) $\int \sqrt{x^3 + 1}\, dx$ or $\int x^2\sqrt{x^3 + 1}\, dx$

(b) $\int \cot 2x\, dx$ or $\int \cot^3 2x \csc^2 2x\, dx$

82. Comparing Methods Find the indefinite integral in two ways. Explain any difference in the forms of the answers.

(a) $\int (2x - 1)^2\, dx$ (b) $\int \sin x \cos x\, dx$

83. Depreciation The rate of depreciation dV/dt of a machine is inversely proportional to the square of $(t + 1)$, where V is the value of the machine t years after it was purchased. The initial value of the machine was \$500,000, and its value decreased \$100,000 in the first year. Estimate its value after 4 years.

84. HOW DO YOU SEE IT? The graph shows the flow rate of water at a pumping station for one day.

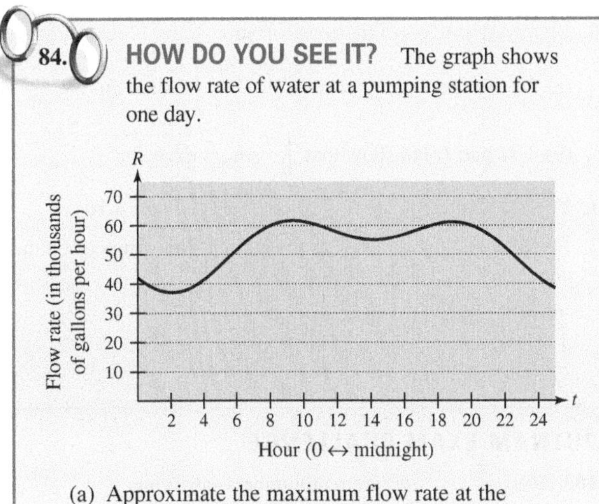

(a) Approximate the maximum flow rate at the pumping station. At what time does this occur?

(b) Explain how you can find the amount of water used during the day.

(c) Approximate the two-hour period when the least amount of water is used. Explain your reasoning.

85. Sales The sales S (in thousands of units) of a seasonal product are given by the model

$$S = 74.50 + 43.75 \sin \frac{\pi t}{6}$$

where t is the time in months, with $t = 1$ corresponding to January. Find the average sales for each time period.

(a) The first quarter $(0 \le t \le 3)$

(b) The second quarter $(3 \le t \le 6)$

(c) The entire year $(0 \le t \le 12)$

86. Electricity

The oscillating current in an electrical circuit is

$I = 2\sin(60\pi t) + \cos(120\pi t)$

where I is measured in amperes and t is measured in seconds. Find the average current for each time interval.

(a) $0 \le t \le \frac{1}{60}$

(b) $0 \le t \le \frac{1}{240}$

(c) $0 \le t \le \frac{1}{30}$

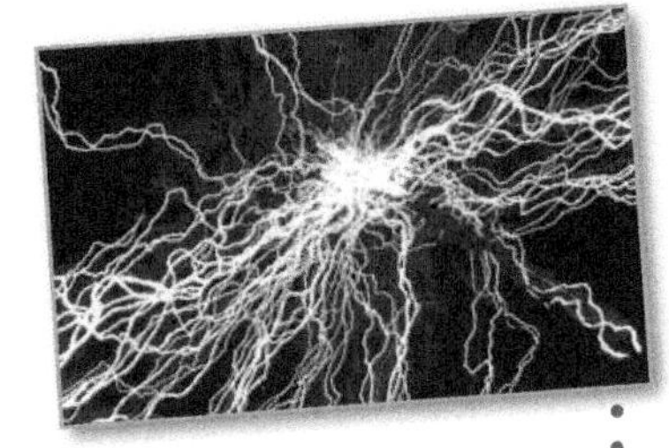

87. Graphical Analysis Consider the functions f and g, where

$$f(x) = 6 \sin x \cos^2 x \quad \text{and} \quad g(t) = \int_0^t f(x)\, dx.$$

(a) Use a graphing utility to graph f and g in the same viewing window.

(b) Explain why g is nonnegative.

(c) Identify the points on the graph of g that correspond to the extrema of f.

(d) Does each of the zeros of f correspond to an extremum of g? Explain.

(e) Consider the function

$$h(t) = \int_{\pi/2}^t f(x)\, dx.$$

Use a graphing utility to graph h. What is the relationship between g and h? Verify your conjecture.

88. Finding a Limit Using a Definite Integral Find

$$\lim_{n\to\infty} \sum_{i=1}^{n} \frac{\sin(i\pi/n)}{n}$$

by evaluating an appropriate definite integral over the interval [0, 1].

89. Rewriting Integrals

(a) Show that $\int_0^1 x^3(1 - x)^8\, dx = \int_0^1 x^8(1 - x)^3\, dx$

(b) Show that $\int_0^1 x^a(1 - x)^b\, dx = \int_0^1 x^b(1 - x)^a\, dx.$

90. Rewriting Integrals

(a) Show that $\int_0^{\pi/2} \sin^2 x\, dx = \int_0^{\pi/2} \cos^2 x\, dx.$

(b) Show that

$$\int_0^{\pi/2} \sin^n x\, dx = \int_0^{\pi/2} \cos^n x\, dx$$

where n is a positive integer.

Probability **In Exercises 91 and 92, the function**

$$f(x) = kx^n(1-x)^m, \quad 0 \le x \le 1$$

where $n > 0$, $m > 0$, and k is a constant, can be used to represent various probability distributions. If k is chosen such that

$$\int_0^1 f(x)\,dx = 1$$

then the probability that x will fall between a and b $(0 \le a \le b \le 1)$ is

$$P_{a,b} = \int_a^b f(x)\,dx.$$

91. The probability that a person will remember between $100a\%$ and $100b\%$ of material learned in an experiment is

$$P_{a,b} = \int_a^b \frac{15}{4}x\sqrt{1-x}\,dx$$

where x represents the proportion remembered. (See figure.)

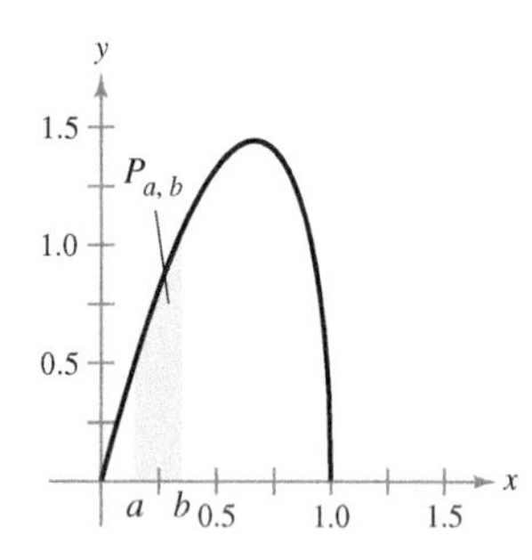

(a) For a randomly chosen individual, what is the probability that he or she will recall between 50% and 75% of the material?

(b) What is the median percent recall? That is, for what value of b is it true that the probability of recalling 0 to b is 0.5?

92. The probability that ore samples taken from a region contain between $100a\%$ and $100b\%$ iron is

$$P_{a,b} = \int_a^b \frac{1155}{32}x^3(1-x)^{3/2}\,dx$$

where x represents the proportion of iron. (See figure.)

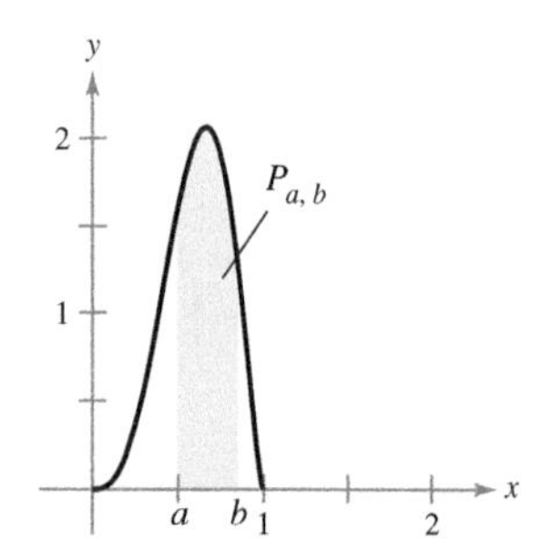

(a) What is the probability that a sample will contain between 0% and 25% iron?

(b) What is the probability that a sample will contain between 50% and 100% iron?

True or False? **In Exercises 93–98, determine whether the statement is true or false. If it is false, explain why or give an example that shows it is false.**

93. $\displaystyle\int 3x^2(x^3+5)^{-2}\,dx = -(x^3+5)^{-1} + C$

94. $\displaystyle\int x(x^2+1)\,dx = \tfrac{1}{2}x^2\left(\tfrac{1}{3}x^3 + x\right) + C$

95. $\displaystyle\int_{-10}^{10} (ax^3 + bx^2 + cx + d)\,dx = 2\int_0^{10} (bx^2 + d)\,dx$

96. $\displaystyle\int_a^b \sin x\,dx = \int_a^{b+2\pi} \sin x\,dx$

97. $\displaystyle 4\int \sin x\cos x\,dx = -\cos 2x + C$

98. $\displaystyle\int \sin^2 2x\cos 2x\,dx = \tfrac{1}{3}\sin^3 2x + C$

99. Rewriting Integrals Assume that f is continuous everywhere and that c is a constant. Show that

$$\int_{ca}^{cb} f(x)\,dx = c\int_a^b f(cx)\,dx.$$

100. Integration and Differentiation

(a) Verify that $\sin u - u\cos u + C = \displaystyle\int u\sin u\,du.$

(b) Use part (a) to show that $\displaystyle\int_0^{\pi^2} \sin\sqrt{x}\,dx = 2\pi.$

101. Proof Prove the second property of Theorem 4.16.

102. Rewriting Integrals Show that if f is continuous on the entire real number line, then

$$\int_a^b f(x+h)\,dx = \int_{a+h}^{b+h} f(x)\,dx.$$

PUTNAM EXAM CHALLENGE

103. If $a_0, a_1, \ldots, a_n$ are real numbers satisfying

$$\frac{a_0}{1} + \frac{a_1}{2} + \cdots + \frac{a_n}{n+1} = 0,$$

show that the equation

$$a_0 + a_1x + a_2x^2 + \cdots + a_nx^n = 0$$

has at least one real root.

104. Find all the continuous positive functions $f(x)$, for $0 \le x \le 1$, such that

$$\int_0^1 f(x)\,dx = 1$$

$$\int_0^1 f(x)x\,dx = \alpha$$

$$\int_0^1 f(x)x^2\,dx = \alpha^2$$

where α is a given real number.

4.6 Numerical Integration

- **Approximate a definite integral using the Trapezoidal Rule.**
- **Approximate a definite integral using Simpson's Rule.**
- **Analyze the approximate errors in the Trapezoidal Rule and Simpson's Rule.**

The Trapezoidal Rule

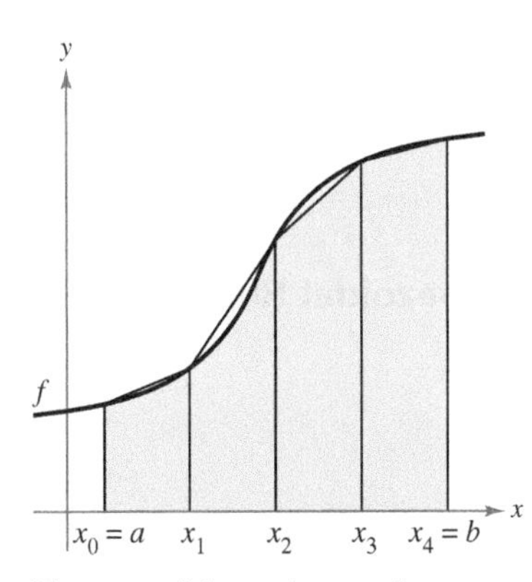

The area of the region can be approximated using four trapezoids.
Figure 4.42

Some elementary functions simply do not have antiderivatives that are elementary functions. For example, there is no elementary function that has any of the following functions as its derivative.

$$\sqrt[3]{x}\sqrt{1-x}, \qquad \sqrt{x}\cos x, \qquad \frac{\cos x}{x}, \qquad \sqrt{1-x^3}, \qquad \sin x^2$$

If you need to evaluate a definite integral involving a function whose antiderivative cannot be found, then while the Fundamental Theorem of Calculus is still true, it cannot be easily applied. In this case, it is easier to resort to an approximation technique. Two such techniques are described in this section.

One way to approximate a definite integral is to use n trapezoids, as shown in Figure 4.42. In the development of this method, assume that f is continuous and positive on the interval $[a, b]$. So, the definite integral

$$\int_a^b f(x)\,dx$$

represents the area of the region bounded by the graph of f and the x-axis, from $x = a$ to $x = b$. First, partition the interval $[a, b]$ into n subintervals, each of width $\Delta x = (b - a)/n$, such that

$$a = x_0 < x_1 < x_2 < \cdot \cdot \cdot < x_n = b.$$

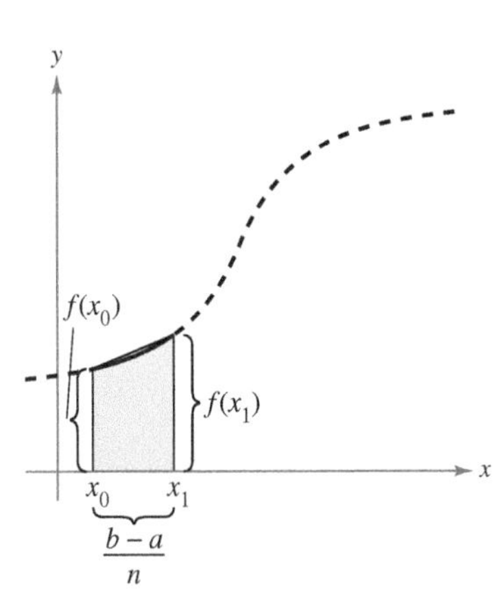

The area of the first trapezoid is
$\left[\frac{f(x_0) + f(x_1)}{2}\right]\left(\frac{b-a}{n}\right)$.
Figure 4.43

Then form a trapezoid for each subinterval (see Figure 4.43). The area of the ith trapezoid is

$$\text{Area of } i\text{th trapezoid} = \left[\frac{f(x_{i-1}) + f(x_i)}{2}\right]\left(\frac{b-a}{n}\right).$$

This implies that the sum of the areas of the n trapezoids is

$$\begin{aligned}
\text{Area} &= \left(\frac{b-a}{n}\right)\left[\frac{f(x_0) + f(x_1)}{2} + \cdot \cdot \cdot + \frac{f(x_{n-1}) + f(x_n)}{2}\right] \\
&= \left(\frac{b-a}{2n}\right)[f(x_0) + f(x_1) + f(x_1) + f(x_2) + \cdot \cdot \cdot + f(x_{n-1}) + f(x_n)] \\
&= \left(\frac{b-a}{2n}\right)[f(x_0) + 2f(x_1) + 2f(x_2) + \cdot \cdot \cdot + 2f(x_{n-1}) + f(x_n)].
\end{aligned}$$

Letting $\Delta x = (b - a)/n$, you can take the limit as $n \to \infty$ to obtain

$$\begin{aligned}
&\lim_{n\to\infty}\left(\frac{b-a}{2n}\right)[f(x_0) + 2f(x_1) + \cdot \cdot \cdot + 2f(x_{n-1}) + f(x_n)] \\
&= \lim_{n\to\infty}\left[\frac{[f(a) - f(b)]\,\Delta x}{2} + \sum_{i=1}^{n} f(x_i)\,\Delta x\right] \\
&= \lim_{n\to\infty}\frac{[f(a) - f(b)](b-a)}{2n} + \lim_{n\to\infty}\sum_{i=1}^{n} f(x_i)\,\Delta x \\
&= 0 + \int_a^b f(x)\,dx.
\end{aligned}$$

The result is summarized in the next theorem.

> **THEOREM 4.17 The Trapezoidal Rule**
>
> Let f be continuous on $[a, b]$. The Trapezoidal Rule for approximating $\int_a^b f(x)\,dx$ is
>
> $$\int_a^b f(x)\,dx \approx \frac{b-a}{2n}[f(x_0) + 2f(x_1) + 2f(x_2) + \cdots + 2f(x_{n-1}) + f(x_n)].$$
>
> Moreover, as $n \to \infty$, the right-hand side approaches $\int_a^b f(x)\,dx$.

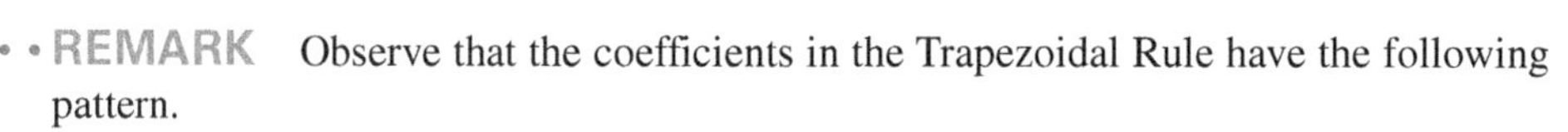

REMARK Observe that the coefficients in the Trapezoidal Rule have the following pattern.

1 2 2 2 . . . 2 2 1

EXAMPLE 1 Approximation with the Trapezoidal Rule

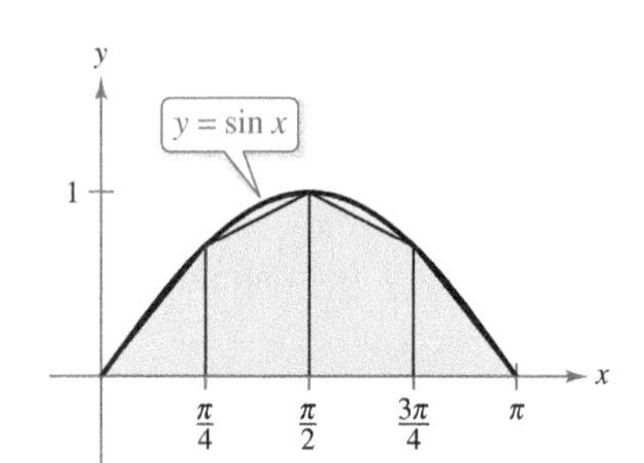

Four subintervals

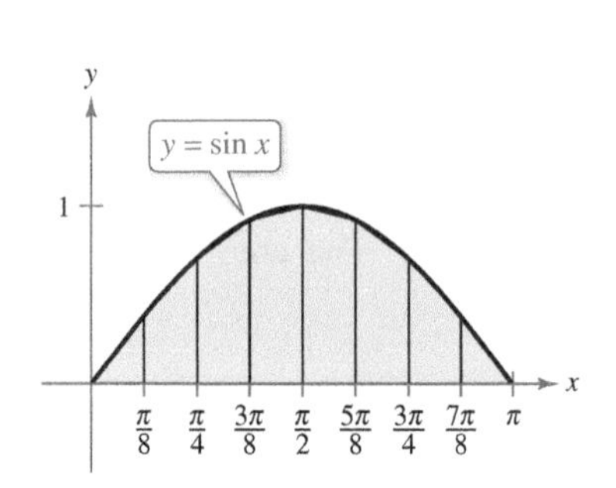

Eight subintervals

Trapezoidal approximations
Figure 4.44

Use the Trapezoidal Rule to approximate

$$\int_0^{\pi} \sin x\,dx.$$

Compare the results for $n = 4$ and $n = 8$, as shown in Figure 4.44.

Solution When $n = 4$, $\Delta x = \pi/4$, and you obtain

$$\begin{aligned}\int_0^{\pi} \sin x\,dx &\approx \frac{\pi}{8}\left(\sin 0 + 2\sin\frac{\pi}{4} + 2\sin\frac{\pi}{2} + 2\sin\frac{3\pi}{4} + \sin\pi\right)\\ &= \frac{\pi}{8}\left(0 + \sqrt{2} + 2 + \sqrt{2} + 0\right)\\ &= \frac{\pi(1+\sqrt{2})}{4}\\ &\approx 1.896.\end{aligned}$$

When $n = 8$, $\Delta x = \pi/8$, and you obtain

$$\begin{aligned}\int_0^{\pi} \sin x\,dx &\approx \frac{\pi}{16}\left(\sin 0 + 2\sin\frac{\pi}{8} + 2\sin\frac{\pi}{4} + 2\sin\frac{3\pi}{8} + 2\sin\frac{\pi}{2}\right.\\ &\qquad \left. + 2\sin\frac{5\pi}{8} + 2\sin\frac{3\pi}{4} + 2\sin\frac{7\pi}{8} + \sin\pi\right)\\ &= \frac{\pi}{16}\left(2 + 2\sqrt{2} + 4\sin\frac{\pi}{8} + 4\sin\frac{3\pi}{8}\right)\\ &\approx 1.974.\end{aligned}$$

For this particular integral, you could have found an antiderivative and determined that the exact area of the region is 2. ■

TECHNOLOGY Most graphing utilities and computer algebra systems have built-in programs that can be used to approximate the value of a definite integral. Try using such a program to approximate the integral in Example 1. How close is your approximation? When you use such a program, you need to be aware of its limitations. Often, you are given no indication of the degree of accuracy of the approximation. Other times, you may be given an approximation that is completely wrong. For instance, try using a built-in numerical integration program to evaluate

$$\int_{-1}^{2} \frac{1}{x}\,dx.$$

Your calculator should give an error message. Does yours?

It is interesting to compare the Trapezoidal Rule with the Midpoint Rule given in Section 4.2. For the Trapezoidal Rule, you average the function values at the endpoints of the subintervals, but for the Midpoint Rule, you take the function values of the subinterval midpoints.

$$\int_a^b f(x)\,dx \approx \sum_{i=1}^{n} f\left(\frac{x_i + x_{i-1}}{2}\right)\Delta x \qquad \text{Midpoint Rule}$$

$$\int_a^b f(x)\,dx \approx \sum_{i=1}^{n} \left(\frac{f(x_i) + f(x_{i-1})}{2}\right)\Delta x \qquad \text{Trapezoidal Rule}$$

There are two important points that should be made concerning the Trapezoidal Rule (or the Midpoint Rule). First, the approximation tends to become more accurate as n increases. For instance, in Example 1, when $n = 16$, the Trapezoidal Rule yields an approximation of 1.994. Second, although you could have used the Fundamental Theorem to evaluate the integral in Example 1, this theorem cannot be used to evaluate an integral as simple as $\int_0^\pi \sin x^2\,dx$ because $\sin x^2$ has no elementary antiderivative. Yet, the Trapezoidal Rule can be applied to estimate this integral.

Simpson's Rule

One way to view the trapezoidal approximation of a definite integral is to say that on each subinterval, you approximate f by a *first*-degree polynomial. In Simpson's Rule, named after the English mathematician Thomas Simpson (1710–1761), you take this procedure one step further and approximate f by *second*-degree polynomials.

Before presenting Simpson's Rule, consider the next theorem for evaluating integrals of polynomials of degree 2 (or less).

THEOREM 4.18 Integral of $p(x) = Ax^2 + Bx + C$

If $p(x) = Ax^2 + Bx + C$, then

$$\int_a^b p(x)\,dx = \left(\frac{b-a}{6}\right)\left[p(a) + 4p\left(\frac{a+b}{2}\right) + p(b)\right].$$

Proof

$$\begin{aligned}\int_a^b p(x)\,dx &= \int_a^b (Ax^2 + Bx + C)\,dx \\ &= \left[\frac{Ax^3}{3} + \frac{Bx^2}{2} + Cx\right]_a^b \\ &= \frac{A(b^3 - a^3)}{3} + \frac{B(b^2 - a^2)}{2} + C(b-a) \\ &= \left(\frac{b-a}{6}\right)[2A(a^2 + ab + b^2) + 3B(b+a) + 6C]\end{aligned}$$

By expansion and collection of terms, the expression inside the brackets becomes

$$\underbrace{(Aa^2 + Ba + C)}_{p(a)} + \underbrace{4\left[A\left(\frac{b+a}{2}\right)^2 + B\left(\frac{b+a}{2}\right) + C\right]}_{4p\left(\frac{a+b}{2}\right)} + \underbrace{(Ab^2 + Bb + C)}_{p(b)}$$

and you can write

$$\int_a^b p(x)\,dx = \left(\frac{b-a}{6}\right)\left[p(a) + 4p\left(\frac{a+b}{2}\right) + p(b)\right].$$

See LarsonCalculus.com for Bruce Edwards's video of this proof. ■

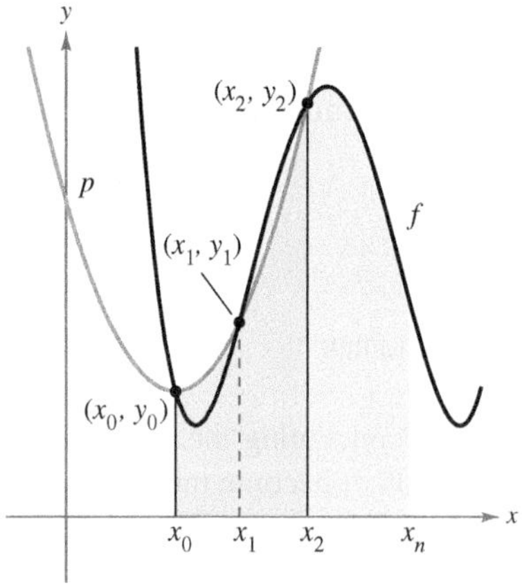

$$\int_{x_0}^{x_2} p(x)\,dx \approx \int_{x_0}^{x_2} f(x)\,dx$$

Figure 4.45

To develop Simpson's Rule for approximating a definite integral, you again partition the interval $[a, b]$ into n subintervals, each of width $\Delta x = (b - a)/n$. This time, however, n is required to be even, and the subintervals are grouped in pairs such that

$$a = \underbrace{x_0 < x_1 < x_2}_{[x_0, x_2]} < \underbrace{x_3 < x_4}_{[x_2, x_4]} < \cdot \cdot \cdot < \underbrace{x_{n-2} < x_{n-1} < x_n}_{[x_{n-2}, x_n]} = b.$$

On each (double) subinterval $[x_{i-2}, x_i]$, you can approximate f by a polynomial p of degree less than or equal to 2. (See Exercise 47.) For example, on the subinterval $[x_0, x_2]$, choose the polynomial of least degree passing through the points (x_0, y_0), (x_1, y_1), and (x_2, y_2), as shown in Figure 4.45. Now, using p as an approximation of f on this subinterval, you have, by Theorem 4.18,

$$\begin{aligned}
\int_{x_0}^{x_2} f(x)\,dx &\approx \int_{x_0}^{x_2} p(x)\,dx \\
&= \frac{x_2 - x_0}{6}\left[p(x_0) + 4p\left(\frac{x_0 + x_2}{2}\right) + p(x_2)\right] \\
&= \frac{2[(b - a)/n]}{6}[p(x_0) + 4p(x_1) + p(x_2)] \\
&= \frac{b - a}{3n}[f(x_0) + 4f(x_1) + f(x_2)].
\end{aligned}$$

Repeating this procedure on the entire interval $[a, b]$ produces the next theorem.

REMARK Observe that the coefficients in Simpson's Rule have the following pattern.

1 4 2 4 2 4 . . . 4 2 4 1

THEOREM 4.19 Simpson's Rule

Let f be continuous on $[a, b]$ and let n be an even integer. Simpson's Rule for approximating $\int_a^b f(x)\,dx$ is

$$\int_a^b f(x)\,dx \approx \frac{b - a}{3n}[f(x_0) + 4f(x_1) + 2f(x_2) + 4f(x_3) + \cdot \cdot \cdot + 4f(x_{n-1}) + f(x_n)].$$

Moreover, as $n \to \infty$, the right-hand side approaches $\int_a^b f(x)\,dx$.

REMARK In Section 4.2, Example 8, the Midpoint Rule with $n = 4$ approximates $\int_0^{\pi} \sin x\,dx$ as 2.052. In Example 1, the Trapezoidal Rule with $n = 4$ gives an approximation of 1.896. In Example 2, Simpson's Rule with $n = 4$ gives an approximation of 2.005. The antiderivative would produce the true value of 2.

In Example 1, the Trapezoidal Rule was used to estimate $\int_0^{\pi} \sin x\,dx$. In the next example, Simpson's Rule is applied to the same integral.

EXAMPLE 2 Approximation with Simpson's Rule

See LarsonCalculus.com for an interactive version of this type of example.

Use Simpson's Rule to approximate

$$\int_0^{\pi} \sin x\,dx.$$

Compare the results for $n = 4$ and $n = 8$.

Solution When $n = 4$, you have

$$\int_0^{\pi} \sin x\,dx \approx \frac{\pi}{12}\left(\sin 0 + 4 \sin\frac{\pi}{4} + 2 \sin\frac{\pi}{2} + 4 \sin\frac{3\pi}{4} + \sin \pi\right) \approx 2.005.$$

When $n = 8$, you have $\int_0^{\pi} \sin x\,dx \approx 2.0003$.

FOR FURTHER INFORMATION For proofs of the formulas used for estimating the errors involved in the use of the Midpoint Rule and Simpson's Rule, see the article "Elementary Proofs of Error Estimates for the Midpoint and Simpson's Rules" by Edward C. Fazekas, Jr. and Peter R. Mercer in *Mathematics Magazine*. To view this article, go to *MathArticles.com*.

Error Analysis

When you use an approximation technique, it is important to know how accurate you can expect the approximation to be. The next theorem, which is listed without proof, gives the formulas for estimating the errors involved in the use of Simpson's Rule and the Trapezoidal Rule. In general, when using an approximation, you can think of the error E as the difference between $\int_a^b f(x)\,dx$ and the approximation.

THEOREM 4.20 Errors in the Trapezoidal Rule and Simpson's Rule

If f has a continuous second derivative on $[a, b]$, then the error E in approximating $\int_a^b f(x)\,dx$ by the Trapezoidal Rule is

$$|E| \le \frac{(b-a)^3}{12n^2}[\max|f''(x)|], \quad a \le x \le b. \qquad \text{Trapezoidal Rule}$$

Moreover, if f has a continuous fourth derivative on $[a, b]$, then the error E in approximating $\int_a^b f(x)\,dx$ by Simpson's Rule is

$$|E| \le \frac{(b-a)^5}{180n^4}[\max|f^{(4)}(x)|], \quad a \le x \le b. \qquad \text{Simpson's Rule}$$

TECHNOLOGY If you have access to a computer algebra system, use it to evaluate the definite integral in Example 3. You should obtain a value of

$$\int_0^1 \sqrt{1+x^2}\,dx = \frac{1}{2}\left[\sqrt{2} + \ln\left(1+\sqrt{2}\right)\right] \approx 1.14779.$$

(The symbol "ln" represents the natural logarithmic function, which you will study in Section 5.1.)

Theorem 4.20 states that the errors generated by the Trapezoidal Rule and Simpson's Rule have upper bounds dependent on the extreme values of $f''(x)$ and $f^{(4)}(x)$ in the interval $[a, b]$. Furthermore, these errors can be made arbitrarily small by *increasing n*, provided that f'' and $f^{(4)}$ are continuous and therefore bounded in $[a, b]$.

EXAMPLE 3 The Approximate Error in the Trapezoidal Rule

Determine a value of n such that the Trapezoidal Rule will approximate the value of

$$\int_0^1 \sqrt{1+x^2}\,dx$$

with an error that is less than or equal to 0.01.

Solution Begin by letting $f(x) = \sqrt{1+x^2}$ and finding the second derivative of f.

$$f'(x) = x(1+x^2)^{-1/2} \quad \text{and} \quad f''(x) = (1+x^2)^{-3/2}$$

The maximum value of $|f''(x)|$ on the interval $[0, 1]$ is $|f''(0)| = 1$. So, by Theorem 4.20, you can write

$$|E| \le \frac{(b-a)^3}{12n^2}|f''(0)| = \frac{1}{12n^2}(1) = \frac{1}{12n^2}.$$

To obtain an error E that is less than 0.01, you must choose n such that $1/(12n^2) \le 1/100$.

$$100 \le 12n^2 \implies n \ge \sqrt{\frac{100}{12}} \approx 2.89$$

So, you can choose $n = 3$ (because n must be greater than or equal to 2.89) and apply the Trapezoidal Rule, as shown in Figure 4.46, to obtain

$$\int_0^1 \sqrt{1+x^2}\,dx \approx \frac{1}{6}\left[\sqrt{1+0^2} + 2\sqrt{1+\left(\tfrac{1}{3}\right)^2} + 2\sqrt{1+\left(\tfrac{2}{3}\right)^2} + \sqrt{1+1^2}\right] \approx 1.154.$$

So, by adding and subtracting the error from this estimate, you know that

$$1.144 \le \int_0^1 \sqrt{1+x^2}\,dx \le 1.164.$$

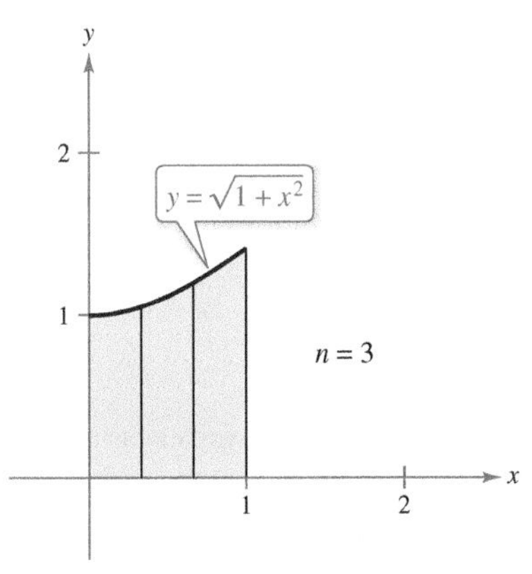

$1.144 \le \int_0^1 \sqrt{1+x^2}\,dx \le 1.164$

Figure 4.46

4.6 Exercises

See CalcChat.com for tutorial help and worked-out solutions to odd-numbered exercises.

Using the Trapezoidal Rule and Simpson's Rule In Exercises 1–10, use the Trapezoidal Rule and Simpson's Rule to approximate the value of the definite integral for the given value of n. Round your answer to four decimal places and compare the results with the exact value of the definite integral.

1. $\int_0^2 x^2\,dx,\quad n = 4$
2. $\int_1^2 \left(\frac{x^2}{4} + 1\right) dx,\quad n = 4$
3. $\int_0^2 x^3\,dx,\quad n = 4$
4. $\int_2^3 \frac{2}{x^2}\,dx,\quad n = 4$
5. $\int_1^3 x^3\,dx,\quad n = 6$
6. $\int_0^8 \sqrt[3]{x}\,dx,\quad n = 8$
7. $\int_4^9 \sqrt{x}\,dx,\quad n = 8$
8. $\int_1^4 (4 - x^2)\,dx,\quad n = 6$
9. $\int_0^1 \frac{2}{(x + 2)^2}\,dx,\quad n = 4$
10. $\int_0^2 x\sqrt{x^2 + 1}\,dx,\quad n = 4$

Using the Trapezoidal Rule and Simpson's Rule In Exercises 11–20, approximate the definite integral using the Trapezoidal Rule and Simpson's Rule with $n = 4$. Compare these results with the approximation of the integral using a graphing utility.

11. $\int_0^2 \sqrt{1 + x^3}\,dx$
12. $\int_0^2 \frac{1}{\sqrt{1 + x^3}}\,dx$
13. $\int_0^1 \sqrt{x}\sqrt{1 - x}\,dx$
14. $\int_{\pi/2}^{\pi} \sqrt{x}\sin x\,dx$
15. $\int_0^{\sqrt{\pi/2}} \sin x^2\,dx$
16. $\int_0^{\sqrt{\pi/4}} \tan x^2\,dx$
17. $\int_3^{3.1} \cos x^2\,dx$
18. $\int_0^{\pi/2} \sqrt{1 + \sin^2 x}\,dx$
19. $\int_0^{\pi/4} x\tan x\,dx$
20. $\int_0^{\pi} f(x)\,dx,\quad f(x) = \begin{cases} \frac{\sin x}{x}, & x > 0 \\ 1, & x = 0 \end{cases}$

WRITING ABOUT CONCEPTS

21. **Polynomial Approximations** The Trapezoidal Rule and Simpson's Rule yield approximations of a definite integral $\int_a^b f(x)\,dx$ based on polynomial approximations of f. What is the degree of the polynomials used for each?

22. **Describing an Error** Describe the size of the error when the Trapezoidal Rule is used to approximate $\int_a^b f(x)\,dx$ when $f(x)$ is a linear function. Use a graph to explain your answer.

Estimating Errors In Exercises 23–26, use the error formulas in Theorem 4.20 to estimate the errors in approximating the integral, with $n = 4$, using (a) the Trapezoidal Rule and (b) Simpson's Rule.

23. $\int_1^3 2x^3\,dx$
24. $\int_3^5 (5x + 2)\,dx$
25. $\int_2^4 \frac{1}{(x - 1)^2}\,dx$
26. $\int_0^{\pi} \cos x\,dx$

Estimating Errors In Exercises 27–30, use the error formulas in Theorem 4.20 to find n such that the error in the approximation of the definite integral is less than or equal to 0.00001 using (a) the Trapezoidal Rule and (b) Simpson's Rule.

27. $\int_1^3 \frac{1}{x}\,dx$
28. $\int_0^1 \frac{1}{1 + x}\,dx$
29. $\int_0^2 \sqrt{x + 2}\,dx$
30. $\int_0^{\pi/2} \sin x\,dx$

Estimating Errors Using Technology In Exercises 31–34, use a computer algebra system and the error formulas to find n such that the error in the approximation of the definite integral is less than or equal to 0.00001 using (a) the Trapezoidal Rule and (b) Simpson's Rule.

31. $\int_0^2 \sqrt{1 + x}\,dx$
32. $\int_0^2 (x + 1)^{2/3}\,dx$
33. $\int_0^1 \tan x^2\,dx$
34. $\int_0^1 \sin x^2\,dx$

35. **Finding the Area of a Region** Approximate the area of the shaded region using
(a) the Trapezoidal Rule with $n = 4$.
(b) Simpson's Rule with $n = 4$.

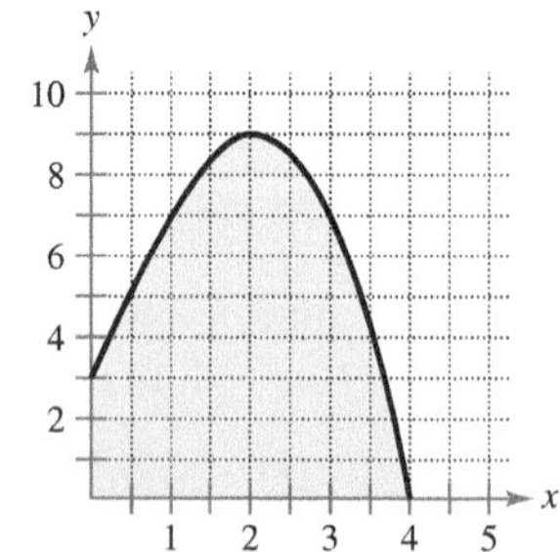

Figure for 35

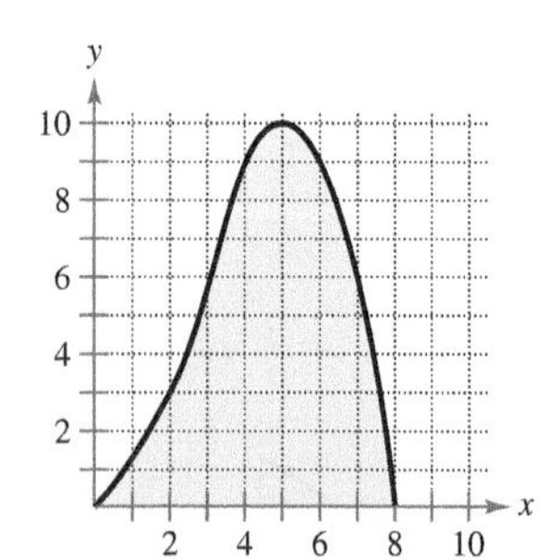

Figure for 36

36. **Finding the Area of a Region** Approximate the area of the shaded region using
(a) the Trapezoidal Rule with $n = 8$.
(b) Simpson's Rule with $n = 8$.

37. **Area** Use Simpson's Rule with $n = 14$ to approximate the area of the region bounded by the graphs of $y = \sqrt{x}\cos x$, $y = 0$, $x = 0$, and $x = \pi/2$.

38. Circumference The **elliptic integral**

$$8\sqrt{3}\int_0^{\pi/2} \sqrt{1 - \frac{2}{3}\sin^2\theta}\, d\theta$$

gives the circumference of an ellipse. Use Simpson's Rule with $n = 8$ to approximate the circumference.

39. Surveying

Use the Trapezoidal Rule to estimate the number of square meters of land, where x and y are measured in meters, as shown in the figure. The land is bounded by a stream and two straight roads that meet at right angles.

x	0	100	200	300	400	500
y	125	125	120	112	90	90

x	600	700	800	900	1000
y	95	88	75	35	0

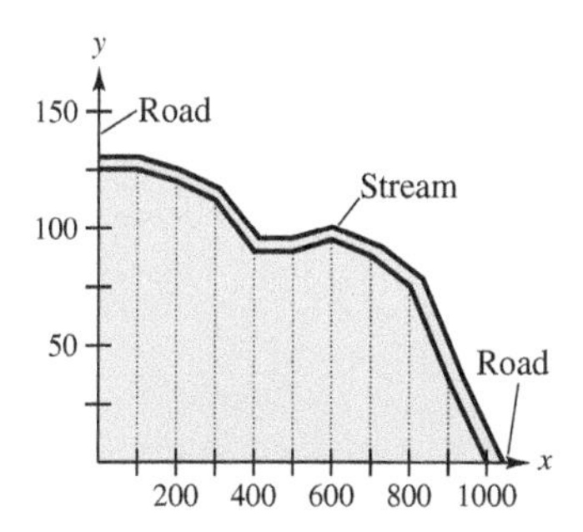

40. HOW DO YOU SEE IT? The function $f(x)$ is concave upward on the interval $[0, 2]$ and the function $g(x)$ is concave downward on the interval $[0, 2]$.

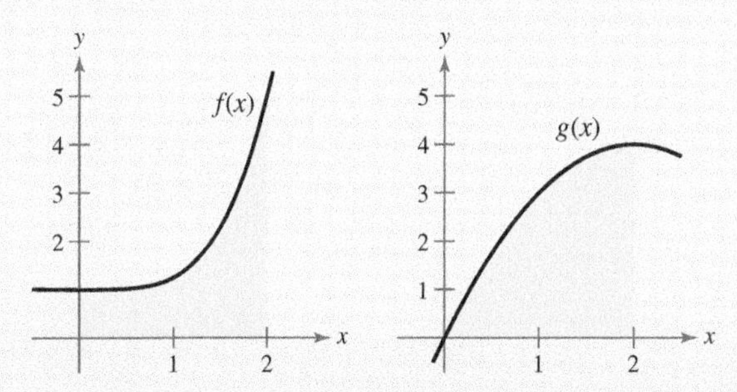

(a) Using the Trapezoidal Rule with $n = 4$, which integral would be overestimated? Which integral would be underestimated? Explain your reasoning.

(b) Which rule would you use for more accurate approximations of $\int_0^2 f(x)\,dx$ and $\int_0^2 g(x)\,dx$, the Trapezoidal Rule or Simpson's Rule? Explain your reasoning.

41. Work To determine the size of the motor required to operate a press, a company must know the amount of work done when the press moves an object linearly 5 feet. The variable force to move the object is

$$F(x) = 100x\sqrt{125 - x^3}$$

where F is given in pounds and x gives the position of the unit in feet. Use Simpson's Rule with $n = 12$ to approximate the work W (in foot-pounds) done through one cycle when

$$W = \int_0^5 F(x)\,dx.$$

42. Approximating a Function The table lists several measurements gathered in an experiment to approximate an unknown continuous function $y = f(x)$.

x	0.00	0.25	0.50	0.75	1.00
y	4.32	4.36	4.58	5.79	6.14

x	1.25	1.50	1.75	2.00
y	7.25	7.64	8.08	8.14

(a) Approximate the integral

$$\int_0^2 f(x)\,dx$$

using the Trapezoidal Rule and Simpson's Rule.

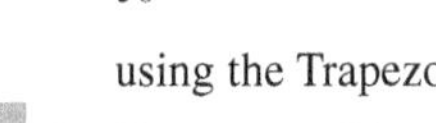

(b) Use a graphing utility to find a model of the form $y = ax^3 + bx^2 + cx + d$ for the data. Integrate the resulting polynomial over $[0, 2]$ and compare the result with the integral from part (a).

Approximation of Pi In Exercises 43 and 44, use Simpson's Rule with $n = 6$ to approximate π using the given equation. (In Section 5.7, you will be able to evaluate the integral using inverse trigonometric functions.)

43. $\pi = \displaystyle\int_0^{1/2} \frac{6}{\sqrt{1 - x^2}}\,dx$ **44.** $\pi = \displaystyle\int_0^1 \frac{4}{1 + x^2}\,dx$

45. Using Simpson's Rule Use Simpson's Rule with $n = 10$ and a computer algebra system to approximate t in the integral equation

$$\int_0^t \sin\sqrt{x}\,dx = 2.$$

46. Proof Prove that Simpson's Rule is exact when approximating the integral of a cubic polynomial function, and demonstrate the result with $n = 4$ for

$$\int_0^1 x^3\,dx.$$

47. Proof Prove that you can find a polynomial

$$p(x) = Ax^2 + Bx + C$$

that passes through any three points (x_1, y_1), (x_2, y_2), and (x_3, y_3), where the x_i's are distinct.

Review Exercises

See CalcChat.com for tutorial help and worked-out solutions to odd-numbered exercises.

Finding an Indefinite Integral **In Exercises 1–8, find the indefinite integral.**

1. $\int (x^3 + 4)\,dx$
2. $\int (x^4 + 3)\,dx$
3. $\int (4x^2 + x + 3)\,dx$
4. $\int \frac{6}{\sqrt[3]{x}}\,dx$
5. $\int \frac{x^4 + 8}{x^3}\,dx$
6. $\int \frac{x^2 + 2x - 6}{x^4}\,dx$
7. $\int (2\csc^2 x - 9\sin x)\,dx$
8. $\int (5\cos x - 2\sec^2 x)\,dx$

Finding a Particular Solution **In Exercises 9–12, find the particular solution of the differential equation that satisfies the initial condition(s).**

9. $f'(x) = -6x,\ f(1) = -2$
10. $f'(x) = 9x^2 + 1,\ f(0) = 7$
11. $f''(x) = 24x,\ f'(-1) = 7,\ f(1) = -4$
12. $f''(x) = 2\cos x,\ f'(0) = 4,\ f(0) = -5$

13. **Vertical Motion** A ball is thrown vertically upward from ground level with an initial velocity of 96 feet per second. Assume the acceleration of the ball is $a(t) = -32$ feet per second per second. (Neglect air resistance.)
    (a) How long will it take the ball to rise to its maximum height? What is the maximum height?
    (b) After how many seconds is the velocity of the ball one-half the initial velocity?
    (c) What is the height of the ball when its velocity is one-half the initial velocity?

14. **Vertical Motion** With what initial velocity must an object be thrown upward (from a height of 3 meters) to reach a maximum height of 150 meters? Assume the acceleration of the object is $a(t) = 9.8$ meters per second per second. (Neglect air resistance.)

Finding a Sum **In Exercises 15 and 16, find the sum by adding each term together. Use the summation capabilities of a graphing utility to verify your result.**

15. $\sum_{i=1}^{5} (5i - 3)$
16. $\sum_{k=0}^{3} (k^2 + 1)$

Using Sigma Notation **In Exercises 17 and 18, use sigma notation to write the sum.**

17. $\frac{1}{5(3)} + \frac{2}{5(4)} + \frac{3}{5(5)} + \cdots + \frac{10}{5(12)}$

18. $\left(\frac{3}{n}\right)\left(\frac{1+1}{n}\right)^2 + \left(\frac{3}{n}\right)\left(\frac{2+1}{n}\right)^2 + \cdots + \left(\frac{3}{n}\right)\left(\frac{n+1}{n}\right)^2$

Evaluating a Sum **In Exercises 19–24, use the properties of summation and Theorem 4.2 to evaluate the sum. Use the summation capabilities of a graphing utility to verify your result.**

19. $\sum_{i=1}^{24} 8$
20. $\sum_{i=1}^{75} 5i$
21. $\sum_{i=1}^{20} 2i$
22. $\sum_{i=1}^{30} (3i - 4)$
23. $\sum_{i=1}^{20} (i + 1)^2$
24. $\sum_{i=1}^{12} i(i^2 - 1)$

Finding Upper and Lower Sums for a Region **In Exercises 25 and 26, use upper and lower sums to approximate the area of the region using the given number of subintervals (of equal width.)**

25. $y = \frac{10}{x^2 + 1}$

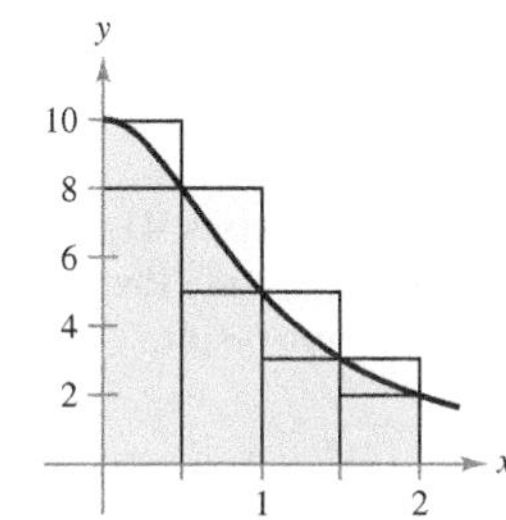

26. $y = 9 - \frac{1}{4}x^2$

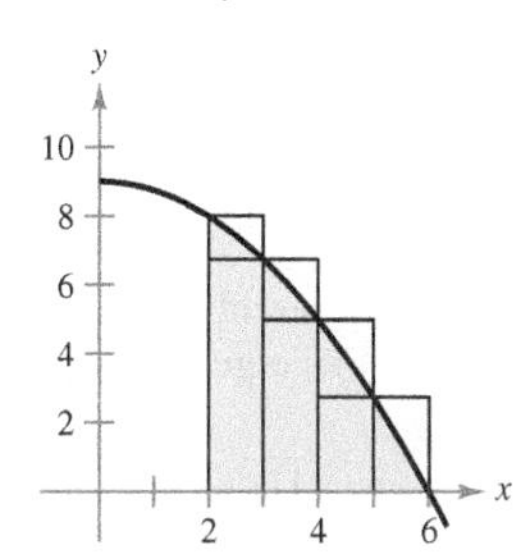

Finding Upper and Lower Sums for a Region **In Exercises 27 and 28, find the upper and lower sums for the region bounded by the graph of the function and the x-axis on the given interval. Leave your answer in terms of n, the number of subintervals.**

	Function	Interval
27.	$f(x) = 4x + 1$	$[2, 3]$
28.	$f(x) = 7x^2$	$[0, 3]$

Finding Area by the Limit Definition **In Exercises 29–32, use the limit process to find the area of the region bounded by the graph of the function and the x-axis over the given interval. Sketch the region.**

29. $y = 8 - 2x,\ [0, 3]$
30. $y = x^2 + 3,\ [0, 2]$
31. $y = 5 - x^2,\ [-2, 1]$
32. $y = \frac{1}{4}x^3,\ [2, 4]$

Approximating Area with the Midpoint Rule **In Exercises 33 and 34, use the Midpoint Rule with $n = 4$ to approximate the area of the region bounded by the graph of the function and the x-axis over the given interval.**

33. $f(x) = 16 - x^2,\ [0, 4]$
34. $f(x) = \sin \pi x,\ [0, 1]$

Evaluating a Definite Integral as a Limit **In Exercises 35 and 36, evaluate the definite integral by the limit definition.**

35. $\int_{-3}^{5} 6x\,dx$
36. $\int_{0}^{3} (1 - 2x^2)\,dx$

Evaluating a Definite Integral Using a Geometric Formula **In Exercises 37 and 38, sketch the region whose area is given by the definite integral. Then use a geometric formula to evaluate the integral.**

37. $\int_{0}^{5} (5 - |x - 5|)\,dx$
38. $\int_{-6}^{6} \sqrt{36 - x^2}\,dx$

39. Using Properties of Definite Integrals Given

$$\int_4^8 f(x)\,dx = 12 \quad \text{and} \quad \int_4^8 g(x)\,dx = 5, \text{ evaluate}$$

(a) $\int_4^8 [f(x) + g(x)]\,dx.$ (b) $\int_4^8 [f(x) - g(x)]\,dx.$

(c) $\int_4^8 [2f(x) - 3g(x)]\,dx.$ (d) $\int_4^8 7f(x)\,dx.$

40. Using Properties of Definite Integrals Given

$$\int_0^2 f(x)\,dx = 2 \quad \text{and} \quad \int_2^5 f(x) = -5, \text{ evaluate}$$

(a) $\int_0^5 f(x)\,dx.$ (b) $\int_5^2 f(x)\,dx.$

(c) $\int_3^3 f(x)\,dx.$ (d) $\int_2^5 -8f(x)\,dx.$

Evaluating a Definite Integral In Exercises 41–46, use the Fundamental Theorem of Calculus to evaluate the definite integral. Use a graphing utility to verify your result.

41. $\int_0^6 (x - 1)\,dx$ **42.** $\int_{-2}^1 (4x^4 - x)\,dx$

43. $\int_4^9 x\sqrt{x}\,dx$ **44.** $\int_1^4 \left(\frac{1}{x^3} + x\right)dx$

45. $\int_0^{3\pi/4} \sin\theta\,d\theta$ **46.** $\int_{-\pi/4}^{\pi/4} \sec^2 t\,dt$

Finding the Area of a Region In Exercises 47 and 48, find the area of the given region.

47. $y = \sin x$ **48.** $y = x + \cos x$

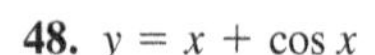

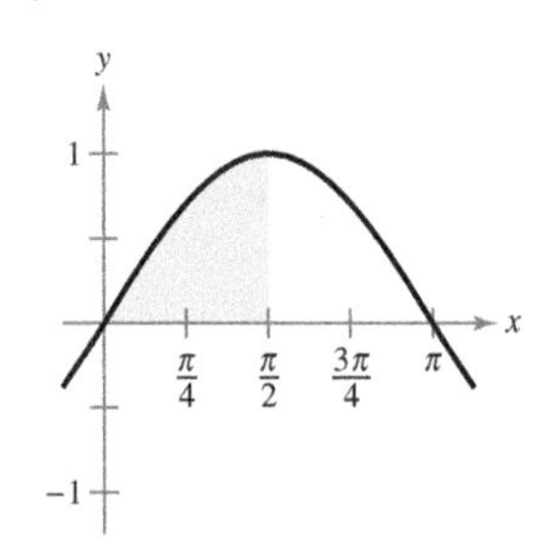

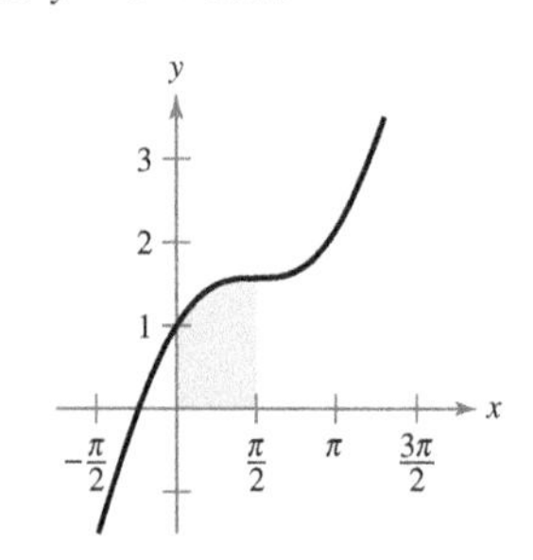

Finding the Area of a Region In Exercises 49–52, find the area of the region bounded by the graphs of the equations.

49. $y = 8 - x, x = 0, x = 6, y = 0$

50. $y = -x^2 + x + 6, y = 0$

51. $y = x - x^3, x = 0, x = 1, y = 0$

52. $y = \sqrt{x}(1 - x), y = 0$

Using the Mean Value Theorem for Integrals In Exercises 53 and 54, find the value(s) of c guaranteed by the Mean Value Theorem for Integrals for the function over the given interval.

53. $f(x) = 3x^2, \quad [1, 3]$

54. $f(x) = \sin x, \quad [0, \pi]$

Finding the Average Value of a Function In Exercises 55 and 56, find the average value of the function over the given interval and all values of x in the interval for which the function equals its average value.

55. $f(x) = \frac{1}{\sqrt{x}}, \quad [4, 9]$ **56.** $f(x) = x^3, \quad [0, 2]$

Using the Second Fundamental Theorem of Calculus In Exercises 57 and 58, use the Second Fundamental Theorem of Calculus to find $F'(x)$.

57. $F(x) = \int_0^x t^2\sqrt{1 + t^3}\,dt$ **58.** $F(x) = \int_1^x \frac{1}{t^2}\,dt$

Finding an Indefinite Integral In Exercises 59–66, find the indefinite integral.

59. $\int x(1 - 3x^2)^4\,dx$ **60.** $\int 6x^3\sqrt{3x^4 + 2}\,dx$

61. $\int \sin^3 x \cos x\,dx$ **62.** $\int x \sin 3x^2\,dx$

63. $\int \frac{\cos\theta}{\sqrt{1 - \sin\theta}}\,d\theta$ **64.** $\int \frac{\sin x}{\sqrt{\cos x}}\,dx$

65. $\int x\sqrt{8 - x}\,dx$ **66.** $\int \sqrt{1 + \sqrt{x}}\,dx$

Evaluating a Definite Integral In Exercises 67–72, evaluate the definite integral. Use a graphing utility to verify your result.

67. $\int_0^1 (3x + 1)^5\,dx$ **68.** $\int_0^1 x^2(x^3 - 2)^3\,dx$

69. $\int_0^3 \frac{1}{\sqrt{1 + x}}\,dx$ **70.** $\int_3^6 \frac{x}{3\sqrt{x^2 - 8}}\,dx$

71. $2\pi\int_0^1 (y + 1)\sqrt{1 - y}\,dy$ **72.** $2\pi\int_{-1}^0 x^2\sqrt{x + 1}\,dx$

Finding the Area of a Region In Exercises 73 and 74, find the area of the region. Use a graphing utility to verify your result.

73. $\int_1^9 x\sqrt[3]{x - 1}\,dx$ **74.** $\int_0^{\pi/2} (\cos x + \sin 2x)\,dx$

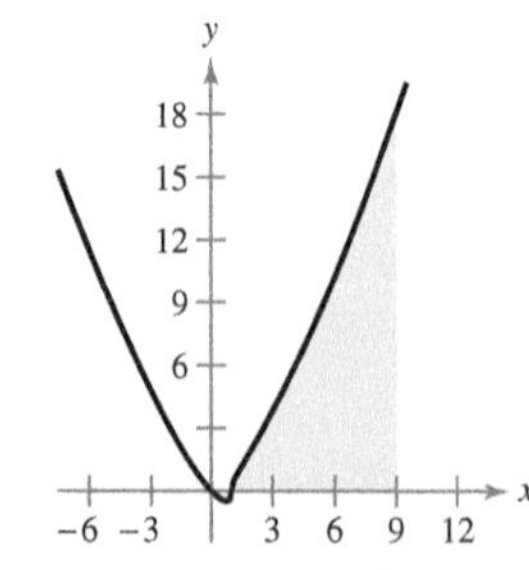

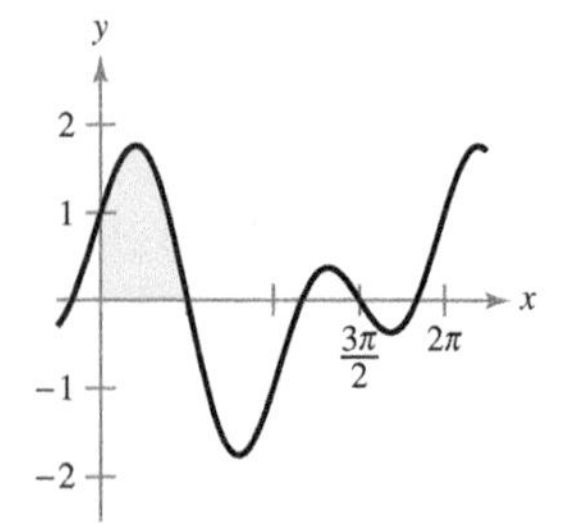

Even and Odd Functions In Exercises 75 and 76, evaluate the integral using the properties of even and odd functions as an aid.

75. $\int_{-2}^2 (x^3 - 2x)\,dx$ **76.** $\int_{-\pi}^{\pi} (\cos x + x^2)\,dx$

P.S. Problem Solving

See CalcChat.com for tutorial help and worked-out solutions to odd-numbered exercises.

1. **Using a Function** Let $L(x) = \int_1^x \frac{1}{t}\,dt,\ x > 0.$

   (a) Find $L(1)$.

   (b) Find $L'(x)$ and $L'(1)$.

   (c) Use a graphing utility to approximate the value of x (to three decimal places) for which $L(x) = 1$.

   (d) Prove that $L(x_1x_2) = L(x_1) + L(x_2)$ for all positive values of x_1 and x_2.

2. **Parabolic Arch** Archimedes showed that the area of a parabolic arch is equal to $\frac{2}{3}$ the product of the base and the height (see figure).

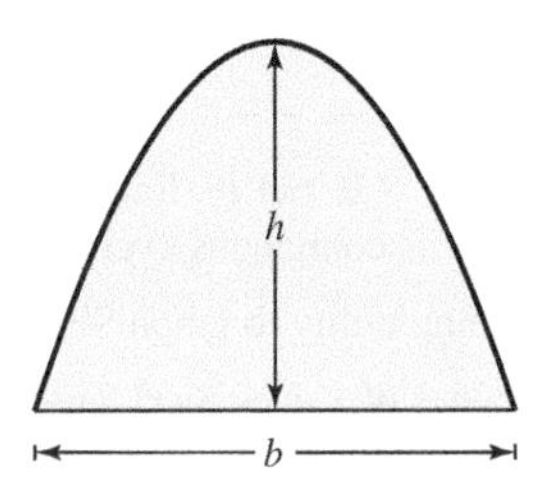

   (a) Graph the parabolic arch bounded by $y = 9 - x^2$ and the x-axis. Use an appropriate integral to find the area A.

   (b) Find the base and height of the arch and verify Archimedes' formula.

   (c) Prove Archimedes' formula for a general parabola.

Evaluating a Sum and a Limit **In Exercises 3 and 4, (a) write the area under the graph of the given function defined on the given interval as a limit. Then (b) evaluate the sum in part (a), and (c) evaluate the limit using the result of part (b).**

3. $y = x^4 - 4x^3 + 4x^2,\quad [0, 2]$

$$\left(\textit{Hint: } \sum_{i=1}^{n} i^4 = \frac{n(n+1)(2n+1)(3n^2+3n-1)}{30}\right)$$

4. $y = \frac{1}{2}x^5 + 2x^3,\quad [0, 2]$

$$\left(\textit{Hint: } \sum_{i=1}^{n} i^5 = \frac{n^2(n+1)^2(2n^2+2n-1)}{12}\right)$$

5. **Fresnel Function** The **Fresnel function** S is defined by the integral

$$S(x) = \int_0^x \sin\frac{\pi t^2}{2}\,dt.$$

   (a) Graph the function $y = \sin\frac{\pi x^2}{2}$ on the interval $[0, 3]$.

   (b) Use the graph in part (a) to sketch the graph of S on the interval $[0, 3]$.

   (c) Locate all relative extrema of S on the interval $(0, 3)$.

   (d) Locate all points of inflection of S on the interval $(0, 3)$.

6. **Approximation** The **Two-Point Gaussian Quadrature Approximation** for f is

$$\int_{-1}^{1} f(x)\,dx \approx f\left(-\frac{1}{\sqrt{3}}\right) + f\left(\frac{1}{\sqrt{3}}\right).$$

   (a) Use this formula to approximate

$$\int_{-1}^{1} \cos x\,dx.$$

   Find the error of the approximation.

   (b) Use this formula to approximate

$$\int_{-1}^{1} \frac{1}{1+x^2}\,dx.$$

   (c) Prove that the Two-Point Gaussian Quadrature Approximation is exact for all polynomials of degree 3 or less.

7. **Extrema and Points of Inflection** The graph of the function f consists of the three line segments joining the points $(0, 0)$, $(2, -2)$, $(6, 2)$, and $(8, 3)$. The function F is defined by the integral

$$F(x) = \int_0^x f(t)\,dt.$$

   (a) Sketch the graph of f.

   (b) Complete the table.

x	0	1	2	3	4	5	6	7	8
$F(x)$									

   (c) Find the extrema of F on the interval $[0, 8]$.

   (d) Determine all points of inflection of F on the interval $(0, 8)$.

8. **Falling Objects** Galileo Galilei (1564–1642) stated the following proposition concerning falling objects:

   The time in which any space is traversed by a uniformly accelerating body is equal to the time in which that same space would be traversed by the same body moving at a uniform speed whose value is the mean of the highest speed of the accelerating body and the speed just before acceleration began.

   Use the techniques of this chapter to verify this proposition.

9. **Proof** Prove $\int_0^x f(t)(x-t)\,dt = \int_0^x \left(\int_0^t f(v)\,dv\right)dt.$

10. **Proof** Prove $\int_a^b f(x)f'(x)\,dx = \frac{1}{2}([f(b)]^2 - [f(a)]^2).$

11. **Riemann Sum** Use an appropriate Riemann sum to evaluate the limit

$$\lim_{n\to\infty} \frac{\sqrt{1} + \sqrt{2} + \sqrt{3} + \cdots + \sqrt{n}}{n^{3/2}}.$$

12. Riemann Sum Use an appropriate Riemann sum to evaluate the limit

$$\lim_{n\to\infty} \frac{1^5 + 2^5 + 3^5 + \cdot \cdot \cdot + n^5}{n^6}.$$

13. Proof Suppose that f is integrable on $[a, b]$ and $0 < m \le f(x) \le M$ for all x in the interval $[a, b]$. Prove that

$$m(a - b) \le \int_a^b f(x)\,dx \le M(b - a).$$

Use this result to estimate $\int_0^1 \sqrt{1 + x^4}\,dx$.

14. Using a Continuous Function Let f be continuous on the interval $[0, b]$, where $f(x) + f(b - x) \ne 0$ on $[0, b]$.

(a) Show that $\displaystyle\int_0^b \frac{f(x)}{f(x) + f(b - x)}\,dx = \frac{b}{2}$.

(b) Use the result in part (a) to evaluate

$$\int_0^1 \frac{\sin x}{\sin(1 - x) + \sin x}\,dx.$$

(c) Use the result in part (a) to evaluate

$$\int_0^3 \frac{\sqrt{x}}{\sqrt{x} + \sqrt{3 - x}}\,dx.$$

15. Velocity and Acceleration A car travels in a straight line for 1 hour. Its velocity v in miles per hour at six-minute intervals is shown in the table.

t (hours)	0	0.1	0.2	0.3	0.4	0.5
v (mi/h)	0	10	20	40	60	50

t (hours)	0.6	0.7	0.8	0.9	1.0
v (mi/h)	40	35	40	50	65

(a) Produce a reasonable graph of the velocity function v by graphing these points and connecting them with a smooth curve.

(b) Find the open intervals over which the acceleration a is positive.

(c) Find the average acceleration of the car (in miles per hour per hour) over the interval $[0, 0.4]$.

(d) What does the integral

$$\int_0^1 v(t)\,dt$$

signify? Approximate this integral using the Midpoint Rule with five subintervals.

(e) Approximate the acceleration at $t = 0.8$.

16. Proof Prove that if f is a continuous function on a closed interval $[a, b]$, then

$$\left|\int_a^b f(x)\,dx\right| \le \int_a^b |f(x)|\,dx.$$

17. Verifying a Sum Verify that

$$\sum_{i=1}^{n} i^2 = \frac{n(n + 1)(2n + 1)}{6}$$

by showing the following.

(a) $(1 + i)^3 - i^3 = 3i^2 + 3i + 1$

(b) $(n + 1)^3 = \displaystyle\sum_{i=1}^{n} (3i^2 + 3i + 1) + 1$

(c) $\displaystyle\sum_{i=1}^{n} i^2 = \frac{n(n + 1)(2n + 1)}{6}$

18. Sine Integral Function The **sine integral function**

$$\text{Si}(x) = \int_0^x \frac{\sin t}{t}\,dt$$

is often used in engineering. The function

$$f(t) = \frac{\sin t}{t}$$

is not defined at $t = 0$, but its limit is 1 as $t \to 0$. So, define $f(0) = 1$. Then f is continuous everywhere.

(a) Use a graphing utility to graph $\text{Si}(x)$.

(b) At what values of x does $\text{Si}(x)$ have relative maxima?

(c) Find the coordinates of the first inflection point where $x > 0$.

(d) Decide whether $\text{Si}(x)$ has any horizontal asymptotes. If so, identify each.

19. Upper and Lower Sums Consider the region bounded by $y = mx$, $y = 0$, $x = 0$, and $x = b$.

(a) Find the upper and lower sums to approximate the area of the region when $\Delta x = b/4$.

(b) Find the upper and lower sums to approximate the area of the region when $\Delta x = b/n$.

(c) Find the area of the region by letting n approach infinity in both sums in part (b). Show that, in each case, you obtain the formula for the area of a triangle.

20. Minimizing an Integral Determine the limits of integration where $a \le b$ such that

$$\int_a^b (x^2 - 16)\,dx$$

has minimal value.

21. Finding a Function The graph of f' is shown. Find and sketch the graph of f given that f is continuous and $f(0) = 1$.

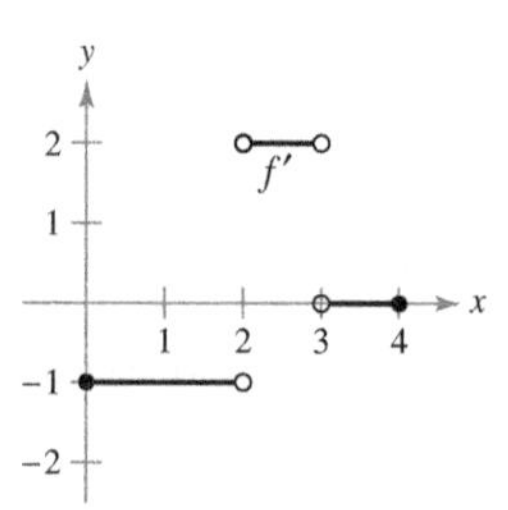

5 Logarithmic, Exponential, and Other Transcendental Functions

Radioactive Half-Life Model (Example 1, p. 352)

Breaking Strength (Exercise 102, p. 360)

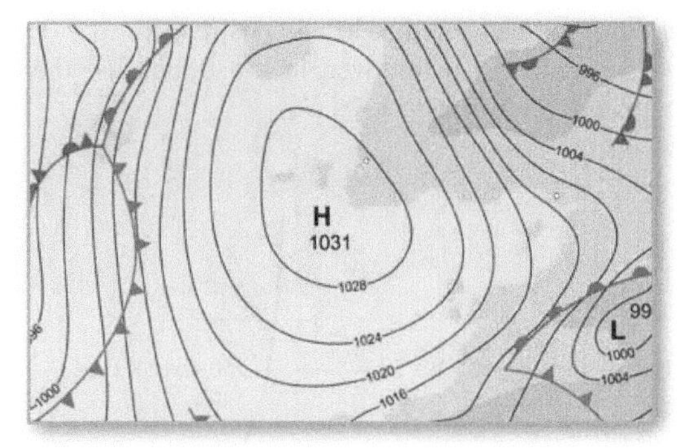

Atmospheric Pressure (Exercise 85, p. 349)

Heat Transfer (Exercise 93, p. 332)

Sound Intensity (Exercise 104, p. 323)

5.1 The Natural Logarithmic Function: Differentiation

- **Develop and use properties of the natural logarithmic function.**
- **Understand the definition of the number e.**
- **Find derivatives of functions involving the natural logarithmic function.**

The Natural Logarithmic Function

Recall that the General Power Rule

$$\int x^n\,dx = \frac{x^{n+1}}{n+1} + C, \quad n \neq -1 \qquad \text{General Power Rule}$$

has an important disclaimer—it does not apply when $n = -1$. Consequently, you have not yet found an antiderivative for the function $f(x) = 1/x$. In this section, you will use the Second Fundamental Theorem of Calculus to *define* such a function. This antiderivative is a function that you have not encountered previously in the text. It is neither algebraic nor trigonometric but falls into a new class of functions called *logarithmic functions.* This particular function is the **natural logarithmic function.**

JOHN NAPIER (1550–1617)

Logarithms were invented by the Scottish mathematician John Napier. Napier coined the term *logarithm*, from the two Greek words *logos* (or ratio) and *arithmos* (or number), to describe the theory that he spent 20 years developing and that first appeared in the book *Mirifici Logarithmorum canonis descriptio* (A Description of the Marvelous Rule of Logarithms). Although he did not introduce the *natural* logarithmic function, it is sometimes called the *Napierian* logarithm.

See LarsonCalculus.com to read more of this biography.

Definition of the Natural Logarithmic Function

The **natural logarithmic function** is defined by

$$\ln x = \int_1^x \frac{1}{t}\,dt, \quad x > 0.$$

The domain of the natural logarithmic function is the set of all positive real numbers.

From this definition, you can see that $\ln x$ is positive for $x > 1$ and negative for $0 < x < 1$, as shown in Figure 5.1. Moreover, $\ln 1 = 0$, because the upper and lower limits of integration are equal when $x = 1$.

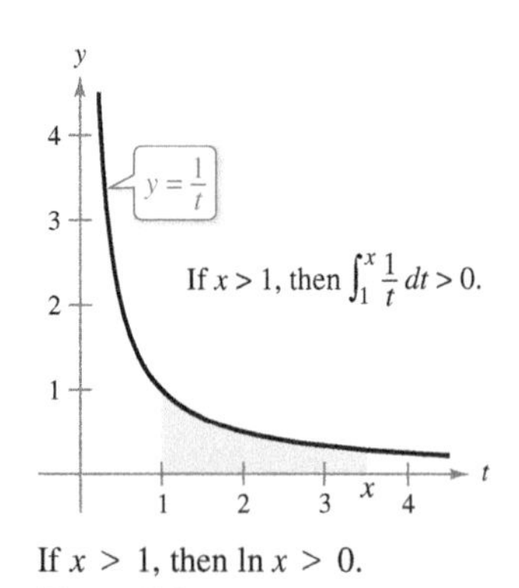

If $x > 1$, then $\ln x > 0$.

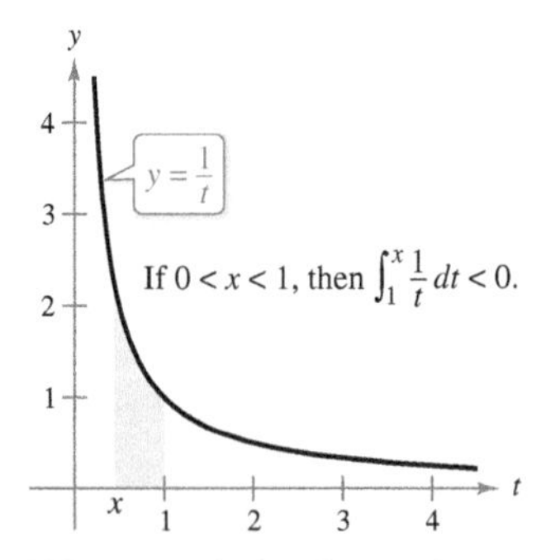

If $0 < x < 1$, then $\ln x < 0$.

Figure 5.1

Exploration

Graphing the Natural Logarithmic Function Using *only* the definition of the natural logarithmic function, sketch a graph of the function. Explain your reasoning.

To sketch the graph of $y = \ln x$, you can think of the natural logarithmic function as an *antiderivative* given by the differential equation

$$\frac{dy}{dx} = \frac{1}{x}.$$

Figure 5.2 is a computer-generated graph, called a *slope field (or direction field)*, showing small line segments of slope $1/x$. The graph of $y = \ln x$ is the solution that passes through the point $(1, 0)$. (You will study slope fields in Section 6.1.)

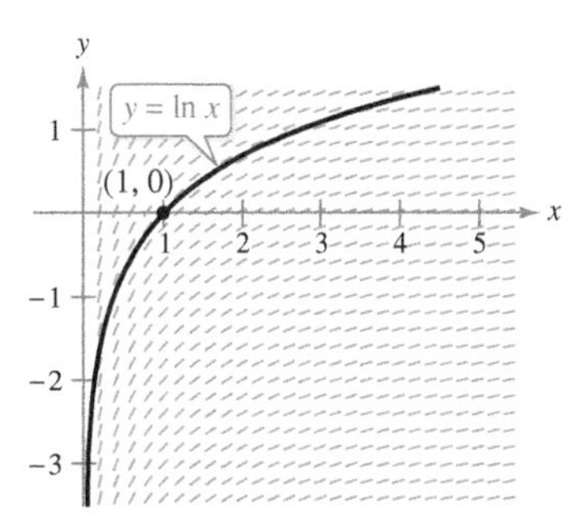

Each small line segment has a slope of $\frac{1}{x}$.

Figure 5.2

THEOREM 5.1 Properties of the Natural Logarithmic Function

The natural logarithmic function has the following properties.

1. The domain is $(0, \infty)$ and the range is $(-\infty, \infty)$.
2. The function is continuous, increasing, and one-to-one.
3. The graph is concave downward.

Proof The domain of $f(x) = \ln x$ is $(0, \infty)$ by definition. Moreover, the function is continuous because it is differentiable. It is increasing because its derivative

$$f'(x) = \frac{1}{x} \qquad \text{First derivative}$$

is positive for $x > 0$, as shown in Figure 5.3. It is concave downward because its second derivative

$$f''(x) = -\frac{1}{x^2} \qquad \text{Second derivative}$$

is negative for $x > 0$. The proof that f is one-to-one is given in Appendix A. The following limits imply that its range is the entire real number line.

$$\lim_{x \to 0^+} \ln x = -\infty$$

and

$$\lim_{x \to \infty} \ln x = \infty$$

Verification of these two limits is given in Appendix A. ■

The natural logarithmic function is increasing, and its graph is concave downward.

Figure 5.3

Using the definition of the natural logarithmic function, you can prove several important properties involving operations with natural logarithms. If you are already familiar with logarithms, you will recognize that the properties listed on the next page are characteristic of all logarithms.

THEOREM 5.2 Logarithmic Properties

If a and b are positive numbers and n is rational, then the following properties are true.

1. $\ln 1 = 0$ **2.** $\ln(ab) = \ln a + \ln b$

3. $\ln(a^n) = n \ln a$ **4.** $\ln\left(\frac{a}{b}\right) = \ln a - \ln b$

Proof The first property has already been discussed. The proof of the second property follows from the fact that two antiderivatives of the same function differ at most by a constant. From the Second Fundamental Theorem of Calculus and the definition of the natural logarithmic function, you know that

$$\frac{d}{dx}[\ln x] = \frac{d}{dx}\left[\int_1^x \frac{1}{t}\,dt\right] = \frac{1}{x}.$$

So, consider the two derivatives

$$\frac{d}{dx}[\ln(ax)] = \frac{a}{ax} = \frac{1}{x}$$

and

$$\frac{d}{dx}[\ln a + \ln x] = 0 + \frac{1}{x} = \frac{1}{x}.$$

Because $\ln(ax)$ and $(\ln a + \ln x)$ are both antiderivatives of $1/x$, they must differ at most by a constant, $\ln(ax) = \ln a + \ln x + C$. By letting $x = 1$, you can see that $C = 0$. The third property can be proved similarly by comparing the derivatives of $\ln(x^n)$ and $n \ln x$. Finally, using the second and third properties, you can prove the fourth property.

$$\ln\left(\frac{a}{b}\right) = \ln[a(b^{-1})] = \ln a + \ln(b^{-1}) = \ln a - \ln b$$

EXAMPLE 1 Expanding Logarithmic Expressions

a. $\ln \frac{10}{9} = \ln 10 - \ln 9$ Property 4

b. $\ln\sqrt{3x+2} = \ln(3x+2)^{1/2}$ Rewrite with rational exponent.

$= \frac{1}{2}\ln(3x+2)$ Property 3

c. $\ln\frac{6x}{5} = \ln(6x) - \ln 5$ Property 4

$= \ln 6 + \ln x - \ln 5$ Property 2

d. $\ln\frac{(x^2+3)^2}{x\sqrt[3]{x^2+1}} = \ln(x^2+3)^2 - \ln\left(x\sqrt[3]{x^2+1}\right)$

$= 2\ln(x^2+3) - [\ln x + \ln(x^2+1)^{1/3}]$

$= 2\ln(x^2+3) - \ln x - \ln(x^2+1)^{1/3}$

$= 2\ln(x^2+3) - \ln x - \frac{1}{3}\ln(x^2+1)$

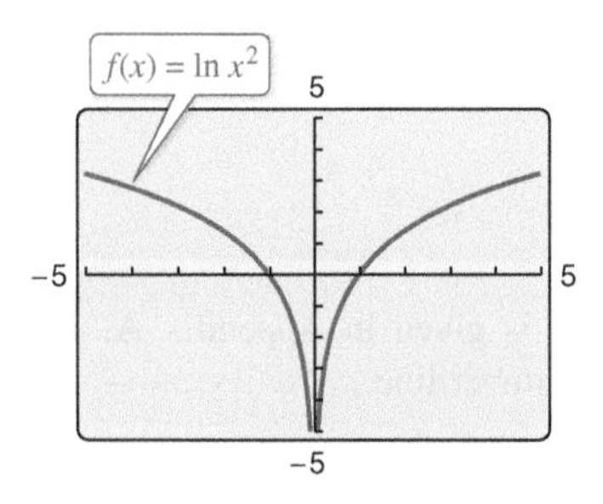

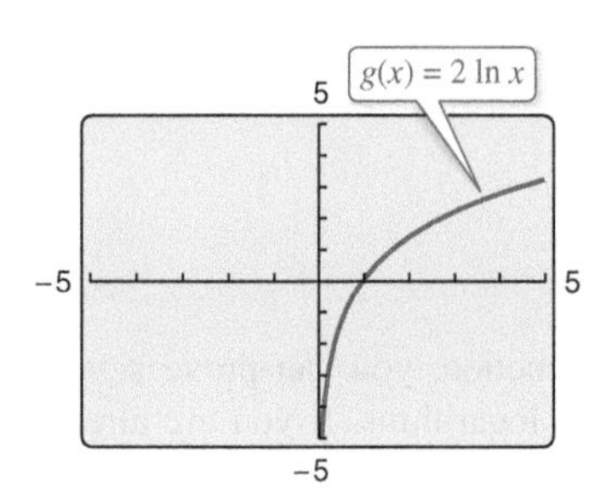

Figure 5.4

When using the properties of logarithms to rewrite logarithmic functions, you must check to see whether the domain of the rewritten function is the same as the domain of the original. For instance, the domain of $f(x) = \ln x^2$ is all real numbers except $x = 0$, and the domain of $g(x) = 2 \ln x$ is all positive real numbers. (See Figure 5.4.)

The Number e

THE NUMBER e

The symbol e was first used by mathematician Leonhard Euler to represent the base of natural logarithms in a letter to another mathematician, Christian Goldbach, in 1731.

It is likely that you have studied logarithms in an algebra course. There, without the benefit of calculus, logarithms would have been defined in terms of a **base** number. For example, common logarithms have a base of 10 and therefore $\log_{10}10 = 1$. (You will learn more about this in Section 5.5.)

The **base for the natural logarithm** is defined using the fact that the natural logarithmic function is continuous, is one-to-one, and has a range of $(-\infty, \infty)$. So, there must be a unique real number x such that $\ln x = 1$, as shown in Figure 5.5. This number is denoted by the letter e. It can be shown that e is irrational and has the following decimal approximation.

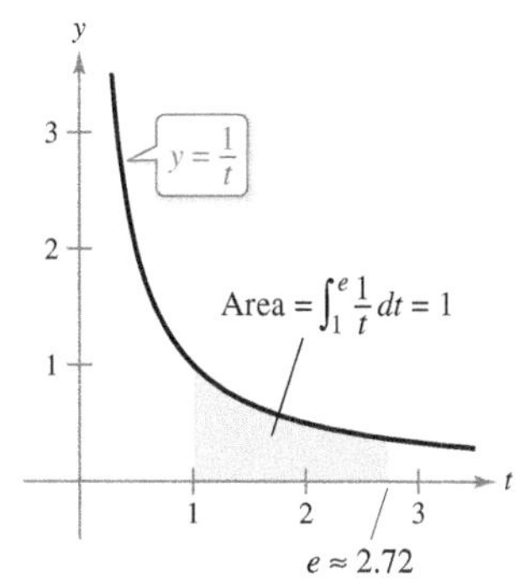

e is the base for the natural logarithm because $\ln e = 1$.
Figure 5.5

$$e \approx 2.71828182846$$

Definition of e

The letter e denotes the positive real number such that

$$\ln e = \int_1^e \frac{1}{t}\,dt = 1.$$

FOR FURTHER INFORMATION To learn more about the number e, see the article "Unexpected Occurrences of the Number e" by Harris S. Shultz and Bill Leonard in *Mathematics Magazine*. To view this article, go to *MathArticles.com.*

Once you know that $\ln e = 1$, you can use logarithmic properties to evaluate the natural logarithms of several other numbers. For example, by using the property

$$\begin{aligned}\ln(e^n) &= n \ln e \\ &= n(1) \\ &= n\end{aligned}$$

you can evaluate $\ln(e^n)$ for various values of n, as shown in the table and in Figure 5.6.

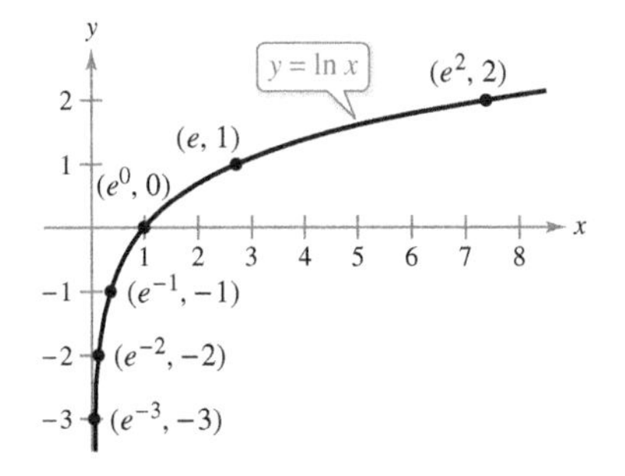

If $x = e^n$, then $\ln x = n$.
Figure 5.6

x	$\frac{1}{e^3} \approx 0.050$	$\frac{1}{e^2} \approx 0.135$	$\frac{1}{e} \approx 0.368$	$e^0 = 1$	$e \approx 2.718$	$e^2 \approx 7.389$
$\ln x$	-3	-2	-1	0	1	2

The logarithms shown in the table above are convenient because the x-values are integer powers of e. Most logarithmic expressions are, however, best evaluated with a calculator.

EXAMPLE 2 **Evaluating Natural Logarithmic Expressions**

a. $\ln 2 \approx 0.693$

b. $\ln 32 \approx 3.466$

c. $\ln 0.1 \approx -2.303$

The Derivative of the Natural Logarithmic Function

The derivative of the natural logarithmic function is given in Theorem 5.3. The first part of the theorem follows from the definition of the natural logarithmic function as an antiderivative. The second part of the theorem is simply the Chain Rule version of the first part.

THEOREM 5.3 Derivative of the Natural Logarithmic Function

Let u be a differentiable function of x.

1. $\frac{d}{dx}[\ln x] = \frac{1}{x}, \quad x > 0$
2. $\frac{d}{dx}[\ln u] = \frac{1}{u}\frac{du}{dx} = \frac{u'}{u}, \quad u > 0$

EXAMPLE 3 Differentiation of Logarithmic Functions

See LarsonCalculus.com for an interactive version of this type of example.

a. $\frac{d}{dx}[\ln 2x] = \frac{u'}{u} = \frac{2}{2x} = \frac{1}{x}$ $\qquad u = 2x$

b. $\frac{d}{dx}[\ln(x^2 + 1)] = \frac{u'}{u} = \frac{2x}{x^2 + 1}$ $\qquad u = x^2 + 1$

c. $\frac{d}{dx}[x \ln x] = x\left(\frac{d}{dx}[\ln x]\right) + (\ln x)\left(\frac{d}{dx}[x]\right)$ $\qquad$ Product Rule

$$= x\left(\frac{1}{x}\right) + (\ln x)(1)$$

$$= 1 + \ln x$$

d. $\frac{d}{dx}[(\ln x)^3] = 3(\ln x)^2 \frac{d}{dx}[\ln x]$ $\qquad$ Chain Rule

$$= 3(\ln x)^2 \frac{1}{x}$$

Napier used logarithmic properties to simplify *calculations* involving products, quotients, and powers. Of course, given the availability of calculators, there is now little need for this particular application of logarithms. However, there is great value in using logarithmic properties to simplify *differentiation* involving products, quotients, and powers.

EXAMPLE 4 Logarithmic Properties as Aids to Differentiation

Differentiate

$$f(x) = \ln\sqrt{x + 1}.$$

Solution Because

$$f(x) = \ln\sqrt{x + 1} = \ln(x + 1)^{1/2} = \frac{1}{2}\ln(x + 1) \qquad \text{Rewrite before differentiating.}$$

you can write

$$f'(x) = \frac{1}{2}\left(\frac{1}{x + 1}\right) = \frac{1}{2(x + 1)}. \qquad \text{Differentiate.}$$

EXAMPLE 5 **Logarithmic Properties as Aids to Differentiation**

Differentiate $f(x) = \ln \dfrac{x(x^2+1)^2}{\sqrt{2x^3-1}}$.

Solution Because

$$f(x) = \ln \frac{x(x^2+1)^2}{\sqrt{2x^3-1}} \qquad \text{Write original function.}$$

$$= \ln x + 2\ln(x^2+1) - \frac{1}{2}\ln(2x^3-1) \qquad \text{Rewrite before differentiating.}$$

you can write

$$f'(x) = \frac{1}{x} + 2\left(\frac{2x}{x^2+1}\right) - \frac{1}{2}\left(\frac{6x^2}{2x^3-1}\right) \qquad \text{Differentiate.}$$

$$= \frac{1}{x} + \frac{4x}{x^2+1} - \frac{3x^2}{2x^3-1}. \qquad \text{Simplify.}$$

In Examples 4 and 5, be sure you see the benefit of applying logarithmic properties *before* differentiating. Consider, for instance, the difficulty of direct differentiation of the function given in Example 5.

On occasion, it is convenient to use logarithms as aids in differentiating *nonlogarithmic* functions. This procedure is called **logarithmic differentiation.** In general, use logarithmic differentiation when differentiating (1) a function involving many factors or (2) a function having both a variable base and a variable exponent [see Section 5.5, Example 5(d)].

EXAMPLE 6 **Logarithmic Differentiation**

Find the derivative of $y = \dfrac{(x-2)^2}{\sqrt{x^2+1}}, \quad x \neq 2.$

Solution Note that $y > 0$ for all $x \neq 2$. So, $\ln y$ is defined. Begin by taking the natural logarithm of each side of the equation. Then apply logarithmic properties and differentiate implicitly. Finally, solve for y'.

$$y = \frac{(x-2)^2}{\sqrt{x^2+1}}, \quad x \neq 2 \qquad \text{Write original equation.}$$

$$\ln y = \ln \frac{(x-2)^2}{\sqrt{x^2+1}} \qquad \text{Take natural log of each side.}$$

$$\ln y = 2\ln(x-2) - \frac{1}{2}\ln(x^2+1) \qquad \text{Logarithmic properties}$$

$$\frac{y'}{y} = 2\left(\frac{1}{x-2}\right) - \frac{1}{2}\left(\frac{2x}{x^2+1}\right) \qquad \text{Differentiate.}$$

$$\frac{y'}{y} = \frac{x^2+2x+2}{(x-2)(x^2+1)} \qquad \text{Simplify.}$$

$$y' = y\left[\frac{x^2+2x+2}{(x-2)(x^2+1)}\right] \qquad \text{Solve for } y'.$$

$$y' = \frac{(x-2)^2}{\sqrt{x^2+1}}\left[\frac{x^2+2x+2}{(x-2)(x^2+1)}\right] \qquad \text{Substitute for } y.$$

$$y' = \frac{(x-2)(x^2+2x+2)}{(x^2+1)^{3/2}} \qquad \text{Simplify.}$$

REMARK You could also solve the problem in Example 6 without using logarithmic differentiation by using the Power and Quotient Rules. Use these rules to find the derivative and show that the result is equivalent to the one in Example 6. Which method do you prefer?

Because the natural logarithm is undefined for negative numbers, you will often encounter expressions of the form $\ln|u|$. The next theorem states that you can differentiate functions of the form $y = \ln|u|$ as though the absolute value notation was not present.

THEOREM 5.4 Derivative Involving Absolute Value

If u is a differentiable function of x such that $u \neq 0$, then

$$\frac{d}{dx}[\ln|u|] = \frac{u'}{u}.$$

Proof If $u > 0$, then $|u| = u$, and the result follows from Theorem 5.3. If $u < 0$, then $|u| = -u$, and you have

$$\begin{aligned}\frac{d}{dx}[\ln|u|] &= \frac{d}{dx}[\ln(-u)]\\ &= \frac{-u'}{-u}\\ &= \frac{u'}{u}.\end{aligned}$$

EXAMPLE 7 Derivative Involving Absolute Value

Find the derivative of

$$f(x) = \ln|\cos x|.$$

Solution Using Theorem 5.4, let $u = \cos x$ and write

$$\begin{aligned}\frac{d}{dx}[\ln|\cos x|] &= \frac{u'}{u} && \frac{d}{dx}[\ln|u|] = \frac{u'}{u}\\ &= \frac{-\sin x}{\cos x} && u = \cos x\\ &= -\tan x. && \text{Simplify.}\end{aligned}$$

EXAMPLE 8 Finding Relative Extrema

Locate the relative extrema of

$$y = \ln(x^2 + 2x + 3).$$

Solution Differentiating y, you obtain

$$\frac{dy}{dx} = \frac{2x + 2}{x^2 + 2x + 3}.$$

Because $dy/dx = 0$ when $x = -1$, you can apply the First Derivative Test and conclude that a relative minimum occurs at the point $(-1, \ln 2)$. Because there are no other critical points, it follows that this is the only relative extremum, as shown in the figure.

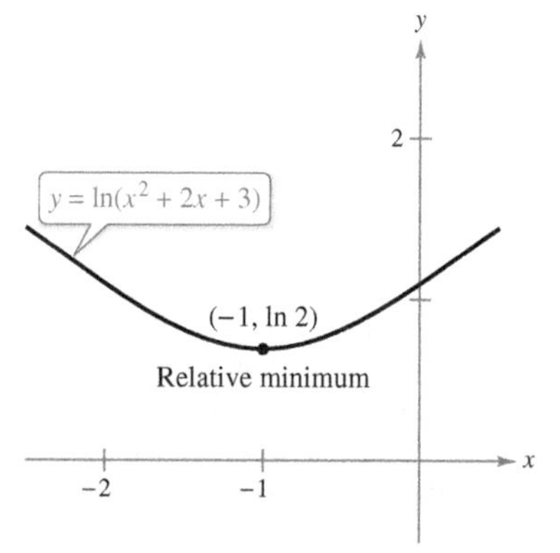

The derivative of y changes from negative to positive at $x = -1$.

5.1 Exercises

See CalcChat.com for tutorial help and worked-out solutions to odd-numbered exercises.

CONCEPT CHECK

1. **Natural Logarithmic Function** Explain why $\ln x$ is positive for $x > 1$ and negative for $0 < x < 1$.
2. **Logarithmic Properties** What is the value of n?

   $\ln 4 + \ln(n^{-1}) = \ln 4 - \ln 7$
3. **The Number e** How is the number e defined?
4. **Differentiation of Logarithmic Functions** State the Chain Rule version of the derivative of the natural logarithmic function in your own words.

Evaluating a Logarithm Using Technology **In Exercises 5–8, use a graphing utility to evaluate the logarithm by (a) using the natural logarithm key and (b) using the integration capabilities to evaluate the integral $\int_1^x (1/t)\,dt$.**

5. $\ln 45$ **6.** $\ln 8.3$

7. $\ln 0.8$ **8.** $\ln 0.6$

Matching **In Exercises 9–12, match the function with its graph. [The graphs are labeled (a), (b), (c), and (d).]**

(a)
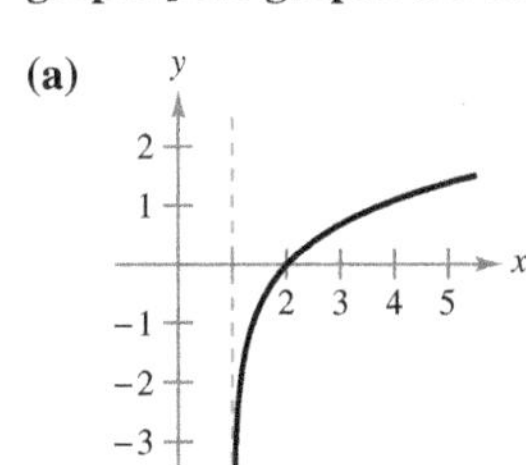

(b)
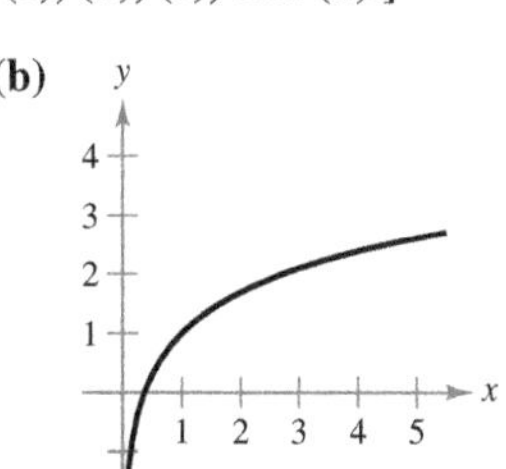

(c)
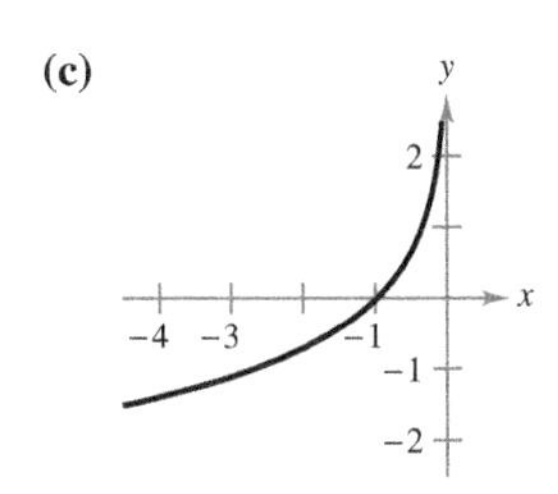

(d)
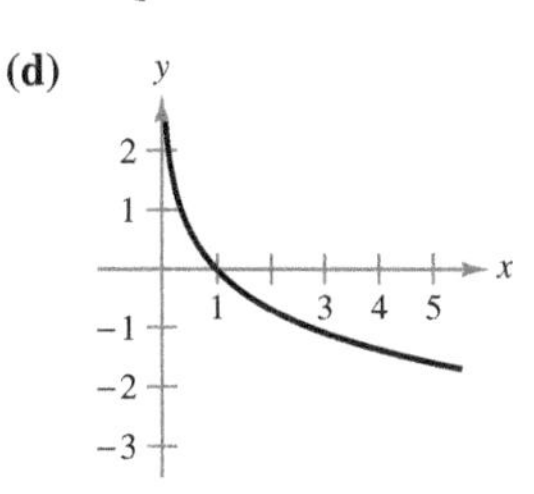

9. $f(x) = \ln x + 1$ **10.** $f(x) = -\ln x$

11. $f(x) = \ln(x - 1)$ **12.** $f(x) = -\ln(-x)$

Sketching a Graph **In Exercises 13–18, sketch the graph of the function and state its domain.**

13. $f(x) = 3 \ln x$

14. $f(x) = -2 \ln x$

15. $f(x) = \ln 2x$

16. $f(x) = \ln|x|$

17. $f(x) = \ln(x - 3)$

18. $f(x) = \ln x - 4$

Using Properties of Logarithms **In Exercises 19 and 20, use the properties of logarithms to approximate the indicated logarithms, given that $\ln 2 \approx 0.6931$ and $\ln 3 \approx 1.0986$.**

19. (a) $\ln 6$ (b) $\ln \frac{2}{3}$ (c) $\ln 81$ (d) $\ln \sqrt{3}$

20. (a) $\ln 0.25$ (b) $\ln 24$ (c) $\ln \sqrt[3]{12}$ (d) $\ln \frac{1}{72}$

Expanding a Logarithmic Expression **In Exercises 21–30, use the properties of logarithms to expand the logarithmic expression.**

21. $\ln \dfrac{x}{4}$ **22.** $\ln \sqrt{x^5}$

23. $\ln \dfrac{xy}{z}$ **24.** $\ln(xyz)$

25. $\ln\left(x\sqrt{x^2 + 5}\right)$ **26.** $x \ln \sqrt{x - 4}$

27. $\ln \sqrt{\dfrac{x - 1}{x}}$ **28.** $\ln(3e^2)$

29. $\ln z(z - 1)^2$ **30.** $\ln \dfrac{z}{e}$

Condensing a Logarithmic Expression **In Exercises 31–36, write the expression as a logarithm of a single quantity.**

31. $\ln(x - 2) - \ln(x + 2)$ **32.** $3 \ln x + 2 \ln y - 4 \ln z$

33. $\frac{1}{3}[2 \ln(x + 3) + \ln x - \ln(x^2 - 1)]$

34. $2[\ln x - \ln(x + 1) - \ln(x - 1)]$

35. $4 \ln 2 - \frac{1}{2} \ln(x^3 + 6x)$

36. $\frac{3}{2}[\ln(x^2 + 1) - \ln(x + 1) - \ln(x - 1)]$

Verifying Properties of Logarithms **In Exercises 37 and 38, (a) verify that $f = g$ by using a graphing utility to graph f and g in the same viewing window and (b) verify that $f = g$ algebraically.**

37. $f(x) = \ln \dfrac{x^2}{4}, \quad x > 0, \quad g(x) = 2 \ln x - \ln 4$

38. $f(x) = \ln \sqrt{x(x^2 + 1)}, \quad g(x) = \frac{1}{2}[\ln x + \ln(x^2 + 1)]$

Finding a Limit **In Exercises 39–42, find the limit.**

39. $\displaystyle\lim_{x \to 3^+} \ln(x - 3)$ **40.** $\displaystyle\lim_{x \to 6^-} \ln(6 - x)$

41. $\displaystyle\lim_{x \to 2^-} \ln[x^2(3 - x)]$ **42.** $\displaystyle\lim_{x \to 5^+} \ln \frac{x}{\sqrt{x - 4}}$

Finding a Derivative **In Exercises 43–66, find the derivative of the function.**

43. $f(x) = \ln 3x$ **44.** $f(x) = \ln(x - 1)$

45. $f(x) = \ln(x^2 + 3)$ **46.** $h(x) = \ln(2x^2 + 1)$

47. $y = (\ln x)^4$ **48.** $y = x^2 \ln x$

49. $y = \ln(t + 1)^2$ **50.** $y = \ln \sqrt{x^2 - 4}$

51. $y = \ln\left(x\sqrt{x^2 - 1}\right)$

52. $y = \ln[t(t^2 + 3)^3]$

53. $f(x) = \ln \dfrac{x}{x^2 + 1}$

54. $f(x) = \ln \dfrac{2x}{x + 3}$

55. $g(t) = \dfrac{\ln t}{t^2}$

56. $h(t) = \dfrac{\ln t}{t^3 + 5}$

57. $y = \ln(\ln x^2)$

58. $y = \ln(\ln x)$

59. $y = \ln \sqrt{\dfrac{x + 1}{x - 1}}$

60. $y = \ln \sqrt[3]{\dfrac{x - 1}{x + 1}}$

61. $f(x) = \ln \dfrac{\sqrt{4 + x^2}}{x}$

62. $f(x) = \ln\left(x + \sqrt{4 + x^2}\right)$

63. $y = \ln|\sin x|$

64. $y = \ln|\csc x|$

65. $y = \ln\left|\dfrac{\cos x}{\cos x - 1}\right|$

66. $y = \ln|\sec x + \tan x|$

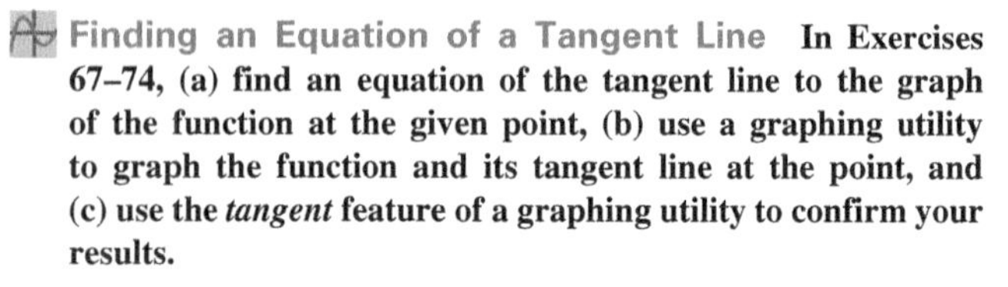

Finding an Equation of a Tangent Line **In Exercises 67–74, (a) find an equation of the tangent line to the graph of the function at the given point, (b) use a graphing utility to graph the function and its tangent line at the point, and (c) use the *tangent* feature of a graphing utility to confirm your results.**

67. $y = \ln x^4, \quad (1, 0)$

68. $y = \ln x^{2/3}, \quad (-1, 0)$

69. $f(x) = 3x^2 - \ln x, \quad (1, 3)$

70. $f(x) = 4 - x^2 - \ln\left(\frac{1}{2}x + 1\right), \quad (0, 4)$

71. $f(x) = \ln\sqrt{1 + \sin^2 x}, \quad \left(\dfrac{\pi}{4}, \ln\sqrt{\dfrac{3}{2}}\right)$

72. $f(x) = \sin 2x \ln x^2, \quad (1, 0)$

73. $y = x^3 \ln x^4, \quad (-1, 0)$

74. $f(x) = \dfrac{1}{2}x \ln x^2, \quad (-1, 0)$

Logarithmic Differentiation **In Exercises 75–80, use logarithmic differentiation to find dy/dx.**

75. $y = x\sqrt{x^2 + 1}, \quad x > 0$

76. $y = \sqrt{x^2(x + 1)(x + 2)}, \quad x > 0$

77. $y = \dfrac{x^2\sqrt{3x - 2}}{(x + 1)^2}, \quad x > \dfrac{2}{3}$

78. $y = \sqrt{\dfrac{x^2 - 1}{x^2 + 1}}, \quad x > 1$

79. $y = \dfrac{x(x - 1)^{3/2}}{\sqrt{x + 1}}, \quad x > 1$

80. $y = \dfrac{(x + 1)(x - 2)}{(x - 1)(x + 2)}, \quad x > 2$

Implicit Differentiation **In Exercises 81–84, use implicit differentiation to find dy/dx.**

81. $x^2 - 3 \ln y + y^2 = 10$

82. $\ln xy + 5x = 30$

83. $4x^3 + \ln y^2 + 2y = 2x$

84. $4xy + \ln x^2 y = 7$

Differential Equation **In Exercises 85 and 86, verify that the function is a solution of the differential equation.**

Function	Differential Equation
85. $y = 2 \ln x + 3$	$xy'' + y' = 0$
86. $y = x \ln x - 4x$	$x + y - xy' = 0$

Relative Extrema and Points of Inflection **In Exercises 87–92, locate any relative extrema and points of inflection. Use a graphing utility to confirm your results.**

87. $y = \dfrac{x^2}{2} - \ln x$

88. $y = 2x - \ln 2x$

89. $y = x \ln x$

90. $y = \dfrac{\ln x}{x}$

91. $y = \dfrac{x}{\ln x}$

92. $y = x^2 \ln \dfrac{x}{4}$

Using Newton's Method **In Exercises 93 and 94, use Newton's Method to approximate, to three decimal places, the x-coordinate of the point of intersection of the graphs of the two equations. Use a graphing utility to verify your result.**

93. $y = \ln x, \quad y = -x$

94. $y = \ln x, \quad y = 3 - x$

EXPLORING CONCEPTS

Comparing Functions **In Exercises 95 and 96, let f be a function that is positive and differentiable on the entire real number line and let $g(x) = \ln f(x)$.**

95. When g is increasing, must f be increasing? Explain.

96. When the graph of f is concave upward, must the graph of g be concave upward? Explain.

97. **Think About It** Is $\ln xy = \ln x \ln y$ a valid property of logarithms, where $x > 0$ and $y > 0$? Explain.

98. **HOW DO YOU SEE IT?** The graph shows the temperature T (in degrees Celsius) of an object h hours after it is removed from a furnace.

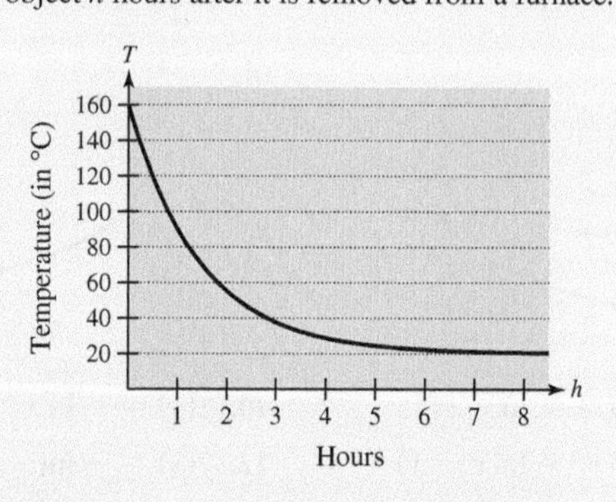

(a) Find $\lim_{h\to\infty} T$. What does this limit represent?

(b) When is the temperature changing most rapidly?

True or False? **In Exercises 99–102, determine whether the statement is true or false. If it is false, explain why or give an example that shows it is false.**

99. $\ln(a^{n+m}) = n \ln a + m \ln a$, where $a > 0$ and m and n are rational.

100. $\dfrac{d}{dx}[\ln(cx)] = \dfrac{d}{dx}[\ln x]$, where $c > 0$

101. If $y = \ln \pi$, then $y' = 1/\pi$.

102. If $y = \ln e$, then $y' = 1$.

103. Home Mortgage The term t (in years) of a \$200,000 home mortgage at 7.5% interest can be approximated by

$$t = 13.375 \ln\left(\frac{x}{x - 1250}\right), \quad x > 1250$$

where x is the monthly payment in dollars.

(a) Use a graphing utility to graph the model.

(b) Use the model to approximate the term of a home mortgage for which the monthly payment is \$1398.43. What is the total amount paid?

(c) Use the model to approximate the term of a home mortgage for which the monthly payment is \$1611.19. What is the total amount paid?

(d) Find the instantaneous rates of change of t with respect to x when $x = \$1398.43$ and $x = \$1611.19$.

(e) Write a short paragraph describing the benefit of the higher monthly payment.

104. Sound Intensity The relationship between the number of decibels β and the intensity of a sound I in watts per centimeter squared is

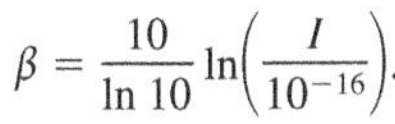

$$\beta = \frac{10}{\ln 10} \ln\left(\frac{I}{10^{-16}}\right).$$

(a) Use the properties of logarithms to write the formula in simpler form.

(b) Determine the number of decibels of a sound with an intensity of 10^{-5} watt per square centimeter.

Aceshot1/Shutterstock.com

105. Modeling Data The table shows the temperatures T (in degrees Fahrenheit) at which water boils at selected pressures p (in pounds per square inch). *(Source: Standard Handbook of Mechanical Engineers)*

p	5	10	14.696 (1 atm)	20
T	162.24	193.21	212.00	227.96

p	30	40	60	80	100
T	250.33	267.25	292.71	312.03	327.81

A model that approximates the data is

$$T = 87.97 + 34.96 \ln p + 7.91\sqrt{p}.$$

(a) Use a graphing utility to plot the data and graph the model.

(b) Find the rates of change of T with respect to p when $p = 10$ and $p = 70$.

(c) Use a graphing utility to graph T'. Find $\lim_{p \to \infty} T'(p)$ and interpret the result in the context of the problem.

106. Modeling Data The atmospheric pressure decreases with increasing altitude. At sea level, the average air pressure is one atmosphere (1.033227 kilograms per square centimeter). The table shows the pressures p (in atmospheres) at selected altitudes h (in kilometers).

h	0	5	10	15	20	25
p	1	0.55	0.25	0.12	0.06	0.02

(a) Use a graphing utility to find a model of the form $p = a + b \ln h$ for the data. Explain why the result is an error message.

(b) Use a graphing utility to find the logarithmic model $h = a + b \ln p$ for the data.

(c) Use a graphing utility to plot the data and graph the model from part (b).

(d) Use the model to estimate the altitude when $p = 0.75$.

(e) Use the model to estimate the pressure when $h = 13$.

(f) Use the model to find the rates of change of pressure when $h = 5$ and $h = 20$. Interpret the results.

107. Tractrix A person walking along a dock drags a boat by a 10-meter rope. The boat travels along a path known as a *tractrix* (see figure). The equation of this path is

$$y = 10 \ln\left(\frac{10 + \sqrt{100 - x^2}}{x}\right) - \sqrt{100 - x^2}.$$

(a) Use a graphing utility to graph the function.

(b) What are the slopes of this path when $x = 5$ and $x = 9$?

(c) What does the slope of the path approach as x approaches 10 from the left?

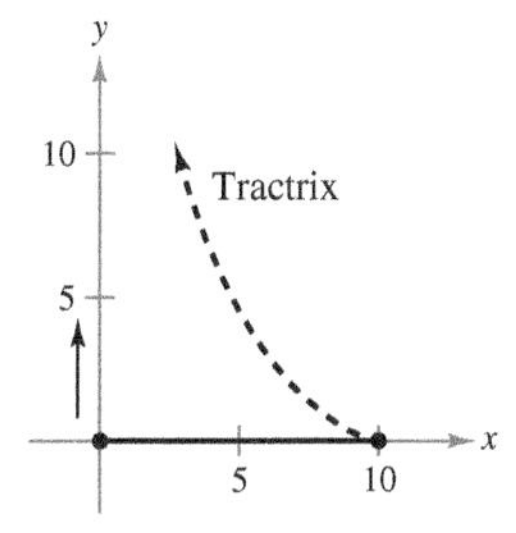

108. Prime Number Theorem There are 25 prime numbers less than 100. The **Prime Number Theorem** states that the number of primes less than x approaches

$$p(x) \approx \frac{x}{\ln x}.$$

Use this approximation to estimate the rate (in primes per 100 integers) at which the prime numbers occur when

(a) $x = 1000$.

(b) $x = 1{,}000{,}000$.

(c) $x = 1{,}000{,}000{,}000$.

109. Conjecture Use a graphing utility to graph f and g in the same viewing window and determine which is increasing at the greater rate for large values of x. What can you conclude about the rate of growth of the natural logarithmic function?

(a) $f(x) = \ln x, \quad g(x) = \sqrt{x}$

(b) $f(x) = \ln x, \quad g(x) = \sqrt[4]{x}$

5.2 The Natural Logarithmic Function: Integration

- **Use the Log Rule for Integration to integrate a rational function.**
- **Integrate trigonometric functions.**

Log Rule for Integration

The differentiation rules

$$\frac{d}{dx}[\ln|x|] = \frac{1}{x} \quad \text{and} \quad \frac{d}{dx}[\ln|u|] = \frac{u'}{u}$$

that you studied in the preceding section produce the following integration rule.

THEOREM 5.5 Log Rule for Integration

Let u be a differentiable function of x.

1. $\displaystyle\int \frac{1}{x}\,dx = \ln|x| + C$ **2.** $\displaystyle\int \frac{1}{u}\,du = \ln|u| + C$

Because $du = u'\,dx$, the second formula can also be written as

$$\int \frac{u'}{u}\,dx = \ln|u| + C. \qquad \text{Alternative form of Log Rule}$$

Exploration

Integrating Rational Functions

Early in Chapter 4, you learned rules that allowed you to integrate *any* polynomial function. The Log Rule presented in this section goes a long way toward enabling you to integrate rational functions. For instance, each of the following functions can be integrated with the Log Rule.

$\frac{2}{x}$	Example 1
$\frac{1}{4x-1}$	Example 2
$\frac{x}{x^2+1}$	Example 3
$\frac{3x^2+1}{x^3+x}$	Example 4(a)
$\frac{x+1}{x^2+2x}$	Example 4(c)
$\frac{1}{3x+2}$	Example 4(d)
$\frac{x^2+x+1}{x^2+1}$	Example 5
$\frac{2x}{(x+1)^2}$	Example 6

There are still some rational functions that cannot be integrated using the Log Rule. Give examples of these functions and explain your reasoning.

EXAMPLE 1 Using the Log Rule for Integration

$$\int \frac{2}{x}\,dx = 2\int \frac{1}{x}\,dx \qquad \text{Constant Multiple Rule}$$
$$= 2\ln|x| + C \qquad \text{Log Rule for Integration}$$
$$= \ln x^2 + C \qquad \text{Property of logarithms}$$

Because x^2 cannot be negative, the absolute value notation is unnecessary in the final form of the antiderivative.

EXAMPLE 2 Using the Log Rule with a Change of Variables

Find $\displaystyle\int \frac{1}{4x-1}\,dx.$

Solution If you let $u = 4x - 1$, then $du = 4\,dx$.

$$\int \frac{1}{4x-1}\,dx = \frac{1}{4}\int\left(\frac{1}{4x-1}\right)4\,dx \qquad \text{Multiply and divide by 4.}$$
$$= \frac{1}{4}\int \frac{1}{u}\,du \qquad \text{Substitute: } u = 4x - 1.$$
$$= \frac{1}{4}\ln|u| + C \qquad \text{Apply Log Rule.}$$
$$= \frac{1}{4}\ln|4x-1| + C \qquad \text{Back-substitute.}$$

Example 3 uses the alternative form of the Log Rule. To apply this rule, look for quotients in which the numerator is the derivative of the denominator.

EXAMPLE 3 Finding Area with the Log Rule

Find the area of the region bounded by the graph of

$$y = \frac{x}{x^2 + 1}$$

the x-axis, and the line $x = 3$.

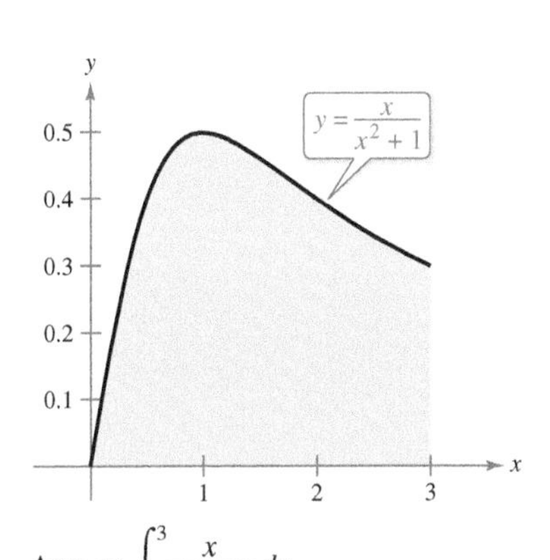

$\text{Area} = \int_0^3 \frac{x}{x^2+1}\,dx$

The area of the region bounded by the graph of y, the x-axis, and $x = 3$ is $\frac{1}{2}\ln 10$.

Figure 5.7

Solution In Figure 5.7, you can see that the area of the region is given by the definite integral

$$\int_0^3 \frac{x}{x^2 + 1}\,dx.$$

If you let $u = x^2 + 1$, then $u' = 2x$. To apply the Log Rule, multiply and divide by 2 as shown.

$$\begin{aligned}
\int_0^3 \frac{x}{x^2+1}\,dx &= \frac{1}{2}\int_0^3 \frac{2x}{x^2+1}\,dx && \text{Multiply and divide by 2.}\\
&= \frac{1}{2}\Big[\ln(x^2+1)\Big]_0^3 && \int \frac{u'}{u}\,dx = \ln|u| + C\\
&= \frac{1}{2}(\ln 10 - \ln 1)\\
&= \frac{1}{2}\ln 10 && \ln 1 = 0\\
&\approx 1.151
\end{aligned}$$

EXAMPLE 4 Recognizing Quotient Forms of the Log Rule

a. $\displaystyle\int \frac{3x^2+1}{x^3+x}\,dx = \ln|x^3+x| + C$ $\qquad u = x^3 + x$

b. $\displaystyle\int \frac{\sec^2 x}{\tan x}\,dx = \ln|\tan x| + C$ $\qquad u = \tan x$

c. $\displaystyle\int \frac{x+1}{x^2+2x}\,dx = \frac{1}{2}\int \frac{2x+2}{x^2+2x}\,dx$ $\qquad u = x^2 + 2x$

$$= \frac{1}{2}\ln|x^2+2x| + C$$

d. $\displaystyle\int \frac{1}{3x+2}\,dx = \frac{1}{3}\int \frac{3}{3x+2}\,dx$ $\qquad u = 3x + 2$

$$= \frac{1}{3}\ln|3x+2| + C$$

With antiderivatives involving logarithms, it is easy to obtain forms that look quite different but are still equivalent. For instance, both

$$\ln|(3x+2)^{1/3}| + C$$

and

$$\ln|3x+2|^{1/3} + C$$

are equivalent to the antiderivative listed in Example 4(d).

Integrals to which the Log Rule can be applied often appear in disguised form. For instance, when a rational function has a *numerator of degree greater than or equal to that of the denominator*, division may reveal a form to which you can apply the Log Rule. This is shown in Example 5.

EXAMPLE 5 Using Long Division Before Integrating

See LarsonCalculus.com for an interactive version of this type of example.

Find the indefinite integral.

$$\int \frac{x^2 + x + 1}{x^2 + 1}\,dx$$

Solution Begin by using long division to rewrite the integrand.

$$\frac{x^2 + x + 1}{x^2 + 1} \quad \Longrightarrow \quad x^2 + 1\overline{)\,x^2 + x + 1} \;\; (\text{quotient } 1;\; x^2 + 1 \text{ subtracted, remainder } x) \quad \Longrightarrow \quad 1 + \frac{x}{x^2 + 1}$$

Now, you can integrate to obtain

$$\int \frac{x^2 + x + 1}{x^2 + 1}\,dx = \int \left(1 + \frac{x}{x^2 + 1}\right) dx \qquad \text{Rewrite using long division.}$$

$$= \int dx + \frac{1}{2}\int \frac{2x}{x^2 + 1}\,dx \qquad \text{Rewrite as two integrals.}$$

$$= x + \frac{1}{2}\ln(x^2 + 1) + C. \qquad \text{Integrate.}$$

Check this result by differentiating to obtain the original integrand. ■

The next example presents another instance in which the use of the Log Rule is disguised. In this case, a change of variables helps you recognize the Log Rule.

EXAMPLE 6 Change of Variables with the Log Rule

Find the indefinite integral.

$$\int \frac{2x}{(x + 1)^2}\,dx$$

Solution If you let $u = x + 1$, then $du = dx$ and $x = u - 1$.

$$\int \frac{2x}{(x + 1)^2}\,dx = \int \frac{2(u - 1)}{u^2}\,du \qquad \text{Substitute.}$$

$$= 2\int \left(\frac{u}{u^2} - \frac{1}{u^2}\right) du \qquad \text{Rewrite as two fractions.}$$

$$= 2\int \frac{du}{u} - 2\int u^{-2}\,du \qquad \text{Rewrite as two integrals.}$$

$$= 2\ln|u| - 2\left(\frac{u^{-1}}{-1}\right) + C \qquad \text{Integrate.}$$

$$= 2\ln|u| + \frac{2}{u} + C \qquad \text{Simplify.}$$

$$= 2\ln|x + 1| + \frac{2}{x + 1} + C \qquad \text{Back-substitute.}$$

Check this result by differentiating to obtain the original integrand. ■

TECHNOLOGY If you have access to a computer algebra system, use it to find the indefinite integrals in Examples 5 and 6. How does the form of the antiderivative that it gives you compare with that given in Examples 5 and 6?

As you study the methods shown in Examples 5 and 6, be aware that both methods involve rewriting a disguised integrand so that it fits one or more of the basic integration formulas. Throughout the remaining sections of Chapter 5 and in Chapter 8, much time will be devoted to integration techniques. To master these techniques, you must recognize the "form-fitting" nature of integration. In this sense, integration is not nearly as straightforward as differentiation. Differentiation takes the form

"Here is the question; what is the answer?"

Integration is more like

"Here is the answer; what is the question?"

Here are some guidelines you can use for integration.

GUIDELINES FOR INTEGRATION

1. Learn a basic list of integration formulas.
2. Find an integration formula that resembles all or part of the integrand and, by trial and error, find a choice of u that will make the integrand conform to the formula.
3. When you cannot find a u-substitution that works, try altering the integrand. You might try a trigonometric identity, multiplication and division by the same quantity, addition and subtraction of the same quantity, or long division. Be creative.
4. If you have access to computer software that will find antiderivatives symbolically, use it.
5. Check your result by differentiating to obtain the original integrand.

EXAMPLE 7 ***u*-Substitution and the Log Rule**

Solve the differential equation

$$\frac{dy}{dx} = \frac{1}{x \ln x}.$$

Solution The solution can be written as an indefinite integral.

$$y = \int \frac{1}{x \ln x}\, dx$$

Because the integrand is a quotient whose denominator is raised to the first power, you should try the Log Rule. There are three basic choices for u. The choices

$$u = x \quad \text{and} \quad u = x \ln x$$

fail to fit the u'/u form of the Log Rule. However, the third choice does fit. Letting $u = \ln x$ produces $u' = 1/x$, and you obtain the following.

$$\begin{aligned} \int \frac{1}{x \ln x}\, dx &= \int \frac{1/x}{\ln x}\, dx && \text{Divide numerator and denominator by } x. \\ &= \int \frac{u'}{u}\, dx && \text{Substitute: } u = \ln x. \\ &= \ln|u| + C && \text{Apply Log Rule.} \\ &= \ln|\ln x| + C && \text{Back-substitute.} \end{aligned}$$

So, the solution is $y = \ln|\ln x| + C$.

REMARK Keep in mind that you can check your answer to an integration problem by differentiating the answer. For instance, in Example 7, the derivative of $y = \ln|\ln x| + C$ is $y' = 1/(x \ln x)$.

Integrals of Trigonometric Functions

In Section 4.1, you looked at six trigonometric integration rules—the six that correspond directly to differentiation rules. With the Log Rule, you can now complete the set of basic trigonometric integration formulas.

EXAMPLE 8 Using a Trigonometric Identity

Find $\int \tan x \, dx$.

Solution This integral does not seem to fit any formulas on our basic list. However, by using a trigonometric identity, you obtain

$$\int \tan x \, dx = \int \frac{\sin x}{\cos x} \, dx.$$

Knowing that $D_x[\cos x] = -\sin x$, you can let $u = \cos x$ and write

$$\int \tan x \, dx = -\int \frac{-\sin x}{\cos x} \, dx$$ Apply trigonometric identity and multiply and divide by -1.

$$= -\int \frac{u'}{u} \, dx$$ Substitute: $u = \cos x$.

$$= -\ln|u| + C$$ Apply Log Rule.

$$= -\ln|\cos x| + C.$$ Back-substitute. ■

Example 8 used a trigonometric identity to derive an integration rule for the tangent function. The next example takes a rather unusual step (multiplying and dividing by the same quantity) to derive an integration rule for the secant function.

EXAMPLE 9 Derivation of the Secant Formula

Find $\int \sec x \, dx$.

Solution Consider the following procedure.

$$\int \sec x \, dx = \int (\sec x)\left(\frac{\sec x + \tan x}{\sec x + \tan x}\right) dx$$ Multiply and divide by $\sec x + \tan x$.

$$= \int \frac{\sec^2 x + \sec x \tan x}{\sec x + \tan x} \, dx$$

Letting u be the denominator of this quotient produces

$$u = \sec x + \tan x$$

and

$$u' = \sec x \tan x + \sec^2 x.$$

So, you can conclude that

$$\int \sec x \, dx = \int \frac{\sec^2 x + \sec x \tan x}{\sec x + \tan x} \, dx$$ Rewrite integrand.

$$= \int \frac{u'}{u} \, dx$$ Substitute: $u = \sec x + \tan x$.

$$= \ln|u| + C$$ Apply Log Rule.

$$= \ln|\sec x + \tan x| + C.$$ Back-substitute. ■

With the results of Examples 8 and 9, you now have integration formulas for $\sin x$, $\cos x$, $\tan x$, and $\sec x$. The integrals of the six basic trigonometric functions are summarized below. (For proofs of $\cot u$ and $\csc u$, see Exercises 85 and 86.)

REMARK Using trigonometric identities and properties of logarithms, you could rewrite these six integration rules in other forms. For instance, you could write

$$\int \csc u\, du = \ln|\csc u - \cot u| + C.$$

(See Exercises 87–90.)

INTEGRALS OF THE SIX BASIC TRIGONOMETRIC FUNCTIONS

$$\int \sin u\, du = -\cos u + C \qquad \int \cos u\, du = \sin u + C$$

$$\int \tan u\, du = -\ln|\cos u| + C \qquad \int \cot u\, du = \ln|\sin u| + C$$

$$\int \sec u\, du = \ln|\sec u + \tan u| + C \qquad \int \csc u\, du = -\ln|\csc u + \cot u| + C$$

EXAMPLE 10 Integrating Trigonometric Functions

Evaluate $\displaystyle\int_0^{\pi/4} \sqrt{1 + \tan^2 x}\, dx$.

Solution Using $1 + \tan^2 x = \sec^2 x$, you can write

$$\begin{aligned}
\int_0^{\pi/4} \sqrt{1 + \tan^2 x}\, dx &= \int_0^{\pi/4} \sqrt{\sec^2 x}\, dx \\
&= \int_0^{\pi/4} \sec x\, dx && \sec x \ge 0 \text{ for } 0 \le x \le \frac{\pi}{4}. \\
&= \ln|\sec x + \tan x| \Big]_0^{\pi/4} \\
&= \ln(\sqrt{2} + 1) - \ln 1 \\
&\approx 0.881.
\end{aligned}$$

EXAMPLE 11 Finding an Average Value

Find the average value of

$$f(x) = \tan x$$

on the interval $[0, \pi/4]$.

Solution

$$\begin{aligned}
\text{Average value} &= \frac{1}{(\pi/4) - 0} \int_0^{\pi/4} \tan x\, dx && \text{Average value} = \frac{1}{b-a}\int_a^b f(x)\, dx \\
&= \frac{4}{\pi}\int_0^{\pi/4} \tan x\, dx && \text{Simplify.} \\
&= \frac{4}{\pi}\Big[-\ln|\cos x|\Big]_0^{\pi/4} && \text{Integrate.} \\
&= -\frac{4}{\pi}\left[\ln\frac{\sqrt{2}}{2} - \ln 1\right] \\
&= -\frac{4}{\pi}\ln\frac{\sqrt{2}}{2} \\
&\approx 0.441
\end{aligned}$$

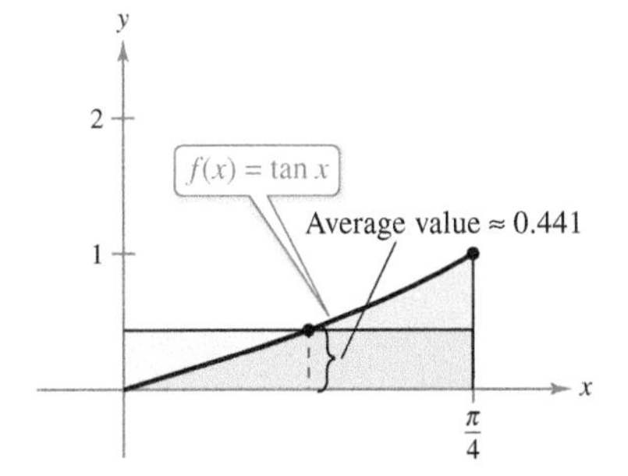

Figure 5.8

The average value is about 0.441, as shown in Figure 5.8.

5.2 Exercises

See CalcChat.com for tutorial help and worked-out solutions to odd-numbered exercises.

CONCEPT CHECK

1. **Log Rule for Integration** Can you use the Log Rule to find the integral below? Explain.

$$\int \frac{x}{(x^2-4)^3}\,dx$$

2. **Long Division** Explain when to use long division before applying the Log Rule.

3. **Guidelines for Integration** Describe two ways to alter an integrand so that it fits an integration formula.

4. **Trigonometric Functions** Integrating which trigonometric function results in $\ln|\sin x| + C$?

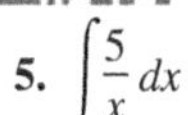

Finding an Indefinite Integral **In Exercises 5–28, find the indefinite integral.**

5. $\displaystyle\int \frac{5}{x}\,dx$

6. $\displaystyle\int \frac{1}{x-5}\,dx$

7. $\displaystyle\int \frac{1}{2x+5}\,dx$

8. $\displaystyle\int \frac{9}{5-4x}\,dx$

9. $\displaystyle\int \frac{x}{x^2-3}\,dx$

10. $\displaystyle\int \frac{x^2}{5-x^3}\,dx$

11. $\displaystyle\int \frac{4x^3+3}{x^4+3x}\,dx$

12. $\displaystyle\int \frac{x^2-2x}{x^3-3x^2}\,dx$

13. $\displaystyle\int \frac{x^2-7}{7x}\,dx$

14. $\displaystyle\int \frac{x^3-8x}{x^2}\,dx$

15. $\displaystyle\int \frac{x^2+2x+3}{x^3+3x^2+9x}\,dx$

16. $\displaystyle\int \frac{x^2+4x}{x^3+6x^2+5}\,dx$

17. $\displaystyle\int \frac{x^2-3x+2}{x+1}\,dx$

18. $\displaystyle\int \frac{2x^2+7x-3}{x-2}\,dx$

19. $\displaystyle\int \frac{x^3-3x^2+5}{x-3}\,dx$

20. $\displaystyle\int \frac{x^3-6x-20}{x+5}\,dx$

21. $\displaystyle\int \frac{x^4+x-4}{x^2+2}\,dx$

22. $\displaystyle\int \frac{x^3-4x^2-4x+20}{x^2-5}\,dx$

23. $\displaystyle\int \frac{(\ln x)^2}{x}\,dx$

24. $\displaystyle\int \frac{dx}{x(\ln x^2)^3}$

25. $\displaystyle\int \frac{1}{\sqrt{x}(1-3\sqrt{x})}\,dx$

26. $\displaystyle\int \frac{1}{x^{2/3}(1+x^{1/3})}\,dx$

27. $\displaystyle\int \frac{6x}{(x-5)^2}\,dx$

28. $\displaystyle\int \frac{x(x-2)}{(x-1)^3}\,dx$

Change of Variables **In Exercises 29–32, find the indefinite integral by making a change of variables (*Hint:* Let *u* be the denominator of the integrand.)**

29. $\displaystyle\int \frac{1}{1+\sqrt{2x}}\,dx$

30. $\displaystyle\int \frac{4}{1+\sqrt{5x}}\,dx$

31. $\displaystyle\int \frac{\sqrt{x}}{\sqrt{x}-3}\,dx$

32. $\displaystyle\int \frac{\sqrt[3]{x}}{\sqrt[3]{x}-1}\,dx$

Finding an Indefinite Integral of a Trigonometric Function **In Exercises 33–42, find the indefinite integral.**

33. $\displaystyle\int \cot\frac{\theta}{3}\,d\theta$

34. $\displaystyle\int \theta\tan 2\theta^2\,d\theta$

35. $\displaystyle\int \csc 2x\,dx$

36. $\displaystyle\int \sec\frac{x}{2}\,dx$

37. $\displaystyle\int (5-\cos 3\theta)\,d\theta$

38. $\displaystyle\int \left(2-\tan\frac{\theta}{4}\right)d\theta$

39. $\displaystyle\int \frac{\cos t}{1+\sin t}\,dt$

40. $\displaystyle\int \frac{\csc^2 t}{\cot t}\,dt$

41. $\displaystyle\int \frac{\sec x\tan x}{\sec x-1}\,dx$

42. $\displaystyle\int (\sec 2x+\tan 2x)\,dx$

Differential Equation **In Exercises 43–46, find the general solution of the differential equation. Use a graphing utility to graph three solutions, one of which passes through the given point.**

43. $\dfrac{dy}{dx} = \dfrac{3}{2-x},\quad (1, 0)$

44. $\dfrac{dy}{dx} = \dfrac{x-2}{x},\quad (-1, 0)$

45. $\dfrac{dy}{dx} = \dfrac{2x}{x^2-9},\quad (0, 4)$

46. $\dfrac{dr}{dt} = \dfrac{\sec^2 t}{\tan t+1},\quad (\pi, 4)$

Finding a Particular Solution **In Exercises 47 and 48, find the particular solution of the differential equation that satisfies the initial conditions.**

47. $f''(x) = \dfrac{2}{x^2},\quad f'(1) = 1,\quad f(1) = 1,\quad x > 0$

48. $f''(x) = -\dfrac{4}{(x-1)^2} - 2,\quad f'(2) = 0,\quad f(2) = 3,\quad x > 1$

Slope Field **In Exercises 49 and 50, a differential equation, a point, and a slope field are given. (a) Sketch two approximate solutions of the differential equation on the slope field, one of which passes through the given point. (To print an enlarged copy of the graph, go to *MathGraphs.com*.) (b) Use integration and the given point to find the particular solution of the differential equation and use a graphing utility to graph the solution. Compare the result with the sketch in part (a) that passes through the given point.**

49. $\dfrac{dy}{dx} = \dfrac{1}{x+2},\quad (0, 1)$

50. $\dfrac{dy}{dx} = \dfrac{\ln x}{x},\quad (1, -2)$

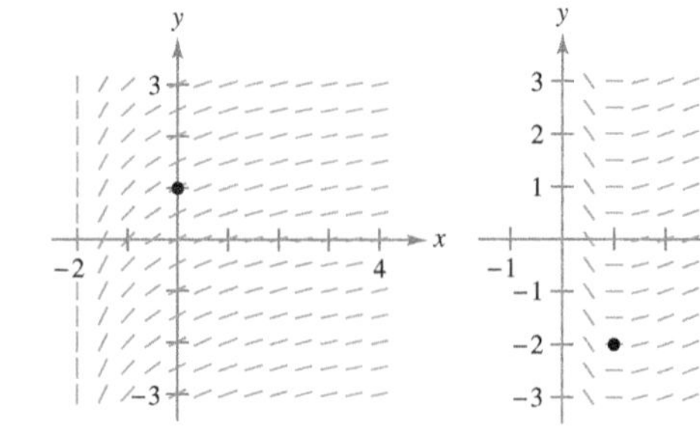

Evaluating a Definite Integral In Exercises 51–58, evaluate the definite integral. Use a graphing utility to verify your result.

51. $\int_0^4 \frac{5}{3x+1}\,dx$

52. $\int_{-1}^1 \frac{1}{2x+3}\,dx$

53. $\int_1^e \frac{(1+\ln x)^2}{x}\,dx$

54. $\int_e^{e^2} \frac{1}{x\ln x}\,dx$

55. $\int_0^2 \frac{x^2-2}{x+1}\,dx$

56. $\int_0^1 \frac{x-1}{x+1}\,dx$

57. $\int_1^2 \frac{1-\cos\theta}{\theta-\sin\theta}\,d\theta$

58. $\int_{\pi/8}^{\pi/4} (\csc 2\theta - \cot 2\theta)\,d\theta$

Finding an Integral Using Technology In Exercises 59 and 60, use a computer algebra system to find or evaluate the integral.

59. $\int \frac{1-\sqrt{x}}{1+\sqrt{x}}\,dx$

60. $\int_{-\pi/4}^{\pi/4} \frac{\sin^2 x - \cos^2 x}{\cos x}\,dx$

Finding a Derivative In Exercises 61–64, find $F'(x)$.

61. $F(x) = \int_1^x \frac{1}{t}\,dt$

62. $F(x) = \int_0^x \tan t\,dt$

63. $F(x) = \int_1^{4x} \cot t\,dt$

64. $F(x) = \int_0^{x^2} \frac{3}{t+1}\,dt$

Area In Exercises 65–68, find the area of the given region. Use a graphing utility to verify your result.

65. $y = \frac{6}{x}$

66. $y = \frac{1+\ln x^3}{x}$

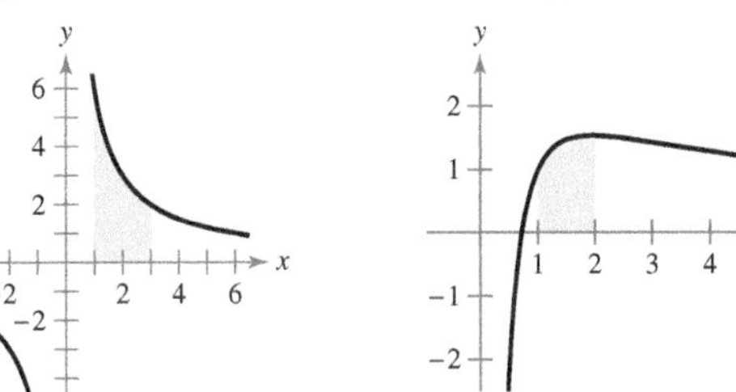

67. $y = \csc(x+1)$

68. $y = \frac{\sin x}{1+\cos x}$

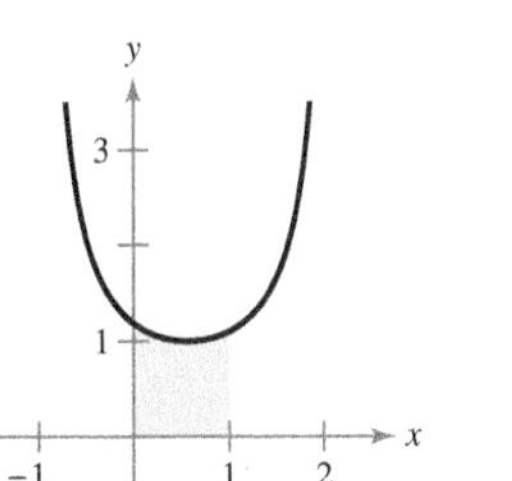

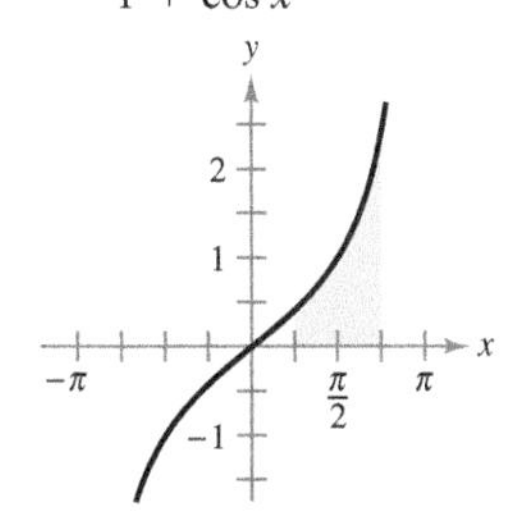

Area In Exercises 69–72, find the area of the region bounded by the graphs of the equations. Use a graphing utility to verify your result.

69. $y = \frac{x^2+4}{x}, \quad x = 1, \quad x = 4, \quad y = 0$

70. $y = \frac{5x}{x^2+2}, \quad x = 1, \quad x = 5, \quad y = 0$

71. $y = 2\sec\frac{\pi x}{6}, \quad x = 0, \quad x = 2, \quad y = 0$

72. $y = 2x - \tan 0.3x, \quad x = 1, \quad x = 4, \quad y = 0$

Finding the Average Value of a Function In Exercises 73–76, find the average value of the function over the given interval.

73. $f(x) = \frac{8}{x^2}, \quad [2, 4]$

74. $f(x) = \frac{4(x+1)}{x^2}, \quad [2, 4]$

75. $f(x) = \frac{2\ln x}{x}, \quad [1, e]$

76. $f(x) = \sec\frac{\pi x}{6}, \quad [0, 2]$

Midpoint Rule In Exercises 77 and 78, use the Midpoint Rule with $n = 4$ to approximate the value of the definite integral. Use a graphing utility to verify your result.

77. $\int_1^3 \frac{12}{x}\,dx$

78. $\int_0^{\pi/4} \sec x\,dx$

EXPLORING CONCEPTS

Approximation In Exercises 79 and 80, determine which value best approximates the area of the region between the x-axis and the graph of the function over the given interval. Make your selection on the basis of a sketch of the region, not by performing calculations.

79. $f(x) = \sec x, \quad [0, 1]$

(a) 6 (b) -6 (c) $\frac{1}{2}$ (d) 1.25 (e) 3

80. $f(x) = \frac{2x}{x^2+1}, \quad [0, 4]$

(a) 3 (b) 7 (c) -2 (d) 5 (e) 1

81. **Napier's Inequality** For $0 < x < y$, use the Mean Value Theorem to show that

$$\frac{1}{y} < \frac{\ln y - \ln x}{y - x} < \frac{1}{x}.$$

82. **Think About It** Is the function

$$F(x) = \int_x^{2x} \frac{1}{t}\,dt$$

constant, increasing, or decreasing on the interval $(0, \infty)$? Explain.

83. **Finding a Value** Find a value of x such that

$$\int_1^x \frac{3}{t}\,dt = \int_{1/4}^x \frac{1}{t}\,dt.$$

84. **Finding a Value** Find a value of x such that

$$\int_1^x \frac{1}{t}\,dt$$

is equal to (a) $\ln 5$ and (b) 1.

85. Proof Prove that

$$\int \cot u\, du = \ln|\sin u| + C.$$

86. Proof Prove that

$$\int \csc u\, du = -\ln|\csc u + \cot u| + C.$$

Using Properties of Logarithms and Trigonometric Identities **In Exercises 87–90, show that the two formulas are equivalent.**

87. $\int \tan x\, dx = -\ln|\cos x| + C$

$\int \tan x\, dx = \ln|\sec x| + C$

88. $\int \cot x\, dx = \ln|\sin x| + C$

$\int \cot x\, dx = -\ln|\csc x| + C$

89. $\int \sec x\, dx = \ln|\sec x + \tan x| + C$

$\int \sec x\, dx = -\ln|\sec x - \tan x| + C$

90. $\int \csc x\, dx = -\ln|\csc x + \cot x| + C$

$\int \csc x\, dx = \ln|\csc x - \cot x| + C$

91. Population Growth A population of bacteria P is changing at a rate of

$$\frac{dP}{dt} = \frac{3000}{1 + 0.25t}$$

where t is the time in days. The initial population (when $t = 0$) is 1000.

(a) Write an equation that gives the population at any time t.

(b) Find the population when $t = 3$ days.

92. Sales The rate of change in sales S is inversely proportional to time t ($t > 1$), measured in weeks. Find S as a function of t when the sales after 2 and 4 weeks are 200 units and 300 units, respectively.

93. Heat Transfer

Find the time required for an object to cool from 300°F to 250°F by evaluating

$$t = \frac{10}{\ln 2}\int_{250}^{300} \frac{1}{T - 100}\, dT$$

where t is time in minutes.

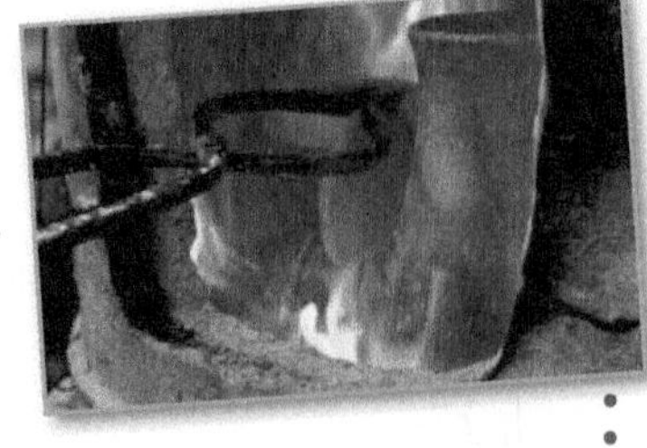

94. Average Price The demand equation for a product is

$$p = \frac{90{,}000}{400 + 3x}$$

where p is the price (in dollars) and x is the number of units (in thousands). Find the average price p on the interval $40 \le x \le 50$.

95. Area and Slope Graph the function

$$f(x) = \frac{x}{1 + x^2}$$

on the interval $[0, \infty)$.

(a) Find the area bounded by the graph of f and the line $y = \frac{1}{2}x$.

(b) Determine the values of the slope m such that the line $y = mx$ and the graph of f enclose a finite region.

(c) Calculate the area of this region as a function of m.

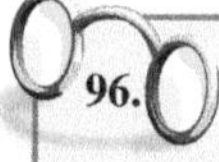

96. HOW DO YOU SEE IT? Use the graph of f' shown in the figure to answer the following.

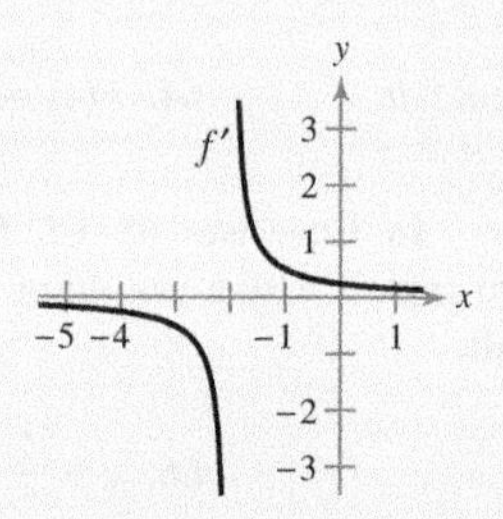

(a) Approximate the slope of f at $x = -1$. Explain.

(b) Approximate any open intervals on which the graph of f is increasing and any open intervals on which it is decreasing. Explain.

True or False? **In Exercises 97–100, determine whether the statement is true or false. If it is false, explain why or give an example that shows it is false.**

97. $\ln|x^4| = \ln x^4$

98. $\ln|\cos \theta^2| = \ln(\cos \theta^2)$

99. $\int \frac{1}{x}\, dx = \ln|cx|, \quad c \ne 0$

100. $\int_{-1}^{2} \frac{1}{x}\, dx = \Big[\ln|x|\Big]_{-1}^{2} = \ln 2 - \ln 1 = \ln 2$

PUTNAM EXAM CHALLENGE

101. Suppose that f is a function on the interval $[1, 3]$ such that $-1 \le f(x) \le 1$ for all x and $\int_1^3 f(x)\, dx = 0$. How large can $\int_1^3 \frac{f(x)}{x}\, dx$ be?

5.3 Inverse Functions

- Verify that one function is the inverse function of another function.
- Determine whether a function has an inverse function.
- Find the derivative of an inverse function.

Inverse Functions

Recall from Section P.3 that a function can be represented by a set of ordered pairs. For instance, the function $f(x) = x + 3$ from $A = \{1, 2, 3, 4\}$ to $B = \{4, 5, 6, 7\}$ can be written as

$$f: \{(1, 4), (2, 5), (3, 6), (4, 7)\}.$$

By interchanging the first and second coordinates of each ordered pair, you can form the **inverse function** of f. This function is denoted by f^{-1}. It is a function from B to A and can be written as

$$f^{-1}: \{(4, 1), (5, 2), (6, 3), (7, 4)\}.$$

Note that the domain of f is equal to the range of f^{-1}, and vice versa, as shown in Figure 5.9. The functions f and f^{-1} have the effect of "undoing" each other. That is, when you form the composition of f with f^{-1} or the composition of f^{-1} with f, you obtain the identity function.

$$f(f^{-1}(x)) = x \quad \text{and} \quad f^{-1}(f(x)) = x$$

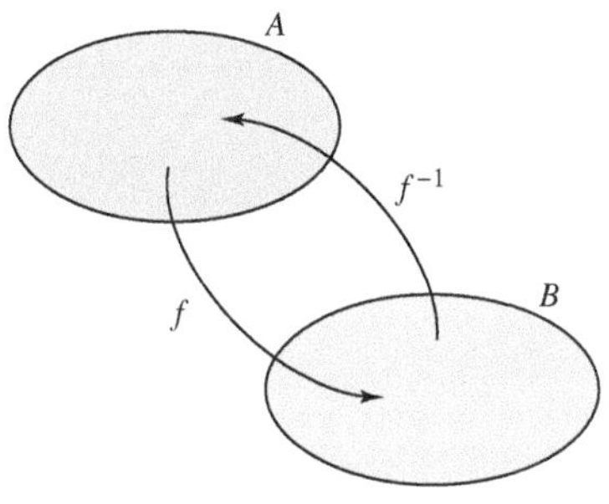

A

f^{-1}

B

f

Domain of f = range of f^{-1}
Domain of f^{-1} = range of f
Figure 5.9

REMARK Although the notation used to denote an inverse function resembles *exponential notation*, it is a different use of -1 as a superscript. That is, in general,

$$f^{-1}(x) \neq \frac{1}{f(x)}.$$

Exploration

Finding Inverse Functions Explain how to "undo" each of the functions below. Then use your explanation to write the inverse function of f.

a. $f(x) = x - 5$

b. $f(x) = 6x$

c. $f(x) = \dfrac{x}{2}$

d. $f(x) = 3x + 2$

e. $f(x) = x^3$

f. $f(x) = 4(x - 2)$

Use a graphing utility to graph each function and its inverse function in the same "square" viewing window. What observation can you make about each pair of graphs?

Definition of Inverse Function

A function g is the **inverse function** of the function f when

$$f(g(x)) = x \text{ for each } x \text{ in the domain of } g$$

and

$$g(f(x)) = x \text{ for each } x \text{ in the domain of } f.$$

The function g is denoted by f^{-1} (read "f inverse").

Here are some important observations about inverse functions.

1. If g is the inverse function of f, then f is the inverse function of g.
2. The domain of f^{-1} is equal to the range of f, and the range of f^{-1} is equal to the domain of f.
3. A function need not have an inverse function, but when it does, the inverse function is unique (see Exercise 94).

You can think of f^{-1} as undoing what has been done by f. For example, subtraction can be used to undo addition, and division can be used to undo multiplication. So,

$$f(x) = x + c \quad \text{and} \quad f^{-1}(x) = x - c$$ Subtraction can be used to undo addition.

are inverse functions of each other and

$$f(x) = cx \quad \text{and} \quad f^{-1}(x) = \frac{x}{c}, \quad c \neq 0$$ Division can be used to undo multiplication.

are inverse functions of each other.

EXAMPLE 1 Verifying Inverse Functions

Show that the functions are inverse functions of each other.

$$f(x) = 2x^3 - 1 \quad \text{and} \quad g(x) = \sqrt[3]{\frac{x+1}{2}}$$

REMARK In Example 1, try comparing the functions f and g verbally.

For f: First cube x, then multiply by 2, then subtract 1.

For g: First add 1, then divide by 2, then take the cube root.

Do you see the "undoing pattern"?

Solution Because the domains and ranges of both f and g consist of all real numbers, you can conclude that both composite functions exist for all x. The composition of f with g is

$$\begin{aligned} f(g(x)) &= 2\left(\sqrt[3]{\frac{x+1}{2}}\right)^3 - 1 \\ &= 2\left(\frac{x+1}{2}\right) - 1 \\ &= x + 1 - 1 \\ &= x. \end{aligned}$$

The composition of g with f is

$$g(f(x)) = \sqrt[3]{\frac{(2x^3 - 1) + 1}{2}} = \sqrt[3]{\frac{2x^3}{2}} = \sqrt[3]{x^3} = x.$$

Because $f(g(x)) = x$ and $g(f(x)) = x$, you can conclude that f and g are inverse functions of each other (see Figure 5.10).

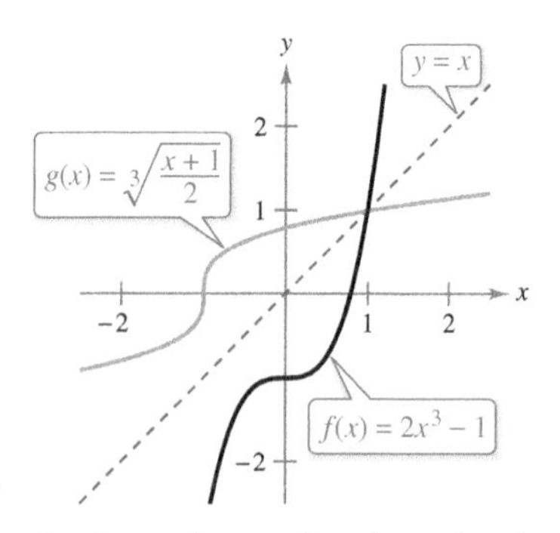

f and g are inverse functions of each other.
Figure 5.10

In Figure 5.10, the graphs of f and $g = f^{-1}$ appear to be mirror images of each other with respect to the line $y = x$. The graph of f^{-1} is a **reflection** of the graph of f in the line $y = x$. This idea is generalized in the next theorem.

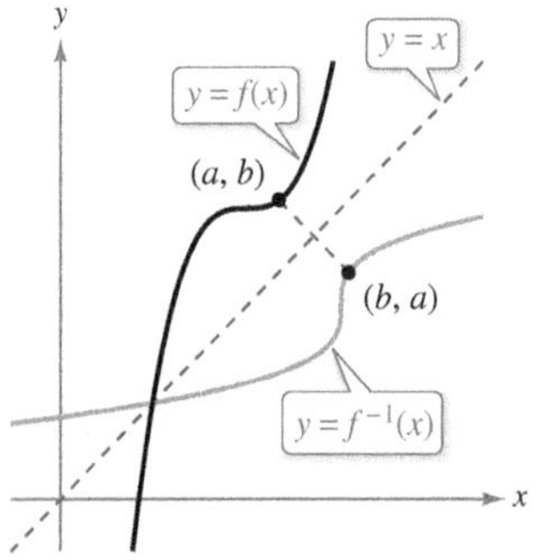

THEOREM 5.6 Reflective Property of Inverse Functions

The graph of f contains the point (a, b) if and only if the graph of f^{-1} contains the point (b, a).

Proof If (a, b) is on the graph of f, then $f(a) = b$, and you can write

$$f^{-1}(b) = f^{-1}(f(a)) = a.$$

So, (b, a) is on the graph of f^{-1}, as shown in Figure 5.11. A similar argument will prove the theorem in the other direction.

The graph of f^{-1} is a reflection of the graph of f in the line $y = x$.
Figure 5.11

Existence of an Inverse Function

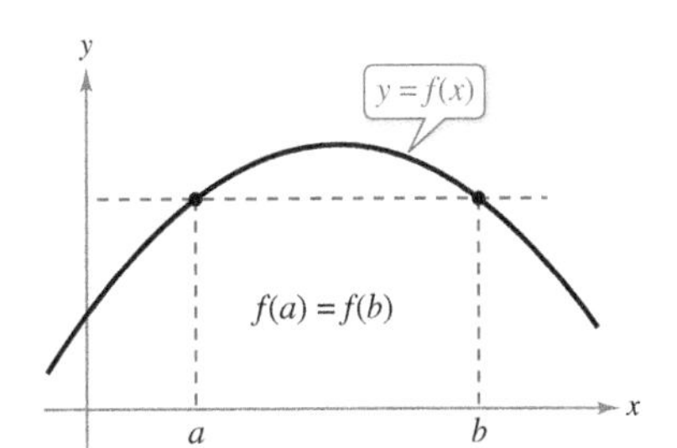

If a horizontal line intersects the graph of f twice, then f is not one-to-one.
Figure 5.12

Not every function has an inverse function, and Theorem 5.6 suggests a graphical test for those that do—the **Horizontal Line Test** for an inverse function. This test states that a function f has an inverse function if and only if every horizontal line intersects the graph of f at most once (see Figure 5.12). The next theorem formally states why the Horizontal Line Test is valid. (Recall from Section 3.3 that a function is *strictly monotonic* when it is either increasing on its entire domain or decreasing on its entire domain.)

THEOREM 5.7 The Existence of an Inverse Function

1. A function has an inverse function if and only if it is one-to-one.
2. If f is strictly monotonic on its entire domain, then it is one-to-one and therefore has an inverse function.

Proof The proof of the first part of the theorem is left as an exercise (see Exercise 95). To prove the second part of the theorem, recall from Section P.3 that f is one-to-one when for x_1 and x_2 in its domain

$$x_1 \neq x_2 \implies f(x_1) \neq f(x_2).$$

Now, choose x_1 and x_2 in the domain of f. If $x_1 \neq x_2$, then, because f is strictly monotonic, it follows that either $f(x_1) < f(x_2)$ or $f(x_1) > f(x_2)$. In either case, $f(x_1) \neq f(x_2)$. So, f is one-to-one on the interval. ■

EXAMPLE 2 The Existence of an Inverse Function

a. From the graph of $f(x) = x^3 + x - 1$ shown in Figure 5.13(a), it appears that f is increasing over its entire domain. To verify this, note that the derivative, $f'(x) = 3x^2 + 1$, is positive for all real values of x. So, f is strictly monotonic, and it must have an inverse function.

b. From the graph of $f(x) = x^3 - x + 1$ shown in Figure 5.13(b), you can see that the function does not pass the Horizontal Line Test. In other words, it is not one-to-one. For instance, f has the same value when $x = -1, 0,$ and 1.

$$f(-1) = f(1) = f(0) = 1 \qquad \text{Not one-to-one}$$

So, by Theorem 5.7, f does not have an inverse function. ■

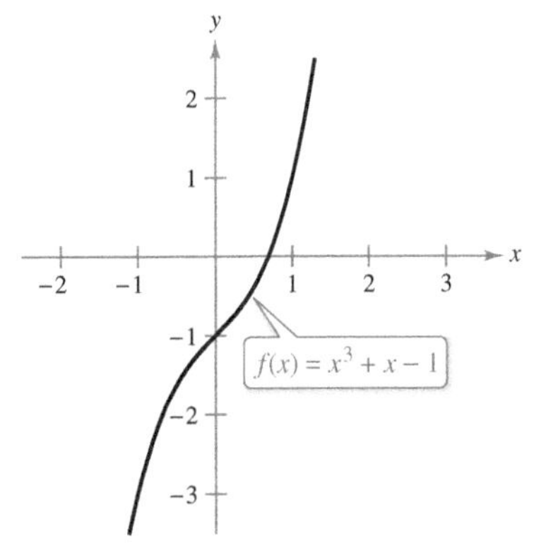

(a) Because f is increasing over its entire domain, it has an inverse function.

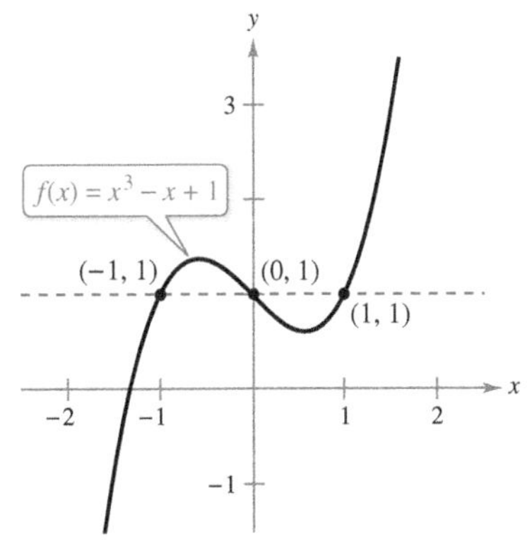

(b) Because f is not one-to-one, it does not have an inverse function.
Figure 5.13

Often, it is easier to prove that a function *has* an inverse function than to find the inverse function. For instance, it would be difficult algebraically to find the inverse function of the function in Example 2(a).

GUIDELINES FOR FINDING AN INVERSE FUNCTION

1. Use Theorem 5.7 to determine whether the function $y = f(x)$ has an inverse function.
2. Solve for x as a function of y: $x = g(y) = f^{-1}(y)$.
3. Interchange x and y. The resulting equation is $y = f^{-1}(x)$.
4. Define the domain of f^{-1} as the range of f.
5. Verify that $f(f^{-1}(x)) = x$ and $f^{-1}(f(x)) = x$.

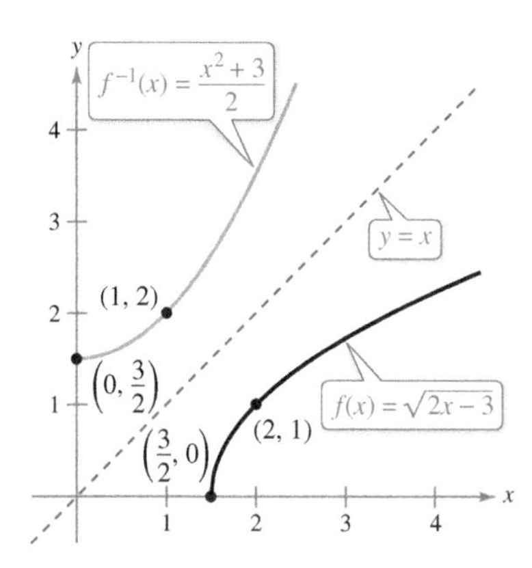

The domain of f^{-1}, $[0, \infty)$, is the range of f.
Figure 5.14

EXAMPLE 3 Finding an Inverse Function

Find the inverse function of $f(x) = \sqrt{2x - 3}$.

Solution From the graph of f in Figure 5.14, it appears that f is increasing over its entire domain, $[3/2, \infty)$. To verify this, note that

$$f'(x) = \frac{1}{\sqrt{2x-3}}$$

is positive on the domain of f. So, f is strictly monotonic, and it must have an inverse function. To find an equation for the inverse function, let $y = f(x)$, and solve for x in terms of y.

$$\sqrt{2x-3} = y \qquad \text{Let } y = f(x).$$

$$2x - 3 = y^2 \qquad \text{Square each side.}$$

$$x = \frac{y^2+3}{2} \qquad \text{Solve for } x.$$

$$y = \frac{x^2+3}{2} \qquad \text{Interchange } x \text{ and } y.$$

$$f^{-1}(x) = \frac{x^2+3}{2} \qquad \text{Replace } y \text{ by } f^{-1}(x).$$

The domain of f^{-1} is the range of f, which is $[0, \infty)$. You can verify this result as shown.

$$f(f^{-1}(x)) = \sqrt{2\left(\frac{x^2+3}{2}\right) - 3} = \sqrt{x^2} = x, \quad x \geq 0$$

$$f^{-1}(f(x)) = \frac{\left(\sqrt{2x-3}\right)^2 + 3}{2} = \frac{2x - 3 + 3}{2} = x, \quad x \geq \frac{3}{2}$$

Theorem 5.7 is useful in the next type of problem. You are given a function that is *not* one-to-one on its domain. By restricting the domain to an interval on which the function is strictly monotonic, you can conclude that the new function *is* one-to-one on the restricted domain.

EXAMPLE 4 Testing Whether a Function Is One-to-One

See LarsonCalculus.com for an interactive version of this type of example.

Show that the sine function

$$f(x) = \sin x$$

is not one-to-one on the entire real number line. Then show that $[-\pi/2, \pi/2]$ is the largest interval, centered at the origin, on which f is strictly monotonic.

Solution It is clear that f is not one-to-one, because many different x-values yield the same y-value. For instance,

$$\sin 0 = 0 = \sin \pi.$$

Moreover, f is increasing on the open interval $(-\pi/2, \pi/2)$, because its derivative

$$f'(x) = \cos x$$

is positive there. Finally, because the left and right endpoints correspond to relative extrema of the sine function, you can conclude that f is increasing on the closed interval $[-\pi/2, \pi/2]$ *and* that on any larger interval the function is not strictly monotonic (see Figure 5.15).

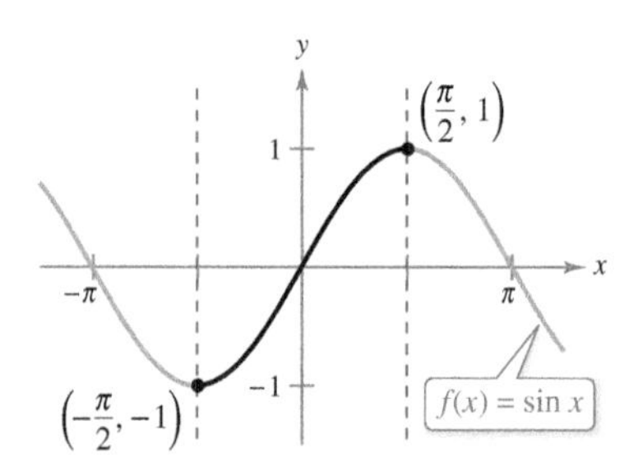

f is one-to-one on the interval $[-\pi/2, \pi/2]$.
Figure 5.15

Derivative of an Inverse Function

The next two theorems discuss the derivative of an inverse function. The reasonableness of Theorem 5.8 follows from the reflective property of inverse functions, as shown in Figure 5.11.

THEOREM 5.8 Continuity and Differentiability of Inverse Functions

Let f be a function whose domain is an interval I. If f has an inverse function, then the following statements are true.

1. If f is continuous on its domain, then f^{-1} is continuous on its domain.
2. If f is increasing on its domain, then f^{-1} is increasing on its domain.
3. If f is decreasing on its domain, then f^{-1} is decreasing on its domain.
4. If f is differentiable on an interval containing c and $f'(c) \neq 0$, then f^{-1} is differentiable at $f(c)$.

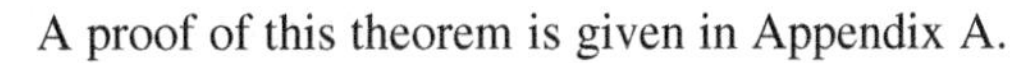

A proof of this theorem is given in Appendix A.

Exploration

Graph the inverse functions $f(x) = x^3$ and $g(x) = x^{1/3}$. Calculate the slopes of f at $(1, 1)$, $(2, 8)$, and $(3, 27)$, and the slopes of g at $(1, 1)$, $(8, 2)$, and $(27, 3)$. What do you observe? What happens at $(0, 0)$?

THEOREM 5.9 The Derivative of an Inverse Function

Let f be a function that is differentiable on an interval I. If f has an inverse function g, then g is differentiable at any x for which $f'(g(x)) \neq 0$. Moreover,

$$g'(x) = \frac{1}{f'(g(x))}, \quad f'(g(x)) \neq 0.$$

A proof of this theorem is given in Appendix A.

EXAMPLE 5 Evaluating the Derivative of an Inverse Function

Let $f(x) = \frac{1}{4}x^3 + x - 1$.

a. What is the value of $f^{-1}(x)$ when $x = 3$?

b. What is the value of $(f^{-1})'(x)$ when $x = 3$?

Solution Notice that f is one-to-one and therefore has an inverse function.

a. Because $f(x) = 3$ when $x = 2$, you know that $f^{-1}(3) = 2$.

b. Because the function f is differentiable and has an inverse function, you can apply Theorem 5.9 (with $g = f^{-1}$) to write

$$(f^{-1})'(3) = \frac{1}{f'(f^{-1}(3))} = \frac{1}{f'(2)}.$$

Moreover, using $f'(x) = \frac{3}{4}x^2 + 1$, you can conclude that

$$(f^{-1})'(3) = \frac{1}{f'(2)} = \frac{1}{\frac{3}{4}(2^2) + 1} = \frac{1}{4}.$$

■

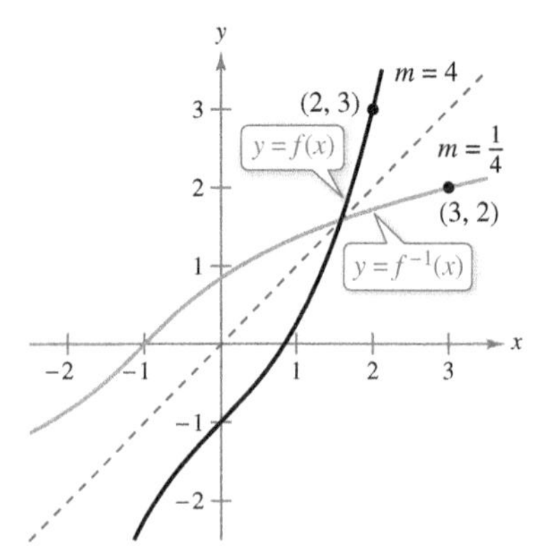

The graphs of the inverse functions f and f^{-1} have reciprocal slopes at points (a, b) and (b, a).
Figure 5.16

In Example 5, note that at the point (2, 3), the slope of the graph of f is $m = 4$, and at the point (3, 2), the slope of the graph of f^{-1} is

$$m = \frac{1}{4}$$

as shown in Figure 5.16. In general, if $y = g(x) = f^{-1}(x)$, then $f(y) = x$ and $f'(y) = \frac{dx}{dy}$. It follows from Theorem 5.9 that

$$g'(x) = \frac{dy}{dx} = \frac{1}{f'(g(x))} = \frac{1}{f'(y)} = \frac{1}{(dx/dy)}.$$

This reciprocal relationship is sometimes written as

$$\frac{dy}{dx} = \frac{1}{dx/dy}.$$

EXAMPLE 6 Graphs of Inverse Functions Have Reciprocal Slopes

Let $f(x) = x^2$ (for $x \geq 0$), and let $f^{-1}(x) = \sqrt{x}$. Show that the slopes of the graphs of f and f^{-1} are reciprocals at each of the following points.

a. (2, 4) and (4, 2) **b.** (3, 9) and (9, 3)

Solution The derivatives of f and f^{-1} are

$$f'(x) = 2x \quad \text{and} \quad (f^{-1})'(x) = \frac{1}{2\sqrt{x}}.$$

a. At (2, 4), the slope of the graph of f is $f'(2) = 2(2) = 4$. At (4, 2), the slope of the graph of f^{-1} is

$$(f^{-1})'(4) = \frac{1}{2\sqrt{4}} = \frac{1}{2(2)} = \frac{1}{4}.$$

b. At (3, 9), the slope of the graph of f is $f'(3) = 2(3) = 6$. At (9, 3), the slope of the graph of f^{-1} is

$$(f^{-1})'(9) = \frac{1}{2\sqrt{9}} = \frac{1}{2(3)} = \frac{1}{6}.$$

So, in both cases, the slopes are reciprocals, as shown in Figure 5.17.

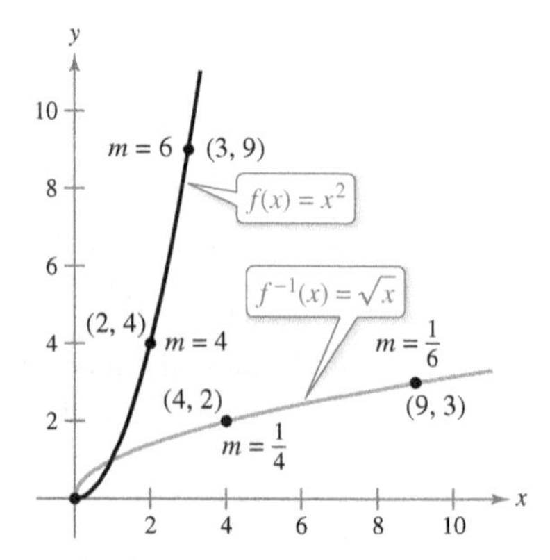

At (0, 0), the derivative of f is 0, and the derivative of f^{-1} does not exist.
Figure 5.17

5.3 Exercises

See CalcChat.com for tutorial help and worked-out solutions to odd-numbered exercises.

CONCEPT CHECK

1. **Inverse Functions** In your own words, describe what it means to say that the function g is the inverse function of the function f.
2. **Reflective Property of Inverse Functions** Describe the relationship between the graph of a function and the graph of its inverse function.
3. **Domain of an Inverse Function** The function f has an inverse function, f^{-1}. Is the domain of f the same as the domain of f^{-1}? Explain.
4. **Behavior of an Inverse Function** The function f is decreasing on its domain and has an inverse function, f^{-1}. Is f^{-1} increasing, decreasing, or constant on its domain?

Matching **In Exercises 5–8, match the graph of the function with the graph of its inverse function. [The graphs of the inverse functions are labeled (a), (b), (c), and (d).]**

(a)

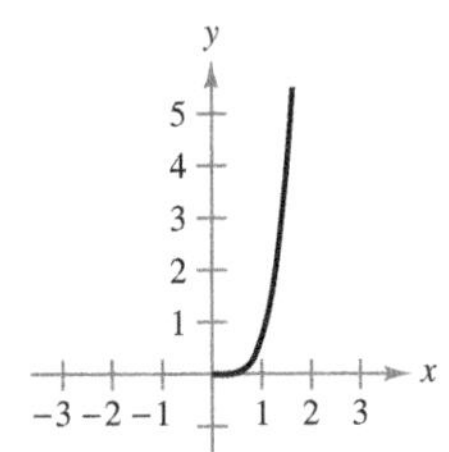

(b)

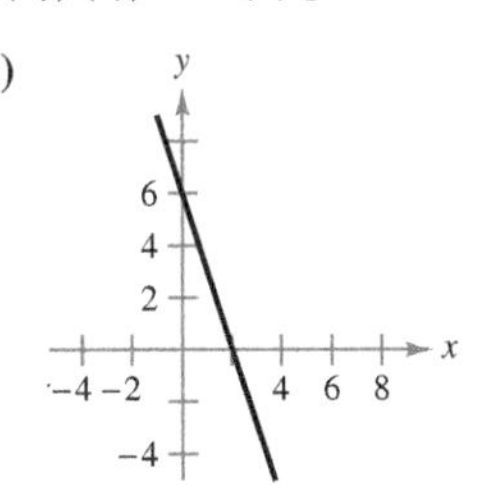

(c)

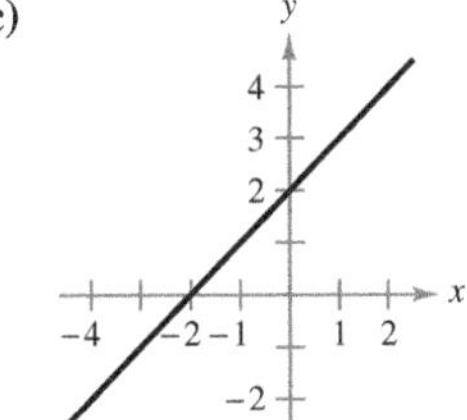

(d)

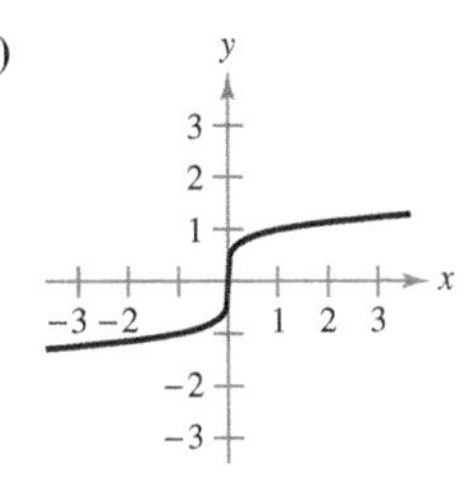

5.

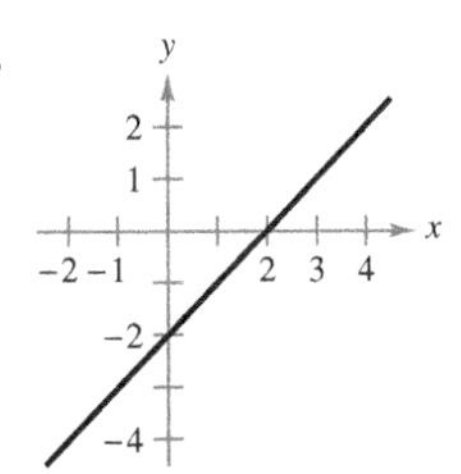

6.

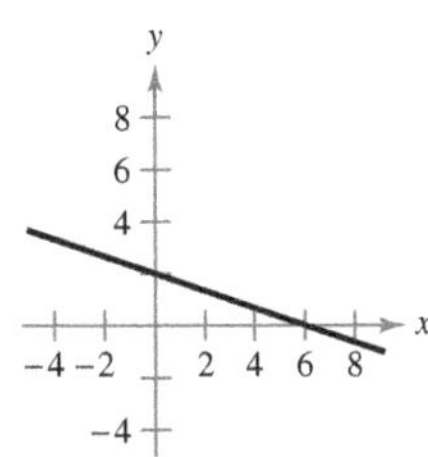

7.

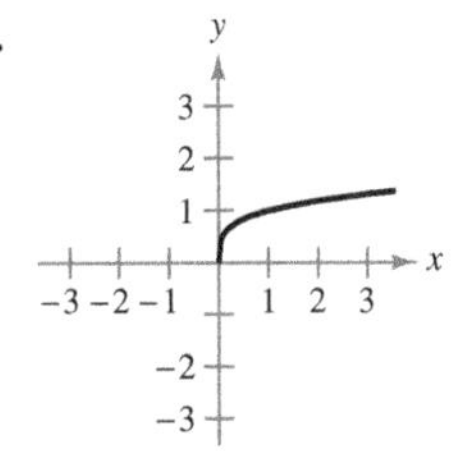

8.

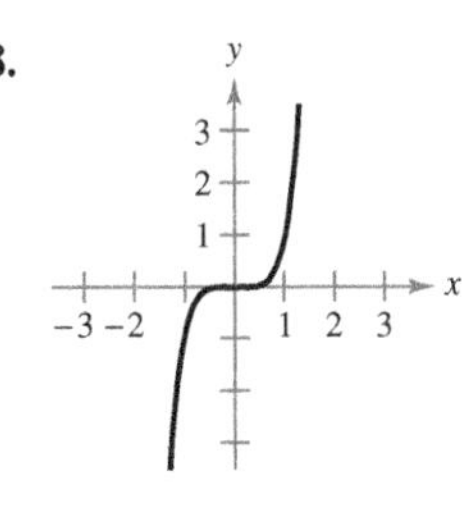

Verifying Inverse Functions **In Exercises 9–16, show that f and g are inverse functions (a) analytically and (b) graphically.**

9. $f(x) = 5x + 1, \quad g(x) = \dfrac{x-1}{5}$
10. $f(x) = 3 - 4x, \quad g(x) = \dfrac{3-x}{4}$
11. $f(x) = x^3, \quad g(x) = \sqrt[3]{x}$
12. $f(x) = 1 - x^3, \quad g(x) = \sqrt[3]{1-x}$
13. $f(x) = \sqrt{x-4}, \quad g(x) = x^2 + 4, \quad x \geq 0$
14. $f(x) = 16 - x^2, \quad x \geq 0, \quad g(x) = \sqrt{16-x}$
15. $f(x) = \dfrac{1}{x}, \quad g(x) = \dfrac{1}{x}$
16. $f(x) = \dfrac{1}{1+x}, \quad x \geq 0, \quad g(x) = \dfrac{1-x}{x}, \quad 0 < x \leq 1$

Using the Horizontal Line Test **In Exercises 17–24, use a graphing utility to graph the function. Then use the Horizontal Line Test to determine whether the function is one-to-one on its entire domain and therefore has an inverse function.**

17. $f(x) = \frac{3}{4}x + 6$
18. $f(x) = 1 - x^3$
19. $f(\theta) = \sin\theta$
20. $f(x) = x\cos x$
21. $h(s) = \dfrac{1}{s-2} - 3$
22. $g(t) = \dfrac{1}{\sqrt{t^2+1}}$
23. $f(x) = \ln x$
24. $h(x) = \ln x^2$

Determining Whether a Function Has an Inverse Function **In Exercises 25–30, use the derivative to determine whether the function is strictly monotonic on its entire domain and therefore has an inverse function.**

25. $f(x) = 2 - x - x^3$
26. $f(x) = x^3 - 6x^2 + 12x$
27. $f(x) = 8x^3 + x^2 - 1$
28. $f(x) = 1 - x^3 - 6x^5$
29. $f(x) = \ln(x-3)$
30. $f(x) = \cos\dfrac{3x}{2}$

Verifying a Function Has an Inverse Function **In Exercises 31–34, show that f is strictly monotonic on the given interval and therefore has an inverse function on that interval.**

31. $f(x) = (x-4)^2, \quad [4, \infty)$
32. $f(x) = |x+2|, \quad [-2, \infty)$
33. $f(x) = \cot x, \quad (0, \pi)$
34. $f(x) = \sec x, \quad \left[0, \dfrac{\pi}{2}\right)$

Finding an Inverse Function **In Exercises 35–46, (a) find the inverse function of f, (b) graph f and f^{-1} on the same set of coordinate axes, (c) describe the relationship between the graphs, and (d) state the domains and ranges of f and f^{-1}.**

35. $f(x) = 2x - 3$
36. $f(x) = 9 - 5x$
37. $f(x) = x^5$
38. $f(x) = x^3 - 1$
39. $f(x) = \sqrt{x}$
40. $f(x) = x^4, \quad x \geq 0$
41. $f(x) = \sqrt{4 - x^2}, \quad 0 \leq x \leq 2$
42. $f(x) = \sqrt{x^2 - 4}, \quad x \geq 2$
43. $f(x) = \sqrt[3]{x - 1}$
44. $f(x) = x^{2/3}, \quad x \geq 0$
45. $f(x) = \dfrac{x}{\sqrt{x^2 + 7}}$
46. $f(x) = \dfrac{x + 2}{x}$

Finding an Inverse Function **In Exercises 47 and 48, use the graph of the function f to make a table of values for the given points. Then make a second table that can be used to find f^{-1} and sketch the graph of f^{-1}. To print an enlarged copy of the graph, go to *MathGraphs.com*.**

47.
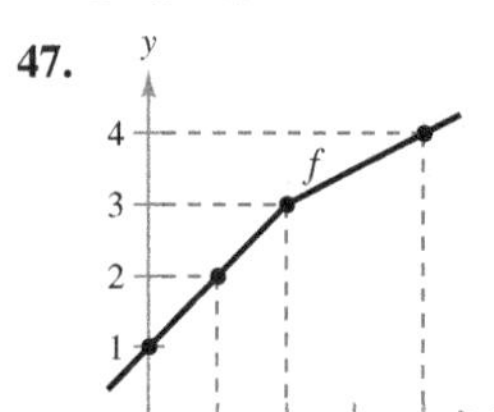

48.
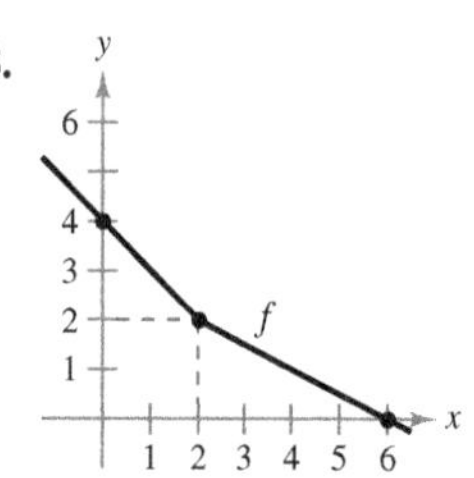

49. **Cost** You need a total of 50 pounds of two commodities costing \$1.25 and \$2.75 per pound.

(a) Verify that the total cost is $y = 1.25x + 2.75(50 - x)$, where x is the number of pounds of the less expensive commodity.

(b) Find the inverse function of the cost function. What does each variable represent in the inverse function?

(c) What is the domain of the inverse function? Validate or explain your answer using the context of the problem.

(d) Determine the number of pounds of the less expensive commodity purchased when the total cost is \$73.

50. **Temperature** The formula $C = \frac{5}{9}(F - 32)$, where $F \geq -459.6$, represents Celsius temperature C as a function of Fahrenheit temperature F.

(a) Find the inverse function of C.

(b) What does the inverse function represent?

(c) What is the domain of the inverse function? Validate or explain your answer using the context of the problem.

(d) The temperature is 22°C. What is the corresponding temperature in degrees Fahrenheit?

Testing Whether a Function Is One-to-One **In Exercises 51–54, determine whether the function is one-to-one. If it is, find its inverse function.**

51. $f(x) = \sqrt{x - 2}$
52. $f(x) = -3$
53. $f(x) = |x - 2|, \quad x \leq 2$
54. $f(x) = ax + b, \quad a \neq 0$

Making a Function One-to-One **In Exercises 55–58, the function is not one-to-one. Delete part of the domain so that the function that remains is one-to-one. Find the inverse function of the remaining function and give the domain of the inverse function. (*Note:* There is more than one correct answer.)**

55. $f(x) = (x - 3)^2$
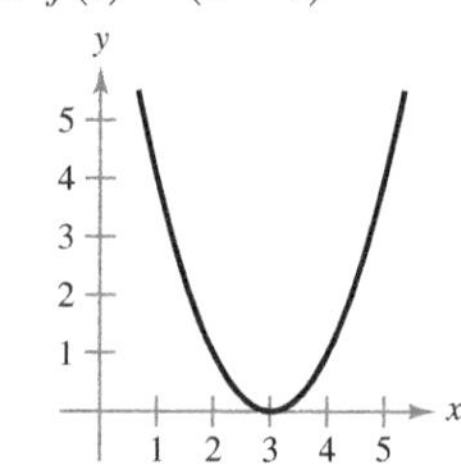

56. $f(x) = |x - 3|$
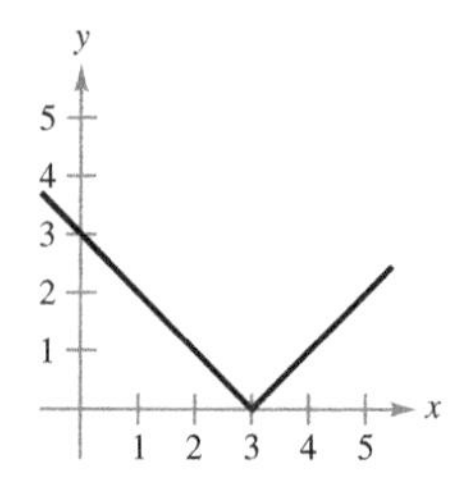

57. $f(x) = |x + 3|$
58. $f(x) = 16 - x^4$

Think About It **In Exercises 59–62, decide whether the function has an inverse function. If so, describe what the inverse function represents.**

59. $g(t)$ is the volume of water that has passed through a water line t minutes after a control valve is opened.

60. $h(t)$ is the height of the tide t hours after midnight, where $0 \leq t < 24$.

61. $C(t)$ is the cost of a long-distance phone call lasting t minutes.

62. $A(r)$ is the area of a circle of radius r.

Evaluating the Derivative of an Inverse Function **In Exercises 63–70, verify that f has an inverse function. Then use the function f and the given real number a to find $(f^{-1})'(a)$. (*Hint:* See Example 5.)**

63. $f(x) = 5 - 2x^3, \; a = 7$
64. $f(x) = x^3 + 3x - 1, \; a = -5$
65. $f(x) = \frac{1}{27}(x^5 + 2x^3), \; a = -11$
66. $f(x) = \sqrt{x - 4}, \; a = 2$
67. $f(x) = \sin x, \; -\dfrac{\pi}{2} \leq x \leq \dfrac{\pi}{2}, \; a = \dfrac{1}{2}$
68. $f(x) = \cos 2x, \; 0 \leq x \leq \dfrac{\pi}{2}, \; a = 1$
69. $f(x) = \dfrac{x + 6}{x - 2}, \; x > 2, \; a = 3$
70. $f(x) = \dfrac{x + 3}{x + 1}, \; x > -1, \; a = 2$

Using Inverse Functions **In Exercises 71–74, (a) find the domains of f and f^{-1}, (b) find the ranges of f and f^{-1}, (c) graph f and f^{-1}, and (d) show that the slopes of the graphs of f and f^{-1} are reciprocals at the given points.**

	Functions	Point
71.	$f(x) = x^3$	$\left(\frac{1}{2}, \frac{1}{8}\right)$
	$f^{-1}(x) = \sqrt[3]{x}$	$\left(\frac{1}{8}, \frac{1}{2}\right)$
72.	$f(x) = 3 - 4x$	$(1, -1)$
	$f^{-1}(x) = \dfrac{3 - x}{4}$	$(-1, 1)$

Functions	Point
73. $f(x) = \sqrt{x - 4}$	$(5, 1)$
$f^{-1}(x) = x^2 + 4, \quad x \geq 0$	$(1, 5)$
74. $f(x) = \dfrac{4}{1 + x^2}, \quad x \geq 0$	$(1, 2)$
$f^{-1}(x) = \sqrt{\dfrac{4 - x}{x}}$	$(2, 1)$

Using Composite and Inverse Functions **In Exercises 75–78, use the functions $f(x) = \frac{1}{8}x - 3$ and $g(x) = x^3$ to find the given value.**

75. $(f^{-1} \circ g^{-1})(1)$

76. $(g^{-1} \circ f^{-1})(-3)$

77. $(f^{-1} \circ f^{-1})(-2)$

78. $(g^{-1} \circ g^{-1})(8)$

Using Composite and Inverse Functions **In Exercises 79–82, use the functions $f(x) = x + 4$ and $g(x) = 2x - 5$ to find the given function.**

79. $g^{-1} \circ f^{-1}$

80. $f^{-1} \circ g^{-1}$

81. $(f \circ g)^{-1}$

82. $(g \circ f)^{-1}$

EXPLORING CONCEPTS

83. Inverse Function Consider the function $f(x) = x^n$, where n is odd. Does f^{-1} exist? Explain.

84. Think About It Does adding a constant term to a function affect the existence of an inverse function? Explain.

Explaining Why a Function Is Not One-to-One
In Exercises 85 and 86, the derivative of the function has the same sign for all x in its domain, but the function is not one-to-one. Explain why the function is not one-to-one.

85. $f(x) = \tan x$

86. $f(x) = \dfrac{x}{x^2 - 4}$

87. Think About It The function $f(x) = k(2 - x - x^3)$ is one-to-one and $f^{-1}(3) = -2$. Find k.

88. HOW DO YOU SEE IT? Use the information in the graph of f below.

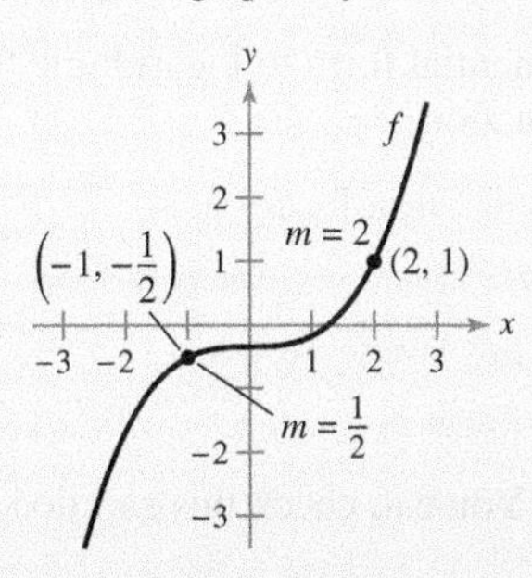

(a) What is the slope of the tangent line to the graph of f^{-1} at the point $\left(-\frac{1}{2}, -1\right)$? Explain.

(b) What is the slope of the tangent line to the graph of f^{-1} at the point $(1, 2)$? Explain.

True or False? **In Exercises 89 and 90, determine whether the statement is true or false. If it is false, explain why or give an example that shows it is false.**

89. If f is an even function, then f^{-1} exists.

90. If the inverse function of f exists, then the y-intercept of f is an x-intercept of f^{-1}.

91. Making a Function One-to-One

(a) Show that $f(x) = 2x^3 + 3x^2 - 36x$ is not one-to-one on $(-\infty, \infty)$.

(b) Determine the greatest value c such that f is one-to-one on $(-c, c)$.

92. Proof Let f and g be one-to-one functions. Prove that

(a) $f \circ g$ is one-to-one.

(b) $(f \circ g)^{-1}(x) = (g^{-1} \circ f^{-1})(x)$.

93. Proof Prove that if f has an inverse function, then $(f^{-1})^{-1} = f$.

94. Proof Prove that if a function has an inverse function, then the inverse function is unique.

95. Proof Prove that a function has an inverse function if and only if it is one-to-one.

96. Using Theorem 5.7 Is the converse of the second part of Theorem 5.7 true? That is, if a function is one-to-one (and therefore has an inverse function), then must the function be strictly monotonic? If so, prove it. If not, give a counterexample.

97. Derivative of an Inverse Function Show that

$$f(x) = \int_2^x \sqrt{1 + t^2}\, dt$$

is one-to-one and find $(f^{-1})'(0)$.

98. Derivative of an Inverse Function Show that

$$f(x) = \int_2^x \frac{dt}{\sqrt{1 + t^4}}$$

is one-to-one and find $(f^{-1})'(0)$.

99. Inverse Function Let

$$f(x) = \frac{x - 2}{x - 1}.$$

Show that f is its own inverse function. What can you conclude about the graph of f? Explain.

100. Using a Function Let $f(x) = \dfrac{ax + b}{cx + d}$.

(a) Show that f is one-to-one if and only if $bc - ad \neq 0$.

(b) Given $bc - ad \neq 0$, find f^{-1}.

(c) Determine the values of a, b, c, and d such that $f = f^{-1}$.

101. Concavity Let f be twice-differentiable and one-to-one on an open interval I. Show that its inverse function g satisfies

$$g''(x) = -\frac{f''(g(x))}{[f'(g(x))]^3}.$$

When f is increasing and concave downward, what is the concavity of g?

5.4 Exponential Functions: Differentiation and Integration

- **Develop properties of the natural exponential function.**
- **Differentiate natural exponential functions.**
- **Integrate natural exponential functions.**

The Natural Exponential Function

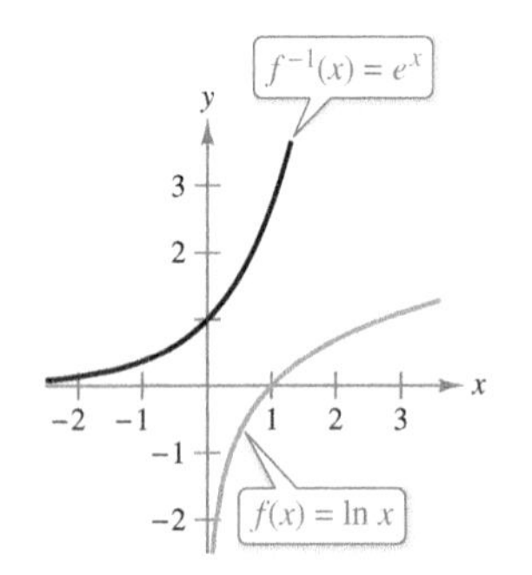

The inverse function of the natural logarithmic function is the natural exponential function.
Figure 5.18

The function $f(x) = \ln x$ is increasing on its entire domain, and therefore it has an inverse function f^{-1}. The domain of f^{-1} is the set of all real numbers, and the range is the set of positive real numbers, as shown in Figure 5.18. So, for any real number x,

$$f(f^{-1}(x)) = \ln[f^{-1}(x)] = x. \qquad x \text{ is any real number.}$$

If x is rational, then

$$\ln(e^x) = x \ln e = x(1) = x. \qquad x \text{ is a rational number.}$$

Because the natural logarithmic function is one-to-one, you can conclude that $f^{-1}(x)$ and e^x agree for *rational* values of x. The next definition extends the meaning of e^x to include *all* real values of x.

Definition of the Natural Exponential Function

The inverse function of the natural logarithmic function $f(x) = \ln x$ is called the **natural exponential function** and is denoted by

$$f^{-1}(x) = e^x.$$

That is,

$$y = e^x \quad \text{if and only if} \quad x = \ln y.$$

The inverse relationship between the natural logarithmic function and the natural exponential function can be summarized as shown.

$$\ln(e^x) = x \quad \text{and} \quad e^{\ln x} = x \qquad \text{Inverse relationship}$$

EXAMPLE 1 Solving an Exponential Equation

Solve $7 = e^{x+1}$.

Solution You can convert from exponential form to logarithmic form by *taking the natural logarithm of each side* of the equation.

$$7 = e^{x+1} \qquad \text{Write original equation.}$$
$$\ln 7 = \ln(e^{x+1}) \qquad \text{Take natural logarithm of each side.}$$
$$\ln 7 = x + 1 \qquad \text{Apply inverse property.}$$
$$-1 + \ln 7 = x \qquad \text{Solve for } x.$$

So, the solution is $-1 + \ln 7 \approx 0.946$. You can check this solution as shown.

$$7 = e^{x+1} \qquad \text{Write original equation.}$$
$$7 \stackrel{?}{=} e^{(-1+\ln 7)+1} \qquad \text{Substitute } -1 + \ln 7 \text{ for } x \text{ in original equation.}$$
$$7 \stackrel{?}{=} e^{\ln 7} \qquad \text{Simplify.}$$
$$7 = 7 \checkmark \qquad \text{Solution checks.}$$

TECHNOLOGY You can use a graphing utility to check a solution of an equation. One way to do this is to graph the left- and right-hand sides of the equation and then use the *intersect* feature. For instance, to check the solution to Example 2, enter $y_1 = \ln(2x - 3)$ and $y_2 = 5$. The solution of the original equation is the x-value of each point of intersection (see figure). So the solution of the original equation is $x \approx 75.707$.

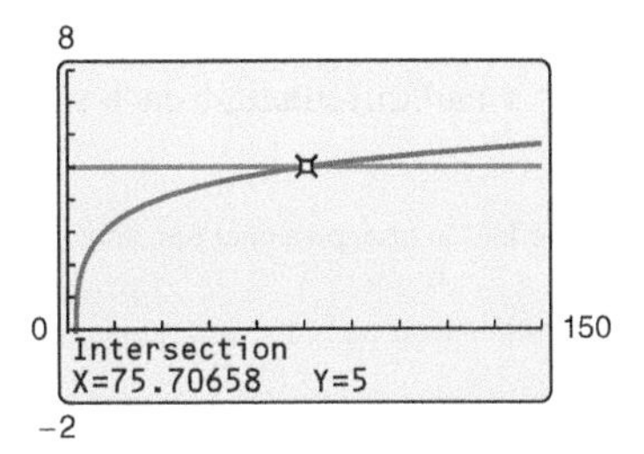

EXAMPLE 2 **Solving a Logarithmic Equation**

Solve $\ln(2x - 3) = 5$.

Solution To convert from logarithmic form to exponential form, you can *exponentiate each side* of the logarithmic equation.

$\ln(2x - 3) = 5$	Write original equation.
$e^{\ln(2x-3)} = e^5$	Exponentiate each side.
$2x - 3 = e^5$	Apply inverse property.
$x = \frac{1}{2}(e^5 + 3)$	Solve for x.
$x \approx 75.707$	Use a calculator.

The familiar rules for operating with rational exponents can be extended to the natural exponential function, as shown in the next theorem.

THEOREM 5.10 Operations with Exponential Functions

Let a and b be any real numbers.

1. $e^a e^b = e^{a+b}$ **2.** $\dfrac{e^a}{e^b} = e^{a-b}$

Proof To prove Property 1, you can write

$$\ln(e^a e^b) = \ln(e^a) + \ln(e^b) = a + b = \ln(e^{a+b}).$$

Because the natural logarithmic function is one-to-one, you can conclude that

$$e^a e^b = e^{a+b}.$$

The proof of the other property is given in Appendix A.

In Section 5.3, you learned that an inverse function f^{-1} shares many properties with f. So, the natural exponential function inherits the properties listed below from the natural logarithmic function.

Properties of the Natural Exponential Function

1. The domain of $f(x) = e^x$ is
$$(-\infty, \infty)$$
and the range is
$$(0, \infty).$$
2. The function $f(x) = e^x$ is continuous, increasing, and one-to-one on its entire domain.
3. The graph of $f(x) = e^x$ is concave upward on its entire domain.
4. $\lim_{x\to-\infty} e^x = 0$
5. $\lim_{x\to\infty} e^x = \infty$

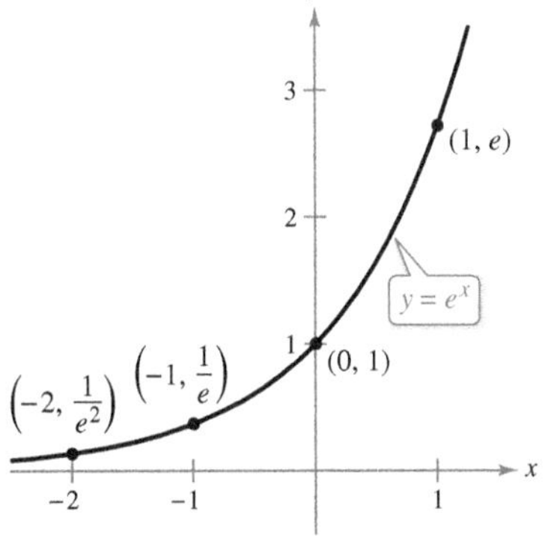

The natural exponential function is increasing, and its graph is concave upward.

Derivatives of Exponential Functions

One of the most intriguing (and useful) characteristics of the natural exponential function is that *it is its own derivative*. In other words, it is a solution of the differential equation $y' = y$. This result is stated in the next theorem.

REMARK You can interpret this theorem geometrically by saying that the slope of the graph of $f(x) = e^x$ at any point (x, e^x) is equal to the y-coordinate of the point.

THEOREM 5.11 Derivatives of the Natural Exponential Function

Let u be a differentiable function of x.

1. $\dfrac{d}{dx}[e^x] = e^x$
2. $\dfrac{d}{dx}[e^u] = e^u \dfrac{du}{dx}$

Proof To prove Property 1, use the fact that $\ln e^x = x$ and differentiate each side of the equation.

$$\ln e^x = x \qquad \text{Definition of exponential function}$$

$$\frac{d}{dx}[\ln e^x] = \frac{d}{dx}[x] \qquad \text{Differentiate each side with respect to } x.$$

$$\frac{1}{e^x}\frac{d}{dx}[e^x] = 1$$

$$\frac{d}{dx}[e^x] = e^x \qquad \text{Multiply each side by } e^x.$$

The derivative of e^u follows from the Chain Rule. ■

EXAMPLE 3 Differentiating Exponential Functions

a. $\dfrac{d}{dx}[e^{2x-1}] = e^u\dfrac{du}{dx} = 2e^{2x-1}$ $\qquad u = 2x - 1$

b. $\dfrac{d}{dx}[e^{-3/x}] = e^u\dfrac{du}{dx} = \left(\dfrac{3}{x^2}\right)e^{-3/x} = \dfrac{3e^{-3/x}}{x^2}$ $\qquad u = -\dfrac{3}{x}$

c. $\dfrac{d}{dx}[x^2e^x] = x^2(e^x) + e^x(2x) = xe^x(x + 2)$ $\qquad$ Product Rule and Theorem 5.11

d. $\dfrac{d}{dx}\left[\dfrac{e^{3x}}{e^x + 1}\right] = \dfrac{(e^x + 1)(3e^{3x}) - e^{3x}(e^x)}{(e^x + 1)^2} = \dfrac{3e^{4x} + 3e^{3x} - e^{4x}}{(e^x + 1)^2} = \dfrac{e^{3x}(2e^x + 3)}{(e^x + 1)^2}$

EXAMPLE 4 Locating Relative Extrema

Find the relative extrema of

$$f(x) = xe^x.$$

Solution The derivative of f is

$$f'(x) = x(e^x) + e^x(1) \qquad \text{Product Rule}$$
$$= e^x(x + 1).$$

Because e^x is never 0, the derivative is 0 only when $x = -1$. Moreover, by the First Derivative Test, you can determine that this corresponds to a relative minimum, as shown in Figure 5.19. Because the derivative $f'(x) = e^x(x + 1)$ is defined for all x, there are no other critical points. ■

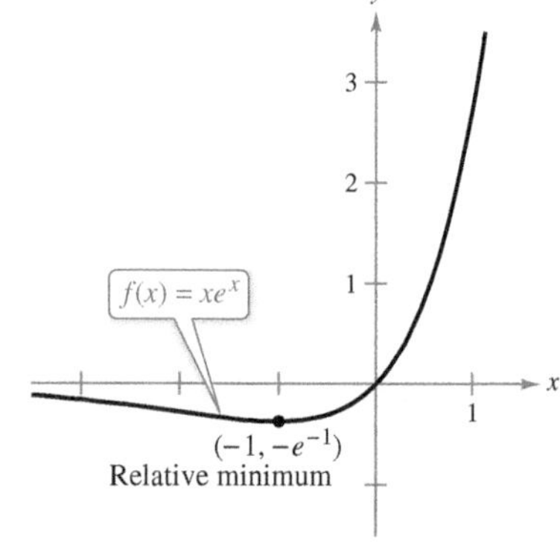

The derivative of f changes from negative to positive at $x = -1$.
Figure 5.19

EXAMPLE 5 **Finding an Equation of a Tangent Line**

Find an equation of the tangent line to the graph of $f(x) = 2 + e^{1-x}$ at the point $(1, 3)$.

Solution Begin by finding $f'(x)$.

$$f(x) = 2 + e^{1-x} \qquad \text{Write original function.}$$
$$f'(x) = e^{1-x}(-1) \qquad u = 1 - x$$
$$= -e^{1-x} \qquad \text{First derivative}$$

To find the slope of the tangent line at $(1, 3)$, evaluate $f'(1)$.

$$f'(1) = -e^{1-1} = -e^0 = -1 \qquad \text{Slope of tangent line at } (1, 3)$$

Now, using the point-slope form of the equation of a line, you can write

$$y - y_1 = m(x - x_1) \qquad \text{Point-slope form}$$
$$y - 3 = -1(x - 1) \qquad \text{Substitute for } y_1, m, \text{ and } x_1.$$
$$y = -x + 4. \qquad \text{Equation of tangent line at } (1, 3)$$

The graph of f and its tangent line at $(1, 3)$ are shown in Figure 5.20.

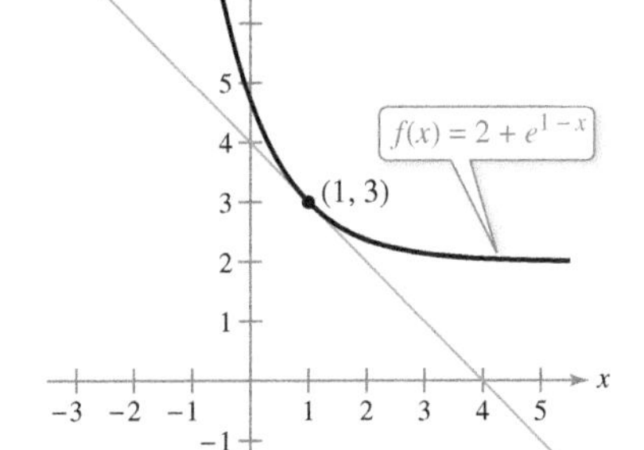

Figure 5.20

EXAMPLE 6 **The Standard Normal Probability Density Function**

See LarsonCalculus.com for an interactive version of this type of example.

Show that the *standard normal probability density function*

$$f(x) = \frac{1}{\sqrt{2\pi}} e^{-x^2/2}$$

has points of inflection when $x = \pm 1$.

Solution To locate possible points of inflection, find the x-values for which the second derivative is 0.

$$f(x) = \frac{1}{\sqrt{2\pi}} e^{-x^2/2} \qquad \text{Write original function.}$$
$$f'(x) = \frac{1}{\sqrt{2\pi}} (-x)e^{-x^2/2} \qquad \text{First derivative}$$
$$f''(x) = \frac{1}{\sqrt{2\pi}} [(-x)(-x)e^{-x^2/2} + (-1)e^{-x^2/2}] \qquad \text{Product Rule}$$
$$= \frac{1}{\sqrt{2\pi}} (e^{-x^2/2})(x^2 - 1) \qquad \text{Second derivative}$$

So, $f''(x) = 0$ when $x = \pm 1$, and you can apply the techniques of Chapter 3 to conclude that these values yield the two points of inflection shown in the figure below.

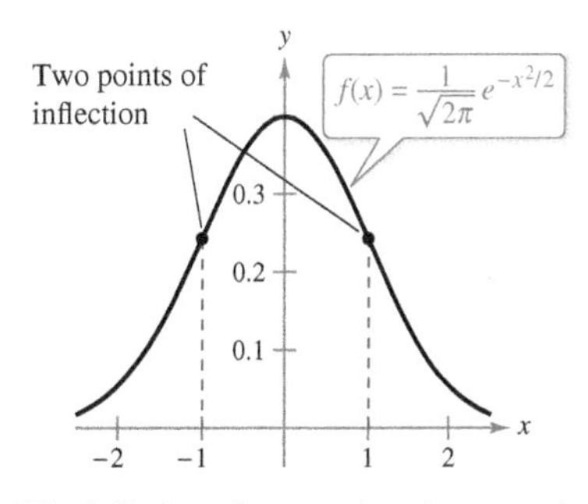

The bell-shaped curve given by a standard normal probability density function

REMARK The general form of a normal probability density function (whose mean is 0) is

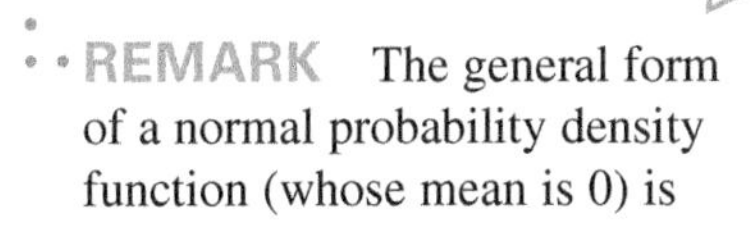

$$f(x) = \frac{1}{\sigma\sqrt{2\pi}} e^{-x^2/(2\sigma^2)}$$

where σ is the standard deviation (σ is the lowercase Greek letter sigma). This "bell-shaped curve" has points of inflection when $x = \pm\sigma$.

FOR FURTHER INFORMATION To learn about derivatives of exponential functions of order 1/2, see the article "A Child's Garden of Fractional Derivatives" by Marcia Kleinz and Thomas J. Osler in *The College Mathematics Journal*. To view this article, go to *MathArticles.com*.

Integrals of Exponential Functions

Each differentiation formula in Theorem 5.11 has a corresponding integration formula.

THEOREM 5.12 Integration Rules for Exponential Functions

Let u be a differentiable function of x.

1. $\int e^x\,dx = e^x + C$ **2.** $\int e^u\,du = e^u + C$

EXAMPLE 7 Integrating Exponential Functions

Find the indefinite integral.

$$\int e^{3x+1}\,dx$$

Solution If you let $u = 3x + 1$, then $du = 3\,dx$.

$$\int e^{3x+1}\,dx = \frac{1}{3}\int e^{3x+1}(3)\,dx \quad \text{Multiply and divide by 3.}$$

$$= \frac{1}{3}\int e^u\,du \quad \text{Substitute: } u = 3x + 1.$$

$$= \frac{1}{3}e^u + C \quad \text{Apply Exponential Rule.}$$

$$= \frac{e^{3x+1}}{3} + C \quad \text{Back-substitute.}$$

REMARK In Example 7, the missing *constant* factor 3 was introduced to create $du = 3\,dx$. However, remember that you cannot introduce a missing *variable* factor in the integrand. For instance,

$$\int e^{-x^2}\,dx \neq \frac{1}{x}\int e^{-x^2}(x\,dx).$$

EXAMPLE 8 Integrating Exponential Functions

Find the indefinite integral.

$$\int 5xe^{-x^2}\,dx$$

Solution If you let $u = -x^2$, then $du = -2x\,dx$ or $x\,dx = -du/2$.

$$\int 5xe^{-x^2}\,dx = \int 5e^{-x^2}(x\,dx) \quad \text{Regroup integrand.}$$

$$= \int 5e^u\left(-\frac{du}{2}\right) \quad \text{Substitute: } u = -x^2.$$

$$= -\frac{5}{2}\int e^u\,du \quad \text{Constant Multiple Rule}$$

$$= -\frac{5}{2}e^u + C \quad \text{Apply Exponential Rule.}$$

$$= -\frac{5}{2}e^{-x^2} + C \quad \text{Back-substitute.}$$

EXAMPLE 9 Integrating Exponential Functions

Find each indefinite integral.

a. $\displaystyle\int \frac{e^{1/x}}{x^2}\,dx$ **b.** $\displaystyle\int \sin x\, e^{\cos x}\,dx$

Solution

a. $\displaystyle\int \frac{e^{1/x}}{x^2}\,dx = -\int \overbrace{e^{1/x}}^{e^u}\overbrace{\left(-\frac{1}{x^2}\right)dx}^{du}$ $\quad u = \frac{1}{x}$

$\displaystyle = -e^{1/x} + C$

b. $\displaystyle\int \sin x\, e^{\cos x}\,dx = -\int \overbrace{e^{\cos x}}^{e^u}\overbrace{(-\sin x)\,dx}^{du}$ $\quad u = \cos x$

$\displaystyle = -e^{\cos x} + C$

EXAMPLE 10 Finding Areas Bounded by Exponential Functions

Evaluate each definite integral.

a. $\displaystyle\int_0^1 e^{-x}\,dx$ **b.** $\displaystyle\int_0^1 \frac{e^x}{1+e^x}\,dx$ **c.** $\displaystyle\int_{-1}^0 e^x \cos(e^x)\,dx$

Solution

a. $\displaystyle\int_0^1 e^{-x}\,dx = -e^{-x}\Big]_0^1$ See Figure 5.21(a).

$\displaystyle = -e^{-1} - (-1)$

$\displaystyle = 1 - \frac{1}{e}$

≈ 0.632

b. $\displaystyle\int_0^1 \frac{e^x}{1+e^x}\,dx = \ln(1+e^x)\Big]_0^1$ See Figure 5.21(b).

$= \ln(1+e) - \ln 2$

≈ 0.620

c. $\displaystyle\int_{-1}^0 e^x\cos(e^x)\,dx = \sin(e^x)\Big]_{-1}^0$ See Figure 5.21(c).

$= \sin 1 - \sin(e^{-1})$

≈ 0.482

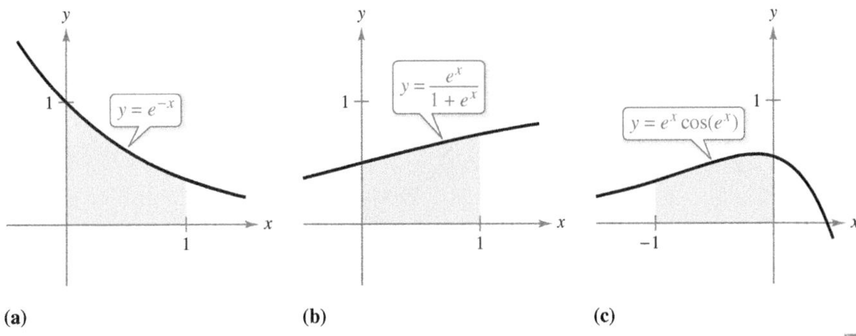

(a) (b) (c)

Figure 5.21

5.4 Exercises

See CalcChat.com for tutorial help and worked-out solutions to odd-numbered exercises.

CONCEPT CHECK

1. **Natural Exponential Function** Describe the graph of $f(x) = e^x$.

2. **A Function and Its Derivative** Which of the following functions are their own derivative?

$y = e^x + 4 \qquad y = e^x \qquad y = e^{4x} \qquad y = 4e^x$

Solving an Exponential or Logarithmic Equation **In Exercises 3–18, solve for x accurate to three decimal places.**

3. $e^{\ln x} = 4$ **4.** $e^{\ln 3x} = 24$

5. $e^x = 12$ **6.** $5e^x = 36$

7. $9 - 2e^x = 7$ **8.** $8e^x - 12 = 7$

9. $50e^{-x} = 30$ **10.** $100e^{-2x} = 35$

11. $\dfrac{800}{100 - e^{x/2}} = 50$ **12.** $\dfrac{5000}{1 + e^{2x}} = 2$

13. $\ln x = 2$ **14.** $\ln x^2 = -8$

15. $\ln(x - 3) = 2$ **16.** $\ln 4x = 1$

17. $\ln\sqrt{x + 2} = 1$ **18.** $\ln(x - 2)^2 = 12$

Sketching a Graph **In Exercises 19–24, sketch the graph of the function.**

19. $y = e^{-x}$ **20.** $y = \frac{1}{3}e^x$

21. $y = e^x + 1$ **22.** $y = -e^{x-1}$

23. $y = e^{-x^2}$ **24.** $y = e^{-x/2}$

Matching **In Exercises 25–28, match the equation with the correct graph. Assume that a and C are positive real numbers. [The graphs are labeled (a), (b), (c), and (d).]**

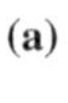

(a)

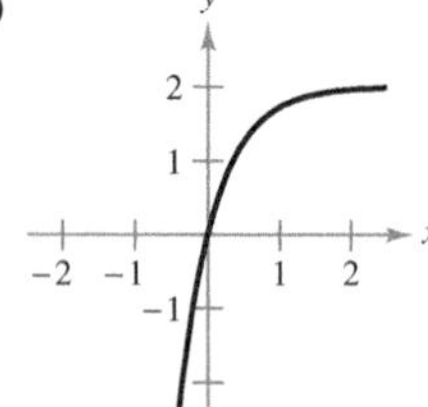

(b)

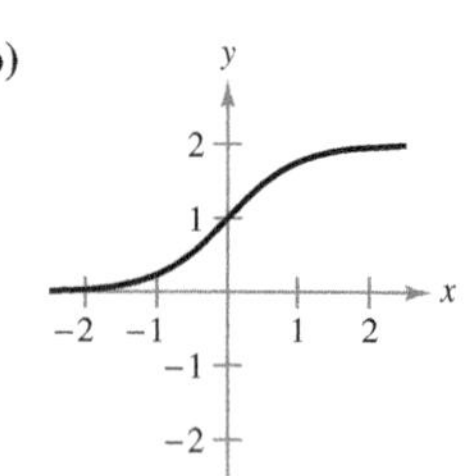

(c)

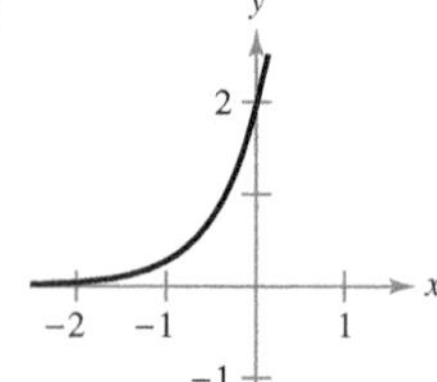

(d)

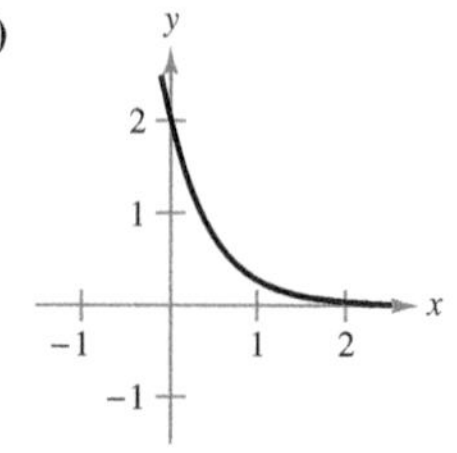

25. $y = Ce^{ax}$ **26.** $y = Ce^{-ax}$

27. $y = C(1 - e^{-ax})$ **28.** $y = \dfrac{C}{1 + e^{-ax}}$

Inverse Functions **In Exercises 29–32, illustrate that the functions are inverse functions of each other by sketching their graphs on the same set of coordinate axes.**

29. $f(x) = e^{2x}$, $g(x) = \ln\sqrt{x}$ **30.** $f(x) = e^{x/3}$, $g(x) = \ln x^3$

31. $f(x) = e^x - 1$, $g(x) = \ln(x + 1)$ **32.** $f(x) = e^{x-1}$, $g(x) = 1 + \ln x$

Finding a Derivative **In Exercises 33–54, find the derivative of the function.**

33. $y = e^{5x}$ **34.** $y = e^{-8x}$

35. $y = e^{\sqrt{x}}$ **36.** $y = e^{-2x^3}$

37. $y = e^{x-4}$ **38.** $y = 5e^{x^2+5}$

39. $y = e^x \ln x$ **40.** $y = xe^{4x}$

41. $y = (x + 1)^2 e^x$ **42.** $y = x^2 e^{-x}$

43. $g(t) = (e^{-t} + e^t)^3$ **44.** $g(t) = e^{-3/t^2}$

45. $y = \ln(2 - e^{5x})$ **46.** $y = \ln\left(\dfrac{1 + e^x}{1 - e^x}\right)$

47. $y = \dfrac{2}{e^x + e^{-x}}$ **48.** $y = \dfrac{e^x - e^{-x}}{2}$

49. $y = \dfrac{e^x + 1}{e^x - 1}$ **50.** $y = \dfrac{e^{2x}}{e^{2x} + 1}$

51. $y = e^x(\sin x + \cos x)$ **52.** $y = e^{2x} \tan 2x$

53. $F(x) = \displaystyle\int_{\pi}^{\ln x} \cos e^t \, dt$ **54.** $F(x) = \displaystyle\int_{0}^{e^{2x}} \ln(t + 1) \, dt$

Finding an Equation of a Tangent Line **In Exercises 55–62, find an equation of the tangent line to the graph of the function at the given point.**

55. $f(x) = e^{3x}$, $(0, 1)$ **56.** $f(x) = e^{-x} - 6$, $(0, -5)$

57. $y = e^{3x - x^2}$, $(3, 1)$ **58.** $y = e^{-2x + x^2}$, $(2, 1)$

59. $f(x) = e^{-x} \ln x$, $(1, 0)$ **60.** $y = \ln \dfrac{e^x + e^{-x}}{2}$, $(0, 0)$

61. $y = x^2 e^x - 2xe^x + 2e^x$, $(1, e)$

62. $y = xe^x - e^x$, $(1, 0)$

Implicit Differentiation **In Exercises 63 and 64, use implicit differentiation to find dy/dx.**

63. $xe^y - 10x + 3y = 0$ **64.** $e^{xy} + x^2 - y^2 = 10$

Finding the Equation of a Tangent Line **In Exercises 65 and 66, use implicit differentiation to find an equation of the tangent line to the graph of the equation at the given point.**

65. $xe^y + ye^x = 1$, $(0, 1)$

66. $1 + \ln xy = e^{x-y}$, $(1, 1)$

Finding a Second Derivative **In Exercises 67 and 68, find the second derivative of the function.**

67. $f(x) = (3 + 2x)e^{-3x}$

68. $g(x) = \sqrt{x} + e^x \ln x$

Differential Equation **In Exercises 69 and 70, show that the function $y = f(x)$ is a solution of the differential equation.**

69. $y = 4e^{-x}$
$y'' - y = 0$

70. $y = e^{3x} + e^{-3x}$
$y'' - 9y = 0$

Relative Extrema and Points of Inflection **In Exercises 71–78, find the relative extrema and the points of inflection (if any exist) of the function. Use a graphing utility to graph the function and confirm your results.**

71. $f(x) = \dfrac{e^x + e^{-x}}{2}$

72. $f(x) = \dfrac{e^x - e^{-x}}{2}$

73. $g(x) = \dfrac{1}{\sqrt{2\pi}}e^{-(x-2)^2/2}$

74. $g(x) = \dfrac{1}{\sqrt{2\pi}}e^{-(x-3)^2/2}$

75. $f(x) = (2 - x)e^x$

76. $f(x) = xe^{-x}$

77. $g(t) = 1 + (2 + t)e^{-t}$

78. $f(x) = -2 + e^{3x}(4 - 2x)$

79. Area Find the area of the largest rectangle that can be inscribed under the curve $y = e^{-x^2}$ in the first and second quadrants.

80. Area Perform the following steps to find the maximum area of the rectangle shown in the figure.

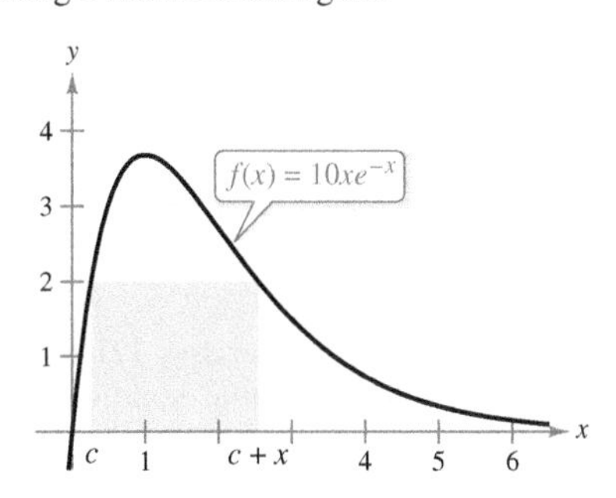

(a) Solve for c in the equation $f(c) = f(c + x)$.

(b) Use the result in part (a) to write the area A as a function of x. [*Hint:* $A = xf(c)$]

(c) Use a graphing utility to graph the area function. Use the graph to approximate the dimensions of the rectangle of maximum area. Determine the maximum area.

(d) Use a graphing utility to graph the expression for c found in part (a). Use the graph to approximate

$$\lim_{x\to 0^+} c \quad \text{and} \quad \lim_{x\to\infty} c.$$

Use this result to describe the changes in dimensions and position of the rectangle for $0 < x < \infty$.

81. Finding an Equation of a Tangent Line Find the point on the graph of the function $f(x) = e^{2x}$ such that the tangent line to the graph at that point passes through the origin. Use a graphing utility to graph f and the tangent line in the same viewing window.

82. HOW DO YOU SEE IT? The figure shows the graphs of f and g, where a is a positive real number. Identify the open interval(s) on which the graphs of f and g are (a) increasing or decreasing and (b) concave upward or concave downward.

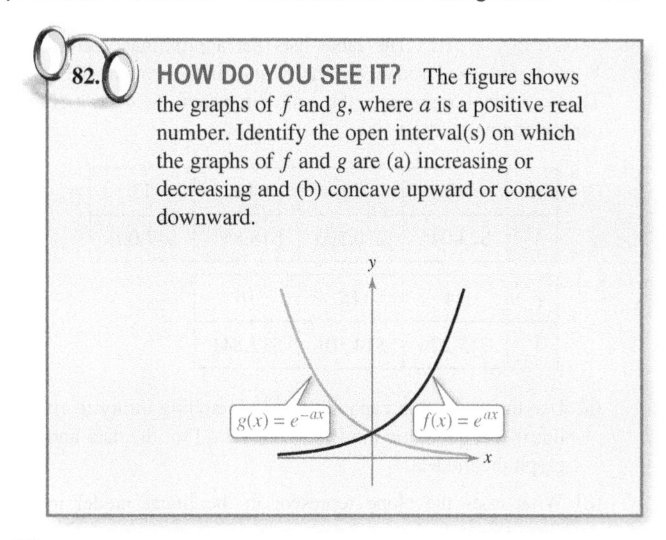

83. Depreciation The value V of an item t years after it is purchased is $V = 15{,}000e^{-0.6286t}$, $0 \le t \le 10$.

(a) Use a graphing utility to graph the function.

(b) Find the rates of change of V with respect to t when $t = 1$ and $t = 5$.

(c) Use a graphing utility to graph the tangent lines to the function when $t = 1$ and $t = 5$.

84. Harmonic Motion The displacement from equilibrium of a mass oscillating on the end of a spring suspended from a ceiling is $y = 1.56e^{-0.22t} \cos 4.9t$, where y is the displacement (in feet) and t is the time (in seconds). Use a graphing utility to graph the displacement function on the interval $[0, 10]$. Find a value of t past which the displacement is less than 3 inches from equilibrium.

85. Atmospheric Pressure

A meteorologist measures the atmospheric pressure P (in millibars) at altitude h (in kilometers). The data are shown below.

h	0	5	10	15	20
P	1013.2	547.5	233.0	121.6	50.7

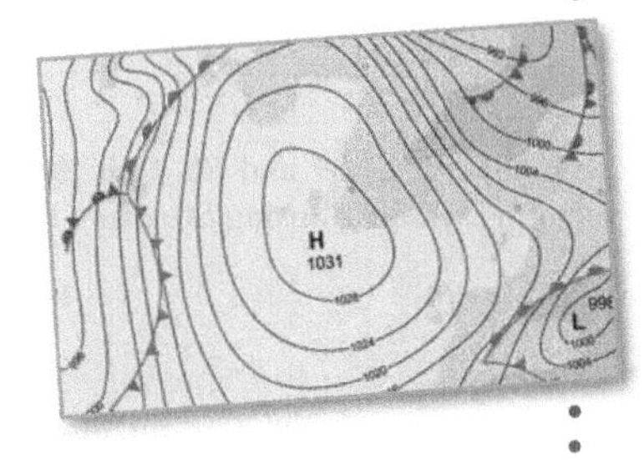

(a) Use a graphing utility to plot the points $(h, \ln P)$. Use the regression capabilities of the graphing utility to find a linear model for the revised data points.

(b) The line in part (a) has the form $\ln P = ah + b$. Write the equation in exponential form.

(c) Use a graphing utility to plot the original data and graph the exponential model in part (b).

(d) Find the rates of change of the pressure when $h = 5$ and $h = 18$.

86. Modeling Data The table lists the approximate values V of a mid-sized sedan for the years 2010 through 2016. The variable t represents the time (in years), with $t = 10$ corresponding to 2010.

t	10	11	12	13	14	15	16
V	\$23,046	\$20,596	\$18,851	\$17,001	\$15,226	\$14,101	\$12,841

(a) Use the regression capabilities of a graphing utility to fit linear and quadratic models to the data. Plot the data and graph the models.

(b) What does the slope represent in the linear model in part (a)?

(c) Use the regression capabilities of a graphing utility to fit an exponential model to the data.

(d) Determine the horizontal asymptote of the exponential model found in part (c). Interpret its meaning in the context of the problem.

(e) Use the exponential model to find the rates of decrease in the value of the sedan when $t = 12$ and $t = 15$.

Linear and Quadratic Approximation **In Exercises 87 and 88, use a graphing utility to graph the function. Then graph**

$$P_1(x) = f(0) + f'(0)(x - 0) \quad \text{and}$$

$$P_2(x) = f(0) + f'(0)(x - 0) + \tfrac{1}{2}f''(0)(x - 0)^2$$

in the same viewing window. Compare the values of f, P_1, P_2, and their first derivatives at $x = 0$.

87. $f(x) = e^x$

88. $f(x) = e^{x/2}$

Stirling's Formula **For large values of n,**

$$n! = 1 \cdot 2 \cdot 3 \cdot 4 \cdot \cdot \cdot (n - 1) \cdot n$$

can be approximated by Stirling's Formula,

$$n! \approx \left(\frac{n}{3}\right)^n \sqrt{2\pi n}.$$

In Exercises 89 and 90, find the exact value of $n!$ and then approximate $n!$ using Stirling's Formula.

89. $n = 12$

90. $n = 15$

Finding an Indefinite Integral **In Exercises 91–108, find the indefinite integral.**

91. $\int e^{5x}(5)\,dx$

92. $\int e^{-x^4}(-4x^3)\,dx$

93. $\int e^{5x-3}\,dx$

94. $\int e^{1-3x}\,dx$

95. $\int (2x + 1)e^{x^2+x}\,dx$

96. $\int e^x(e^x + 1)^2\,dx$

97. $\int \frac{e^{\sqrt{x}}}{\sqrt{x}}\,dx$

98. $\int \frac{e^{1/x^2}}{x^3}\,dx$

99. $\int \frac{e^{-x}}{1 + e^{-x}}\,dx$

100. $\int \frac{e^{2x}}{1 + e^{2x}}\,dx$

101. $\int e^x\sqrt{1 - e^x}\,dx$

102. $\int \frac{e^x - e^{-x}}{e^x + e^{-x}}\,dx$

103. $\int \frac{e^x + e^{-x}}{e^x - e^{-x}}\,dx$

104. $\int \frac{2e^x - 2e^{-x}}{(e^x + e^{-x})^2}\,dx$

105. $\int \frac{5 - e^x}{e^{2x}}\,dx$

106. $\int \frac{e^{-3x} + 2e^{2x} + 3}{e^x}\,dx$

107. $\int e^{-x}\tan(e^{-x})\,dx$

108. $\int e^{2x}\csc(e^{2x})\,dx$

Evaluating a Definite Integral **In Exercises 109–118, evaluate the definite integral. Use a graphing utility to verify your result.**

109. $\int_0^1 e^{-2x}\,dx$

110. $\int_{-1}^1 e^{1+4x}\,dx$

111. $\int_0^1 xe^{-x^2}\,dx$

112. $\int_{-2}^0 x^2e^{x^3/2}\,dx$

113. $\int_1^3 \frac{e^{3/x}}{x^2}\,dx$

114. $\int_0^{\sqrt{2}} xe^{-x^2/2}\,dx$

115. $\int_0^2 \frac{e^{4x}}{1 + e^{4x}}\,dx$

116. $\int_{-2}^0 \frac{e^{x+1}}{7 - e^{x+1}}\,dx$

117. $\int_0^{\pi/2} e^{\sin \pi x}\cos \pi x\,dx$

118. $\int_{\pi/3}^{\pi/2} e^{\sec 2x}\sec 2x \tan 2x\,dx$

Differential Equation **In Exercises 119 and 120, find the general solution of the differential equation.**

119. $\frac{dy}{dx} = xe^{9x^2}$

120. $\frac{dy}{dx} = (e^x - e^{-x})^2$

Differential Equation **In Exercises 121 and 122, find the particular solution of the differential equation that satisfies the initial conditions.**

121. $f''(x) = \frac{1}{2}(e^x + e^{-x}),\ f(0) = 1,\ f'(0) = 0$

122. $f''(x) = \sin x + e^{2x},\ f(0) = \frac{1}{4},\ f'(0) = \frac{1}{2}$

Area **In Exercises 123–126, find the area of the region bounded by the graphs of the equations. Use a graphing utility to verify your result.**

123. $y = e^x,\ y = 0,\ x = 0,\ x = 6$

124. $y = e^{-2x},\ y = 0,\ x = -1,\ x = 3$

125. $y = xe^{-x^2/4},\ y = 0,\ x = 0,\ x = \sqrt{6}$

126. $y = e^{-2x} + 2,\ y = 0,\ x = 0,\ x = 2$

Midpoint Rule **In Exercises 127 and 128, use the Midpoint Rule with $n = 12$ to approximate the value of the definite integral. Use a graphing utility to verify your result.**

127. $\int_0^4 \sqrt{x}\,e^x\,dx$

128. $\int_0^2 2xe^{-x}\,dx$

EXPLORING CONCEPTS

129. Asymptotes Compare the asymptotes of the natural exponential function with those of the natural logarithmic function.

130. Comparing Graphs Use a graphing utility to graph $f(x) = e^x$ and the given function in the same viewing window. How are the two graphs related?

(a) $g(x) = e^{x-2}$ (b) $h(x) = -\frac{1}{2}e^x$ (c) $q(x) = e^{-x} + 3$

True or False? **In Exercises 131–134, determine whether the statement is true or false. If it is false, explain why or give an example that shows it is false.**

131. If $f(x) = g(x)e^x$, then $f'(x) = g'(x)e^x$.

132. If $f(x) = \ln x$, then $f(e^{n+1}) - f(e^n) = 1$ for any value of n.

133. The graphs of $f(x) = e^x$ and $g(x) = e^{-x}$ meet at right angles.

134. If $f(x) = g(x)e^x$, then the only zeros of f are the zeros of g.

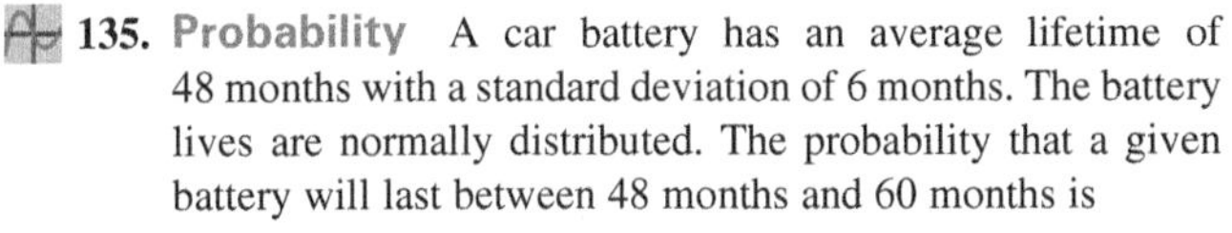

135. Probability A car battery has an average lifetime of 48 months with a standard deviation of 6 months. The battery lives are normally distributed. The probability that a given battery will last between 48 months and 60 months is

$$0.0065 \int_{48}^{60} e^{-0.0139(t-48)^2}\, dt.$$

Use the integration capabilities of a graphing utility to approximate the integral. Interpret the resulting probability.

136. Probability The median waiting time (in minutes) for people waiting for service in a convenience store is given by the solution of the equation

$$\int_0^x 0.3e^{-0.3t}\, dt = \frac{1}{2}.$$

What is the median waiting time?

137. Modeling Data A valve on a storage tank is opened for 4 hours to release a chemical in a manufacturing process. The flow rate R (in liters per hour) at time t (in hours) is given in the table.

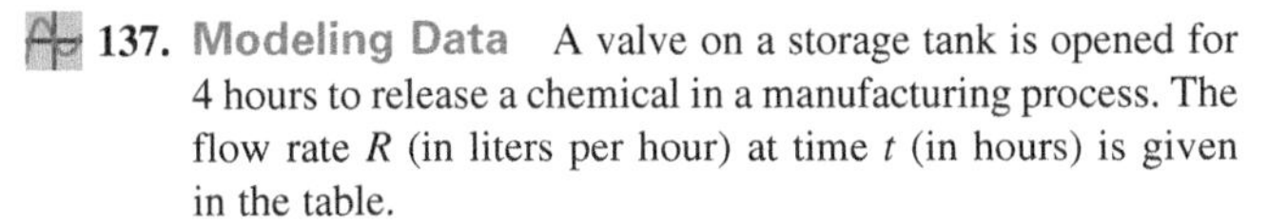

t	0	1	2	3	4
R	425	240	118	71	36

(a) Use the regression capabilities of a graphing utility to find a linear model for the points $(t, \ln R)$. Write the resulting equation of the form $\ln R = at + b$ in exponential form.

(b) Use a graphing utility to plot the data and graph the exponential model.

(c) Use a definite integral to approximate the number of liters of chemical released during the 4 hours.

138. Using the Area of a Region Find the value of a such that the area bounded by $y = e^{-x}$, the x-axis, $x = -a$, and $x = a$ is $\frac{8}{3}$.

139. Analyzing a Graph Consider the function

$$f(x) = \frac{2}{1 + e^{1/x}}.$$

(a) Use a graphing utility to graph f.

(b) Write a short paragraph explaining why the graph has a horizontal asymptote at $y = 1$ and why the function has a nonremovable discontinuity at $x = 0$.

140. Analyzing a Function Let $f(x) = \dfrac{\ln x}{x}$.

(a) Graph f on $(0, \infty)$ and show that f is strictly decreasing on (e, ∞).

(b) Show that if $e \le A < B$, then $A^B > B^A$.

(c) Use part (b) to show that $e^\pi > \pi^e$.

141. Deriving an Inequality Given $e^x \ge 1$ for $x \ge 0$, it follows that

$$\int_0^x e^t\, dt \ge \int_0^x 1\, dt.$$

Perform this integration to derive the inequality

$$e^x \ge 1 + x$$

for $x \ge 0$.

142. Solving an Equation Find, to three decimal places, the value of x such that $e^{-x} = x$. (Use Newton's Method or the *zero* or *root* feature of a graphing utility.)

143. Analyzing a Graph Consider

$$f(x) = xe^{-kx}$$

for $k > 0$. Find the relative extrema and the points of inflection of the function.

144. Finding the Maximum Rate of Change Verify that the function

$$y = \frac{L}{1 + ae^{-x/b}}, \quad a > 0, \quad b > 0, \quad L > 0$$

increases at a maximum rate when $y = \dfrac{L}{2}$.

PUTNAM EXAM CHALLENGE

145. Let S be a class of functions from $[0, \infty)$ to $[0, \infty)$ that satisfies:

(i) The functions $f_1(x) = e^x - 1$ and $f_2(x) = \ln(x + 1)$ are in S;

(ii) If $f(x)$ and $g(x)$ are in S, the functions $f(x) + g(x)$ and $f(g(x))$ are in S;

(iii) If $f(x)$ and $g(x)$ are in S and $f(x) \ge g(x)$ for all $x \ge 0$, then the function $f(x) - g(x)$ is in S.

Prove that if $f(x)$ and $g(x)$ are in S, then the function $f(x)g(x)$ is also in S.

This problem was composed by the Committee on the Putnam Prize Competition.

5.5 Bases Other than e and Applications

- **Define exponential functions that have bases other than e.**
- **Differentiate and integrate exponential functions that have bases other than e.**
- **Use exponential functions to model compound interest and exponential growth.**

Bases Other than e

The **base** of the natural exponential function is e. This "natural" base can be used to assign a meaning to a general base a.

> **Definition of Exponential Function to Base *a***
>
> If a is a positive real number ($a \neq 1$) and x is any real number, then the **exponential function to the base *a*** is denoted by a^x and is defined by
>
> $$a^x = e^{(\ln a)x}.$$
>
> If $a = 1$, then $y = 1^x = 1$ is a constant function.

Exponential functions obey the usual laws of exponents. For instance, here are some familiar properties.

1. $a^0 = 1$ **2.** $a^x a^y = a^{x+y}$ **3.** $\dfrac{a^x}{a^y} = a^{x-y}$ **4.** $(a^x)^y = a^{xy}$

When modeling the half-life of a radioactive sample, it is convenient to use $\frac{1}{2}$ as the base of the exponential model. (*Half-life* is the number of years required for half of the atoms in a sample of radioactive material to decay.)

Carbon dating uses the radioactive isotope carbon-14 to estimate the age of dead organic materials. The method is based on the decay rate of carbon-14 (see Example 1), a compound organisms take in when they are alive.

EXAMPLE 1 Radioactive Half-Life Model

The half-life of carbon-14 is about 5715 years. A sample contains 1 gram of carbon-14. How much will be present in 10,000 years?

Solution Let $t = 0$ represent the present time and let y represent the amount (in grams) of carbon-14 in the sample. Using a base of $\frac{1}{2}$, you can model y by the equation

$$y = \left(\frac{1}{2}\right)^{t/5715}.$$

Notice that when $t = 5715$, the amount is reduced to half of the original amount.

$$y = \left(\frac{1}{2}\right)^{5715/5715} = \frac{1}{2} \text{ gram}$$

When $t = 11{,}430$, the amount is reduced to a quarter of the original amount and so on. To find the amount of carbon-14 after 10,000 years, substitute 10,000 for t.

$$y = \left(\frac{1}{2}\right)^{10{,}000/5715}$$

$$\approx 0.30 \text{ gram}$$

The graph of y is shown at the right.

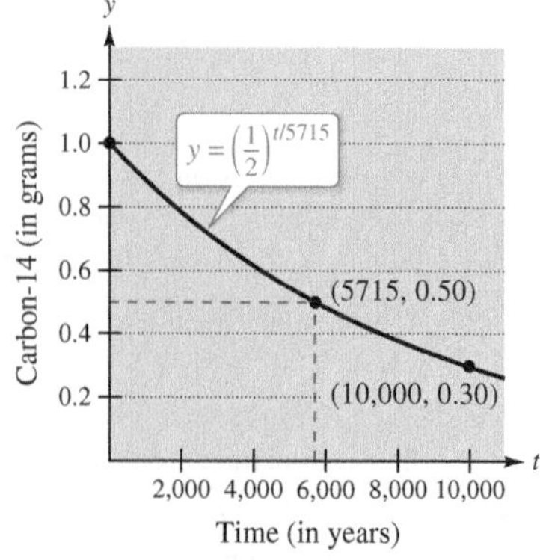

The half-life of carbon-14 is about 5715 years.

Logarithmic functions to bases other than e can be defined in much the same way as exponential functions to other bases are defined.

REMARK In precalculus, you learned that $\log_a x$ is the value to which a must be raised to produce x. This agrees with the definition at the right because

$$\begin{aligned} a^{\log_a x} &= a^{(1/\ln a)\ln x} \\ &= (e^{\ln a})^{(1/\ln a)\ln x} \\ &= e^{(\ln a/\ln a)\ln x} \\ &= e^{\ln x} \\ &= x. \end{aligned}$$

Definition of Logarithmic Function to Base *a*

If a is a positive real number $(a \neq 1)$ and x is any positive real number, then the **logarithmic function to the base *a*** is denoted by $\log_a x$ and is defined as

$$\log_a x = \frac{1}{\ln a} \ln x.$$

Logarithmic functions to the base a have properties similar to those of the natural logarithmic function given in Theorem 5.2. (Assume x and y are positive numbers and n is rational.)

1. $\log_a 1 = 0$ — Log of 1

2. $\log_a xy = \log_a x + \log_a y$ — Log of a product

3. $\log_a x^n = n \log_a x$ — Log of a power

4. $\log_a \frac{x}{y} = \log_a x - \log_a y$ — Log of a quotient

From the definitions of the exponential and logarithmic functions to the base a, it follows that $f(x) = a^x$ and $g(x) = \log_a x$ are inverse functions of each other.

Properties of Inverse Functions

1. $y = a^x$ if and only if $x = \log_a y$

2. $a^{\log_a x} = x$, for $x > 0$

3. $\log_a a^x = x$, for all x

The logarithmic function to the base 10 is called the **common logarithmic function.** So, for common logarithms,

$$y = 10^x \quad \text{if and only if} \quad x = \log_{10} y.$$ Property of inverse functions

EXAMPLE 2 **Bases Other than e**

Solve for x in each equation.

a. $3^x = \frac{1}{81}$ **b.** $\log_2 x = -4$

Solution

a. To solve this equation, you can apply the logarithmic function to the base 3 to each side of the equation.

$$\begin{aligned} 3^x &= \frac{1}{81} \\ \log_3 3^x &= \log_3 \frac{1}{81} \\ x &= \log_3 3^{-4} \\ x &= -4 \end{aligned}$$

b. To solve this equation, you can apply the exponential function to the base 2 to each side of the equation.

$$\begin{aligned} \log_2 x &= -4 \\ 2^{\log_2 x} &= 2^{-4} \\ x &= \frac{1}{2^4} \\ x &= \frac{1}{16} \end{aligned}$$

■

Differentiation and Integration

To differentiate exponential and logarithmic functions to other bases, you have three options: (1) use the definitions of a^x and $\log_a x$ and differentiate using the rules for the natural exponential and logarithmic functions, (2) use logarithmic differentiation, or (3) use the differentiation rules for bases other than e given in the next theorem.

REMARK These differentiation rules are similar to those for the natural exponential function and the natural logarithmic function. In fact, they differ only by the constant factors $\ln a$ and $1/\ln a$. This points out one reason why, for calculus, e is the most convenient base.

THEOREM 5.13 Derivatives for Bases Other than *e*

Let a be a positive real number $(a \neq 1)$, and let u be a differentiable function of x.

1. $\dfrac{d}{dx}[a^x] = (\ln a)a^x$
2. $\dfrac{d}{dx}[a^u] = (\ln a)a^u \dfrac{du}{dx}$
3. $\dfrac{d}{dx}[\log_a x] = \dfrac{1}{(\ln a)x}$
4. $\dfrac{d}{dx}[\log_a u] = \dfrac{1}{(\ln a)u}\dfrac{du}{dx}$

Proof By definition, $a^x = e^{(\ln a)x}$. So, you can prove the first rule by letting $u = (\ln a)x$ and differentiating with base e to obtain

$$\frac{d}{dx}[a^x] = \frac{d}{dx}[e^{(\ln a)x}] = e^u\frac{du}{dx} = e^{(\ln a)x}(\ln a) = (\ln a)a^x.$$

To prove the third rule, you can write

$$\frac{d}{dx}[\log_a x] = \frac{d}{dx}\left[\frac{1}{\ln a}\ln x\right] = \frac{1}{\ln a}\left(\frac{1}{x}\right) = \frac{1}{(\ln a)x}.$$

The second and fourth rules are simply the Chain Rule versions of the first and third rules. ■

EXAMPLE 3 Differentiating Functions to Other Bases

Find the derivative of each function.

a. $y = 2^x$ **b.** $y = 2^{3x}$ **c.** $y = \log_{10} \cos x$ **d.** $y = \log_3 \dfrac{\sqrt{x}}{x+5}$

Solution

a. $y' = \dfrac{d}{dx}[2^x] = (\ln 2)2^x$

b. $y' = \dfrac{d}{dx}[2^{3x}] = (\ln 2)2^{3x}(3) = (3 \ln 2)2^{3x}$

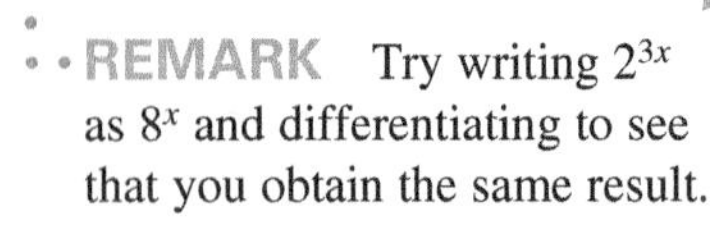

REMARK Try writing 2^{3x} as 8^x and differentiating to see that you obtain the same result.

c. $y' = \dfrac{d}{dx}[\log_{10} \cos x] = \dfrac{-\sin x}{(\ln 10)\cos x} = -\dfrac{1}{\ln 10}\tan x$

d. Before differentiating, rewrite the function using logarithmic properties.

$$y = \log_3 \frac{\sqrt{x}}{x+5} = \frac{1}{2}\log_3 x - \log_3(x+5)$$

Next, apply Theorem 5.13 to differentiate the function.

$$\begin{aligned} y' &= \frac{d}{dx}\left[\frac{1}{2}\log_3 x - \log_3(x+5)\right] \\ &= \frac{1}{2(\ln 3)x} - \frac{1}{(\ln 3)(x+5)} \\ &= \frac{5-x}{2(\ln 3)x(x+5)} \end{aligned}$$

■

Occasionally, an integrand involves an exponential function to a base other than e. When this occurs, there are two options: (1) convert to base e using the formula $a^x = e^{(\ln a)x}$ and then integrate, or (2) integrate directly, using the integration formula

$$\int a^x\,dx = \left(\frac{1}{\ln a}\right)a^x + C$$

which follows from Theorem 5.13.

EXAMPLE 4 Integrating an Exponential Function to Another Base

Find $\int 2^x\,dx$.

Solution

$$\int 2^x\,dx = \frac{1}{\ln 2}2^x + C$$

When the Power Rule, $D_x[x^n] = nx^{n-1}$, was introduced in Chapter 2, the exponent n was required to be a rational number. Now the rule is extended to cover any real value of n. Try to prove this theorem using logarithmic differentiation.

THEOREM 5.14 The Power Rule for Real Exponents

Let n be any real number, and let u be a differentiable function of x.

1. $\dfrac{d}{dx}[x^n] = nx^{n-1}$
2. $\dfrac{d}{dx}[u^n] = nu^{n-1}\dfrac{du}{dx}$

The next example compares the derivatives of four types of functions. Each function uses a different differentiation formula, depending on whether the base and the exponent are constants or variables.

EXAMPLE 5 Comparing Variables and Constants

a. $\dfrac{d}{dx}[e^e] = 0$ — Constant Rule

b. $\dfrac{d}{dx}[e^x] = e^x$ — Exponential Rule

c. $\dfrac{d}{dx}[x^e] = ex^{e-1}$ — Power Rule

d. Use logarithmic differentiation.

$$y = x^x$$
$$\ln y = \ln x^x$$
$$\ln y = x\ln x$$
$$\frac{y'}{y} = x\left(\frac{1}{x}\right) + (\ln x)(1)$$
$$\frac{y'}{y} = 1 + \ln x$$
$$y' = y(1 + \ln x)$$
$$y' = x^x(1 + \ln x)$$

REMARK Be sure you see that there is no simple differentiation rule for calculating the derivative of $y = x^x$. In general, when $y = u(x)^{v(x)}$, you need to use logarithmic differentiation.

Applications of Exponential Functions

An amount of P dollars is deposited in an account at an annual interest rate r (in decimal form). What is the balance in the account at the end of 1 year? The answer depends on the number of times n the interest is compounded according to the formula

$$A = P\left(1 + \frac{r}{n}\right)^n.$$

For instance, the result for a deposit of \$1000 at 8% interest compounded n times a year is shown in the table at the right.

n	A
1	\$1080.00
2	\$1081.60
4	\$1082.43
12	\$1083.00
365	\$1083.28

As n increases, the balance A approaches a limit. To develop this limit, use the next theorem. To test the reasonableness of this theorem, try evaluating

$$\left(\frac{x+1}{x}\right)^x$$

for several values of x, as shown in the table at the left.

x	$\left(\frac{x+1}{x}\right)^x$
10	2.59374
100	2.70481
1000	2.71692
10,000	2.71815
100,000	2.71827
1,000,000	2.71828

THEOREM 5.15 A Limit Involving *e*

$$\lim_{x\to\infty}\left(1 + \frac{1}{x}\right)^x = \lim_{x\to\infty}\left(\frac{x+1}{x}\right)^x = e$$

A proof of this theorem is given in Appendix A.

Given Theorem 5.15, take another look at the formula for the balance A in an account in which the interest is compounded n times per year. By taking the limit as n approaches infinity, you obtain

$$A = \lim_{n\to\infty} P\left(1 + \frac{r}{n}\right)^n \qquad \text{Take limit as } n\to\infty.$$

$$= P \lim_{n\to\infty}\left[\left(1 + \frac{1}{n/r}\right)^{n/r}\right]^r \qquad \text{Rewrite.}$$

$$= P\left[\lim_{x\to\infty}\left(1 + \frac{1}{x}\right)^x\right]^r \qquad \text{Let } x = n/r. \text{ Then } x\to\infty \text{ as } n\to\infty.$$

$$= Pe^r. \qquad \text{Apply Theorem 5.15.}$$

This limit produces the balance after 1 year of **continuous compounding.** So, for a deposit of \$1000 at 8% interest compounded continuously, the balance at the end of 1 year would be

$$A = 1000e^{0.08} \approx \$1083.29.$$

SUMMARY OF COMPOUND INTEREST FORMULAS

Let P = amount of deposit, t = number of years, A = balance after t years, r = annual interest rate (in decimal form), and n = number of compoundings per year.

1. Compounded n times per year: $A = P\left(1 + \dfrac{r}{n}\right)^{nt}$

2. Compounded continuously: $A = Pe^{rt}$

EXAMPLE 6 Continuous, Quarterly, and Monthly Compounding

See LarsonCalculus.com for an interactive version of this type of example.

A deposit of \$2500 is made in an account that pays an annual interest rate of 5%. Find the balance in the account at the end of 5 years when the interest is compounded (a) quarterly, (b) monthly, and (c) continuously.

Solution

a. $A = P\left(1 + \frac{r}{n}\right)^{nt}$ Compounded quarterly

$= 2500\left(1 + \frac{0.05}{4}\right)^{4(5)}$

$= 2500(1.0125)^{20}$

$= \$3205.09$

b. $A = P\left(1 + \frac{r}{n}\right)^{nt}$ Compounded monthly

$= 2500\left(1 + \frac{0.05}{12}\right)^{12(5)}$

$\approx 2500(1.0041667)^{60}$

$= \$3208.40$

c. $A = Pe^{rt}$ Compounded continuously

$= 2500\left[e^{0.05(5)}\right]$

$= 2500e^{0.25}$

$= \$3210.06$

EXAMPLE 7 Bacterial Culture Growth

A bacterial culture is growing according to the *logistic growth function*

$$y = \frac{1.25}{1 + 0.25e^{-0.4t}}, \quad t \geq 0$$

where y is the weight of the culture in grams and t is the time in hours. Find the weight of the culture after (a) 0 hours, (b) 1 hour, and (c) 10 hours. (d) What is the limit as t approaches infinity?

Solution

a. When $t = 0$, $y = \frac{1.25}{1 + 0.25e^{-0.4(0)}}$

$= 1$ gram.

b. When $t = 1$, $y = \frac{1.25}{1 + 0.25e^{-0.4(1)}}$

≈ 1.071 grams.

c. When $t = 10$, $y = \frac{1.25}{1 + 0.25e^{-0.4(10)}}$

≈ 1.244 grams.

d. Taking the limit as t approaches infinity, you obtain

$$\lim_{t \to \infty} \frac{1.25}{1 + 0.25e^{-0.4t}} = \frac{1.25}{1 + 0} = 1.25 \text{ grams.}$$

The graph of the function is shown in Figure 5.22.

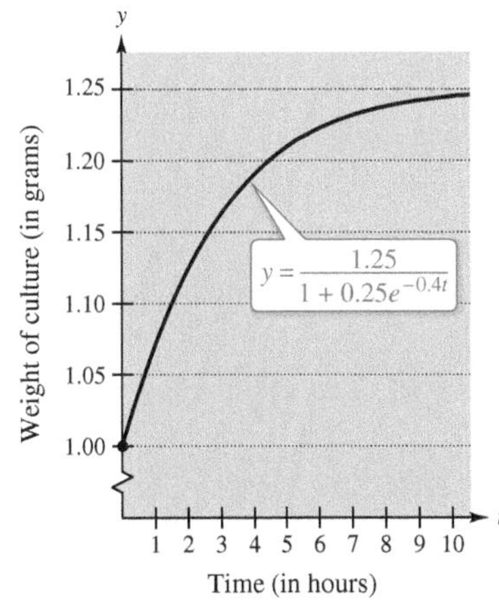

The limit of the weight of the culture as $t \to \infty$ is 1.25 grams.
Figure 5.22

5.5 Exercises

See CalcChat.com for tutorial help and worked-out solutions to odd-numbered exercises.

CONCEPT CHECK

1. **Derivatives for Bases Other than e** What are the values of a and b?

$$\frac{d}{dx}[6^{4x}] = a(\ln b)6^{4x}$$

2. **Integration for Bases Other than e** What are two options for finding the indefinite integral below?

$$\int 5^t\,dt$$

3. **Logarithmic Differentiation** Explain when it is necessary to use logarithmic differentiation to find the derivative of an exponential function.

4. **Compound Interest Formulas** Explain how to choose which compound interest formula to use to find the balance of a deposit.

Evaluating a Logarithmic Expression **In Exercises 5–10, evaluate the expression without using a calculator.**

5. $\log_2 \frac{1}{8}$ **6.** $\log_3 81$

7. $\log_7 1$ **8.** $\log_a \dfrac{1}{a}$

9. $\log_{64} 32$ **10.** $\log_{27} \frac{1}{9}$

Exponential and Logarithmic Forms of Equations **In Exercises 11–14, write the exponential equation as a logarithmic equation or vice versa.**

11. (a) $2^3 = 8$ (b) $3^{-1} = \frac{1}{3}$

12. (a) $27^{2/3} = 9$ (b) $16^{3/4} = 8$

13. (a) $\log_{10} 0.01 = -2$ (b) $\log_{0.5} 8 = -3$

14. (a) $\log_3 \frac{1}{9} = -2$ (b) $49^{1/2} = 7$

Sketching a Graph **In Exercises 15–20, sketch the graph of the function.**

15. $y = 2^x$ **16.** $y = 4^{x-1}$

17. $y = \left(\frac{1}{3}\right)^x$ **18.** $y = 2^{x^2}$

19. $h(x) = 5^{x-2}$ **20.** $y = 3^{-|x|}$

Solving an Equation **In Exercises 21–26, solve for x.**

21. (a) $\log_{10} 1000 = x$ (b) $\log_{10} 0.1 = x$

22. (a) $\log_3 \frac{1}{81} = x$ (b) $\log_6 36 = x$

23. (a) $\log_3 x = -1$ (b) $\log_2 x = -4$

24. (a) $\log_4 x = -2$ (b) $\log_5 x = 3$

25. (a) $x^2 - x = \log_5 25$ (b) $3x + 5 = \log_2 64$

26. (a) $\log_3 x + \log_3(x - 2) = 1$ (b) $\log_{10}(x + 3) - \log_{10} x = 1$

Solving an Equation **In Exercises 27–36, solve the equation accurate to three decimal places.**

27. $3^{2x} = 75$ **28.** $6^{-2x} = 74$

29. $2^{3-z} = 625$ **30.** $3(5^{x-1}) = 86$

31. $\left(1 + \dfrac{0.09}{12}\right)^{12t} = 3$ **32.** $\left(1 + \dfrac{0.10}{365}\right)^{365t} = 2$

33. $\log_2(x - 1) = 5$ **34.** $\log_{10}(t - 3) = 2.6$

35. $\log_7 x^3 = 1.9$ **36.** $\log_5 \sqrt{x - 4} = 3.2$

Inverse Functions **In Exercises 37 and 38, illustrate that the functions are inverse functions of each other by sketching their graphs on the same set of coordinate axes.**

37. $f(x) = 4^x$, $g(x) = \log_4 x$ **38.** $f(x) = 3^x$, $g(x) = \log_3 x$

Finding a Derivative **In Exercises 39–60, find the derivative of the function.**

39. $f(x) = 4^x$ **40.** $f(x) = 3^{4x}$

41. $y = 5^{-4x}$ **42.** $y = 6^{3x-4}$

43. $f(x) = x\,9^x$ **44.** $y = -7x(8^{-2x})$

45. $f(t) = \dfrac{-2t^2}{8^t}$ **46.** $f(t) = \dfrac{3^{2t}}{t}$

47. $h(\theta) = 2^{-\theta}\cos \pi\theta$ **48.** $g(\alpha) = 5^{-\alpha/2}\sin 2\alpha$

49. $y = \log_4 (6x + 1)$ **50.** $y = \log_3(x^2 - 3x)$

51. $h(t) = \log_5(4 - t)^2$ **52.** $g(t) = \log_2(t^2 + 7)^3$

53. $y = \log_5 \sqrt{x^2 - 1}$ **54.** $f(x) = \log_2 \sqrt[3]{2x + 1}$

55. $f(x) = \log_2 \dfrac{x^2}{x - 1}$ **56.** $y = \log_{10} \dfrac{x^2 - 1}{x}$

57. $h(x) = \log_3 \dfrac{x\sqrt{x - 1}}{2}$ **58.** $g(x) = \log_5 \dfrac{4}{x^2\sqrt{1 - x}}$

59. $g(t) = \dfrac{10 \log_4 t}{t}$

60. $f(t) = t^{3/2} \log_2 \sqrt{t + 1}$

Finding an Equation of a Tangent Line **In Exercises 61–64, find an equation of the tangent line to the graph of the function at the given point.**

61. $y = 2^{-x}$, $(-1, 2)$ **62.** $y = 5^{x-2}$, $(2, 1)$

63. $y = \log_3 x$, $(27, 3)$ **64.** $y = \log_{10} 2x$, $(5, 1)$

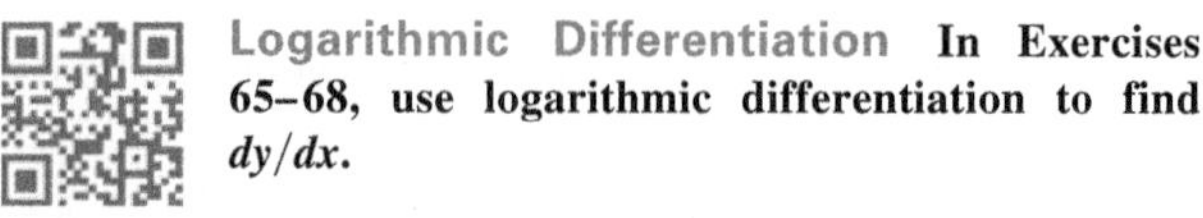

Logarithmic Differentiation **In Exercises 65–68, use logarithmic differentiation to find dy/dx.**

65. $y = x^{2/x}$ **66.** $y = x^{x-1}$

67. $y = (x - 2)^{x+1}$ **68.** $y = (1 + x)^{1/x}$

Finding an Indefinite Integral In Exercises 69–76, find the indefinite integral.

69. $\int 3^x\,dx$

70. $\int 2^{-x}\,dx$

71. $\int (x^2 + 2^{-x})\,dx$

72. $\int (x^4 + 5^x)\,dx$

73. $\int x(5^{-x^2})\,dx$

74. $\int (4 - x)6^{(4-x)^2}\,dx$

75. $\int \frac{3^{2x}}{1 + 3^{2x}}\,dx$

76. $\int 2^{\sin x} \cos x\,dx$

Evaluating a Definite Integral In Exercises 77–80, evaluate the definite integral. Use a graphing utility to verify your result.

77. $\int_{-1}^{2} 2^x\,dx$

78. $\int_{-4}^{4} 3^{x/4}\,dx$

79. $\int_{0}^{1} (5^x - 3^x)\,dx$

80. $\int_{1}^{3} (4^{x+1} + 2^x)\,dx$

Area In Exercises 81 and 82, find the area of the region bounded by the graphs of the equations. Use a graphing utility to verify your result.

81. $y = \frac{\log_4 x}{x}$, $y = 0$, $x = 1$, $x = 5$

82. $y = 3^{\cos x} \sin x$, $y = 0$, $x = 0$, $x = \pi$

EXPLORING CONCEPTS

83. **Exponential Function** What happens to the rate of change of the exponential function $y = a^x$ as a becomes larger?

84. **Logarithmic Function** What happens to the rate of change of the logarithmic function $y = \log_a x$ as a becomes larger?

85. **Analyzing a Logarithmic Equation** Consider the function $f(x) = \log_{10} x$.

(a) What is the domain of f?

(b) Find f^{-1}.

(c) Let x be a real number between 1000 and 10,000. Determine the interval in which $f(x)$ will be found.

(d) Determine the interval in which x will be found if $f(x)$ is negative.

(e) When $f(x)$ is increased by one unit, x must have been increased by what factor?

(f) Find the ratio of x_1 to x_2 given that $f(x_1) = 3n$ and $f(x_2) = n$.

86. **Comparing Rates of Growth** Order the functions

$$f(x) = \log_2 x,\ g(x) = x^x,\ h(x) = x^2, \text{ and } k(x) = 2^x$$

from the one with the greatest rate of growth to the one with the least rate of growth for large values of x.

87. **Inflation** When the annual rate of inflation averages 5% over the next 10 years, the approximate cost C of goods or services during any year in that decade is

$$C(t) = P(1.05)^t$$

where t is the time in years and P is the present cost.

(a) The price of an oil change for your car is presently \$24.95. Estimate the price 10 years from now.

(b) Find the rates of change of C with respect to t when $t = 1$ and $t = 8$.

(c) Verify that the rate of change of C is proportional to C. What is the constant of proportionality?

88. **Depreciation** After t years, the value of a car purchased for \$25,000 is

$$V(t) = 25{,}000\left(\tfrac{3}{4}\right)^t.$$

(a) Use a graphing utility to graph the function and determine the value of the car 2 years after it was purchased.

(b) Find the rates of change of V with respect to t when $t = 1$ and $t = 4$.

(c) Use a graphing utility to graph $V'(t)$ and determine the horizontal asymptote of $V'(t)$. Interpret its meaning in the context of the problem.

Compound Interest In Exercises 89–92, complete the table by determining the balance A for P dollars invested at rate r for t years and compounded n times per year.

n	1	2	4	12	365	Continuous Compounding
A						

89. $P = \$1000$, $r = 3\frac{1}{2}\%$, $t = 10$ years

90. $P = \$2500$, $r = 6\%$, $t = 20$ years

91. $P = \$7500$, $r = 4.8\%$, $t = 30$ years

92. $P = \$4000$, $r = 4\%$, $t = 15$ years

Compound Interest In Exercises 93–96, complete the table by determining the amount of money P (present value) that should be invested at rate r to produce a balance of \$100,000 in t years.

t	1	10	20	30	40	50
P						

93. $r = 4\%$, Compounded continuously

94. $r = 0.6\%$, Compounded continuously

95. $r = 5\%$, Compounded monthly

96. $r = 2\%$, Compounded daily

97. Compound Interest Assume that you can earn 6% on an investment, compounded daily. Which of the following options would yield the greatest balance after 8 years?

(a) \$20,000 now (b) \$30,000 after 8 years

(c) \$8000 now and \$20,000 after 4 years

(d) \$9000 now, \$9000 after 4 years, and \$9000 after 8 years

98. Compound Interest Consider a deposit of \$100 placed in an account for 20 years at $r\%$ compounded continuously. Use a graphing utility to graph the exponential functions describing the growth of the investment over the 20 years for the following interest rates. Compare the ending balances.

(a) $r = 3\%$ (b) $r = 5\%$ (c) $r = 6\%$

99. Timber Yield The yield V (in millions of cubic feet per acre) for a stand of timber at age t is $V = 6.7e^{-48.1/t}$, where t is measured in years.

(a) Find the limiting volume of wood per acre as t approaches infinity.

(b) Find the rates at which the yield is changing when $t = 20$ and $t = 60$.

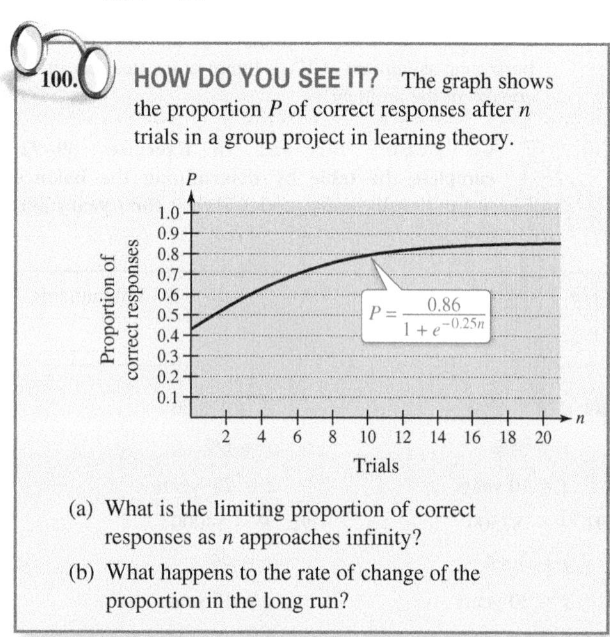

(a) What is the limiting proportion of correct responses as n approaches infinity?

(b) What happens to the rate of change of the proportion in the long run?

101. Population Growth A lake is stocked with 500 fish, and the population p is growing according to the logistic curve

$$p(t) = \frac{10,000}{1 + 19e^{-t/5}}$$

where t is measured in months.

 (a) Use a graphing utility to graph the function.

(b) Find the fish populations after 6 months, 12 months, 24 months, 36 months, and 48 months. What is the limiting size of the fish population?

(c) Find the rates at which the fish population is changing after 1 month and after 10 months.

(d) After how many months is the population increasing most rapidly?

 **102. Modeling Data**

The breaking strengths B (in tons) of steel cables of various diameters d (in inches) are shown in the table.

d	0.50	0.75	1.00	1.25	1.50	1.75
B	9.85	21.8	38.3	59.2	84.4	114.0

(a) Use the regression capabilities of a graphing utility to fit an exponential model to the data.

(b) Use a graphing utility to plot the data and graph the model.

(c) Find the rates of growth of the model when $d = 0.8$ and $d = 1.5$.

103. Comparing Models The total numbers y of AIDS cases by year of diagnosis in Canada for the years 2005 through 2014 are shown in the table, with $x = 5$ corresponding to 2005. *(Source: Public Health Agency of Canada)*

x	5	6	7	8	9
y	434	398	371	367	296

x	10	11	12	13	14
y	276	234	223	226	188

(a) Use the regression capabilities of a graphing utility to find the following models for the data.

$y_1 = ax + b$

$y_2 = a + b \ln x$

$y_3 = ab^x$

$y_4 = ax^b$

(b) Use a graphing utility to plot the data and graph each of the models. Which model do you think best fits the data?

(c) Find the rate of change of each of the models in part (a) for the year 2012. Which model is decreasing at the greatest rate in 2012?

104. An Approximation of e Complete the table to demonstrate that e can also be defined as

$$\lim_{x \to 0^+} (1 + x)^{1/x}.$$

x	1	10^{-1}	10^{-2}	10^{-4}	10^{-6}
$(1 + x)^{1/x}$					

Modeling Data In Exercises 105 and 106, find an exponential function that fits the experimental data collected over time t.

105.

t	0	1	2	3	4
y	1200.00	720.00	432.00	259.20	155.52

106.

t	0	1	2	3	4
y	600.00	630.00	661.50	694.58	729.30

Using Properties of Exponents In Exercises 107–110, find the exact value of the expression.

107. $5^{1/\ln 5}$ **108.** $6^{(\ln 10)/\ln 6}$

109. $9^{1/\ln 3}$ **110.** $32^{1/\ln 2}$

111. Comparing Functions

(a) Show that $(2^3)^2 \neq 2^{(3^2)}$.

(b) Are

$$f(x) = (x^x)^x \text{ and } g(x) = x^{(x^x)}$$

the same function? Why or why not?

(c) Find $f'(x)$ and $g'(x)$.

112. Finding an Inverse Function Let

$$f(x) = \frac{a^x - 1}{a^x + 1}$$

for $a > 0$, $a \neq 1$. Show that f has an inverse function. Then find f^{-1}.

113. Logistic Differential Equation Show that solving the logistic differential equation

$$\frac{dy}{dt} = \frac{8}{25}y\left(\frac{5}{4} - y\right), \quad y(0) = 1$$

results in the logistic growth function in Example 7.

$$\left[\textit{Hint: } \frac{1}{y\left(\frac{5}{4} - y\right)} = \frac{4}{5}\left(\frac{1}{y} + \frac{1}{\frac{5}{4} - y}\right)\right]$$

114. Using Properties of Exponents Given the exponential function $f(x) = a^x$, show that

(a) $f(u + v) = f(u) \cdot f(v)$.

(b) $f(2x) = [f(x)]^2$.

115. Tangent Lines

(a) Determine y' given $y^x = x^y$.

(b) Find the slope of the tangent line to the graph of $y^x = x^y$ at each of the following points.

(i) (c, c) (ii) $(2, 4)$ (iii) $(4, 2)$

(c) At what points on the graph of $y^x = x^y$ does the tangent line not exist?

PUTNAM EXAM CHALLENGE

116. Which is greater

$$\left(\sqrt{n}\right)^{\sqrt{n+1}} \quad \text{or} \quad \left(\sqrt{n+1}\right)^{\sqrt{n}}$$

where $n > 8$?

117. Show that if x is positive, then

$$\log_e\left(1 + \frac{1}{x}\right) > \frac{1}{1 + x}.$$

SECTION PROJECT

Using Graphing Utilities to Estimate Slope

Let $f(x) = \begin{cases} |x|^x, & x \neq 0 \\ 1, & x = 0. \end{cases}$

(a) Use a graphing utility to graph f in the viewing window $-3 \le x \le 3$, $-2 \le y \le 2$. What is the domain of f?

(b) Use the *zoom* and *trace* features of a graphing utility to estimate

$$\lim_{x \to 0} f(x).$$

(c) Write a short paragraph explaining why the function f is continuous for all real numbers.

(d) Visually estimate the slope of f at the point $(0, 1)$.

(e) Explain why the derivative of a function can be approximated by the formula

$$\frac{f(x + \Delta x) - f(x - \Delta x)}{2\Delta x}$$

for small values of Δx. Use this formula to approximate the slope of f at the point $(0, 1)$.

$$f'(0) \approx \frac{f(0 + \Delta x) - f(0 - \Delta x)}{2\Delta x} = \frac{f(\Delta x) - f(-\Delta x)}{2\Delta x}$$

What do you think the slope of the graph of f is at $(0, 1)$?

(f) Find a formula for the derivative of f and determine $f'(0)$. Write a short paragraph explaining how a graphing utility might lead you to approximate the slope of a graph incorrectly.

(g) Use your formula for the derivative of f to find the relative extrema of f. Verify your answer using a graphing utility.

FOR FURTHER INFORMATION For more information on using graphing utilities to estimate slope, see the article "Computer-Aided Delusions" by Richard L. Hall in *The College Mathematics Journal*. To view this article, go to *MathArticles.com*.

5.7 Inverse Trigonometric Functions: Differentiation

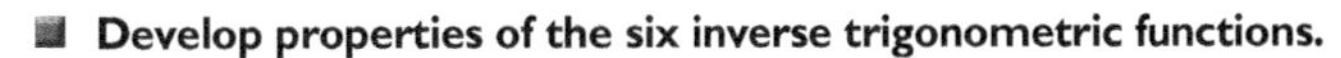

- **Develop properties of the six inverse trigonometric functions.**
- **Differentiate an inverse trigonometric function.**
- **Review the basic differentiation rules for elementary functions.**

Inverse Trigonometric Functions

This section begins with a rather surprising statement: *None of the six basic trigonometric functions has an inverse function.* This statement is true because all six trigonometric functions are periodic and therefore are not one-to-one. In this section, you will examine these six functions to see whether their domains can be redefined in such a way that they will have inverse functions on the *restricted domains.*

In Example 4 of Section 5.3, you saw that the sine function is increasing (and therefore is one-to-one) on the interval

$$\left[-\frac{\pi}{2}, \frac{\pi}{2}\right]$$

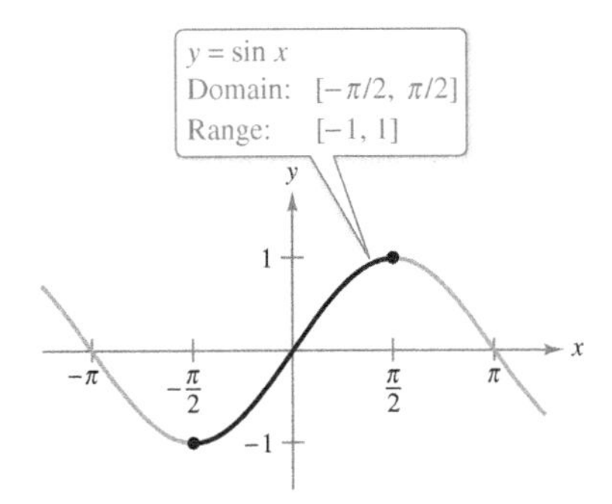

The sine function is one-to-one on $[-\pi/2, \pi/2]$.
Figure 5.25

as shown in Figure 5.25. On this interval, you can define the inverse of the *restricted* sine function as

$$y = \arcsin x \quad \text{if and only if} \quad \sin y = x$$

where $-1 \le x \le 1$ and $-\pi/2 \le \arcsin x \le \pi/2$.

Under suitable restrictions, each of the six trigonometric functions is one-to-one and so has an inverse function, as shown in the next definition. (Note that the term "iff" is used to represent the phrase "if and only if.")

REMARK The term "arcsin x" is read as "the arcsine of x" or sometimes "the angle whose sine is x." An alternative notation for the inverse sine function is "$\sin^{-1} x$."

Definitions of Inverse Trigonometric Functions

Function	Domain	Range
$y = \arcsin x$ iff $\sin y = x$	$-1 \le x \le 1$	$-\frac{\pi}{2} \le y \le \frac{\pi}{2}$
$y = \arccos x$ iff $\cos y = x$	$-1 \le x \le 1$	$0 \le y \le \pi$
$y = \arctan x$ iff $\tan y = x$	$-\infty < x < \infty$	$-\frac{\pi}{2} < y < \frac{\pi}{2}$
$y = \text{arccot } x$ iff $\cot y = x$	$-\infty < x < \infty$	$0 < y < \pi$
$y = \text{arcsec } x$ iff $\sec y = x$	$\lvert x \rvert \ge 1$	$0 \le y \le \pi,\ \ y \ne \frac{\pi}{2}$
$y = \text{arccsc } x$ iff $\csc y = x$	$\lvert x \rvert \ge 1$	$-\frac{\pi}{2} \le y \le \frac{\pi}{2},\ \ y \ne 0$

Exploration

The Inverse Secant Function In the definitions of the inverse trigonometric functions, the inverse secant function is defined by restricting the domain of the secant function to the intervals $[0, \pi/2) \cup (\pi/2, \pi]$. Most other texts and reference books agree with this, but some disagree. What other domains might make sense? Explain your reasoning graphically. Most calculators do not have a key for the inverse secant function. How can you use a calculator to evaluate the inverse secant function?

The graphs of the six inverse trigonometric functions are shown in Figure 5.26.

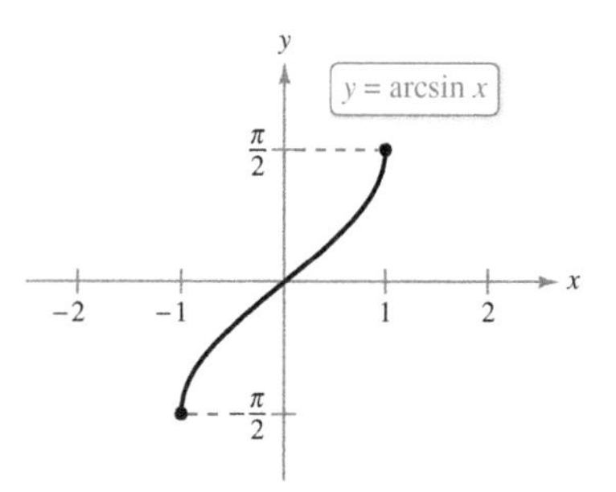

Domain: $[-1, 1]$
Range: $[-\pi/2, \pi/2]$

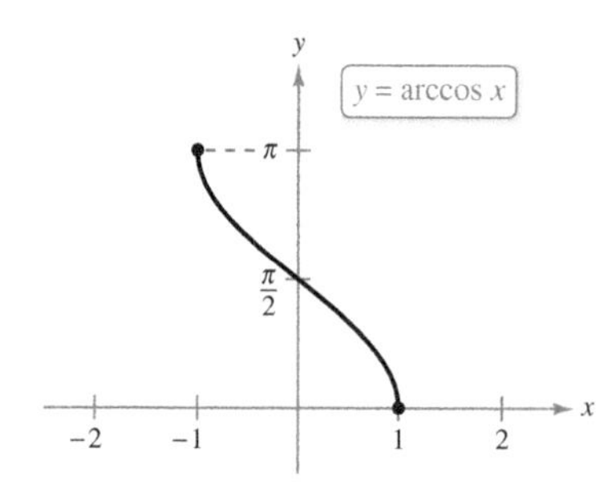

Domain: $[-1, 1]$
Range: $[0, \pi]$

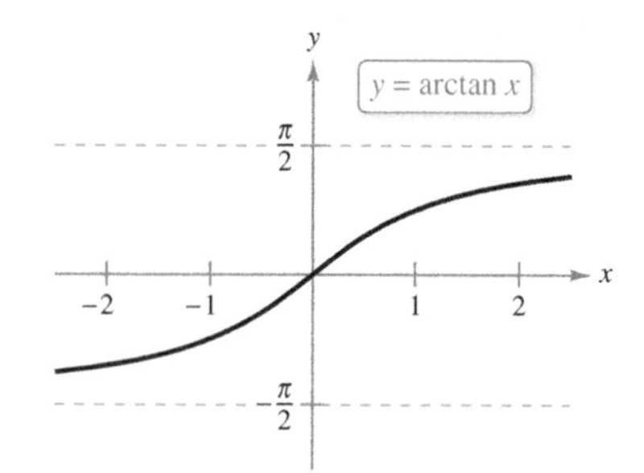

Domain: $(-\infty, \infty)$
Range: $(-\pi/2, \pi/2)$

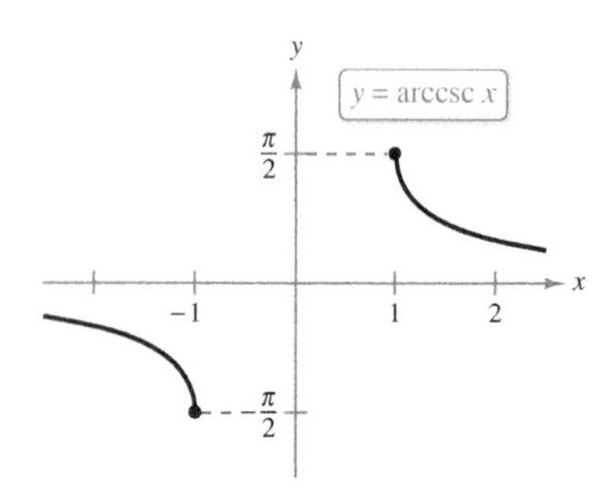

Domain: $(-\infty, -1] \cup [1, \infty)$
Range: $[-\pi/2, 0) \cup (0, \pi/2]$

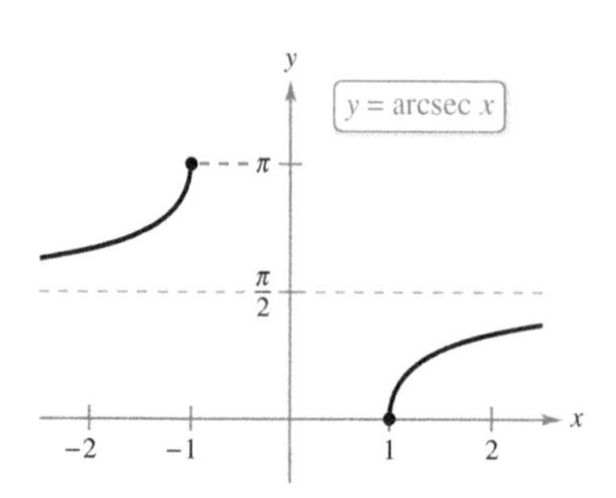

Domain: $(-\infty, -1] \cup [1, \infty)$
Range: $[0, \pi/2) \cup (\pi/2, \pi]$

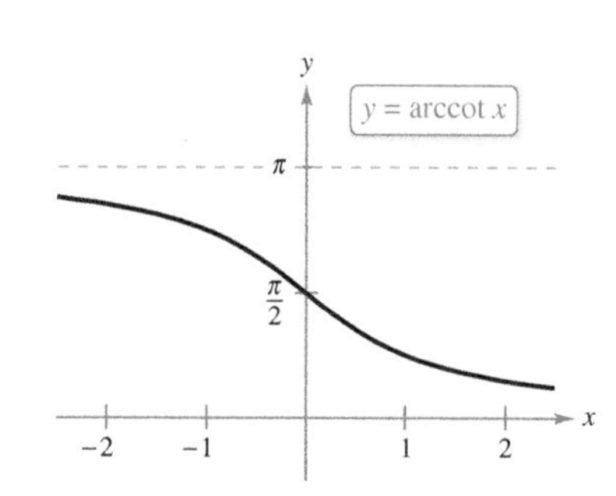

Domain: $(-\infty, \infty)$
Range: $(0, \pi)$

Figure 5.26

When evaluating inverse trigonometric functions, remember that they denote angles in *radian measure*.

EXAMPLE 1 Evaluating Inverse Trigonometric Functions

Evaluate each function.

a. $\arcsin\left(-\dfrac{1}{2}\right)$ **b.** $\arccos 0$ **c.** $\arctan \sqrt{3}$ **d.** $\arcsin(0.3)$

Solution

a. By definition, $y = \arcsin\left(-\frac{1}{2}\right)$ implies that $\sin y = -\frac{1}{2}$. In the interval $[-\pi/2, \pi/2]$, the correct value of y is $-\pi/6$.

$$\arcsin\left(-\frac{1}{2}\right) = -\frac{\pi}{6}$$

b. By definition, $y = \arccos 0$ implies that $\cos y = 0$. In the interval $[0, \pi]$, you have $y = \pi/2$.

$$\arccos 0 = \frac{\pi}{2}$$

c. By definition, $y = \arctan \sqrt{3}$ implies that $\tan y = \sqrt{3}$. In the interval $(-\pi/2, \pi/2)$, you have $y = \pi/3$.

$$\arctan \sqrt{3} = \frac{\pi}{3}$$

d. Using a calculator set in *radian* mode produces

$$\arcsin(0.3) \approx 0.305.$$

Inverse functions have the properties $f(f^{-1}(x)) = x$ and $f^{-1}(f(x)) = x$. When applying these properties to inverse trigonometric functions, remember that the trigonometric functions have inverse functions only in restricted domains. For x-values outside these domains, these two properties do not hold. For example, $\arcsin(\sin \pi)$ is equal to 0, not π.

Properties of Inverse Trigonometric Functions

If $-1 \le x \le 1$ and $-\pi/2 \le y \le \pi/2$, then

$$\sin(\arcsin x) = x \quad \text{and} \quad \arcsin(\sin y) = y.$$

If $-\pi/2 < y < \pi/2$, then

$$\tan(\arctan x) = x \quad \text{and} \quad \arctan(\tan y) = y.$$

If $|x| \ge 1$ and $0 \le y < \pi/2$ or $\pi/2 < y \le \pi$, then

$$\sec(\operatorname{arcsec} x) = x \quad \text{and} \quad \operatorname{arcsec}(\sec y) = y.$$

Similar properties hold for the other inverse trigonometric functions.

EXAMPLE 2 Solving an Equation

$$\arctan(2x - 3) = \frac{\pi}{4} \qquad \text{Original equation}$$

$$\tan[\arctan(2x - 3)] = \tan\frac{\pi}{4} \qquad \text{Take tangent of each side.}$$

$$2x - 3 = 1 \qquad \tan(\arctan x) = x$$

$$x = 2 \qquad \text{Solve for } x.$$

Some problems in calculus require that you evaluate expressions such as $\cos(\arcsin x)$, as shown in Example 3.

EXAMPLE 3 Using Right Triangles

a. Given $y = \arcsin x$, where $0 < y < \pi/2$, find $\cos y$.

b. Given $y = \operatorname{arcsec}(\sqrt{5}/2)$, find $\tan y$.

Solution

a. Because $y = \arcsin x$, you know that $\sin y = x$. This relationship between x and y can be represented by a right triangle, as shown in the figure at the right.

$$\cos y = \cos(\arcsin x) = \frac{\text{adj.}}{\text{hyp.}} = \sqrt{1 - x^2}$$

(This result is also valid for $-\pi/2 < y < 0$.)

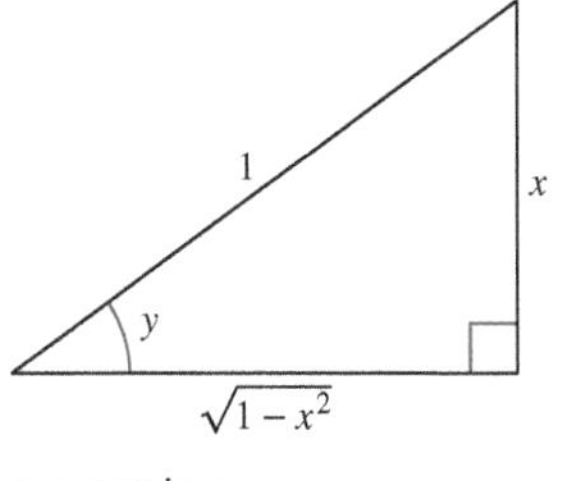

$y = \arcsin x$

b. Use the right triangle shown in the figure at the left.

$$\tan y = \tan\left(\operatorname{arcsec}\frac{\sqrt{5}}{2}\right) = \frac{\text{opp.}}{\text{adj.}} = \frac{1}{2}$$

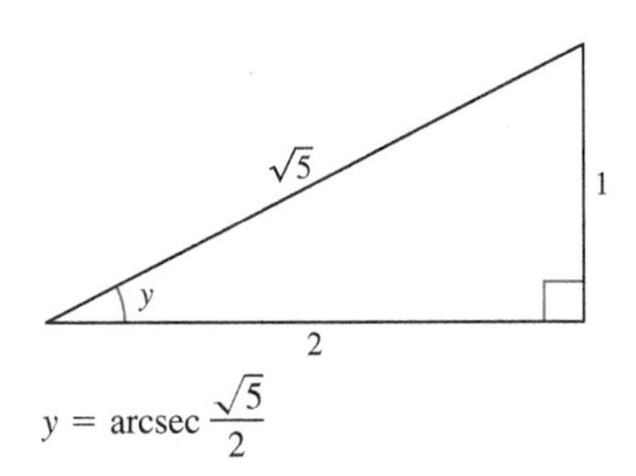

$y = \operatorname{arcsec}\dfrac{\sqrt{5}}{2}$

Derivatives of Inverse Trigonometric Functions

In Section 5.1, you saw that the derivative of the *transcendental* function $f(x) = \ln x$ is the *algebraic* function $f'(x) = 1/x$. You will now see that the derivatives of the inverse trigonometric functions also are algebraic (even though the inverse trigonometric functions are themselves transcendental).

The next theorem lists the derivatives of the six inverse trigonometric functions. Note that the derivatives of arccos u, arccot u, and arccsc u are the *negatives* of the derivatives of arcsin u, arctan u, and arcsec u, respectively.

REMARK There is no common agreement on the definition of arcsec x (or arccsc x) for negative values of x. When we defined the range of the arcsecant, we chose to preserve the reciprocal identity

$$\text{arcsec } x = \arccos \frac{1}{x}.$$

One consequence of this definition is that its graph has a positive slope at every x-value in its domain. (See Figure 5.26.) This accounts for the absolute value sign in the formula for the derivative of arcsec x.

THEOREM 5.18 Derivatives of Inverse Trigonometric Functions

Let u be a differentiable function of x.

$$\frac{d}{dx}[\arcsin u] = \frac{u'}{\sqrt{1-u^2}} \qquad \frac{d}{dx}[\arccos u] = \frac{-u'}{\sqrt{1-u^2}}$$

$$\frac{d}{dx}[\arctan u] = \frac{u'}{1+u^2} \qquad \frac{d}{dx}[\text{arccot } u] = \frac{-u'}{1+u^2}$$

$$\frac{d}{dx}[\text{arcsec } u] = \frac{u'}{|u|\sqrt{u^2-1}} \qquad \frac{d}{dx}[\text{arccsc } u] = \frac{-u'}{|u|\sqrt{u^2-1}}$$

Proofs for arcsin u and arccos u are given in Appendix A. [The proofs for the other rules are left as an exercise (see Exercise 94).]

TECHNOLOGY If your graphing utility does not have the arcsecant function, you can obtain its graph using

$$f(x) = \text{arcsec } x = \arccos \frac{1}{x}.$$

EXAMPLE 4 Differentiating Inverse Trigonometric Functions

a. $\dfrac{d}{dx}[\arcsin(2x)] = \dfrac{2}{\sqrt{1-(2x)^2}} = \dfrac{2}{\sqrt{1-4x^2}}$

b. $\dfrac{d}{dx}[\arctan(3x)] = \dfrac{3}{1+(3x)^2} = \dfrac{3}{1+9x^2}$

c. $\dfrac{d}{dx}[\arcsin \sqrt{x}] = \dfrac{(1/2)\,x^{-1/2}}{\sqrt{1-x}} = \dfrac{1}{2\sqrt{x}\sqrt{1-x}} = \dfrac{1}{2\sqrt{x-x^2}}$

d. $\dfrac{d}{dx}[\text{arcsec } e^{2x}] = \dfrac{2e^{2x}}{e^{2x}\sqrt{(e^{2x})^2-1}} = \dfrac{2}{\sqrt{e^{4x}-1}}$

The absolute value sign is not necessary because $e^{2x} > 0$.

EXAMPLE 5 A Derivative That Can Be Simplified

$$y = \arcsin x + x\sqrt{1-x^2}$$

$$\begin{aligned} y' &= \frac{1}{\sqrt{1-x^2}} + x\left(\frac{1}{2}\right)(-2x)(1-x^2)^{-1/2} + \sqrt{1-x^2} \\ &= \frac{1}{\sqrt{1-x^2}} - \frac{x^2}{\sqrt{1-x^2}} + \sqrt{1-x^2} \\ &= \sqrt{1-x^2} + \sqrt{1-x^2} \\ &= 2\sqrt{1-x^2} \end{aligned}$$

From Example 5, you can see one of the benefits of inverse trigonometric functions—they can be used to integrate common algebraic functions. For instance, from the result shown in the example, it follows that

$$\int \sqrt{1-x^2}\,dx = \frac{1}{2}\left(\arcsin x + x\sqrt{1-x^2}\right).$$

EXAMPLE 6 Analyzing an Inverse Trigonometric Graph

Analyze the graph of $y = (\arctan x)^2$.

Solution From the derivative

$$y' = 2(\arctan x)\left(\frac{1}{1 + x^2}\right) = \frac{2 \arctan x}{1 + x^2}$$

you can see that the only critical number is $x = 0$. By the First Derivative Test, this value corresponds to a relative minimum. From the second derivative

$$y'' = \frac{(1 + x^2)\left(\frac{2}{1 + x^2}\right) - (2 \arctan x)(2x)}{(1 + x^2)^2} = \frac{2(1 - 2x \arctan x)}{(1 + x^2)^2}$$

it follows that points of inflection occur when $2x \arctan x = 1$. Using Newton's Method, these points occur when $x \approx \pm 0.765$. Finally, because

$$\lim_{x \to \pm\infty} (\arctan x)^2 = \frac{\pi^2}{4}$$

it follows that the graph has a horizontal asymptote at $y = \pi^2/4$. The graph is shown in Figure 5.27.

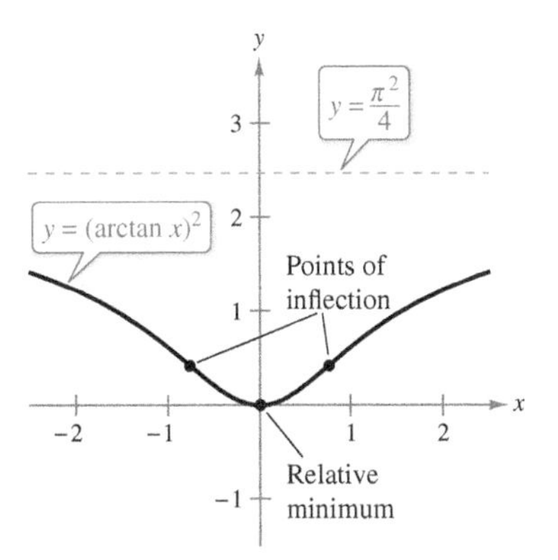

The graph of $y = (\arctan x)^2$ has a horizontal asymptote at $y = \pi^2/4$.
Figure 5.27

EXAMPLE 7 Maximizing an Angle

▷ *See LarsonCalculus.com for an interactive version of this type of example.*

A photographer is taking a picture of a painting hung in an art gallery. The height of the painting is 4 feet. The camera lens is 1 foot below the lower edge of the painting, as shown in the figure at the right. How far should the camera be from the painting to maximize the angle subtended by the camera lens?

Not drawn to scale

The camera should be 2.236 feet from the painting to maximize the angle β.

Solution In the figure, let β be the angle to be maximized.

$$\beta = \theta - \alpha = \operatorname{arccot}\frac{x}{5} - \operatorname{arccot} x$$

Differentiating produces

$$\frac{d\beta}{dx} = \frac{-1/5}{1 + (x^2/25)} - \frac{-1}{1 + x^2} = \frac{-5}{25 + x^2} + \frac{1}{1 + x^2} = \frac{4(5 - x^2)}{(25 + x^2)(1 + x^2)}.$$

Because $d\beta/dx = 0$ when $x = \sqrt{5}$, you can conclude from the First Derivative Test that this distance yields a maximum value of β. So, the distance is $x \approx 2.236$ feet and the angle is $\beta \approx 0.7297$ radian $\approx 41.81°$. ■

REMARK In Example 7, you could also let $\theta = \arctan(5/x)$ and $\alpha = \arctan(1/x)$. Although these expressions are more difficult to use than those in Example 7, you should obtain the same answer. Try verifying this.

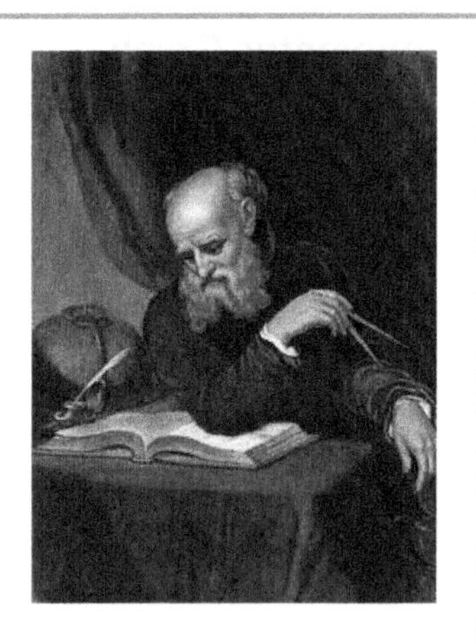

GALILEO GALILEI (1564–1642)

Galileo's approach to science departed from the accepted Aristotelian view that nature had describable *qualities*, such as "fluidity" and "potentiality." He chose to describe the physical world in terms of measurable *quantities*, such as time, distance, force, and mass.
See LarsonCalculus.com to read more of this biography.

Review of Basic Differentiation Rules

In the 1600s, Europe was ushered into the scientific age by such great thinkers as Descartes, Galileo, Huygens, Newton, and Kepler. These men believed that nature is governed by basic laws—laws that can, for the most part, be written in terms of mathematical equations. One of the most influential publications of this period—*Dialogue on the Great World Systems*, by Galileo Galilei—has become a classic description of modern scientific thought.

As mathematics has developed during the past few hundred years, a small number of elementary functions have proven sufficient for modeling most* phenomena in physics, chemistry, biology, engineering, economics, and a variety of other fields. An **elementary function** is a function from the following list or one that can be formed as the sum, product, quotient, or composition of functions in the list.

Algebraic Functions	**Transcendental Functions**
Polynomial functions	Logarithmic functions
Rational functions	Exponential functions
Functions involving radicals	Trigonometric functions
	Inverse trigonometric functions

With the differentiation rules introduced so far in the text, you can differentiate *any* elementary function. For convenience, these differentiation rules are summarized below.

BASIC DIFFERENTIATION RULES FOR ELEMENTARY FUNCTIONS

1. $\frac{d}{dx}[cu] = cu'$

2. $\frac{d}{dx}[u \pm v] = u' \pm v'$

3. $\frac{d}{dx}[uv] = uv' + vu'$

4. $\frac{d}{dx}\left[\frac{u}{v}\right] = \frac{vu' - uv'}{v^2}$

5. $\frac{d}{dx}[c] = 0$

6. $\frac{d}{dx}[u^n] = nu^{n-1}u'$

7. $\frac{d}{dx}[x] = 1$

8. $\frac{d}{dx}[|u|] = \frac{u}{|u|}(u'),\ u \neq 0$

9. $\frac{d}{dx}[\ln u] = \frac{u'}{u}$

10. $\frac{d}{dx}[e^u] = e^u u'$

11. $\frac{d}{dx}[\log_a u] = \frac{u'}{(\ln a)u}$

12. $\frac{d}{dx}[a^u] = (\ln a)a^u u'$

13. $\frac{d}{dx}[\sin u] = (\cos u)u'$

14. $\frac{d}{dx}[\cos u] = -(\sin u)u'$

15. $\frac{d}{dx}[\tan u] = (\sec^2 u)u'$

16. $\frac{d}{dx}[\cot u] = -(\csc^2 u)u'$

17. $\frac{d}{dx}[\sec u] = (\sec u \tan u)u'$

18. $\frac{d}{dx}[\csc u] = -(\csc u \cot u)u'$

19. $\frac{d}{dx}[\arcsin u] = \frac{u'}{\sqrt{1 - u^2}}$

20. $\frac{d}{dx}[\arccos u] = \frac{-u'}{\sqrt{1 - u^2}}$

21. $\frac{d}{dx}[\arctan u] = \frac{u'}{1 + u^2}$

22. $\frac{d}{dx}[\text{arccot } u] = \frac{-u'}{1 + u^2}$

23. $\frac{d}{dx}[\text{arcsec } u] = \frac{u'}{|u|\sqrt{u^2 - 1}}$

24. $\frac{d}{dx}[\text{arccsc } u] = \frac{-u'}{|u|\sqrt{u^2 - 1}}$

FOR FURTHER INFORMATION For more on the derivative of the arctangent function, see the article "Differentiating the Arctangent Directly" by Eric Key in *The College Mathematics Journal*. To view this article, go to *MathArticles.com*.

* Some important functions used in engineering and science (such as Bessel functions and gamma functions) are not elementary functions.

5.7 Exercises

See CalcChat.com for tutorial help and worked-out solutions to odd-numbered exercises.

CONCEPT CHECK

1. **Inverse Trigonometric Function** Describe the meaning of $\arccos x$ in your own words.
2. **Restricted Domain** What is a restricted domain? Why are restricted domains necessary to define inverse trigonometric functions?
3. **Inverse Trigonometric Functions** Which inverse trigonometric function has a range of $0 < y < \pi$?
4. **Finding a Derivative** What is the missing value?

$$\frac{d}{dx}[\text{arccsc } x^3] = \frac{\square}{|x^3|\sqrt{x^6 - 1}}$$

Finding Coordinates In Exercises 5 and 6, determine the missing coordinates of the points on the graph of the function.

5.
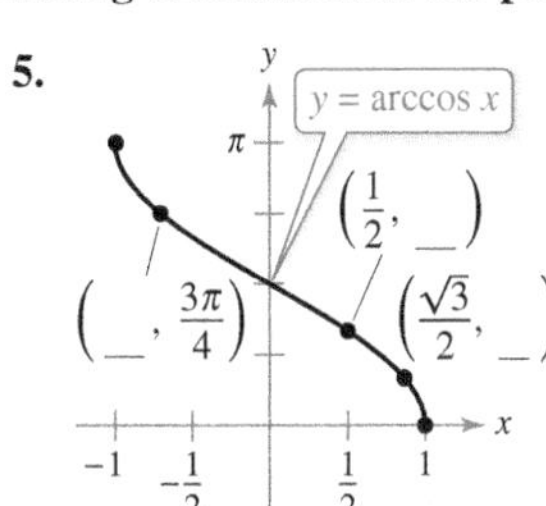

6.
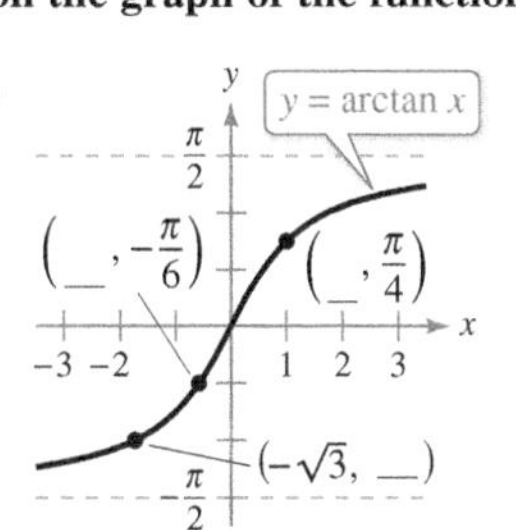

Evaluating Inverse Trigonometric Functions In Exercises 7–14, evaluate the expression without using a calculator.

7. $\arcsin \frac{1}{2}$
8. $\arcsin 0$
9. $\arccos \frac{1}{2}$
10. $\arccos(-1)$
11. $\arctan \frac{\sqrt{3}}{3}$
12. $\text{arccot}(-\sqrt{3})$
13. $\text{arccsc}(-\sqrt{2})$
14. $\text{arcsec } 2$

Approximating Inverse Trigonometric Functions In Exercises 15–18, use a calculator to approximate the value. Round your answer to two decimal places.

15. $\arccos(0.051)$
16. $\arcsin(-0.39)$
17. $\text{arcsec } 1.269$
18. $\text{arccsc}(-4.487)$

Using a Right Triangle In Exercises 19–24, use the figure to write the expression in algebraic form given $y = \arccos x$, where $0 < y < \pi/2$.

19. $\cos y$
20. $\sin y$
21. $\tan y$
22. $\cot y$
23. $\sec y$
24. $\csc y$

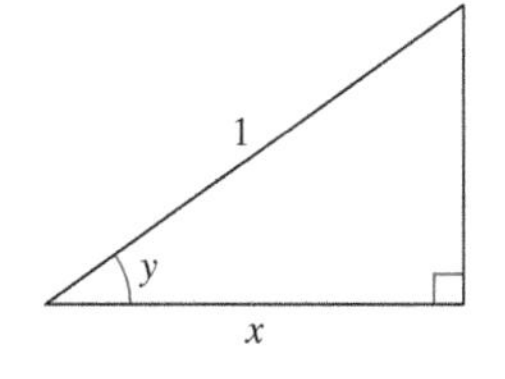

Evaluating an Expression In Exercises 25–28, evaluate each expression without using a calculator. (*Hint:* Sketch a right triangle, as demonstrated in Example 3.)

25. (a) $\sin\left(\arctan \frac{3}{4}\right)$
    (b) $\sec\left(\arcsin \frac{4}{5}\right)$
26. (a) $\tan\left(\arccos \frac{\sqrt{2}}{2}\right)$
    (b) $\cos\left(\arcsin \frac{5}{13}\right)$
27. (a) $\cot\left[\arcsin\left(-\frac{1}{2}\right)\right]$
    (b) $\csc\left[\arctan\left(-\frac{5}{12}\right)\right]$
28. (a) $\sec\left[\arctan\left(-\frac{3}{5}\right)\right]$
    (b) $\tan\left[\arcsin\left(-\frac{5}{6}\right)\right]$

Simplifying an Expression Using a Right Triangle In Exercises 29–36, write the expression in algebraic form. (*Hint:* Sketch a right triangle, as demonstrated in Example 3.)

29. $\cos(\arcsin 2x)$
30. $\sec(\arctan 6x)$
31. $\sin(\text{arcsec } x)$
32. $\cos(\text{arccot } x)$
33. $\tan\left(\text{arcsec } \frac{x}{3}\right)$
34. $\sec[\arcsin(x - 1)]$
35. $\csc\left(\arctan \frac{x}{\sqrt{2}}\right)$
36. $\cos\left(\arcsin \frac{x - h}{r}\right)$

Solving an Equation In Exercises 37–40, solve the equation for x.

37. $\arcsin(3x - \pi) = \frac{1}{2}$
38. $\arctan(2x - 5) = -1$
39. $\arcsin \sqrt{2x} = \arccos \sqrt{x}$
40. $\arccos x = \text{arcsec } x$

Finding a Derivative In Exercises 41–56, find the derivative of the function.

41. $f(x) = \arcsin(x - 1)$
42. $f(t) = \text{arccsc}(-t^2)$
43. $g(x) = 3 \arccos \frac{x}{2}$
44. $f(x) = \text{arcsec } 2x$
45. $f(x) = \arctan e^x$
46. $f(x) = \text{arccot } \sqrt{x}$
47. $g(x) = \frac{\arcsin 3x}{x}$
48. $h(x) = x^2 \arctan 5x$
49. $h(t) = \sin(\arccos t)$
50. $f(x) = \arcsin x + \arccos x$
51. $y = 2x \arccos x - 2\sqrt{1 - x^2}$
52. $y = x \arctan 2x - \frac{1}{4}\ln(1 + 4x^2)$
53. $y = \frac{1}{2}\left(\frac{1}{2}\ln\frac{x + 1}{x - 1} + \arctan x\right)$
54. $y = \frac{1}{2}\left[x\sqrt{4 - x^2} + 4 \arcsin \frac{x}{2}\right]$
55. $y = 8 \arcsin \frac{x}{4} - \frac{x\sqrt{16 - x^2}}{2}$
56. $y = \arctan x + \frac{x}{1 + x^2}$

Finding an Equation of a Tangent Line In Exercises 57–62, find an equation of the tangent line to the graph of the function at the given point.

57. $y = 2 \arcsin x, \quad \left(\frac{1}{2}, \frac{\pi}{3}\right)$

58. $y = -\frac{1}{4} \arccos x, \quad \left(-\frac{1}{2}, -\frac{\pi}{6}\right)$

59. $y = \arctan \frac{x}{2}, \quad \left(2, \frac{\pi}{4}\right)$

60. $y = \operatorname{arcsec} 4x, \quad \left(\frac{\sqrt{2}}{4}, \frac{\pi}{4}\right)$

61. $y = 4x \arccos(x - 1), \quad (1, 2\pi)$

62. $y = 3x \arcsin x, \quad \left(\frac{1}{2}, \frac{\pi}{4}\right)$

Finding Relative Extrema In Exercises 63–66, find any relative extrema of the function.

63. $f(x) = \operatorname{arcsec} x - x$

64. $f(x) = \arcsin x - 2x$

65. $f(x) = \arctan x - \arctan(x - 4)$

66. $h(x) = \arcsin x - 2 \arctan x$

Analyzing an Inverse Trigonometric Graph In Exercises 67–70, analyze and sketch a graph of the function. Identify any relative extrema, points of inflection, and asymptotes. Use a graphing utility to verify your results.

67. $f(x) = \arcsin(x - 1)$

68. $f(x) = \arctan x + \frac{\pi}{2}$

69. $f(x) = \operatorname{arcsec} 2x$

70. $f(x) = \arccos \frac{x}{4}$

Implicit Differentiation In Exercises 71–74, use implicit differentiation to find an equation of the tangent line to the graph of the equation at the given point.

71. $x^2 + x \arctan y = y - 1, \quad \left(-\frac{\pi}{4}, 1\right)$

72. $\arctan(xy) = \arcsin(x + y), \quad (0, 0)$

73. $\arcsin x + \arcsin y = \frac{\pi}{2}, \quad \left(\frac{\sqrt{2}}{2}, \frac{\sqrt{2}}{2}\right)$

74. $\arctan(x + y) = y^2 + \frac{\pi}{4}, \quad (1, 0)$

75. **Finding Values**

(a) Use a graphing utility to evaluate arcsin(arcsin 0.5) and arcsin(arcsin 1).

(b) Let

$f(x) = \arcsin(\arcsin x)$.

Find the values of x in the interval $-1 \le x \le 1$ such that $f(x)$ is a real number.

76. **HOW DO YOU SEE IT?** The graph of $g(x) = \cos x$ is shown below. Explain whether the points

$$\left(-\frac{1}{2}, \frac{2\pi}{3}\right), \quad \left(0, \frac{\pi}{2}\right), \quad \text{and} \quad \left(\frac{1}{2}, -\frac{\pi}{3}\right)$$

lie on the graph of $y = \arccos x$.

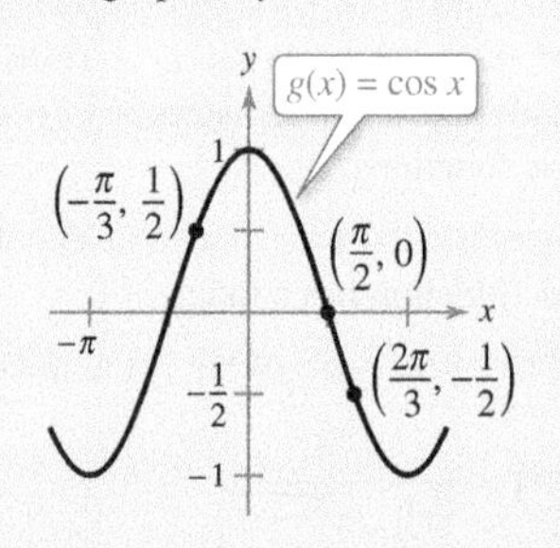

EXPLORING CONCEPTS

77. **Inverse Trigonometric Functions** Determine whether

$$\frac{\arcsin x}{\arccos x} = \arctan x.$$

78. **Inverse Trigonometric Functions** Determine whether each inverse trigonometric function can be defined as shown. Explain.

(a) $y = \operatorname{arcsec} x$, Domain: $x > 1$, Range: $-\frac{\pi}{2} < y < \frac{\pi}{2}$

(b) $y = \operatorname{arccsc} x$, Domain: $x > 1$, Range: $0 < y < \pi$

79. **Inverse Trigonometric Functions** Explain why $\sin 2\pi = 0$ does not imply that $\arcsin 0 = 2\pi$.

80. **Inverse Trigonometric Functions** Explain why $\tan \pi = 0$ does not imply that $\arctan 0 = \pi$.

Verifying Identities In Exercises 81 and 82, verify each identity.

81. (a) $\operatorname{arccsc} x = \arcsin \frac{1}{x}, \quad |x| \ge 1$

(b) $\arctan x + \arctan \frac{1}{x} = \frac{\pi}{2}, \quad x > 0$

82. (a) $\arcsin(-x) = -\arcsin x, \quad |x| \le 1$

(b) $\arccos(-x) = \pi - \arccos x, \quad |x| \le 1$

True or False? In Exercises 83–86, determine whether the statement is true or false. If it is false, explain why or give an example that shows it is false.

83. The slope of the graph of the inverse tangent function is positive for all x.

84. The range of $y = \arcsin x$ is $[0, \pi]$.

85. $\frac{d}{dx}[\arctan(\tan x)] = 1$ for all x in the domain.

86. $\arcsin^2 x + \arccos^2 x = 1$

87. Angular Rate of Change An airplane flies at an altitude of 5 miles toward a point directly over an observer. Consider θ and x as shown in the figure.

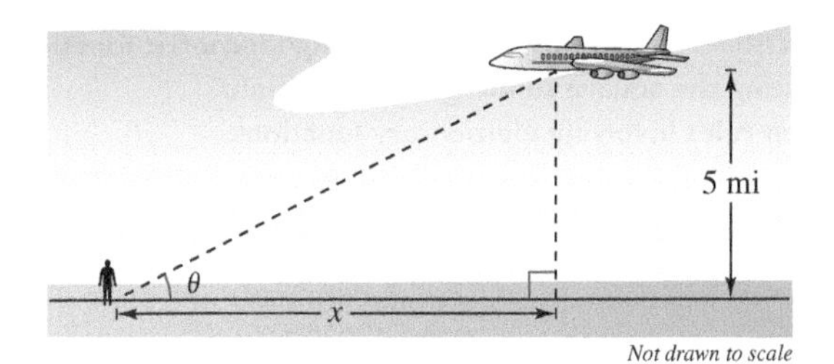

(a) Write θ as a function of x.

(b) The speed of the plane is 400 miles per hour. Find $d\theta/dt$ when $x = 10$ miles and $x = 3$ miles.

88. Writing Repeat Exercise 87 for an altitude of 3 miles and describe how the altitude affects the rate of change of θ.

89. Angular Rate of Change In a free-fall experiment, an object is dropped from a height of 256 feet. A camera on the ground 500 feet from the point of impact records the fall of the object (see figure).

(a) Find the position function that yields the height of the object at time t, assuming the object is released at time $t = 0$. At what time will the object reach ground level?

(b) Find the rates of change of the angle of elevation of the camera when $t = 1$ and $t = 2$.

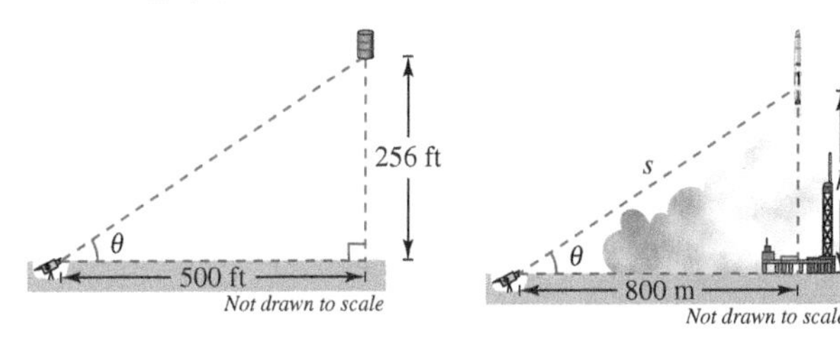

Figure for 89 Figure for 90

90. Angular Rate of Change A television camera at ground level is filming the lift-off of a rocket at a point 800 meters from the launch pad. Let θ be the angle of elevation of the rocket and let s be the distance between the camera and the rocket (see figure). Write θ as a function of s for the period of time when the rocket is moving vertically. Differentiate the result to find $d\theta/dt$ in terms of s and ds/dt.

91. Maximizing an Angle A billboard 85 feet wide is perpendicular to a straight road and is 40 feet from the road (see figure). Find the point on the road at which the angle θ subtended by the billboard is a maximum.

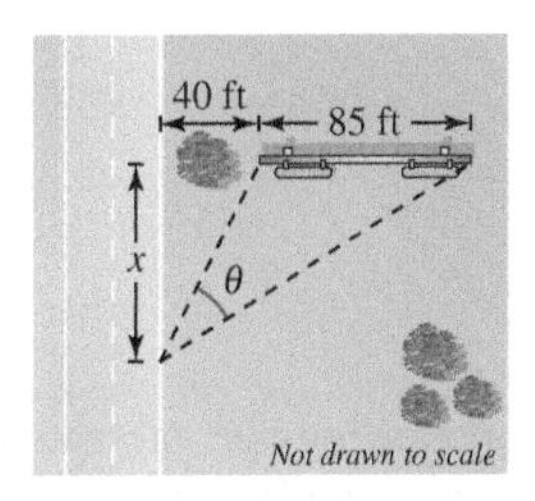

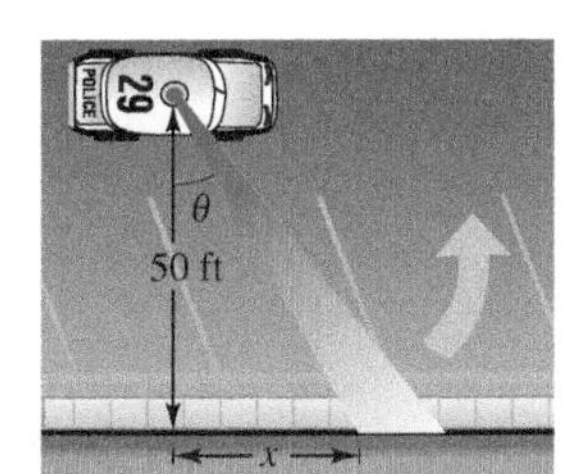

Figure for 91 Figure for 92

92. Angular Speed A patrol car is parked 50 feet from a long warehouse (see figure). The revolving light on top of the car turns at a rate of 30 revolutions per minute. Write θ as a function of x. How fast is the light beam moving along the wall when the beam makes an angle of $\theta = 45°$ with the line perpendicular from the light to the wall?

93. Proof

(a) Prove that $\arctan x + \arctan y = \arctan \dfrac{x + y}{1 - xy}$, $xy \neq 1$.

(b) Use the formula in part (a) to show that

$$\arctan \frac{1}{2} + \arctan \frac{1}{3} = \frac{\pi}{4}.$$

94. Proof Prove each differentiation formula.

(a) $\dfrac{d}{dx}[\arctan u] = \dfrac{u'}{1 + u^2}$

(b) $\dfrac{d}{dx}[\operatorname{arccot} u] = \dfrac{-u'}{1 + u^2}$

(c) $\dfrac{d}{dx}[\operatorname{arcsec} u] = \dfrac{u'}{|u|\sqrt{u^2 - 1}}$

(d) $\dfrac{d}{dx}[\operatorname{arccsc} u] = \dfrac{-u'}{|u|\sqrt{u^2 - 1}}$

95. Describing a Graph Use a graphing utility to graph the function $f(x) = \arccos x + \arcsin x$ on the interval $[-1, 1]$.

(a) Describe the graph of f.

(b) Verify the result of part (a) analytically.

96. Think About It Use a graphing utility to graph $f(x) = \sin x$ and $g(x) = \arcsin(\sin x)$.

(a) Explain why the graph of g is not the line $y = x$.

(b) Determine the extrema of g.

97. Maximizing an Angle In the figure, find the value of c in the interval $[0, 4]$ on the x-axis that maximizes angle θ.

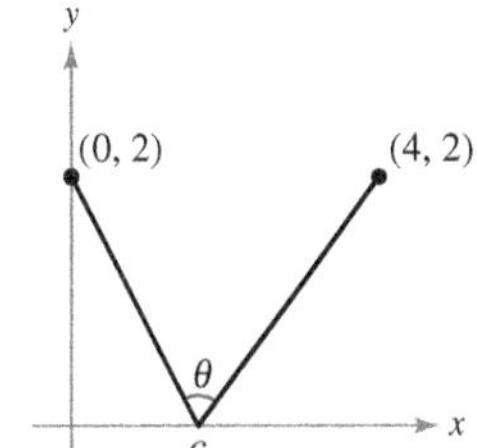

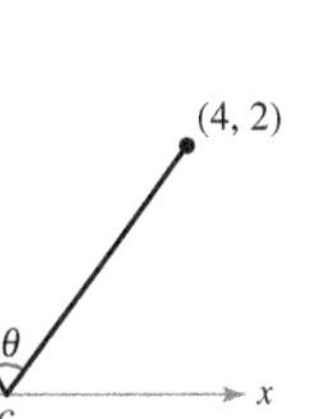

Figure for 97 Figure for 98

98. Finding a Distance In the figure, find PR such that $0 \leq PR \leq 3$ and $m\angle\theta$ is a maximum.

99. Proof Prove that $\arcsin x = \arctan\left(\dfrac{x}{\sqrt{1 - x^2}}\right)$, $|x| < 1$.

100. Inverse Secant Function Some calculus textbooks define the inverse secant function using the range $[0, \pi/2) \cup [\pi, 3\pi/2)$.

(a) Sketch the graph $y = \operatorname{arcsec} x$ using this range.

(b) Show that $y' = \dfrac{1}{x\sqrt{x^2 - 1}}$.

5.8 Inverse Trigonometric Functions: Integration

- **Integrate functions whose antiderivatives involve inverse trigonometric functions.**
- **Use the method of completing the square to integrate a function.**
- **Review the basic integration rules involving elementary functions.**

Integrals Involving Inverse Trigonometric Functions

The derivatives of the six inverse trigonometric functions fall into three pairs. In each pair, the derivative of one function is the negative of the other. For example,

$$\frac{d}{dx}[\arcsin x] = \frac{1}{\sqrt{1 - x^2}}$$

and

$$\frac{d}{dx}[\arccos x] = -\frac{1}{\sqrt{1 - x^2}}.$$

When listing the *antiderivative* that corresponds to each of the inverse trigonometric functions, you need to use only one member from each pair. It is conventional to use $\arcsin x$ as the antiderivative of $1/\sqrt{1 - x^2}$, rather than $-\arccos x$. The next theorem gives one antiderivative formula for each of the three pairs. The proofs of these integration rules are left to you (see Exercises 73–75).

FOR FURTHER INFORMATION For a detailed proof of rule 2 of Theorem 5.19, see the article "A Direct Proof of the Integral Formula for Arctangent" by Arnold J. Insel in *The College Mathematics Journal*. To view this article, go to *MathArticles.com*.

THEOREM 5.19 Integrals Involving Inverse Trigonometric Functions

Let u be a differentiable function of x, and let $a > 0$.

1. $\displaystyle\int \frac{du}{\sqrt{a^2 - u^2}} = \arcsin \frac{u}{a} + C$

2. $\displaystyle\int \frac{du}{a^2 + u^2} = \frac{1}{a} \arctan \frac{u}{a} + C$

3. $\displaystyle\int \frac{du}{u\sqrt{u^2 - a^2}} = \frac{1}{a} \operatorname{arcsec} \frac{|u|}{a} + C$

EXAMPLE 1 Integration with Inverse Trigonometric Functions

a. $\displaystyle\int \frac{dx}{\sqrt{4 - x^2}} = \arcsin \frac{x}{2} + C$ $\qquad u = x, a = 2$

b. $\displaystyle\int \frac{dx}{2 + 9x^2} = \frac{1}{3}\int \frac{3\,dx}{(\sqrt{2})^2 + (3x)^2}$ $\qquad u = 3x, a = \sqrt{2}$

$$= \frac{1}{3\sqrt{2}} \arctan \frac{3x}{\sqrt{2}} + C$$

c. $\displaystyle\int \frac{dx}{x\sqrt{4x^2 - 9}} = \int \frac{2\,dx}{2x\sqrt{(2x)^2 - 3^2}}$ $\qquad u = 2x, a = 3$

$$= \frac{1}{3} \operatorname{arcsec} \frac{|2x|}{3} + C$$

The integrals in Example 1 are fairly straightforward applications of integration formulas. Unfortunately, this is not typical. The integration formulas for inverse trigonometric functions can be disguised in many ways.

EXAMPLE 2 Integration by Substitution

Find $\int \frac{dx}{\sqrt{e^{2x} - 1}}$.

Solution As it stands, this integral does not fit any of the three inverse trigonometric formulas. Using the substitution $u = e^x$, however, produces

$$u = e^x \implies du = e^x\,dx \implies dx = \frac{du}{e^x} = \frac{du}{u}.$$

With this substitution, you can integrate as shown.

$$\int \frac{dx}{\sqrt{e^{2x} - 1}} = \int \frac{dx}{\sqrt{(e^x)^2 - 1}} \qquad \text{Write } e^{2x} \text{ as } (e^x)^2.$$

$$= \int \frac{du/u}{\sqrt{u^2 - 1}} \qquad \text{Substitute.}$$

$$= \int \frac{du}{u\sqrt{u^2 - 1}} \qquad \text{Rewrite to fit Arcsecant Rule.}$$

$$= \text{arcsec}\,\frac{|u|}{1} + C \qquad \text{Apply Arcsecant Rule.}$$

$$= \text{arcsec}\,e^x + C \qquad \text{Back-substitute.}$$

TECHNOLOGY PITFALL A symbolic integration utility can be useful for integrating functions such as the one in Example 2. In some cases, however, the utility may fail to find an antiderivative for two reasons. First, some elementary functions do not have antiderivatives that are elementary functions. Second, every utility has limitations—you might have entered a function that the utility was not programmed to handle. You should also remember that antiderivatives involving trigonometric functions or logarithmic functions can be written in many different forms. For instance, one utility found the integral in Example 2 to be

$$\int \frac{dx}{\sqrt{e^{2x} - 1}} = \arctan\sqrt{e^{2x} - 1} + C.$$

Try showing that this antiderivative is equivalent to the one found in Example 2.

EXAMPLE 3 Rewriting as the Sum of Two Quotients

Find $\int \frac{x + 2}{\sqrt{4 - x^2}}\,dx$.

Solution This integral does not appear to fit any of the basic integration formulas. By splitting the integrand into two parts, however, you can see that the first part can be found with the Power Rule and the second part yields an inverse sine function.

$$\int \frac{x + 2}{\sqrt{4 - x^2}}\,dx = \int \frac{x}{\sqrt{4 - x^2}}\,dx + \int \frac{2}{\sqrt{4 - x^2}}\,dx$$

$$= -\frac{1}{2}\int (4 - x^2)^{-1/2}(-2x)\,dx + 2\int \frac{1}{\sqrt{4 - x^2}}\,dx$$

$$= -\frac{1}{2}\left[\frac{(4 - x^2)^{1/2}}{1/2}\right] + 2\arcsin\frac{x}{2} + C$$

$$= -\sqrt{4 - x^2} + 2\arcsin\frac{x}{2} + C$$

Completing the Square

Completing the square helps when quadratic functions are involved in the integrand. For example, the quadratic $x^2 + bx + c$ can be written as the difference of two squares by adding and subtracting $(b/2)^2$.

$$x^2 + bx + c = x^2 + bx + \left(\frac{b}{2}\right)^2 - \left(\frac{b}{2}\right)^2 + c = \left(x + \frac{b}{2}\right)^2 - \left(\frac{b}{2}\right)^2 + c$$

EXAMPLE 4 Completing the Square

See LarsonCalculus.com for an interactive version of this type of example.

Find $\displaystyle\int \frac{dx}{x^2 - 4x + 7}$.

Solution You can write the denominator as the sum of two squares, as shown.

$$x^2 - 4x + 7 = (x^2 - 4x + 4) - 4 + 7 = (x - 2)^2 + 3 = u^2 + a^2$$

Now, in this completed square form, let $u = x - 2$ and $a = \sqrt{3}$.

$$\int \frac{dx}{x^2 - 4x + 7} = \int \frac{dx}{(x - 2)^2 + 3} = \frac{1}{\sqrt{3}} \arctan \frac{x - 2}{\sqrt{3}} + C$$

When the leading coefficient is not 1, it helps to factor before completing the square. For instance, you can complete the square of $2x^2 - 8x + 10$ by factoring first.

$$\begin{aligned} 2x^2 - 8x + 10 &= 2(x^2 - 4x + 5) \\ &= 2(x^2 - 4x + 4 - 4 + 5) \\ &= 2[(x - 2)^2 + 1] \end{aligned}$$

To complete the square when the coefficient of x^2 is negative, use the same factoring process shown above. For instance, you can complete the square for $3x - x^2$ as shown.

$$3x - x^2 = -(x^2 - 3x) = -\left[x^2 - 3x + \left(\tfrac{3}{2}\right)^2 - \left(\tfrac{3}{2}\right)^2\right] = \left(\tfrac{3}{2}\right)^2 - \left(x - \tfrac{3}{2}\right)^2$$

EXAMPLE 5 Completing the Square

Find the area of the region bounded by the graph of

$$f(x) = \frac{1}{\sqrt{3x - x^2}}$$

the x-axis, and the lines $x = \frac{3}{2}$ and $x = \frac{9}{4}$.

Solution In Figure 5.28, you can see that the area is

$$\begin{aligned} \text{Area} &= \int_{3/2}^{9/4} \frac{1}{\sqrt{3x - x^2}}\,dx \\ &= \int_{3/2}^{9/4} \frac{dx}{\sqrt{(3/2)^2 - [x - (3/2)]^2}} \qquad \text{Use completed square form derived above.} \\ &= \arcsin \frac{x - (3/2)}{3/2} \Bigg]_{3/2}^{9/4} \\ &= \arcsin \frac{1}{2} - \arcsin 0 \\ &= \frac{\pi}{6} \\ &\approx 0.524. \end{aligned}$$

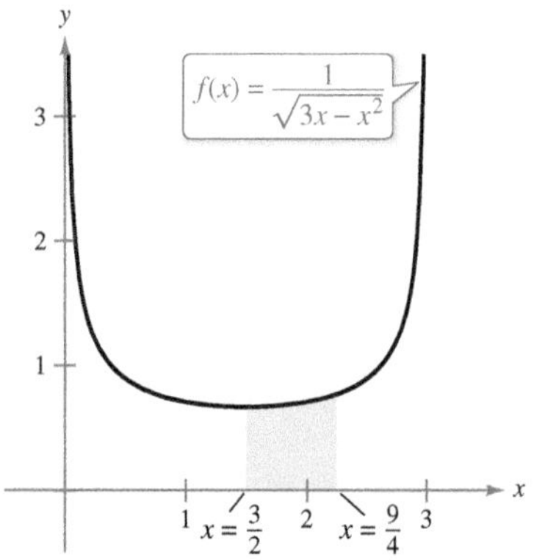

The area of the region bounded by the graph of f, the x-axis, $x = \frac{3}{2}$, and $x = \frac{9}{4}$ is $\pi/6$.

Figure 5.28

Review of Basic Integration Rules

You have now completed the introduction of the **basic integration rules.** To be efficient at applying these rules, you should have practiced enough so that each rule is committed to memory.

BASIC INTEGRATION RULES ($a > 0$)

1. $\int kf(u)\,du = k\int f(u)\,du$

2. $\int [f(u) \pm g(u)]\,du = \int f(u)\,du \pm \int g(u)\,du$

3. $\int du = u + C$

4. $\int u^n\,du = \dfrac{u^{n+1}}{n+1} + C, \quad n \neq -1$

5. $\int \dfrac{du}{u} = \ln|u| + C$

6. $\int e^u\,du = e^u + C$

7. $\int a^u\,du = \left(\dfrac{1}{\ln a}\right)a^u + C$

8. $\int \sin u\,du = -\cos u + C$

9. $\int \cos u\,du = \sin u + C$

10. $\int \tan u\,du = -\ln|\cos u| + C$

11. $\int \cot u\,du = \ln|\sin u| + C$

12. $\int \sec u\,du = \ln|\sec u + \tan u| + C$

13. $\int \csc u\,du = -\ln|\csc u + \cot u| + C$

14. $\int \sec^2 u\,du = \tan u + C$

15. $\int \csc^2 u\,du = -\cot u + C$

16. $\int \sec u \tan u\,du = \sec u + C$

17. $\int \csc u \cot u\,du = -\csc u + C$

18. $\int \dfrac{du}{\sqrt{a^2 - u^2}} = \arcsin\dfrac{u}{a} + C$

19. $\int \dfrac{du}{a^2 + u^2} = \dfrac{1}{a}\arctan\dfrac{u}{a} + C$

20. $\int \dfrac{du}{u\sqrt{u^2 - a^2}} = \dfrac{1}{a}\operatorname{arcsec}\dfrac{|u|}{a} + C$

You can learn a lot about the nature of integration by comparing this list with the summary of differentiation rules given in the preceding section. For differentiation, you now have rules that allow you to differentiate *any* elementary function. For integration, this is far from true.

The integration rules listed above are primarily those that were happened on during the development of differentiation rules. So far, you have not learned any rules or techniques for finding the antiderivative of a general product or quotient, the natural logarithmic function, or the inverse trigonometric functions. More important, you cannot apply any of the rules in this list unless you can create the proper du corresponding to the u in the formula. The point is that you need to work more on integration techniques, which you will do in Chapter 8. The next two examples should give you a better feeling for the integration problems that you *can* and *cannot* solve with the techniques and rules you now know.

EXAMPLE 6 Comparing Integration Problems

Find as many of the following integrals as you can using the formulas and techniques you have studied so far in the text.

a. $\displaystyle\int \frac{dx}{x\sqrt{x^2 - 1}}$

b. $\displaystyle\int \frac{x\,dx}{\sqrt{x^2 - 1}}$

c. $\displaystyle\int \frac{dx}{\sqrt{x^2 - 1}}$

Solution

a. You *can* find this integral (it fits the Arcsecant Rule).

$$\int \frac{dx}{x\sqrt{x^2 - 1}} = \operatorname{arcsec}|x| + C$$

b. You *can* find this integral (it fits the Power Rule).

$$\int \frac{x\,dx}{\sqrt{x^2 - 1}} = \frac{1}{2}\int (x^2 - 1)^{-1/2}(2x)\,dx$$
$$= \frac{1}{2}\left[\frac{(x^2 - 1)^{1/2}}{1/2}\right] + C$$
$$= \sqrt{x^2 - 1} + C$$

c. You *cannot* find this integral using the techniques you have studied so far. (You should scan the list of basic integration rules to verify this conclusion.)

EXAMPLE 7 Comparing Integration Problems

Find as many of the following integrals as you can using the formulas and techniques you have studied so far in the text.

a. $\displaystyle\int \frac{dx}{x \ln x}$

b. $\displaystyle\int \frac{\ln x\,dx}{x}$

c. $\displaystyle\int \ln x\,dx$

Solution

a. You *can* find this integral (it fits the Log Rule).

$$\int \frac{dx}{x \ln x} = \int \frac{1/x}{\ln x}\,dx$$
$$= \ln|\ln x| + C$$

b. You *can* find this integral (it fits the Power Rule).

$$\int \frac{\ln x\,dx}{x} = \int \left(\frac{1}{x}\right)(\ln x)^1\,dx$$
$$= \frac{(\ln x)^2}{2} + C$$

c. You *cannot* find this integral using the techniques you have studied so far. ■

REMARK Note in Examples 6 and 7 that the *simplest* functions are the ones that you cannot yet integrate.

5.8 Exercises

CONCEPT CHECK

1. **Integration Rules** Decide whether you can find each integral using the formulas and techniques you have studied so far. Explain.

(a) $\int \frac{2\,dx}{\sqrt{x^2+4}}$ (b) $\int \frac{dx}{x\sqrt{x^2-9}}$

2. **Completing the Square** In your own words, describe the process of completing the square of a quadratic function. Explain when completing the square is useful for finding an integral.

Finding an Indefinite Integral **In Exercises 3–22, find the indefinite integral.**

3. $\int \frac{dx}{\sqrt{9-x^2}}$

4. $\int \frac{dx}{\sqrt{1-4x^2}}$

5. $\int \frac{1}{x\sqrt{4x^2-1}}\,dx$

6. $\int \frac{12}{1+9x^2}\,dx$

7. $\int \frac{1}{\sqrt{1-(x+1)^2}}\,dx$

8. $\int \frac{7}{4+(3-x)^2}\,dx$

9. $\int \frac{t}{\sqrt{1-t^4}}\,dt$

10. $\int \frac{1}{x\sqrt{x^4-4}}\,dx$

11. $\int \frac{t}{t^4+25}\,dt$

12. $\int \frac{1}{x\sqrt{1-(\ln x)^2}}\,dx$

13. $\int \frac{e^{2x}}{4+e^{4x}}\,dx$

14. $\int \frac{5}{x\sqrt{9x^2-11}}\,dx$

15. $\int \frac{-\csc x \cot x}{\sqrt{25-\csc^2 x}}\,dx$

16. $\int \frac{\sin x}{7+\cos^2 x}\,dx$

17. $\int \frac{1}{\sqrt{x}\sqrt{1-x}}\,dx$

18. $\int \frac{3}{2\sqrt{x}(1+x)}\,dx$

19. $\int \frac{x-3}{x^2+1}\,dx$

20. $\int \frac{x^2+8}{x\sqrt{x^2-4}}\,dx$

21. $\int \frac{x+5}{\sqrt{9-(x-3)^2}}\,dx$

22. $\int \frac{x-2}{(x+1)^2+4}\,dx$

Evaluating a Definite Integral **In Exercises 23–34, evaluate the definite integral.**

23. $\int_0^{1/6} \frac{3}{\sqrt{1-9x^2}}\,dx$

24. $\int_0^{\sqrt{2}} \frac{1}{\sqrt{4-x^2}}\,dx$

25. $\int_0^{\sqrt{3}/2} \frac{1}{1+4x^2}\,dx$

26. $\int_{\sqrt{3}}^{3} \frac{1}{x\sqrt{4x^2-9}}\,dx$

27. $\int_1^7 \frac{1}{9+(x+2)^2}\,dx$

28. $\int_1^4 \frac{1}{x\sqrt{16x^2-5}}\,dx$

29. $\int_0^{\ln 5} \frac{e^x}{1+e^{2x}}\,dx$

30. $\int_{\ln 2}^{\ln 4} \frac{e^{-x}}{\sqrt{1-e^{-2x}}}\,dx$

31. $\int_{\pi/2}^{\pi} \frac{\sin x}{1+\cos^2 x}\,dx$

32. $\int_0^{\pi/2} \frac{\cos x}{1+\sin^2 x}\,dx$

33. $\int_0^{1/\sqrt{2}} \frac{\arcsin x}{\sqrt{1-x^2}}\,dx$

34. $\int_0^{1/\sqrt{2}} \frac{\arccos x}{\sqrt{1-x^2}}\,dx$

Completing the Square **In Exercises 35–42, find or evaluate the integral by completing the square.**

35. $\int_0^2 \frac{dx}{x^2-2x+2}$

36. $\int_{-2}^{3} \frac{dx}{x^2+4x+8}$

37. $\int \frac{dx}{\sqrt{-2x^2+8x+4}}$

38. $\int \frac{dx}{3x^2-6x+12}$

39. $\int \frac{1}{\sqrt{-x^2-4x}}\,dx$

40. $\int \frac{2}{\sqrt{-x^2+4x}}\,dx$

41. $\int_2^3 \frac{2x-3}{\sqrt{4x-x^2}}\,dx$

42. $\int_3^4 \frac{1}{(x-1)\sqrt{x^2-2x}}\,dx$

Integration by Substitution **In Exercises 43–46, use the specified substitution to find or evaluate the integral.**

43. $\int \sqrt{e^t-3}\,dt$, $u=\sqrt{e^t-3}$

44. $\int \frac{\sqrt{x-2}}{x+1}\,dx$, $u=\sqrt{x-2}$

45. $\int_1^3 \frac{dx}{\sqrt{x}(1+x)}$, $u=\sqrt{x}$

46. $\int_0^1 \frac{dx}{2\sqrt{3-x}\sqrt{x+1}}$, $u=\sqrt{x+1}$

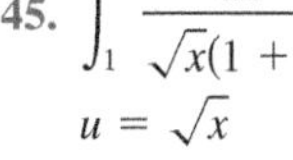

Comparing Integration Problems **In Exercises 47–50, find the indefinite integrals, if possible, using the formulas and techniques you have studied so far in the text.**

47. (a) $\int \frac{1}{\sqrt{1-x^2}}\,dx$ (b) $\int \frac{x}{\sqrt{1-x^2}}\,dx$ (c) $\int \frac{1}{x\sqrt{1-x^2}}\,dx$

48. (a) $\int e^{x^2}\,dx$ (b) $\int xe^{x^2}\,dx$ (c) $\int \frac{1}{x^2}e^{1/x}\,dx$

49. (a) $\int \sqrt{x-1}\,dx$ (b) $\int x\sqrt{x-1}\,dx$ (c) $\int \frac{x}{\sqrt{x-1}}\,dx$

50. (a) $\int \frac{1}{1+x^4}\,dx$ (b) $\int \frac{x}{1+x^4}\,dx$ (c) $\int \frac{x^3}{1+x^4}\,dx$

EXPLORING CONCEPTS

Comparing Antiderivatives **In Exercises 51 and 52, show that the antiderivatives are equivalent.**

51. $\displaystyle\int \frac{3x^2}{\sqrt{1-x^6}}\,dx = \arcsin x^3 + C \text{ or } \arccos\sqrt{1-x^6} + C$

52. $\displaystyle\int \frac{6}{4+9x^2}\,dx = \arctan\frac{3x}{2} + C \text{ or } \operatorname{arccsc}\frac{\sqrt{4+9x^2}}{3x} + C$

53. Inverse Trigonometric Functions The antiderivative of

$$\int \frac{1}{\sqrt{1-x^2}}\,dx$$

can be either $\arcsin x + C$ or $-\arccos x + C$. Does this mean that $\arcsin x = -\arccos x$? Explain.

54. HOW DO YOU SEE IT? Using the graph, which value best approximates the area of the region between the x-axis and the function over the interval $\left[-\frac{1}{2}, \frac{1}{2}\right]$? Explain.

$f(x) = \dfrac{1}{\sqrt{1-x^2}}$

(a) -3 (b) $\frac{1}{2}$ (c) 1 (d) 2 (e) 4

Slope Field **In Exercises 55 and 56, a differential equation, a point, and a slope field are given. (a) Sketch two approximate solutions of the differential equation on the slope field, one of which passes through the given point. (To print an enlarged copy of the graph, go to *MathGraphs.com*.) (b) Use integration and the given point to find the particular solution of the differential equation and use a graphing utility to graph the solution. Compare the result with the sketch in part (a) that passes through the given point.**

55. $\dfrac{dy}{dx} = \dfrac{2}{9+x^2}$, $(0, 2)$

56. $\dfrac{dy}{dx} = \dfrac{2}{\sqrt{25-x^2}}$, $(5, \pi)$

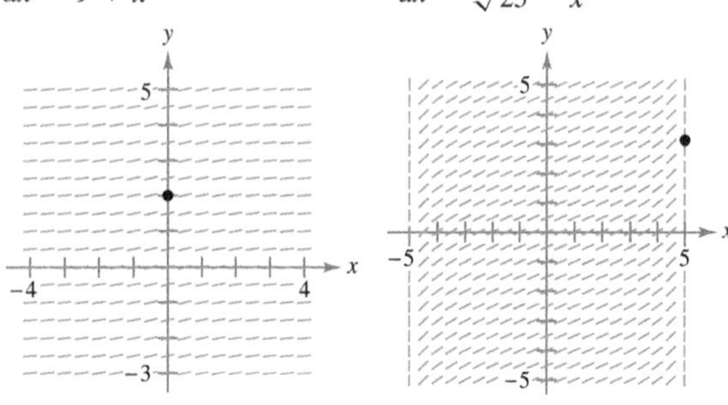

Slope Field **In Exercises 57–60, use a graphing utility to graph the slope field for the differential equation and graph the particular solution satisfying the specified initial condition.**

57. $\dfrac{dy}{dx} = \dfrac{10}{x\sqrt{x^2-1}}$

$y(3) = 0$

58. $\dfrac{dy}{dx} = \dfrac{1}{12+x^2}$

$y(4) = 2$

59. $\dfrac{dy}{dx} = \dfrac{2y}{\sqrt{16-x^2}}$

$y(0) = 2$

60. $\dfrac{dy}{dx} = \dfrac{\sqrt{y}}{1+x^2}$

$y(0) = 4$

Differential Equation **In Exercises 61 and 62, find the particular solution of the differential equation that satisfies the initial condition.**

61. $\dfrac{dy}{dx} = \dfrac{1}{\sqrt{4-x^2}}$

$y(0) = \pi$

62. $\dfrac{dy}{dx} = \dfrac{1}{4+x^2}$

$y(2) = \pi$

Area **In Exercises 63–66, find the area of the given region. Use a graphing utility to verify your result.**

63. $y = \dfrac{2}{\sqrt{4-x^2}}$

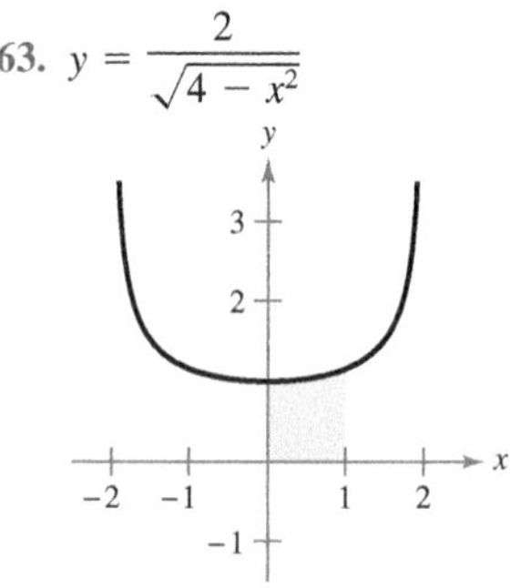

64. $y = \dfrac{1}{x\sqrt{x^2-1}}$

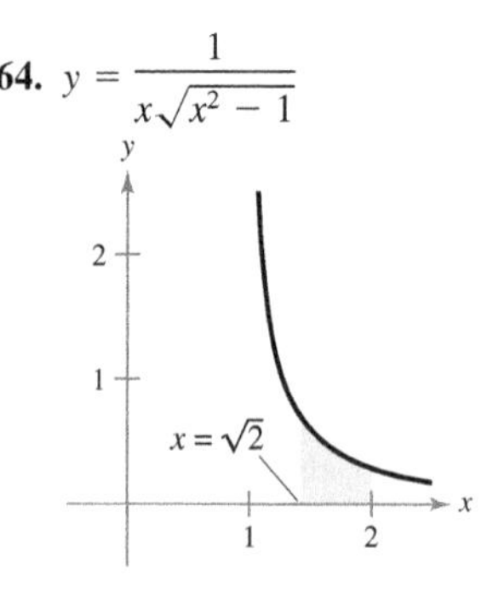

65. $y = \dfrac{3\cos x}{1+\sin^2 x}$

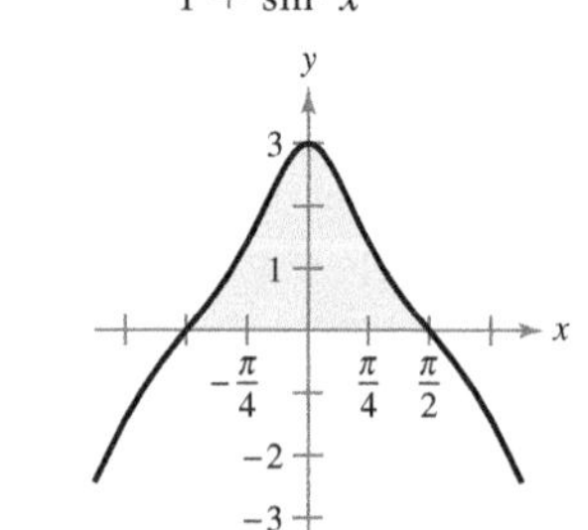

66. $y = \dfrac{4e^x}{1+e^{2x}}$

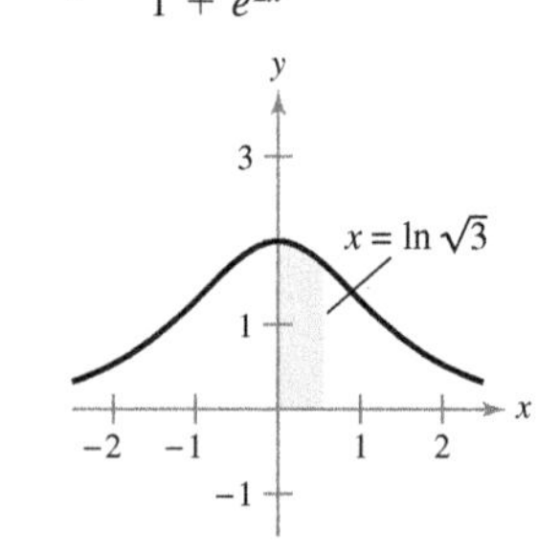

67. Area

(a) Sketch the region whose area is represented by

$$\int_0^1 \arcsin x\,dx.$$

(b) Use the integration capabilities of a graphing utility to approximate the area.

(c) Find the exact area analytically.

68. Approximating Pi

(a) Show that

$$\int_0^1 \frac{4}{1+x^2}\,dx = \pi.$$

(b) Approximate the number π by using the integration capabilities of a graphing utility.

69. Investigation Consider the function

$$F(x) = \frac{1}{2}\int_x^{x+2} \frac{2}{t^2+1}\,dt.$$

(a) Write a short paragraph giving a geometric interpretation of the function $F(x)$ relative to the function

$$f(x) = \frac{2}{x^2+1}.$$

Use what you have written to guess the value of x that will make F maximum.

(b) Perform the specified integration to find an alternative form of $F(x)$. Use calculus to locate the value of x that will make F maximum and compare the result with your guess in part (a).

70. Comparing Integrals Consider the integral

$$\int \frac{1}{\sqrt{6x - x^2}}\,dx.$$

(a) Find the integral by completing the square of the radicand.

(b) Find the integral by making the substitution $u = \sqrt{x}$.

(c) The antiderivatives in parts (a) and (b) appear to be significantly different. Use a graphing utility to graph each antiderivative in the same viewing window and determine the relationship between them. Find the domain of each.

True or False? **In Exercises 71 and 72, determine whether the statement is true or false. If it is false, explain why or give an example that shows it is false.**

71. $\displaystyle\int \frac{dx}{3x\sqrt{9x^2-16}} = \frac{1}{4}\operatorname{arcsec}\frac{3x}{4} + C$

72. $\displaystyle\int \frac{dx}{25+x^2} = \frac{1}{25}\arctan\frac{x}{25} + C$

Verifying an Integration Rule **In Exercises 73–75, verify the rule by differentiating. Let $a > 0$.**

73. $\displaystyle\int \frac{du}{\sqrt{a^2-u^2}} = \arcsin\frac{u}{a} + C$

74. $\displaystyle\int \frac{du}{a^2+u^2} = \frac{1}{a}\arctan\frac{u}{a} + C$

75. $\displaystyle\int \frac{du}{u\sqrt{u^2-a^2}} = \frac{1}{a}\operatorname{arcsec}\frac{|u|}{a} + C$

76. Proof Graph $y_1 = \dfrac{x}{1+x^2}$, $y_2 = \arctan x$, and $y_3 = x$ on $[0, 10]$. Prove that $\dfrac{x}{1+x^2} < \arctan x < x$ for $x > 0$.

77. Numerical Integration

(a) Write an integral that represents the area of the region in the figure.

(b) Use the Midpoint Rule with $n = 8$ to estimate the area of the region.

(c) Explain how you can use the results of parts (a) and (b) to estimate π.

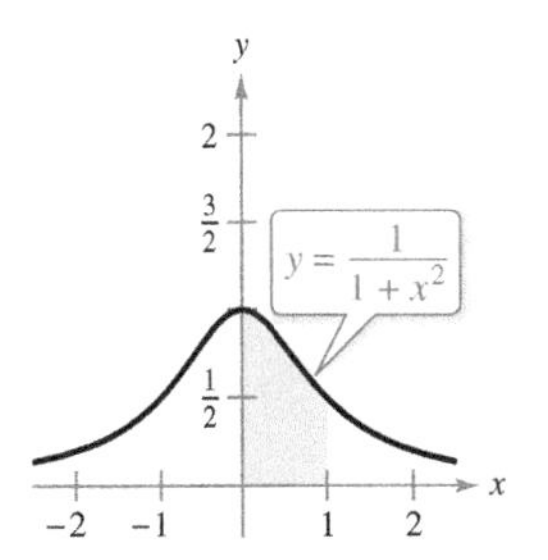

78. Vertical Motion An object is projected upward from ground level with an initial velocity of 500 feet per second. In this exercise, the goal is to analyze the motion of the object during its upward flight.

(a) If air resistance is neglected, find the velocity of the object as a function of time. Use a graphing utility to graph this function.

(b) Use the result of part (a) to find the position function and determine the maximum height attained by the object.

(c) If the air resistance is proportional to the square of the velocity, you obtain the equation

$$\frac{dv}{dt} = -(32 + kv^2)$$

where 32 feet per second per second is the acceleration due to gravity and k is a constant. Find the velocity as a function of time by solving the equation

$$\int \frac{dv}{32+kv^2} = -\int dt.$$

(d) Use a graphing utility to graph the velocity function $v(t)$ in part (c) for $k = 0.001$. Use the graph to approximate the time t_0 at which the object reaches its maximum height.

(e) Use the integration capabilities of a graphing utility to approximate the integral

$$\int_0^{t_0} v(t)\,dt$$

where $v(t)$ and t_0 are those found in part (d). This is the approximation of the maximum height of the object.

(f) Explain the difference between the results in parts (b) and (e).

FOR FURTHER INFORMATION For more information on this topic, see the article "What Goes Up Must Come Down; Will Air Resistance Make It Return Sooner, or Later?" by John Lekner in *Mathematics Magazine*. To view this article, go to *MathArticles.com*.

5.9 Hyperbolic Functions

- **Develop properties of hyperbolic functions.**
- **Differentiate and integrate hyperbolic functions.**
- **Develop properties of inverse hyperbolic functions.**
- **Differentiate and integrate functions involving inverse hyperbolic functions.**

Hyperbolic Functions

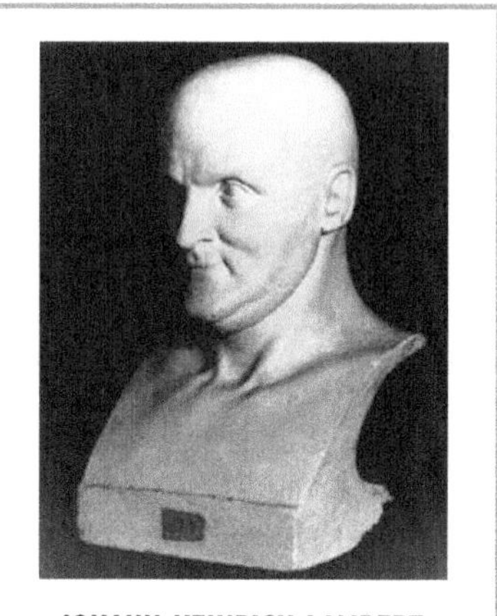

JOHANN HEINRICH LAMBERT (1728–1777)

The first person to publish a comprehensive study on hyperbolic functions was Johann Heinrich Lambert, a Swiss-German mathematician and colleague of Euler.

See LarsonCalculus.com to read more of this biography.

In this section, you will look briefly at a special class of exponential functions called **hyperbolic functions.** The name *hyperbolic function* arose from comparison of the area of a semicircular region, as shown in Figure 5.29, with the area of a region under a hyperbola, as shown in Figure 5.30.

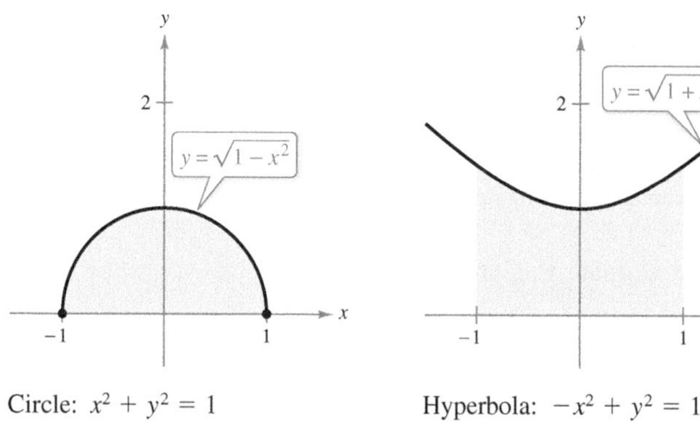

Circle: $x^2 + y^2 = 1$
Figure 5.29

Hyperbola: $-x^2 + y^2 = 1$
Figure 5.30

The integral for the semicircular region involves an inverse trigonometric (circular) function:

$$\int_{-1}^{1} \sqrt{1-x^2}\,dx = \frac{1}{2}\left[x\sqrt{1-x^2} + \arcsin x\right]_{-1}^{1} = \frac{\pi}{2} \approx 1.571.$$

The integral for the hyperbolic region involves an inverse hyperbolic function:

$$\int_{-1}^{1} \sqrt{1+x^2}\,dx = \frac{1}{2}\left[x\sqrt{1+x^2} + \sinh^{-1} x\right]_{-1}^{1} \approx 2.296.$$

This is only one of many ways in which the hyperbolic functions are similar to the trigonometric functions.

REMARK The notation $\sinh x$ is read as "the hyperbolic sine of x," $\cosh x$ as "the hyperbolic cosine of x," and so on.

Definitions of the Hyperbolic Functions

$$\sinh x = \frac{e^x - e^{-x}}{2} \qquad \operatorname{csch} x = \frac{1}{\sinh x}, \quad x \neq 0$$

$$\cosh x = \frac{e^x + e^{-x}}{2} \qquad \operatorname{sech} x = \frac{1}{\cosh x}$$

$$\tanh x = \frac{\sinh x}{\cosh x} \qquad \coth x = \frac{1}{\tanh x}, \quad x \neq 0$$

FOR FURTHER INFORMATION For more information on the development of hyperbolic functions, see the article "An Introduction to Hyperbolic Functions in Elementary Calculus" by Jerome Rosenthal in *Mathematics Teacher.* To view this article, go to *MathArticles.com.*

The graphs of the six hyperbolic functions and their domains and ranges are shown in Figure 5.31. Note that the graph of $\sinh x$ can be obtained by adding the corresponding y-coordinates of the exponential functions $f(x) = \frac{1}{2}e^x$ and $g(x) = -\frac{1}{2}e^{-x}$. Likewise, the graph of $\cosh x$ can be obtained by adding the corresponding y-coordinates of the exponential functions $f(x) = \frac{1}{2}e^x$ and $h(x) = \frac{1}{2}e^{-x}$.

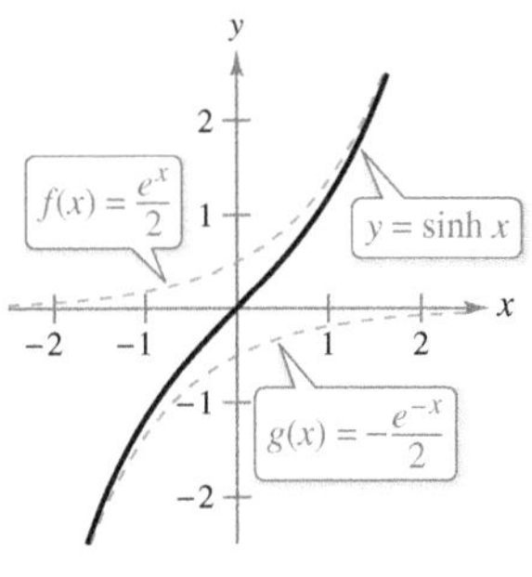

Domain: $(-\infty, \infty)$
Range: $(-\infty, \infty)$

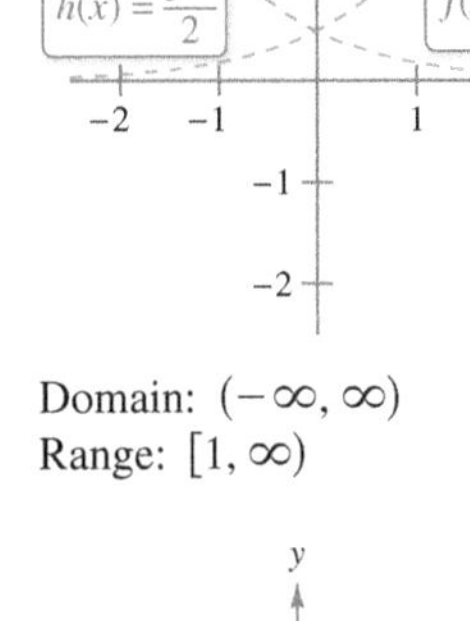

Domain: $(-\infty, \infty)$
Range: $[1, \infty)$

Domain: $(-\infty, \infty)$
Range: $(-1, 1)$

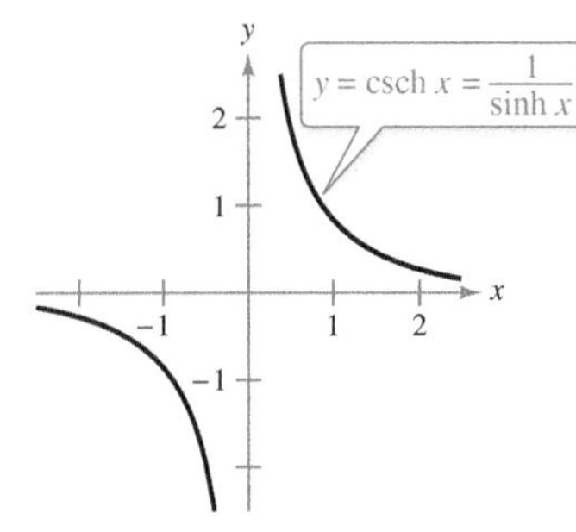

Domain: $(-\infty, 0) \cup (0, \infty)$
Range: $(-\infty, 0) \cup (0, \infty)$

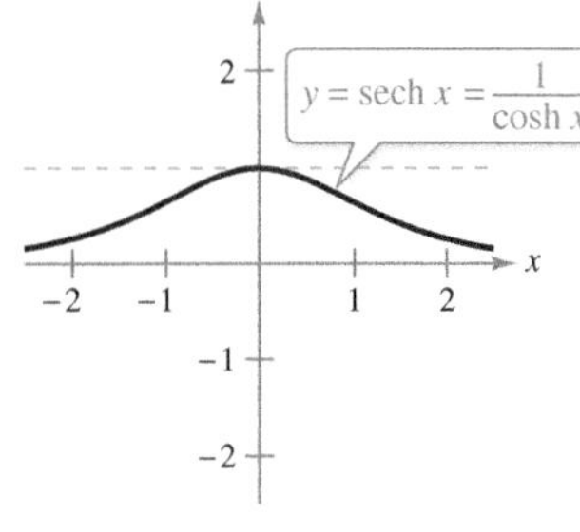

Domain: $(-\infty, \infty)$
Range: $(0, 1]$

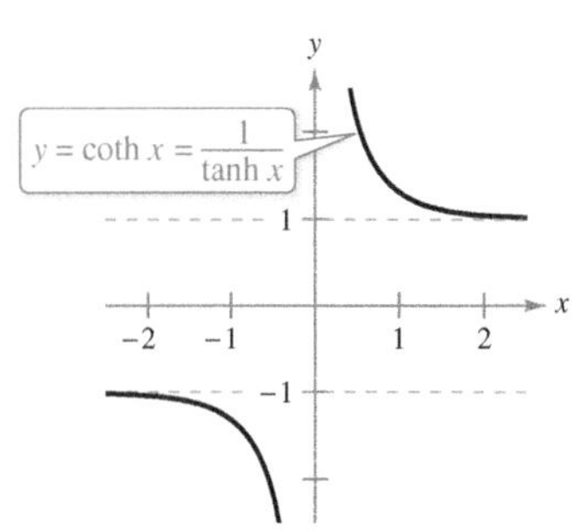

Domain: $(-\infty, 0) \cup (0, \infty)$
Range: $(-\infty, -1) \cup (1, \infty)$

Figure 5.31

Many of the trigonometric identities have corresponding *hyperbolic identities*. For instance,

$$\begin{aligned}\cosh^2 x - \sinh^2 x &= \left(\frac{e^x + e^{-x}}{2}\right)^2 - \left(\frac{e^x - e^{-x}}{2}\right)^2 \\ &= \frac{e^{2x} + 2 + e^{-2x}}{4} - \frac{e^{2x} - 2 + e^{-2x}}{4} \\ &= \frac{4}{4} \\ &= 1.\end{aligned}$$

FOR FURTHER INFORMATION To understand geometrically the relationship between the hyperbolic and exponential functions, see the article "A Short Proof Linking the Hyperbolic and Exponential Functions" by Michael J. Seery in *The AMATYC Review*.

HYPERBOLIC IDENTITIES

$\cosh^2 x - \sinh^2 x = 1$
$\tanh^2 x + \operatorname{sech}^2 x = 1$
$\coth^2 x - \operatorname{csch}^2 x = 1$

$\sinh(x + y) = \sinh x \cosh y + \cosh x \sinh y$
$\sinh(x - y) = \sinh x \cosh y - \cosh x \sinh y$
$\cosh(x + y) = \cosh x \cosh y + \sinh x \sinh y$
$\cosh(x - y) = \cosh x \cosh y - \sinh x \sinh y$

$\sinh^2 x = \dfrac{-1 + \cosh 2x}{2}$

$\cosh^2 x = \dfrac{1 + \cosh 2x}{2}$

$\sinh 2x = 2 \sinh x \cosh x$

$\cosh 2x = \cosh^2 x + \sinh^2 x$

Differentiation and Integration of Hyperbolic Functions

Because the hyperbolic functions are written in terms of e^x and e^{-x}, you can easily derive rules for their derivatives. The next theorem lists these derivatives with the corresponding integration rules.

THEOREM 5.20 Derivatives and Integrals of Hyperbolic Functions

Let u be a differentiable function of x.

$$\frac{d}{dx}[\sinh u] = (\cosh u)u' \qquad \int \cosh u\, du = \sinh u + C$$

$$\frac{d}{dx}[\cosh u] = (\sinh u)u' \qquad \int \sinh u\, du = \cosh u + C$$

$$\frac{d}{dx}[\tanh u] = (\text{sech}^2\, u)u' \qquad \int \text{sech}^2\, u\, du = \tanh u + C$$

$$\frac{d}{dx}[\coth u] = -(\text{csch}^2\, u)u' \qquad \int \text{csch}^2\, u\, du = -\coth u + C$$

$$\frac{d}{dx}[\text{sech}\, u] = -(\text{sech}\, u \tanh u)u' \qquad \int \text{sech}\, u \tanh u\, du = -\text{sech}\, u + C$$

$$\frac{d}{dx}[\text{csch}\, u] = -(\text{csch}\, u \coth u)u' \qquad \int \text{csch}\, u \coth u\, du = -\text{csch}\, u + C$$

Proof Here is a proof of two of the differentiation rules. (You are asked to prove some of the other differentiation rules in Exercises 99–101.)

$$\begin{aligned}\frac{d}{dx}[\sinh x] &= \frac{d}{dx}\left[\frac{e^x - e^{-x}}{2}\right]\\ &= \frac{e^x + e^{-x}}{2}\\ &= \cosh x\end{aligned}$$

$$\begin{aligned}\frac{d}{dx}[\tanh x] &= \frac{d}{dx}\left[\frac{\sinh x}{\cosh x}\right]\\ &= \frac{(\cosh x)(\cosh x) - (\sinh x)(\sinh x)}{\cosh^2 x}\\ &= \frac{1}{\cosh^2 x}\\ &= \text{sech}^2\, x\end{aligned}$$

EXAMPLE 1 Differentiation of Hyperbolic Functions

a. $\frac{d}{dx}[\sinh(x^2 - 3)] = 2x\cosh(x^2 - 3)$

b. $\frac{d}{dx}[\ln(\cosh x)] = \frac{\sinh x}{\cosh x} = \tanh x$

c. $\frac{d}{dx}[x \sinh x - \cosh x] = x\cosh x + \sinh x - \sinh x = x\cosh x$

d. $\frac{d}{dx}[(x - 1)\cosh x - \sinh x] = (x - 1)\sinh x + \cosh x - \cosh x = (x - 1)\sinh x$

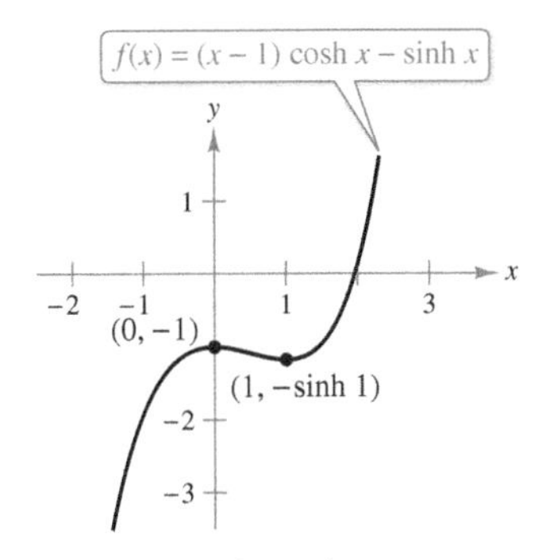

$f''(0) < 0$, so $(0, -1)$ is a relative maximum. $f''(1) > 0$, so $(1, -\sinh 1)$ is a relative minimum.
Figure 5.32

EXAMPLE 2 Finding Relative Extrema

Find the relative extrema of

$$f(x) = (x - 1)\cosh x - \sinh x.$$

Solution Using the result of Example 1(d), set the first derivative of f equal to 0.

$$(x - 1)\sinh x = 0$$

So, the critical numbers are $x = 1$ and $x = 0$. Using the Second Derivative Test, you can verify that the point $(0, -1)$ yields a relative maximum and the point $(1, -\sinh 1)$ yields a relative minimum, as shown in Figure 5.32. Try using a graphing utility to confirm this result. If your graphing utility does not have hyperbolic functions, you can use exponential functions, as shown.

$$\begin{aligned} f(x) &= (x - 1)\left(\frac{1}{2}\right)(e^x + e^{-x}) - \frac{1}{2}(e^x - e^{-x}) \\ &= \frac{1}{2}(xe^x + xe^{-x} - e^x - e^{-x} - e^x + e^{-x}) \\ &= \frac{1}{2}(xe^x + xe^{-x} - 2e^x) \end{aligned}$$

When a uniform flexible cable, such as a telephone wire, is suspended from two points, it takes the shape of a *catenary*, as discussed in Example 3.

EXAMPLE 3 Hanging Power Cables

See LarsonCalculus.com for an interactive version of this type of example.

Power cables are suspended between two towers, forming the catenary shown in Figure 5.33. The equation for this catenary is

$$y = a\cosh\frac{x}{a}.$$

The distance between the two towers is $2b$. Find the slope of the catenary at the point where the cable meets the right-hand tower.

Solution Differentiating produces

$$y' = a\left(\frac{1}{a}\right)\sinh\frac{x}{a} = \sinh\frac{x}{a}.$$

At the point $(b, a\cosh(b/a))$, the slope (from the left) is $m = \sinh\dfrac{b}{a}$.

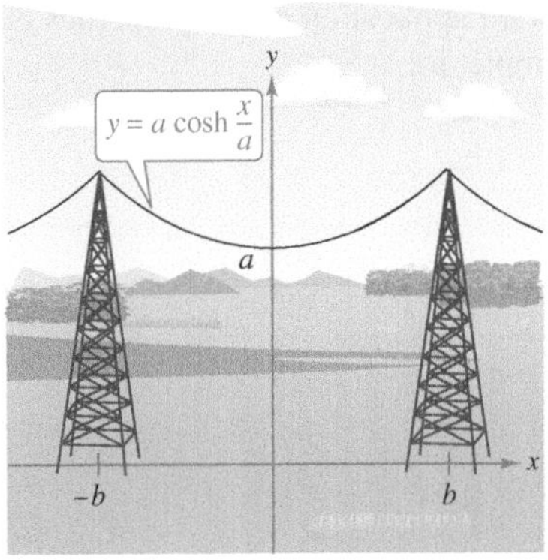

Catenary
Figure 5.33

FOR FURTHER INFORMATION In Example 3, the cable is a catenary between two supports at the same height. To learn about the shape of a cable hanging between supports of different heights, see the article "Reexamining the Catenary" by Paul Cella in *The College Mathematics Journal*. To view this article, go to *MathArticles.com*.

EXAMPLE 4 Integrating a Hyperbolic Function

Find $\displaystyle\int \cosh 2x \sinh^2 2x \, dx$.

Solution

$$\begin{aligned} \int \cosh 2x \sinh^2 2x \, dx &= \frac{1}{2}\int (\sinh 2x)^2(2\cosh 2x)\, dx \qquad u = \sinh 2x \\ &= \frac{1}{2}\left[\frac{(\sinh 2x)^3}{3}\right] + C \\ &= \frac{\sinh^3 2x}{6} + C \end{aligned}$$

Inverse Hyperbolic Functions

Unlike trigonometric functions, hyperbolic functions are not periodic. In fact, by looking back at Figure 5.31, you can see that four of the six hyperbolic functions are actually one-to-one (the hyperbolic sine, tangent, cosecant, and cotangent). So, you can apply Theorem 5.7 to conclude that these four functions have inverse functions. The other two (the hyperbolic cosine and secant) are one-to-one when their domains are restricted to the positive real numbers, and for this restricted domain they also have inverse functions. Because the hyperbolic functions are defined in terms of exponential functions, it is not surprising to find that the inverse hyperbolic functions can be written in terms of logarithmic functions, as shown in the next theorem.

THEOREM 5.21 Inverse Hyperbolic Functions

Function	Domain
$\sinh^{-1} x = \ln\left(x + \sqrt{x^2 + 1}\right)$	$(-\infty, \infty)$
$\cosh^{-1} x = \ln\left(x + \sqrt{x^2 - 1}\right)$	$[1, \infty)$
$\tanh^{-1} x = \dfrac{1}{2} \ln \dfrac{1 + x}{1 - x}$	$(-1, 1)$
$\coth^{-1} x = \dfrac{1}{2} \ln \dfrac{x + 1}{x - 1}$	$(-\infty, -1) \cup (1, \infty)$
$\operatorname{sech}^{-1} x = \ln \dfrac{1 + \sqrt{1 - x^2}}{x}$	$(0, 1]$
$\operatorname{csch}^{-1} x = \ln\left(\dfrac{1}{x} + \dfrac{\sqrt{1 + x^2}}{\lvert x \rvert}\right)$	$(-\infty, 0) \cup (0, \infty)$

Proof The proof of this theorem is a straightforward application of the properties of the exponential and logarithmic functions. For example, for

$$f(x) = \sinh x = \frac{e^x - e^{-x}}{2}$$

and

$$g(x) = \ln\left(x + \sqrt{x^2 + 1}\right)$$

you can show that

$$f(g(x)) = x \quad \text{and} \quad g(f(x)) = x$$

which implies that g is the inverse function of f. ■

TECHNOLOGY You can use a graphing utility to confirm graphically the results of Theorem 5.21. For instance, graph the following functions.

$y_1 = \tanh x$ — Hyperbolic tangent

$y_2 = \dfrac{e^x - e^{-x}}{e^x + e^{-x}}$ — Definition of hyperbolic tangent

$y_3 = \tanh^{-1} x$ — Inverse hyperbolic tangent

$y_4 = \dfrac{1}{2} \ln \dfrac{1 + x}{1 - x}$ — Definition of inverse hyperbolic tangent

The resulting display is shown in Figure 5.34. As you watch the graphs being traced out, notice that $y_1 = y_2$ and $y_3 = y_4$. Also notice that the graph of y_1 is the reflection of the graph of y_3 in the line $y = x$.

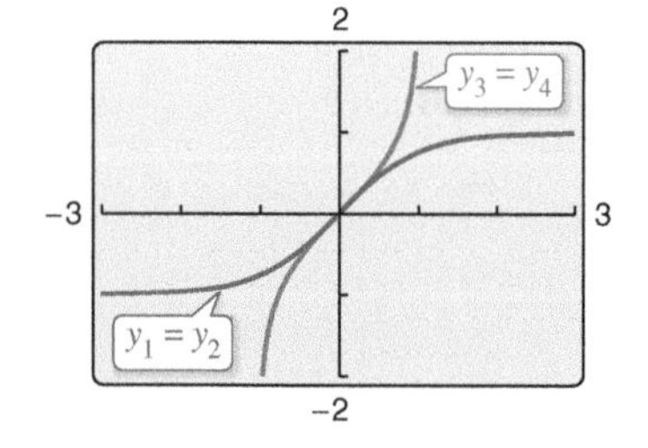

Graphs of the hyperbolic tangent function and the inverse hyperbolic tangent function
Figure 5.34

The graphs of the inverse hyperbolic functions are shown in Figure 5.35.

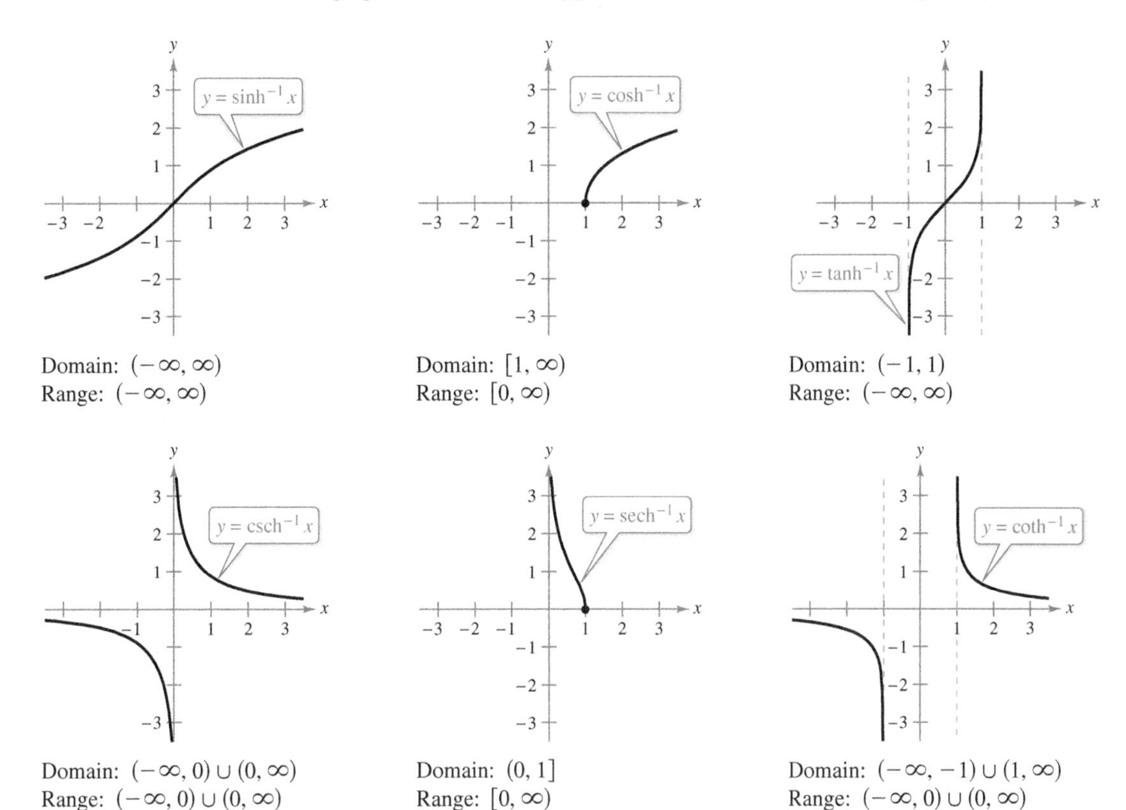

Figure 5.35

The inverse hyperbolic secant can be used to define a curve called a *tractrix* or *pursuit curve*, as discussed in Example 5.

EXAMPLE 5 A Tractrix

A person is holding a rope that is tied to a boat, as shown in Figure 5.36. As the person walks along the dock, the boat travels along a **tractrix,** given by the equation

$$y = a \, \text{sech}^{-1} \frac{x}{a} - \sqrt{a^2 - x^2}$$

where a is the length of the rope. For $a = 20$ feet, find the distance the person must walk to bring the boat to a position 5 feet from the dock.

Solution In Figure 5.36, notice that the distance the person has walked is

$$\begin{aligned} y_1 &= y + \sqrt{20^2 - x^2} \\ &= \left(20 \, \text{sech}^{-1} \frac{x}{20} - \sqrt{20^2 - x^2}\right) + \sqrt{20^2 - x^2} \\ &= 20 \, \text{sech}^{-1} \frac{x}{20}. \end{aligned}$$

When $x = 5$, this distance is

$$y_1 = 20 \, \text{sech}^{-1} \frac{5}{20} = 20 \ln \frac{1 + \sqrt{1 - (1/4)^2}}{1/4} = 20 \ln\left(4 + \sqrt{15}\right) \approx 41.27 \text{ feet.}$$

So, the person must walk about 41.27 feet to bring the boat to a position 5 feet from the dock.

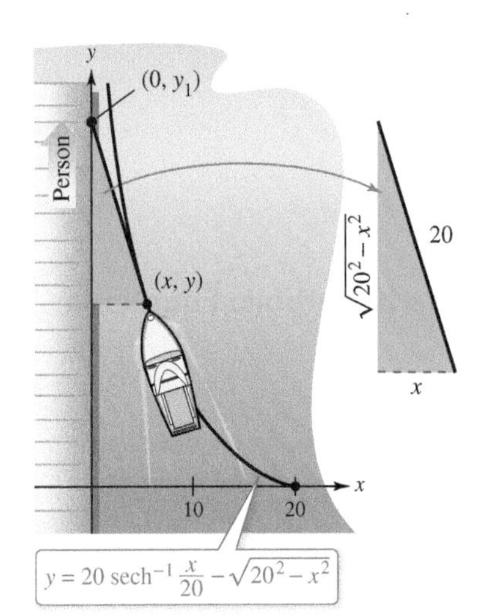

A person must walk about 41.27 feet to bring the boat to a position 5 feet from the dock.
Figure 5.36

Inverse Hyperbolic Functions: Differentiation and Integration

The derivatives of the inverse hyperbolic functions, which resemble the derivatives of the inverse trigonometric functions, are listed in Theorem 5.22 with the corresponding integration formulas (in logarithmic form). You can verify each of these formulas by applying the logarithmic definitions of the inverse hyperbolic functions. (See Exercises 102–104.)

THEOREM 5.22 Differentiation and Integration Involving Inverse Hyperbolic Functions

Let u be a differentiable function of x.

$$\frac{d}{dx}[\sinh^{-1} u] = \frac{u'}{\sqrt{u^2+1}} \qquad \frac{d}{dx}[\cosh^{-1} u] = \frac{u'}{\sqrt{u^2-1}}$$

$$\frac{d}{dx}[\tanh^{-1} u] = \frac{u'}{1-u^2} \qquad \frac{d}{dx}[\coth^{-1} u] = \frac{u'}{1-u^2}$$

$$\frac{d}{dx}[\operatorname{sech}^{-1} u] = \frac{-u'}{u\sqrt{1-u^2}} \qquad \frac{d}{dx}[\operatorname{csch}^{-1} u] = \frac{-u'}{|u|\sqrt{1+u^2}}$$

$$\int \frac{du}{\sqrt{u^2 \pm a^2}} = \ln\left(u + \sqrt{u^2 \pm a^2}\right) + C$$

$$\int \frac{du}{a^2 - u^2} = \frac{1}{2a}\ln\left|\frac{a+u}{a-u}\right| + C$$

$$\int \frac{du}{u\sqrt{a^2 \pm u^2}} = -\frac{1}{a}\ln\frac{a + \sqrt{a^2 \pm u^2}}{|u|} + C$$

EXAMPLE 6 Differentiation of Inverse Hyperbolic Functions

a. $$\frac{d}{dx}[\sinh^{-1}(2x)] = \frac{2}{\sqrt{(2x)^2+1}} = \frac{2}{\sqrt{4x^2+1}}$$

b. $$\frac{d}{dx}[\tanh^{-1}(x^3)] = \frac{3x^2}{1-(x^3)^2} = \frac{3x^2}{1-x^6}$$

EXAMPLE 7 Integration Using Inverse Hyperbolic Functions

a. $$\int \frac{dx}{x\sqrt{4-9x^2}} = \int \frac{3\,dx}{(3x)\sqrt{2^2-(3x)^2}} \qquad \int \frac{du}{u\sqrt{a^2-u^2}}$$

$$= -\frac{1}{2}\ln\frac{2+\sqrt{4-9x^2}}{|3x|} + C \qquad -\frac{1}{a}\ln\frac{a+\sqrt{a^2-u^2}}{|u|} + C$$

REMARK Let $a = 2$ and $u = 3x$.

b. $$\int \frac{dx}{5-4x^2} = \frac{1}{2}\int \frac{2\,dx}{(\sqrt{5})^2-(2x)^2} \qquad \int \frac{du}{a^2-u^2}$$

$$= \frac{1}{2}\left(\frac{1}{2\sqrt{5}}\ln\left|\frac{\sqrt{5}+2x}{\sqrt{5}-2x}\right|\right) + C \qquad \frac{1}{2a}\ln\left|\frac{a+u}{a-u}\right| + C$$

$$= \frac{1}{4\sqrt{5}}\ln\left|\frac{\sqrt{5}+2x}{\sqrt{5}-2x}\right| + C$$

REMARK Let $a = \sqrt{5}$ and $u = 2x$.

5.9 Exercises

See CalcChat.com for tutorial help and worked-out solutions to odd-numbered exercises.

CONCEPT CHECK

1. **Hyperbolic Functions** Describe how the name *hyperbolic function* arose.
2. **Domains of Hyperbolic Functions** Which hyperbolic functions have domains that are not all real numbers?
3. **Hyperbolic Identities** Which hyperbolic identity corresponds to the trigonometric identity

$$\sin^2 x = \frac{1 - \cos 2x}{2}?$$

4. **Derivatives of Inverse Hyperbolic Functions** What is the missing value?

$$\frac{d}{dx}[\operatorname{sech}^{-1}(3x)] = \frac{\square}{3x\sqrt{1 - 9x^2}}$$

Evaluating a Function **In Exercises 5–10, evaluate the function. If the value is not a rational number, round your answer to three decimal places.**

5. (a) $\sinh 3$ (b) $\tanh(-2)$
6. (a) $\cosh 0$ (b) $\operatorname{sech} 1$
7. (a) $\operatorname{csch}(\ln 2)$ (b) $\coth(\ln 5)$
8. (a) $\sinh^{-1} 0$ (b) $\tanh^{-1} 0$
9. (a) $\cosh^{-1} 2$ (b) $\operatorname{sech}^{-1} \frac{2}{3}$
10. (a) $\operatorname{csch}^{-1} 2$ (b) $\coth^{-1} 3$

Verifying an Identity **In Exercises 11–18, verify the identity.**

11. $\sinh x + \cosh x = e^x$
12. $\cosh x - \sinh x = e^{-x}$
13. $\tanh^2 x + \operatorname{sech}^2 x = 1$
14. $\coth^2 x - \operatorname{csch}^2 x = 1$
15. $\cosh^2 x = \dfrac{1 + \cosh 2x}{2}$
16. $\sinh^2 x = \dfrac{-1 + \cosh 2x}{2}$
17. $\sinh 2x = 2 \sinh x \cosh x$
18. $\sinh(x + y) = \sinh x \cosh y + \cosh x \sinh y$

Finding Values of Hyperbolic Functions **In Exercises 19 and 20, use the value of the given hyperbolic function to find the values of the other hyperbolic functions.**

19. $\sinh x = \dfrac{3}{2}$
20. $\tanh x = \dfrac{1}{2}$

Finding a Limit **In Exercises 21–24, find the limit.**

21. $\lim_{x \to \infty} \sinh x$
22. $\lim_{x \to -\infty} \tanh x$
23. $\lim_{x \to 0} \dfrac{\sinh x}{x}$
24. $\lim_{x \to 0^-} \coth x$

Finding a Derivative **In Exercises 25–34, find the derivative of the function.**

25. $f(x) = \sinh 9x$
26. $f(x) = \cosh(8x + 1)$
27. $y = \operatorname{sech} 5x^2$
28. $f(x) = \tanh(4x^2 + 3x)$
29. $f(x) = \ln(\sinh x)$
30. $y = \ln\left(\tanh \dfrac{x}{2}\right)$
31. $h(t) = \dfrac{t}{6} \sinh(-3t)$
32. $y = (x^2 + 1) \coth \dfrac{x}{3}$
33. $f(t) = \arctan(\sinh t)$
34. $g(x) = \operatorname{sech}^2 3x$

Finding an Equation of a Tangent Line **In Exercises 35–38, find an equation of the tangent line to the graph of the function at the given point.**

35. $y = \sinh(1 - x^2), \quad (1, 0)$
36. $y = x^{\cosh x}, \quad (1, 1)$
37. $y = (\cosh x - \sinh x)^2, \quad (0, 1)$
38. $y = e^{\sinh x}, \quad (0, 1)$

Finding Relative Extrema **In Exercises 39–42, find the relative extrema of the function. Use a graphing utility to confirm your result.**

39. $g(x) = x \operatorname{sech} x$
40. $h(x) = 2 \tanh x - x$
41. $f(x) = \sin x \sinh x - \cos x \cosh x, \quad -4 \le x \le 4$
42. $f(x) = x \sinh(x - 1) - \cosh(x - 1)$

Catenary **In Exercises 43 and 44, a model for a power cable suspended between two towers is given. (a) Graph the model. (b) Find the heights of the cable at the towers and at the midpoint between the towers. (c) Find the slope of the cable at the point where the cable meets the right-hand tower.**

43. $y = 10 + 15 \cosh \dfrac{x}{15}, \quad -15 \le x \le 15$
44. $y = 18 + 25 \cosh \dfrac{x}{25}, \quad -25 \le x \le 25$

Finding an Indefinite Integral **In Exercises 45–54, find the indefinite integral.**

45. $\displaystyle\int \cosh 4x \, dx$
46. $\displaystyle\int \operatorname{sech}^2 3x \, dx$
47. $\displaystyle\int \sinh(1 - 2x) \, dx$
48. $\displaystyle\int \frac{\cosh \sqrt{x}}{\sqrt{x}} \, dx$
49. $\displaystyle\int \cosh^2(x - 1) \sinh(x - 1) \, dx$
50. $\displaystyle\int \frac{\sinh x}{1 + \sinh^2 x} \, dx$

51. $\int \frac{\cosh x}{\sinh x}\,dx$

52. $\int \frac{\operatorname{csch}(1/x)\coth(1/x)}{x^2}\,dx$

53. $\int x \operatorname{csch}^2 \frac{x^2}{2}\,dx$

54. $\int \operatorname{sech}^3 x \tanh x\,dx$

Evaluating a Definite Integral **In Exercises 55–60, evaluate the definite integral.**

55. $\int_0^{\ln 2} \tanh x\,dx$

56. $\int_0^1 \cosh^2 x\,dx$

57. $\int_3^4 \operatorname{csch}^2 (x - 2)\,dx$

58. $\int_{1/2}^1 \operatorname{sech}^2 (2x - 1)\,dx$

59. $\int_{5/3}^2 \operatorname{csch}(3x - 4)\coth(3x - 4)\,dx$

60. $\int_0^{\ln 2} 2e^{-x}\cosh x\,dx$

EXPLORING CONCEPTS

61. **Using a Graph** Explain graphically why there is no solution to $\cosh x = \sinh x$.

62. **Hyperbolic Functions** Use the graphs on page 391 to determine whether each hyperbolic function is even, odd, or neither.

63. **Think About It** Verify the results of Exercise 62 algebraically.

64. **HOW DO YOU SEE IT?** Use the graphs of f and g shown in the figures to answer the following.

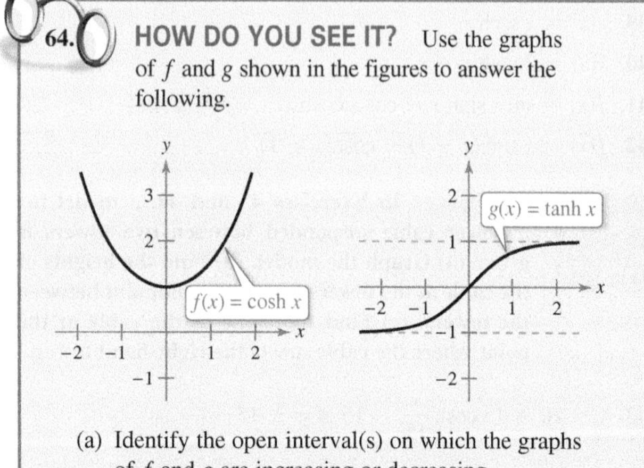

(a) Identify the open interval(s) on which the graphs of f and g are increasing or decreasing.

(b) Identify the open interval(s) on which the graphs of f and g are concave upward or concave downward.

Finding a Derivative **In Exercises 65–74, find the derivative of the function.**

65. $y = \cosh^{-1}(3x)$

66. $y = \operatorname{csch}^{-1}(1 - x)$

67. $y = \tanh^{-1}\sqrt{x}$

68. $f(x) = \coth^{-1}(x^2)$

69. $y = \sinh^{-1}(\tan x)$

70. $y = \tanh^{-1}(\sin 2x)$

71. $y = \operatorname{sech}^{-1}(\sin x),\ 0 < x < \pi/2$

72. $y = \coth^{-1}(e^{2x})$

73. $y = 2x \sinh^{-1}(2x) - \sqrt{1 + 4x^2}$

74. $y = x \tanh^{-1} x + \ln\sqrt{1 - x^2}$

Finding an Indefinite Integral **In Exercises 75–82, find the indefinite integral using the formulas from Theorem 5.22.**

75. $\int \frac{1}{3 - 9x^2}\,dx$

76. $\int \frac{1}{2x\sqrt{1 - 4x^2}}\,dx$

77. $\int \frac{1}{\sqrt{1 + e^{2x}}}\,dx$

78. $\int \frac{x}{9 - x^4}\,dx$

79. $\int \frac{1}{\sqrt{x}\sqrt{1 + x}}\,dx$

80. $\int \frac{\sqrt{x}}{\sqrt{1 + x^3}}\,dx$

81. $\int \frac{-1}{4x - x^2}\,dx$

82. $\int \frac{dx}{(x + 2)\sqrt{x^2 + 4x + 8}}$

Evaluating a Definite Integral **In Exercises 83–86, evaluate the definite integral using the formulas from Theorem 5.22.**

83. $\int_3^7 \frac{1}{\sqrt{x^2 - 4}}\,dx$

84. $\int_1^3 \frac{1}{x\sqrt{4 + x^2}}\,dx$

85. $\int_{-1}^1 \frac{1}{16 - 9x^2}\,dx$

86. $\int_0^1 \frac{1}{\sqrt{25x^2 + 1}}\,dx$

Differential Equation **In Exercises 87 and 88, find the general solution of the differential equation.**

87. $\frac{dy}{dx} = \frac{x^3 - 21x}{5 + 4x - x^2}$

88. $\frac{dy}{dx} = \frac{1 - 2x}{4x - x^2}$

Area **In Exercises 89–92, find the area of the given region.**

89. $y = \operatorname{sech} \frac{x}{2}$

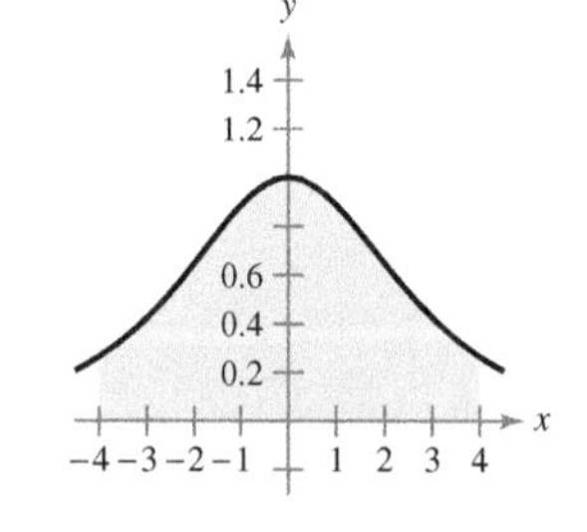

90. $y = \tanh 2x$

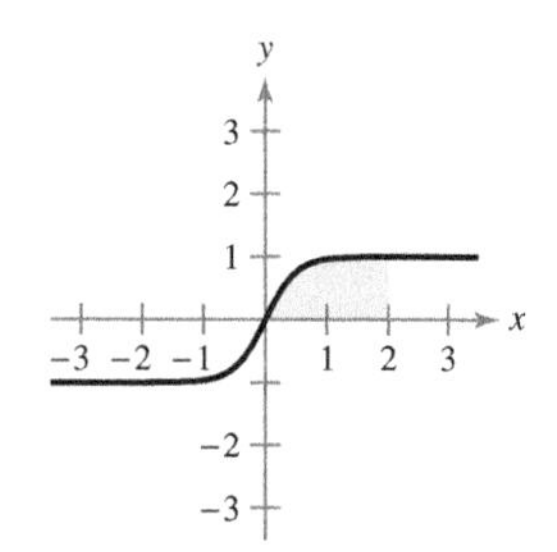

91. $y = \frac{5x}{\sqrt{x^4 + 1}}$

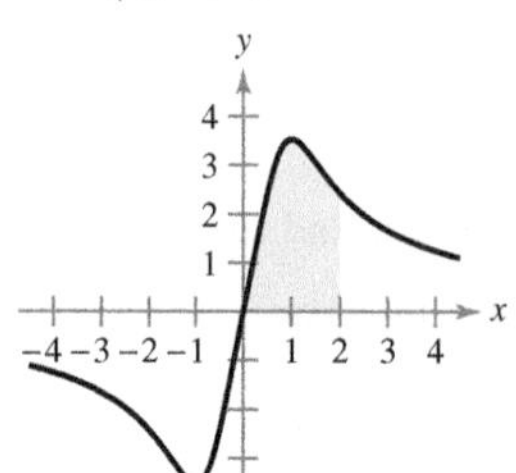

92. $y = \frac{6}{x\sqrt{9 - x^2}}$

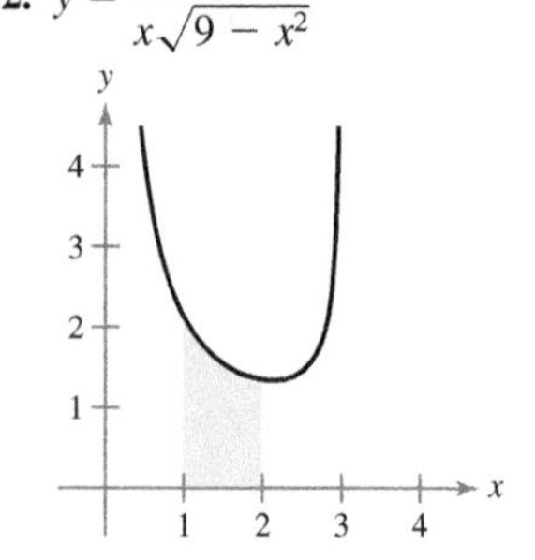

93. Tractrix Consider the equation of a tractrix

$$y = a\,\text{sech}^{-1}\left(\frac{x}{a}\right) - \sqrt{a^2 - x^2}, \quad a > 0.$$

(a) Find dy/dx.

(b) Let L be the tangent line to the tractrix at the point P. When L intersects the y-axis at the point Q, show that the distance between P and Q is a.

94. Tractrix Show that the boat in Example 5 is always pointing toward the person.

95. Proof Prove that

$$\tanh^{-1} x = \frac{1}{2}\ln\left(\frac{1 + x}{1 - x}\right), \quad -1 < x < 1.$$

96. Proof Prove that

$$\sinh^{-1} t = \ln\left(t + \sqrt{t^2 + 1}\right).$$

97. Using a Right Triangle Show that

$$\arctan(\sinh x) = \arcsin(\tanh x).$$

98. Integration Let $x > 0$ and $b > 0$. Show that

$$\int_{-b}^{b} e^{xt}\,dt = \frac{2 \sinh bx}{x}.$$

Proof In Exercises 99–101, prove the differentiation formula.

99. $\dfrac{d}{dx}[\cosh x] = \sinh x$

100. $\dfrac{d}{dx}[\coth x] = -\text{csch}^2 x$

101. $\dfrac{d}{dx}[\text{sech}\, x] = -\text{sech}\, x \tanh x$

Verifying a Differentiation Formula In Exercises 102–104, verify the differentiation formula.

102. $\dfrac{d}{dx}[\cosh^{-1} x] = \dfrac{1}{\sqrt{x^2 - 1}}$

103. $\dfrac{d}{dx}[\sinh^{-1} x] = \dfrac{1}{\sqrt{x^2 + 1}}$

104. $\dfrac{d}{dx}[\text{sech}^{-1} x] = \dfrac{-1}{x\sqrt{1 - x^2}}$

PUTNAM EXAM CHALLENGE

105. From the vertex $(0, c)$ of the catenary $y = c\cosh(x/c)$ a line L is drawn perpendicular to the tangent to the catenary at point P. Prove that the length of L intercepted by the axes is equal to the ordinate y of the point P.

106. Prove or disprove: there is at least one straight line normal to the graph of $y = \cosh x$ at a point $(a, \cosh a)$ and also normal to the graph of $y = \sinh x$ at a point $(c, \sinh c)$.

[At a point on a graph, the normal line is the perpendicular to the tangent at that point. Also, $\cosh x = (e^x + e^{-x})/2$ and $\sinh x = (e^x - e^{-x})/2$.]

These problems were composed by the Committee on the Putnam Prize Competition.

SECTION PROJECT

Mercator Map

When flying or sailing, pilots expect to be given a steady compass course to follow. On a standard flat map, this is difficult because a steady compass course results in a curved line, as shown below.

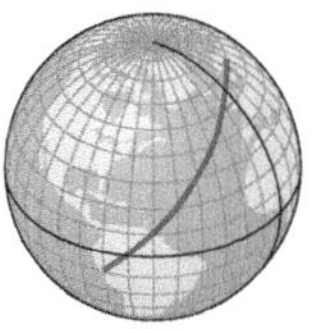

Globe: flight with constant 45° bearing

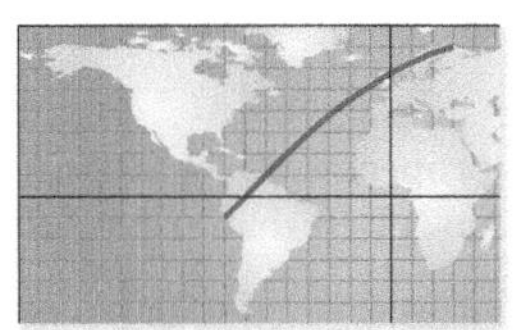

Standard flat map: flight with constant 45° bearing

For curved lines to appear as straight lines on a flat map, Flemish geographer Gerardus Mercator (1512-1594) realized that latitude lines must be stretched horizontally by a scaling factor of sec ϕ, where ϕ is the angle (in radians) of the latitude line. The Mercator map has latitude lines that are not equidistant, as shown at the right.

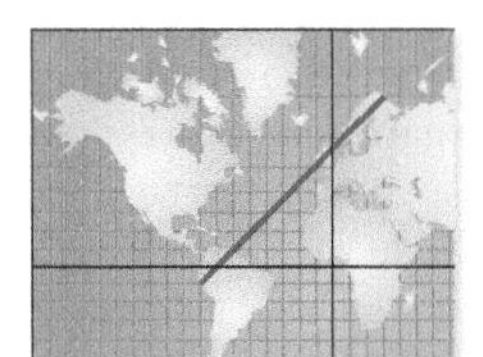

Mercator map: flight with constant 45° bearing

To calculate these vertical lengths, imagine a globe with radius R and latitude lines marked at angles of every $\Delta\phi$ radians, with $\Delta\phi = \phi_i - \phi_{i-1}$, as shown in the figure on the left below. The arc length of consecutive latitude lines is $R\Delta\phi$. On the corresponding Mercator map, the vertical distance between the ith and $(i - 1)$st latitude lines is $R\Delta\phi \sec \phi_i$, and the total vertical distance from the equator to the nth latitude line is approximately $\sum_{i=1}^{n} R\Delta\phi \sec \phi_i$, as shown in the figure on the right below.

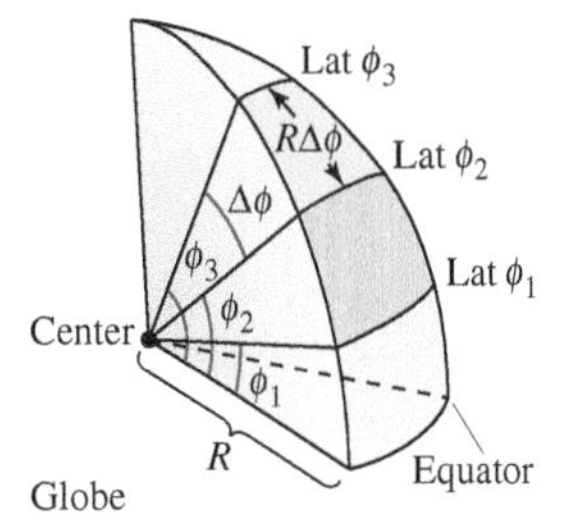

Globe

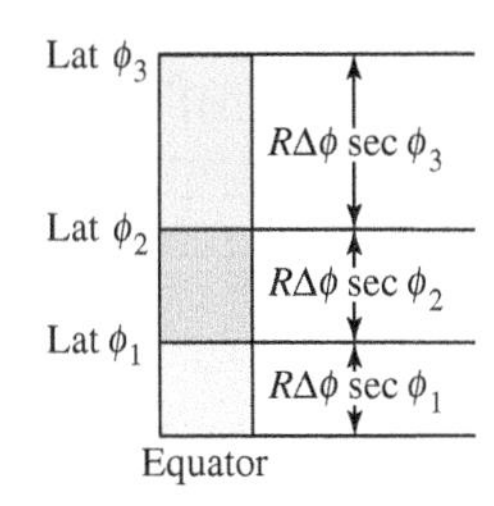

Mercator map

Mercator maps are still used by websites to display the world.

(a) Explain how to calculate the total vertical distance on a Mercator map from the equator to the nth latitude line using calculus.

(b) Using a globe radius of $R = 6$ inches, find the total vertical distances on a Mercator map from the equator to the latitude lines whose angles are 30°, 45°, and 60°.

(c) Explain what happens when you attempt to find the total vertical distance on a Mercator map from the equator to the North Pole.

(d) The *Gudermannian function* $\text{gd}(y) = \displaystyle\int_0^y \frac{dt}{\cosh t}$ expresses the latitude $\phi(y) = \text{gd}(y)$ in terms of the vertical position y on a Mercator map. Show that $\text{gd}(y) = \arctan(\sinh y)$.

Review Exercises

See CalcChat.com for tutorial help and worked-out solutions to odd-numbered exercises.

Sketching a Graph In Exercises 1 and 2, sketch the graph of the function and state its domain.

1. $f(x) = \ln x - 3$
2. $f(x) = \ln(x + 3)$

Using Properties of Logarithms In Exercises 3 and 4, use the properties of logarithms to approximate the indicated logarithms, given that $\ln 4 \approx 1.3863$ and $\ln 5 \approx 1.6094$.

3. (a) $\ln 20$ (b) $\ln \frac{4}{5}$ (c) $\ln 625$ (d) $\ln \sqrt{5}$
4. (a) $\ln 0.0625$ (b) $\ln \frac{5}{4}$ (c) $\ln 16$ (d) $\ln \sqrt[3]{80}$

Expanding a Logarithmic Expression In Exercises 5 and 6, use the properties of logarithms to expand the logarithmic expression.

5. $\ln \sqrt[5]{\dfrac{4x^2 - 1}{4x^2 + 1}}$
6. $\ln[(x^2 + 1)(x - 1)]$

Condensing a Logarithmic Expression In Exercises 7 and 8, write the expression as the logarithm of a single quantity.

7. $\ln 3 + \frac{1}{3}\ln(4 - x^2) - \ln x$
8. $3[\ln x - 2\ln(x^2 + 1)] + 2\ln 5$

Finding a Derivative In Exercises 9–16, find the derivative of the function.

9. $g(x) = \ln \sqrt{2x}$
10. $f(x) = \ln(3x^2 + 2x)$
11. $f(x) = x\sqrt{\ln x}$
12. $f(x) = [\ln(2x)]^3$
13. $y = \ln \sqrt{\dfrac{x^2 + 4}{x^2 - 4}}$
14. $y = \ln \dfrac{4x}{x - 6}$
15. $y = \dfrac{1}{\ln(1 - 7x)}$
16. $y = \dfrac{\ln 5x}{1 - x}$

Finding an Equation of a Tangent Line In Exercises 17 and 18, find an equation of the tangent line to the graph of the function at the given point.

17. $y = \ln(2 + x) + \dfrac{2}{2 + x}$, $(-1, 2)$
18. $y = 2x^2 + \ln x^2$, $(1, 2)$

Logarithmic Differentiation In Exercises 19 and 20, use logarithmic differentiation to find dy/dx.

19. $y = x^2\sqrt{x - 1}$, $x > 1$
20. $y = \dfrac{x + 2}{\sqrt{3x - 2}}$, $x > \dfrac{2}{3}$

Finding an Indefinite Integral In Exercises 21–26, find the indefinite integral.

21. $\displaystyle\int \frac{1}{7x - 2}\,dx$
22. $\displaystyle\int \frac{x^2}{x^3 + 1}\,dx$
23. $\displaystyle\int \frac{\sin x}{1 + \cos x}\,dx$
24. $\displaystyle\int \frac{\ln \sqrt{x}}{x}\,dx$
25. $\displaystyle\int \frac{x^2 - 6x + 1}{x^2 + 1}\,dx$
26. $\displaystyle\int \frac{dx}{\sqrt{x}(2\sqrt{x} + 5)}$

Evaluating a Definite Integral In Exercises 27–30, evaluate the definite integral.

27. $\displaystyle\int_1^4 \frac{2x + 1}{2x}\,dx$
28. $\displaystyle\int_1^e \frac{\ln x}{x}\,dx$
29. $\displaystyle\int_0^{\pi/3} \sec\theta\,d\theta$
30. $\displaystyle\int_0^{\pi} \tan\frac{\theta}{3}\,d\theta$

Area In Exercises 31 and 32, find the area of the region bounded by the graphs of the equations. Use a graphing utility to verify your result.

31. $y = \dfrac{6x^2}{x^3 - 2}$, $x = 3$, $x = 5$, $y = 0$
32. $y = x + \csc\dfrac{\pi x}{12}$, $x = 2$, $x = 6$, $y = 0$

Finding an Inverse Function In Exercises 33–38, (a) find the inverse function of f, (b) graph f and f^{-1} on the same set of coordinate axes, (c) verify that $f^{-1}(f(x)) = x$ and $f(f^{-1}(x)) = x$, and (d) state the domains and ranges of f and f^{-1}.

33. $f(x) = \frac{1}{2}x - 3$
34. $f(x) = 5x - 7$
35. $f(x) = \sqrt{x + 1}$
36. $f(x) = x^3 + 2$
37. $f(x) = \sqrt[3]{x + 1}$
38. $f(x) = x^2 - 5$, $x \ge 0$

Evaluating the Derivative of an Inverse Function In Exercises 39–42, verify that f has an inverse function. Then use the function f and the given real number a to find $(f^{-1})'(a)$. (*Hint:* Use Theorem 5.9.)

39. $f(x) = x^3 + 2$, $a = -1$
40. $f(x) = x\sqrt{x - 3}$, $a = 4$
41. $f(x) = \tan x$, $-\dfrac{\pi}{4} \le x \le \dfrac{\pi}{4}$, $a = \dfrac{\sqrt{3}}{3}$
42. $f(x) = \cos x$, $0 \le x \le \pi$, $a = 0$

Solving an Exponential or Logarithmic Equation **In Exercises 43–46, solve for x accurate to three decimal places.**

43. $e^{3x} = 30$

44. $-4 + 3e^{-2x} = 6$

45. $\ln \sqrt{x+1} = 2$

46. $\ln x + \ln(x - 3) = 0$

Finding a Derivative **In Exercises 47–52, find the derivative of the function.**

47. $g(t) = t^2 e^t$

48. $g(x) = \ln \dfrac{e^x}{1 + e^x}$

49. $y = \sqrt{e^{2x} + e^{-2x}}$

50. $h(z) = e^{-z^2/2}$

51. $g(x) = \dfrac{x^3}{e^{2x}}$

52. $y = 3e^{-3/t}$

Finding an Equation of a Tangent Line **In Exercises 53 and 54, find an equation of the tangent line to the graph of the function at the given point.**

53. $f(x) = e^{6x}$, $(0, 1)$

54. $h(x) = -xe^{2-x}$, $(2, -2)$

Finding Extrema and Points of Inflection **In Exercises 55 and 56, find the extrema and points of inflection (if any exist) of the function. Use a graphing utility to graph the function and confirm your results.**

55. $f(x) = (x + 1)e^{-x}$

56. $g(x) = \dfrac{1}{\sqrt{2\pi}} e^{-(x-5)^2/2}$

Finding an Indefinite Integral **In Exercises 57–60, find the indefinite integral.**

57. $\displaystyle\int xe^{1-x^2}\,dx$

58. $\displaystyle\int x^2 e^{x^3+1}\,dx$

59. $\displaystyle\int \frac{e^{4x} - e^{2x} + 1}{e^x}\,dx$

60. $\displaystyle\int \frac{e^{2x} - e^{-2x}}{e^{2x} + e^{-2x}}\,dx$

Evaluating a Definite Integral **In Exercises 61–64, evaluate the definite integral.**

61. $\displaystyle\int_0^1 xe^{-3x^2}\,dx$

62. $\displaystyle\int_{1/2}^2 \frac{e^{1/x}}{x^2}\,dx$

63. $\displaystyle\int_1^3 \frac{e^x}{e^x - 1}\,dx$

64. $\displaystyle\int_{1/4}^5 \frac{e^{4x} + 1}{4x + e^{4x}}\,dx$

65. **Area** Find the area of the region bounded by the graphs of

$$y = 2e^{-x},\quad y = 0,\quad x = 0,\quad \text{and}\quad x = 2.$$

66. **Depreciation** The value V of an item t years after it is purchased is $V = 9000e^{-0.6t}$, $0 \le t \le 5$.
   (a) Use a graphing utility to graph the function.
   (b) Find the rates of change of V with respect to t when $t = 1$ and $t = 4$.
   (c) Use a graphing utility to graph the tangent lines to the function when $t = 1$ and $t = 4$.

Sketching a Graph **In Exercises 67 and 68, sketch the graph of the function.**

67. $y = 3^{x/2}$

68. $y = \left(\dfrac{1}{4}\right)^x$

Solving an Equation **In Exercises 69–74, solve the equation accurate to three decimal places.**

69. $4^{1-x} = 52$

70. $2(3^{x+2}) = 17$

71. $\left(1 + \dfrac{0.03}{12}\right)^{12t} = 3$

72. $\left(1 + \dfrac{0.06}{365}\right)^{365t} = 2$

73. $\log_6(x + 1) = 2$

74. $\log_5 x^2 = 4.1$

Finding a Derivative **In Exercises 75–82, find the derivative of the function.**

75. $f(x) = 3^{x-1}$

76. $f(x) = 5^{3x}$

77. $g(t) = \dfrac{2^{3t}}{t^2}$

78. $f(x) = x(4^{-3x})$

79. $g(x) = \log_3 \sqrt{1 - x}$

80. $h(x) = \log_5 \dfrac{x}{x - 1}$

81. $y = x^{2x+1}$

82. $y = (3x + 5)^x$

Finding an Indefinite Integral **In Exercises 83 and 84, find the indefinite integral.**

83. $\displaystyle\int (x + 1)5^{(x+1)^2}\,dx$

84. $\displaystyle\int \frac{2^{-1/t}}{t^2}\,dt$

Evaluating a Definite Integral **In Exercises 85 and 86, evaluate the definite integral.**

85. $\displaystyle\int_1^2 6^x\,dx$

86. $\displaystyle\int_{-4}^0 9^{x/2}\,dx$

87. **Compound Interest**
   (a) A deposit of \$550 is made in a savings account that pays an annual interest rate of 1% compounded monthly. What is the balance after 11 years?
   (b) How large a deposit, at 5% interest compounded continuously, must be made to obtain a balance of \$10,000 in 15 years?
   (c) A deposit earns interest at a rate of r percent compounded continuously and doubles in value in 10 years. Find r.

88. **Climb Rate** The time t (in minutes) for a small plane to climb to an altitude of h feet is

$$t = 50 \log_{10} \frac{18{,}000}{18{,}000 - h}$$

where 18,000 feet is as high as the plane can fly.
   (a) Determine the domain of the function appropriate for the context of the problem.
   (b) Use a graphing utility to graph the function and identify any asymptotes.
   (c) Find the time when the altitude is increasing at the greatest rate.

Evaluating a Limit In Exercises 89–96, use L'Hôpital's Rule to evaluate the limit.

89. $\lim_{x\to 1} \frac{(\ln x)^2}{x-1}$

90. $\lim_{x\to 0} \frac{\sin \pi x}{\sin 5\pi x}$

91. $\lim_{x\to \infty} \frac{e^{2x}}{x^2}$

92. $\lim_{x\to \infty} xe^{-x^2}$

93. $\lim_{x\to \infty} (\ln x)^{2/x}$

94. $\lim_{x\to 1^+} (x-1)^{\ln x}$

95. $\lim_{n\to \infty} 1000\left(1+\frac{0.09}{n}\right)^n$

96. $\lim_{x\to \infty} \left(1+\frac{4}{x}\right)^x$

Evaluating an Expression In Exercises 97 and 98, evaluate each expression without using a calculator. (*Hint:* Make a sketch of a right triangle.)

97. (a) $\sin\left(\arcsin \frac{1}{2}\right)$
(b) $\cos\left(\arcsin \frac{1}{2}\right)$

98. (a) $\tan(\text{arccot } 2)$
(b) $\cos\left(\text{arcsec } \sqrt{5}\right)$

Finding a Derivative In Exercises 99–104, find the derivative of the function.

99. $y = \text{arccsc } 2x^2$

100. $y = \frac{1}{2} \arctan e^{2x}$

101. $y = x \text{ arcsec } x$

102. $y = \sqrt{x^2-4} - 2 \text{ arcsec } \frac{x}{2}, \quad 2 < x < 4$

103. $y = x(\arcsin x)^2 - 2x + 2\sqrt{1-x^2} \arcsin x$

104. $y = \tan(\arcsin x)$

Finding an Indefinite Integral In Exercises 105–110, find the indefinite integral.

105. $\int \frac{1}{e^{2x}+e^{-2x}}\,dx$

106. $\int \frac{1}{3+25x^2}\,dx$

107. $\int \frac{x}{\sqrt{1-x^4}}\,dx$

108. $\int \frac{1}{x\sqrt{9x^2-49}}\,dx$

109. $\int \frac{\arctan(x/2)}{4+x^2}\,dx$

110. $\int \frac{\arcsin 2x}{\sqrt{1-4x^2}}\,dx$

Evaluating a Definite Integral In Exercises 111–114, evaluate the definite integral.

111. $\int_0^{1/7} \frac{dx}{\sqrt{1-49x^2}}$

112. $\int_0^1 \frac{2x^2}{\sqrt{4-x^6}}\,dx$

113. $\int_{-1}^2 \frac{10e^{2x}}{25+e^{4x}}\,dx$

114. $\int_{\pi/3}^{\pi/2} \frac{\cos x}{(\sin x)\sqrt{\sin^2 x - (1/4)}}\,dx$

Area In Exercises 115 and 116, find the area of the given region.

115. $y = \frac{4-x}{\sqrt{4-x^2}}$

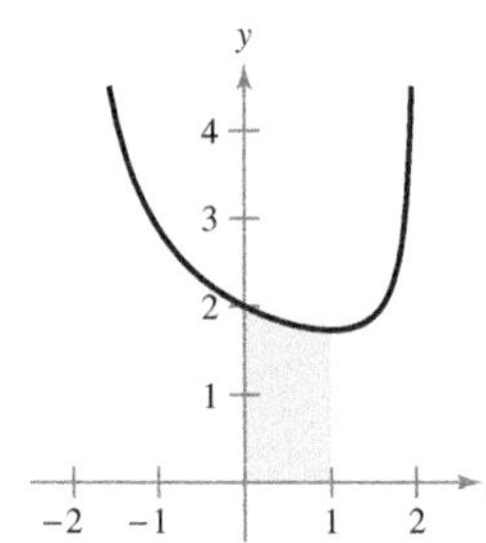

116. 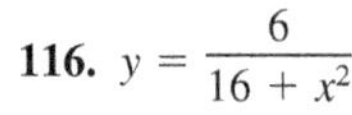 $y = \frac{6}{16+x^2}$

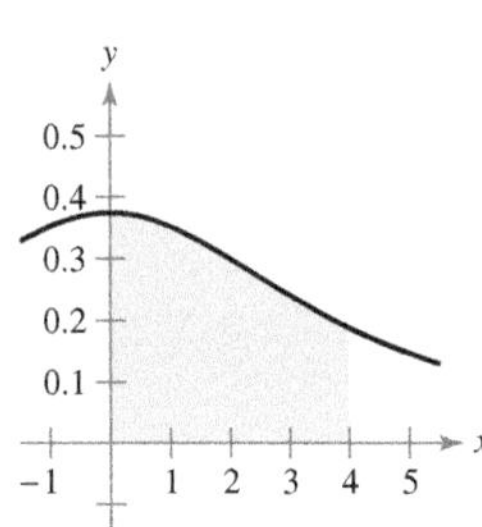

Verifying an Identity In Exercises 117 and 118, verify the identity.

117. $\cosh 2x = \cosh^2 x + \sinh^2 x$

118. $\cosh(x-y) = \cosh x \cosh y - \sinh x \sinh y$

Finding a Derivative In Exercises 119–124, find the derivative of the function.

119. $y = \text{sech}(4x-1)$

120. $y = 2x - \cosh\sqrt{x}$

121. $y = \coth 8x^2$

122. $y = \ln(\coth x)$

123. $y = \sinh^{-1}(4x)$

124. $y = x \tanh^{-1}(2x)$

Finding an Indefinite Integral In Exercises 125–130, find the indefinite integral.

125. $\int x^2 \,\text{sech}^2\, x^3\,dx$

126. $\int \sinh 6x\,dx$

127. $\int \frac{\text{sech}^2 x}{\tanh x}\,dx$

128. $\int \text{csch}^4\, 3x \coth 3x\,dx$

129. $\int \frac{1}{9-4x^2}\,dx$

130. $\int \frac{x}{\sqrt{x^4-1}}\,dx$

Evaluating a Definite Integral In Exercises 131–134, evaluate the definite integral.

131. $\int_1^2 \text{sech } 2x \tanh 2x\,dx$

132. $\int_0^1 \sinh^2 x\,dx$

133. $\int_0^1 \frac{3}{\sqrt{9x^2+16}}\,dx$

134. $\int_{-1}^0 \frac{2}{49-4x^2}\,dx$

P.S. Problem Solving

See CalcChat.com for tutorial help and worked-out solutions to odd-numbered exercises.

1. Approximation To approximate e^x, you can use a function of the form

$$f(x) = \frac{a + bx}{1 + cx}.$$

(This function is known as a **Padé approximation.**) The values of $f(0)$, $f'(0)$, and $f''(0)$ are equal to the corresponding values of e^x. Show that these values are equal to 1 and find the values of a, b, and c such that $f(0) = f'(0) = f''(0) = 1$. Then use a graphing utility to compare the graphs of f and e^x.

2. Symmetry Recall that the graph of a function $y = f(x)$ is symmetric with respect to the origin if, whenever (x, y) is a point on the graph, $(-x, -y)$ is also a point on the graph. The graph of the function $y = f(x)$ is **symmetric with respect to the point (a, b)** if, whenever $(a - x, b - y)$ is a point on the graph, $(a + x, b + y)$ is also a point on the graph, as shown in the figure.

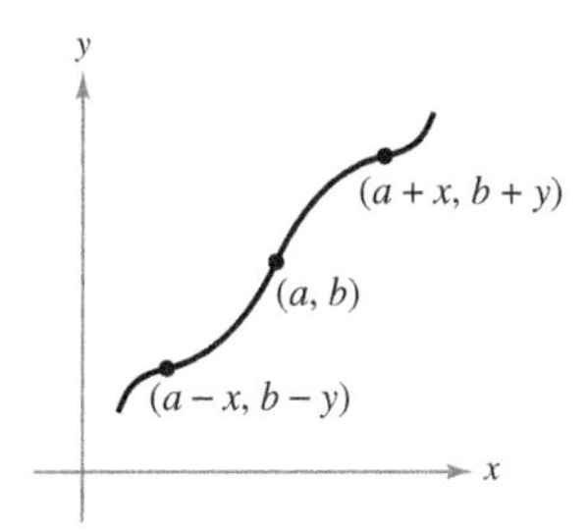

(a) Sketch the graph of $y = \sin x$ on the interval $[0, 2\pi]$. Write a short paragraph explaining how the symmetry of the graph with respect to the point $(\pi, 0)$ allows you to conclude that

$$\int_0^{2\pi} \sin x \, dx = 0.$$

(b) Sketch the graph of $y = \sin x + 2$ on the interval $[0, 2\pi]$. Use the symmetry of the graph with respect to the point $(\pi, 2)$ to evaluate the integral

$$\int_0^{2\pi} (\sin x + 2) \, dx.$$

(c) Sketch the graph of $y = \arccos x$ on the interval $[-1, 1]$. Use the symmetry of the graph to evaluate the integral

$$\int_{-1}^{1} \arccos x \, dx.$$

(d) Evaluate the integral $\displaystyle\int_0^{\pi/2} \frac{1}{1 + (\tan x)^{\sqrt{2}}} \, dx.$

3. Finding a Value Find the value of the positive constant c such that

$$\lim_{x\to\infty} \left(\frac{x + c}{x - c}\right)^x = 9.$$

4. Finding a Value Find the value of the positive constant c such that

$$\lim_{x\to\infty} \left(\frac{x - c}{x + c}\right)^x = \frac{1}{4}.$$

5. Finding Limits Use a graphing utility to estimate each limit. Then calculate each limit using L'Hôpital's Rule. What can you conclude about the form $0 \cdot \infty$?

(a) $\displaystyle\lim_{x\to 0^+} \left(\cot x + \frac{1}{x}\right)$ (b) $\displaystyle\lim_{x\to 0^+} \left(\cot x - \frac{1}{x}\right)$

(c) $\displaystyle\lim_{x\to 0^+} \left[\left(\cot x + \frac{1}{x}\right)\left(\cot x - \frac{1}{x}\right)\right]$

6. Areas and Angles

(a) Let $P(\cos t, \sin t)$ be a point on the unit circle $x^2 + y^2 = 1$ in the first quadrant (see figure). Show that t is equal to twice the area of the shaded circular sector AOP.

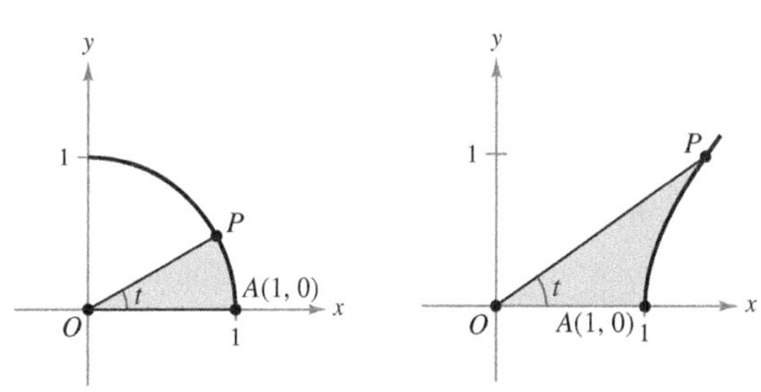

Figure for part (a) Figure for part (b)

(b) Let $P(\cosh t, \sinh t)$ be a point on the unit hyperbola $x^2 - y^2 = 1$ in the first quadrant (see figure). Show that t is equal to twice the area of the shaded region AOP. Begin by showing that the area of the shaded region AOP is given by the formula

$$A(t) = \frac{1}{2} \cosh t \sinh t - \int_1^{\cosh t} \sqrt{x^2 - 1} \, dx.$$

7. Intersection Graph the exponential function $y = a^x$ for $a = 0.5$, 1.2, and 2.0. Which of these curves intersects the line $y = x$? Determine all positive numbers a for which the curve $y = a^x$ intersects the line $y = x$.

8. Length The line $x = 1$ is tangent to the unit circle at A. The length of segment QA equals the length of the circular arc $\widehat{PA}$ (see figure). Show that the length of segment OR approaches 2 as P approaches A.

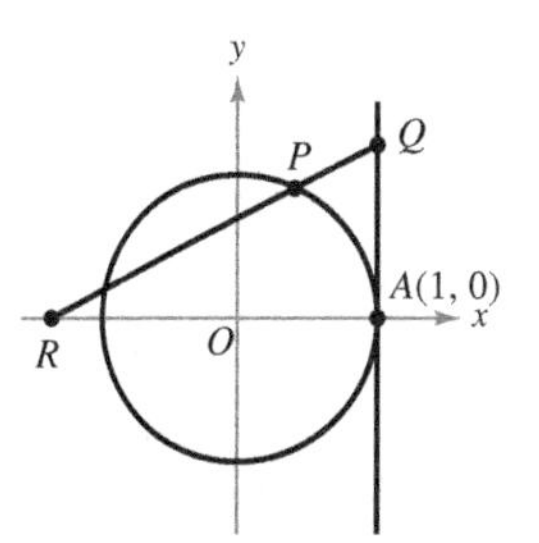

9. Area Consider the three regions A, B, and C determined by the graph of $f(x) = \arcsin x$, as shown in the figure.

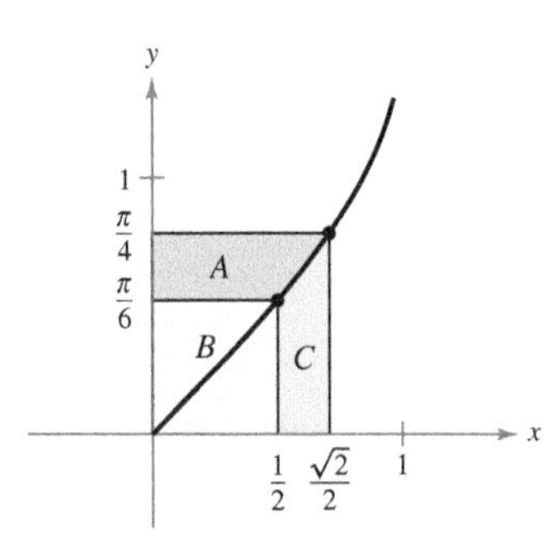

(a) Calculate the areas of regions A and B.

(b) Use your answers in part (a) to evaluate the integral

$$\int_{1/2}^{\sqrt{2}/2} \arcsin x \, dx.$$

(c) Use the methods in part (a) to evaluate the integral

$$\int_{1}^{3} \ln x \, dx.$$

(d) Use the methods in part (a) to evaluate the integral

$$\int_{1}^{\sqrt{3}} \arctan x \, dx.$$

10. Distance Let L be the tangent line to the graph of the function $y = \ln x$ at the point (a, b), where c is the y-intercept of the tangent line, as shown in the figure. Show that the distance between b and c is always equal to 1.

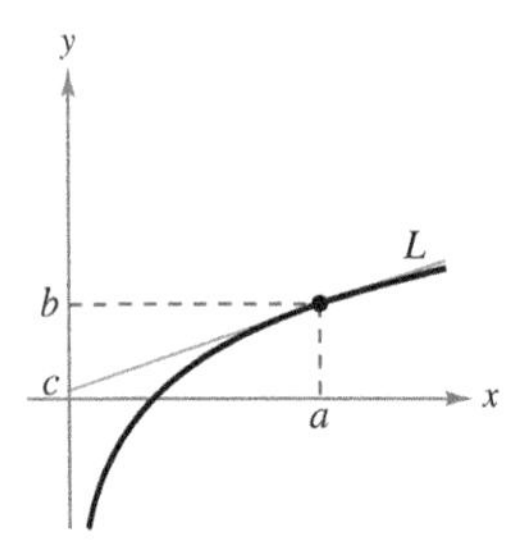

Figure for 10

Figure for 11

11. Distance Let L be the tangent line to the graph of the function $y = e^x$ at the point (a, b), where c is the y-intercept of the tangent line, as shown in the figure. Show that the distance between a and c is always equal to 1.

12. Gudermannian Function The **Gudermannian function** of x is $\text{gd}(x) = \arctan(\sinh x)$.

(a) Graph gd using a graphing utility.

(b) Show that gd is an odd function.

(c) Show that gd is monotonic and therefore has an inverse.

(d) Find the point of inflection of gd.

(e) Verify that $\text{gd}(x) = \arcsin(\tanh x)$.

13. Decreasing Function Show that $f(x) = \dfrac{\ln x^n}{x}$ is a decreasing function for $x > e$ and $n > 0$.

14. Area Use integration by substitution to find the area under the curve

$$y = \frac{1}{\sin^2 x + 4\cos^2 x}$$

between $x = 0$ and $x = \dfrac{\pi}{4}$.

15. Area Use integration by substitution to find the area under the curve

$$y = \frac{1}{\sqrt{x} + x}$$

between $x = 1$ and $x = 4$.

16. Mortgage A \$120,000 home mortgage for 35 years at $9\frac{1}{2}\%$ has a monthly payment of \$985.93. Part of the monthly payment goes for the interest charge on the unpaid balance, and the remainder of the payment is used to reduce the principal. The amount that goes for interest is

$$u = M - \left(M - \frac{Pr}{12}\right)\left(1 + \frac{r}{12}\right)^{12t}$$

and the amount that goes toward reduction of the principal is

$$v = \left(M - \frac{Pr}{12}\right)\left(1 + \frac{r}{12}\right)^{12t}.$$

In these formulas, P is the amount of the mortgage, r is the interest rate (in decimal form), M is the monthly payment, and t is the time in years.

(a) Use a graphing utility to graph each function in the same viewing window. (The viewing window should show all 35 years of mortgage payments.)

(b) In the early years of the mortgage, the larger part of the monthly payment goes for what purpose? Approximate the time when the monthly payment is evenly divided between interest and principal reduction.

(c) Use the graphs in part (a) to make a conjecture about the relationship between the slopes of the tangent lines to the two curves for a specified value of t. Give an analytical argument to verify your conjecture. Find $u'(15)$ and $v'(15)$.

(d) Repeat parts (a) and (b) for a repayment period of 20 years ($M = \$1118.56$). What can you conclude?

17. Approximating a Function

(a) Use a graphing utility to compare the graph of the function $y = e^x$ with the graph of each given function.

(i) $y_1 = 1 + \dfrac{x}{1!}$

(ii) $y_2 = 1 + \dfrac{x}{1!} + \dfrac{x^2}{2!}$

(iii) $y_3 = 1 + \dfrac{x}{1!} + \dfrac{x^2}{2!} + \dfrac{x^3}{3!}$

(b) Identify the pattern of successive polynomials in part (a), extend the pattern one more term, and compare the graph of the resulting polynomial function with the graph of $y = e^x$.

(c) What do you think this pattern implies?

Appendices

A Proofs of Selected Theorems

The text version of Appendix A, Proofs of Selected Theorems, is available at *CengageBrain.com*. Also, to enhance your study of calculus, each proof is available in video format at *LarsonCalculus.com*. At this website, you can watch videos of Bruce Edwards explaining each proof in the text and in Appendix A. To access a video, visit the website at *LarsonCalculus.com* or scan the code near the proof or the proof's reference.

Sample Video: Bruce Edwards's Proof of the Power Rule at *LarsonCalculus.com*

2.2 Basic Differentiation Rules and Rates of Change 111

The Power Rule

Before proving the next rule, it is important to review the procedure for expanding a binomial.

$$(x + \Delta x)^2 = x^2 + 2x\Delta x + (\Delta x)^2$$
$$(x + \Delta x)^3 = x^3 + 3x^2\Delta x + 3x(\Delta x)^2 + (\Delta x)^3$$
$$(x + \Delta x)^4 = x^4 + 4x^3\Delta x + 6x^2(\Delta x)^2 + 4x(\Delta x)^3 + (\Delta x)^4$$
$$(x + \Delta x)^5 = x^5 + 5x^4\Delta x + 10x^3(\Delta x)^2 + 10x^2(\Delta x)^3 + 5x(\Delta x)^4 + (\Delta x)^5$$

The general binomial expansion for a positive integer n is

$$(x + \Delta x)^n = x^n + nx^{n-1}(\Delta x) + \underbrace{\frac{n(n-1)x^{n-2}}{2}(\Delta x)^2 + \cdots + (\Delta x)^n}_{(\Delta x)^2 \text{ is a factor of these terms.}}.$$

This binomial expansion is used in proving a special case of the Power Rule.

THEOREM 2.3 The Power Rule

If n is a rational number, then the function $f(x) = x^n$ is differentiable and

$$\frac{d}{dx}[x^n] = nx^{n-1}.$$

For f to be differentiable at $x = 0$, n must be a number such that x^{n-1} is defined on an interval containing 0.

REMARK From Example 7 in Section 2.1, you know that the function $f(x) = x^{1/3}$ is defined at $x = 0$ but is not differentiable at $x = 0$. This is because $x^{-2/3}$ is not defined on an interval containing 0.

Proof If n is a positive integer greater than 1, then the binomial expansion produces

$$\frac{d}{dx}[x^n] = \lim_{\Delta x \to 0} \frac{(x + \Delta x)^n - x^n}{\Delta x}$$
$$= \lim_{\Delta x \to 0} \frac{x^n + nx^{n-1}(\Delta x) + \frac{n(n-}{}\ldots}{}$$
$$= \lim_{\Delta x \to 0} \left[nx^{n-1} + \frac{n(n-1)x^{n-2}}{2} \ldots \right.$$
$$= nx^{n-1} + 0 + \cdots + 0$$
$$= nx^{n-1}.$$

This proves the case for which n is a positive inte… the case for $n = 1$. Example 7 in Section 2.3 … integer. In Exercise 73 in Section 2.5, you are … rational. (In Section 5.5, the Power Rule will be…

When using the Power Rule, the case f… separate differentiation rule. That is,

$$\frac{d}{dx}[x] = 1.$$

This rule is consistent with the fact that the s… Figure 2.15.

The slope of the line $y = x$ is 1.
Figure 2.15

B Integration Tables

Forms Involving u^n

1. $$\int u^n\,du = \frac{u^{n+1}}{n+1} + C, \quad n \neq -1$$

2. $$\int \frac{1}{u}\,du = \ln|u| + C$$

Forms Involving $a + bu$

3. $$\int \frac{u}{a+bu}\,du = \frac{1}{b^2}(bu - a\ln|a+bu|) + C$$

4. $$\int \frac{u}{(a+bu)^2}\,du = \frac{1}{b^2}\left(\frac{a}{a+bu} + \ln|a+bu|\right) + C$$

5. $$\int \frac{u}{(a+bu)^n}\,du = \frac{1}{b^2}\left[\frac{-1}{(n-2)(a+bu)^{n-2}} + \frac{a}{(n-1)(a+bu)^{n-1}}\right] + C, \quad n \neq 1, 2$$

6. $$\int \frac{u^2}{a+bu}\,du = \frac{1}{b^3}\left[-\frac{bu}{2}(2a - bu) + a^2\ln|a+bu|\right] + C$$

7. $$\int \frac{u^2}{(a+bu)^2}\,du = \frac{1}{b^3}\left(bu - \frac{a^2}{a+bu} - 2a\ln|a+bu|\right) + C$$

8. $$\int \frac{u^2}{(a+bu)^3}\,du = \frac{1}{b^3}\left[\frac{2a}{a+bu} - \frac{a^2}{2(a+bu)^2} + \ln|a+bu|\right] + C$$

9. $$\int \frac{u^2}{(a+bu)^n}\,du = \frac{1}{b^3}\left[\frac{-1}{(n-3)(a+bu)^{n-3}} + \frac{2a}{(n-2)(a+bu)^{n-2}} - \frac{a^2}{(n-1)(a+bu)^{n-1}}\right] + C, \quad n \neq 1, 2, 3$$

10. $$\int \frac{1}{u(a+bu)}\,du = \frac{1}{a}\ln\left|\frac{u}{a+bu}\right| + C$$

11. $$\int \frac{1}{u(a+bu)^2}\,du = \frac{1}{a}\left(\frac{1}{a+bu} + \frac{1}{a}\ln\left|\frac{u}{a+bu}\right|\right) + C$$

12. $$\int \frac{1}{u^2(a+bu)}\,du = -\frac{1}{a}\left(\frac{1}{u} + \frac{b}{a}\ln\left|\frac{u}{a+bu}\right|\right) + C$$

13. $$\int \frac{1}{u^2(a+bu)^2}\,du = -\frac{1}{a^2}\left[\frac{a+2bu}{u(a+bu)} + \frac{2b}{a}\ln\left|\frac{u}{a+bu}\right|\right] + C$$

Forms Involving $a + bu + cu^2$, $b^2 \neq 4ac$

14. $$\int \frac{1}{a+bu+cu^2}\,du = \begin{cases} \dfrac{2}{\sqrt{4ac-b^2}}\arctan\dfrac{2cu+b}{\sqrt{4ac-b^2}} + C, & b^2 < 4ac \\[2ex] \dfrac{1}{\sqrt{b^2-4ac}}\ln\left|\dfrac{2cu+b-\sqrt{b^2-4ac}}{2cu+b+\sqrt{b^2-4ac}}\right| + C, & b^2 > 4ac \end{cases}$$

15. $$\int \frac{u}{a+bu+cu^2}\,du = \frac{1}{2c}\left(\ln|a+bu+cu^2| - b\int \frac{1}{a+bu+cu^2}\,du\right)$$

Forms Involving $\sqrt{a + bu}$

16. $$\int u^n\sqrt{a+bu}\,du = \frac{2}{b(2n+3)}\left[u^n(a+bu)^{3/2} - na\int u^{n-1}\sqrt{a+bu}\,du\right]$$

17. $$\int \frac{1}{u\sqrt{a+bu}}\,du = \begin{cases} \dfrac{1}{\sqrt{a}}\ln\left|\dfrac{\sqrt{a+bu}-\sqrt{a}}{\sqrt{a+bu}+\sqrt{a}}\right| + C, & a > 0 \\[2ex] \dfrac{2}{\sqrt{-a}}\arctan\sqrt{\dfrac{a+bu}{-a}} + C, & a < 0 \end{cases}$$

18. $$\int \frac{1}{u^n\sqrt{a+bu}}\,du = \frac{-1}{a(n-1)}\left[\frac{\sqrt{a+bu}}{u^{n-1}} + \frac{(2n-3)b}{2}\int \frac{1}{u^{n-1}\sqrt{a+bu}}\,du\right], \quad n \neq 1$$

19. $\displaystyle\int \frac{\sqrt{a+bu}}{u}\,du = 2\sqrt{a+bu} + a\int \frac{1}{u\sqrt{a+bu}}\,du$

20. $\displaystyle\int \frac{\sqrt{a+bu}}{u^n}\,du = \frac{-1}{a(n-1)}\left[\frac{(a+bu)^{3/2}}{u^{n-1}} + \frac{(2n-5)b}{2}\int \frac{\sqrt{a+bu}}{u^{n-1}}\,du\right],\ n \neq 1$

21. $\displaystyle\int \frac{u}{\sqrt{a+bu}}\,du = \frac{-2(2a-bu)}{3b^2}\sqrt{a+bu} + C$

22. $\displaystyle\int \frac{u^n}{\sqrt{a+bu}}\,du = \frac{2}{(2n+1)b}\left(u^n\sqrt{a+bu} - na\int \frac{u^{n-1}}{\sqrt{a+bu}}\,du\right)$

Forms Involving $a^2 \pm u^2$, $a > 0$

23. $\displaystyle\int \frac{1}{a^2+u^2}\,du = \frac{1}{a}\arctan\frac{u}{a} + C$

24. $\displaystyle\int \frac{1}{u^2-a^2}\,du = -\int \frac{1}{a^2-u^2}\,du = \frac{1}{2a}\ln\left|\frac{u-a}{u+a}\right| + C$

25. $\displaystyle\int \frac{1}{(a^2 \pm u^2)^n}\,du = \frac{1}{2a^2(n-1)}\left[\frac{u}{(a^2 \pm u^2)^{n-1}} + (2n-3)\int \frac{1}{(a^2 \pm u^2)^{n-1}}\,du\right],\ n \neq 1$

Forms Involving $\sqrt{u^2 \pm a^2}$, $a > 0$

26. $\displaystyle\int \sqrt{u^2 \pm a^2}\,du = \frac{1}{2}\left(u\sqrt{u^2 \pm a^2} \pm a^2\ln\left|u + \sqrt{u^2 \pm a^2}\right|\right) + C$

27. $\displaystyle\int u^2\sqrt{u^2 \pm a^2}\,du = \frac{1}{8}\left[u(2u^2 \pm a^2)\sqrt{u^2 \pm a^2} - a^4\ln\left|u + \sqrt{u^2 \pm a^2}\right|\right] + C$

28. $\displaystyle\int \frac{\sqrt{u^2+a^2}}{u}\,du = \sqrt{u^2+a^2} - a\ln\left|\frac{a + \sqrt{u^2+a^2}}{u}\right| + C$

29. $\displaystyle\int \frac{\sqrt{u^2-a^2}}{u}\,du = \sqrt{u^2-a^2} - a\,\mathrm{arcsec}\,\frac{|u|}{a} + C$

30. $\displaystyle\int \frac{\sqrt{u^2 \pm a^2}}{u^2}\,du = \frac{-\sqrt{u^2 \pm a^2}}{u} + \ln\left|u + \sqrt{u^2 \pm a^2}\right| + C$

31. $\displaystyle\int \frac{1}{\sqrt{u^2 \pm a^2}}\,du = \ln\left|u + \sqrt{u^2 \pm a^2}\right| + C$

32. $\displaystyle\int \frac{1}{u\sqrt{u^2+a^2}}\,du = \frac{-1}{a}\ln\left|\frac{a + \sqrt{u^2+a^2}}{u}\right| + C$

33. $\displaystyle\int \frac{1}{u\sqrt{u^2-a^2}}\,du = \frac{1}{a}\,\mathrm{arcsec}\,\frac{|u|}{a} + C$

34. $\displaystyle\int \frac{u^2}{\sqrt{u^2 \pm a^2}}\,du = \frac{1}{2}\left(u\sqrt{u^2 \pm a^2} \mp a^2\ln\left|u + \sqrt{u^2 \pm a^2}\right|\right) + C$

35. $\displaystyle\int \frac{1}{u^2\sqrt{u^2 \pm a^2}}\,du = \mp\frac{\sqrt{u^2 \pm a^2}}{a^2u} + C$

36. $\displaystyle\int \frac{1}{(u^2 \pm a^2)^{3/2}}\,du = \frac{\pm u}{a^2\sqrt{u^2 \pm a^2}} + C$

Forms Involving $\sqrt{a^2 - u^2}$, $a > 0$

37. $\displaystyle\int \sqrt{a^2-u^2}\,du = \frac{1}{2}\left(u\sqrt{a^2-u^2} + a^2\arcsin\frac{u}{a}\right) + C$

38. $\displaystyle\int u^2\sqrt{a^2-u^2}\,du = \frac{1}{8}\left[u(2u^2-a^2)\sqrt{a^2-u^2} + a^4\arcsin\frac{u}{a}\right] + C$

39. $\displaystyle\int \frac{\sqrt{a^2-u^2}}{u}\,du = \sqrt{a^2-u^2} - a\ln\left|\frac{a+\sqrt{a^2-u^2}}{u}\right| + C$

40. $\displaystyle\int \frac{\sqrt{a^2-u^2}}{u^2}\,du = \frac{-\sqrt{a^2-u^2}}{u} - \arcsin\frac{u}{a} + C$

41. $\displaystyle\int \frac{1}{\sqrt{a^2-u^2}}\,du = \arcsin\frac{u}{a} + C$

42. $\displaystyle\int \frac{1}{u\sqrt{a^2-u^2}}\,du = \frac{-1}{a}\ln\left|\frac{a+\sqrt{a^2-u^2}}{u}\right| + C$

43. $\displaystyle\int \frac{u^2}{\sqrt{a^2-u^2}}\,du = \frac{1}{2}\left(-u\sqrt{a^2-u^2} + a^2\arcsin\frac{u}{a}\right) + C$

44. $\displaystyle\int \frac{1}{u^2\sqrt{a^2-u^2}}\,du = \frac{-\sqrt{a^2-u^2}}{a^2u} + C$

45. $\displaystyle\int \frac{1}{(a^2-u^2)^{3/2}}\,du = \frac{u}{a^2\sqrt{a^2-u^2}} + C$

Forms Involving sin *u* or cos *u*

46. $\displaystyle\int \sin u\,du = -\cos u + C$

47. $\displaystyle\int \cos u\,du = \sin u + C$

48. $\displaystyle\int \sin^2 u\,du = \frac{1}{2}(u - \sin u\cos u) + C$

49. $\displaystyle\int \cos^2 u\,du = \frac{1}{2}(u + \sin u\cos u) + C$

50. $\displaystyle\int \sin^n u\,du = -\frac{\sin^{n-1}u\cos u}{n} + \frac{n-1}{n}\int \sin^{n-2}u\,du$

51. $\displaystyle\int \cos^n u\,du = \frac{\cos^{n-1}u\sin u}{n} + \frac{n-1}{n}\int \cos^{n-2}u\,du$

52. $\displaystyle\int u\sin u\,du = \sin u - u\cos u + C$

53. $\displaystyle\int u\cos u\,du = \cos u + u\sin u + C$

54. $\displaystyle\int u^n\sin u\,du = -u^n\cos u + n\int u^{n-1}\cos u\,du$

55. $\displaystyle\int u^n\cos u\,du = u^n\sin u - n\int u^{n-1}\sin u\,du$

56. $\displaystyle\int \frac{1}{1\pm\sin u}\,du = \tan u \mp \sec u + C$

57. $\displaystyle\int \frac{1}{1\pm\cos u}\,du = -\cot u \pm \csc u + C$

58. $\displaystyle\int \frac{1}{\sin u\cos u}\,du = \ln|\tan u| + C$

Forms Involving tan *u*, cot *u*, sec *u*, or csc *u*

59. $\displaystyle\int \tan u\,du = -\ln|\cos u| + C$

60. $\displaystyle\int \cot u\,du = \ln|\sin u| + C$

61. $\displaystyle\int \sec u\,du = \ln|\sec u + \tan u| + C$

62. $\displaystyle\int \csc u\,du = \ln|\csc u - \cot u| + C$ or $\displaystyle\int \csc u\,du = -\ln|\csc u + \cot u| + C$

63. $\displaystyle\int \tan^2 u\,du = -u + \tan u + C$

64. $\displaystyle\int \cot^2 u\,du = -u - \cot u + C$

65. $\displaystyle\int \sec^2 u\,du = \tan u + C$

66. $\displaystyle\int \csc^2 u\,du = -\cot u + C$

67. $\displaystyle\int \tan^n u\,du = \frac{\tan^{n-1}u}{n-1} - \int \tan^{n-2}u\,du,\ n \neq 1$

68. $\displaystyle\int \cot^n u\,du = -\frac{\cot^{n-1}u}{n-1} - \int \cot^{n-2}u\,du,\ n \neq 1$

69. $\displaystyle\int \sec^n u\,du = \frac{\sec^{n-2}u\tan u}{n-1} + \frac{n-2}{n-1}\int \sec^{n-2}u\,du,\ n \neq 1$

70. $\displaystyle\int \csc^n u\,du = -\frac{\csc^{n-2}u\cot u}{n-1} + \frac{n-2}{n-1}\int \csc^{n-2}u\,du,\ n \neq 1$

71. $\displaystyle\int \frac{1}{1 \pm \tan u}\, du = \frac{1}{2}(u \pm \ln|\cos u \pm \sin u|) + C$

72. $\displaystyle\int \frac{1}{1 \pm \cot u}\, du = \frac{1}{2}(u \mp \ln|\sin u \pm \cos u|) + C$

73. $\displaystyle\int \frac{1}{1 \pm \sec u}\, du = u + \cot u \mp \csc u + C$

74. $\displaystyle\int \frac{1}{1 \pm \csc u}\, du = u - \tan u \pm \sec u + C$

Forms Involving Inverse Trigonometric Functions

75. $\displaystyle\int \arcsin u\, du = u \arcsin u + \sqrt{1 - u^2} + C$

76. $\displaystyle\int \arccos u\, du = u \arccos u - \sqrt{1 - u^2} + C$

77. $\displaystyle\int \arctan u\, du = u \arctan u - \ln\sqrt{1 + u^2} + C$

78. $\displaystyle\int \operatorname{arccot} u\, du = u \operatorname{arccot} u + \ln\sqrt{1 + u^2} + C$

79. $\displaystyle\int \operatorname{arcsec} u\, du = u \operatorname{arcsec} u - \ln\left|u + \sqrt{u^2 - 1}\right| + C$

80. $\displaystyle\int \operatorname{arccsc} u\, du = u \operatorname{arccsc} u + \ln\left|u + \sqrt{u^2 - 1}\right| + C$

Forms Involving e^u

81. $\displaystyle\int e^u\, du = e^u + C$

82. $\displaystyle\int ue^u\, du = (u - 1)e^u + C$

83. $\displaystyle\int u^n e^u\, du = u^n e^u - n\int u^{n-1} e^u\, du$

84. $\displaystyle\int \frac{1}{1 + e^u}\, du = u - \ln(1 + e^u) + C$

85. $\displaystyle\int e^{au} \sin bu\, du = \frac{e^{au}}{a^2 + b^2}(a \sin bu - b \cos bu) + C$

86. $\displaystyle\int e^{au} \cos bu\, du = \frac{e^{au}}{a^2 + b^2}(a \cos bu + b \sin bu) + C$

Forms Involving $\ln u$

87. $\displaystyle\int \ln u\, du = u(-1 + \ln u) + C$

88. $\displaystyle\int u \ln u\, du = \frac{u^2}{4}(-1 + 2 \ln u) + C$

89. $\displaystyle\int u^n \ln u\, du = \frac{u^{n+1}}{(n + 1)^2}[-1 + (n + 1)\ln u] + C,\ n \neq -1$

90. $\displaystyle\int (\ln u)^2\, du = u[2 - 2 \ln u + (\ln u)^2] + C$

91. $\displaystyle\int (\ln u)^n\, du = u(\ln u)^n - n\int (\ln u)^{n-1}\, du$

Forms Involving Hyperbolic Functions

92. $\displaystyle\int \cosh u\, du = \sinh u + C$

93. $\displaystyle\int \sinh u\, du = \cosh u + C$

94. $\displaystyle\int \operatorname{sech}^2 u\, du = \tanh u + C$

95. $\displaystyle\int \operatorname{csch}^2 u\, du = -\coth u + C$

96. $\displaystyle\int \operatorname{sech} u \tanh u\, du = -\operatorname{sech} u + C$

97. $\displaystyle\int \operatorname{csch} u \coth u\, du = -\operatorname{csch} u + C$

Forms Involving Inverse Hyperbolic Functions (in logarithmic form)

98. $\displaystyle\int \frac{du}{\sqrt{u^2 \pm a^2}} = \ln\left(u + \sqrt{u^2 \pm a^2}\right) + C$

99. $\displaystyle\int \frac{du}{a^2 - u^2} = \frac{1}{2a}\ln\left|\frac{a + u}{a - u}\right| + C$

100. $\displaystyle\int \frac{du}{u\sqrt{a^2 \pm u^2}} = -\frac{1}{a}\ln\frac{a + \sqrt{a^2 \pm u^2}}{|u|} + C$

Chapter 1

Section 1.1 (page 51)

1. Calculus is the mathematics of change. Precalculus mathematics is more static.

Answers will vary. Sample answer:

Precalculus	**Calculus**
Area of a rectangle	Area under a curve
Work done by a constant force	Work done by a variable force
Center of a rectangle	Centroid of a region

3. Precalculus: 300 ft

5. Calculus: Slope of the tangent line at $x = 2$ is 0.16.

7. (a)

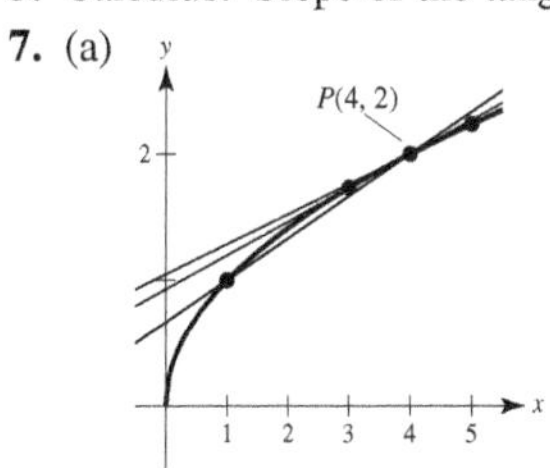

(b) $x = 1$: $m = \frac{1}{3}$

$x = 3$: $m = \dfrac{1}{\sqrt{3} + 2} \approx 0.2679$

$x = 5$: $m = \dfrac{1}{\sqrt{5} + 2} \approx 0.2361$

(c) $\frac{1}{4}$; You can improve your approximation of the slope at $x = 4$ by considering x-values very close to 4.

9. Area $\approx$ 10.417; Area $\approx$ 9.145; Use more rectangles.

11. (a) About 5.66 (b) About 6.11

(c) Increase the number of line segments.

Section 1.2 (page 59)

1. As the graph of the function approaches 8 on the horizontal axis, the graph approaches 25 on the vertical axis.

3.

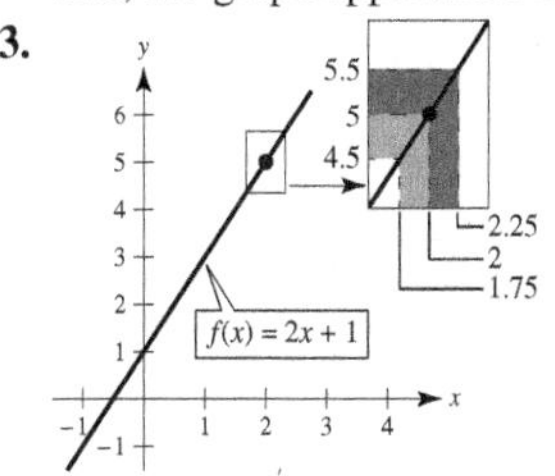

5.

x	3.9	3.99	3.999	4
$f(x)$	0.3448	0.3344	0.3334	?

x	4.001	4.01	4.1
$f(x)$	0.3332	0.3322	0.3226

$\lim_{x\to 4} \frac{x-4}{x^2-5x+4} \approx 0.3333 \left(\text{Actual limit is } \frac{1}{3}.\right)$

7.

x	−0.1	−0.01	−0.001	0
$f(x)$	0.5132	0.5013	0.5001	?

x	0.001	0.01	0.1
$f(x)$	0.4999	0.4988	0.4881

$\lim_{x\to 0} \frac{\sqrt{x+1}-1}{x} \approx 0.5000 \left(\text{Actual limit is } \frac{1}{2}.\right)$

9.

x	−0.1	−0.01	−0.001	0
$f(x)$	0.9983	0.99998	1.0000	?

x	0.001	0.01	0.1
$f(x)$	1.0000	0.99998	0.9983

$\lim_{x\to 0} \frac{\sin x}{x} \approx 1.0000$ (Actual limit is 1.)

11.

x	0.9	0.99	0.999	1
$f(x)$	0.2564	0.2506	0.2501	?

x	1.001	1.01	1.1
$f(x)$	0.2499	0.2494	0.2439

$\lim_{x\to 1} \frac{x-2}{x^2+x-6} \approx 0.2500 \left(\text{Actual limit is } \frac{1}{4}.\right)$

13.

x	0.9	0.99	0.999	1
$f(x)$	0.7340	0.6733	0.6673	?

x	1.001	1.01	1.1
$f(x)$	0.6660	0.6600	0.6015

$\lim_{x\to 1} \frac{x^4-1}{x^6-1} \approx 0.6666 \left(\text{Actual limit is } \frac{2}{3}.\right)$

15.

x	−6.1	−6.01	−6.001	−6
$f(x)$	−0.1248	−0.1250	−0.1250	?

x	−5.999	−5.99	−5.9
$f(x)$	−0.1250	−0.1250	−0.1252

$\lim_{x\to -6} \frac{\sqrt{10-x}-4}{x+6} \approx -0.1250 \left(\text{Actual limit is } -\frac{1}{8}.\right)$

17.

x	−0.1	−0.01	−0.001	0
$f(x)$	1.9867	1.9999	2.0000	?

x	0.001	0.01	0.1
$f(x)$	2.0000	1.9999	1.9867

$\lim_{x\to 0} \frac{\sin 2x}{x} \approx 2.0000$ (Actual limit is 2.)

19.

x	−0.1	−0.01	−0.001	0
$f(x)$	−2000	-2×10^6	-2×10^9	?

x	0.001	0.01	0.1
$f(x)$	2×10^9	2×10^6	2000

As x approaches 0 from the left, the function decreases without bound. As x approaches 0 from the right, the function increases without bound.

21. 1 **23.** 2

25. Limit does not exist. The function approaches 1 from the right side of 2, but it approaches −1 from the left side of 2.

27. Limit does not exist. The function oscillates between 1 and −1 as x approaches 0.

29. (a) 2
(b) Limit does not exist. The function approaches 1 from the right side of 1, but it approaches 3.5 from the left side of 1.
(c) Value does not exist. The function is undefined at $x = 4$.
(d) 2

31.

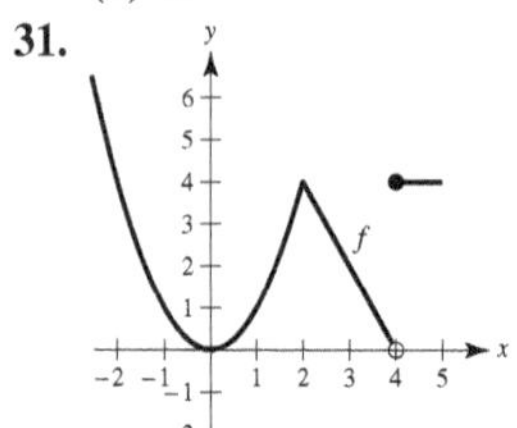

33.

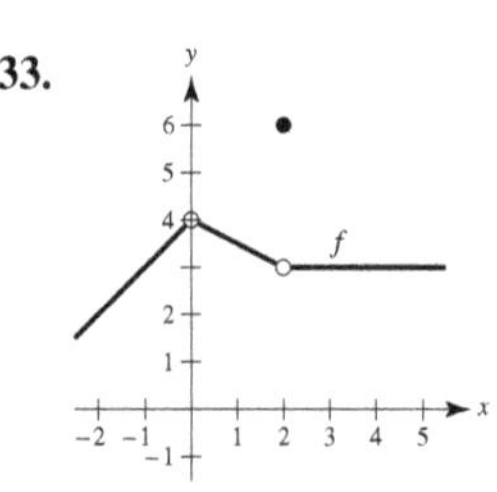

$\lim_{x\to c} f(x)$ exists for all points on the graph except where $c = 4$.

35. $\delta = 0.4$ **37.** $\delta = \frac{1}{11} \approx 0.091$

39. $L = 8$

Answers will vary. Sample answers:
(a) $\delta \approx 0.0033$ (b) $\delta \approx 0.00167$

41. $L = 1$
Answers will vary.
Sample answers:
(a) $\delta = 0.002$
(b) $\delta = 0.001$

43. $L = 12$
Answers will vary.
Sample answers:
(a) $\delta = 0.00125$
(b) $\delta = 0.000625$

45. 6 **47.** −3 **49.** 3 **51.** 0 **53.** 10

55. 2 **57.** 4

59.

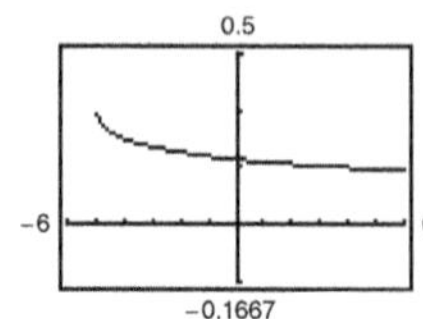

$\lim_{x\to 4} f(x) = \frac{1}{6}$
Domain: $[-5, 4) \cup (4, \infty)$
The graph has a hole at $x = 4$.

61. (a) \$17.89; the cost of a 10-minute, 45-second phone call

(b) 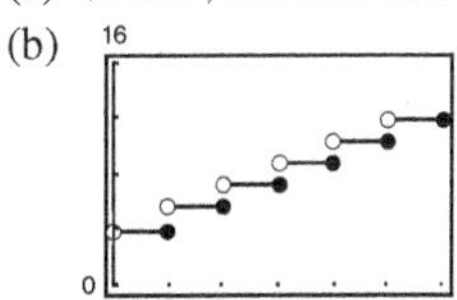

The limit does not exist because the limits from the right and left are not equal.

63. Choosing a smaller positive value of δ will still satisfy the inequality $|f(x) - L| < \varepsilon$.

65. No. The fact that $f(2) = 4$ has no bearing on the existence of the limit of $f(x)$ as x approaches 2.

67. (a) $r = \dfrac{3}{\pi} \approx 0.9549$ cm

(b) $\dfrac{5.5}{2\pi} \le r \le \dfrac{6.5}{2\pi}$, or approximately $0.8754 < r < 1.0345$

(c) $\lim\limits_{r\to 3/\pi} 2\pi r = 6$; $\varepsilon = 0.5$; $\delta \approx 0.0796$

69.

x	-0.001	-0.0001	-0.00001
$f(x)$	2.7196	2.7184	2.7183

x	0.00001	0.0001	0.001
$f(x)$	2.7183	2.7181	2.7169

$\lim\limits_{x\to 0} f(x) \approx 2.7183$

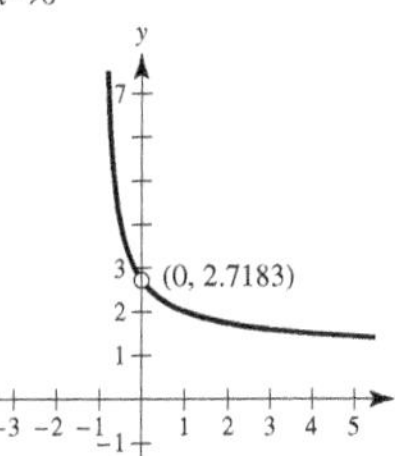

71. 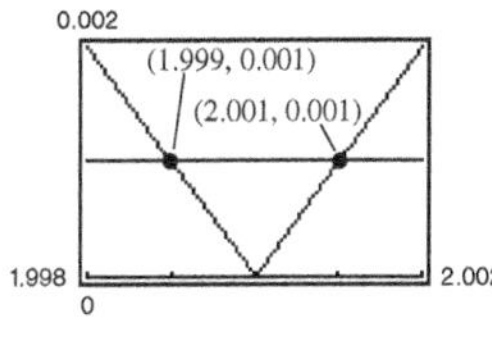

$\delta = 0.001$, $(1.999, 2.001)$

73. False. The existence or nonexistence of $f(x)$ at $x = c$ has no bearing on the existence of the limit of $f(x)$ as $x \to c$.

75. False. See Exercise 23.

77. Yes. As x approaches 0.25 from either side, $\sqrt{x}$ becomes arbitrarily close to 0.5.

79. $\lim\limits_{x\to 0} \dfrac{\sin nx}{x} = n$ **81–83.** Proofs

85. Putnam Problem B1, 1986

Section 1.3 *(page 71)*

1. Substitute c for x and simplify.

3. If a function f is squeezed between two functions h and g, $h(x) \le f(x) \le g(x)$, and h and g have the same limit L as $x \to c$, then $\lim\limits_{x\to c} f(x)$ exists and equals L.

5. 8 **7.** -1 **9.** 0 **11.** 7 **13.** $\sqrt{11}$ **15.** 125

17. $\frac{3}{5}$ **19.** $\frac{1}{5}$ **21.** 7 **23.** (a) 4 (b) 64 (c) 64

25. (a) 3 (b) 2 (c) 2 **27.** 1 **29.** $\frac{1}{2}$ **31.** 1

33. $\frac{1}{2}$ **35.** -1 **37.** (a) 10 (b) $\frac{12}{5}$ (c) $\frac{4}{5}$ (d) $\frac{1}{5}$

39. (a) 256 (b) 4 (c) 48 (d) 64

41. $f(x) = \dfrac{x^2 + 3x}{x}$ and $g(x) = x + 3$ agree except at $x = 0$.

$\lim\limits_{x\to 0} f(x) = \lim\limits_{x\to 0} g(x) = 3$

43. $f(x) = \dfrac{x^2 - 1}{x + 1}$ and $g(x) = x - 1$ agree except at $x = -1$.

$\lim\limits_{x\to -1} f(x) = \lim\limits_{x\to -1} g(x) = -2$

45. $f(x) = \dfrac{x^3 - 8}{x - 2}$ and $g(x) = x^2 + 2x + 4$ agree except at $x = 2$.

$\lim\limits_{x\to 2} f(x) = \lim\limits_{x\to 2} g(x) = 12$

47. -1 **49.** $\dfrac{1}{8}$ **51.** $\dfrac{5}{6}$ **53.** $\dfrac{1}{6}$ **55.** $\dfrac{\sqrt{5}}{10}$

57. $-\frac{1}{9}$ **59.** 2 **61.** $2x - 2$ **63.** $\frac{1}{5}$ **65.** 0

67. 0 **69.** 0 **71.** 0 **73.** $\frac{3}{2}$

75. 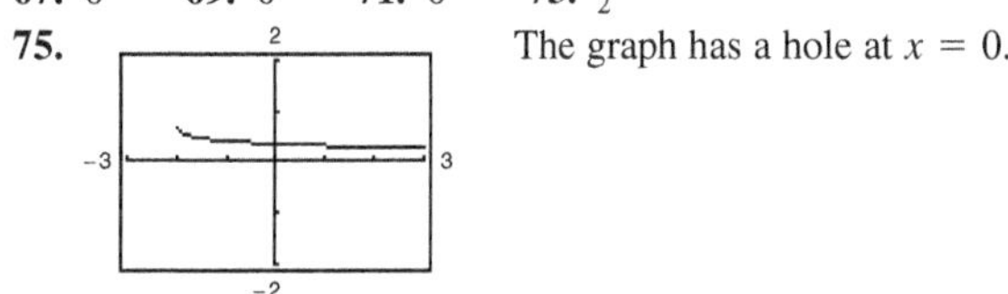

The graph has a hole at $x = 0$.

Answers will vary. Sample answer:

x	-0.1	-0.01	-0.001	0.001	0.01	0.1
$f(x)$	0.358	0.354	0.354	0.354	0.353	0.349

$\lim\limits_{x\to 0} \dfrac{\sqrt{x+2} - \sqrt{2}}{x} \approx 0.354$; Actual limit is $\dfrac{1}{2\sqrt{2}} = \dfrac{\sqrt{2}}{4}$.

77. 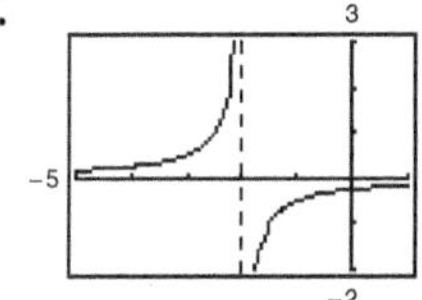

The graph has a hole at $x = 0$.

Answers will vary. Sample answer:

x	-0.1	-0.01	-0.001
$f(x)$	-0.263	-0.251	-0.250

x	0.001	0.01	0.1
$f(x)$	-0.250	-0.249	-0.238

$\lim\limits_{x\to 0} \dfrac{[1/(2 + x)] - (1/2)}{x} \approx -0.250$; Actual limit is $-\dfrac{1}{4}$.

79.

The graph has a hole at $t = 0$.

Answers will vary. Sample answer:

t	-0.1	-0.01	0	0.01	0.1
$f(t)$	2.96	2.9996	?	2.9996	2.96

$\lim\limits_{t\to 0} \dfrac{\sin 3t}{t} \approx 3.0000$; Actual limit is 3.

81. 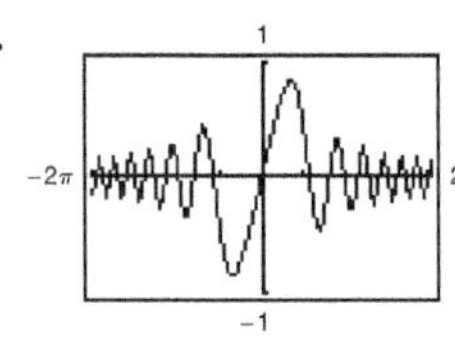

The graph has a hole at $x = 0$.

Answers will vary. Sample answer:

x	-0.1	-0.01	-0.001	0	0.001	0.01	0.1
$f(x)$	-0.1	-0.01	-0.001	?	0.001	0.01	0.1

$\lim_{x\to 0} \frac{\sin x^2}{x} = 0$; Actual limit is 0.

83. 3 **85.** $2x - 4$ **87.** $x^{-1/2}$

89. $-1/(x + 3)^2$ **91.** 4

93. 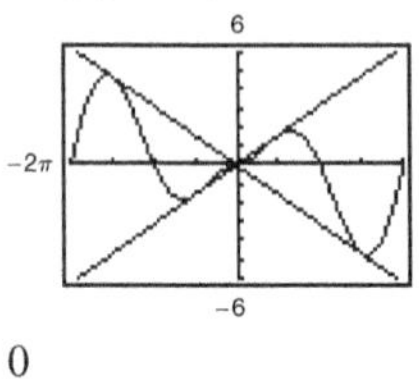

0

95. 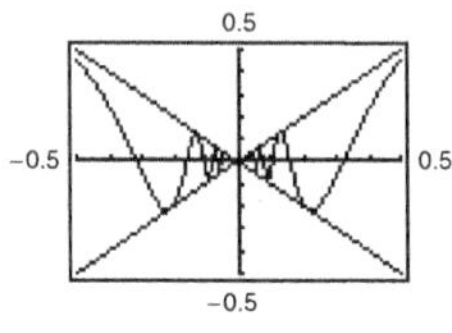

0

The graph has a hole at $x = 0$.

97. (a) f and g agree at all but one point if c is a real number such that $f(x) = g(x)$ for all $x \neq c$.

(b) Sample answer: $f(x) = \frac{x^2 - 1}{x - 1}$ and $g(x) = x + 1$ agree at all points except $x = 1$.

99. 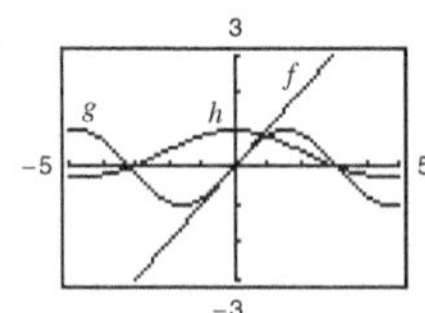

The magnitudes of $f(x)$ and $g(x)$ are approximately equal when x is close to 0. Therefore, their ratio is approximately 1.

101. -64 ft/sec (speed = 64 ft/sec) **103.** -29.4 m/sec

105. Let $f(x) = 1/x$ and $g(x) = -1/x$.

$\lim_{x\to 0} f(x)$ and $\lim_{x\to 0} g(x)$ do not exist. However,

$$\lim_{x\to 0} [f(x) + g(x)] = \lim_{x\to 0} \left[\frac{1}{x} + \left(-\frac{1}{x}\right)\right] = \lim_{x\to 0} 0 = 0$$

and therefore does exist.

107–111. Proofs

113. Let $f(x) = \begin{cases} 4, & x \geq 0 \\ -4, & x < 0 \end{cases}$.

$\lim_{x\to 0} |f(x)| = \lim_{x\to 0} 4 = 4$

$\lim_{x\to 0} f(x)$ does not exist because for $x < 0$, $f(x) = -4$ and for $x \geq 0$, $f(x) = 4$.

115. False. The limit does not exist because the function approaches 1 from the right side of 0 and approaches -1 from the left side of 0.

117. True.

119. False. The limit does not exist because $f(x)$ approaches 3 from the left side of 2 and approaches 0 from the right side of 2.

121. Proof

123. (a) All $x \neq 0, \frac{\pi}{2} + n\pi$

(b) 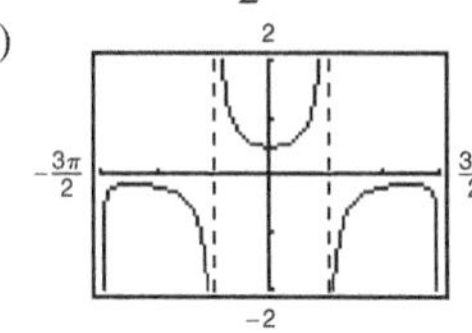

The domain is not obvious. The hole at $x = 0$ is not apparent from the graph.

(c) $\frac{1}{2}$ (d) $\frac{1}{2}$

Section 1.4 *(page 83)*

1. A function is continuous at a point c if there is no interruption of the graph at c.

3. The limit exists because the limit from the left and the limit from the right are equivalent.

5. (a) 3 (b) 3 (c) 3; $f(x)$ is continuous on $(-\infty, \infty)$.

7. (a) 0 (b) 0 (c) 0; Discontinuity at $x = 3$

9. (a) -3 (b) 3 (c) Limit does not exist. Discontinuity at $x = 2$

11. $\frac{1}{16}$ **13.** $\frac{1}{10}$

15. Limit does not exist. The function decreases without bound as x approaches -3 from the left.

17. -1 **19.** $-\frac{1}{x^2}$ **21.** $\frac{5}{2}$ **23.** 2

25. Limit does not exist. The function decreases without bound as x approaches π from the left and increases without bound as x approaches π from the right.

27. 8 **29.** 2

31. Discontinuities at $x = -2$ and $x = 2$

33. Discontinuities at every integer

35. Continuous on $[-7, 7]$ **37.** Continuous on $[-1, 4]$

39. Nonremovable discontinuity at $x = 0$

41. Nonremovable discontinuities at $x = -2$ and $x = 2$

43. Continuous for all real x

45. Nonremovable discontinuity at $x = 1$
Removable discontinuity at $x = 0$

47. Removable discontinuity at $x = -2$
Nonremovable discontinuity at $x = 5$

49. Nonremovable discontinuity at $x = -7$

51. Nonremovable discontinuity at $x = 2$

53. Continuous for all real x

55. Nonremovable discontinuities at integer multiples of $\frac{\pi}{2}$

57. Nonremovable discontinuities at each integer

59. $a = 7$ **61.** $a = 2$ **63.** $a = -1, b = 1$

65. Continuous for all real x

67. Nonremovable discontinuities at $x = 1$ and $x = -1$

69. Continuous on the open intervals $\ldots, (-3\pi, -\pi), (-\pi, \pi), (\pi, 3\pi), \ldots$

71. 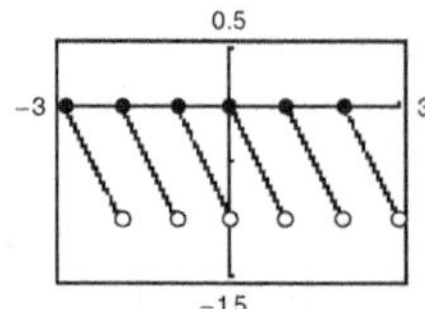

Nonremovable discontinuity at each integer

73. 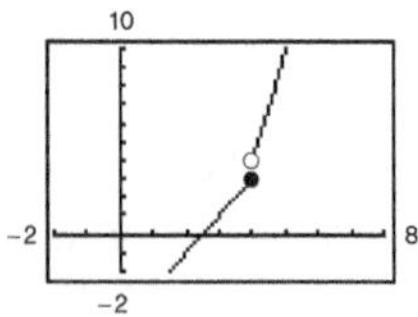

Nonremovable discontinuity at $x = 4$

75. Continuous on $(-\infty, \infty)$ **77.** Continuous on $[0, \infty)$

79. Continuous on the open intervals . . . , $(-6, -2)$, $(-2, 2)$, $(2, 6)$, . . .

81. Continuous on $(-\infty, \infty)$

83. Because $f(x)$ is continuous on the interval $[1, 2]$ and $f(1) = \frac{37}{12}$ and $f(2) = -\frac{8}{3}$, by the Intermediate Value Theorem there exists a real number c in $[1, 2]$ such that $f(c) = 0$.

85. Because $f(x)$ is continuous on the interval $[0, \pi]$ and $f(0) = -3$ and $f(\pi) \approx 8.87$, by the Intermediate Value Theorem there exists a real number c in $[0, \pi]$ such that $f(c) = 0$.

87. Consider the intervals $[1, 3]$ and $[3, 5]$.
$f(1) = 2 > 0$ and $f(3) = -2 < 0$. So, there is at least one zero in the interval $[1, 3]$.
$f(3) = -2 < 0$ and $f(5) = 2 > 0$. So, there is at least one zero in the interval $[3, 5]$.

89. 0.68, 0.6823 **91.** 0.95, 0.9472 **93.** 0.56, 0.5636

95. $f(3) = 11$; $c = 3$

97. $f(0) \approx 0.6458$, $f(5) \approx 1.464$; $c = 2$

99. $f(1) = 0$, $f(3) = 24$; $c = 2$

101. Answers will vary. Sample answer:
$$f(x) = \frac{1}{(x - a)(x - b)}$$

103. If f and g are continuous for all real x, then so is $f + g$ (Theorem 1.11, part 2). However, $\frac{f}{g}$ might not be continuous if $g(x) = 0$. For example, let $f(x) = x$ and $g(x) = x^2 - 1$. Then f and g are continuous for all real x, but $\frac{f}{g}$ is not continuous at $x = \pm 1$.

105. True

107. False. $f(x) = \cos x$ has two zeros in $[0, 2\pi]$. However, $f(0)$ and $f(2\pi)$ have the same sign.

109. False. A rational function can be written as $\frac{P(x)}{Q(x)}$, where P and Q are polynomials of degree m and n, respectively. It can have, at most, n discontinuities.

111. The functions differ by 1 for non-integer values of x.

113.

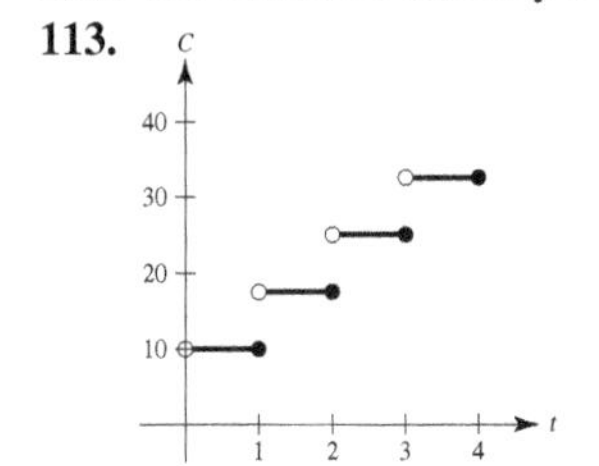

There is a jump discontinuity every gigabyte.

115–117. Proofs **119.** Answers will vary.

121. (a)

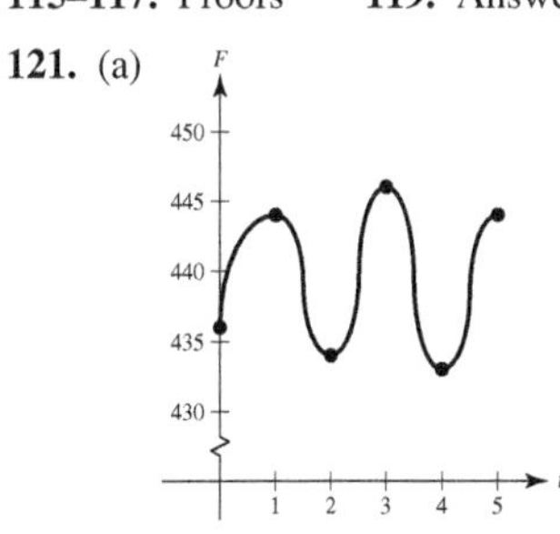

(b) No. The frequency is oscillating.

123. $c = \frac{-1 \pm \sqrt{5}}{2}$

125. Domain: $[-c^2, 0) \cup (0, \infty)$; Let $f(0) = \frac{1}{2c}$.

127. $h(x)$ has a nonremovable discontinuity at every integer except 0.

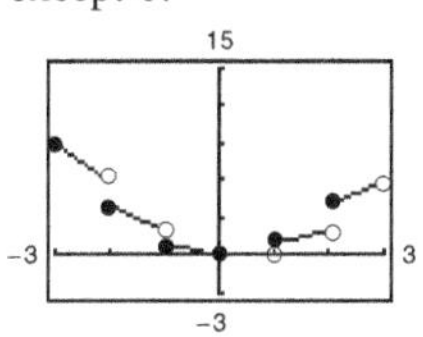

129. Putnam Problem B2, 1988

Section 1.5 *(page 92)*

1. A limit in which $f(x)$ increases or decreases without bound as x approaches c is called an infinite limit. ∞ is not a number. Rather, the symbol $\lim_{x\to c} f(x) = \infty$ says how the limit fails to exist.

3. $\lim_{x\to -2^+} 2\left|\frac{x}{x^2 - 4}\right| = \infty$, $\lim_{x\to -2^-} 2\left|\frac{x}{x^2 - 4}\right| = \infty$

5. $\lim_{x\to -2^+} \tan\frac{\pi x}{4} = -\infty$, $\lim_{x\to -2^-} \tan\frac{\pi x}{4} = \infty$

7. $\lim_{x\to 4^+} \frac{1}{x - 4} = \infty$, $\lim_{x\to 4^-} \frac{1}{x - 4} = -\infty$

9. $\lim_{x\to 4^+} \frac{1}{(x - 4)^2} = \infty$, $\lim_{x\to 4^-} \frac{1}{(x - 4)^2} = \infty$

11.

x	-3.5	-3.1	-3.01	-3.001	-3
$f(x)$	0.31	1.64	16.6	167	?

x	-2.999	-2.99	-2.9	-2.5
$f(x)$	-167	-16.7	-1.69	-0.36

$\lim_{x\to -3^+} f(x) = -\infty$; $\lim_{x\to -3^-} f(x) = \infty$

13.

x	-3.5	-3.1	-3.01	-3.001	-3
$f(x)$	3.8	16	151	1501	?

x	-2.999	-2.99	-2.9	-2.5
$f(x)$	-1499	-149	-14	-2.3

$\lim_{x\to -3^+} f(x) = -\infty$; $\lim_{x\to -3^-} f(x) = \infty$

15.

x	-3.5	-3.1	-3.01	-3.001	-3
$f(x)$	-1.7321	-9.514	-95.49	-954.9	?

x	-2.999	-2.99	-2.9	-2.5
$f(x)$	954.9	95.49	9.514	1.7321

$\lim_{x\to -3^-} f(x) = -\infty$; $\lim_{x\to -3^+} f(x) = \infty$

17. $x = 0$ **19.** $x = \pm 2$ **21.** No vertical asymptote

23. $x = -2, x = 1$ **25.** $x = 0, x = 3$

27. No vertical asymptote **29.** $x = n$, n is an integer

31. $t = n\pi$, n is a nonzero integer

33. Removable discontinuity at $x = -1$

35. Vertical asymptote at $x = -1$ **37.** ∞ **39.** $-\frac{1}{5}$

41. $-\infty$ **43.** $-\infty$ **45.** ∞ **47.** 0 **49.** ∞

51.

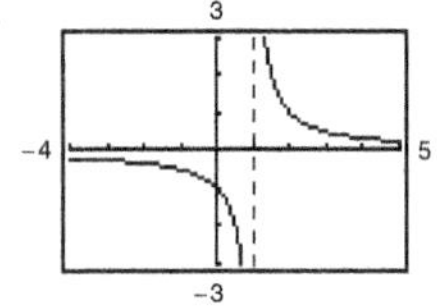

$\lim_{x \to 1^+} f(x) = \infty$

53. (a) ∞ (b) $-\infty$ (c) 0

55. Answers will vary. Sample answer: $f(x) = \dfrac{x - 3}{x^2 - 4x - 12}$

57.

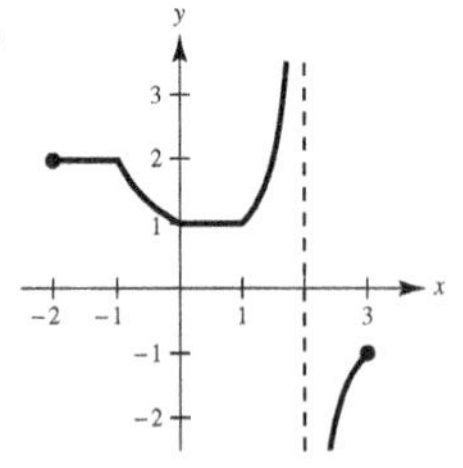

59. (a)

x	1	0.5	0.2	0.1
$f(x)$	0.1585	0.0411	0.0067	0.0017

x	0.01	0.001	0.0001
$f(x)$	≈ 0	≈ 0	≈ 0

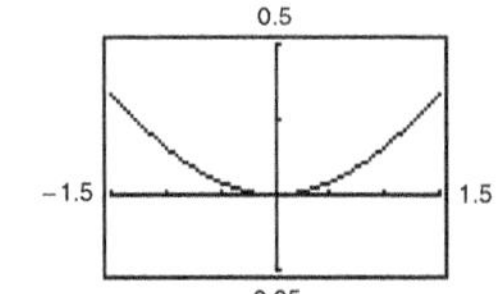

$\lim_{x \to 0^+} \dfrac{x - \sin x}{x} = 0$

(b)

x	1	0.5	0.2	0.1
$f(x)$	0.1585	0.0823	0.0333	0.0167

x	0.01	0.001	0.0001
$f(x)$	0.0017	≈ 0	≈ 0

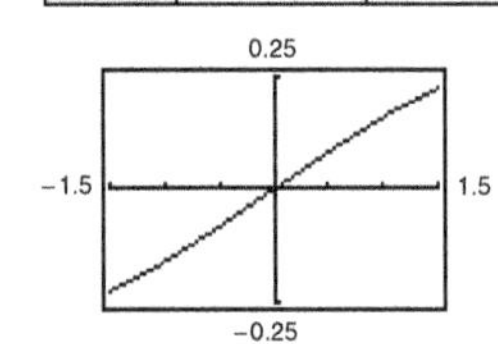

$\lim_{x \to 0^+} \dfrac{x - \sin x}{x^2} = 0$

(c)

x	1	0.5	0.2	0.1
$f(x)$	0.1585	0.1646	0.1663	0.1666

x	0.01	0.001	0.0001
$f(x)$	0.1667	0.1667	0.1667

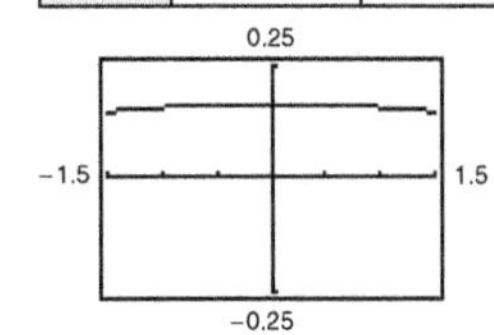

$\lim_{x \to 0^+} \dfrac{x - \sin x}{x^3} = 0.1\overline{6} \text{ or } \dfrac{1}{6}$

(d)

x	1	0.5	0.2	0.1
$f(x)$	0.1585	0.3292	0.8317	1.6658

x	0.01	0.001	0.0001
$f(x)$	16.67	166.7	1667.0

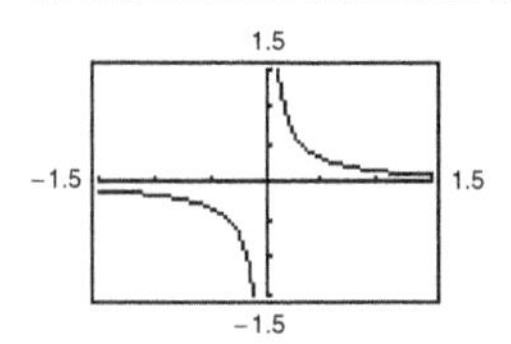

$\lim_{x \to 0^+} \dfrac{x - \sin x}{x^4} = \infty$

For $n > 3$, $\lim_{x \to 0^+} \dfrac{x - \sin x}{x^n} = \infty$.

61. (a) $\frac{7}{12}$ ft/sec (b) $\frac{3}{2}$ ft/sec

(c) $\lim_{x \to 25^-} \dfrac{2x}{\sqrt{625 - x^2}} = \infty$

63. (a) $A = 50 \tan \theta - 50\theta$; Domain: $\left(0, \dfrac{\pi}{2}\right)$

(b)

θ	0.3	0.6	0.9	1.2	1.5
$f(\theta)$	0.47	4.21	18.0	68.6	630.1

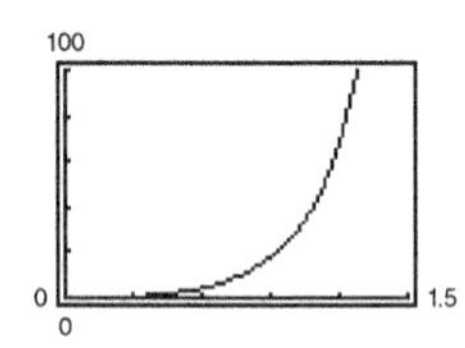

(c) $\lim_{\theta \to \pi/2} A = \infty$

65. True. **67.** False; let $f(x) = \tan x$

69. Let $f(x) = \dfrac{1}{x^2}$ and $g(x) = \dfrac{1}{x^4}$, and let $c = 0$. $\lim_{x \to 0} \dfrac{1}{x^2} = \infty$ and $\lim_{x \to 0} \dfrac{1}{x^4} = \infty$, but $\lim_{x \to 0} \left(\dfrac{1}{x^2} - \dfrac{1}{x^4}\right) = \lim_{x \to 0} \left(\dfrac{x^2 - 1}{x^4}\right) = -\infty \neq 0$.

71. Given $\lim_{x \to c} f(x) = \infty$, let $g(x) = 1$. Then $\lim_{x \to c} \dfrac{g(x)}{f(x)} = 0$ by Theorem 1.15.

73–75. Proofs

Review Exercises for Chapter 1 *(page 95)*

1. Calculus

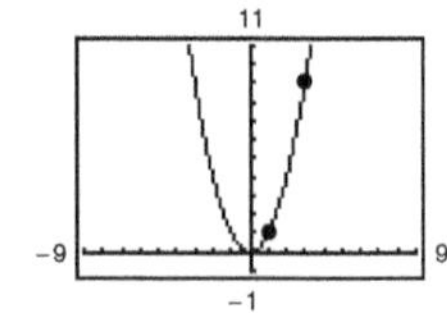

Estimate: 8.3

3.

x	2.9	2.99	2.999	3
$f(x)$	−0.9091	−0.9901	−0.9990	?

x	3.001	3.01	3.1
$f(x)$	−1.0010	−1.0101	−1.1111

$\lim_{x\to 0} \frac{x-3}{x^2-7x+12} \approx -1.0000$

5. (a) Limit does not exist. The function approaches 3 from the left side of 2, but it approaches 2 from the right side of 2.
(b) 0

7. 5; Proof **9.** −3; Proof **11.** 36 **13.** $\sqrt{6} \approx 2.45$
15. 16 **17.** $\frac{4}{3}$ **19.** −1 **21.** $\frac{1}{2}$ **23.** −1
25. 0 **27.** $\sqrt{3}/2$ **29.** −3 **31.** −5

33. The graph has a hole at $x = 0$.

x	−0.1	−0.01	−0.001	0
$f(x)$	0.3352	0.3335	0.3334	?

x	0.001	0.01	0.1
$f(x)$	0.3333	0.3331	0.3315

$\lim_{x\to 0} \frac{\sqrt{2x+9}-3}{x} \approx 0.3333$; Actual limit is $\frac{1}{3}$.

35.

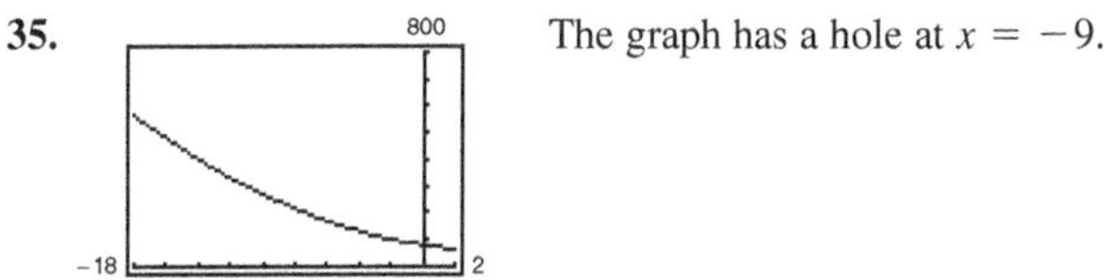

The graph has a hole at $x = -9$.

x	−9.1	−9.01	−9.001	−9
$f(x)$	245.7100	243.2701	243.0270	?

x	−8.999	−8.99	−8.9
$f(x)$	242.9730	242.7301	240.3100

$\lim_{x\to -9} \frac{x^3+729}{x+9} \approx 243.00$; Actual limit is 243.

37. −39.2 m/sec **39.** $\frac{1}{6}$ **41.** $\frac{1}{10}$ **43.** 0
45. Limit does not exist. The function approaches 2 from the left side of 1, but it approaches 1 from the right side of 1.
47. 3 **49.** −4 **51.** Continuous on $[-2, 2]$
53. No discontinuities
55. Nonremovable discontinuity at $x = 5$
57. Nonremovable discontinuities at $x = -1$ and $x = 1$
Removable discontinuity at $x = 0$
59. $c = -\frac{1}{2}$ **61.** Continuous for all real x
63. Continuous on $[0, \infty)$
65. Removable discontinuity at $x = 1$
Continuous on $(-\infty, 1) \cup (1, \infty)$
67. Proof
69. $f(-1) = -8$, $f(2) = 10$
Because f is continuous on the closed interval $[-1, 2]$ and $-8 < 2 < 10$, there is at least one number c in $[-1, 2]$ such that $f(c) = 2$; $c = 1$
71. From the left: $-\infty$
From the right: ∞
73. $x = 0$ **75.** $x = \pm 3$
77. $x = 2n + 1$, where n is an integer **79.** $-\infty$ **81.** $\frac{1}{3}$
83. $-\infty$ **85.** $\frac{4}{5}$ **87.** ∞
89. (a) \$80,000.00 (b) \$720,000.00 (c) ∞

P.S. Problem Solving *(page 97)*

1. (a) Perimeter $\triangle PAO = 1 + \sqrt{(x^2-1)^2+x^2} + \sqrt{x^4+x^2}$
Perimeter $\triangle PBO = 1 + \sqrt{x^4+(x-1)^2} + \sqrt{x^4+x^2}$

(b)

x	4	2	1
Perimeter $\triangle PAO$	33.0166	9.0777	3.4142
Perimeter $\triangle PBO$	33.7712	9.5952	3.4142
$r(x)$	0.9777	0.9461	1.0000

x	0.1	0.01
Perimeter $\triangle PAO$	2.0955	2.0100
Perimeter $\triangle PBO$	2.0006	2.0000
$r(x)$	1.0475	1.0050

1

3. (a) Area (hexagon) $= (3\sqrt{3})/2 \approx 2.5981$
Area (circle) $= \pi \approx 3.1416$
Area (circle) − Area (hexagon) ≈ 0.5435
(b) $A_n = (n/2)\sin(2\pi/n)$
(c)

n	6	12	24	48	96
A_n	2.5981	3.0000	3.1058	3.1326	3.1394

3.1416 or π

5. (a) $m = -\frac{12}{5}$ (b) $y = \frac{5}{12}x - \frac{169}{12}$
(c) $m_x = \frac{-\sqrt{169-x^2}+12}{x-5}$
(d) $\frac{5}{12}$; It is the same as the slope of the tangent line found in part (b).

7. (a) Domain: $[-27, 1) \cup (1, \infty)$
(b)

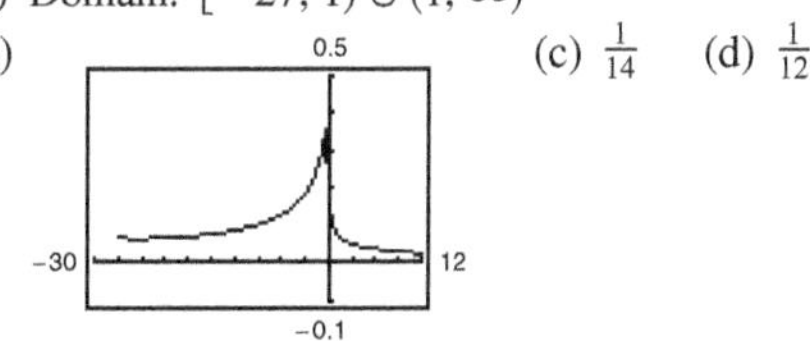

(c) $\frac{1}{14}$ (d) $\frac{1}{12}$

The graph has a hole at $x = 1$.

9. (a) g_1, g_4 (b) g_1 (c) g_1, g_3, g_4

11. 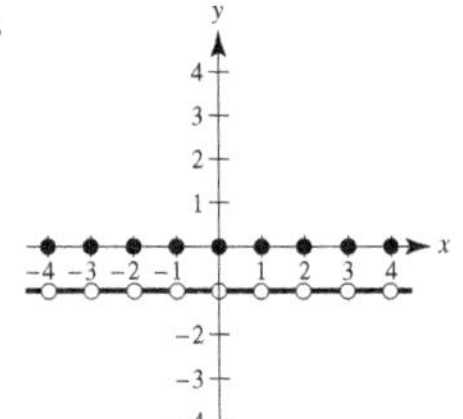

The graph jumps at every integer.

(a) $f(1) = 0,\quad f(0) = 0,\quad f(\frac{1}{2}) = -1,\quad f(-2.7) = -1$

(b) $\lim\limits_{x \to 1^-} f(x) = -1,\quad \lim\limits_{x \to 1^+} f(x) = -1,\quad \lim\limits_{x \to 1/2} f(x) = -1$

(c) There is a discontinuity at each integer.

13. (a) 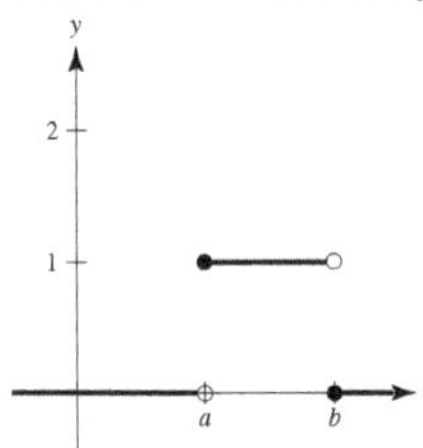

(b) (i) $\lim\limits_{x \to a^+} P_{a,b}(x) = 1$

(ii) $\lim\limits_{x \to a^-} P_{a,b}(x) = 0$

(iii) $\lim\limits_{x \to b^+} P_{a,b}(x) = 0$

(iv) $\lim\limits_{x \to b^-} P_{a,b}(x) = 1$

(c) Continuous for all positive real numbers except a and b

(d) The area under the graph of U and above the x-axis is 1.

Chapter 2

Section 2.1 *(page 107)*

1. Let $(c, f(c))$ represent an arbitrary point on the graph of f. Then the slope of the tangent line at $(c, f(c))$ is

$$m = \lim_{\Delta x \to 0} \frac{f(c + \Delta x) - f(c)}{\Delta x}.$$

3. The limit used to define the slope of a tangent line is also used to define differentiation. The key is to rewrite the difference quotient so that Δx does not occur as a factor of the denominator.

5. $m_1 = 0, m_2 = 5/2$

7. (a)–(d) 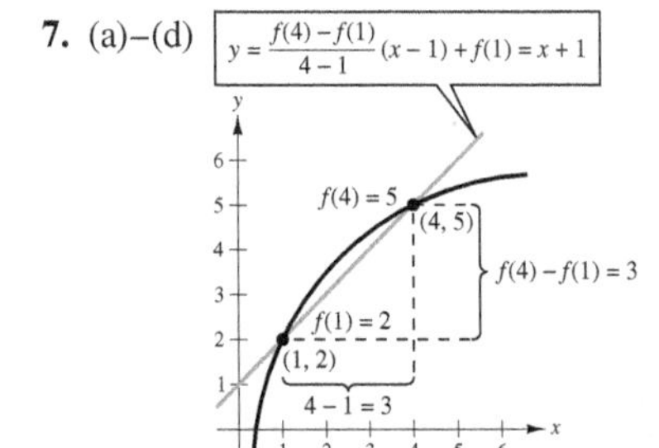

9. $m = -5$

11. $m = 8$ **13.** $m = 3$ **15.** $f'(x) = 0$

17. $f'(x) = -5$ **19.** $h'(s) = \frac{2}{3}$ **21.** $f'(x) = 2x + 1$

23. $f'(x) = 3x^2 - 12$ **25.** $f'(x) = \dfrac{-1}{(x - 1)^2}$

27. $f'(x) = \dfrac{1}{2\sqrt{x + 4}}$

29. (a) Tangent line: $y = -2x + 2$

(b) 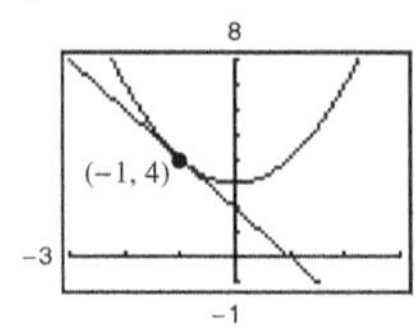

31. (a) Tangent line: $y = 12x - 16$

(b) 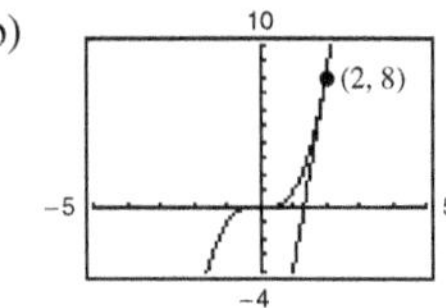

33. (a) Tangent line: $y = \frac{1}{2}x + \frac{1}{2}$

(b) 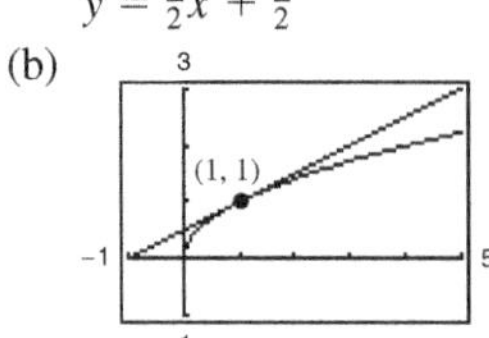

35. (a) Tangent line: $y = \frac{3}{4}x - 2$

(b) 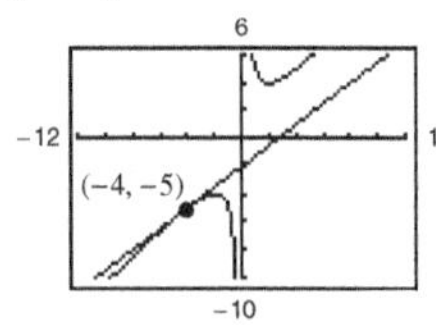

37. $y = -x + 1$ **39.** $y = 3x - 2;\ y = 3x + 2$

41. $y = -\frac{1}{2}x + \frac{3}{2}$

43. 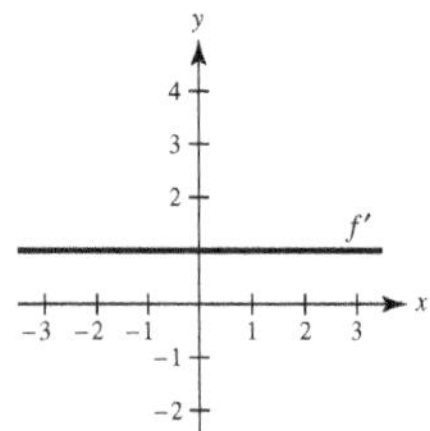

The slope of the graph of f is 1 for all x-values.

45. 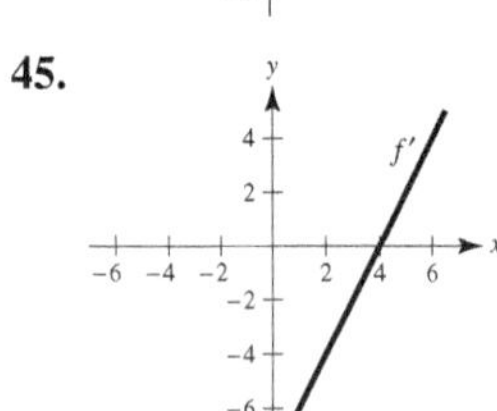

The slope of the graph of f is negative for $x < 4$, positive for $x > 4$, and 0 at $x = 4$.

47. 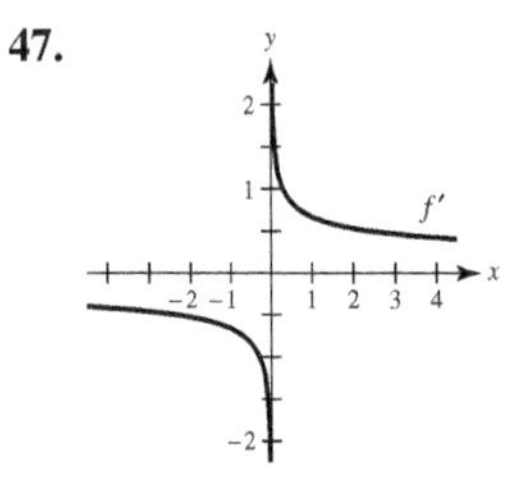

The slope of the graph of f is negative for $x < 0$ and positive for $x > 0$. The slope is undefined at $x = 0$.

49. Answers will vary.
Sample answer: $y = -x$

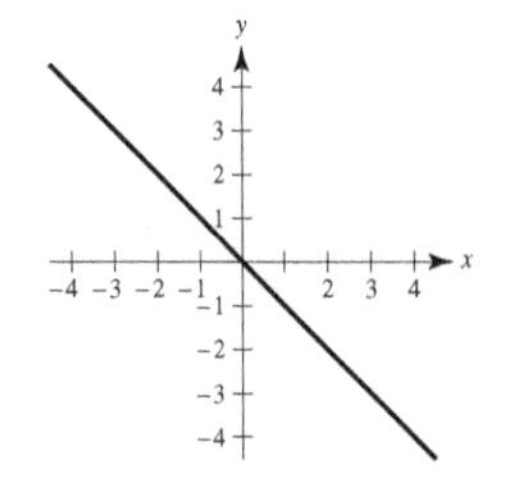

51. No. Consider $f(x) = \sqrt{x}$ and its derivative.

53. $g(4) = 5;\ g'(4) = -\frac{5}{3}$

55. $f(x) = 5 - 3x$
$c = 1$

57. $f(x) = -x^2$
$c = 6$

59. $f(x) = -3x + 2$

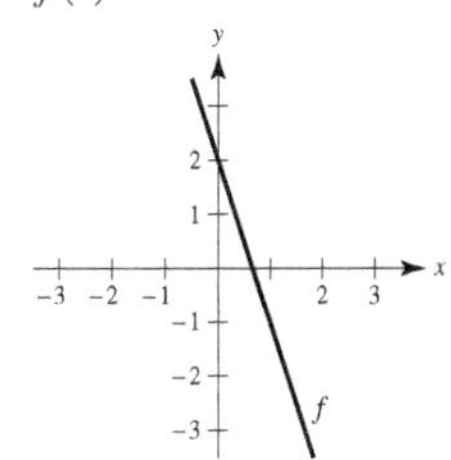

61. $y = 2x + 1,\ y = -2x + 9$

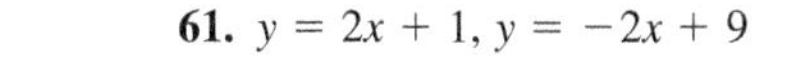

63. (a)

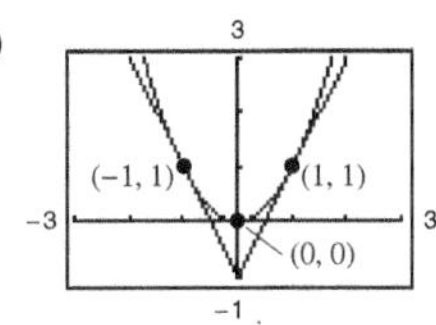

For this function, the slopes of the tangent lines are always distinct for different values of x.

(b)

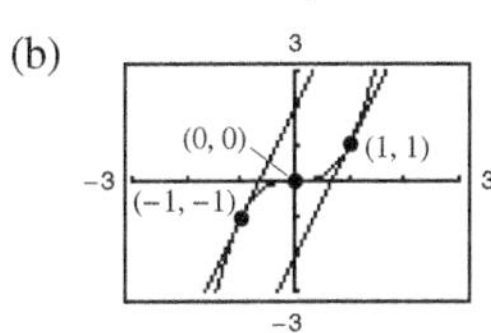

For this function, the slopes of the tangent lines are sometimes the same.

65. (a)

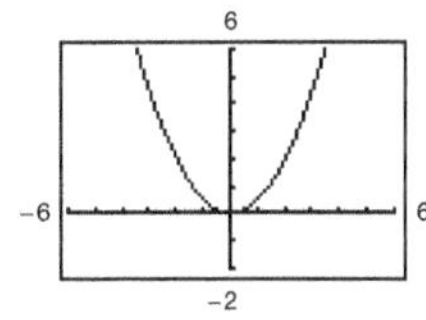

$f'(0) = 0,\ f'(\frac{1}{2}) = \frac{1}{2},\ f'(1) = 1,\ f'(2) = 2$

(b) $f'(-\frac{1}{2}) = -\frac{1}{2},\ f'(-1) = -1,\ f'(-2) = -2$

(c)

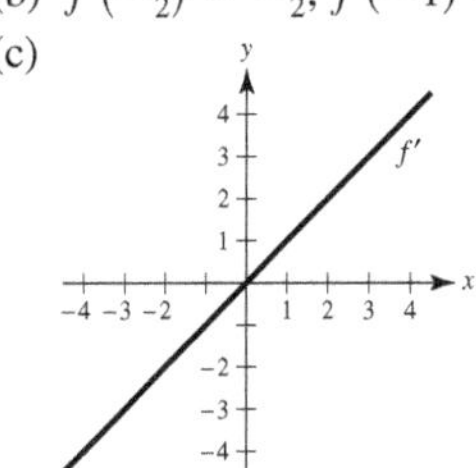

(d) $f'(x) = x$

67. $f(2) = 4,\ f(2.1) = 3.99,\ f'(2) \approx -0.1$ **69.** 4

71. $g(x)$ is not differentiable at $x = 0$.

73. $f(x)$ is not differentiable at $x = 6$.

75. $h(x)$ is not differentiable at $x = -7$.

77. $(-\infty, -4) \cup (-4, \infty)$ **79.** $(-1, \infty)$

81.

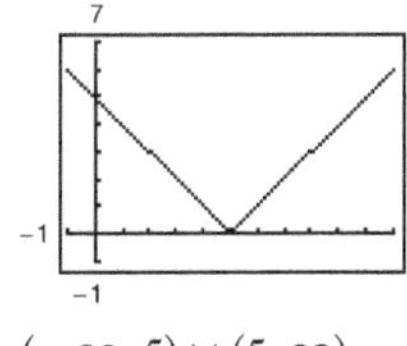

$(-\infty, 5) \cup (5, \infty)$

83.

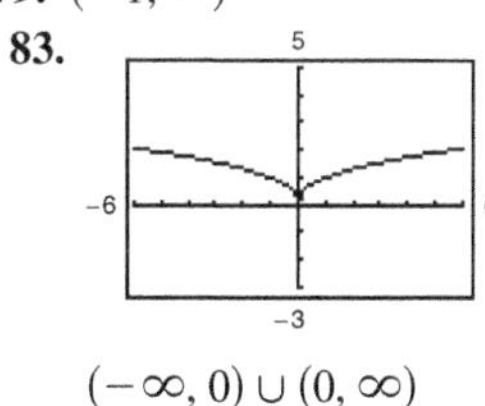

$(-\infty, 0) \cup (0, \infty)$

85. The derivative from the left is -1 and the derivative from the right is 1, so f is not differentiable at $x = 1$.

87. The derivatives from both the right and the left are 0, so $f'(1) = 0$.

89. f is differentiable at $x = 2$.

91. (a) $d = \dfrac{3|m + 1|}{\sqrt{m^2 + 1}}$

(b)

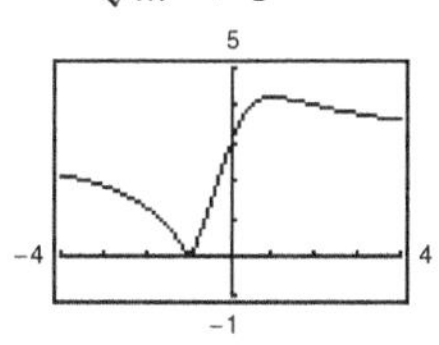

Not differentiable at $m = -1$

93. False. The slope is $\lim\limits_{\Delta x \to 0} \dfrac{f(2 + \Delta x) - f(2)}{\Delta x}$.

95. False. For example, $f(x) = |x|$. The derivative from the left and the derivative from the right both exist but are not equal.

97. Proof

Section 2.2 *(page 118)*

1. 0

3. The derivative of the sine function is the cosine function. The derivative of the cosine function is the negative of the sine function.

5. (a) $\frac{1}{2}$ (b) 3 **7.** 0 **9.** $7x^6$ **11.** $-5/x^6$

13. $1/(9x^{8/9})$ **15.** 1 **17.** $-6t + 2$ **19.** $2x + 12x^2$

21. $3t^2 + 10t - 3$ **23.** $\dfrac{\pi}{2} \cos \theta$ **25.** $2x + \dfrac{1}{2} \sin x$

	Function	Rewrite	Differentiate	Simplify
27.	$y = \dfrac{2}{7x^4}$	$y = \dfrac{2}{7}x^{-4}$	$y' = -\dfrac{8}{7}x^{-5}$	$y' = -\dfrac{8}{7x^5}$
29.	$y = \dfrac{6}{(5x)^3}$	$y = \dfrac{6}{125}x^{-3}$	$y' = -\dfrac{18}{125}x^{-4}$	$y' = -\dfrac{18}{125x^4}$

31. -2 **33.** 0 **35.** 8 **37.** 3 **39.** $\dfrac{2x + 6}{x^3}$

41. $\dfrac{2t + 12}{t^4}$ **43.** $\dfrac{x^3 - 8}{x^3}$ **45.** $\dfrac{3t^2 - 4t + 24}{2t^{5/2}}$

47. $3x^2 + 1$ **49.** $\dfrac{1}{2\sqrt{x}} - \dfrac{2}{x^{2/3}}$ **51.** $\dfrac{3}{\sqrt{x}} - 5 \sin x$

53. $18x + 5 \sin x$

55. (a) $y = 2x - 2$

(b)

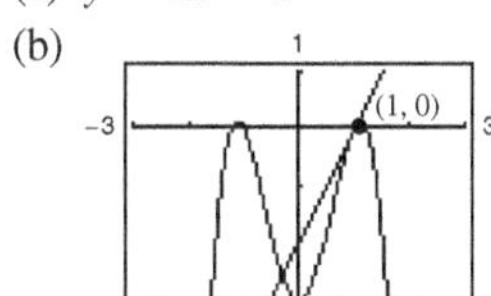

57. (a) $3x + 2y - 7 = 0$

(b)

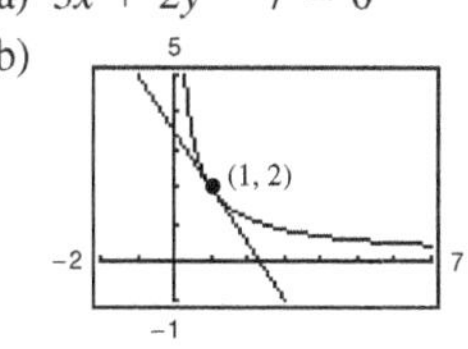

59. $(-1, 2), (0, 3), (1, 2)$ **61.** No horizontal tangents

63. (π, π) **65.** $k = -8$ **67.** $k = 3$

69. $g'(x) = f'(x)$ **71.** $g'(x) = -5f'(x)$

73.

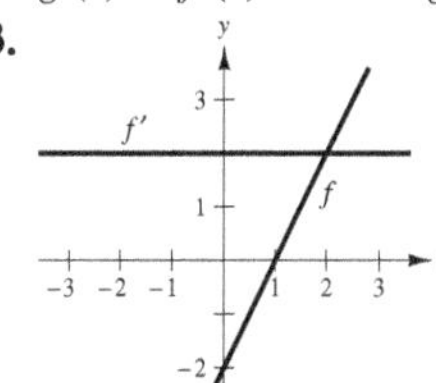

The rate of change of f is constant, and therefore f' is a constant function.

75.

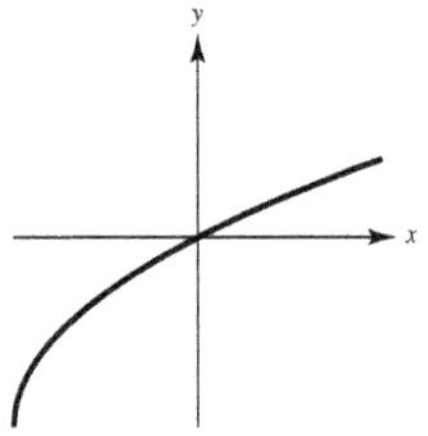

77. $y = 2x - 1$ $y = 4x - 4$

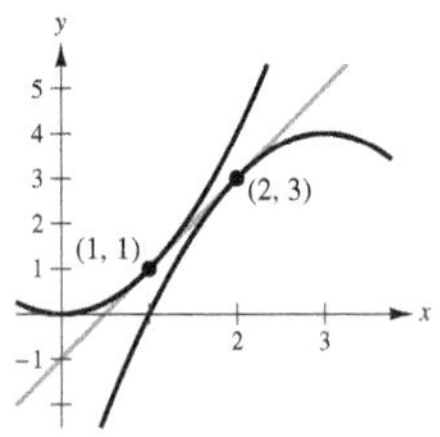

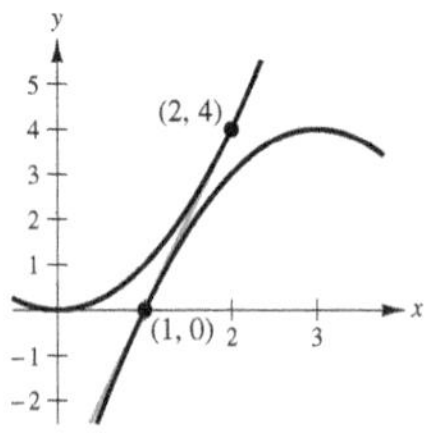

79. $f'(x) = 3 + \cos x \neq 0$ for all x.   **81.** $x - 4y + 4 = 0$

83. (a)

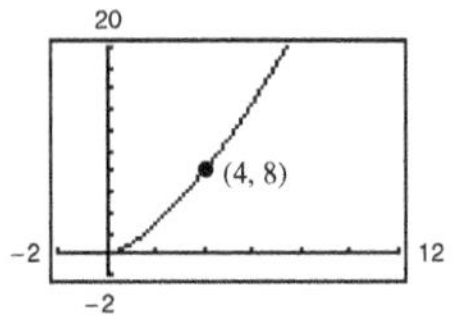

$(3.9, 7.7019)$,
$S(x) = 2.981x - 3.924$

(b) $T(x) = 3(x - 4) + 8 = 3x - 4$
The slope (and equation) of the secant line approaches that of the tangent line at $(4, 8)$ as you choose points closer and closer to $(4, 8)$.

(c)

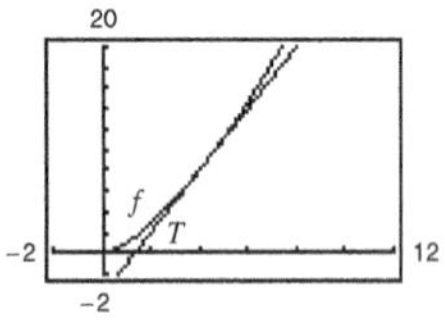

The approximation becomes less accurate.

(d)

Δx	-3	-2	-1	-0.5	-0.1	0
$f(4 + \Delta x)$	1	2.828	5.196	6.548	7.702	8
$T(4 + \Delta x)$	-1	2	5	6.5	7.7	8

Δx	0.1	0.5	1	2	3
$f(4 + \Delta x)$	8.302	9.546	11.180	14.697	18.520
$T(4 + \Delta x)$	8.3	9.5	11	14	17

85. False. Let $f(x) = x$ and $g(x) = x + 1$.

87. False. $\dfrac{dy}{dx} = 0$ **89.** False. $f'(x) = 0$

91. Average rate: 3
Instantaneous rates:
$f'(1) = 3, f'(2) = 3$

93. Average rate: $\frac{1}{2}$
Instantaneous rates:
$f'(1) = 1, f'(2) = \frac{1}{4}$

95. (a) $s(t) = -16t^2 + 1362$, $v(t) = -32t$ (b) -48 ft/sec
(c) $s'(1) = -32$ ft/sec, $s'(2) = -64$ ft/sec
(d) $t = \dfrac{\sqrt{1362}}{4} \approx 9.226$ sec   (e) -295.242 ft/sec

97. $v(5) = 71$ m/sec; $v(10) = 22$ m/sec

99.

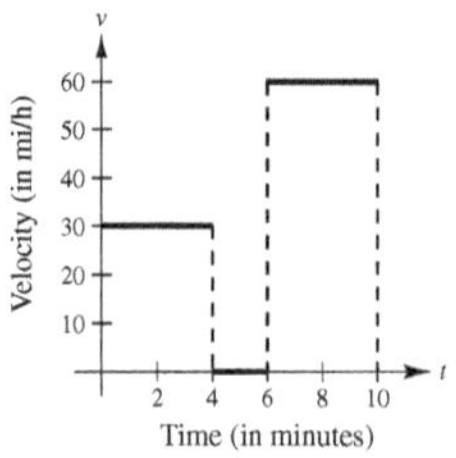

101. $V'(6) = 108\text{ cm}^3/\text{cm}$

103. (a) $R(v) = 0.417v - 0.02$
(b) $B(v) = 0.0056v^2 + 0.001v + 0.04$
(c) $T(v) = 0.0056v^2 + 0.418v + 0.02$
(d)

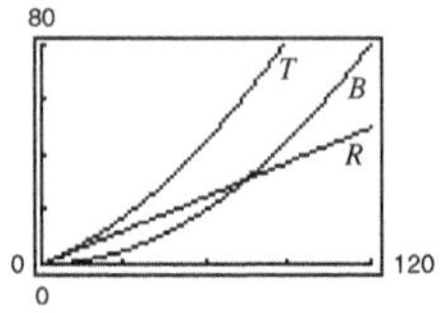

(e) $T'(v) = 0.0112v + 0.418$
$T'(40) = 0.866$
$T'(80) = 1.314$
$T'(100) = 1.538$

(f) Stopping distance increases at an increasing rate.

105. Proof   **107.** $y = 2x^2 - 3x + 1$

109. $9x + y = 0$, $9x + 4y + 27 = 0$ **111.** $a = \frac{1}{3}, b = -\frac{4}{3}$

113. $f_1(x) = |\sin x|$ is differentiable for all $x \neq n\pi$, n an integer.
$f_2(x) = \sin|x|$ is differentiable for all $x \neq 0$.

115. Putnam Problem A2, 2010

Section 2.3 *(page 129)*

1. To find the derivative of the product of two differentiable functions f and g, multiply the first function f by the derivative of the second function g, and then add the second function g times the derivative of the first function f.

3. $\dfrac{d}{dx}\tan x = \sec^2 x$

$\dfrac{d}{dx}\cot x = -\csc^2 x$

$\dfrac{d}{dx}\sec x = \sec x \tan x$

$\dfrac{d}{dx}\csc x = -\csc x \cot x$

5. $-20x + 17$ **7.** $\dfrac{1 - 5t^2}{2\sqrt{t}}$ **9.** $x^2(3\cos x - x\sin x)$

11. $-\dfrac{5}{(x - 5)^2}$ **13.** $\dfrac{1 - 5x^3}{2\sqrt{x}(x^3 + 1)^2}$ **15.** $\dfrac{x\cos x - 2\sin x}{x^3}$

17. $f'(x) = (x^3 + 4x)(6x + 2) + (3x^2 + 2x - 5)(3x^2 + 4)$
$= 15x^4 + 8x^3 + 21x^2 + 16x - 20$
$f'(0) = -20$

19. $f'(x) = \dfrac{x^2 - 6x + 4}{(x - 3)^2}$
$f'(1) = -\dfrac{1}{4}$

21. $f'(x) = \cos x - x\sin x$
$f'\left(\dfrac{\pi}{4}\right) = \dfrac{\sqrt{2}}{8}(4 - \pi)$

	Function	Rewrite	Differentiate	Simplify
23.	$y = \dfrac{x^3 + 6x}{3}$	$y = \frac{1}{3}x^3 + 2x$	$y' = \frac{1}{3}(3x^2) + 2$	$y' = x^2 + 2$
25.	$y = \dfrac{6}{7x^2}$	$y = \frac{6}{7}x^{-2}$	$y' = -\frac{12}{7}x^{-3}$	$y' = -\dfrac{12}{7x^3}$
27.	$y = \dfrac{4x^{3/2}}{x}$	$y = 4x^{1/2}$, $x > 0$	$y' = 2x^{-1/2}$	$y' = \dfrac{2}{\sqrt{x}}$, $x > 0$

29. $\dfrac{3}{(x + 1)^2}$, $x \neq 1$ **31.** $\dfrac{x^2 + 6x - 3}{(x + 3)^2}$ **33.** $\dfrac{3x + 1}{2x^{3/2}}$

35. $-\dfrac{2x^2 - 2x + 3}{x^2(x - 3)^2}$ **37.** $\dfrac{4s^2(3s^2 + 13s + 15)}{(s + 2)^2}$

39. $10x^4 - 8x^3 - 21x^2 - 10x - 30$ **41.** $t(t \cos t + 2 \sin t)$

43. $\dfrac{-(t \sin t + \cos t)}{t^2}$ **45.** $-1 + \sec^2 x$, or $\tan^2 x$

47. $\dfrac{1}{4t^{3/4}} - 6 \csc t \cot t$ **49.** $\dfrac{3}{2} \sec x(\tan x - \sec x)$

51. $\cos x \cot^2 x$ **53.** $x(x \sec^2 x + 2 \tan x)$

55. $4x \cos x + (2 - x^2) \sin x$ **57.** $\dfrac{2x^2 + 8x - 1}{(x + 2)^2}$

59. $-4\sqrt{3}$ **61.** $\dfrac{1}{\pi^2}$

63. (a) $y = -3x - 1$

(b)

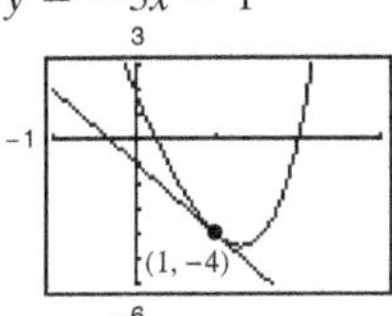

65. (a) $y = 4x + 25$

(b)

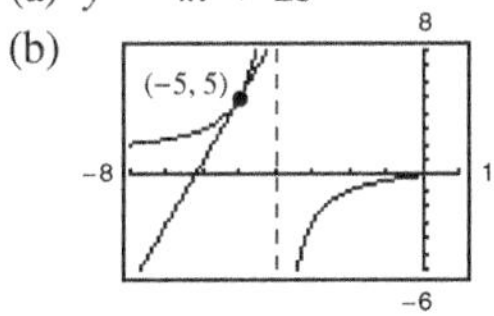

67. (a) $4x - 2y - \pi + 2 = 0$ **69.** $2y + x - 4 = 0$

(b)

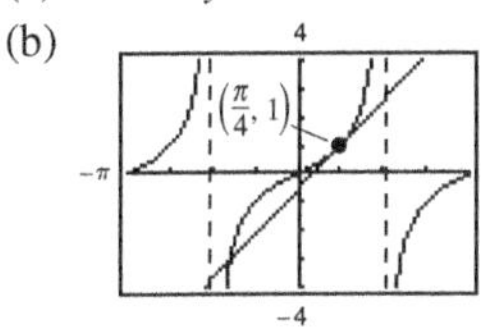

71. $25y - 12x + 16 = 0$ **73.** $(1, 1)$ **75.** $(0, 0), (2, 4)$

77. Tangent lines: $2y + x = 7$, $2y + x = -1$

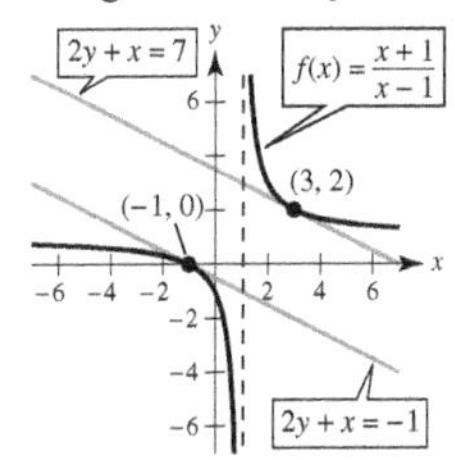

79. $f(x) + 2 = g(x)$ **81.** (a) $p'(1) = 1$ (b) $q'(4) = -\frac{1}{3}$

83. $\dfrac{18t + 5}{2\sqrt{t}}$ cm²/sec

85. (a) −$38.13 thousand/100 components

(b) −$10.37 thousand/100 components

(c) −$3.80 thousand/100 components

The cost decreases with increasing order size.

87. Proof

89. (a) $h(t) = 101.7t + 1593$

$p(t) = 2.1t + 287$

(b) 3000 / 7 / 14 / 2300   320 / 7 / 14 / 300

(c) $A = \dfrac{101.7t + 1593}{2.1t + 287}$

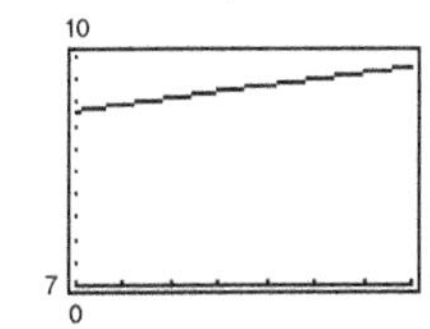

A represents the average health care expenditures per person (in thousands of dollars).

(d) $A'(t) = \dfrac{25{,}842.6}{4.41t^2 + 1205.4t + 82{,}369}$

$A'(t)$ represents the rate of change of the average health care expenditures per person for the given year t.

91. 2 **93.** $\dfrac{3}{\sqrt{x}}$ **95.** $\dfrac{2}{(x - 1)^3}$ **97.** $2 \cos x - x \sin x$

99. $\csc^3 x + \csc x \cot^2 x$ **101.** $6x + \dfrac{6}{25x^{8/5}}$ **103.** $\sin x$

105. 0 **107.** −10

109. $n - 1$ or lower; Answers will vary. Sample answer:
$f(x) = x^3$, $f'(x) = 3x^2$, $f''(x) = 6x$, $f'''(x) = 6$, $f^{(4)}(x) = 0$

111.

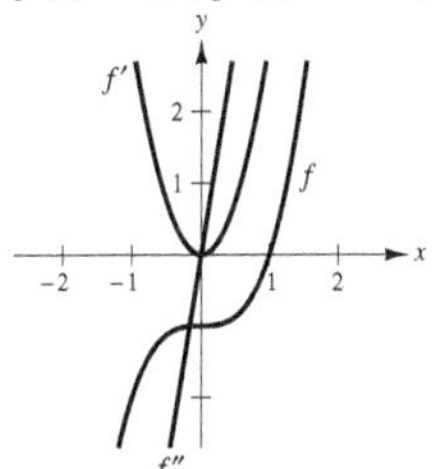

It appears that f is cubic, so f' would be quadratic and f'' would be linear.

113.

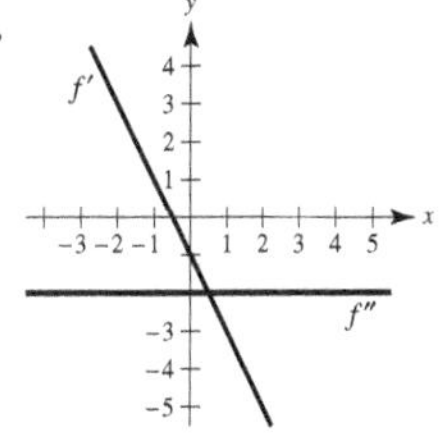

115. Answers will vary.

Sample answer: $f(x) = (x - 2)^2$

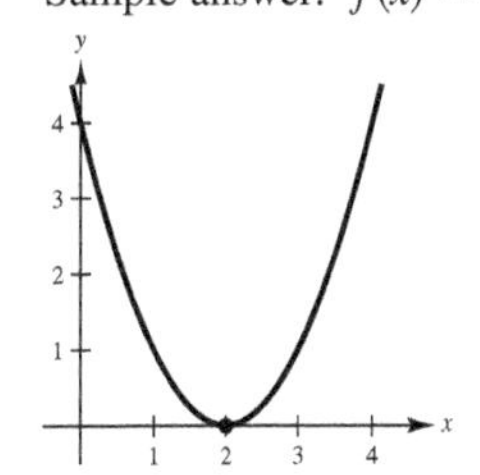

117. $v(3) = 27$ m/sec

$a(3) = -6$ m/sec²

The speed of the object is decreasing.

119.

t	0	1	2	3	4
$s(t)$	0	57.75	99	123.75	132
$v(t)$	66	49.5	33	16.5	0
$a(t)$	−16.5	−16.5	−16.5	−16.5	−16.5

The average velocity on [0, 1] is 57.75, on [1, 2] is 41.25, on [2, 3] is 24.75, and on [3, 4] is 8.25.

121. $f^{(n)}(x) = n(n - 1)(n - 2) \cdot \cdot \cdot (2)(1) = n!$

123. (a) $f''(x) = g(x)h''(x) + 2g'(x)h'(x) + g''(x)h(x)$
$f'''(x) = g(x)h'''(x) + 3g'(x)h''(x) + 3g''(x)h'(x) + g'''(x)h(x)$
$f^{(4)}(x) = g(x)h^{(4)}(x) + 4g'(x)h'''(x) + 6g''(x)h''(x) + 4g'''(x)h'(x) + g^{(4)}(x)h(x)$

(b) $f^{(n)}(x) = g(x)h^{(n)}(x) + \dfrac{n!}{1!(n-1)!}g'(x)h^{(n-1)}(x) + \dfrac{n!}{2!(n-2)!}g''(x)h^{(n-2)}(x) + \cdots + \dfrac{n!}{(n-2)!1!}g^{(n-1)}(x)h'(x) + g^{(n)}(x)h(x)$

125. $n = 1$: $f'(x) = x\cos x + \sin x$
$n = 2$: $f'(x) = x^2\cos x + 2x\sin x$
$n = 3$: $f'(x) = x^3\cos x + 3x^2\sin x$
$n = 4$: $f'(x) = x^4\cos x + 4x^3\sin x$
General rule: $f'(x) = x^n\cos x + nx^{(n-1)}\sin x$

127. $y' = -\dfrac{1}{x^2}$, $y'' = \dfrac{2}{x^3}$,
$$x^3y'' + 2x^2y' = x^3\left(\frac{2}{x^3}\right) + 2x^2\left(\frac{-1}{x^2}\right) = 2 - 2 = 0$$

129. $y' = 2\cos x$, $y'' = -2\sin x$,
$y'' + y = -2\sin x + 2\sin x + 3 = 3$

131. False. $\dfrac{dy}{dx} = f(x)g'(x) + g(x)f'(x)$ **133.** True

135. True **137.** Proof

Section 2.4 *(page 140)*

1. To find the derivative of the composition of two differentiable functions, take the derivative of the outer function and keep the inner function the same. Then multiply by the derivative of the inner function.

	$y = f(g(x))$	$u = g(x)$	$y = f(u)$
3.	$y = (6x-5)^4$	$u = 6x - 5$	$y = u^4$
5.	$y = \dfrac{1}{3x+5}$	$u = 3x + 5$	$y = \dfrac{1}{u}$
7.	$y = \csc^3 x$	$u = \csc x$	$y = u^3$

9. $6(2x-7)^2$ **11.** $-\dfrac{45}{2(4-9x)^{1/6}}$ **13.** $-\dfrac{10s}{\sqrt{5s^2+3}}$

15. $\dfrac{4x}{\sqrt[3]{(6x^2+1)^2}}$ **17.** $-\dfrac{1}{(x-2)^2}$ **19.** $-\dfrac{54s^2}{(s^3-2)^4}$

21. $-\dfrac{3}{2\sqrt{(3x+5)^3}}$ **23.** $x(x-2)^6(9x-4)$

25. $\dfrac{1-2x^2}{\sqrt{1-x^2}}$ **27.** $\dfrac{1}{\sqrt{(x^2+1)^3}}$

29. $\dfrac{-2(x+5)(x^2+10x-2)}{(x^2+2)^3}$ **31.** $\dfrac{8(t+1)^3}{(t+3)^5}$

33. $20x(x^2+3)^9 + 2(x^2+3)^5 + 20x^2(x^2+3)^4 + 2x$

35. $-4\sin 4x$ **37.** $15\sec^2 3x$ **39.** $2\pi^2 x\cos(\pi x)^2$

41. $2\cos 4x$ **43.** $\dfrac{-1-\cos^2 x}{\sin^3 x}$ **45.** $8\sec^2 x\tan x$

47. $\sin 2\theta\cos 2\theta$, or $\dfrac{1}{2}\sin 4\theta$

49. $6\pi(\pi t - 1)\sec(\pi t - 1)^2\tan(\pi t - 1)^2$

51. $(6x - \sin x)\cos(3x^2 + \cos x)$

53. $-\dfrac{3\pi\cos\sqrt{\cot 3\pi x}\csc^2(3\pi x)}{2\sqrt{\cot 3\pi x}}$

55. $\dfrac{1 - 3x^2 - 4x^{3/2}}{2\sqrt{x}(x^2+1)^2}$

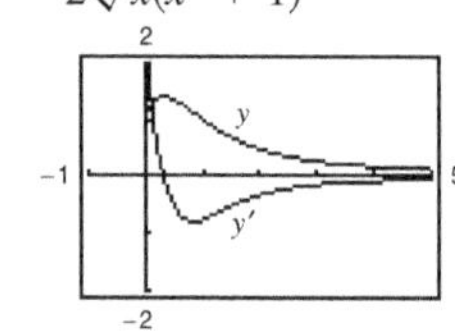

The zero of y' corresponds to the point on the graph of the function where the tangent line is horizontal.

57. $-\dfrac{\sqrt{\dfrac{x+1}{x}}}{2x(x+1)}$

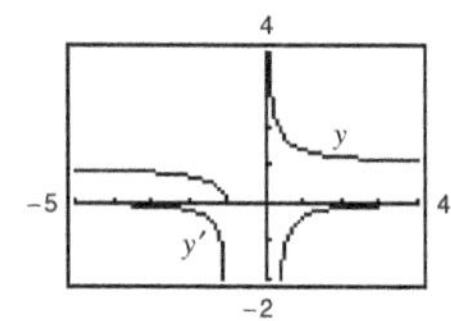

y' has no zeros.

59. $-\dfrac{\pi x\sin(\pi x) + \cos(\pi x) + 1}{x^2}$

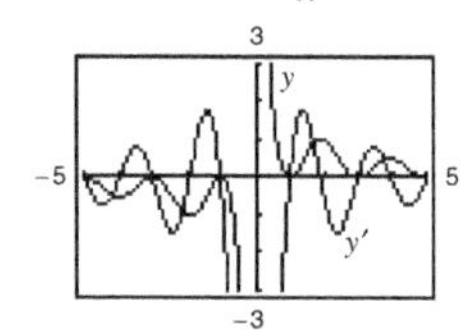

The zeros of y' correspond to the points on the graph of the function where the tangent lines are horizontal.

61. 3; 3 cycles in $[0, 2\pi]$ **63.** $\frac{5}{3}$

65. $-\frac{3}{5}$ **67.** -1 **69.** 0

71. (a) $8x - 5y - 7 = 0$
(b)

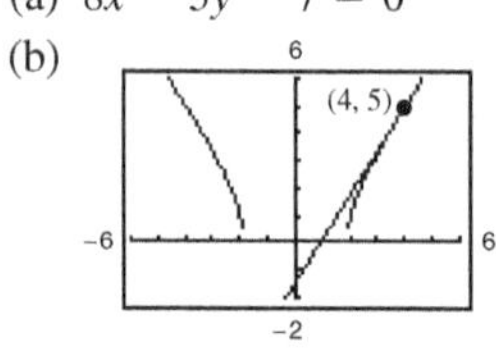

73. (a) $24x + y + 23 = 0$
(b)

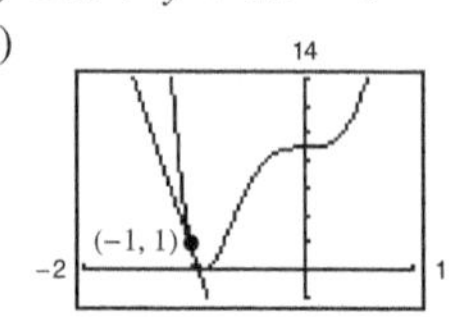

75. (a) $y = 8x - 8\pi$
(b)

77. (a) $4x - y + (1 - \pi) = 0$
(b)

79. $3x + 4y - 25 = 0$

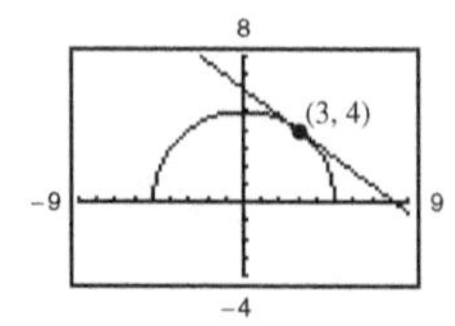

81. $\left(\dfrac{\pi}{6}, \dfrac{3\sqrt{3}}{2}\right), \left(\dfrac{5\pi}{6}, -\dfrac{3\sqrt{3}}{2}\right), \left(\dfrac{3\pi}{2}, 0\right)$ **83.** $2940(2-7x)^2$

85. $\dfrac{242}{(11x-6)^3}$ **87.** $2(\cos x^2 - 2x^2\sin x^2)$

89. $h''(x) = 18x + 6$, 24

91. $f''(x) = -4x^2\cos x^2 - 2\sin x^2$, 0

93. 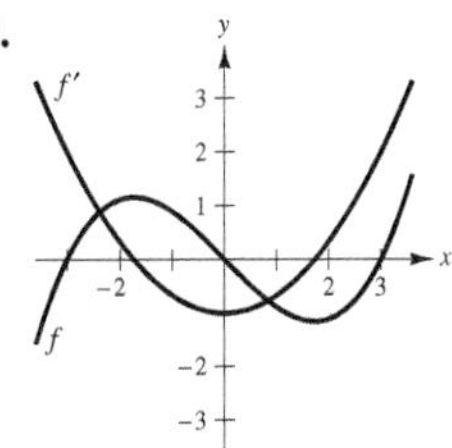

The zeros of f' correspond to the points where the graph of f has horizontal tangents.

95. (a) The rate of change of g is three times as fast as the rate of change of f.
(b) The rate of change of g is $2x$ times as fast as the rate of change of f.

97. (a) $g'(x) = f'(x)$ (b) $h'(x) = 2f'(x)$
(c) $r'(x) = -3f'(-3x)$ (d) $s'(x) = f'(x + 2)$

x	-2	-1	0	1	2	3
$f'(x)$	4	$\frac{2}{3}$	$-\frac{1}{3}$	-1	-2	-4
$g'(x)$	4	$\frac{2}{3}$	$-\frac{1}{3}$	-1	-2	-4
$h'(x)$	8	$\frac{4}{3}$	$-\frac{2}{3}$	-2	-4	-8
$r'(x)$		12	1			
$s'(x)$	$-\frac{1}{3}$	-1	-2	-4		

99. (a) $\frac{1}{2}$
(b) $s'(5)$ does not exist because g is not differentiable at 6.

101. (a) 1.461 (b) -1.016 **103.** 0.2 rad, 1.45 rad/sec

105. (a) 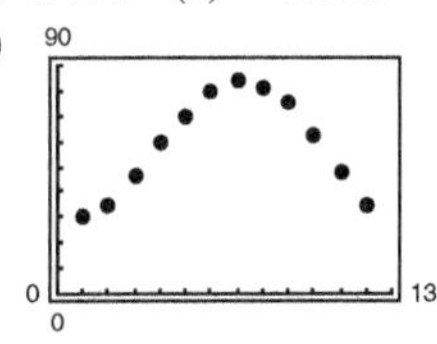

$T(t) = 27.3 \sin(0.49t - 1.90) + 57.1$

(b) 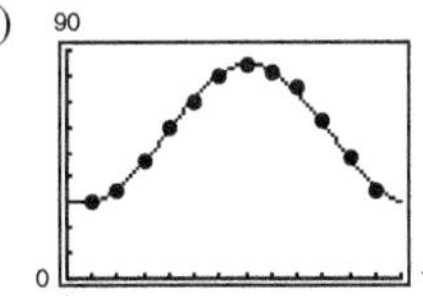

The model is a good fit.

(c) $T'(t) = 13.377 \cos(0.49t - 1.90)$

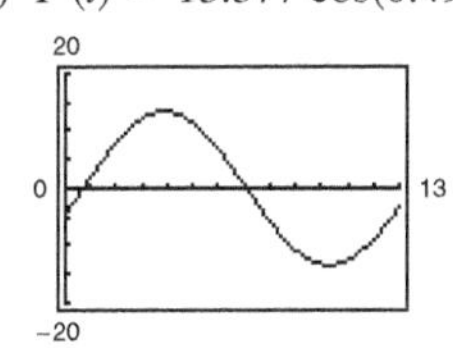

(d) The temperature changes most rapidly around spring (March–May) and fall (Oct.–Nov.)
The temperature changes most slowly around winter (Dec.–Feb.) and summer (Jun.–Aug.)
Yes. Explanations will vary.

107. (a) 0 bacteria per day (b) 177.8 bacteria per day
(c) 44.4 bacteria per day (d) 10.8 bacteria per day
(e) 3.3 bacteria per day
(f) The rate of change of the population is decreasing as time passes.

109. (a) $f'(x) = \beta \cos \beta x$
$f''(x) = -\beta^2 \sin \beta x$
$f'''(x) = -\beta^3 \cos \beta x$
$f^{(4)}(x) = \beta^4 \sin \beta x$
(b) $f''(x) + \beta^2 f(x) = -\beta^2 \sin \beta x + \beta^2(\sin \beta x) = 0$
(c) $f^{(2k)}(x) = (-1)^k \beta^{2k} \sin \beta x$
$f^{(2k-1)}(x) = (-1)^{k+1}\beta^{2k-1} \cos \beta x$

111. (a) $r'(1) = 0$ (b) $s'(4) = \frac{5}{8}$

113. (a) and (b) Proofs

115. $g'(x) = 3\left(\dfrac{3x - 5}{|3x - 5|}\right), \quad x \neq \dfrac{5}{3}$

117. $h'(x) = -|x|\sin x + \dfrac{x}{|x|}\cos x, \quad x \neq 0$

119. (a) $P_1(x) = 2\left(x - \dfrac{\pi}{4}\right) + 1$
$P_2(x) = 2\left(x - \dfrac{\pi}{4}\right)^2 + 2\left(x - \dfrac{\pi}{4}\right) + 1$

(b) 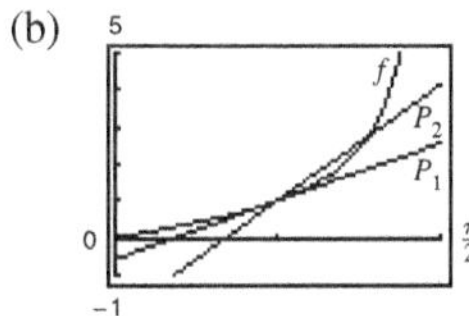

(c) P_2

(d) The accuracy worsens as you move away from $x = \dfrac{\pi}{4}$.

121. True. **123.** True **125.** Putnam Problem A1, 1967

Section 2.5 *(page 149)*

1. Answers will vary. Sample answer: In the explicit form of a function, the dependent variable y is explicitly written as a function of the independent variable x $[y = f(x)]$. In an implicit equation, the dependent variable y is not necessarily written in the form $y = f(x)$. An example of an implicit function is $x^2 + xy = 5$. In explicit form, it would be
$y = \dfrac{5 - x^2}{x}$.

3. You use implicit differentiation to find the derivative in cases where it is difficult to express y as a function of x explicitly.

5. $-\dfrac{x}{y}$ **7.** $-\dfrac{x^4}{y^4}$ **9.** $\dfrac{y - 3x^2}{2y - x}$

11. $\dfrac{1 - 3x^2y^3}{3x^3y^2 - 1}$ **13.** $\dfrac{6xy - 3x^2 - 2y^2}{4xy - 3x^2}$ **15.** $\dfrac{\cos x}{4 \sin 2y}$

17. $-\dfrac{\cot x \csc x + \tan y + 1}{x \sec^2 y}$ **19.** $\dfrac{y \cos xy}{1 - x \cos xy}$

21. (a) $y_1 = \sqrt{64 - x^2}$, $y_2 = -\sqrt{64 - x^2}$
(b) 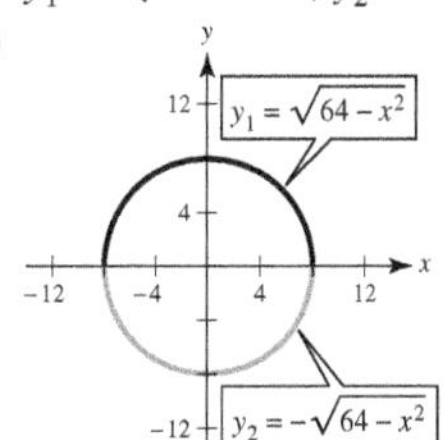

(c) $y' = \mp\dfrac{x}{\sqrt{64 - x^2}} = -\dfrac{x}{y}$ (d) $y' = -\dfrac{x}{y}$

23. (a) $y_1 = \dfrac{\sqrt{x^2+16}}{4}$, $y_2 = \dfrac{-\sqrt{x^2+16}}{4}$

(b)

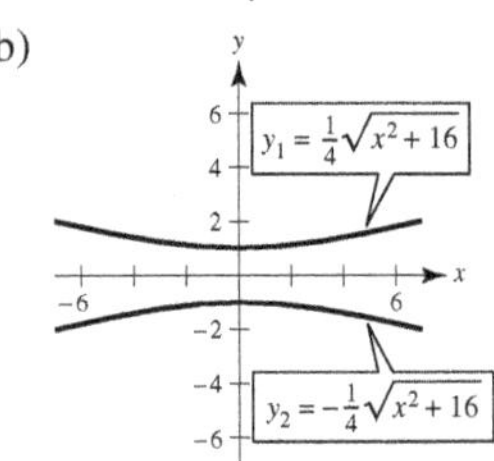

(c) $y' = \dfrac{\pm x}{4\sqrt{x^2+16}} = \dfrac{x}{16y}$ (d) $y' = \dfrac{x}{16y}$

25. $-\dfrac{y}{x}; -\dfrac{1}{6}$ **27.** $\dfrac{98x}{y(x^2+49)^2}$; Undefined

29. $-\dfrac{y(y+2x)}{x(x+2y)}; -1$ **31.** $-\sin^2(x+y)$ or $-\dfrac{x^2}{x^2+1}; 0$

33. $-\frac{1}{2}$ **35.** 0 **37.** $y = -x + 7$

39. $y = \dfrac{\sqrt{3}x}{6} + \dfrac{8\sqrt{3}}{3}$ **41.** $y = -\dfrac{2}{11}x + \dfrac{30}{11}$

43. Answers will vary. Sample answers:
$xy = 2$, $yx^2 + x = 2$; $x^2 + y^2 + y = 4$, $xy + y^2 = 2$

45. (a) $y = -2x + 4$ (b) Answers will vary.

47. $\cos^2 y$, $-\dfrac{\pi}{2} < y < \dfrac{\pi}{2}$, $\dfrac{1}{1+x^2}$ **49.** $-\dfrac{4}{y^3}$

51. $\dfrac{6x^2y + 2y - 20x}{(x^2-1)^2}$ **53.** $\dfrac{x\sin x + 2\cos x + 14y}{7x^2}$

55. $2x + 3y - 30 = 0$

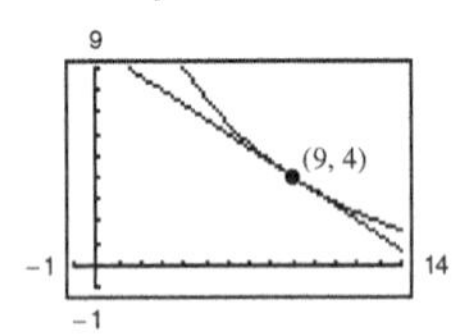

57. At (4, 3):
Tangent line: $4x + 3y - 25 = 0$
Normal line: $3x - 4y = 0$

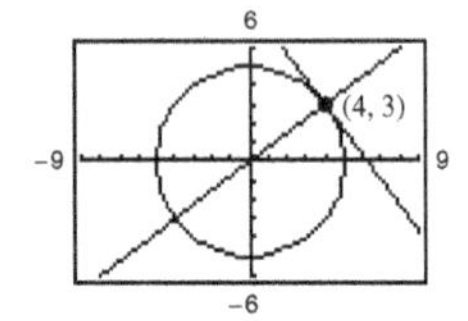

At $(-3, 4)$
Tangent line: $3x - 4y + 25 = 0$
Normal line: $4x + 3y = 0$

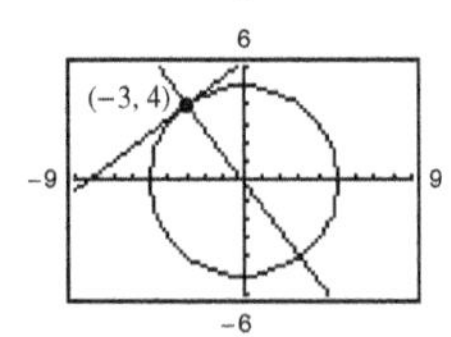

59. $x^2 + y^2 = r^2 \Rightarrow y' = -\dfrac{x}{y} \Rightarrow \dfrac{y}{x}$ = slope of normal line.
Then for (x_0, y_0) on the circle, $x_0 \neq 0$, an equation of the normal line is $y = \left(\dfrac{y_0}{x_0}\right)x$, which passes through the origin. If $x_0 = 0$, the normal line is vertical and passes through the origin.

61. Horizontal tangents: $(-4, 0)$, $(-4, 10)$
Vertical tangents: $(0, 5)$, $(-8, 5)$

63.

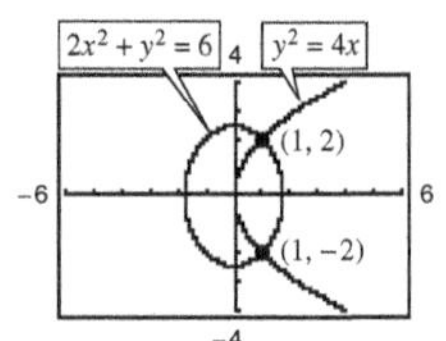

At (1, 2):
Slope of ellipse: -1
Slope of parabola: 1
At $(1, -2)$:
Slope of ellipse: 1
Slope of parabola: -1

65.
At (0, 0):
Slope of line: -1
Slope of sine curve: 1

67. Derivatives: $\dfrac{dy}{dx} = -\dfrac{y}{x}, \dfrac{dy}{dx} = \dfrac{x}{y}$

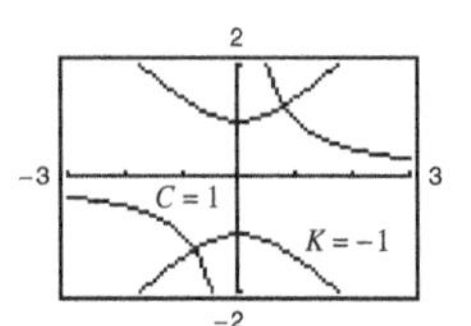

69.

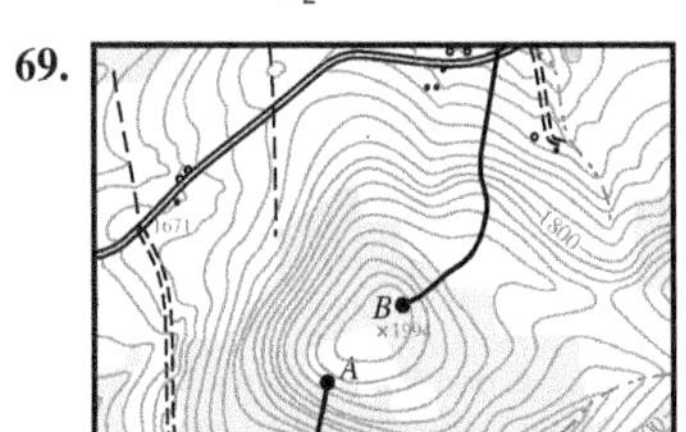

Use starting point B.

71. (a)

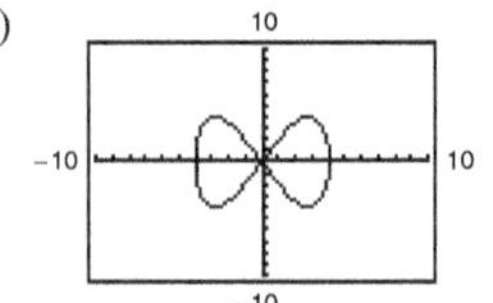

(b)

$y_1 = \frac{1}{3}\left[\left(\sqrt{7}+7\right)x + \left(8\sqrt{7}+23\right)\right]$
$y_2 = -\frac{1}{3}\left[\left(-\sqrt{7}+7\right)x - \left(23 - 8\sqrt{7}\right)\right]$
$y_3 = -\frac{1}{3}\left[\left(\sqrt{7}-7\right)x - \left(23 - 8\sqrt{7}\right)\right]$
$y_4 = -\frac{1}{3}\left[\left(\sqrt{7}+7\right)x - \left(8\sqrt{7}+23\right)\right]$

(c) $\left(\dfrac{8\sqrt{7}}{7}, 5\right)$

73. Proof **75.** $y = -\dfrac{\sqrt{3}}{2}x + 2\sqrt{3}$, $y = \dfrac{\sqrt{3}}{2}x - 2\sqrt{3}$

77. (a) $y = 2x - 6$
(b)

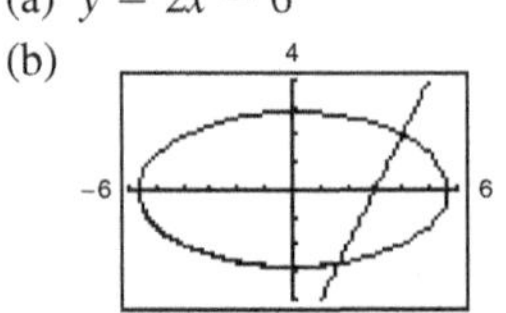

(c) $\left(\frac{28}{17}, -\frac{46}{17}\right)$

Section 2.6 *(page 157)*

1. A related-rate equation is an equation that relates the rates of change of various quantities.

3. (a) $\frac{3}{4}$ (b) 20 **5.** (a) $-\frac{5}{8}$ (b) $\frac{3}{2}$

7. (a) -8 cm/sec (b) 0 cm/sec (c) 8 cm/sec

9. (a) 12 ft/sec (b) 6 ft/sec (c) 3 ft/sec

11. 296π cm²/min

13. (a) 972π in.^{3}/min, $15{,}552\pi$ in.^{3}/min
(b) If $\frac{dr}{dt}$ is constant, $\frac{dV}{dt}$ is proportional to r^2.
15. (a) 72 cm^3/sec (b) 1800 cm^3/sec
17. $\frac{8}{405\pi}$ ft/min **19.** (a) 12.5% (b) $\frac{1}{144}$ m/min
21. (a) $-\frac{7}{12}$ ft/sec, $-\frac{3}{2}$ ft/sec, $-\frac{48}{7}$ ft/sec
(b) $\frac{527}{24}$ ft^2/sec (c) $\frac{1}{12}$ rad/sec
23. Rate of vertical change: $\frac{1}{5}$ m/sec
Rate of horizontal change: $-\frac{\sqrt{3}}{15}$ m/sec
25. (a) -750 mi/h (b) 30 min
27. $-\frac{50}{\sqrt{85}} \approx -5.42$ ft/sec
29. (a) $\frac{25}{3}$ ft/sec (b) $\frac{10}{3}$ ft/sec
31. (a) 12 sec (b) $\frac{1}{2}\sqrt{3}$ m (c) $\frac{\sqrt{5}\pi}{120}$ m/sec
33. Evaporation rate proportional to $S \Rightarrow \frac{dV}{dt} = k(4\pi r^2)$
$V = \left(\frac{4}{3}\right)\pi r^3 \Rightarrow \frac{dV}{dt} = 4\pi r^2 \frac{dr}{dt}$. So $k = \frac{dr}{dt}$.
35. (a) $\frac{dy}{dt} = 3\frac{dx}{dt}$ means that y changes three times as fast as x changes.
(b) y changes slowly when $x \approx 0$ or $x \approx L$. y changes more rapidly when x is near the middle of the interval.
37. 0.6 ohm/sec **39.** About 84.9797 mi/h
41. $\frac{2\sqrt{21}}{525} \approx 0.017$ rad/sec
43. (a) $\frac{200\pi}{3}$ ft/sec (b) 200π ft/sec
(c) About 427.43π ft/sec
45. (a) Proof (b) $\frac{\sqrt{3}s^2}{8}, \frac{s^2}{8}$
47. (a) $r(f) = 0.0096f^3 - 0.559f^2 + 10.54f - 61.5$
(b) $\frac{dr}{dt} = (0.0288f^2 - 1.118f + 10.54)\frac{df}{dt}$;
-0.039 million participants/yr
49. -0.1808 ft/sec^2

Review Exercises for Chapter 2 *(page 161)*

1. 0 **3.** $3x^2 - 2$ **5.** 5
7. f is differentiable at all $x \neq 3$. **9.** 0 **11.** $3x^2 - 22x$
13. $\frac{3}{\sqrt{x}} + \frac{1}{\sqrt[3]{x^2}}$ **15.** $-\frac{4}{3t^3}$ **17.** $4 - 5\cos\theta$
19. $-3\sin\theta - \frac{\cos\theta}{4}$ **21.** -1 **23.** 2
25. (a) 50 vibrations/sec/lb (b) 33.33 vibrations/sec/lb
27. (a) $s(t) = -16t^2 - 30t + 600$
$v(t) = -32t - 30$
(b) -94 ft/sec
(c) $v'(1) = -62$ ft/sec, $v'(3) = -126$ ft/sec
(d) About 5.258 sec (e) About -198.256 ft/sec
29. $4(5x^3 - 15x^2 - 11x - 8)$ **31.** $9x\cos x - \cos x + 9\sin x$
33. $\frac{-(x^2+1)}{(x^2-1)^2}$ **35.** $\frac{4x^3\cos x + x^4\sin x}{\cos^2 x}$
37. $3x^2\sec x\tan x + 6x\sec x$ **39.** $-x\sin x$
41. $y = 4x + 10$ **43.** $y = -8x + 1$ **45.** $-48t$
47. $\frac{225}{4}\sqrt{x}$ **49.** $6\sec^2\theta\tan\theta$ **51.** $8\cot x\csc^2 x$
53. $v(3) = 11$ m/sec, $a(3) = -6$ m/sec^2 **55.** $28(7x+3)^3$
57. $-\frac{6x}{(x^2+5)^4}$ **59.** $-45\sin(9x+1)$
61. $\frac{1}{2}(1 - \cos 2x)$, or $\sin^2 x$ **63.** $(36x+1)(6x+1)^4$
65. $\frac{3x^2(x+10)}{2(x+5)^{5/2}}$ **67.** -2 **69.** -11 **71.** 0
73. $384(8x+5)$ **75.** $2\csc^2 x\cot x$
77. (a) $-18.667°$/h (b) $-7.284°$/h
(c) $-3.240°$/h (d) $-0.747°$/h
79. $-\frac{x}{y}$ **81.** $\frac{y(y^2 - 3x^2)}{x(x^2 - 3y^2)}$ **83.** $\frac{y\sin x + \sin y}{\cos x - x\cos y}$
85. Tangent line: $3x + y - 10 = 0$
Normal line: $x - 3y = 0$

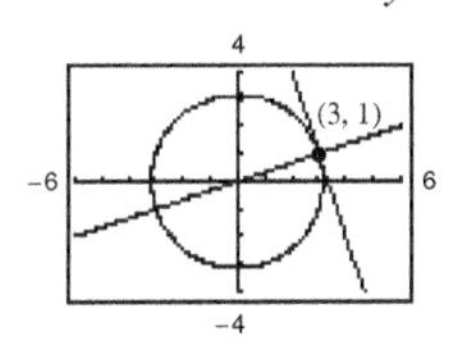

87. (a) $2\sqrt{2}$ units/sec (b) 4 units/sec (c) 8 units/sec
89. 450π km/h

P.S. Problem Solving *(page 163)*

1. (a) $r = \frac{1}{2}$; $x^2 + \left(y - \frac{1}{2}\right)^2 = \frac{1}{4}$
(b) Center: $\left(0, \frac{5}{4}\right)$; $x^2 + \left(y - \frac{5}{4}\right)^2 = 1$
3. $p(x) = 2x^3 + 4x^2 - 5$
5. (a) $y = 4x - 4$ (b) $y = -\frac{1}{4}x + \frac{9}{2}$; $\left(-\frac{9}{4}, \frac{81}{16}\right)$
(c) Tangent line: $y = 0$ (d) Proof
Normal line: $x = 0$
7. (a) Graph $\begin{cases} y_1 = \frac{1}{a}\sqrt{x^2(a^2 - x^2)} \\ y_2 = -\frac{1}{a}\sqrt{x^2(a^2 - x^2)} \end{cases}$ as separate equations.
(b) Answers will vary. Sample answer:

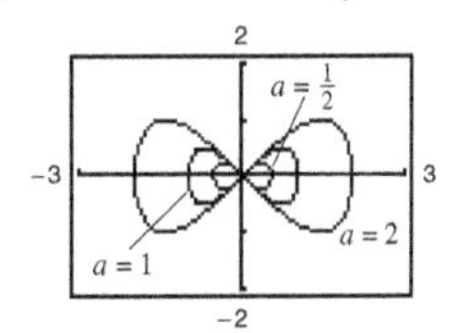

The intercepts will always be $(0, 0)$, $(a, 0)$, and $(-a, 0)$, and the maximum and minimum y-values appear to be $\pm\frac{1}{2}a$.
(c) $\left(\frac{a\sqrt{2}}{2}, \frac{a}{2}\right)$, $\left(\frac{a\sqrt{2}}{2}, -\frac{a}{2}\right)$, $\left(-\frac{a\sqrt{2}}{2}, \frac{a}{2}\right)$, $\left(-\frac{a\sqrt{2}}{2}, -\frac{a}{2}\right)$

9. (a) When the man is 90 ft from the light, the tip of his shadow is $112\frac{1}{2}$ ft from the light. The tip of the child's shadow is $111\frac{1}{9}$ ft from the light, so the man's shadow extends $1\frac{7}{18}$ ft beyond the child's shadow.
(b) When the man is 60 ft from the light, the tip of his shadow is 75 ft from the light. The tip of the child's shadow is $77\frac{7}{9}$ ft from the light, so the child's shadow extends $2\frac{7}{9}$ ft beyond the man's shadow.
(c) $d = 80$ ft
(d) Let x be the distance of the man from the light, and let s be the distance from the light to the tip of the shadow.
If $0 < x < 80$, then $\dfrac{ds}{dt} = -\dfrac{50}{9}$.
If $x > 80$, then $\dfrac{ds}{dt} = -\dfrac{25}{4}$.
There is a discontinuity at $x = 80$.

11. (a) $v(t) = -\frac{27}{5}t + 27$ ft/sec (b) 5 sec; 73.5 ft
$a(t) = -\frac{27}{5}$ ft/sec^2
(c) The acceleration due to gravity on Earth is greater in magnitude than that on the moon.

13. Proof; The graph of L is a line passing through the origin $(0, 0)$.

15. (a) j would be the rate of change of acceleration.
(b) $j = 0$. Acceleration is constant, so there is no change in acceleration.
(c) a: position function, d: velocity function, b: acceleration function, c: jerk function

Chapter 3

Section 3.1 *(page 171)*

1. $f(c)$ is the low point of the graph of f on the interval I.

3. A relative maximum is a peak of the graph. An absolute maximum is the greatest value on the interval I.

5. Find all x-values for which $f'(x) = 0$ and all x-values for which $f'(x)$ does not exist.

7. $f'(0) = 0$ **9.** $f'(2) = 0$ **11.** $f'(-2)$ is undefined.

13. 2, absolute maximum (and relative maximum)

15. 1, absolute maximum (and relative maximum);
2, absolute minimum (and relative minimum);
3, absolute maximum (and relative maximum)

17. $x = \dfrac{3}{4}$ **19.** $t = \dfrac{8}{3}$ **21.** $x = \dfrac{\pi}{3}, \pi, \dfrac{5\pi}{3}$

23. Minimum: $(2, 1)$
Maximum: $(-1, 4)$

25. Minimum: $(-3, -13)$
Maximum: $(0, 5)$

27. Minimum: $\left(-1, -\frac{5}{2}\right)$
Maximum: $(2, 2)$

29. Minimum: $(0, 0)$
Maximum: $(-1, 5)$

31. Minimum: $(1, -6)$ and $(-2, -6)$
Maximum: $(0, 0)$

33. Minimum: $(-1, -1)$
Maximum: $(3, 3)$

35. Minimum value is -2 for $-2 \le x < -1$.
Maximum: $(2, 2)$

37. Minimum: $\left(\dfrac{3\pi}{2}, -1\right)$
Maximum: $\left(\dfrac{5\pi}{6}, \dfrac{1}{2}\right)$

39. Minimum: $(\pi, -3)$
Maxima: $(0, 3)$ and $(2\pi, 3)$

41. (a) Minimum: $(0, -3)$
Maximum: $(2, 1)$
(b) Minimum: $(0, -3)$
(c) Maximum: $(2, 1)$
(d) No extrema

43. (a) Minimum: $(1, -1)$
Maximum: $(-1, 3)$
(b) Maximum: $(3, 3)$
(c) Minimum: $(1, -1)$
(d) Minimum: $(1, -1)$

45.

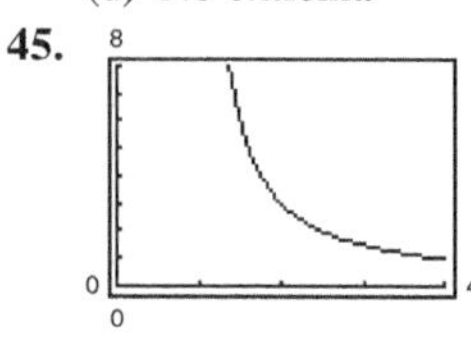

Minimum: $(4, 1)$

47.

Minimum: $(0, 0)$
Maximum: $(2.7149, 1.7856)$

49. (a)

(b) Minimum: $(0.4398, -1.0613)$

51. Maximum: $\left|f''\left(\sqrt[3]{-10 + \sqrt{108}}\right)\right| = f''\left(\sqrt{3} - 1\right) \approx 1.47$

53. Maximum: $\left|f^{(4)}(0)\right| = \frac{56}{81}$

55. Answers will vary. Sample answer: Let $f(x) = \dfrac{1}{x}$. f is continuous on $(0, 1)$ but does not have a maximum or minimum.

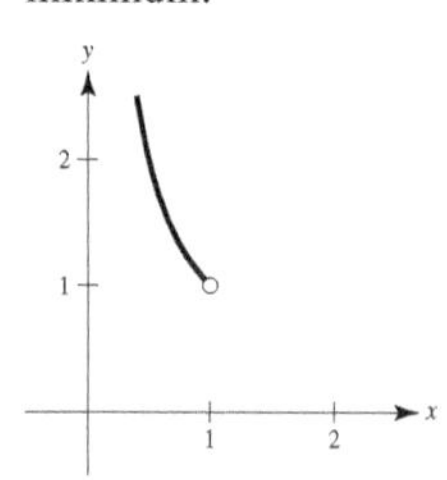

57. (a) Yes. The value is defined.
(b) No. The value is undefined.

59. No. The function is not defined at $x = -2$.

61. Maximum: $P(12) = 72$; No. P is decreasing for $I > 12$.

63. $\theta = \operatorname{arcsec}\sqrt{3} \approx 0.9553$ rad

65. False. The maximum would be 9 if the interval was closed.

67. True **69.** Proof **71.** Putnam Problem B3, 2004

Section 3.2 *(page 178)*

1. Rolle's Theorem gives conditions that guarantee the existence of an extreme value in the interior of a closed interval.

3. $f(-1) = f(1) = 1$; f is not continuous on $[-1, 1]$.

5. $f(0) = f(2) = 0$; f is not differentiable on $(0, 2)$.

7. $(2, 0), (-1, 0)$; $f'\left(\frac{1}{2}\right) = 0$ **9.** $(0, 0), (-4, 0)$; $f'\left(-\frac{8}{3}\right) = 0$

11. $f'\left(\dfrac{3}{2}\right) = 0$ **13.** $f'\left(\dfrac{6 - \sqrt{3}}{3}\right) = 0$; $f'\left(\dfrac{6 + \sqrt{3}}{3}\right) = 0$

15. Not differentiable at $x = 0$ **17.** $f'\left(-2 + \sqrt{5}\right) = 0$

19. $f'\left(\dfrac{\pi}{2}\right) = 0$; $f'\left(\dfrac{3\pi}{2}\right) = 0$ **21.** $f'(1) = 0$

23. Not continuous on $[0, \pi]$

25.

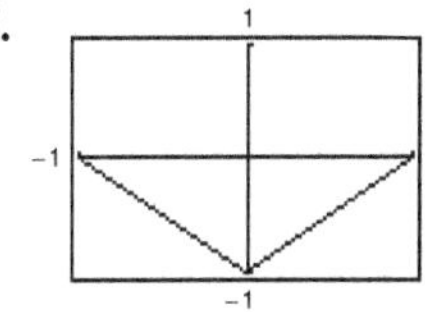

Rolle's Theorem does not apply.

27.

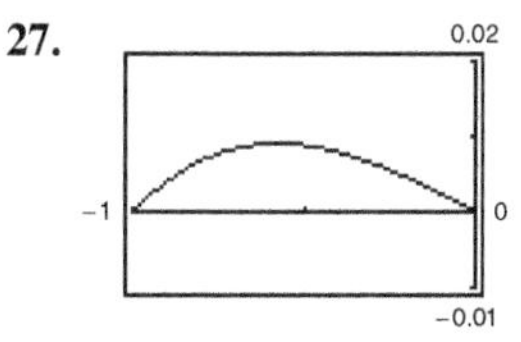

$f'\left(-\frac{6}{\pi}\arccos\frac{3}{\pi}\right) = 0$

29. (a) $f(1) = f(2) = 38$
(b) Velocity $= 0$ for some t in $(1, 2)$; $t = \frac{3}{2}$ sec

31.

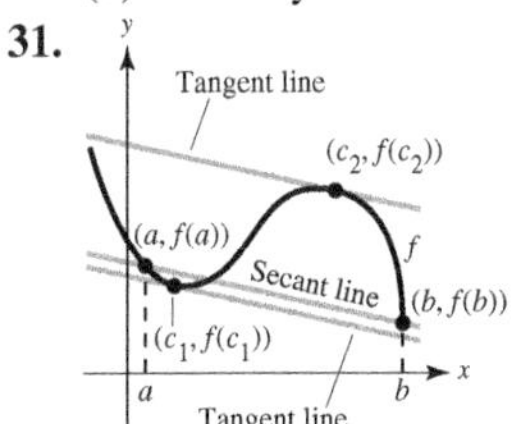

33. The function is not continuous on $[0, 6]$.
35. The function is not continuous on $[0, 6]$.
37. (a) Secant line: $x + y - 3 = 0$ (b) $c = \frac{1}{2}$
(c) Tangent line: $4x + 4y - 21 = 0$
(d)

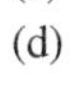

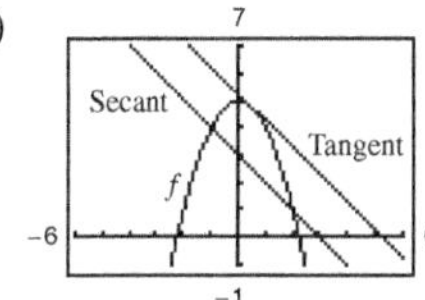

39. $f'\left(\frac{\sqrt{21}}{3}\right) = 42$ **41.** $f'\left(-\frac{\sqrt{3}}{3}\right) = 3$
43. f is not continuous at $x = 1$.
45. f is not differentiable at $x = -\frac{1}{2}$. **47.** $f'\left(\frac{\pi}{2}\right) = 0$
49. (a)–(c)

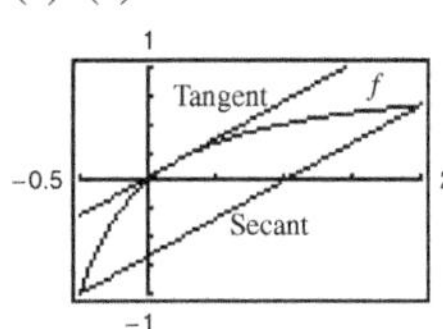

(b) $y = \frac{2}{3}(x - 1)$
(c) $y = \frac{1}{3}(2x + 5 - 2\sqrt{6})$

51. (a)–(c)

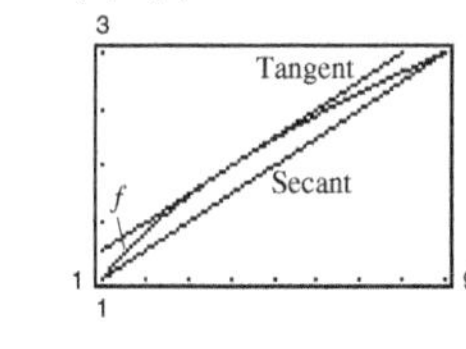

(b) $y = \frac{1}{4}x + \frac{3}{4}$
(c) $y = \frac{1}{4}x + 1$

53. (a) -14.7 m/sec (b) 1.5 sec
55. No. Let $f(x) = x^2$ on $[-1, 2]$.
57. No. $f(x)$ is not continuous on $[0, 1]$. So it does not satisfy the hypothesis of Rolle's Theorem.
59. By the Mean Value Theorem, there is a time when the speed of the plane must equal the average speed of 454.5 miles/hour. The speed was 400 miles/hour when the plane was accelerating to 454.5 miles/hour and decelerating from 454.5 miles/hour.
61. Proof
63. (a)

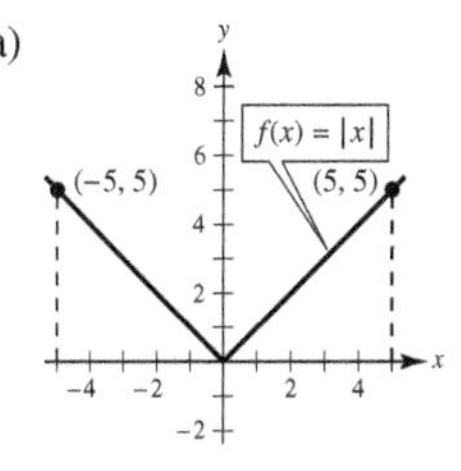

(b)

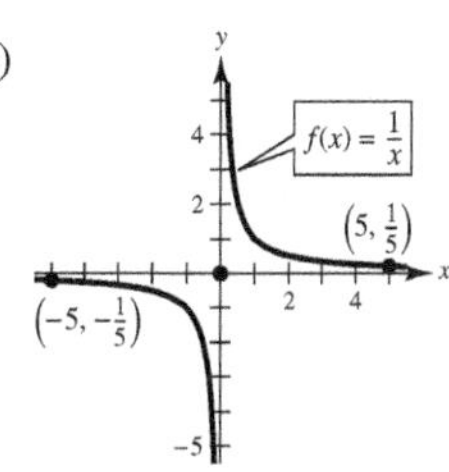

65–67. Proofs **69.** $f(x) = 5$; $f(x) = c$ and $f(2) = 5$.
71. $f(x) = x^2 - 1$; $f(x) = x^2 + c$ and $f(1) = 0$, so $c = -1$.
73. False. f is not continuous on $[-1, 1]$. **75.** True
77–85. Proofs

Section 3.3 *(page 187)*

1. A positive derivative of a function on an open interval implies that the function is increasing on the interval. A negative derivative implies that the function is decreasing. A zero derivative implies that the function is constant.
3. (a) $(0, 6)$ (b) $(6, 8)$
5. Increasing on $(-\infty, -1)$; Decreasing on $(-1, \infty)$
7. Increasing on $(-\infty, -2)$ and $(2, \infty)$; Decreasing on $(-2, 2)$
9. Increasing on $(-\infty, -1)$; Decreasing on $(-1, \infty)$
11. Increasing on $(1, \infty)$; Decreasing on $(-\infty, 1)$
13. Increasing on $(-2\sqrt{2}, 2\sqrt{2})$;
Decreasing on $(-4, -2\sqrt{2})$ and $(2\sqrt{2}, 4)$
15. Increasing on $\left(0, \frac{\pi}{2}\right)$ and $\left(\frac{3\pi}{2}, 2\pi\right)$;
Decreasing on $\left(\frac{\pi}{2}, \frac{3\pi}{2}\right)$
17. Increasing on $\left(0, \frac{7\pi}{6}\right)$ and $\left(\frac{11\pi}{6}, 2\pi\right)$;
Decreasing on $\left(\frac{7\pi}{6}, \frac{11\pi}{6}\right)$
19. (a) Critical number: $x = 4$
(b) Increasing on $(4, \infty)$; Decreasing on $(-\infty, 4)$
(c) Relative minimum: $(4, -16)$
21. (a) Critical number: $x = 1$
(b) Increasing on $(-\infty, 1)$; Decreasing on $(1, \infty)$
(c) Relative maximum: $(1, 5)$
23. (a) Critical numbers: $x = -1, 1$
(b) Increasing on $(-1, 1)$;
Decreasing on $(-\infty, -1)$ and $(1, \infty)$
(c) Relative maximum: $(1, 17)$;
Relative minimum: $(-1, -11)$
25. (a) Critical numbers: $x = -\frac{5}{3}, 1$
(b) Increasing on $\left(-\infty, -\frac{5}{3}\right)$, $(1, \infty)$;
Decreasing on $\left(-\frac{5}{3}, 1\right)$
(c) Relative maximum: $\left(-\frac{5}{3}, \frac{256}{27}\right)$;
Relative minimum: $(1, 0)$
27. (a) Critical numbers: $x = \pm 1$
(b) Increasing on $(-\infty, -1)$ and $(1, \infty)$;
Decreasing on $(-1, 1)$
(c) Relative maximum: $\left(-1, \frac{4}{5}\right)$; Relative minimum: $\left(1, -\frac{4}{5}\right)$
29. (a) Critical number: $x = 0$
(b) Increasing on $(-\infty, \infty)$
(c) No relative extrema
31. (a) Critical number: $x = -2$
(b) Increasing on $(-2, \infty)$; Decreasing on $(-\infty, -2)$
(c) Relative minimum: $(-2, 0)$
33. (a) Critical number: $x = 5$
(b) Increasing on $(-\infty, 5)$; Decreasing on $(5, \infty)$
(c) Relative maximum: $(5, 5)$

35. (a) Critical numbers: $x = \pm\dfrac{\sqrt{2}}{2}$; Discontinuity: $x = 0$

(b) Increasing on $\left(-\infty, -\dfrac{\sqrt{2}}{2}\right)$ and $\left(\dfrac{\sqrt{2}}{2}, \infty\right)$;

Decreasing on $\left(-\dfrac{\sqrt{2}}{2}, 0\right)$ and $\left(0, \dfrac{\sqrt{2}}{2}\right)$

(c) Relative maximum: $\left(-\dfrac{\sqrt{2}}{2}, -2\sqrt{2}\right)$;

Relative minimum: $\left(\dfrac{\sqrt{2}}{2}, 2\sqrt{2}\right)$

37. (a) Critical number: $x = 0$; Discontinuities: $x = \pm 3$

(b) Increasing on $(-\infty, -3)$ and $(-3, 0)$;
Decreasing on $(0, 3)$ and $(3, \infty)$

(c) Relative maximum: $(0, 0)$

39. (a) Critical number: $x = 0$

(b) Increasing on $(-\infty, 0)$; Decreasing on $(0, \infty)$

(c) Relative maximum: $(0, 4)$

41. (a) Critical numbers: $x = \dfrac{\pi}{3}, \dfrac{5\pi}{3}$; Increasing on $\left(\dfrac{\pi}{3}, \dfrac{5\pi}{3}\right)$;

Decreasing on $\left(0, \dfrac{\pi}{3}\right)$ and $\left(\dfrac{5\pi}{3}, 2\pi\right)$

(b) Relative maximum: $\left(\dfrac{5\pi}{3}, \dfrac{5\pi}{3} + \sqrt{3}\right)$;

Relative minimum: $\left(\dfrac{\pi}{3}, \dfrac{\pi}{3} - \sqrt{3}\right)$

43. (a) Critical numbers: $x = \dfrac{\pi}{4}, \dfrac{5\pi}{4}$;

Increasing on $\left(0, \dfrac{\pi}{4}\right), \left(\dfrac{5\pi}{4}, 2\pi\right)$;

Decreasing on $\left(\dfrac{\pi}{4}, \dfrac{5\pi}{4}\right)$

(b) Relative maximum: $\left(\dfrac{\pi}{4}, \sqrt{2}\right)$;

Relative minimum: $\left(\dfrac{5\pi}{4}, -\sqrt{2}\right)$

45. (a) Critical numbers:

$x = \dfrac{\pi}{4}, \dfrac{\pi}{2}, \dfrac{3\pi}{4}, \pi, \dfrac{5\pi}{4}, \dfrac{3\pi}{2}, \dfrac{7\pi}{4}$;

Increasing on $\left(\dfrac{\pi}{4}, \dfrac{\pi}{2}\right), \left(\dfrac{3\pi}{4}, \pi\right), \left(\dfrac{5\pi}{4}, \dfrac{3\pi}{2}\right), \left(\dfrac{7\pi}{4}, 2\pi\right)$;

Decreasing on $\left(0, \dfrac{\pi}{4}\right), \left(\dfrac{\pi}{2}, \dfrac{3\pi}{4}\right), \left(\pi, \dfrac{5\pi}{4}\right), \left(\dfrac{3\pi}{2}, \dfrac{7\pi}{4}\right)$

(b) Relative maxima: $\left(\dfrac{\pi}{2}, 1\right), (\pi, 1), \left(\dfrac{3\pi}{2}, 1\right)$;

Relative minima: $\left(\dfrac{\pi}{4}, 0\right), \left(\dfrac{3\pi}{4}, 0\right), \left(\dfrac{5\pi}{4}, 0\right), \left(\dfrac{7\pi}{4}, 0\right)$

47. (a) Critical numbers: $\dfrac{\pi}{2}, \dfrac{7\pi}{6}, \dfrac{3\pi}{2}, \dfrac{11\pi}{6}$;

Increasing on $\left(0, \dfrac{\pi}{2}\right), \left(\dfrac{7\pi}{6}, \dfrac{3\pi}{2}\right), \left(\dfrac{11\pi}{6}, 2\pi\right)$;

Decreasing on $\left(\dfrac{\pi}{2}, \dfrac{7\pi}{6}\right), \left(\dfrac{3\pi}{2}, \dfrac{11\pi}{6}\right)$

(b) Relative maxima: $\left(\dfrac{\pi}{2}, 2\right), \left(\dfrac{3\pi}{2}, 0\right)$;

Relative minima: $\left(\dfrac{7\pi}{6}, -\dfrac{1}{4}\right), \left(\dfrac{11\pi}{6}, -\dfrac{1}{4}\right)$

49. (a) $f'(x) = \dfrac{2(9 - 2x^2)}{\sqrt{9 - x^2}}$

(b)

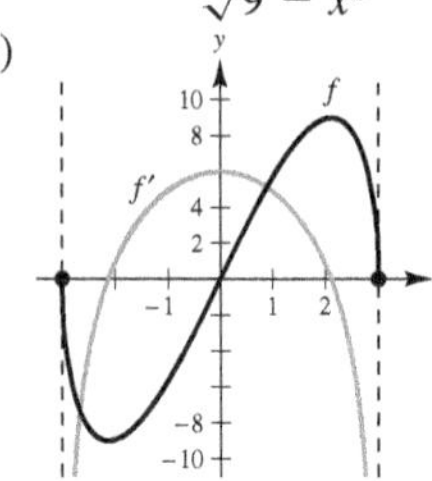

(c) Critical numbers:
$x = \pm\dfrac{3\sqrt{2}}{2}$

(d) $f' > 0$ on $\left(-\dfrac{3\sqrt{2}}{2}, \dfrac{3\sqrt{2}}{2}\right)$;

$f' < 0$ on $\left(-3, -\dfrac{3\sqrt{2}}{2}\right), \left(\dfrac{3\sqrt{2}}{2}, 3\right)$;

f is increasing when f' is positive and decreasing when f' is negative.

51. (a) $f'(t) = t(t \cos t + 2 \sin t)$

(b)

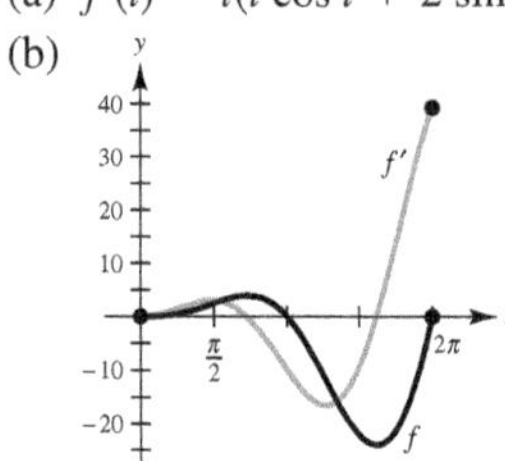

(c) Critical numbers:
$t = 2.2889, 5.0870$

(d) $f' > 0$ on $(0, 2.2889)$, $(5.0870, 2\pi)$;
$f' < 0$ on $(2.2889, 5.0870)$;
f is increasing when f' is positive and decreasing when f' is negative.

53. (a) $f'(x) = -\cos\dfrac{x}{3}$

(b)

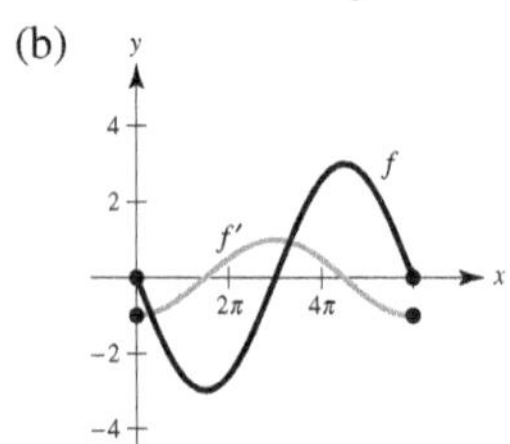

(c) Critical numbers: $x = \dfrac{3\pi}{2}, \dfrac{9\pi}{2}$

(d) $f' > 0$ on $\left(\dfrac{3\pi}{2}, \dfrac{9\pi}{2}\right)$; $f' < 0$ on $\left(0, \dfrac{3\pi}{2}\right), \left(\dfrac{9\pi}{2}, 6\pi\right)$;

f is increasing when f' is positive and decreasing when f' is negative.

55. $f(x)$ is symmetric with respect to the origin.
Zeros: $(0, 0), (\pm\sqrt{3}, 0)$

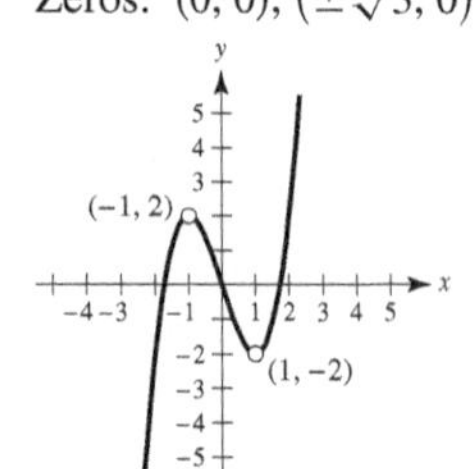

$g(x)$ is continuous on $(-\infty, \infty)$, and $f(x)$ has holes at $x = 1$ and $x = -1$.

57.

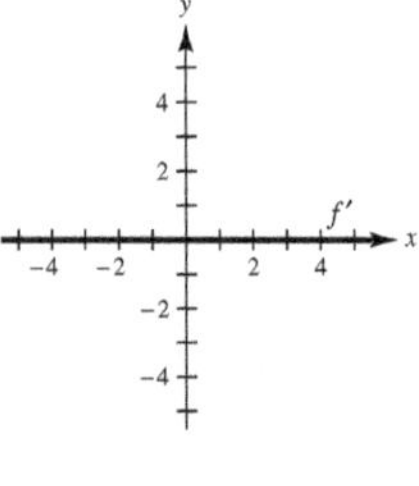

59.

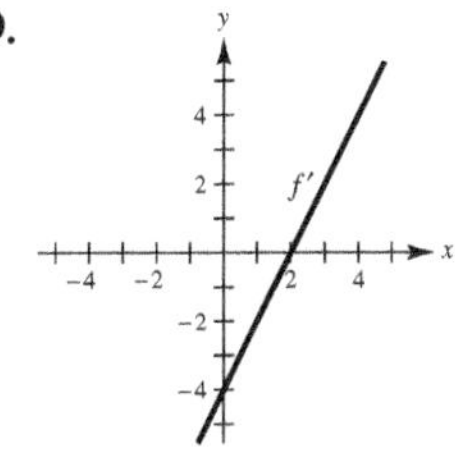

61.

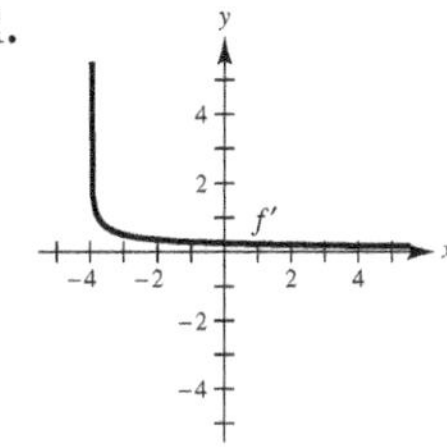

63. $g'(0) < 0$ **65.** $g'(-6) < 0$

67. Answers will vary. Sample answer:

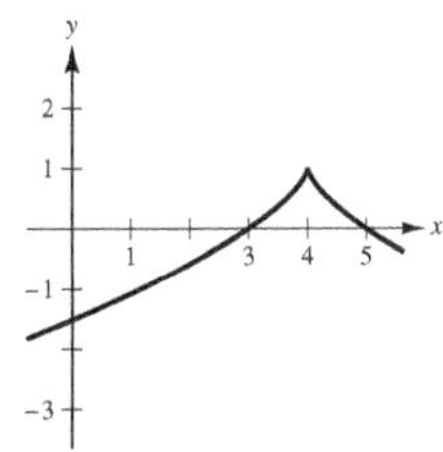

69. No. For example, the product of $f(x) = x$ and $g(x) = x$ is $f(x) \cdot g(x) = x^2$, which is decreasing on $(-\infty, 0)$ and increasing on $(0, \infty)$.

71. $(5, f(5))$ is a relative minimum.

73. (a)

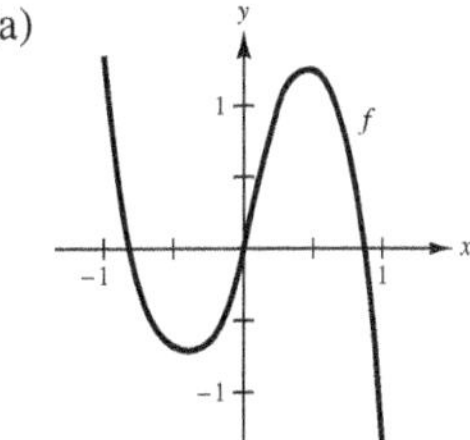

(b) Critical numbers: $x \approx -0.40$ and $x \approx 0.48$

(c) Relative maximum: $(0.48, 1.25)$;
Relative minimum: $(-0.40, 0.75)$

75. (a) $s'(t) = 9.8(\sin\theta)t$; speed $= |9.8(\sin\theta)t|$

(b)

θ	0	$\frac{\pi}{4}$	$\frac{\pi}{3}$	$\frac{\pi}{2}$	$\frac{2\pi}{3}$	$\frac{3\pi}{4}$	π
$s'(t)$	0	$4.9\sqrt{2}t$	$4.9\sqrt{3}t$	$9.8t$	$4.9\sqrt{3}t$	$4.9\sqrt{2}t$	0

The speed is maximum at $\theta = \frac{\pi}{2}$.

77. (a)

t	0	0.5	1	1.5	2	2.5	3
$C(t)$	0	0.055	0.107	0.148	0.171	0.176	0.167

$t = 2.5$ h

(b)

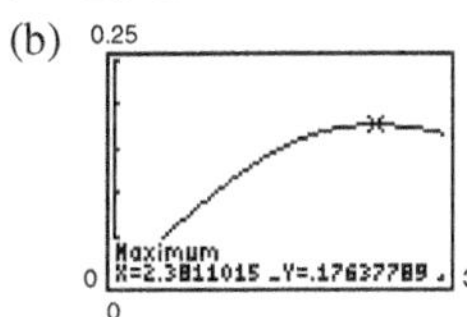

$t \approx 2.38$ h (c) $t \approx 2.38$ h

79. $r = \frac{2R}{3}$

81. (a) $v(t) = 6 - 2t$ (b) $[0, 3)$ (c) $(3, \infty)$ (d) $t = 3$

83. (a) $v(t) = 3t^2 - 10t + 4$

(b) $\left[0, \frac{5 - \sqrt{13}}{3}\right)$ and $\left(\frac{5 + \sqrt{13}}{3}, \infty\right)$

(c) $\left(\frac{5 - \sqrt{13}}{3}, \frac{5 + \sqrt{13}}{3}\right)$ (d) $t = \frac{5 \pm \sqrt{13}}{3}$

85. Answers will vary.

87. (a) Minimum degree: 3

(b) $a_3(0)^3 + a_2(0)^2 + a_1(0) + a_0 = 0$
$a_3(2)^3 + a_2(2)^2 + a_1(2) + a_0 = 2$
$3a_3(0)^2 + 2a_2(0) + a_1 = 0$
$3a_3(2)^2 + 2a_2(2) + a_1 = 0$

(c) $f(x) = -\frac{1}{2}x^3 + \frac{3}{2}x^2$

89. (a) Minimum degree: 4

(b) $a_4(0)^4 + a_3(0)^3 + a_2(0)^2 + a_1(0) + a_0 = 0$
$a_4(2)^4 + a_3(2)^3 + a_2(2)^2 + a_1(2) + a_0 = 4$
$a_4(4)^4 + a_3(4)^3 + a_2(4)^2 + a_1(4) + a_0 = 0$
$4a_4(0)^3 + 3a_3(0)^2 + 2a_2(0) + a_1 = 0$
$4a_4(2)^3 + 3a_3(2)^2 + 2a_2(2) + a_1 = 0$
$4a_4(4)^3 + 3a_3(4)^2 + 2a_2(4) + a_1 = 0$

(c) $f(x) = \frac{1}{4}x^4 - 2x^3 + 4x^2$

91. False. Let $f(x) = \sin x$. **93.** False. Let $f(x) = x^3$.

95. False. Let $f(x) = x^3$. There is a critical number at $x = 0$, but not a relative extremum.

97–99. Proofs **101.** Putnam Problem A3, 2003

Section 3.4 *(page 196)*

1. Find the second derivative of a function and form test intervals by using the values for which the second derivative is zero or does not exist and the values at which the function is not continuous. Determine the sign of the second derivative on these test intervals. If the second derivative is positive, then the graph is concave upward. If the second derivative is negative, then the graph is concave downward.

3. $f' > 0, f'' < 0$ **5.** Concave upward: $(-\infty, \infty)$

7. Concave upward: $(-\infty, 0), \left(\frac{3}{2}, \infty\right)$;
Concave downward: $\left(0, \frac{3}{2}\right)$

9. Concave upward: $(-\infty, -2), (2, \infty)$;
Concave downward: $(-2, 2)$

11. Concave upward: $\left(-\infty, -\frac{1}{6}\right)$;
Concave downward: $\left(-\frac{1}{6}, \infty\right)$

13. Concave upward: $(-\infty, -1), (1, \infty)$;
Concave downward: $(-1, 1)$

15. Concave upward: $\left(-\frac{\pi}{2}, 0\right)$; Concave downward: $\left(0, \frac{\pi}{2}\right)$

17. Point of inflection: $(3, 0)$; Concave downward: $(-\infty, 3)$;
Concave upward: $(3, \infty)$

19. Points of inflection: None; Concave downward: $(-\infty, \infty)$

21. Points of inflection: $(2, -16), (4, 0)$;
Concave upward: $(-\infty, 2), (4, \infty)$;
Concave downward: $(2, 4)$

23. Points of inflection: None; Concave upward: $(-3, \infty)$

25. Points of inflection: None; Concave upward: $(0, \infty)$

27. Point of inflection: $(2\pi, 0)$;
Concave upward: $(2\pi, 4\pi)$; Concave downward: $(0, 2\pi)$

29. Concave upward: $(0, \pi), (2\pi, 3\pi)$;
Concave downward: $(\pi, 2\pi), (3\pi, 4\pi)$

31. Points of inflection: $(\pi, 0)$, $(1.823, 1.452)$, $(4.46, -1.452)$;
Concave upward: $(1.823, \pi)$, $(4.46, 2\pi)$;
Concave downward: $(0, 1.823)$, $(\pi, 4.46)$

33. Relative maximum: $(3, 9)$

35. Relative maximum: $(0, 3)$; Relative minimum: $(2, -1)$

37. Relative minimum: $(3, -25)$

39. Relative minimum: $(0, -3)$

41. Relative maximum: $(-2, -4)$; Relative minimum: $(2, 4)$

43. No relative extrema, because f is nonincreasing.

45. (a) $f'(x) = 0.2x(x-3)^2(5x-6)$;
$f''(x) = 0.4(x-3)(10x^2 - 24x + 9)$

(b) Relative maximum: $(0, 0)$;
Relative minimum: $(1.2, -1.6796)$;
Points of inflection: $(0.4652, -0.7048)$, $(1.9348, -0.9048)$, $(3, 0)$

(c)

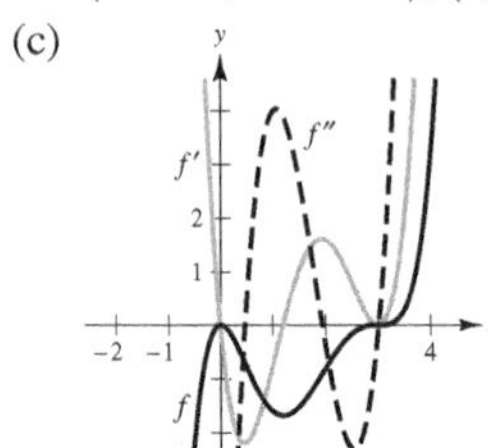

f is increasing when f' is positive and decreasing when f' is negative. f is concave upward when f'' is positive and concave downward when f'' is negative.

47. (a) $f'(x) = \cos x - \cos 3x + \cos 5x$;
$f''(x) = -\sin x + 3\sin 3x - 5\sin 5x$

(b) Relative maximum: $\left(\frac{\pi}{2}, 1.53333\right)$;

Points of inflection: $\left(\frac{\pi}{6}, 0.2667\right)$, $(1.1731, 0.9637)$, $(1.9685, 0.9637)$, $\left(\frac{5\pi}{6}, 0.2667\right)$

(c)

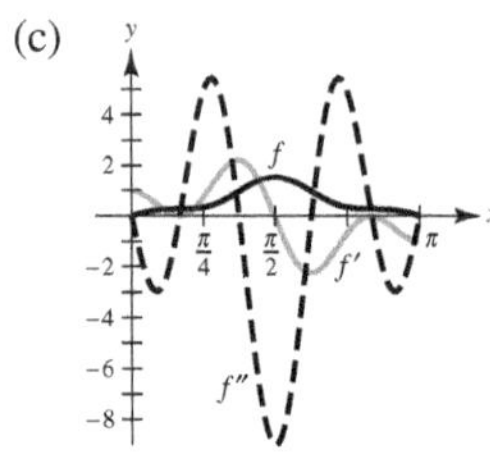

f is increasing when f' is positive and decreasing when f' is negative. f is concave upward when f'' is positive and concave downward when f'' is negative.

49. (a)

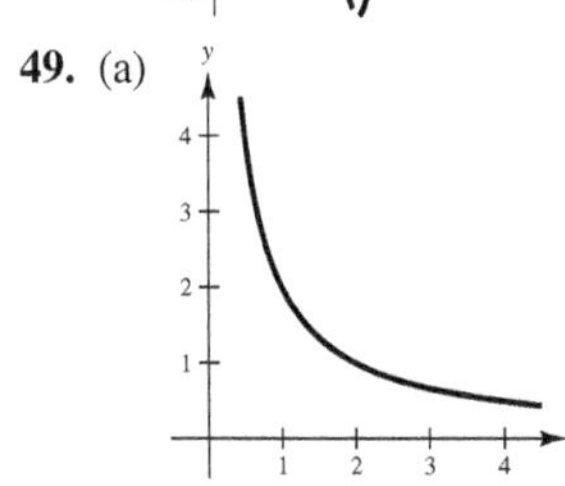

(b)

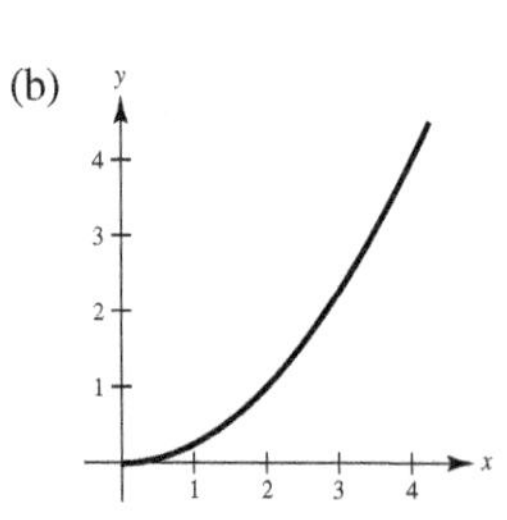

51.

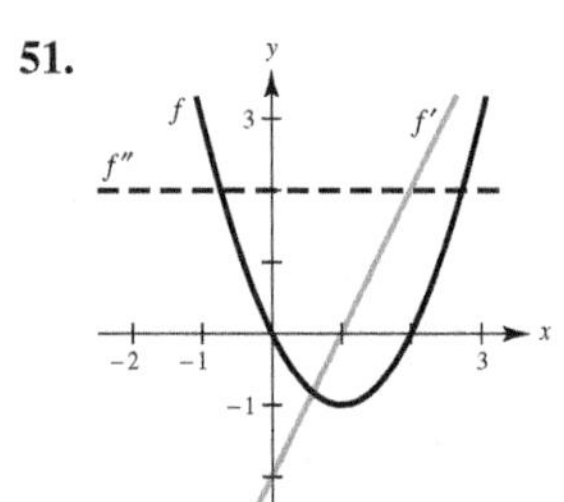

53.

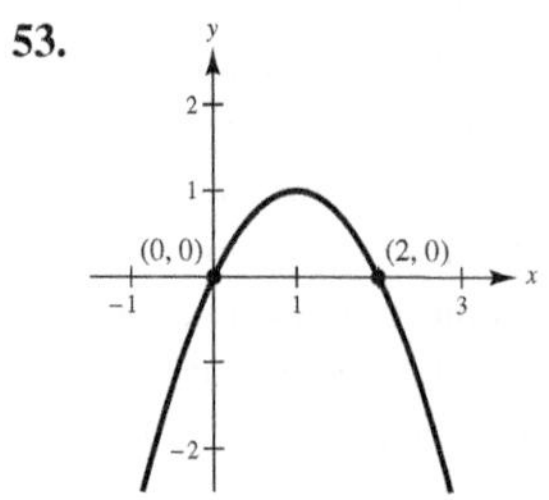

55.

57. Sample answer:

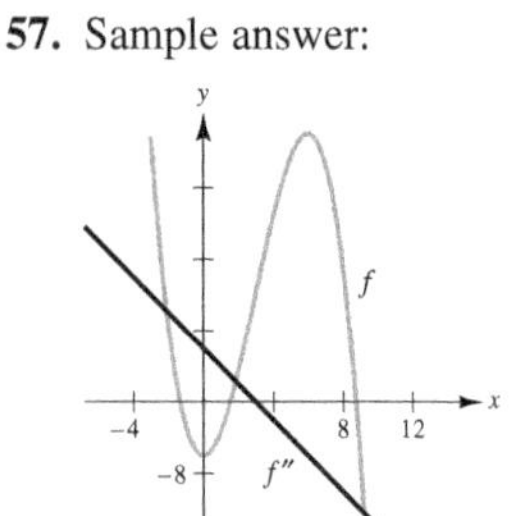

59. (a) $f(x) = (x-2)^n$ has a point of inflection at $(2, 0)$ if n is odd and $n \geq 3$.

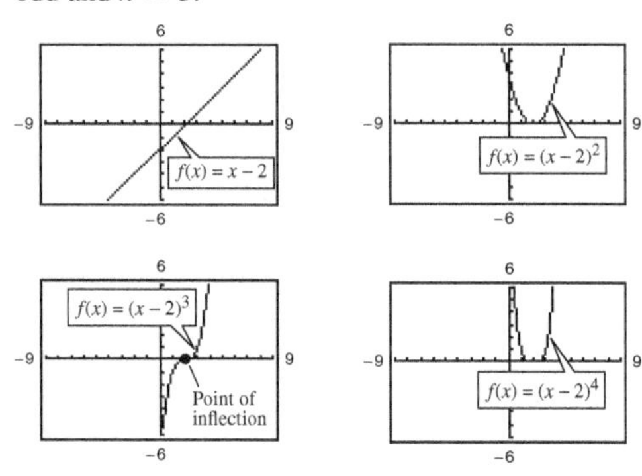

(b) Proof

61. $f(x) = \frac{1}{2}x^3 - 6x^2 + \frac{45}{2}x - 24$

63. (a) $f(x) = \frac{1}{32}x^3 + \frac{3}{16}x^2$ (b) Two miles from touchdown

65. $x = 100$ units

67. (a)

t	0.5	1	1.5	2	2.5	3
S	151.5	555.6	1097.6	1666.7	2193.0	2647.1

$1.5 < t < 2$

(b)

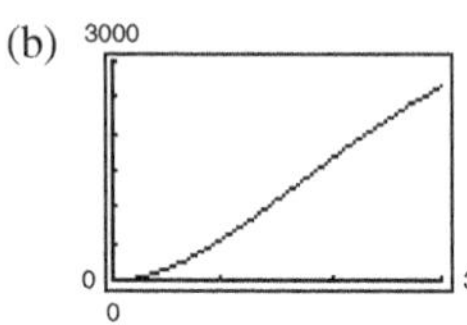

$t \approx 1.5$

(c) About 1.633 yr

69. $P_1(x) = 2\sqrt{2}$

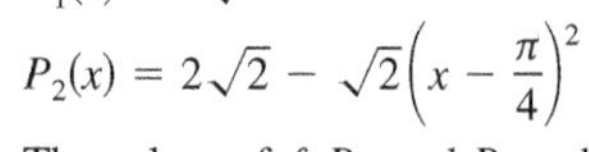

$P_2(x) = 2\sqrt{2} - \sqrt{2}\left(x - \frac{\pi}{4}\right)^2$

The values of f, P_1, and P_2 and their first derivatives are equal when $x = \frac{\pi}{4}$. The approximations worsen as you move away from $x = \frac{\pi}{4}$.

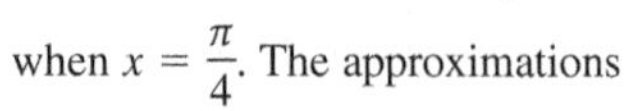

71. $P_1(x) = 1 - \frac{x}{2}$

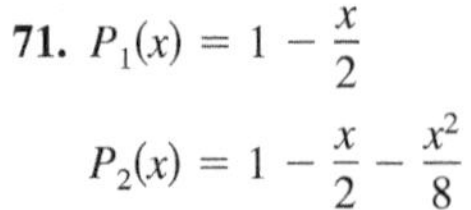

$P_2(x) = 1 - \frac{x}{2} - \frac{x^2}{8}$

The values of f, P_1, and P_2 and their first derivatives are equal when $x = 0$. The approximations worsen as you move away from $x = 0$.

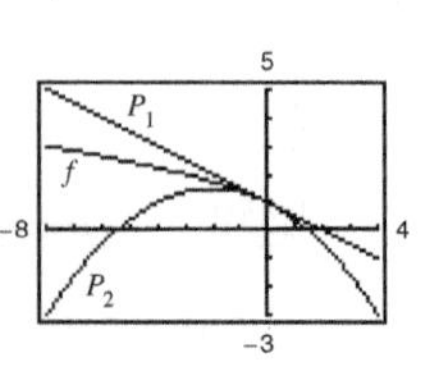

73. 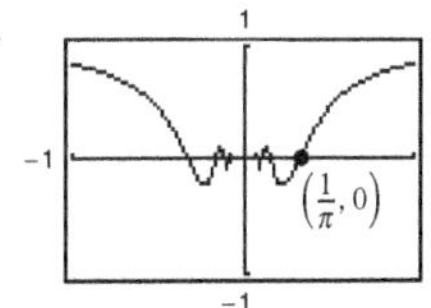

75. True

77. False. f is concave upward at $x = c$ if $f''(c) > 0$.
79. Proof

Section 3.5 *(page 206)*

1. (a) As x increases without bound, $f(x)$ approaches -5.
(b) As x decreases without bound, $f(x)$ approaches 3.
3. 2; one from the left and one from the right
5. f **6.** c **7.** d **8.** a **9.** b **10.** e
11. (a) ∞ (b) 5 (c) 0 **13.** (a) 0 (b) 1 (c) ∞
15. (a) 0 (b) $-\frac{2}{3}$ (c) $-\infty$ **17.** 4 **19.** $\frac{7}{9}$ **21.** 0
23. $-\infty$ **25.** -1 **27.** -2 **29.** $\frac{1}{2}$ **31.** ∞
33. 0 **35.** 0
37.  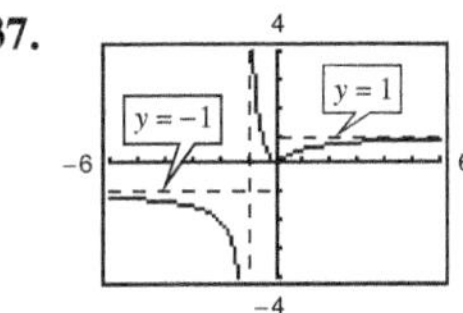

39.

41. 1 **43.** 0 **45.** $\frac{1}{6}$
47.

x	10^0	10^1	10^2	10^3	10^4	10^5	10^6
$f(x)$	1.000	0.513	0.501	0.500	0.500	0.500	0.500

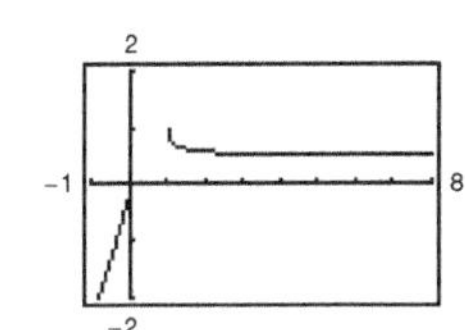

$$\lim_{x\to\infty}\left[x - \sqrt{x(x-1)}\right] = \frac{1}{2}$$

49.

x	10^0	10^1	10^2	10^3	10^4	10^5	10^6
$f(x)$	0.479	0.500	0.500	0.500	0.500	0.500	0.500

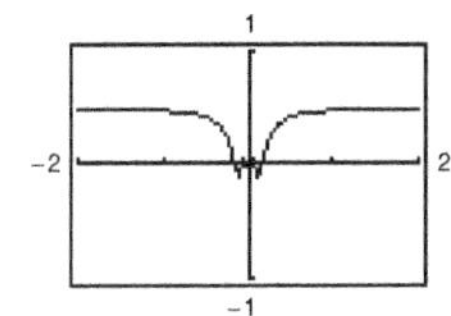

The graph has a hole at $x = 0$.

$$\lim_{x\to\infty} x \sin\frac{1}{2x} = \frac{1}{2}$$

51. 100%
53. An infinite limit is a description of how a limit fails to exist. A limit at infinity deals with the end behavior of a function.
55. (a) 5 (b) -5
57. (a) 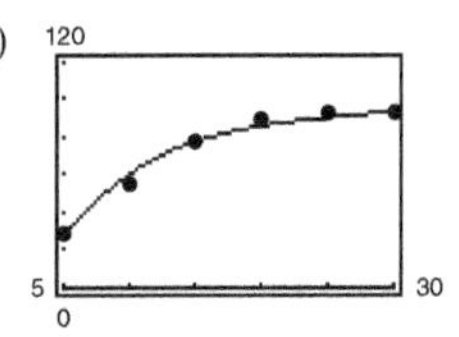

(b) Yes. $\lim_{t\to\infty} S = \frac{100}{1} = 100$

59. (a) $\lim_{x\to\infty} f(x) = 2$
(b) $x_1 = \sqrt{\frac{4-2\varepsilon}{\varepsilon}}, x_2 = -\sqrt{\frac{4-2\varepsilon}{\varepsilon}}$
(c) $M = \sqrt{\frac{4-2\varepsilon}{\varepsilon}}$ (d) $N = -\sqrt{\frac{4-2\varepsilon}{\varepsilon}}$
61. (a) Answers will vary. $M = \frac{5\sqrt{33}}{11}$
(b) Answers will vary. $M = \frac{29\sqrt{177}}{59}$
63–65. Proofs
67. (a) $d(m) = \frac{|3m+3|}{\sqrt{m^2+1}}$
(b) 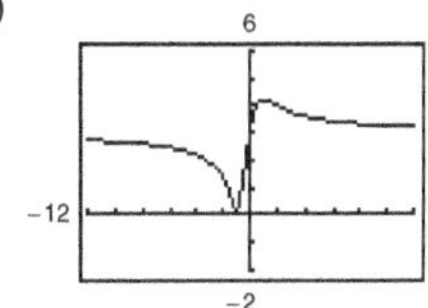

(c) $\lim_{m\to\infty} d(m) = 3$;
$\lim_{m\to-\infty} d(m) = 3$;
As m approaches $\pm\infty$, the distance approaches 3.
69. Proof

Section 3.6 *(page 215)*

1. Domain, range, intercepts, asymptotes, symmetry, end behavior, differentiability, relative extrema, points of inflection, concavity, increasing and decreasing, infinite limits at infinity
3. Rational function; Use long division to rewrite the rational function as the sum of a first-degree polynomial and another rational function.
5. d **6.** c **7.** a **8.** b
9. 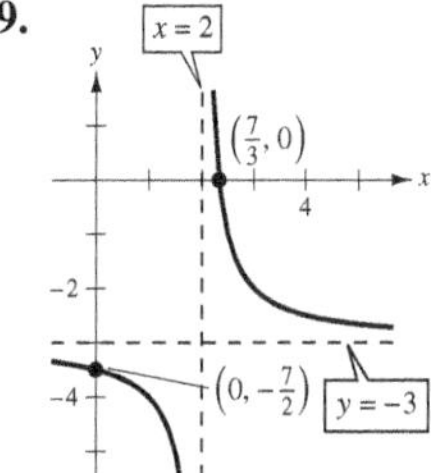

11. 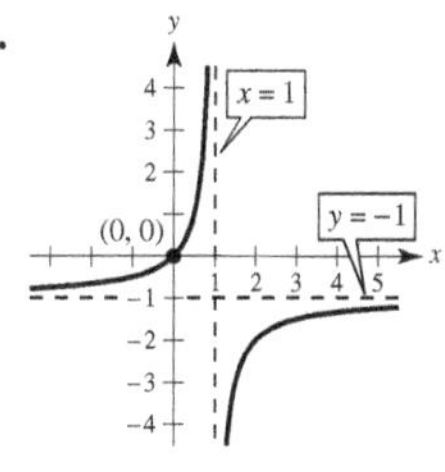

13.

15. 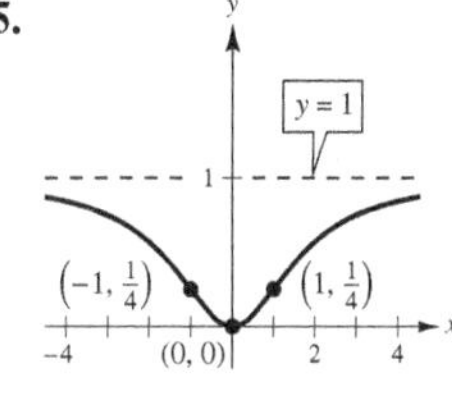

17.

19. 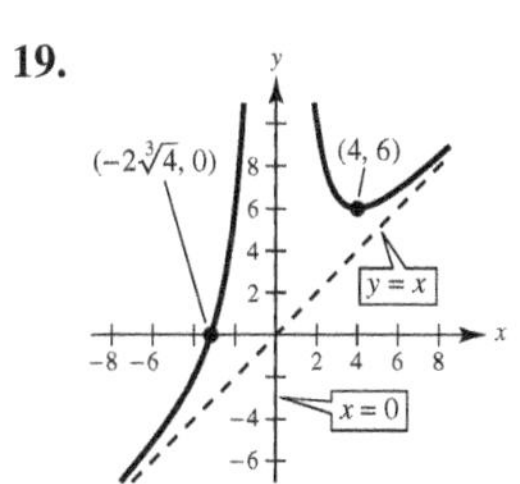

21. 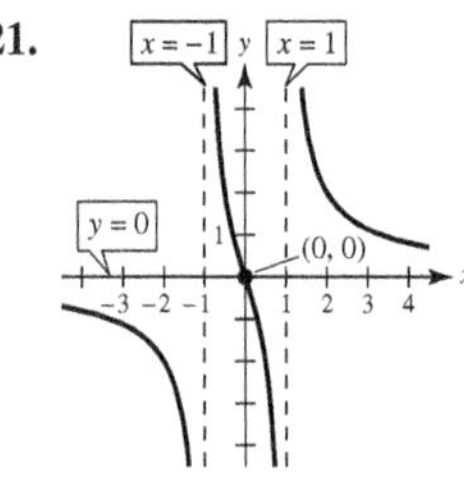

23. 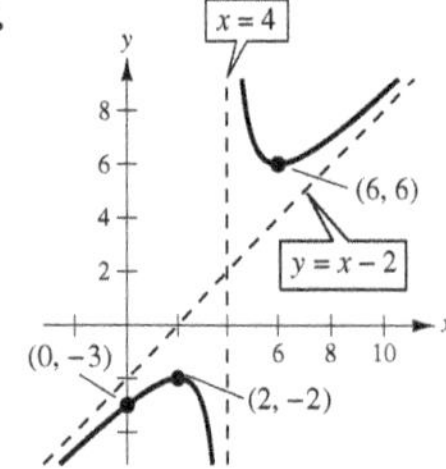

25. 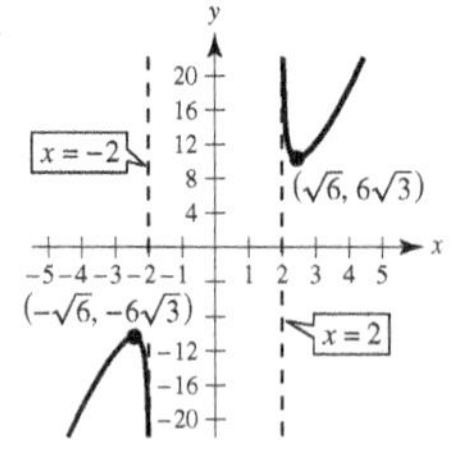

27. 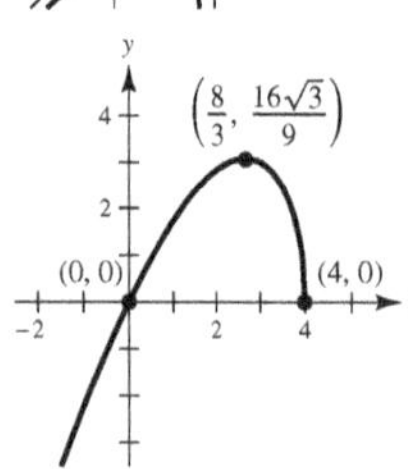

29. 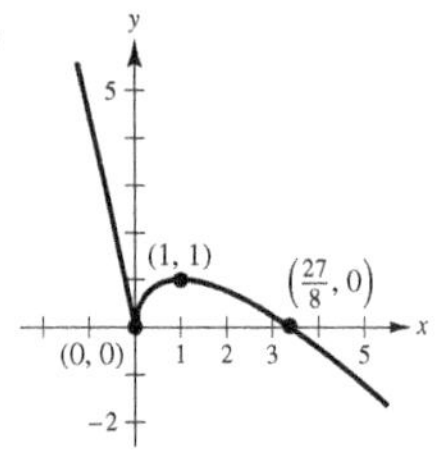

31. 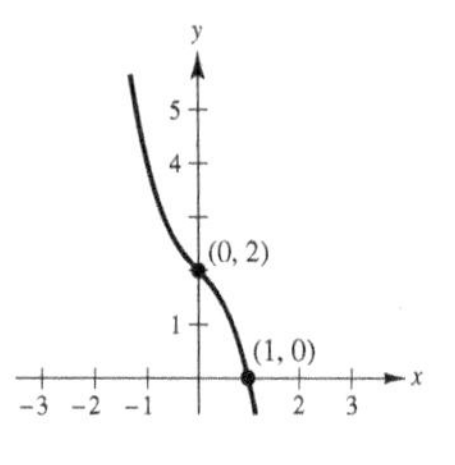

33. 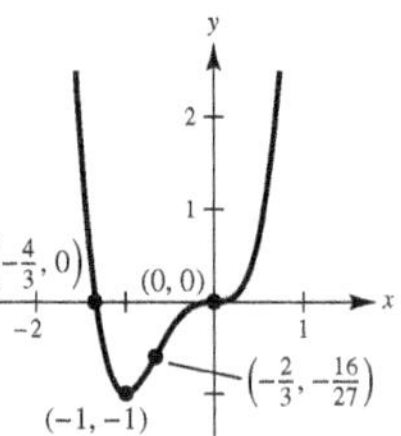

35. 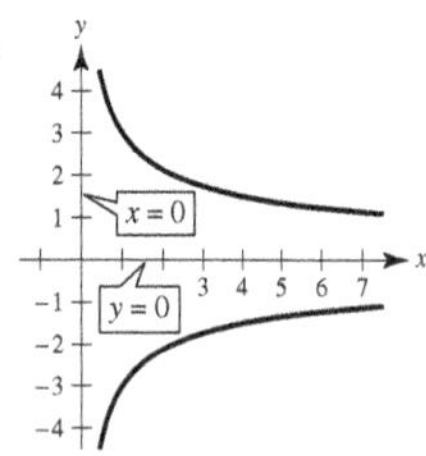

37. 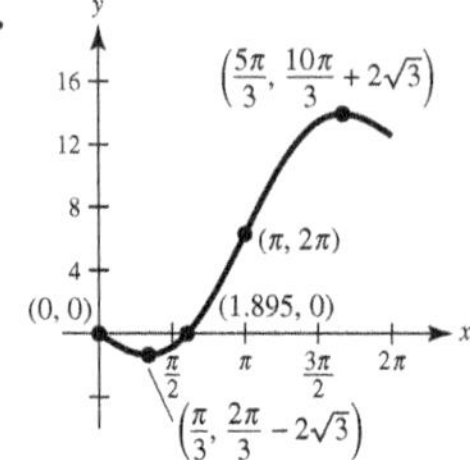

39. 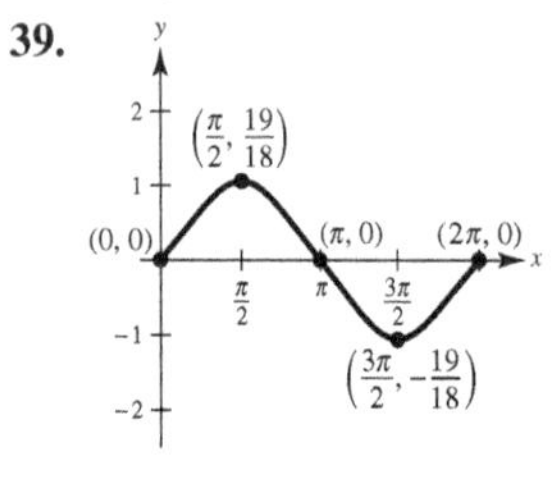

41. 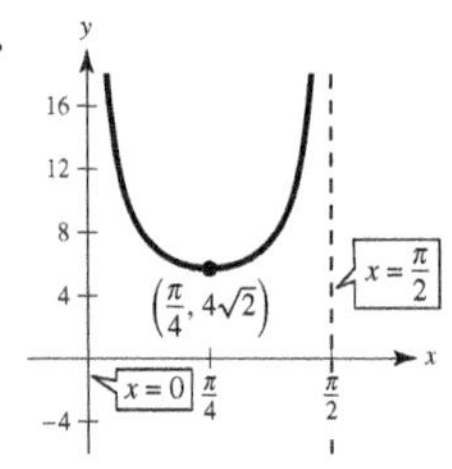

43. 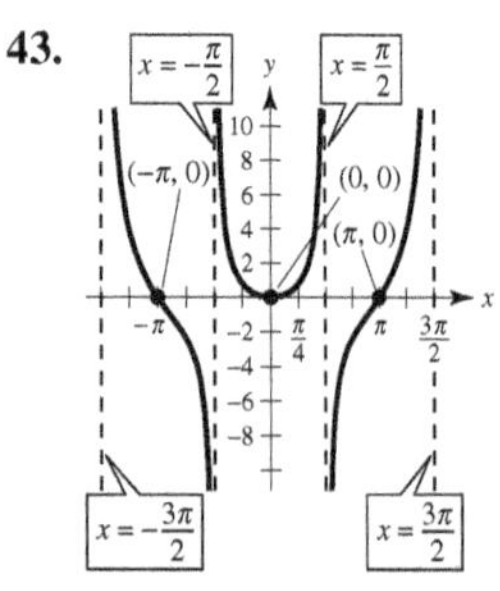

45. 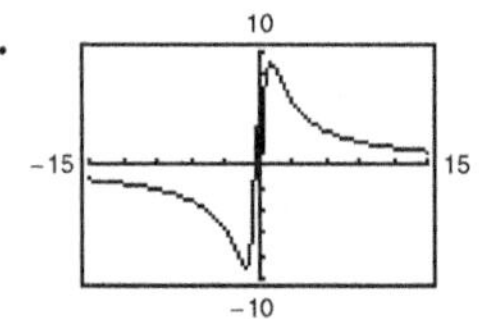

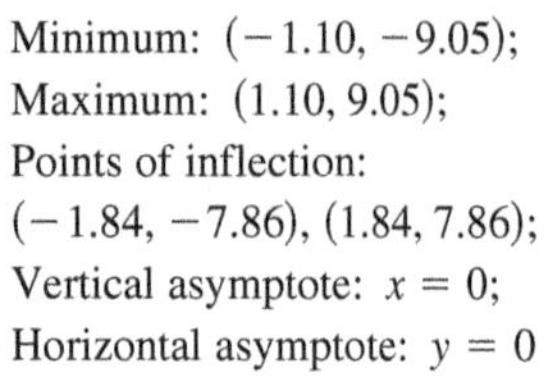

Minimum: $(-1.10, -9.05)$;
Maximum: $(1.10, 9.05)$;
Points of inflection:
$(-1.84, -7.86)$, $(1.84, 7.86)$;
Vertical asymptote: $x = 0$;
Horizontal asymptote: $y = 0$

47. 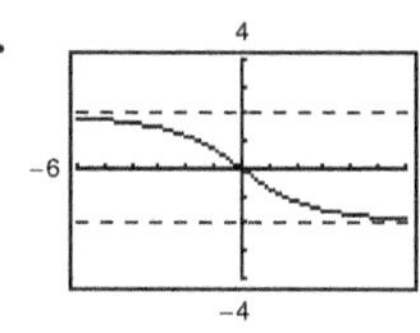

Point of inflection: $(0, 0)$;
Horizontal asymptotes: $y = \pm 2$

49. 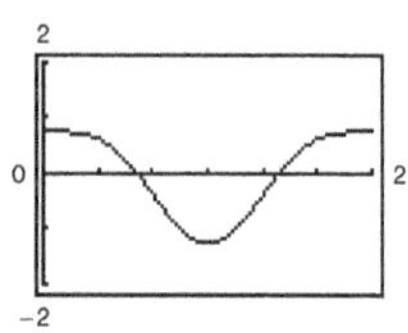

Relative minimum: $\left(\pi, -\frac{5}{4}\right)$;
Points of inflection: $\left(\frac{2\pi}{3}, -\frac{3}{8}\right), \left(\frac{4\pi}{3}, -\frac{3}{8}\right)$

51. 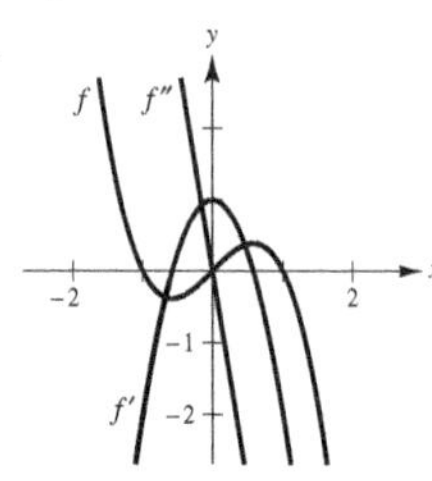

The zeros of f' correspond to the points where the graph of f has horizontal tangents. The zero of f'' corresponds to the point where the graph of f' has a horizontal tangent.

53. 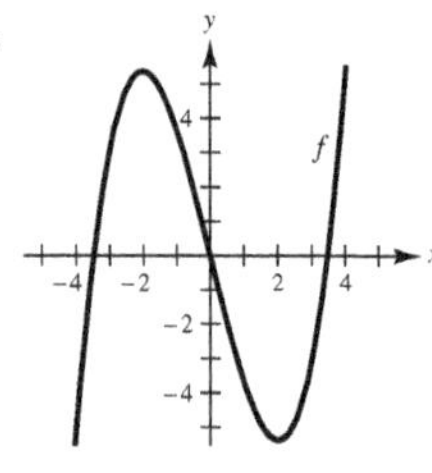

55. 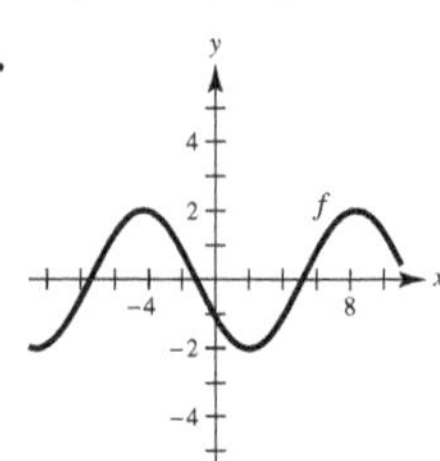

57. (a) 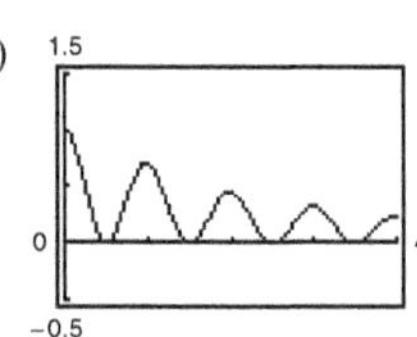

The graph has holes at $x = 0$ and at $x = 4$.
Visually approximated critical numbers: $\frac{1}{2}, 1, \frac{3}{2}, 2, \frac{5}{2}, 3, \frac{7}{2}$

(b) $f'(x) = \dfrac{-x\cos^2 \pi x}{(x^2+1)^{3/2}} - \dfrac{2\pi \sin \pi x \cos \pi x}{\sqrt{x^2+1}}$;

Approximate critical numbers: $\frac{1}{2}$, 0.97, $\frac{3}{2}$, 1.98, $\frac{5}{2}$, 2.98, $\frac{7}{2}$;
The critical numbers where maxima occur appear to be integers in part (a), but by approximating them using f', you can see that they are not integers.

59. Answers will vary. Sample answer: Let

$$f(x) = \frac{-6}{0.1(x-2)^2 + 1} + 6.$$

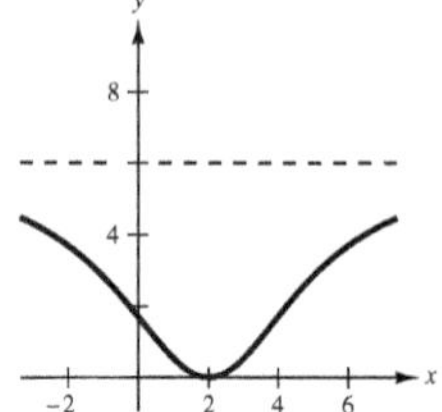

61. f is decreasing on $(2, 8)$, and therefore $f(3) > f(5)$.

63. (a)

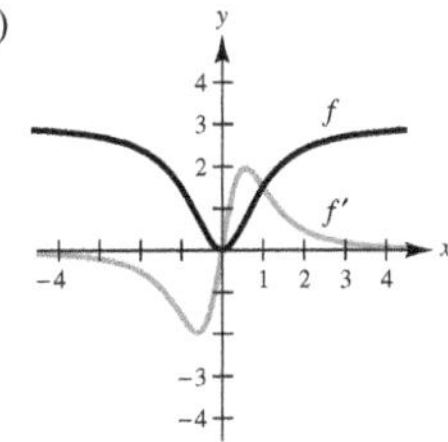

(b) $\lim_{x\to\infty} f(x) = 3, \quad \lim_{x\to\infty} f'(x) = 0$

(c) Because $\lim_{x\to\infty} f(x) = 3$, the graph approaches that of a horizontal line, $\lim_{x\to\infty} f'(x) = 0$.

65.

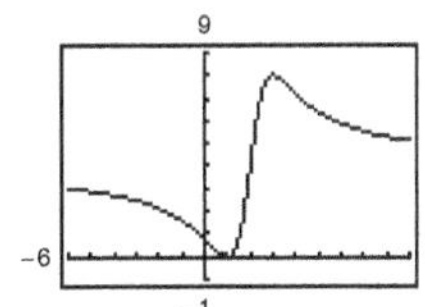

The graph crosses the horizontal asymptote $y = 4$.
The graph of a function f does not cross its vertical asymptote $x = c$ because $f(c)$ does not exist.

67.

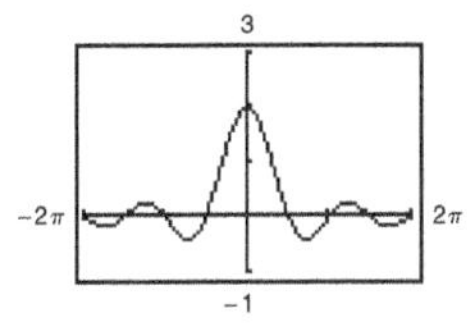

The graph has a hole at $x = 0$.
The graph crosses the horizontal asymptote $y = 0$.
The graph of a function f does not cross its vertical asymptote $x = c$ because $f(c)$ does not exist.

69.

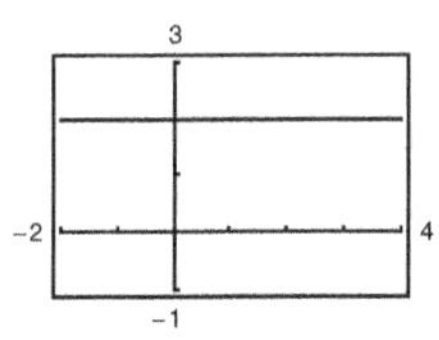

The graph has a hole at $x = 3$. The rational function is not reduced to lowest terms.

71.

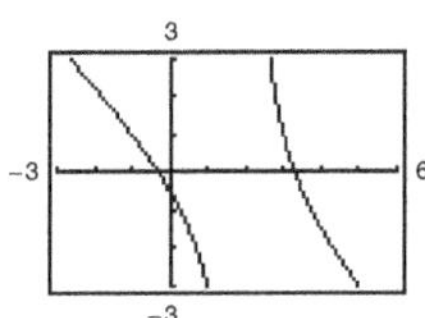

The graph appears to approach the line $y = -x + 1$, which is the slant asymptote.

73.

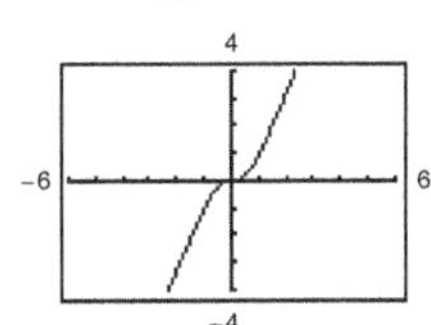

The graph appears to approach the line $y = 2x$, which is the slant asymptote.

75.

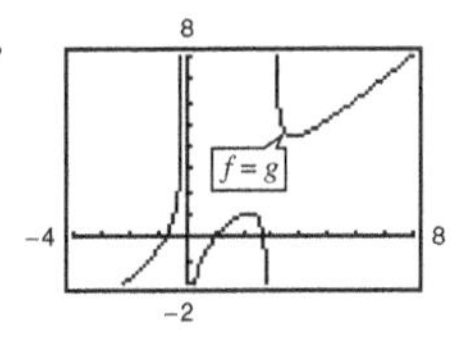

The graph appears to approach the line $y = x$, which is the slant asymptote.

77. (a)–(h) Proofs

79. Answers will vary. Sample answer: $y = \dfrac{1}{x - 3}$

81. Answers will vary.
Sample answer: $y = \dfrac{3x^2 - 7x - 5}{x - 3}$

83. False. Let $f(x) = \dfrac{2x}{\sqrt{x^2 + 2}}$, $f'(x) > 0$ for all real numbers.

85. False. For example,
$$y = \frac{x^3 - 1}{x}$$
does not have a slant asymptote.

87. (a) $(-3, 1)$ (b) $(-7, -1)$
(c) Relative maximum at $x = -3$, relative minimum at $x = 1$
(d) $x = -1$

89. Answers will vary. Sample answer: The graph has a vertical asymptote at $x = b$. If a and b are both positive or both negative, then the graph of f approaches ∞ as x approaches b, and the graph has a minimum at $x = -b$. If a and b have opposite signs, then the graph of f approaches $-\infty$ as x approaches b, and the graph has a maximum at $x = -b$.

91. $y = 4x$, $y = -4x$

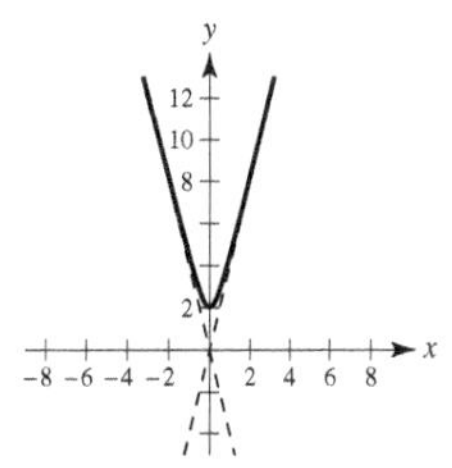

93. (a) When n is even, f is symmetric about the y-axis. When n is odd, f is symmetric about the origin.
(b) $n = 0, 1, 2, 3$ (c) $n = 4$ (d) $y = 2x$
(e)

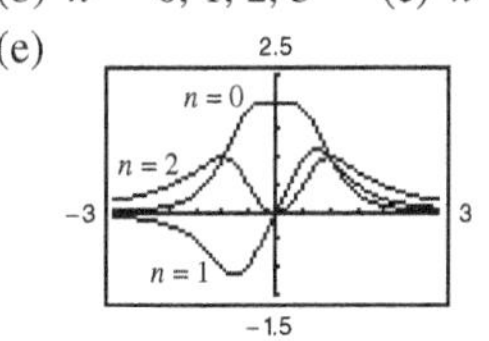

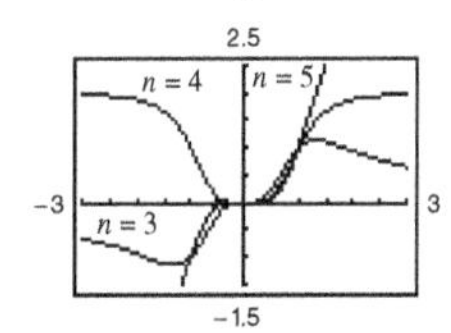

n	0	1	2	3	4	5
M	1	2	3	2	1	0
N	2	3	4	5	2	3

Section 3.7 *(page 224)*

1. A primary equation is a formula for the quantity to be optimized. A secondary equation can be solved for a variable and then substituted into the primary equation to obtain a function of just one variable. A feasible domain is the set of input values that makes sense in an optimization problem.

3. (a)

First Number, x	Second Number	Product, P
10	$110 - 10$	$10(110 - 10) = 1000$
20	$110 - 20$	$20(110 - 20) = 1800$
30	$110 - 30$	$30(110 - 30) = 2400$
40	$110 - 40$	$40(110 - 40) = 2800$
50	$110 - 50$	$50(110 - 50) = 3000$
60	$110 - 60$	$60(110 - 60) = 3000$
70	$110 - 70$	$70(110 - 70) = 2800$
80	$110 - 80$	$80(110 - 80) = 2400$
90	$110 - 90$	$90(110 - 90) = 1800$
100	$110 - 100$	$100(110 - 100) = 1000$

The maximum is attained near $x = 50$ and 60.

(b) $P = x(110 - x)$ (c) 55 and 55

(d) 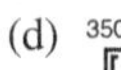

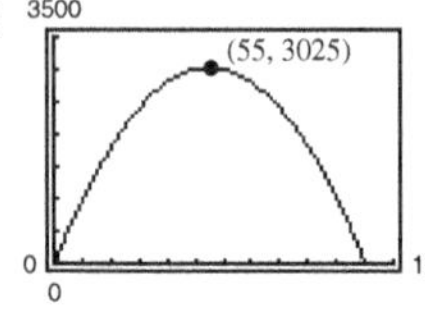

5. $\frac{S}{2}$ and $\frac{S}{2}$ **7.** 21 and 7 **9.** 54 and 27

11. $\ell = w = 20$ m **13.** $\ell = w = 7$ ft

15. $\left(-\sqrt{\frac{5}{2}}, \frac{5}{2}\right), \left(\sqrt{\frac{5}{2}}, \frac{5}{2}\right)$ **17.** 40 in. × 20 in.

19. 900 m × 450 m

21. Rectangular portion: $\frac{16}{\pi + 4} \times \frac{32}{\pi + 4}$ ft

23. (a) $L = \sqrt{x^2 + 4 + \frac{8}{x - 1} + \frac{4}{(x - 1)^2}}, \quad x > 1$

(b) 

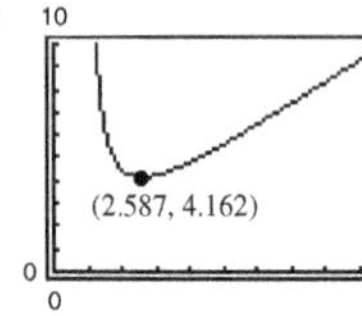

Minimum when $x \approx 2.587$

(c) $(0, 0)$, $(2, 0)$, $(0, 4)$

25. Width: $\frac{5\sqrt{2}}{2}$; Length: $5\sqrt{2}$

27. (a) 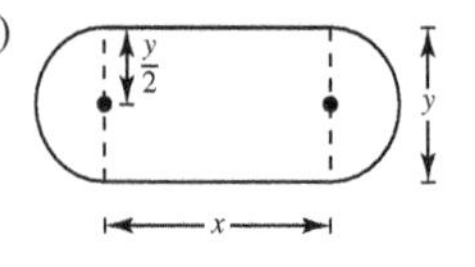

(b)

Length, x	Width, y	Area, xy
10	$(2/\pi)(100 - 10)$	$(10)(2/\pi)(100 - 10) \approx 573$
20	$(2/\pi)(100 - 20)$	$(20)(2/\pi)(100 - 20) \approx 1019$
30	$(2/\pi)(100 - 30)$	$(30)(2/\pi)(100 - 30) \approx 1337$
40	$(2/\pi)(100 - 40)$	$(40)(2/\pi)(100 - 40) \approx 1528$
50	$(2/\pi)(100 - 50)$	$(50)(2/\pi)(100 - 50) \approx 1592$
60	$(2/\pi)(100 - 60)$	$(60)(2/\pi)(100 - 60) \approx 1528$

The maximum area of the rectangle is approximately 1592 m^2.

(c) $A = \frac{2}{\pi}(100x - x^2)$, $0 < x < 100$

(d) $\frac{dA}{dx} = \frac{2}{\pi}(100 - 2x)$
$= 0$ when $x = 50$;
The maximum value is approximately 1592 when $x = 50$.

(e) 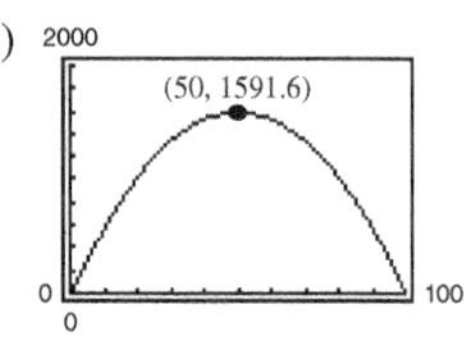

29. 18 in. × 18 in. × 36 in.

31. No. The volume changes because the shape of the container changes when it is squeezed.

33. $r = \sqrt[3]{\frac{21}{2\pi}} \approx 1.50$ ($h = 0$, so the solid is a sphere.)

35. Side of square: $\frac{10\sqrt{3}}{9 + 4\sqrt{3}}$; Side of triangle: $\frac{30}{9 + 4\sqrt{3}}$

37. $w = \frac{20\sqrt{3}}{3}$ in., $h = \frac{20\sqrt{6}}{3}$ in.

39. 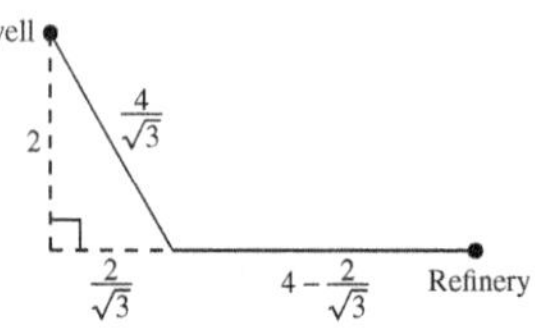

The path of the pipe should go underwater from the oil well to the coast following the hypotenuse of a right triangle with leg lengths of 2 miles and $\frac{2}{\sqrt{3}}$ miles for a distance of $\frac{4}{\sqrt{3}}$ miles. Then the pipe should go down the coast to the refinery for a distance of $\left(4 - \frac{2}{\sqrt{3}}\right)$ miles.

41. One mile from the nearest point on the coast

43. (a) Origin to y-intercept: 2;
Origin to x-intercept: $\frac{\pi}{2}$

(b) $d = \sqrt{x^2 + (2 - 2\sin x)^2}$

(c) Minimum distance is 0.9795 when $x \approx 0.7967$.

45. About 1.153 radians or 66° **47.** $5.\overline{3}\%$

49. $y = \frac{64}{141}x$, $S \approx 6.1$ mi **51.** $y = \frac{3}{10}x$, $S_3 \approx 4.50$ mi

53. $(0, 0)$ **55.** Putnam Problem A1, 1986

Section 3.8 (page 233)

1. Answers will vary. Sample answer: If f is a function continuous on $[a, b]$ and differentiable on (a, b), where $c \in [a, b]$ and $f(c) = 0$, then Newton's Method uses tangent lines to approximate c. First, estimate an initial x_1 close to c. (See graph.) Then determine x_2 using $x_2 = x_1 - \dfrac{f(x_1)}{f'(x_1)}$. Calculate a third estimate x_3 using $x_3 = x_2 - \dfrac{f(x_2)}{f'(x_2)}$. Continue this process until $|x_n - x_{n+1}|$ is within the desired accuracy, and let x_{n+1} be the final approximation of c.

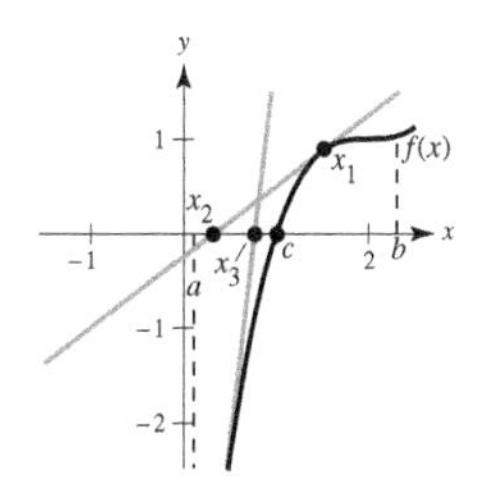

In the answers for Exercises 3 and 5, the values in the tables have been rounded for convenience. Because a calculator and a computer program calculate internally using more digits than they display, you may produce slightly different values from those shown in the tables.

3.

n	x_n	$f(x_n)$	$f'(x_n)$	$\dfrac{f(x_n)}{f'(x_n)}$	$x_n - \dfrac{f(x_n)}{f'(x_n)}$
1	2	−1	4	−0.25	2.25
2	2.25	0.0625	4.5	0.0139	2.2361

5.

n	x_n	$f(x_n)$	$f'(x_n)$	$\dfrac{f(x_n)}{f'(x_n)}$	$x_n - \dfrac{f(x_n)}{f'(x_n)}$
1	1.6	−0.0292	−0.9996	0.0292	1.5708
2	1.5708	0	−1	0	1.5708

7. −1.587 **9.** 0.682 **11.** 1.250, 5.000
13. 0.900, 1.100, 1.900 **15.** 1.935 **17.** 0.569
19. 4.493
21. (a)

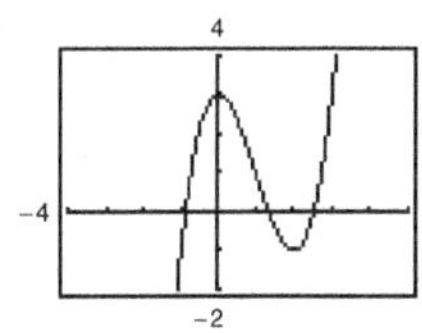

(b) 1.347 (c) 2.532

(d)

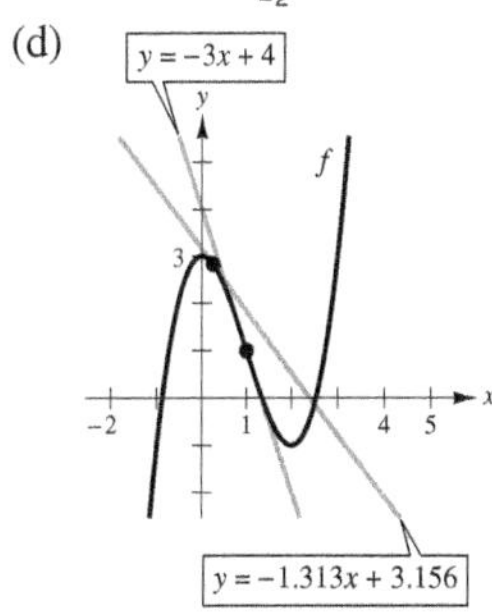

If the initial estimate $x = x_1$ is not sufficiently close to the desired zero of a function, then the x-intercept of the corresponding tangent line to the function may approximate a second zero of the function.

23. $f'(x_1) = 0$ **25.** 0.74
27. The values would be identical. **29.** (1.939, 0.240)

31. $x \approx 1.563$ mi
33. (a) Proof (b) $\sqrt{5} \approx 2.236$, $\sqrt{7} \approx 2.646$ **35.** Proof
37. False; Let $f(x) = \dfrac{x^2 - 1}{x - 1}$. **39.** True **41.** 0.217

Section 3.9 (page 240)

1. $y = f(c) + f'(c)(x - c)$
3. Propagated error $= f(x + \Delta x) - f(x)$, relative error $= \left|\dfrac{dy}{y}\right|$, percent error $= \left|\dfrac{dy}{y}\right| \cdot 100$
5. $T(x) = 4x - 4$

x	1.9	1.99	2	2.01	2.1
$f(x)$	3.610	3.960	4	4.040	4.410
$T(x)$	3.600	3.960	4	4.040	4.400

7. $T(x) = 80x - 128$

x	1.9	1.99	2	2.01	2.1
$f(x)$	24.761	31.208	32	32.808	40.841
$T(x)$	24.000	31.200	32	32.800	40.000

9. $T(x) = (\cos 2)(x - 2) + \sin 2$

x	1.9	1.99	2	2.01	2.1
$f(x)$	0.946	0.913	0.909	0.905	0.863
$T(x)$	0.951	0.913	0.909	0.905	0.868

11. $y - f(0) = f'(0)(x - 0)$
$y - 2 = \frac{1}{4}x$
$y = 2 + \dfrac{x}{4}$

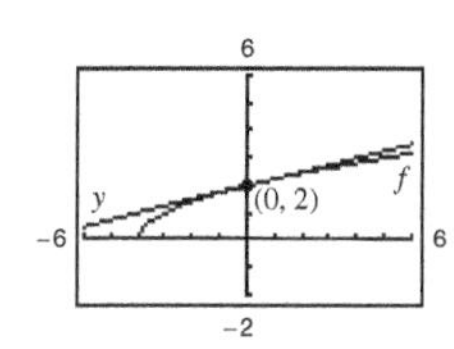

13. $\Delta y = 0.1655$, $dy = 0.15$
15. $\Delta y = -0.039$, $dy = -0.040$
17. $\Delta y \approx -0.053018$, $dy = -0.053$ **19.** $6x\,dx$
21. $(x \sec^2 x + \tan x)\,dx$ **23.** $-\dfrac{3}{(2x - 1)^2}\,dx$
25. $-\dfrac{x}{\sqrt{9 - x^2}}\,dx$ **27.** $(3 - \sin 2x)\,dx$
29. (a) 0.9 (b) 1.04 **31.** (a) 8.035 (b) 7.95
33. (a) $\pm\frac{5}{8}$ in.^{2} (b) 0.625%
35. (a) ± 20.25 in.^{3} (b) ± 5.4 in.^{2} (c) 0.6%, 0.4%
37. 27.5 mi, about 7.3% **39.** (a) $\frac{1}{4}$% (b) 216 sec = 3.6 min
41. 6407 ft
43. $f(x) = \sqrt{x}$, $dy = \dfrac{1}{2\sqrt{x}}\,dx$

$f(99.4) \approx \sqrt{100} + \dfrac{1}{2\sqrt{100}}(-0.6) = 9.97$

Calculator: 9.97

45. $f(x) = \sqrt[4]{x},\ dy = \dfrac{1}{4x^{3/4}}\,dx$

$f(624) \approx \sqrt[4]{625} + \dfrac{1}{4(625)^{3/4}}(-1) = 4.998$

Calculator: 4.998

47. The value of dy becomes closer to the value of Δy as Δx approaches 0. Graphs will vary.

49. True **51.** True **53.** True

Review Exercises for Chapter 3 *(page 242)*

1. Maximum: $(0, 0)$
Minimum: $\left(-\frac{5}{2}, -\frac{25}{4}\right)$

3. Maximum: $(4, 0)$
Minimum: $(0, -2)$

5. Maximum: $\left(3, \frac{2}{3}\right)$
Minimum: $\left(-3, -\frac{2}{3}\right)$

7. Maximum: $(2\pi, 17.57)$
Minimum: $(2.73, 0.88)$

9. $f'(1) = 0$ **11.** Not continuous on $[-2, 2]$

13. $f'\left(\dfrac{2744}{729}\right) = \dfrac{3}{7}$ **15.** f is not differentiable at $x = 5$.

17. $f'(0) = 1$

19. No; The function has a discontinuity at $x = 0$, which is in the interval $[-2, 1]$.

21. Increasing on $\left(-\frac{3}{2}, \infty\right)$; Decreasing on $\left(-\infty, -\frac{3}{2}\right)$

23. Increasing on $(-\infty, 1)$, $(2, \infty)$; Decreasing on $(1, 2)$

25. Increasing on $(1, \infty)$; Decreasing on $(0, 1)$

27. (a) Critical number: $x = 3$
(b) Increasing on $(3, \infty)$; Decreasing on $(-\infty, 3)$
(c) Relative minimum: $(3, -4)$

29. (a) Critical number: $t = 2$
(b) Increasing on $(2, \infty)$; Decreasing on $(-\infty, 2)$
(c) Relative minimum: $(2, -12)$

31. (a) Critical number: $x = -8$; Discontinuity: $x = 0$
(b) Increasing on $(-8, 0)$;
Decreasing on $(-\infty, -8)$ and $(0, \infty)$
(c) Relative minimum: $\left(-8, -\frac{1}{16}\right)$

33. (a) Critical numbers: $x = \dfrac{3\pi}{4}, \dfrac{7\pi}{4}$

(b) Increasing on $\left(\dfrac{3\pi}{4}, \dfrac{7\pi}{4}\right)$;

Decreasing on $\left(0, \dfrac{3\pi}{4}\right)$ and $\left(\dfrac{7\pi}{4}, 2\pi\right)$

(c) Relative minimum: $\left(\dfrac{3\pi}{4}, -\sqrt{2}\right)$;

Relative maximum: $\left(\dfrac{7\pi}{4}, \sqrt{2}\right)$

35. (a) $v(t) = 3 - 4t$ (b) $\left[0, \frac{3}{4}\right)$ (c) $\left(\frac{3}{4}, \infty\right)$ (d) $t = \frac{3}{4}$

37. Point of inflection: $(3, -54)$; Concave upward: $(3, \infty)$; Concave downward: $(-\infty, 3)$

39. Points of inflection: None; Concave upward: $(-5, \infty)$

41. Points of inflection: $\left(\dfrac{\pi}{2}, \dfrac{\pi}{2}\right), \left(\dfrac{3\pi}{2}, \dfrac{3\pi}{2}\right)$;

Concave upward: $\left(\dfrac{\pi}{2}, \dfrac{3\pi}{2}\right)$;

Concave downward: $\left(0, \dfrac{\pi}{2}\right), \left(\dfrac{3\pi}{2}, 2\pi\right)$

43. Relative minimum: $(-9, 0)$

45. Relative maxima: $\left(\dfrac{\sqrt{2}}{2}, \dfrac{1}{2}\right), \left(-\dfrac{\sqrt{2}}{2}, \dfrac{1}{2}\right)$;

Relative minimum: $(0, 0)$

47. Relative maximum: $(-3, -12)$; Relative minimum: $(3, 12)$

49.

51. Increasing and concave downward

53. (a) $D = 0.41489t^4 - 17.1307t^3 + 249.888t^2 - 1499.45t + 3684.8$

(b)

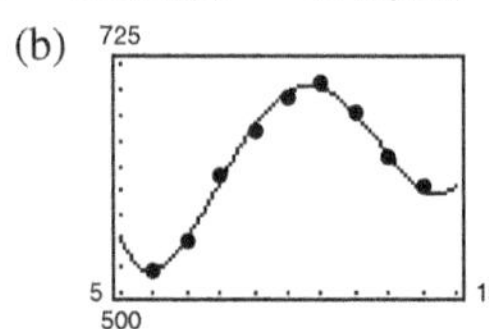

(c) 2011; 2006 (d) 2008

55. 8 **57.** $-\frac{1}{8}$ **59.** $-\infty$ **61.** 0 **63.** 6

65.

67.

69.

71.

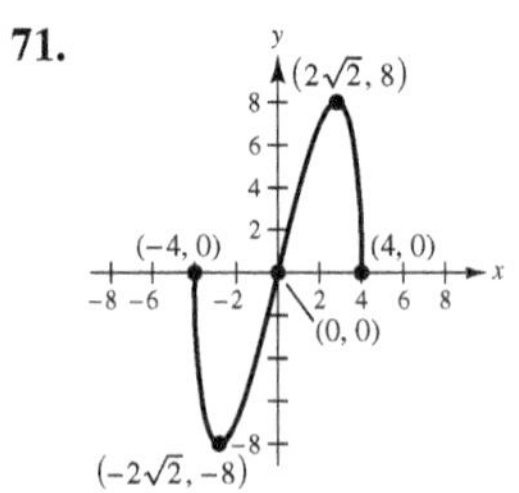

73.

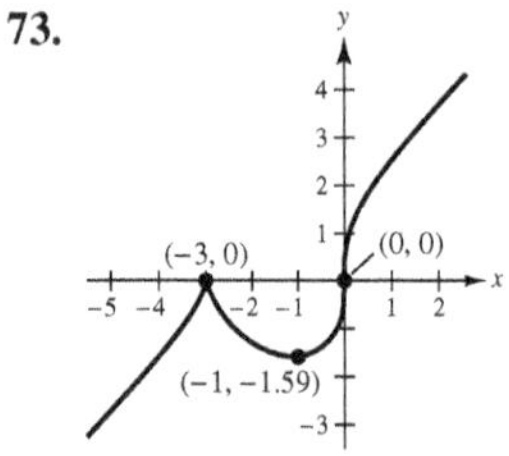

75.

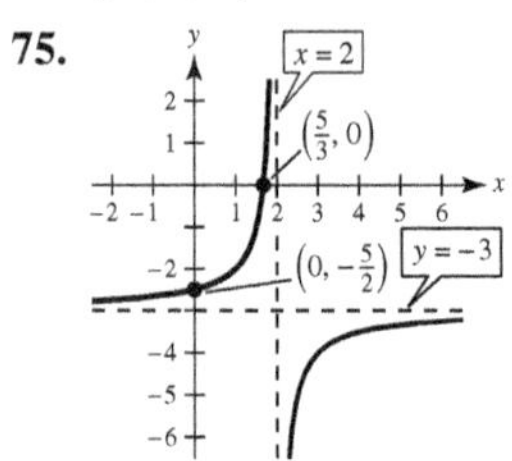

77.

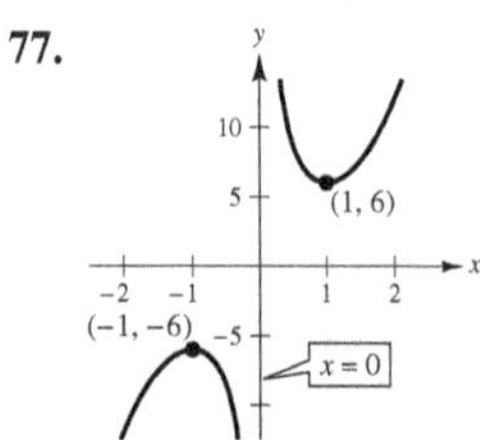

79. 54, 36 **81.** $x = 50$ ft and $y = \frac{200}{3}$ ft

83. $(0, 0), (5, 0), (0, 10)$ **85.** 14.05 ft **87.** $\dfrac{32\pi r^3}{81}$

89. $-1.532, -0.347, 1.879$ **91.** $-2.182, -0.795$

93. -0.755 **95.** $\Delta y = 5.044,\ dy = 4.8$

97. $dy = (1 - \cos x + x \sin x)\,dx$

99. $f(x) = \sqrt{x}$, $dy = \dfrac{1}{2\sqrt{x}}$

$f(63.9) \approx \sqrt{64} + \dfrac{1}{2\sqrt{64}}(-0.1) = 7.99375$

Calculator: 7.99375

101. (a) $\pm 8.1\pi$ cm³ (b) $\pm 1.8\pi$ cm²

(c) About 0.83%, about 0.56%

P.S. Problem Solving *(page 245)*

1. Choices of a may vary.

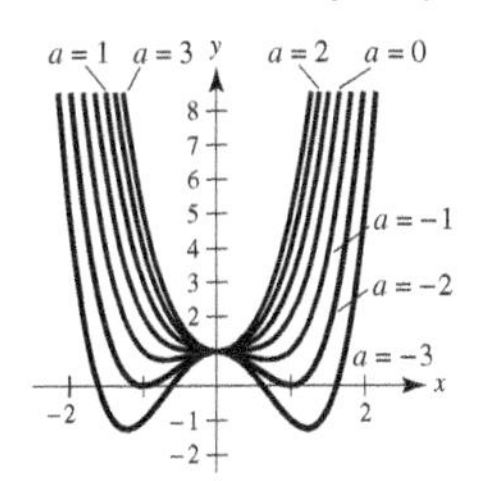

(a) One relative minimum at $(0, 1)$ for $a \geq 0$

(b) One relative maximum at $(0, 1)$ for $a < 0$

(c) Two relative minima for $a < 0$ when $x = \pm\sqrt{-\dfrac{a}{2}}$

(d) If $a < 0$, then there are three critical points. If $a \geq 0$, then there is only one critical point.

3. All c, where c is a real number **5.** Proof

7. The bug should head toward the midpoint of the opposite side. Without calculus, imagine opening up the cube. The shortest distance is the line PQ, passing through the midpoint as shown.

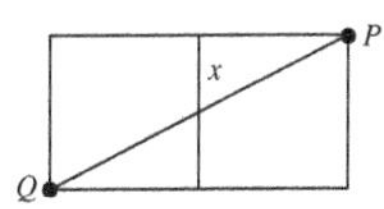

9. $a = 6, b = 1, c = 2$ **11.** Proof

13. Greatest slope: $\left(-\dfrac{\sqrt{3}}{3}, \dfrac{3}{4}\right)$, Least slope: $\left(\dfrac{\sqrt{3}}{3}, \dfrac{3}{4}\right)$

15. Proof **17.** Proof; Point of inflection: $(1, 0)$

19. (a) $P(x) = x - x^2$

(b)

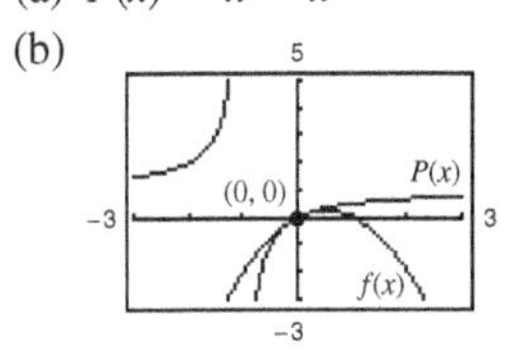

Chapter 4

Section 4.1 *(page 255)*

1. A function F is an antiderivative of f on an interval I when $F'(x) = f(x)$ for all x in I.

3. The particular solution results from knowing the value of $y = F(x)$ for one value of x. Using the initial condition in the general solution, you can solve for C to obtain the particular solution.

5. Proof **7.** $y = 3t^3 + C$ **9.** $y = \frac{2}{5}x^{5/2} + C$

	Original Integral	Rewrite	Integrate	Simplify
11.	$\int \sqrt[3]{x}\,dx$	$\int x^{1/3}\,dx$	$\dfrac{x^{4/3}}{4/3} + C$	$\dfrac{3}{4}x^{4/3} + C$
13.	$\int \dfrac{1}{x\sqrt{x}}\,dx$	$\int x^{-3/2}\,dx$	$\dfrac{x^{-1/2}}{-1/2} + C$	$-\dfrac{2}{\sqrt{x}} + C$

15. $\frac{1}{2}x^2 + 7x + C$ **17.** $\frac{1}{6}x^6 + x + C$

19. $\frac{2}{5}x^{5/2} + x^2 + x + C$ **21.** $\frac{3}{5}x^{5/3} + C$

23. $-\dfrac{1}{4x^4} + C$ **25.** $\dfrac{2}{3}x^{3/2} + 12x^{1/2} + C$

27. $x^3 + \frac{1}{2}x^2 - 2x + C$ **29.** $5 \sin x - 4 \cos x + C$

31. $-\csc x - x^2 + C$ **33.** $\tan\theta + \cos\theta + C$

35. $\tan y + C$ **37.** $f(x) = 3x^2 + 8$

39. $h(x) = x^7 + 5x - 7$ **41.** $f(x) = x^2 + x + 4$

43. $f(x) = -4\sqrt{x} + 3x$

45. (a) Answers will vary. (b) $y = \dfrac{x^3}{3} - x + \dfrac{7}{3}$

Sample answer:

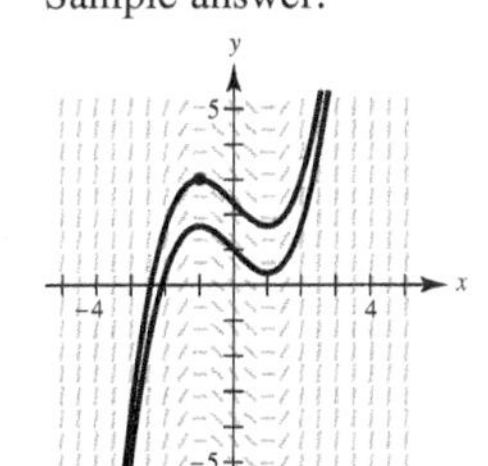

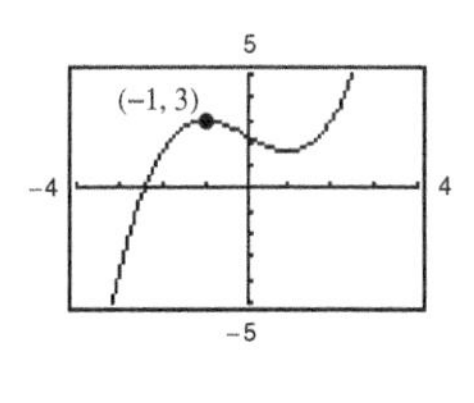

47. (a)

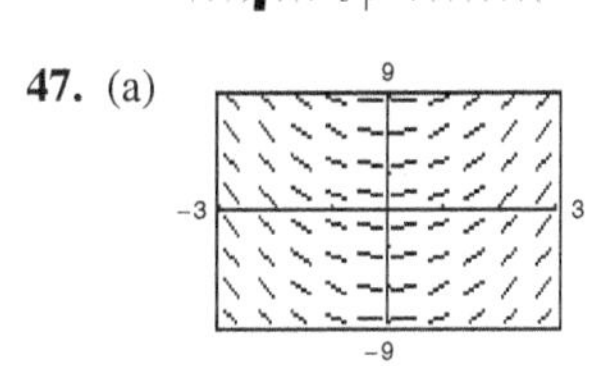

(b) $y = x^2 - 6$

(c)

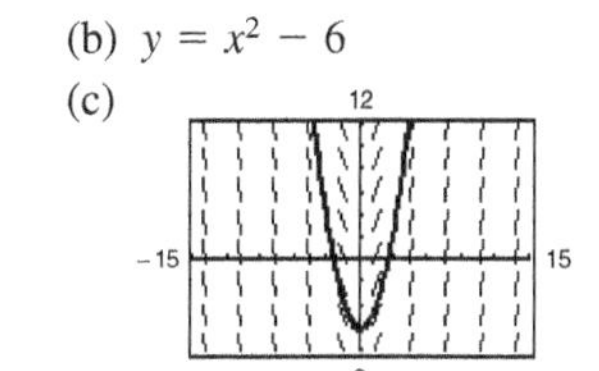

49. Answers will vary. Sample answer:

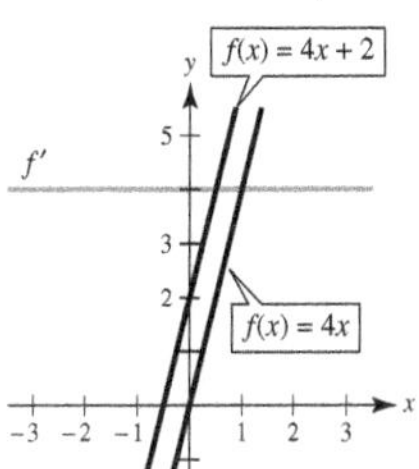

51. $f(x) = \tan^2 x \Rightarrow f'(x) = 2\tan x \cdot \sec^2 x$

$g(x) = \sec^2 x \Rightarrow g'(x) = 2\sec x \cdot \sec x \tan x = f'(x)$

The derivatives are the same, so f and g differ by a constant.

53. $f(x) = \dfrac{x^3}{3} - 4x + \dfrac{16}{3}$

55. (a) $h(t) = \frac{3}{4}t^2 + 5t + 12$ (b) 69 cm **57.** 62.25 ft

59. (a) $t \approx 2.562$ sec (b) $v(t) \approx -65.970$ ft/sec

61. $v_0 \approx 62.3$ m/sec **63.** 320 m; -32 m/sec

65. (a) $v(t) = 3t^2 - 12t + 9$, $a(t) = 6t - 12$

(b) $(0, 1), (3, 5)$ (c) -3

67. $a(t) = -\dfrac{1}{2t^{3/2}}$, $x(t) = 2\sqrt{t} + 2$

69. (a) 1.18 m/sec² (b) 190 m

71. (a) 300 ft (b) 60 ft/sec $\approx$ 41 mi/h

73. False. f has an infinite number of antiderivatives, each differing by a constant.

75. True **77.** True **79.** Proof

81. Putnam Problem B2, 1991

Section 4.2 *(page 267)*

1. The index of summation is i, the upper bound of summation is 8, and the lower bound of summation is 3.

3. You can use the line $y = x$ bounded by $x = a$ and $x = b$. The sum of the areas of the inscribed rectangles in the figure below is the lower sum.

The sum of the areas of the circumscribed rectangles in the figure below is the upper sum.

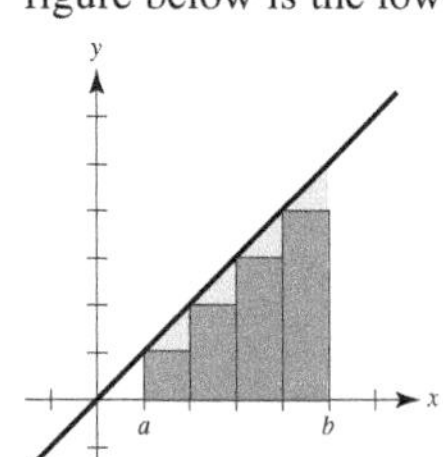

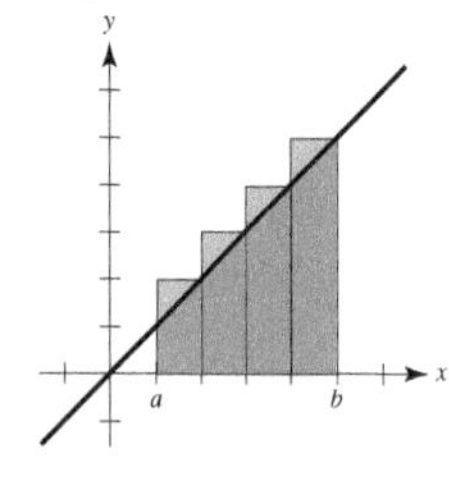

The rectangles in the first graph do not contain all of the area of the region, and the rectangles in the second graph cover more than the area of the region. The exact value of the area lies between these two sums.

5. 75 **7.** $\dfrac{158}{85}$ **9.** $8c$ **11.** $\displaystyle\sum_{i=1}^{11} \frac{1}{5i}$

13. $\displaystyle\sum_{j=1}^{6}\left[7\left(\frac{j}{6}\right)+5\right]$ **15.** $\displaystyle\frac{2}{n}\sum_{i=1}^{n}\left[\left(\frac{2i}{n}\right)^3-\left(\frac{2i}{n}\right)\right]$ **17.** 84

19. 1200 **21.** 2470 **23.** 1876

25. $\dfrac{n+2}{n}$

$n = 10$: $S = 1.2$
$n = 100$: $S = 1.02$
$n = 1000$: $S = 1.002$
$n = 10{,}000$: $S = 1.0002$

27. $\dfrac{2(n+1)(n-1)}{n^2}$

$n = 10$: $S = 1.98$
$n = 100$: $S = 1.9998$
$n = 1000$: $S = 1.999998$
$n = 10{,}000$: $S = 1.99999998$

29. $13 < (\text{Area of region}) < 15$

31. $55 < (\text{Area of region}) < 74.5$

33. $0.7908 < (\text{Area of region}) < 1.1835$

35. The area of the shaded region falls between 12.5 square units and 16.5 square units.

37. $A \approx S \approx 0.768$
$A \approx s \approx 0.518$

39. $A \approx S \approx 0.746$
$A \approx s \approx 0.646$

41. $s(n) = 24 - \dfrac{24}{n}$, $S(n) = 24 + \dfrac{24}{n}$

43. $s(n) = \dfrac{5(2n^2 - 3n + 1)}{6n^2}$, $S(n) = \dfrac{5(2n^2 + 3n + 1)}{6n^2}$

45. (a)

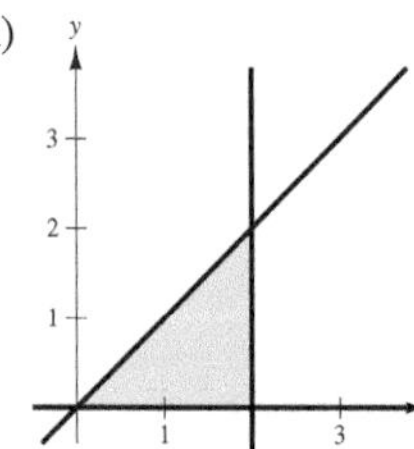

(b) $\Delta x = \dfrac{2-0}{n} = \dfrac{2}{n}$

(c) $s(n) = \displaystyle\sum_{i=1}^{n} f(x_{i-1})\,\Delta x = \sum_{i=1}^{n}\left[(i-1)\left(\frac{2}{n}\right)\right]\left(\frac{2}{n}\right)$

(d) $S(n) = \displaystyle\sum_{i=1}^{n} f(x_i)\,\Delta x = \sum_{i=1}^{n}\left[i\left(\frac{2}{n}\right)\right]\left(\frac{2}{n}\right)$

(e)

n	5	10	50	100
$s(n)$	1.6	1.8	1.96	1.98
$S(n)$	2.4	2.2	2.04	2.02

(f) $\displaystyle\lim_{n\to\infty}\sum_{i=1}^{n}\left[(i-1)\left(\frac{2}{n}\right)\right]\left(\frac{2}{n}\right) = 2$

$\displaystyle\lim_{n\to\infty}\sum_{i=1}^{n}\left[i\left(\frac{2}{n}\right)\right]\left(\frac{2}{n}\right) = 2$

47. $A = 3$

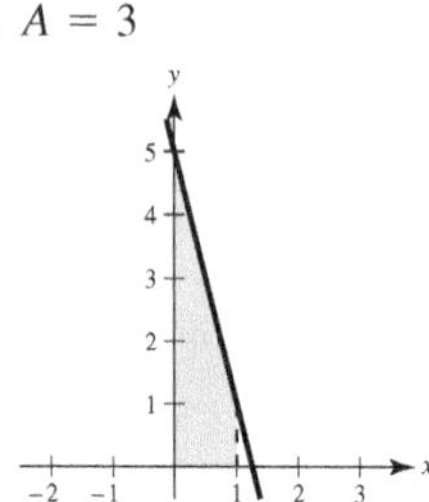

49. $A = \frac{7}{3}$

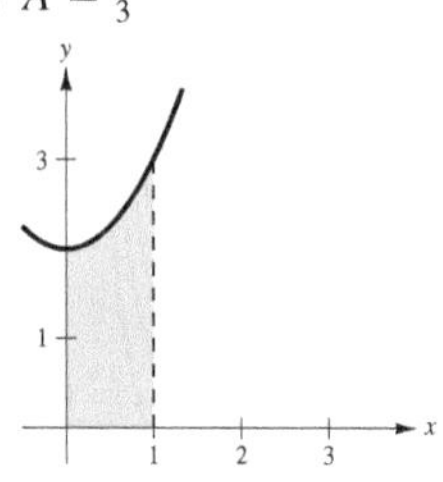

51. $A = 54$

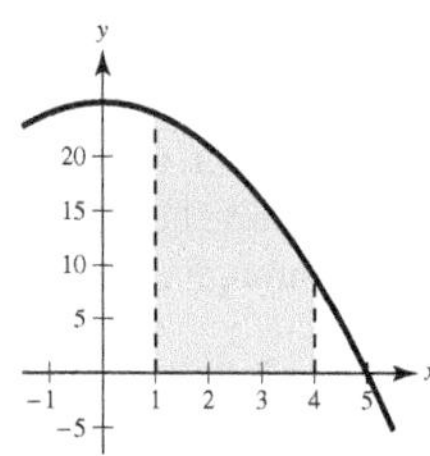

53. $A = 34$

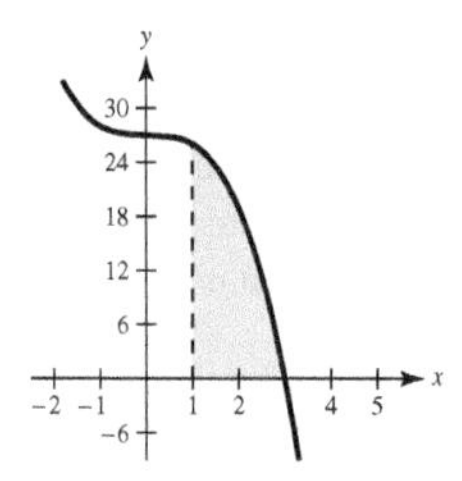

55. $A = \frac{2}{3}$

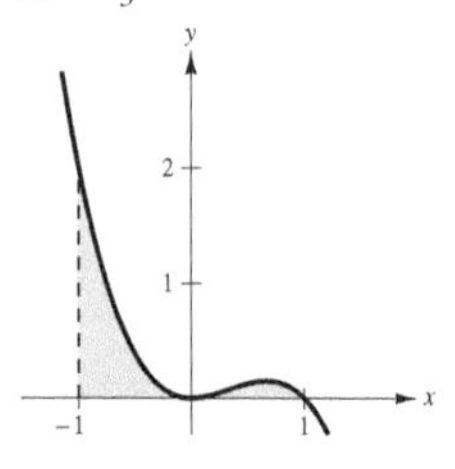

57. $A = 8$

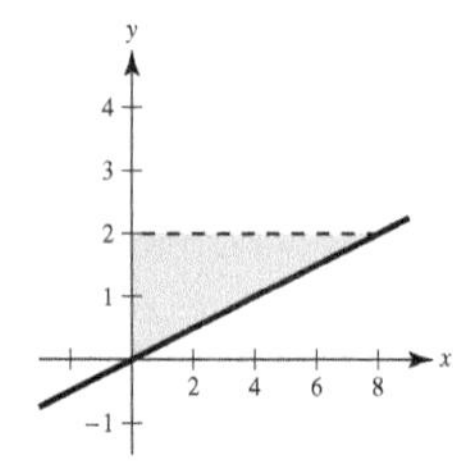

59. $A = \frac{125}{3}$

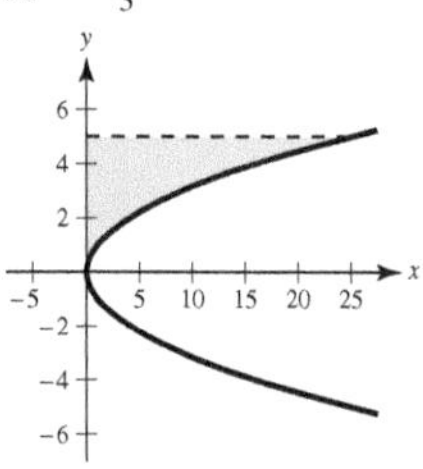

61. $A = \frac{44}{3}$

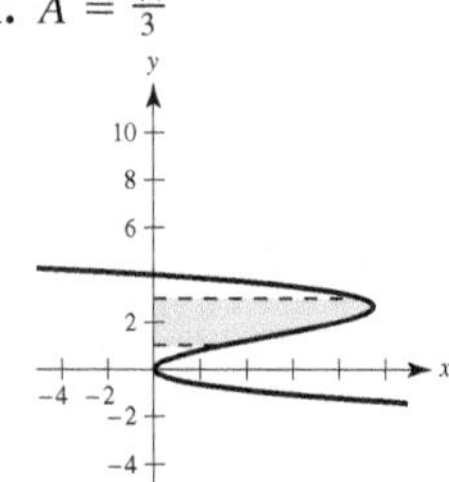

63. $\frac{69}{8}$ **65.** 0.345 **67.** b

69. An overestimate on one side of the midpoint compensates for an underestimate on the other side of the midpoint.

71. (a)

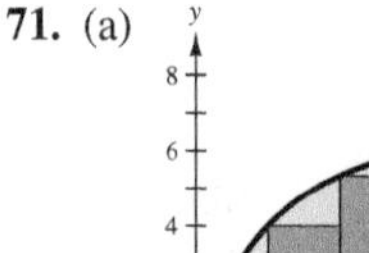

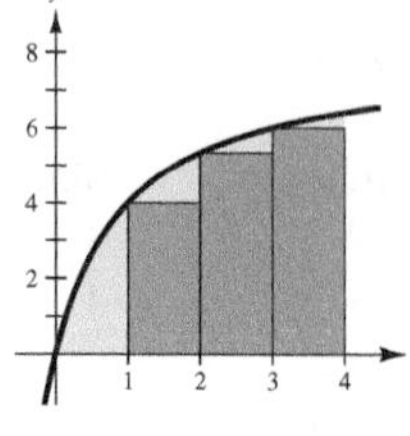

$s(4) = \frac{46}{3}$

(b)

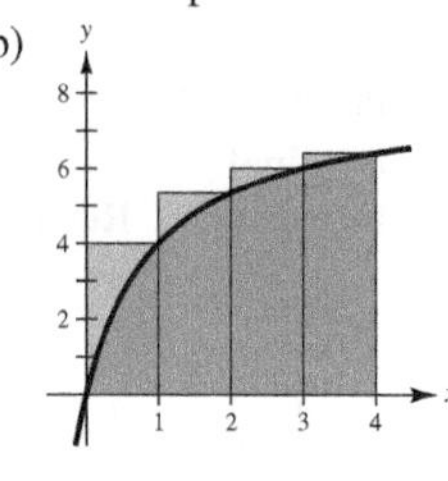

$S(4) = \frac{326}{15}$

(c)

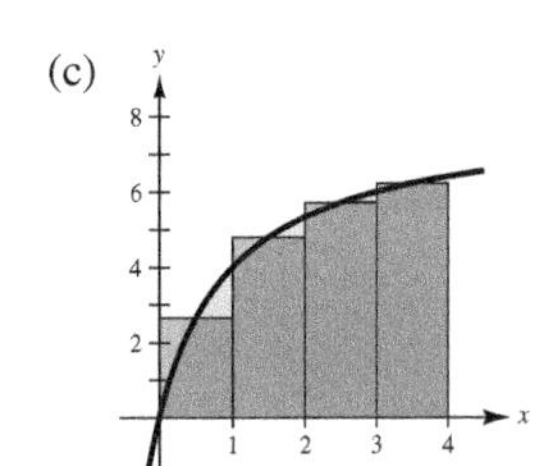

$M(4) = \frac{6112}{315}$

(d) Proof

(e)

n	4	8	20	100	200
$s(n)$	15.333	17.368	18.459	18.995	19.060
$S(n)$	21.733	20.568	19.739	19.251	19.188
$M(n)$	19.403	19.201	19.137	19.125	19.125

(f) Because f is an increasing function, $s(n)$ is always increasing and $S(n)$ is always decreasing.

73. True

75. Suppose there are n rows and $n + 1$ columns. The stars on the left total $1 + 2 + \cdots + n$, as do the stars on the right. There are $n(n + 1)$ stars in total. So, $2[1 + 2 + \cdots + n] = n(n + 1)$ and $1 + 2 + \cdots + n = \dfrac{n(n + 1)}{2}$.

77. When n is odd, there are $\left(\dfrac{n + 1}{2}\right)^2$ seats. When n is even, there are $\dfrac{n^2 + 2n}{4}$ seats.

79. Putnam Problem B1, 1989

Section 4.3 *(page 277)*

1. A Riemann sum represents the addition of all of the subregions for a function f on an interval $[a, b]$.

3. $2\sqrt{3} \approx 3.464$ **5.** 32 **7.** 0 **9.** $\frac{10}{3}$

11. $\displaystyle\int_{-1}^{5} (3x + 10)\, dx$ **13.** $\displaystyle\int_{0}^{4} 5\, dx$ **15.** $\displaystyle\int_{-4}^{4} (4 - |x|)\, dx$

17. $\displaystyle\int_{-5}^{5} (25 - x^2)\, dx$ **19.** $\displaystyle\int_{0}^{\pi/2} \cos x\, dx$ **21.** $\displaystyle\int_{0}^{2} y^3\, dy$

23.

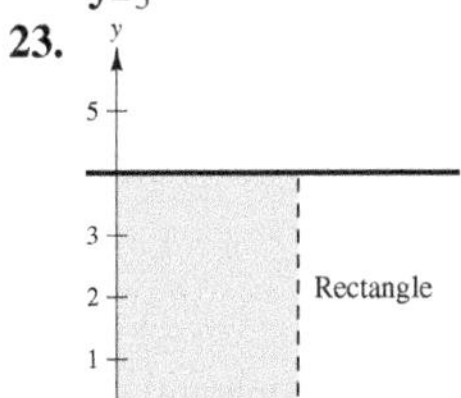

$A = 12$

25.

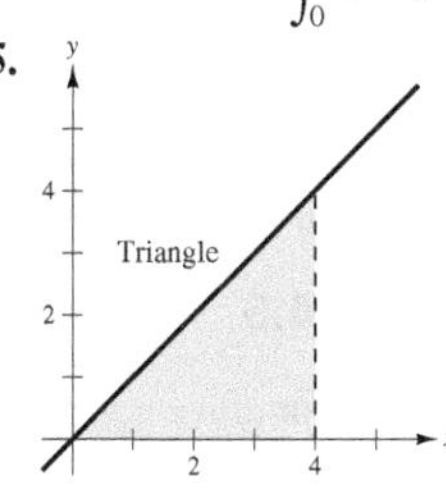

$A = 8$

27.

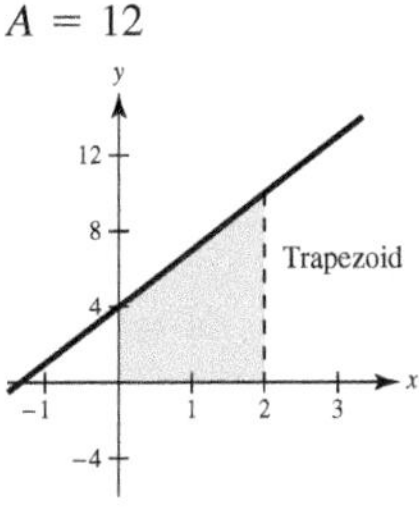

$A = 14$

29.

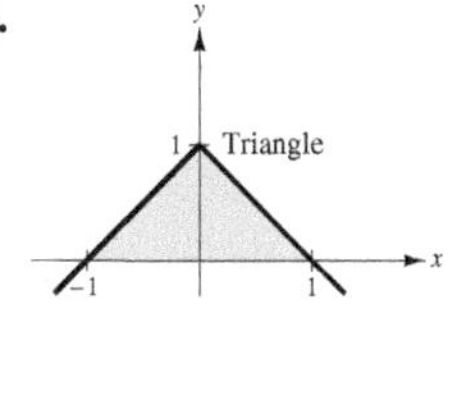

$A = 1$

31.

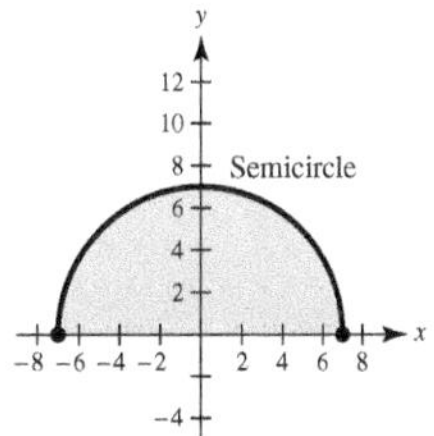

$A = \dfrac{49\pi}{2}$

33. -320 **35.** 80 **37.** -40 **39.** 508

41. (a) 13 (b) -10 (c) 0 (d) 30

43. (a) 8 (b) -12 (c) -4 (d) 30 **45.** $-48, 88$

47. (a) $-\pi$ (b) 4 (c) $-(1 + 2\pi)$ (d) $3 - 2\pi$ (e) $5 + 2\pi$ (f) $23 - 2\pi$

49. (a) 14 (b) 4 (c) 8 (d) 0 **51.** 40 **53.** a

55. Answers will vary. Sample answer:

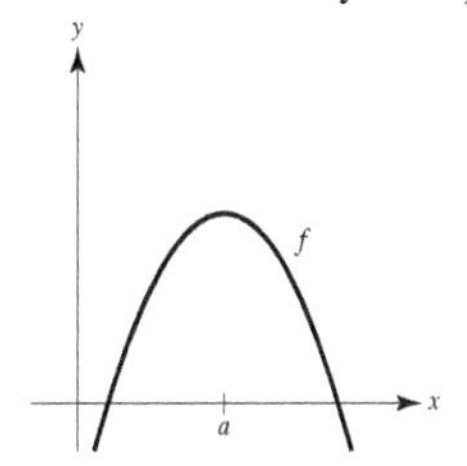

There is no region.

57. Geometric method:

$$\int_{-1}^{3} (x + 2)\, dx = \text{Area of large triangle} - \text{Area of small triangle}$$

$$= \frac{25}{2} - \frac{1}{2} = 12$$

Limit definition:

$$\int_{-1}^{3} (x + 2)\, dx = \lim_{n\to\infty} \sum_{i=1}^{n} \left[\left(-1 + \frac{4i}{n} + 2\right)\left(\frac{4}{n}\right)\right] = 12$$

59. $a = -2, b = 5$

61. Answers will vary. Sample answer: $a = \pi$, $b = 2\pi$

$$\int_{\pi}^{2\pi} \sin x\, dx < 0$$

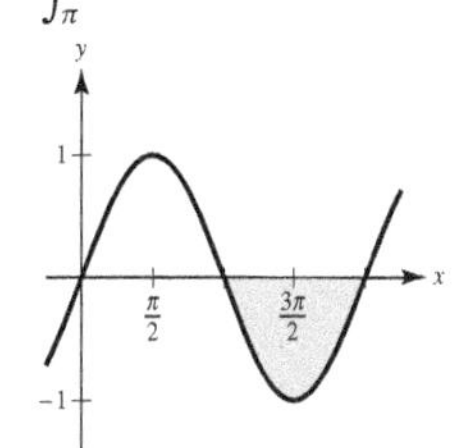

63. True **65.** True **67.** False. $\displaystyle\int_{0}^{2} (-x)\, dx = -2$

69. 272 **71.** Proof

73. No. No matter how small the subintervals, the number of both rational and irrational numbers within each subinterval is infinite, and $f(c_i) = 0$ or $f(c_i) = 1$.

75. $a = -1$ and $b = 1$ maximize the integral.

77. Answers will vary. Sample answer:

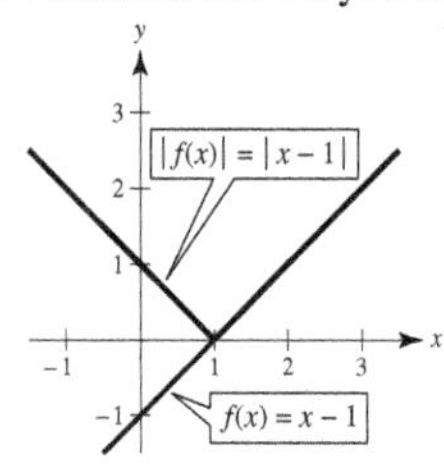

The integrals are equal when f is always greater than or equal to 0 on $[a, b]$.

79. $\frac{1}{3}$

Section 4.4 *(page 292)*

1. Find an antiderivative of the function and evaluate the difference of the antiderivative at the upper limit of integration and the lower limit of integration.

3. The average value of a function on an interval is the integral of the function on $[a, b]$ times $\frac{1}{b-a}$.

5.

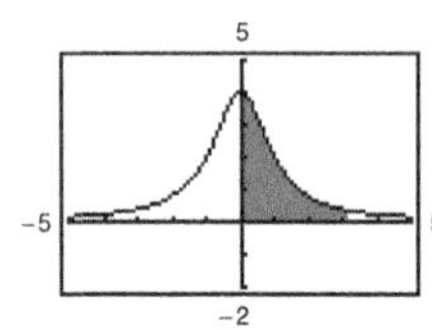

Positive

7.

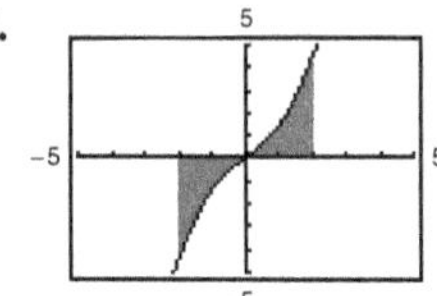

Zero

9. -2 **11.** $-\frac{28}{3}$ **13.** $\frac{1}{3}$ **15.** $\frac{1}{2}$ **17.** $\frac{2}{3}$

19. -4 **21.** $-\frac{1}{18}$ **23.** $-\frac{27}{20}$ **25.** $\frac{25}{2}$ **27.** $\frac{64}{3}$

29. $2 - 7\pi$ **31.** $\frac{\pi}{4}$ **33.** $\frac{2\sqrt{3}}{3}$ **35.** 0 **37.** $\frac{1}{6}$

39. 1 **41.** $\frac{52}{3}$ **43.** 20 **45.** $\frac{32}{3}$ **47.** $\frac{3\sqrt[3]{2}}{2} \approx 1.8899$

49. $2\sqrt{3} \approx 3.4641$ **51.** $\pm\arccos\frac{\sqrt{\pi}}{2} \approx \pm 0.4817$

53. Average value $= \frac{8}{3}$

$x = \pm\frac{2\sqrt{3}}{3}$

55. Average value $= 10.2$

$x \approx 1.3375$

57. Average value $= \frac{2}{\pi}$

$x \approx 0.690,\ x \approx 2.451$

59. (a) $F(x) = 500\sec^2 x$

(b) $\frac{1500\sqrt{3}}{\pi} \approx 827$ N

61. $\frac{2}{\pi} \approx 63.7\%$

63. $F(x) = -\frac{20}{x} + 20$

$F(2) = 10$

$F(5) = 16$

$F(8) = \frac{35}{2}$

65. $F(x) = \sin x$

$F(0) = 0$

$F\left(\frac{\pi}{4}\right) = \frac{\sqrt{2}}{2}$

$F\left(\frac{\pi}{2}\right) = 1$

67. (a) $g(0) = 0,\ g(2) \approx 7,\ g(4) \approx 9,\ g(6) \approx 8,\ g(8) \approx 5$

(b) Increasing: $(0, 4)$; Decreasing: $(4, 8)$

(c) A maximum occurs at $x = 4$.

(d)

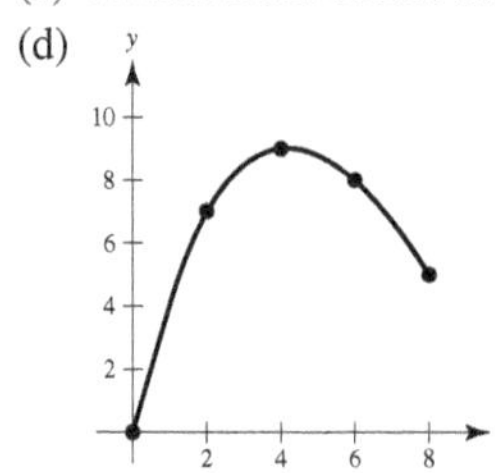

69. $\frac{1}{2}x^2 + 2x$ **71.** $\frac{3}{4}x^{4/3} - 12$ **73.** $\tan x - 1$

75. $x^2 - 2x$ **77.** $\sqrt{x^4 + 1}$ **79.** $\sqrt{x}\csc x$ **81.** 8

83. $\cos x\sqrt{\sin x}$ **85.** $3x^2 \sin x^6$

87.

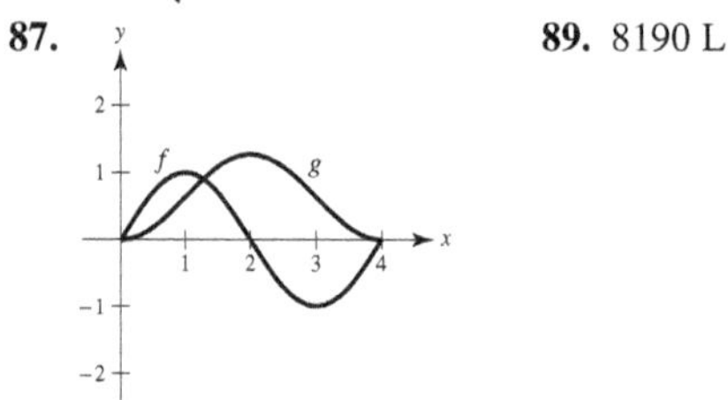

An extremum of g occurs at $x = 2$.

89. 8190 L

91. About 540 ft **93.** (a) $\frac{3}{2}$ ft to the right (b) $\frac{113}{10}$ ft

95. (a) 0 ft (b) $\frac{63}{2}$ ft **97.** (a) 2 ft to the right (b) 2 ft

99. The displacement and total distance traveled are equal when the particle is always moving in the same direction on an interval.

101. The Fundamental Theorem of Calculus requires that f be continuous on $[a, b]$ and that F be an antiderivative for f on the entire interval. On an interval containing c, the function $f(x) = \frac{1}{x - c}$ is not continuous at c.

103. 28 units

105. $f(x) = x^{-2}$ has a nonremovable discontinuity at $x = 0$.

107. $f(x) = \sec^2 x$ has a nonremovable discontinuity at $x = \frac{\pi}{2}$.

109. True

111. $f'(x) = \frac{1}{(1/x)^2 + 1}\left(-\frac{1}{x^2}\right) + \frac{1}{x^2 + 1} = 0$

Because $f'(x) = 0$, $f(x)$ is constant.

113. (a) 0 (b) 0 (c) $xf(x) + \int_0^x f(t)\,dt$ (d) 0

115. Putnam Problem B5, 2006

Section 4.5 *(page 305)*

1. You can move constant multiples outside the integral sign.

$$\int kf(x)\,dx = k\int f(x)\,dx$$

3. The integral of $[g(x)]^n g'(x)$ is $\frac{[g(x)]^{n+1}}{n+1} + C,\ n \neq -1$.

Recall the power rule for polynomials.

	$\int f(g(x))g'(x)\,dx$	$u = g(x)$	$du = g'(x)\,dx$
5.	$\int (5x^2 + 1)^2(10x)\,dx$	$5x^2 + 1$	$10x\,dx$
7.	$\int \tan^2 x \sec^2 x\,dx$	$\tan x$	$\sec^2 x\,dx$

9. $\frac{1}{5}(1 + 6x)^5 + C$ **11.** $\frac{2}{3}(25 - x^2)^{3/2} + C$

13. $\frac{1}{12}(x^4 + 3)^3 + C$ **15.** $\frac{1}{30}(2x^3 - 1)^5 + C$
17. $\frac{1}{3}(t^2 + 2)^{3/2} + C$ **19.** $-\frac{15}{8}(1 - x^2)^{4/3} + C$
21. $\dfrac{7}{4(1 - x^2)^2} + C$ **23.** $-\dfrac{1}{3(1 + x^3)} + C$
25. $-\sqrt{1 - x^2} + C$ **27.** $-\dfrac{1}{4}\left(1 + \dfrac{1}{t}\right)^4 + C$
29. $\sqrt{2x} + C$ **31.** $2x^2 - 4\sqrt{16 - x^2} + C$
33. $-\dfrac{1}{2(x^2 + 2x - 3)} + C$
35. (a) Answers will vary. (b) $y = -\frac{1}{3}(4 - x^2)^{3/2} + 2$
Sample answer:

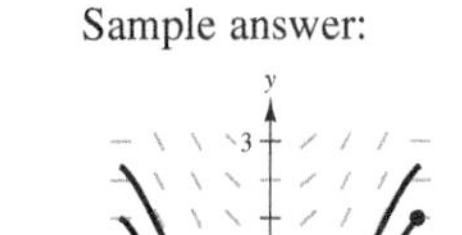

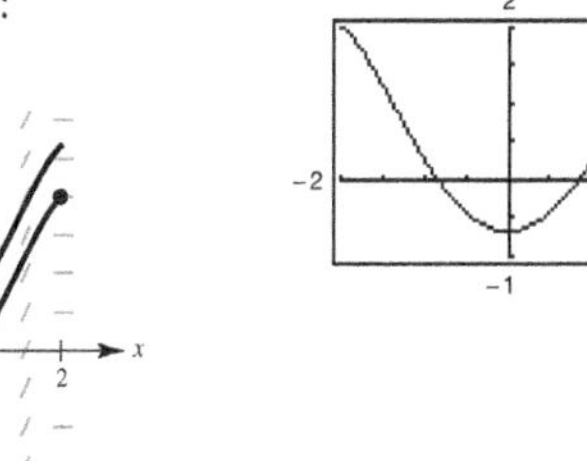

37. $f(x) = (2x^3 + 1)^3 + 3$ **39.** $-\cos \pi x + C$
41. $\dfrac{1}{6}\sin 6x + C$ **43.** $-\sin \dfrac{1}{\theta} + C$
45. $\frac{1}{4}\sin^2 2x + C$ or $-\frac{1}{4}\cos^2 2x + C_1$ or $-\frac{1}{8}\cos 4x + C_2$
47. $\dfrac{1}{2}\tan^2 x + C$ or $\dfrac{1}{2}\sec^2 x + C_1$ **49.** $f(x) = 2\cos\dfrac{x}{2} + 4$
51. $f(x) = \frac{1}{12}(4x^2 - 10)^3 - 8$
53. $\frac{2}{5}(x + 6)^{5/2} - 4(x + 6)^{3/2} + C = \frac{2}{5}(x + 6)^{3/2}(x - 4) + C$
55. $-\left[\frac{2}{3}(1 - x)^{3/2} - \frac{4}{5}(1 - x)^{5/2} + \frac{2}{7}(1 - x)^{7/2}\right] + C =$
$-\frac{2}{105}(1 - x)^{3/2}(15x^2 + 12x + 8) + C$
57. $\frac{1}{8}\left[\frac{2}{5}(2x - 1)^{5/2} + \frac{4}{3}(2x - 1)^{3/2} - 6(2x - 1)^{1/2}\right] + C =$
$\dfrac{\sqrt{2x - 1}}{15}(3x^2 + 2x - 13) + C$
59. $-\frac{1}{8}\cos^4 2x + C$ **61.** 0 **63.** $12 - \frac{8}{9}\sqrt{2}$
65. 2 **67.** $\frac{1}{2}$ **69.** $\frac{1209}{28}$ **71.** $2(\sqrt{3} - 1)$ **73.** $\frac{272}{15}$
75. 0 **77.** (a) 144 (b) 72 (c) -144 (d) 432
79. $2\int_0^3 (4x^2 - 6)\, dx = 36$
81. (a) $\int x^2\sqrt{x^3 + 1}\, dx$; Use substitution with $u = x^3 + 1$.
(b) $\int \cot^3(2x)\csc^2(2x)\, dx$; Use substitution with $u = \cot 2x$.
83. \$340,000
85. (a) 102.532 thousand units (b) 102.352 thousand units
(c) 74.5 thousand units
87. (a)
(b) g is nonnegative, because the graph of f is positive at the beginning and generally has more positive sections than negative ones.
(c) The points on g that correspond to the extrema of f are points of inflection of g.
(d) No, some zeros of f, such as $x = \dfrac{\pi}{2}$, do not correspond to extrema of g. The graph of g continues to increase after $x = \dfrac{\pi}{2}$, because f remains above the x-axis.
(e)

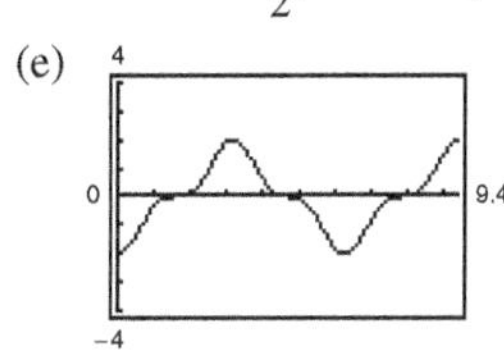

The graph of h is that of g shifted 2 units downward.

89. (a) and (b) Proofs
91. (a) $P_{0.50,\, 0.75} \approx 35.3\%$ (b) $b \approx 58.6\%$
93. True **95.** True **97.** True **99–101.** Proofs
103. Putnam Problem A1, 1958

Review Exercises for Chapter 4 *(page 309)*

1. $\dfrac{x^4}{4} + 4x + C$ **3.** $\dfrac{4}{3}x^3 + \dfrac{1}{2}x^2 + 3x + C$
5. $\dfrac{x^2}{2} - \dfrac{4}{x^2} + C$ **7.** $9\cos x - 2\cot x + C$
9. $y = 1 - 3x^2$ **11.** $f(x) = 4x^3 - 5x - 3$
13. (a) 3 sec; 144 ft (b) $\frac{3}{2}$ sec (c) 108 ft
15. 60 **17.** $\displaystyle\sum_{i=1}^{10} \frac{i}{5(i + 2)}$ **19.** 192 **21.** 420
23. 3310 **25.** 9.038 < (Area of region) < 13.038
27. $s(n) = 11 - \dfrac{2}{n}$, $S(n) = 11 + \dfrac{2}{n}$
29. $A = 15$ **31.** $A = 12$

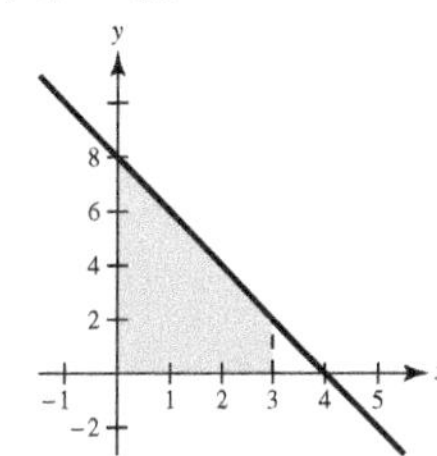

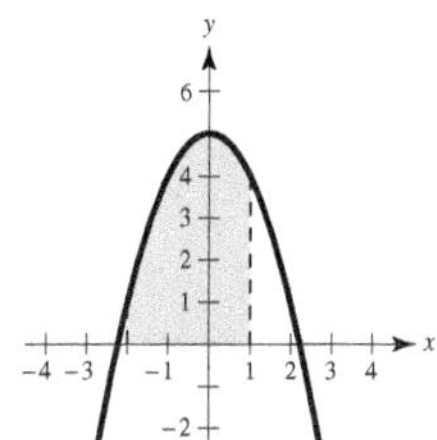

33. 43 **35.** 48
37.

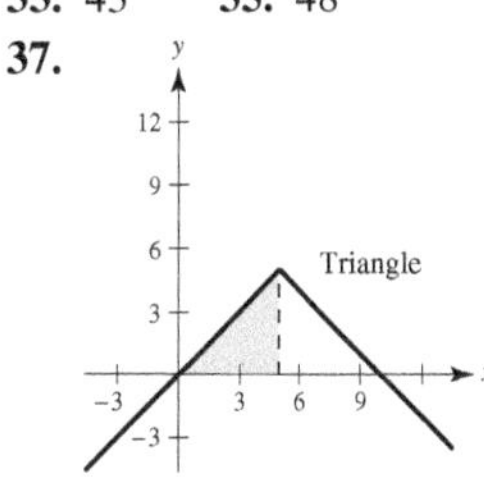

$A = \frac{25}{2}$

39. (a) 17 (b) 7 (c) 9 (d) 84
41. 12 **43.** $\dfrac{422}{5}$ **45.** $\dfrac{\sqrt{2} + 2}{2}$ **47.** 1 **49.** 30
51. $\frac{1}{4}$ **53.** $\sqrt{\frac{13}{3}}$
55. Average value $= \frac{2}{5}$
$x = \frac{25}{4}$
57. $x^2\sqrt{1 + x^3}$ **59.** $-\frac{1}{30}(1 - 3x^2)^5 + C = \frac{1}{30}(3x^2 - 1)^5 + C$
61. $\frac{1}{4}\sin^4 x + C$ **63.** $-2\sqrt{1 - \sin\theta} + C$
65. $\frac{2}{5}(8 - x)^{5/2} - \frac{16}{3}(8 - x)^{3/2} + C$

Section 4.6 *(page 310)*

	Trapezoidal	Simpson's	Exact
1.	2.7500	2.6667	2.6667
3.	4.2500	4.0000	4.0000
5.	20.2222	20.0000	20.0000
7.	12.6640	12.6667	12.6667
9.	0.3352	0.3334	0.3333
	Trapezoidal	**Simpson's**	**Graphing Utility**
11.	3.2833	3.2396	3.2413
13.	0.3415	0.3720	0.3927
15.	0.5495	0.5483	0.5493
17.	-0.0975	-0.0977	-0.0977
19.	0.1940	0.1860	0.1858

21. Trapezoidal: Linear (1st-degree) polynomials
Simpson's: Quadratic (2nd-degree) polynomials

23. (a) 1.500 (b) 0.000 **25.** (a) $\frac{1}{4}$ (b) $\frac{1}{12}$

27. (a) $n = 366$ (b) $n = 26$ **29.** (a) $n = 77$ (b) $n = 8$

31. (a) $n = 130$ (b) $n = 12$ **33.** (a) $n = 643$ (b) $n = 48$

35. (a) 24.5 (b) 25.67 **37.** 0.701 **39.** 89,250 m^2

41. 10,233.58 ft-lb **43.** 3.1416 **45.** 2.477 **47.** Proof

67. $\frac{455}{2}$ **69.** 2 **71.** $\frac{28\pi}{15}$ **73.** $\frac{468}{7}$ **75.** 0

P.S. Problem Solving *(page 311)*

1. (a) $L(1) = 0$ (b) $L'(x) - = \frac{1}{x}, L'(1) = 1$

(c) $x \approx 2.718$ (d) Proof

3. (a) $\lim_{n\to\infty}\left[\frac{32}{n^5}\sum_{i=1}^{n} i^4 - \frac{64}{n^4}\sum_{i=1}^{n} i^3 + \frac{32}{n^3}\sum_{i=1}^{n} i^2\right]$

(b) $\frac{16n^4 - 16}{15n^4}$ (c) $\frac{16}{15}$

5. (a)

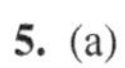

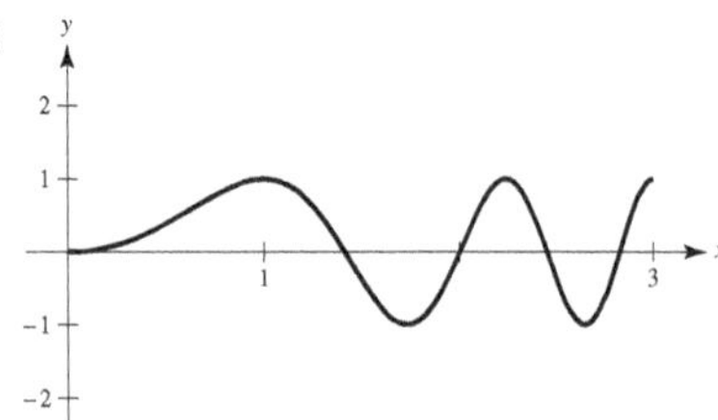

(b)

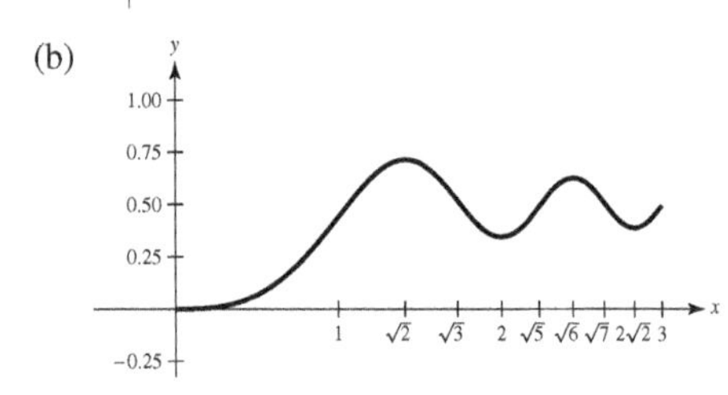

(c) Relative maxima at $x = \sqrt{2}, \sqrt{6}$
Relative minima at $x = 2, 2\sqrt{2}$

(d) Points of inflection at $x = 1, \sqrt{3}, \sqrt{5}, \sqrt{7}$

7. (a)

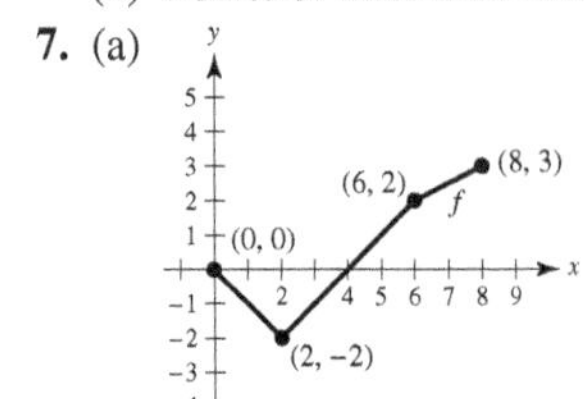

(b)

x	0	1	2	3	4	5	6	7	8
$F(x)$	0	$-\frac{1}{2}$	-2	$-\frac{7}{2}$	-4	$-\frac{7}{2}$	-2	$\frac{1}{4}$	3

(c) $x = 4, 8$ (d) $x = 2$

9. Proof **11.** $\frac{2}{3}$ **13.** Proof; $1 \leq \int_0^1 \sqrt{1 + x^4}\,dx \leq \sqrt{2}$

15. (a)

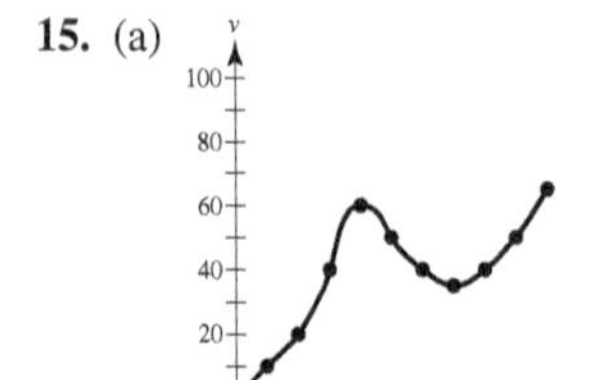

(b) (0, 0.4) and (0.7, 1.0) (c) 150 mi/h^2

(d) Total distance traveled in miles; 37 mi

(e) Answers will vary. Sample answer: 100 mi/h^2

17. (a)–(c) Proofs

19. (a) $S = \frac{5mb^2}{8}, s = \frac{3mb^2}{8}$

(b) $S(n) = \frac{mb^2(n + 1)}{2n}, s(n) = \frac{mb^2(n - 1)}{2n}$

(c) Area $= \frac{1}{2}(b)(mb) = \frac{1}{2}$(base)(height)

21. $f(x) = \begin{cases} -x + 1, & 0 \leq x < 2 \\ 2x - 5, & 2 \leq x < 3 \\ 1, & 3 \leq x \leq 4 \end{cases}$

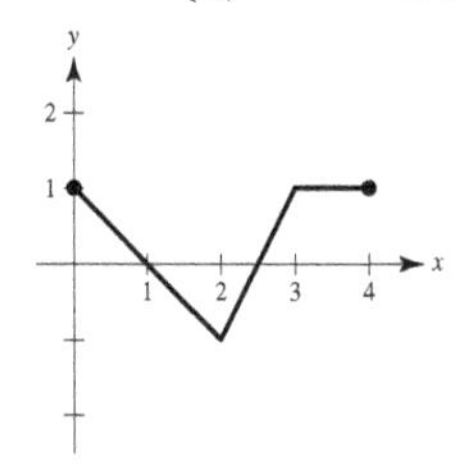

Chapter 5

Section 5.1 *(page 321)*

1. For $x > 1$, $\ln x = \int_1^x \frac{1}{t}\,dt > 0$. For $0 < x < 1$,
$\ln x = \int_1^x \frac{1}{t}\,dt = -\int_x^1 \frac{1}{t}\,dt.$

3. The number e is the base for the natural logarithm:
$\ln e = \int_1^e \frac{1}{t}\,dt = 1.$

5. (a) 3.8067 (b) $\ln 45 = \int_1^{45} \frac{1}{t}\,dt \approx 3.8067$

7. (a) -0.2231 (b) $\ln 0.8 = \int_1^{0.8} \frac{1}{t}\,dt \approx -0.2231$

9. b **10.** d **11.** a **12.** c

13.

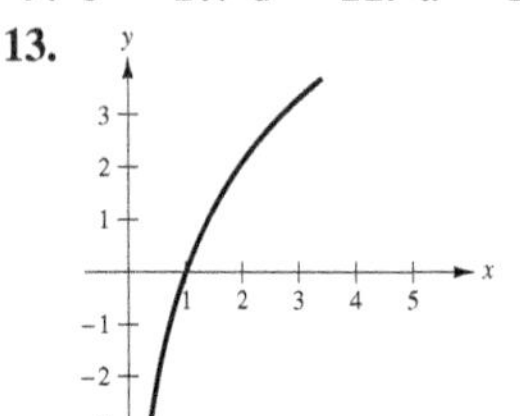

Domain: $x > 0$

15.

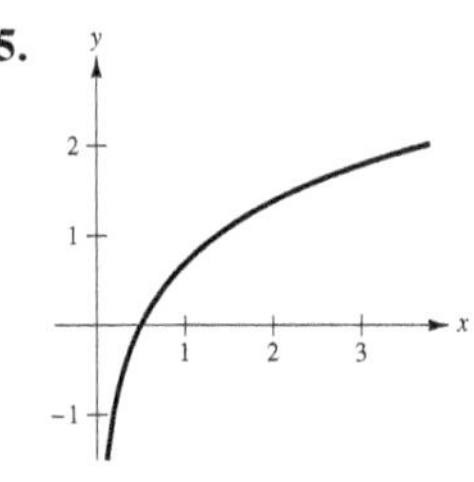

Domain: $x > 0$

17.

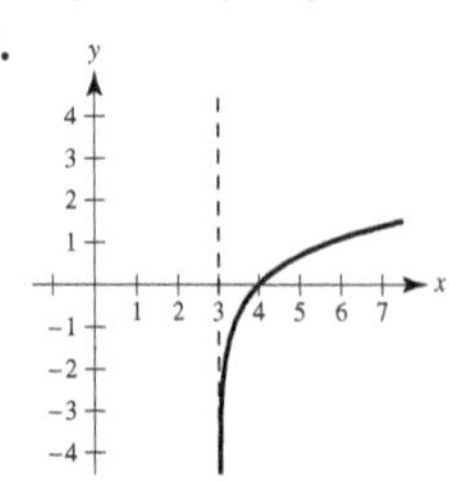

Domain: $x > 3$

19. (a) 1.7917 (b) -0.4055 (c) 4.3944 (d) 0.5493

21. $\ln x - \ln 4$ **23.** $\ln x + \ln y - \ln z$

25. $\ln x + \frac{1}{2}\ln(x^2 + 5)$ **27.** $\frac{1}{2}[\ln(x - 1) - \ln x]$

29. $\ln z + 2\ln(z - 1)$ **31.** $\ln \frac{x - 2}{x + 2}$

33. $\ln \sqrt[3]{\frac{x(x + 3)^2}{x^2 - 1}}$ **35.** $\ln \frac{16}{\sqrt{x^3 + 6x}}$

37. (a)

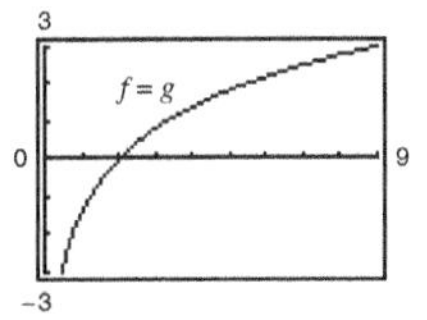

(b) $f(x) = \ln\dfrac{x^2}{4} = \ln x^2 - \ln 4$
$= 2\ln x - \ln 4$
$= g(x)$

39. $-\infty$ **41.** $\ln 4 \approx 1.3863$ **43.** $\dfrac{1}{x}$ **45.** $\dfrac{2x}{x^2+3}$

47. $\dfrac{4(\ln x)^3}{x}$ **49.** $\dfrac{2}{t+1}$ **51.** $\dfrac{2x^2-1}{x(x^2-1)}$

53. $\dfrac{1-x^2}{x(x^2+1)}$ **55.** $\dfrac{1-2\ln t}{t^3}$ **57.** $\dfrac{2}{x\ln x^2} = \dfrac{1}{x\ln x}$

59. $\dfrac{1}{1-x^2}$ **61.** $\dfrac{-4}{x(x^2+4)}$ **63.** $\cot x$

65. $-\tan x + \dfrac{\sin x}{\cos x - 1}$

67. (a) $y = 4x - 4$
(b)

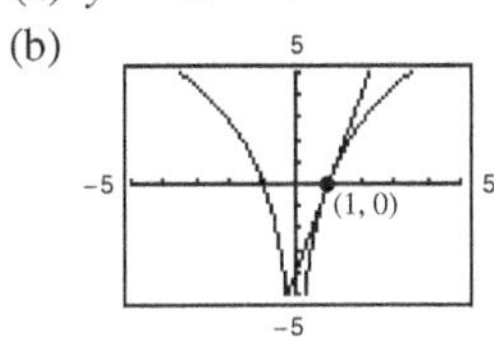

69. (a) $5x - y - 2 = 0$
(b)

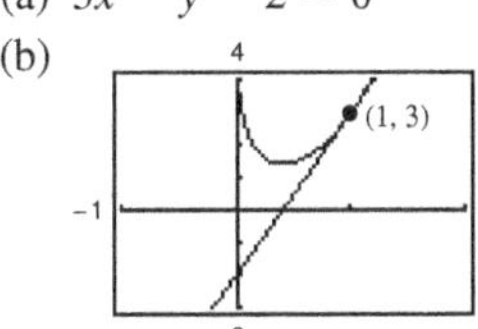

71. (a) $y = \frac{1}{3}x - \frac{1}{12}\pi + \frac{1}{2}\ln\frac{3}{2}$
(b)

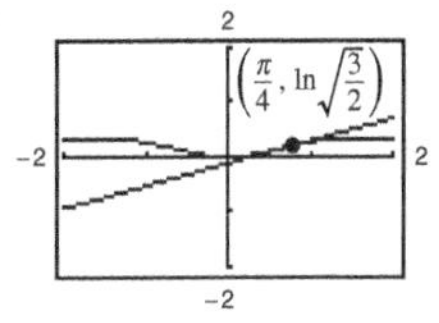

73. (a) $y = 4x + 4$
(b)

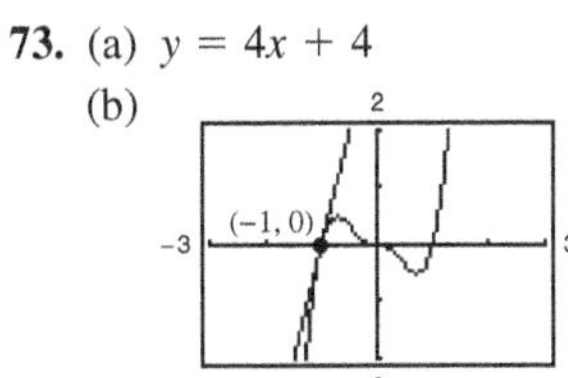

75. $\dfrac{2x^2+1}{\sqrt{x^2+1}}$ **77.** $\dfrac{3x^3+15x^2-8x}{2(x+1)^3\sqrt{3x-2}}$

79. $\dfrac{(2x^2+2x-1)\sqrt{x-1}}{(x+1)^{3/2}}$ **81.** $\dfrac{2xy}{3-2y^2}$

83. $\dfrac{y(1-6x^2)}{1+y}$ **85.** $xy'' + y' = x\left(-\dfrac{2}{x^2}\right) + \dfrac{2}{x} = 0$

87. Relative minimum: $\left(1, \frac{1}{2}\right)$

89. Relative minimum: $(e^{-1}, -e^{-1})$

91. Relative minimum: (e, e); Point of inflection: $\left(e^2, \dfrac{e^2}{2}\right)$

93. $x \approx 0.567$

95. Yes. If the graph of g is increasing, then $g'(x) > 0$. Because $f(x) > 0$, you know that $f'(x) = g'(x)f(x)$ and thus $f'(x) > 0$. Therefore, the graph of f is increasing.

97. No. For example,
$(\ln 2)(\ln 3) \approx 0.76 \neq 1.79 \approx \ln(2 \cdot 3) = \ln 6$.

99. True. **101.** False. π is a constant, so $\dfrac{d}{dx}[\ln \pi] = 0$.

103. (a)

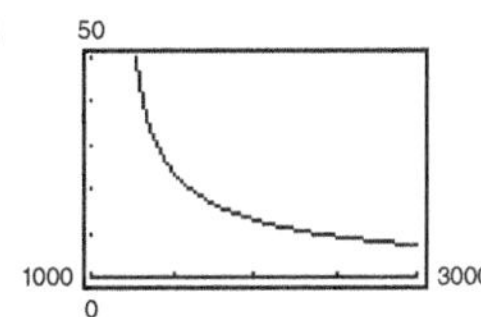

(b) 30 yr; \$503,434.80
(c) 20 yr; \$386,685.60

(d) When $x = 1398.43$, $\dfrac{dt}{dx} \approx -0.0805$. When $x = 1611.19$, $\dfrac{dt}{dx} \approx -0.0287$.

(e) Two benefits of a higher monthly payment are a shorter term and a lower total amount paid.

105. (a)

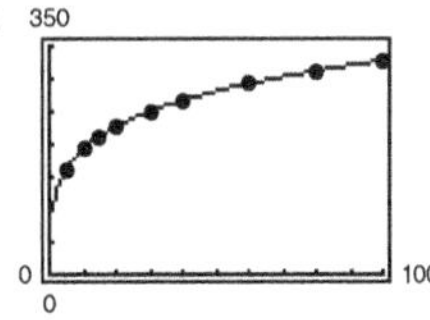

(c)

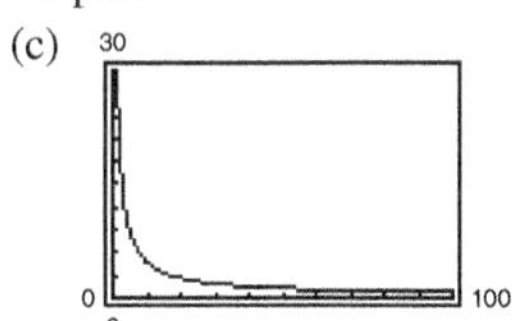

(b) $T'(10) \approx 4.75°/\text{lb/in.}^2$
$T'(70) \approx 0.97°/\text{lb/in.}^2$

$\lim_{p\to\infty} T'(p) = 0$
Answers will vary.

107. (a)

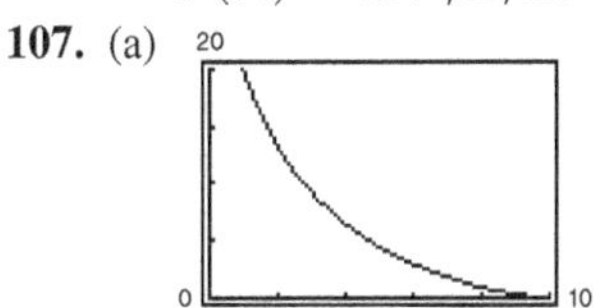

(b) When $x = 5$,
$\dfrac{dy}{dx} = -\sqrt{3}$.
When $x = 9$,
$\dfrac{dy}{dx} = -\dfrac{\sqrt{19}}{9}$.

(c) $\lim_{x\to 10^-} \dfrac{dy}{dx} = 0$

109. (a)

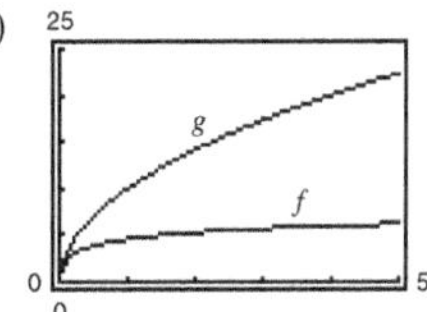

For $x > 4$, $g'(x) > f'(x)$. g is increasing at a faster rate than f for large values of x.

(b)

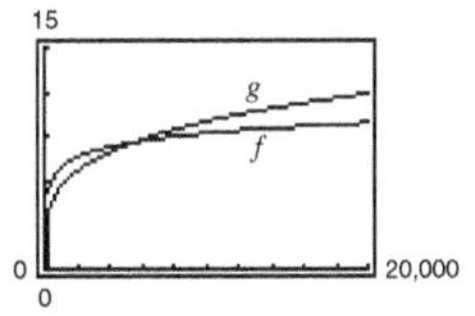

For $x > 256$, $g'(x) > f'(x)$. g is increasing at a faster rate than f for large values of x.

$f(x) = \ln x$ increases very slowly for large values of x.

Section 5.2 *(page 330)*

1. No. To use the Log Rule, look for quotients in which the numerator is the derivative of the denominator, with rewriting in mind.

3. Ways to alter an integrand are to rewrite using a trigonometric identity, multiply and divide by the same quantity, add and subtract the same quantity, or use long division.

5. $5\ln|x| + C$ **7.** $\frac{1}{2}\ln|2x+5| + C$

9. $\frac{1}{2}\ln|x^2-3| + C$ **11.** $\ln|x^4+3x| + C$

13. $\dfrac{x^2}{14} - \ln|x| + C$ **15.** $\dfrac{1}{3}\ln|x^3+3x^2+9x| + C$

17. $\frac{1}{2}x^2 - 4x + 6\ln|x+1| + C$

19. $\frac{1}{3}x^3 + 5\ln|x-3| + C$

21. $\frac{1}{3}x^3 - 2x + \ln\sqrt{x^2+2} + C$ **23.** $\frac{1}{3}(\ln x)^3 + C$

25. $-\frac{2}{3}\ln\left|1 - 3\sqrt{x}\right| + C$ **27.** $6\ln|x-5| - \dfrac{30}{x-5} + C$

29. $\sqrt{2x} - \ln\left|1 + \sqrt{2x}\right| + C$

31. $x + 6\sqrt{x} + 18\ln\left|\sqrt{x} - 3\right| + C$ **33.** $3\ln\left|\sin\dfrac{\theta}{3}\right| + C$

35. $-\frac{1}{2}\ln|\csc 2x + \cot 2x| + C$ **37.** $5\theta - \frac{1}{3}\sin 3\theta + C$

39. $\ln|1 + \sin t| + C$ **41.** $\ln|\sec x - 1| + C$

43. $y = -3\ln|2 - x| + C$ **45.** $y = \ln|x^2 - 9| + C$

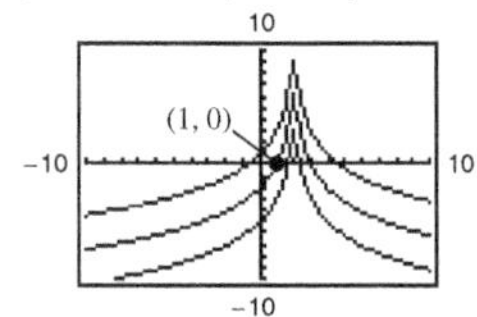

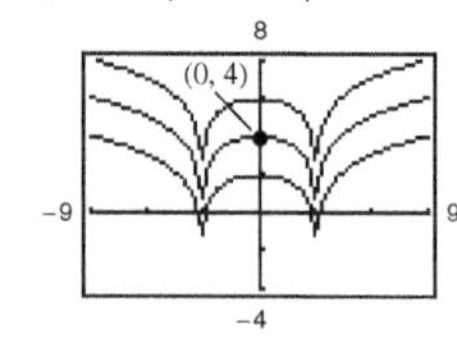

47. $f(x) = -2\ln x + 3x - 2$

49. (a)

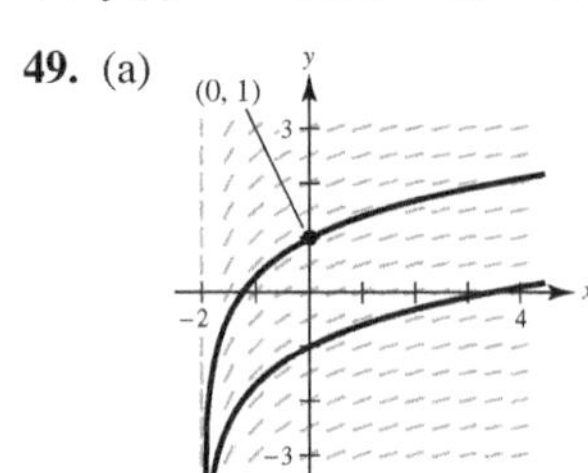

(b) $y = \ln\left(\dfrac{x+2}{2}\right) + 1$

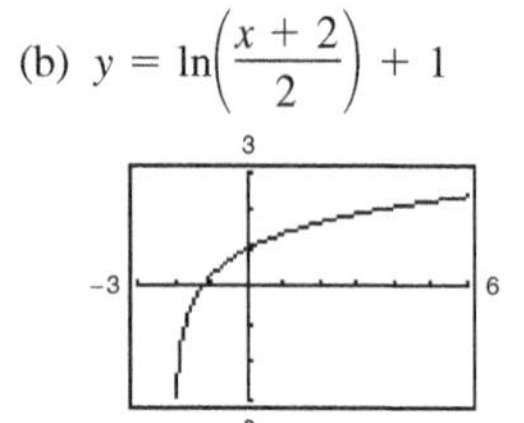

51. $\frac{5}{3}\ln 13 \approx 4.275$ **53.** $\frac{7}{3}$ **55.** $-\ln 3 \approx 1.099$

57. $\ln\left|\dfrac{2 - \sin 2}{1 - \sin 1}\right| \approx 1.929$

59. $4\sqrt{x} - x - 4\ln\left(1 + \sqrt{x}\right) + C$ **61.** $\dfrac{1}{x}$

63. $4\cot 4x$ **65.** $6\ln 3 \approx 6.592$

67. $\ln|\csc 1 + \cot 1| - \ln|\csc 2 + \cot 2| \approx 1.048$

69. $\dfrac{15}{2} + 8\ln 2 \approx 13.045$ **71.** $\dfrac{12}{\pi}\ln\left(2 + \sqrt{3}\right) \approx 5.03$

73. 1 **75.** $\dfrac{1}{e-1} \approx 0.582$ **77.** About 13.077

79. d **81.** Proof **83.** $x = 2$ **85.** Proof

87. $-\ln|\cos x| + C = \ln\left|\dfrac{1}{\cos x}\right| + C = \ln|\sec x| + C$

89. $\ln|\sec x + \tan x| + C = \ln\left|\dfrac{\sec^2 x - \tan^2 x}{\sec x - \tan x}\right| + C$

$= -\ln|\sec x - \tan x| + C$

91. (a) $P(t) = 1000(12\ln|1 + 0.25t| + 1)$ (b) $P(3) \approx 7715$

93. About 4.15 min

95.

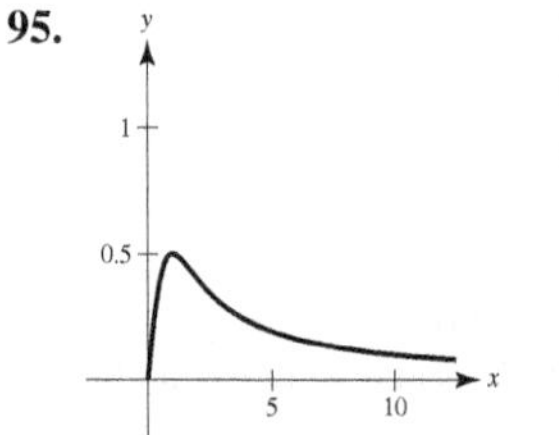

(a) $A = \frac{1}{2}\ln 2 - \frac{1}{4}$
(b) $0 < m < 1$
(c) $A = \frac{1}{2}(m - \ln m - 1)$

97. True **99.** True **101.** Putnam Problem B2, 2014

Section 5.3 *(page 339)*

1. The functions f and g have the effect of "undoing" each other.
3. No. The domain of f^{-1} is the range of f.

5. c **6.** b **7.** a **8.** d

9. (a) $f(g(x)) = 5\left(\dfrac{x-1}{5}\right) + 1 = x$

$g(f(x)) = \dfrac{(5x+1) - 1}{5} = x$

(b)

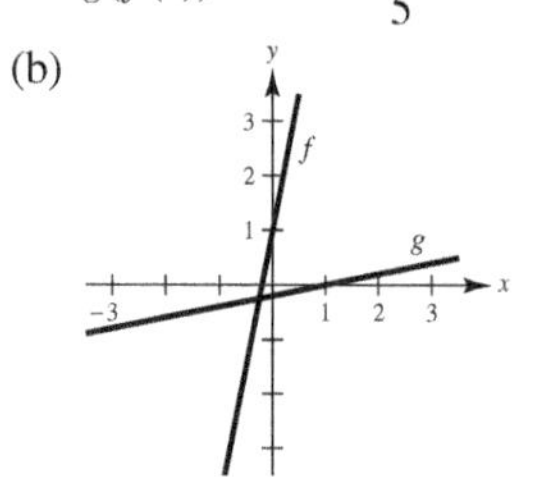

11. (a) $f(g(x)) = \left(\sqrt[3]{x}\right)^3 = x$

$g(f(x)) = \sqrt[3]{x^3} = x$

(b)

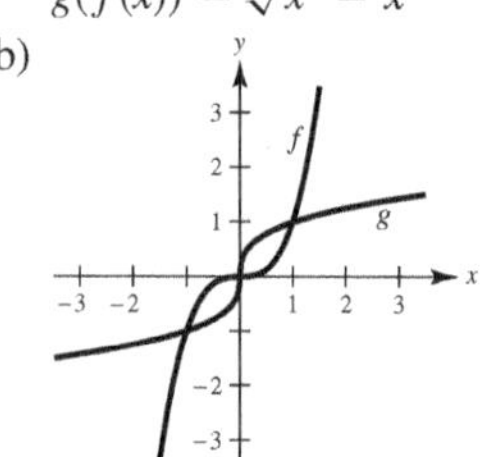

13. (a) $f(g(x)) = \sqrt{x^2 + 4 - 4} = x$

$g(f(x)) = \left(\sqrt{x-4}\right)^2 + 4 = x$

(b)

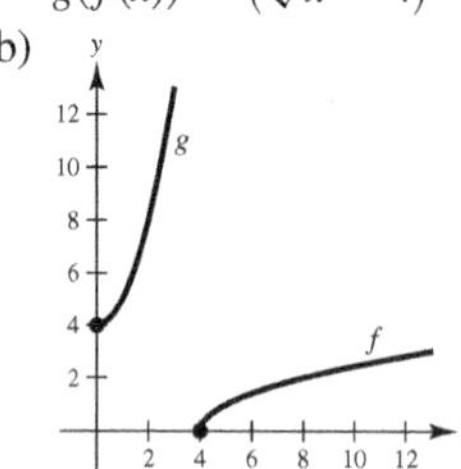

15. (a) $f(g(x)) = \dfrac{1}{1/x} = x$

$g(f(x)) = \dfrac{1}{1/x} = x$

(b)

17.

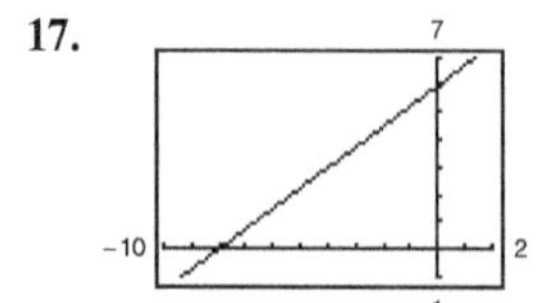

One-to-one, inverse exists

19.

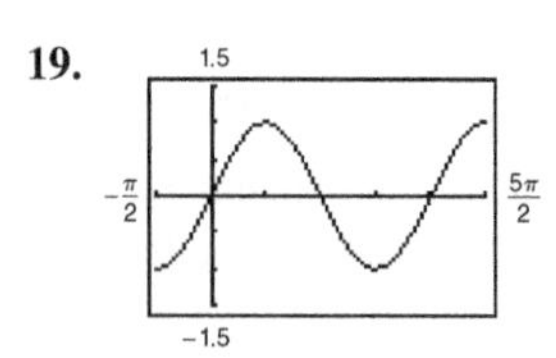

Not one-to-one, inverse does not exist

21.

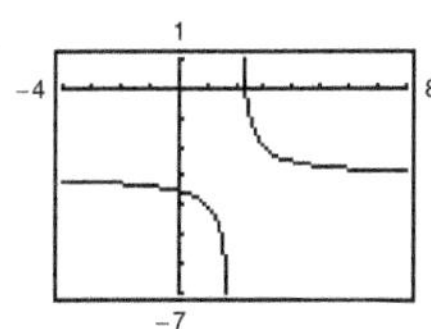

One-to-one, inverse exists

23.

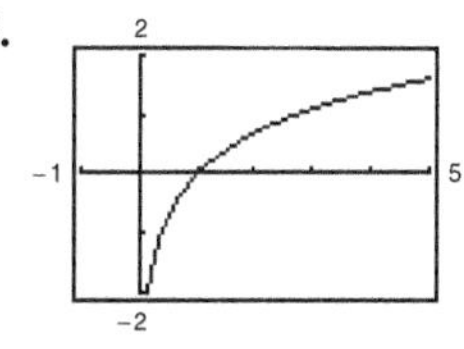

One-to-one, inverse exists

25. Strictly monotonic, inverse exists

27. Not strictly monotonic, inverse does not exist

29. Strictly monotonic, inverse exists

31. $f'(x) = 2(x - 4) > 0$ on $(4, \infty)$

33. $f'(x) = -\csc^2 x < 0$ on $(0, \pi)$

35. (a) $f^{-1}(x) = \dfrac{x + 3}{2}$

(b)

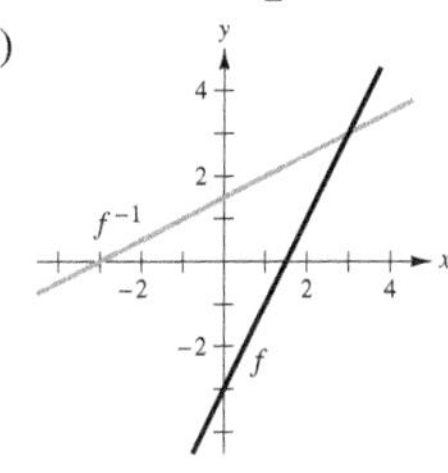

(c) f and f^{-1} are symmetric about $y = x$.

(d) Domain of f and f^{-1}: all real numbers

Range of f and f^{-1}: all real numbers

37. (a) $f^{-1}(x) = x^{1/5}$

(b)

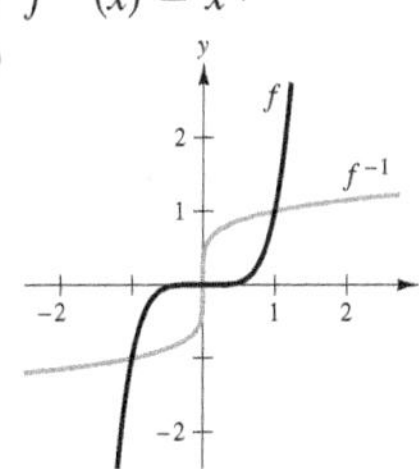

(c) f and f^{-1} are symmetric about $y = x$.

(d) Domain of f and f^{-1}: all real numbers

Range of f and f^{-1}: all real numbers

39. (a) $f^{-1}(x) = x^2, \quad x \geq 0$

(b)

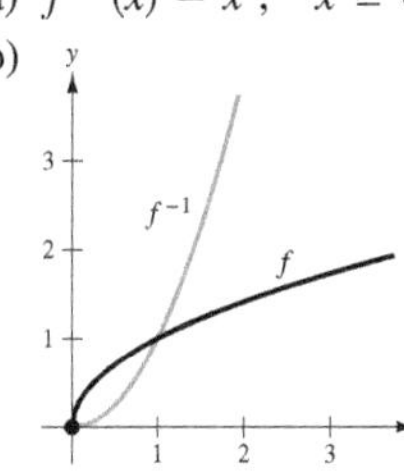

(c) f and f^{-1} are symmetric about $y = x$.

(d) Domain of f and f^{-1}: $x \geq 0$

Range of f and f^{-1}: $y \geq 0$

41. (a) $f^{-1}(x) = \sqrt{4 - x^2}, \quad 0 \leq x \leq 2$

(b)

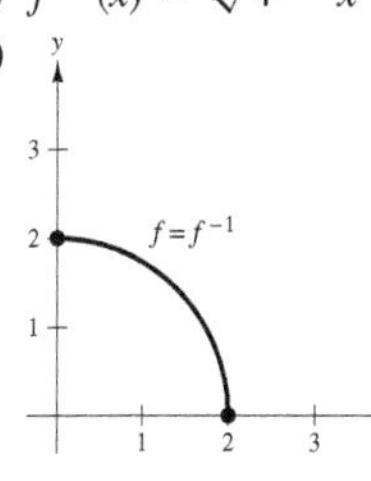

(c) f and f^{-1} are symmetric about $y = x$.

(d) Domain of f and f^{-1}: $0 \leq x \leq 2$

Range of f and f^{-1}: $0 \leq y \leq 2$

43. (a) $f^{-1}(x) = x^3 + 1$

(b)

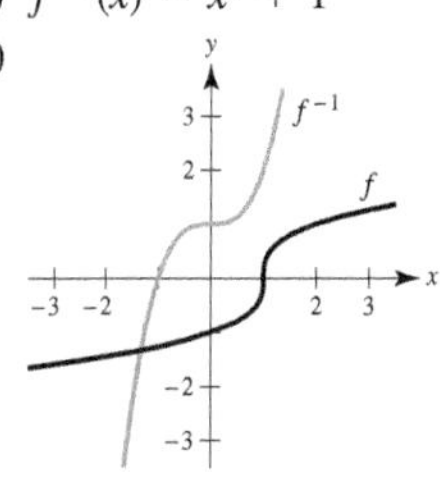

(c) f and f^{-1} are symmetric about $y = x$.

(d) Domain of f and f^{-1}: all real numbers

Range of f and f^{-1}: all real numbers

45. (a) $f^{-1}(x) = \dfrac{\sqrt{7}x}{\sqrt{1 - x^2}}, \quad -1 < x < 1$

(b)

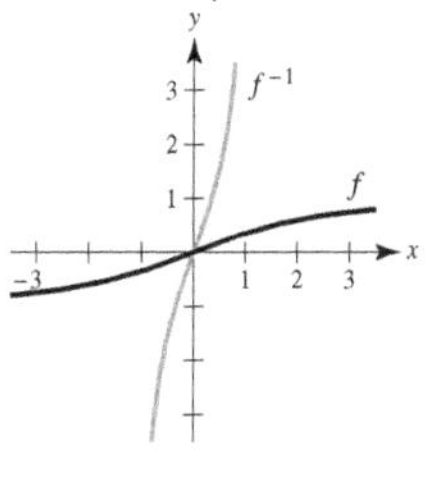

(c) f and f^{-1} are symmetric about $y = x$.

(d) Domain of f: all real numbers

Domain of f^{-1}: $-1 < x < 1$

Range of f: $-1 < y < 1$

Range of f^{-1}: all real numbers

47.

x	0	1	2	4
$f(x)$	1	2	3	4

x	1	2	3	4
$f^{-1}(x)$	0	1	2	4

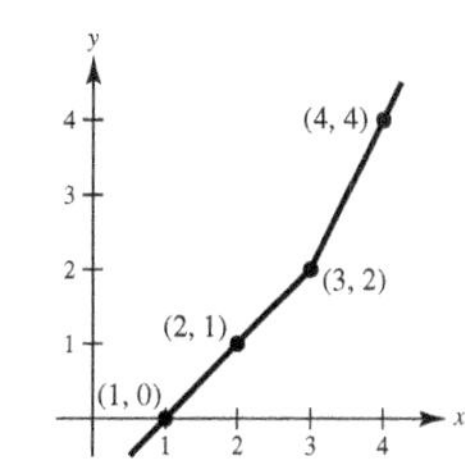

49. (a) Proof

(b) $y = \frac{2}{3}(137.5 - x)$

x: total cost

y: number of pounds of the less expensive commodity

(c) $[62.5, 137.5]$; $50(1.25) = 62.5$ gives the total cost when purchasing 50 pounds of the less expensive commodity, and $50(2.75) = 137.5$ gives the total cost when purchasing 50 pounds of the more expensive commodity.

(d) 43 lb

51. One-to-one

$f^{-1}(x) = x^2 + 2, \quad x \geq 0$

53. One-to-one

$f^{-1}(x) = 2 - x, \quad x \geq 0$

55. Sample answer: $f^{-1}(x) = \sqrt{x} + 3, \quad x \geq 0$

57. Sample answer: $f^{-1}(x) = x - 3, \quad x \geq 0$

59. Inverse exists. Volume is an increasing function and therefore is one-to-one. The inverse function gives the time t corresponding to the volume V.

61. Inverse does not exist. **63.** $-\frac{1}{6}$ **65.** $\frac{1}{17}$

67. $\dfrac{2\sqrt{3}}{3}$ **69.** -2

71. (a) Domain of f: $(-\infty, \infty)$

Domain of f^{-1}: $(-\infty, \infty)$

(b) Range of f: $(-\infty, \infty)$

Range of f^{-1}: $(-\infty, \infty)$

(c)

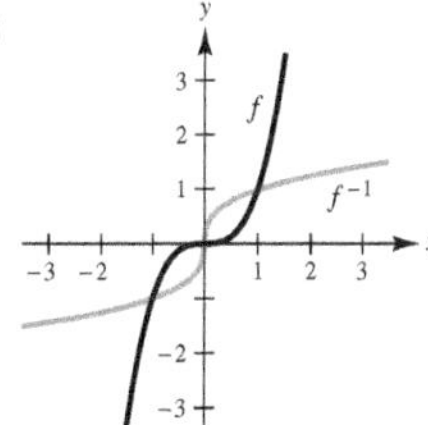

(d) $f'(\frac{1}{2}) = \frac{3}{4}, (f^{-1})'(\frac{1}{8}) = \frac{4}{3}$

73. (a) Domain of f: $[4, \infty)$

Domain of f^{-1}: $[0, \infty)$

(b) Range of f: $[0, \infty)$

Range of f^{-1}: $[4, \infty)$

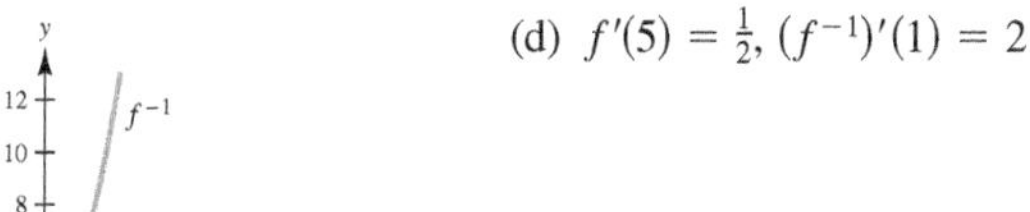

(c)

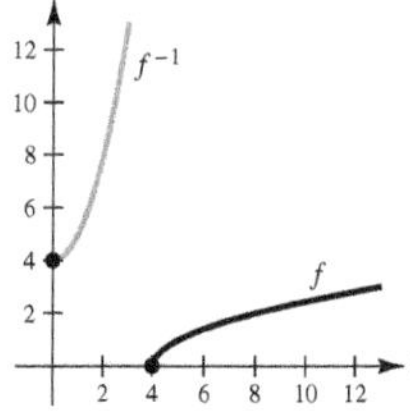

(d) $f'(5) = \frac{1}{2}, (f^{-1})'(1) = 2$

75. 32 **77.** 88 **79.** $(g^{-1} \circ f^{-1})(x) = \dfrac{x+1}{2}$

81. $(f \circ g)^{-1}(x) = \dfrac{x+1}{2}$

83. Yes. Functions of the form $f(x) = x^n$, n is odd, are always increasing or always decreasing. So, it is one-to-one and therefore has an inverse function.

85. Many x-values yield the same y-value. For example, $f(\pi) = 0 = f(0)$. The graph is not continuous at $\dfrac{(2n-1)\pi}{2}$, where n is an integer.

87. $k = \frac{1}{4}$ **89.** False. Let $f(x) = x^2$.

91. (a)

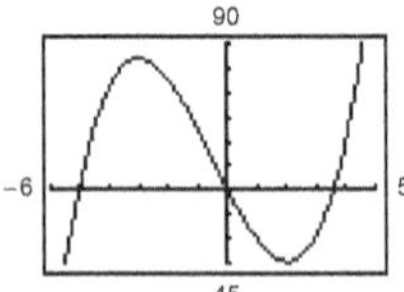

(b) $c = 2$

f does not pass the Horizontal Line Test.

93–95. Proofs **97.** Proof; $\dfrac{\sqrt{5}}{5}$

99. Proof; The graph of f is symmetric about the line $y = x$.

101. Proof; concave upward

Section 5.4 *(page 348)*

1. The graph of $f(x) = e^x$ is concave upward and increasing on the entire domain.

3. $x = 4$ **5.** $x \approx 2.485$ **7.** $x = 0$ **9.** $x \approx 0.511$

11. $x \approx 8.862$ **13.** $x \approx 7.389$ **15.** $x \approx 10.389$

17. $x \approx 5.389$

19.

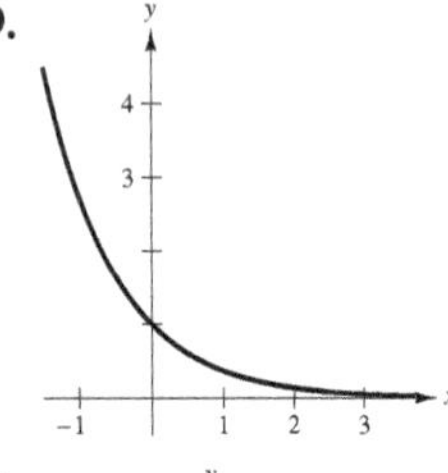

21.

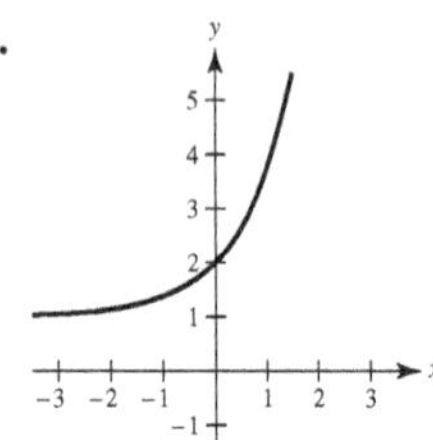

23.

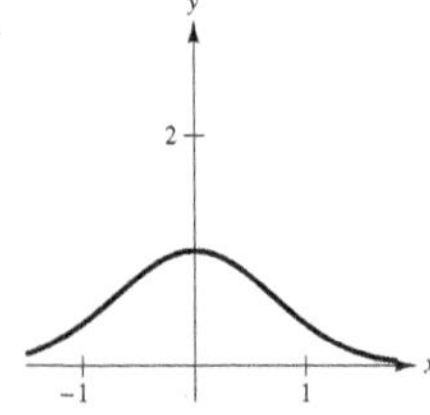

25. c **26.** d **27.** a **28.** b

29.

31.

33. $5e^{5x}$ **35.** $\dfrac{e^{\sqrt{x}}}{2\sqrt{x}}$ **37.** e^{x-4} **39.** $e^x\left(\dfrac{1}{x} + \ln x\right)$

41. $e^x(x+1)(x+3)$ **43.** $3(e^{-t} + e^t)^2(e^t - e^{-t})$

45. $-\dfrac{5e^{5x}}{2 - e^{5x}}$ **47.** $\dfrac{-2(e^x - e^{-x})}{(e^x + e^{-x})^2}$ **49.** $-\dfrac{2e^x}{(e^x - 1)^2}$

51. $2e^x \cos x$ **53.** $\dfrac{\cos x}{x}$ **55.** $y = 3x + 1$

57. $y = -3x + 10$ **59.** $y = \left(\dfrac{1}{e}\right)x - \dfrac{1}{e}$

61. $y = ex$ **63.** $\dfrac{10 - e^y}{xe^y + 3}$ **65.** $y = (-e - 1)x + 1$

67. $3(6x + 5)e^{-3x}$

69. $y'' - y = 0$

$4e^{-x} - 4e^{-x} = 0$

71. Relative minimum: $(0, 1)$

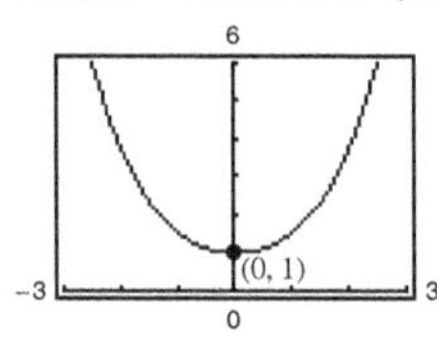

73. Relative maximum: $\left(2, \dfrac{1}{\sqrt{2\pi}}\right)$

Points of inflection:

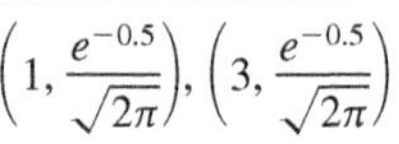

$\left(1, \dfrac{e^{-0.5}}{\sqrt{2\pi}}\right), \left(3, \dfrac{e^{-0.5}}{\sqrt{2\pi}}\right)$

75. Relative maximum: $(1, e)$

Point of inflection: $(0, 2)$

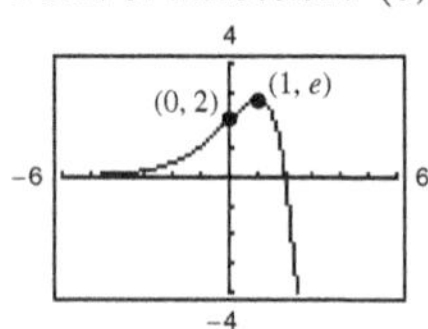

77. Relative maximum: $(-1, 1 + e)$

Point of inflection: $(0, 3)$

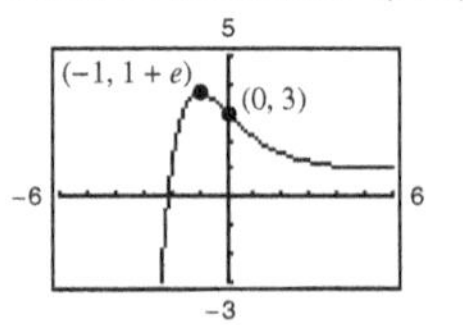

79. $A = \sqrt{2}e^{-1/2}$

81. $\left(\frac{1}{2}, e\right)$

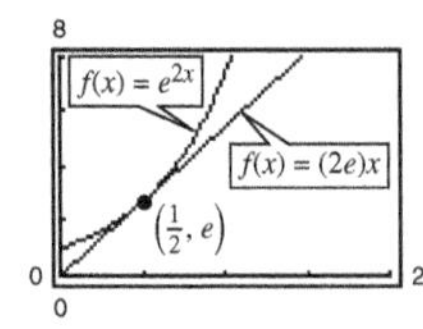

83. (a)

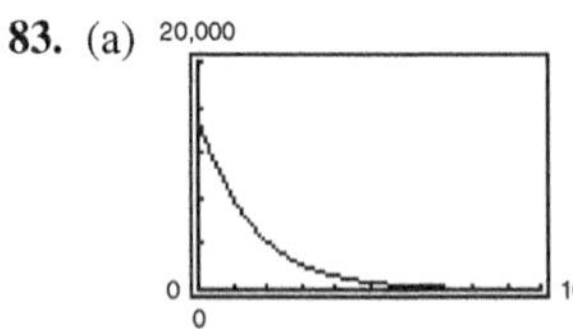

(b) When $t = 1$, $\dfrac{dV}{dt} \approx -5028.84$.

When $t = 5$, $\dfrac{dV}{dt} \approx -406.89$.

(c)

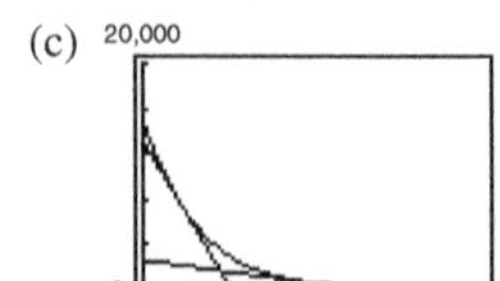

85. (a) 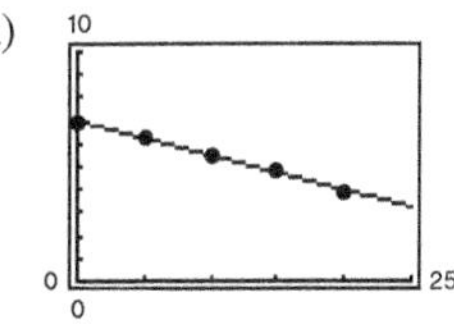

$\ln P = -0.1499h + 6.9797$

(b) $P = 1074.6e^{-0.1499h}$

(c) 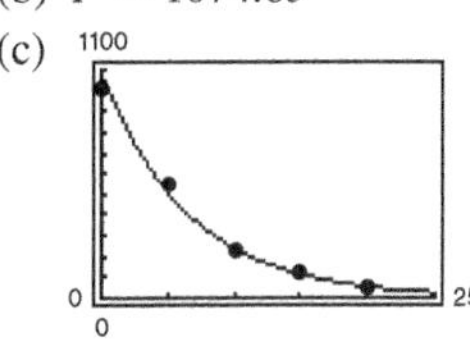

(d) $h = 5$: -76.13 millibars/km
$h = 18$: -10.84 millibars/km

87. $P_1 = 1 + x$; $P_2 = 1 + x + \frac{1}{2}x^2$

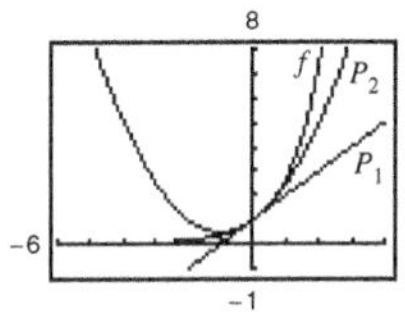

The values of f, P_1, and P_2 and their first derivatives agree at $x = 0$.

89. $12! = 479{,}001{,}600$
Stirling's Formula: $12! \approx 475{,}687{,}487$

91. $e^{5x} + C$ **93.** $\frac{1}{5}e^{5x-3} + C$ **95.** $e^{x^2+x} + C$

97. $2e^{\sqrt{x}} + C$

99. $x - \ln(e^x + 1) + C_1$ or $-\ln(1 + e^{-x}) + C_2$

101. $-\frac{2}{3}(1 - e^x)^{3/2} + C$ **103.** $\ln|e^x - e^{-x}| + C$

105. $-\frac{5}{2}e^{-2x} + e^{-x} + C$ **107.** $\ln|\cos e^{-x}| + C$

109. $\dfrac{e^2 - 1}{2e^2}$ **111.** $\dfrac{e - 1}{2e}$ **113.** $\dfrac{e}{3}(e^2 - 1)$

115. $\dfrac{1}{4}\ln\dfrac{1 + e^8}{2}$ **117.** $\dfrac{1}{\pi}[e^{\sin(\pi^2/2)} - 1]$

119. $y = \frac{1}{18}e^{9x^2} + C$ **121.** $f(x) = \frac{1}{2}(e^x + e^{-x})$

123. $e^6 - 1 \approx 402.4$ **125.** $2(1 - e^{-3/2}) \approx 1.554$

127. 92.190

129. The natural exponential function has a horizontal asymptote $y = 0$ to the left and the natural logarithmic function has a vertical asymptote $x = 0$ from the right.

131. False. The derivative is $e^x(g'(x) + g(x))$.

133. True

135. The probability that a given battery will last between 48 months and 60 months is approximately 47.72%.

137. (a) $R = 428.78e^{-0.6155t}$

(b) 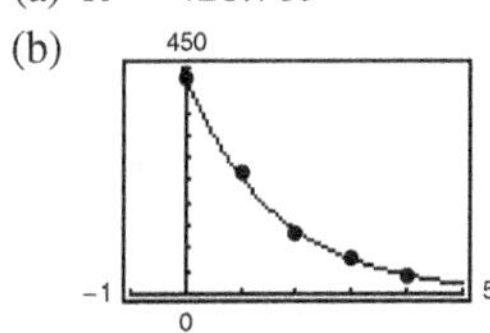

(c) About 637.2 L

139. (a) 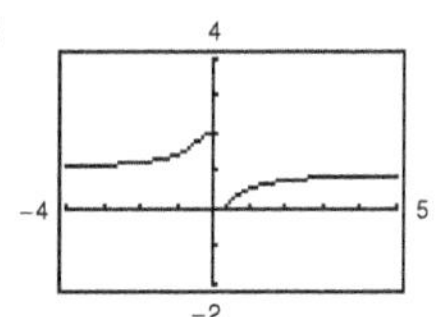

(b) When x increases without bound, $1/x$ approaches zero, and $e^{1/x}$ approaches 1. Therefore, $f(x)$ approaches $2/(1 + 1) = 1$. So, $f(x)$ has a horizontal asymptote at $y = 1$. As x approaches zero from the right, $1/x$ approaches ∞, $e^{1/x}$ approaches ∞, and $f(x)$ approaches zero. As x approaches zero from the left, $1/x$ approaches $-\infty$, $e^{1/x}$ approaches zero, and $f(x)$ approaches 2. The limit does not exist because the left limit does not equal the right limit. Therefore, $x = 0$ is a nonremovable discontinuity.

141. $\displaystyle\int_0^x e^t\,dt \ge \int_0^x 1\,dt$; $e^x - 1 \ge x$; $e^x \ge x + 1$ for $x \ge 0$

143. Relative maximum: $\left(\dfrac{1}{k}, \dfrac{1}{ke}\right)$

Point of inflection: $\left(\dfrac{2}{k}, \dfrac{2}{ke^2}\right)$

145. Putnam Problem B1, 2012

Section 5.5 *(page 358)*

1. $a = 4$, $b = 6$

3. It is necessary when you have a function of the form $y = u(x)^{v(x)}$

5. -3 **7.** 0 **9.** $\frac{5}{6}$

11. (a) $\log_2 8 = 3$ (b) $\log_3\left(\frac{1}{3}\right) = -1$

13. (a) $10^{-2} = 0.01$ (b) $\left(\frac{1}{2}\right)^{-3} = 8$

15. 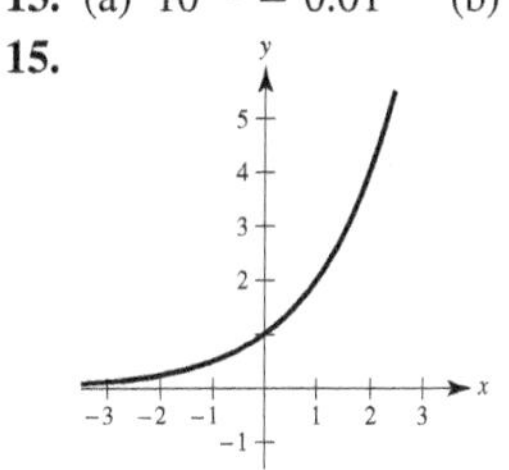

17. 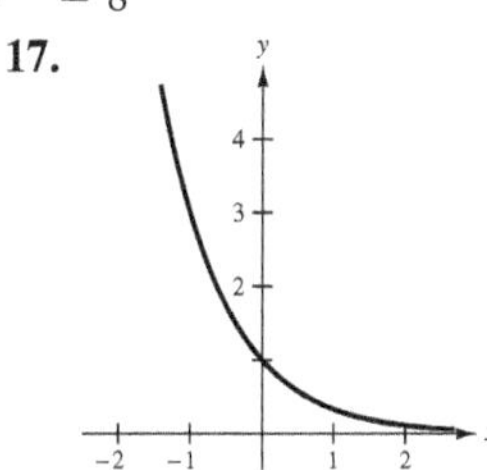

19. 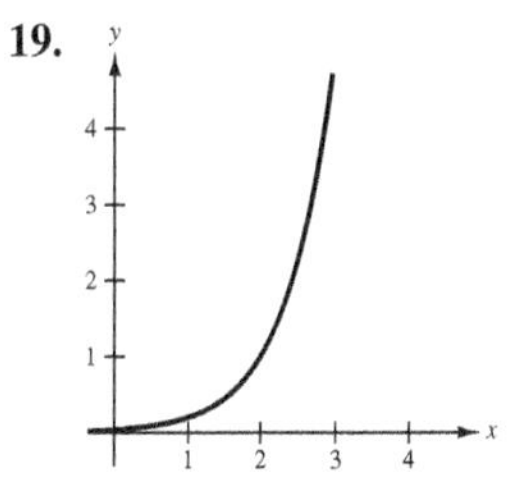

21. (a) $x = 3$ (b) $x = -1$ **23.** (a) $x = \frac{1}{3}$ (b) $x = \frac{1}{16}$

25. (a) $x = -1, 2$ (b) $x = \frac{1}{3}$ **27.** 1.965 **29.** -6.288

31. 12.253 **33.** 33.000 **35.** 3.429

37.

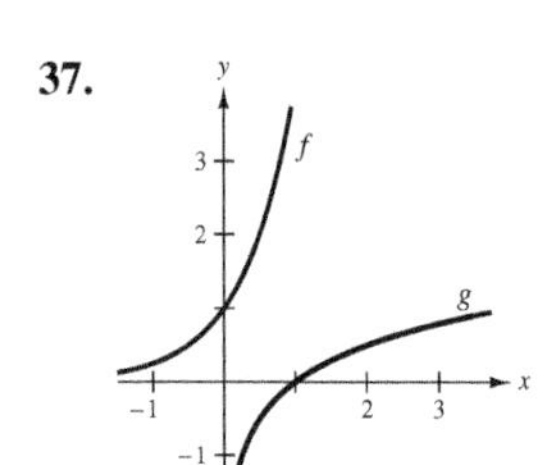

39. $(\ln 4)4^x$ **41.** $(-4\ln 5)5^{-4x}$ **43.** $9^x(x\ln 9 + 1)$

45. $\dfrac{2t^2\ln 8 - 4t}{8^t}$ **47.** $-2^{-\theta}[(\ln 2)\cos \pi\theta + \pi \sin \pi\theta]$

49. $\dfrac{6}{(\ln 4)(6x+1)}$ **51.** $\dfrac{2}{(\ln 5)(t-4)}$ **53.** $\dfrac{x}{(\ln 5)(x^2-1)}$

55. $\dfrac{x-2}{(\ln 2)x(x-1)}$ **57.** $\dfrac{3x-2}{(2x\ln 3)(x-1)}$ **59.** $\dfrac{5(1-\ln t)}{t^2\ln 2}$

61. $y = -2x\ln 2 - 2\ln 2 + 2$ **63.** $y = \dfrac{1}{27\ln 3}x + 3 - \dfrac{1}{\ln 3}$

65. $2(1-\ln x)x^{(2/x)-2}$ **67.** $(x-2)^{x+1}\left[\dfrac{x+1}{x-2} + \ln(x-2)\right]$

69. $\dfrac{3^x}{\ln 3} + C$ **71.** $\dfrac{1}{3}x^3 - \dfrac{2^{-x}}{\ln 2} + C$

73. $-\dfrac{1}{2\ln 5}(5^{-x^2}) + C$ **75.** $\dfrac{\ln(3^{2x}+1)}{2\ln 3} + C$ **77.** $\dfrac{7}{2\ln 2}$

79. $\dfrac{4}{\ln 5} - \dfrac{2}{\ln 3}$ **81.** $\dfrac{(\ln 5)^2}{2\ln 4} \approx 0.934$

83. The exponential function grows more rapidly as a becomes larger.

85. (a) $x > 0$ (b) 10^x (c) $3 \le f(x) \le 4$
(d) $0 < x < 1$ (e) 10 (f) 100^n

87. (a) \$40.64 (b) $C'(1) \approx 0.051P$, $C'(8) \approx 0.072P$
(c) $\ln 1.05$

89.

n	1	2	4	12
A	\$1410.60	\$1414.78	\$1416.91	\$1418.34

n	365	Continuous
A	\$1419.04	\$1419.07

91.

n	1	2	4	12
A	\$30,612.57	\$31,121.37	\$31,385.05	\$31,564.42

n	365	Continuous
A	\$31,652.22	\$31.655.22

93.

t	1	10	20	30
P	\$96,078.94	\$67,032.00	\$44,932.90	\$30,119.42

t	40	50
P	\$20,189.65	\$13,533.53

95.

t	1	10	20	30
P	\$95,132.82	\$60,716.10	\$36,864.45	\$22,382.66

t	40	50
P	\$13,589.88	\$8251.24

97. c

99. (a) 6.7 million ft³/acre
(b) $t = 20$: $\dfrac{dV}{dt} = 0.073$, $t = 60$: $\dfrac{dV}{dt} = 0.040$

101. (a) 12,000 0 0 40
(b) 6 months: 1487 fish
12 months: 3672 fish
24 months: 8648 fish
36 months: 9860 fish
48 months: 9987 fish
Limiting size: 10,000 fish
(c) 1 month: About 114 fish/mo
10 months: About 403 fish/mo
(d) About 15 mo

103. (a) $y_1 = -27.7x + 565$, $y_2 = 843 - 246.3\ln x$,
$y_3 = 706.995(0.9106)^x$, $y_4 = 1765.4563x^{-0.8200}$
(b)

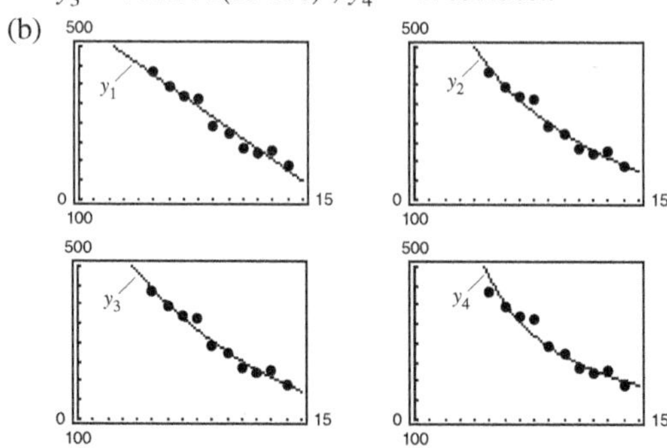

Answers will vary.
(c) $y_1'(12) = -27.7$, $y_2'(12) \approx -20.53$, $y_3'(12) \approx -21.52$,
$y_4'(12) \approx -15.72$; y_1 is decreasing at the greatest rate.

105. $y = 1200(0.6^t)$ **107.** e **109.** e^2

111. (a) $(2^3)^2 = 2^6 = 64$
$2^{(3^2)} = 2^9 = 512$
(b) No. $f(x) = (x^x)^x = x^{(x^2)}$ and $g(x) = x^{(x^x)}$
(c) $f'(x) = x^{x^2}(x + 2x\ln x)$
$g'(x) = x^{x^x + x - 1}[x(\ln x)^2 + x\ln x + 1]$

113. Proof

115. (a) $\dfrac{dy}{dx} = \dfrac{y^2 - yx\ln y}{x^2 - xy\ln x}$
(b) (i) 1 when $c \ne 0$, $c \ne e$ (ii) -3.1774
(iii) -0.3147
(c) (e, e)

117. Putnam Problem B3, 1951

Section 5.7 (page 379)

1. arccos x is the angle, $0 \le \theta \le \pi$, whose cosine is x.

3. arccot **5.** $\left(-\frac{\sqrt{2}}{2}, \frac{3\pi}{4}\right), \left(\frac{1}{2}, \frac{\pi}{3}\right), \left(\frac{\sqrt{3}}{2}, \frac{\pi}{6}\right)$

7. $\frac{\pi}{6}$ **9.** $\frac{\pi}{3}$ **11.** $\frac{\pi}{6}$ **13.** $-\frac{\pi}{4}$ **15.** 1.52

17. $\operatorname{arccos}\frac{1}{1.269} \approx 0.66$ **19.** x **21.** $\frac{\sqrt{1-x^2}}{x}$

23. $\frac{1}{x}$ **25.** (a) $\frac{3}{5}$ (b) $\frac{5}{3}$ **27.** (a) $-\sqrt{3}$ (b) $-\frac{13}{5}$

29. $\sqrt{1-4x^2}$ **31.** $\frac{\sqrt{x^2-1}}{|x|}$ **33.** $\frac{\sqrt{x^2-9}}{3}$

35. $\frac{\sqrt{x^2+2}}{x}$ **37.** $x = \frac{1}{3}\left(\sin\frac{1}{2} + \pi\right) \approx 1.207$ **39.** $x = \frac{1}{3}$

41. $\frac{1}{\sqrt{2x-x^2}}$ **43.** $-\frac{3}{\sqrt{4-x^2}}$ **45.** $\frac{e^x}{1+e^{2x}}$

47. $\frac{3x - \sqrt{1-9x^2}\arcsin 3x}{x^2\sqrt{1-9x^2}}$ **49.** $-\frac{t}{\sqrt{1-t^2}}$

51. $2\arccos x$ **53.** $\frac{1}{1-x^4}$ **55.** $\frac{x^2}{\sqrt{16-x^2}}$

57. $y = \frac{1}{3}\left(4\sqrt{3}x - 2\sqrt{3} + \pi\right)$ **59.** $y = \frac{1}{4}x + \frac{\pi-2}{4}$

61. $y = (2\pi - 4)x + 4$

63. Relative maximum: $(1.272, -0.606)$
Relative minimum: $(-1.272, 3.747)$

65. Relative maximum: $(2, 2.214)$

67. **69.**

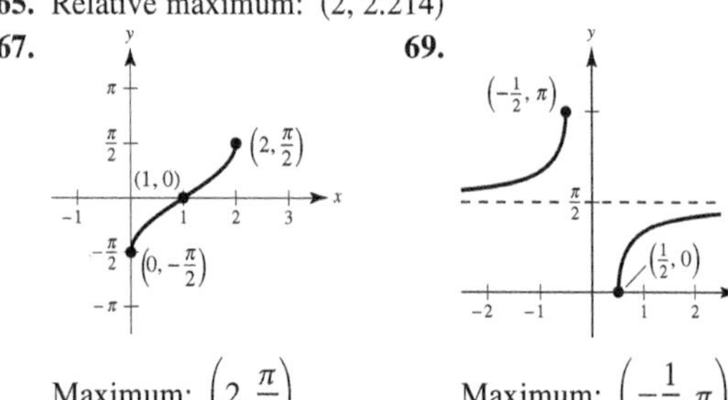

67. Maximum: $\left(2, \frac{\pi}{2}\right)$
Minimum: $\left(0, -\frac{\pi}{2}\right)$
Point of inflection: $(1, 0)$

69. Maximum: $\left(-\frac{1}{2}, \pi\right)$
Minimum: $\left(\frac{1}{2}, 0\right)$
Asymptote: $y = \frac{\pi}{2}$

71. $y = -\frac{2\pi x}{\pi + 8} + 1 - \frac{\pi^2}{2\pi + 16}$ **73.** $y = -x + \sqrt{2}$

75. (a) $\arcsin(\arcsin 0.5) \approx 0.551$
$\arcsin(\arcsin 1)$ does not exist.
(b) $\sin(-1) \le x \le \sin 1$

77. No

79. In order to have a true inverse function, the domain of sine must be restricted. As a result, 2π is not in the range of the arcsine function.

81. (a) and (b) Proofs **83.** True **85.** True

87. (a) $\theta = \operatorname{arccot}\frac{x}{5}$
(b) $x = 10$: 16 rad/h
$x = 3$: 58.824 rad/h

89. (a) $h(t) = -16t^2 + 256$; $t = 4$ sec
(b) $t = 1$: -0.0520 rad/sec
$t = 2$: -0.1116 rad/sec

91. $50\sqrt{2} \approx 70.71$ ft **93.** (a) and (b) Proofs

95.

(a) The graph is a horizontal line at $\frac{\pi}{2}$.
(b) Proof

97. $c = 2$ **99.** Proof

Section 5.8 (page 387)

1. (a) No
(b) Yes. Use the rule involving the arcsecant function.

3. $\arcsin\frac{x}{3} + C$ **5.** $\operatorname{arcsec}|2x| + C$

7. $\arcsin(x+1) + C$ **9.** $\frac{1}{2}\arcsin t^2 + C$

11. $\frac{1}{10}\arctan\frac{t^2}{5} + C$ **13.** $\frac{1}{4}\arctan\frac{e^{2x}}{2} + C$

15. $\arcsin\frac{\csc x}{5} + C$ **17.** $2\arcsin\sqrt{x} + C$

19. $\frac{1}{2}\ln(x^2+1) - 3\arctan x + C$

21. $8\arcsin\frac{x-3}{3} - \sqrt{6x-x^2} + C$ **23.** $\frac{\pi}{6}$ **25.** $\frac{\pi}{6}$

27. $\frac{1}{3}\left(\arctan 3 - \frac{\pi}{4}\right) \approx 0.155$ **29.** $\arctan 5 - \frac{\pi}{4} \approx 0.588$

31. $\frac{\pi}{4}$ **33.** $\frac{1}{32}\pi^2 \approx 0.308$ **35.** $\frac{\pi}{2}$

37. $\frac{\sqrt{2}}{2}\arcsin\left[\frac{\sqrt{6}}{6}(x-2)\right] + C$ **39.** $\arcsin\frac{x+2}{2} + C$

41. $4 - 2\sqrt{3} + \frac{1}{6}\pi \approx 1.059$

43. $2\sqrt{e^t - 3} - 2\sqrt{3}\arctan\frac{\sqrt{e^t-3}}{\sqrt{3}} + C$ **45.** $\frac{\pi}{6}$

47. (a) $\arcsin x + C$ (b) $-\sqrt{1-x^2} + C$ (c) Not possible

49. (a) $\frac{2}{3}(x-1)^{3/2} + C$ (b) $\frac{2}{15}(x-1)^{3/2}(3x+2) + C$
(c) $\frac{2}{3}\sqrt{x-1}(x+2) + C$

51. Proof

53. No. Graphing $f(x) = \arcsin x$ and $g(x) = -\arccos x$, you can see that the graph of f is the graph of g shifted vertically.

55. (a) 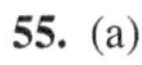

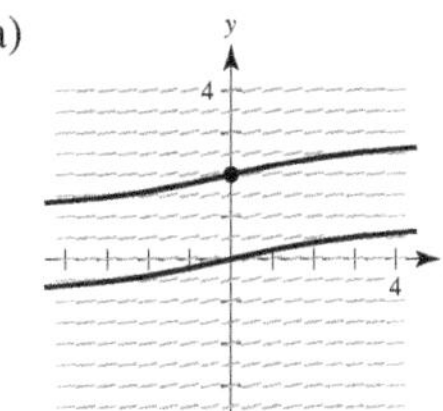

(b) $y = \frac{2}{3}\arctan\frac{x}{3} + 2$

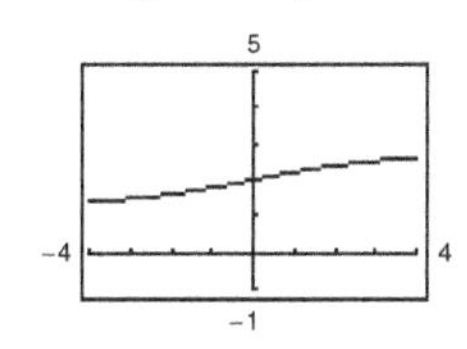

57.

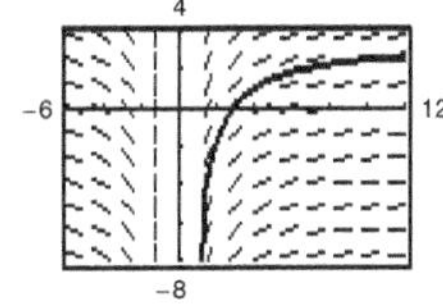

59.

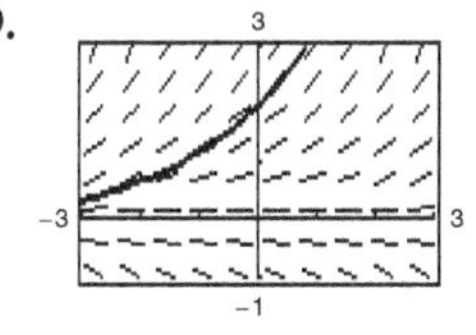

61. $y = \arcsin\frac{x}{2} + \pi$ **63.** $\frac{\pi}{3}$ **65.** $\frac{3\pi}{2}$

67. (a)

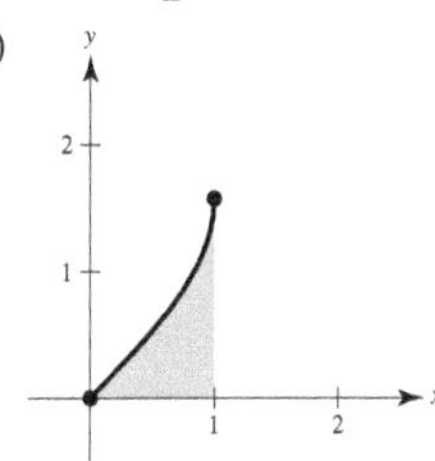

(b) 0.5708 (c) $\frac{\pi - 2}{2}$

69. (a) $F(x)$ represents the average value of $f(x)$ over the interval $[x, x + 2]$; Maximum at $x = -1$

(b) Maximum at $x = -1$

71. False. $\int \frac{dx}{3x\sqrt{9x^2 - 16}} = \frac{1}{12}\operatorname{arcsec}\frac{|3x|}{4} + C$

73–75. Proofs

77. (a) $\int_0^1 \frac{1}{1 + x^2}\,dx$ (b) About 0.7857

(c) Because $\int_0^1 \frac{1}{1 + x^2}\,dx = \frac{\pi}{4}$, you can use the Midpoint Rule to approximate $\frac{\pi}{4}$. Multiplying the result by 4 gives an estimation of π.

Section 5.9 *(page 397)*

1. Hyperbolic function came from the comparison of the area of a semicircular region with the area of a region under a hyperbola.

3. $\sinh^2 x = \frac{-1 + \cosh 2x}{2}$ **5.** (a) 10.018 (b) -0.964

7. (a) $\frac{4}{3}$ (b) $\frac{13}{12}$ **9.** (a) 1.317 (b) 0.962

11–17. Proofs

19. $\cosh x = \frac{\sqrt{13}}{2}$, $\tanh x = \frac{3\sqrt{13}}{13}$, $\operatorname{csch} x = \frac{2}{3}$, $\operatorname{sech} x = \frac{2\sqrt{13}}{13}$, $\coth x = \frac{\sqrt{13}}{3}$

21. ∞ **23.** 1 **25.** $9\cosh 9x$

27. $-10x(\operatorname{sech} 5x^2 \tanh 5x^2)$ **29.** $\coth x$

31. $-\frac{t}{2}\cosh(-3t) + \frac{\sinh(-3t)}{6}$ **33.** $\operatorname{sech} t$

35. $y = -2x + 2$ **37.** $y = 1 - 2x$

39. Relative maximum: $(1.20, 0.66)$
Relative minimum: $(-1.20, -0.66)$

41. Relative maxima: $(\pm\pi, \cosh\pi)$, Relative minimum: $(0, -1)$

43. (a) (b) 33.146 units, 25 units (c) $m = \sinh 1 \approx 1.175$

45. $\frac{1}{4}\sinh 4x + C$ **47.** $-\frac{1}{2}\cosh(1 - 2x) + C$

49. $\frac{1}{3}\cosh^3(x - 1) + C$ **51.** $\ln|\sinh x| + C$

53. $-\coth\frac{x^2}{2} + C$ **55.** $\ln\frac{5}{4}$ **57.** $\coth 1 - \coth 2$

59. $-\frac{1}{3}(\operatorname{csch} 2 - \operatorname{csch} 1)$

61.

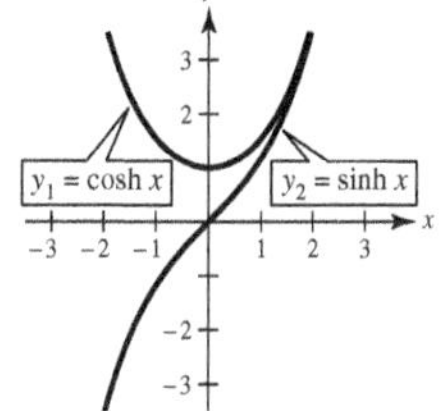

The graphs do not intersect.

63. Proof **65.** $\frac{3}{\sqrt{9x^2 - 1}}$ **67.** $\frac{1}{2\sqrt{x}(1 - x)}$

69. $|\sec x|$ **71.** $-\csc x$ **73.** $2\sinh^{-1}(2x)$

75. $\frac{\sqrt{3}}{18}\ln\left|\frac{1 + \sqrt{3}x}{1 - \sqrt{3}x}\right| + C$ **77.** $\ln\left(\sqrt{e^{2x} + 1} - 1\right) - x + C$

79. $2\sinh^{-1}\sqrt{x} + C = 2\ln\left(\sqrt{x} + \sqrt{1 + x}\right) + C$

81. $\frac{1}{4}\ln\left|\frac{x - 4}{4}\right| + C$ **83.** $\ln\left(\frac{3 + \sqrt{5}}{2}\right)$ **85.** $\frac{\ln 7}{12}$

87. $-\frac{x^2}{2} - 4x - \frac{10}{3}\ln\left|\frac{x - 5}{x + 1}\right| + C$

89. $8\arctan e^2 - 2\pi \approx 5.207$ **91.** $\frac{5}{2}\ln\left(\sqrt{17} + 4\right) \approx 5.237$

93. (a) $-\frac{\sqrt{a^2 - x^2}}{x}$ (b) Proof

95–103. Proofs **105.** Putnam Problem 8, 1939

Review Exercises for Chapter 5 *(page 400)*

1.

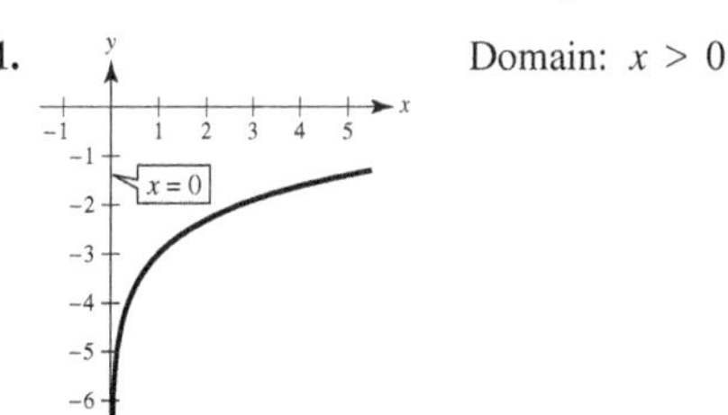

Domain: $x > 0$

3. (a) 2.9957 (b) -0.2231 (c) 6.4376 (d) 0.8047

5. $\frac{1}{5}[\ln(2x + 1) + \ln(2x - 1) - \ln(4x^2 + 1)]$

7. $\ln\frac{3\sqrt[3]{4 - x^2}}{x}$ **9.** $\frac{1}{2x}$ **11.** $\frac{1 + 2\ln x}{2\sqrt{\ln x}}$

13. $-\frac{8x}{x^4 - 16}$ **15.** $\frac{7}{(1 - 7x)[\ln(1 - 7x)]^2}$

17. $y = -x + 1$ **19.** $\dfrac{5x^2 - 4x}{2\sqrt{x-1}}$ **21.** $\frac{1}{7}\ln|7x - 2| + C$

23. $-\ln|1 + \cos x| + C$ **25.** $x - 3\ln(x^2 + 1) + C$

27. $3 + \ln 2$ **29.** $\ln(2 + \sqrt{3})$ **31.** $2 \ln \frac{123}{25} \approx 3.187$

33. (a) $f^{-1}(x) = 2x + 6$

(b)

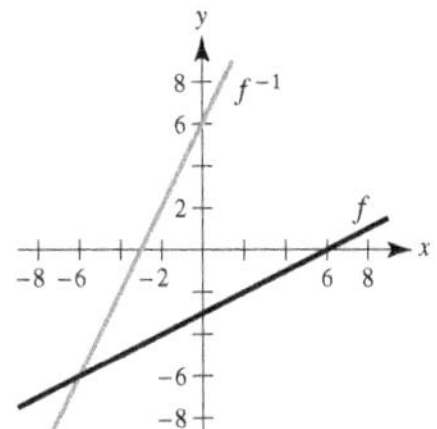

(c) Proof

(d) Domain of f and f^{-1}: all real numbers
Range of f and f^{-1}: all real numbers

35. (a) $f^{-1}(x) = x^2 - 1, \quad x \geq 0$

(b)

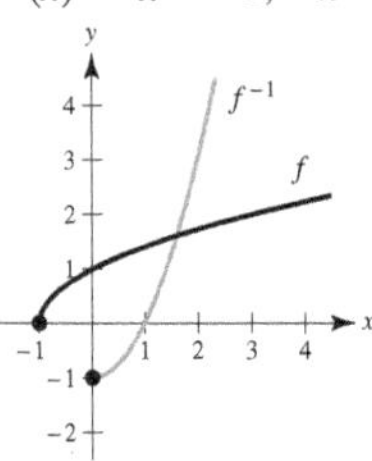

(c) Proof

(d) Domain of f: $x \geq -1$, Domain of f^{-1}: $x \geq 0$
Range of f: $y \geq 0$, Range of f^{-1}: $y \geq -1$

37. (a) $f^{-1}(x) = x^3 - 1$

(b)

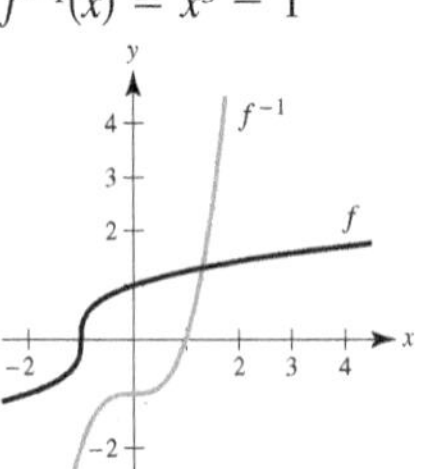

(c) Proof

(d) Domain of f and f^{-1}: all real numbers
Range of f and f^{-1}: all real numbers

39. $\dfrac{1}{3(\sqrt[3]{-3})^2} \approx 0.160$ **41.** $\dfrac{3}{4}$ **43.** $x \approx 1.134$

45. $e^4 - 1 \approx 53.598$ **47.** $te^t(t + 2)$ **49.** $\dfrac{e^{2x} - e^{-2x}}{\sqrt{e^{2x} + e^{-2x}}}$

51. $\dfrac{3x^2 - 2x^3}{e^{2x}}$ **53.** $y = 6x + 1$

55. Relative maximum: $(0, 1)$

Point of inflection: $\left(1, \dfrac{2}{e}\right)$

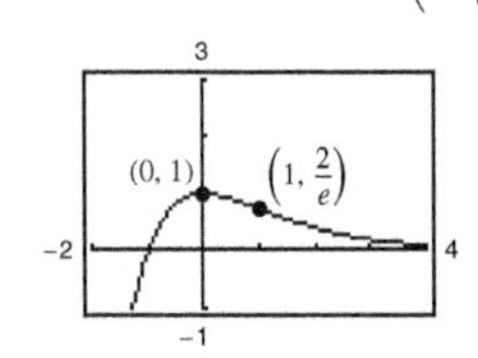

57. $-\dfrac{1}{2}e^{1-x^2} + C$ **59.** $\dfrac{e^{4x} - 3e^{2x} - 3}{3e^x} + C$

61. $\dfrac{1 - e^{-3}}{6}$ **63.** $\ln(e^2 + e + 1)$ **65.** About 1.729

67.

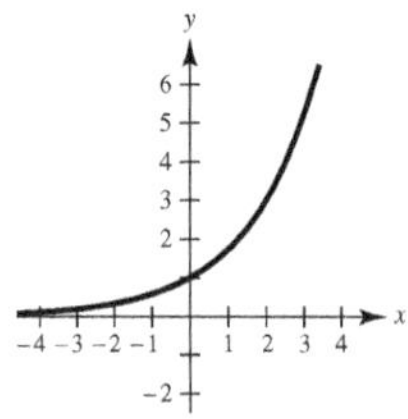

69. $1 - \dfrac{\ln 52}{\ln 4} \approx -1.850$ **71.** $\dfrac{\ln 3}{12 \ln 1.0025} \approx 36.666$

73. 35 **75.** $3^{x-1} \ln 3$ **77.** $\dfrac{8^t(t \ln 8 - 2)}{t^3}$

79. $-\dfrac{1}{(2 - 2x)\ln 3}$ **81.** $x^{2x+1}\left(2 \ln x + 2 + \dfrac{1}{x}\right)$

83. $\dfrac{5^{(x+1)^2}}{2 \ln 5} + C$ **85.** $\dfrac{30}{\ln 6}$

87. (a) \$613.92 (b) \$4723.67 (c) 6.93%

89. 0 **91.** ∞ **93.** 1 **95.** $1000e^{0.09} \approx 1094.17$

97. (a) $\dfrac{1}{2}$ (b) $\dfrac{\sqrt{3}}{2}$ **99.** $-\dfrac{2}{x\sqrt{4x^4 - 1}}$

101. $\dfrac{x}{|x|\sqrt{x^2 - 1}} + \operatorname{arcsec} x$ **103.** $(\arcsin x)^2$

105. $\frac{1}{2} \arctan e^{2x} + C$ **107.** $\frac{1}{2} \arcsin x^2 + C$

109. $\dfrac{1}{4}\left(\arctan \dfrac{x}{2}\right)^2 + C$ **111.** $\dfrac{\pi}{14}$

113. $\arctan \dfrac{e^4}{5} - \arctan \dfrac{e^{-2}}{5}$ **115.** $\dfrac{2}{3}\pi + \sqrt{3} - 2 \approx 1.826$

117. Proof **119.** $y' = -4 \operatorname{sech}(4x - 1) \tanh(4x - 1)$

121. $y' = -16x \operatorname{csch}^2(8x^2)$ **123.** $y' = \dfrac{4}{\sqrt{16x^2 + 1}}$

125. $\frac{1}{3} \tanh x^3 + C$ **127.** $\ln|\tanh x| + C$

129. $\dfrac{1}{12} \ln\left|\dfrac{3 + 2x}{3 - 2x}\right| + C$ **131.** $-\dfrac{1}{2} \operatorname{sech} 4 + \dfrac{1}{2} \operatorname{sech} 2$

133. $\ln 2$

P.S. Problem Solving *(page 403)*

1. $a = 1, b = \frac{1}{2}, c = -\frac{1}{2}$

$f(x) = \dfrac{1 + x/2}{1 - x/2}$

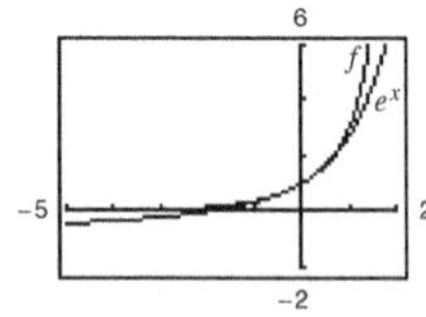

3. $\ln 3$

5. (a) ∞ (b) 0 (c) $-\frac{2}{3}$

The form $0 \cdot \infty$ is indeterminant.

7.

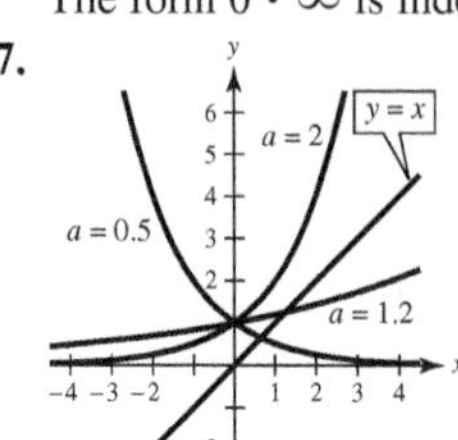

$y = 0.5^x$ and $y = 1.2^x$ intersect the line $y = x$; $0 < a < e^{1/e}$

9. (a) Area of region $A = \dfrac{\sqrt{3} - \sqrt{2}}{2} \approx 0.1589$

Area of region $B = \dfrac{\pi}{12} \approx 0.2618$

(b) $\frac{1}{24}\left[3\pi\sqrt{2} - 12\left(\sqrt{3} - \sqrt{2}\right) - 2\pi\right] \approx 0.1346$

(c) 1.2958 (d) 0.6818

11–13. Proof **15.** $2 \ln \frac{3}{2} \approx 0.8109$

17. (a) (i)

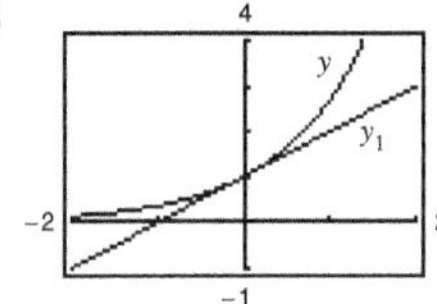

(ii)

(iii)

(b) Pattern: $y_n = 1 + \dfrac{x}{1!} + \dfrac{x^2}{2!} + \cdots + \dfrac{x^n}{n!} + \cdots$

$y_4 = 1 + \dfrac{x}{1!} + \dfrac{x^2}{2!} + \dfrac{x^3}{3!} + \dfrac{x^4}{4!}$

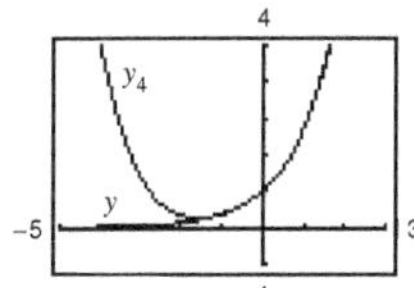

(c) The pattern implies that $e^x = 1 + \dfrac{x}{1!} + \dfrac{x^2}{2!} + \dfrac{x^3}{3} + \cdots$.

Index

C

D

G

J

K

L

M

N

Q

T

Index of Applications (continued from front inside cover)

Business and Economics

Social and Behavioral Sciences

Life Sciences

General

Factors and Zeros of Polynomials

Let $p(x) = a_nx^n + a_{n-1}x^{n-1} + \cdots + a_1x + a_0$ be a polynomial. If $p(a) = 0$, then a is a *zero* of the polynomial and a solution of the equation $p(x) = 0$. Furthermore, $(x - a)$ is a *factor* of the polynomial.

Fundamental Theorem of Algebra

An nth degree polynomial has n (not necessarily distinct) zeros. Although all of these zeros may be imaginary, a real polynomial of odd degree must have at least one real zero.

Quadratic Formula

If $p(x) = ax^2 + bx + c$, and $0 \le b^2 - 4ac$, then the real zeros of p are $x = \left(-b \pm \sqrt{b^2 - 4ac}\right)/2a$.

Special Factors

$x^2 - a^2 = (x - a)(x + a)$

$x^3 - a^3 = (x - a)(x^2 + ax + a^2)$

$x^3 + a^3 = (x + a)(x^2 - ax + a^2)$

$x^4 - a^4 = (x - a)(x + a)(x^2 + a^2)$

Binomial Theorem

$(x + y)^2 = x^2 + 2xy + y^2$

$(x - y)^2 = x^2 - 2xy + y^2$

$(x + y)^3 = x^3 + 3x^2y + 3xy^2 + y^3$

$(x - y)^3 = x^3 - 3x^2y + 3xy^2 - y^3$

$(x + y)^4 = x^4 + 4x^3y + 6x^2y^2 + 4xy^3 + y^4$

$(x - y)^4 = x^4 - 4x^3y + 6x^2y^2 - 4xy^3 + y^4$

$$(x + y)^n = x^n + nx^{n-1}y + \frac{n(n-1)}{2!}x^{n-2}y^2 + \cdots + nxy^{n-1} + y^n$$

$$(x - y)^n = x^n - nx^{n-1}y + \frac{n(n-1)}{2!}x^{n-2}y^2 - \cdots \pm nxy^{n-1} \mp y^n$$

Rational Zero Theorem

If $p(x) = a_nx^n + a_{n-1}x^{n-1} + \cdots + a_1x + a_0$ has integer coefficients, then every *rational zero* of p is of the form $x = r/s$, where r is a factor of a_0 and s is a factor of a_n.

Factoring by Grouping

$acx^3 + adx^2 + bcx + bd = ax^2(cx + d) + b(cx + d) = (ax^2 + b)(cx + d)$

Arithmetic Operations

$ab + ac = a(b + c)$

$$\frac{a}{b} + \frac{c}{d} = \frac{ad + bc}{bd}$$

$$\frac{a + b}{c} = \frac{a}{c} + \frac{b}{c}$$

$$\frac{\left(\frac{a}{b}\right)}{\left(\frac{c}{d}\right)} = \left(\frac{a}{b}\right)\left(\frac{d}{c}\right) = \frac{ad}{bc}$$

$$\frac{\left(\frac{a}{b}\right)}{c} = \frac{a}{bc}$$

$$\frac{a}{\left(\frac{b}{c}\right)} = \frac{ac}{b}$$

$$a\left(\frac{b}{c}\right) = \frac{ab}{c}$$

$$\frac{a - b}{c - d} = \frac{b - a}{d - c}$$

$$\frac{ab + ac}{a} = b + c$$

Exponents and Radicals

$a^0 = 1, \quad a \ne 0$

$(ab)^x = a^xb^x$

$a^xa^y = a^{x+y}$

$\sqrt{a} = a^{1/2}$

$\frac{a^x}{a^y} = a^{x-y}$

$\sqrt[n]{a} = a^{1/n}$

$\left(\frac{a}{b}\right)^x = \frac{a^x}{b^x}$

$\sqrt[n]{a^m} = a^{m/n}$

$a^{-x} = \frac{1}{a^x}$

$\sqrt[n]{ab} = \sqrt[n]{a}\sqrt[n]{b}$

$(a^x)^y = a^{xy}$

$\sqrt[n]{\frac{a}{b}} = \frac{\sqrt[n]{a}}{\sqrt[n]{b}}$

FORMULAS FROM GEOMETRY

Triangle

$h = a \sin \theta$

Area $= \frac{1}{2}bh$

(Law of Cosines)

$c^2 = a^2 + b^2 - 2ab \cos \theta$

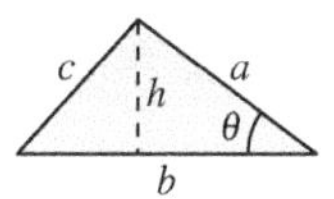

Right Triangle

(Pythagorean Theorem)

$c^2 = a^2 + b^2$

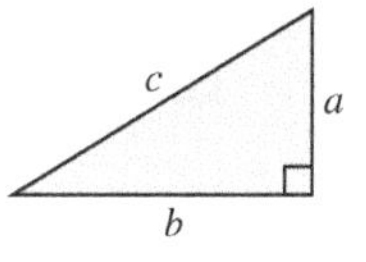

Equilateral Triangle

$h = \frac{\sqrt{3}s}{2}$

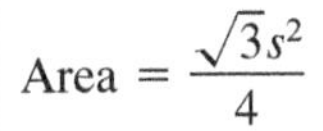

Area $= \frac{\sqrt{3}s^2}{4}$

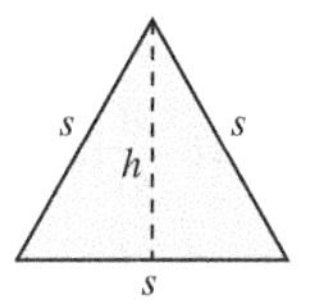

Parallelogram

Area $= bh$

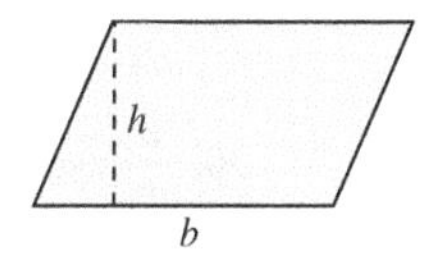

Trapezoid

Area $= \frac{h}{2}(a + b)$

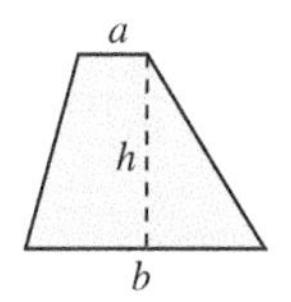

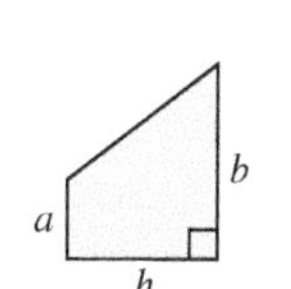

Circle

Area $= \pi r^2$

Circumference $= 2\pi r$

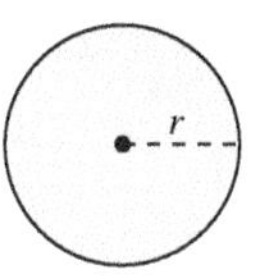

Sector of Circle

(θ in radians)

Area $= \frac{\theta r^2}{2}$

$s = r\theta$

Circular Ring

(p = average radius,

w = width of ring)

Area $= \pi(R^2 - r^2)$

$= 2\pi pw$

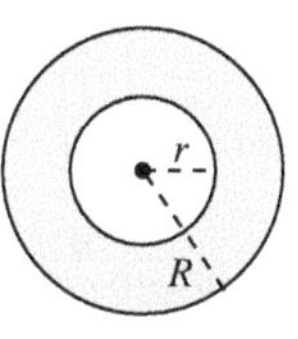

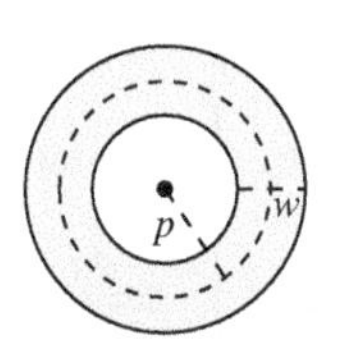

Sector of Circular Ring

(p = average radius,

w = width of ring,

θ in radians)

Area $= \theta pw$

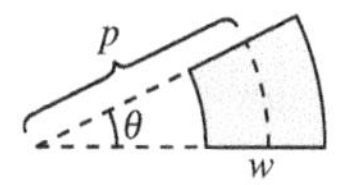

Ellipse

Area $= \pi ab$

Circumference $\approx 2\pi\sqrt{\frac{a^2 + b^2}{2}}$

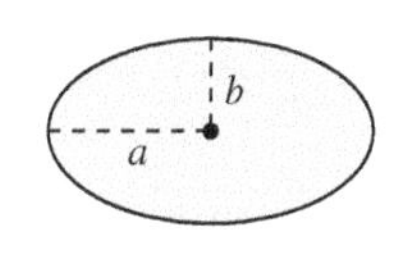

Cone

(A = area of base)

Volume $= \frac{Ah}{3}$

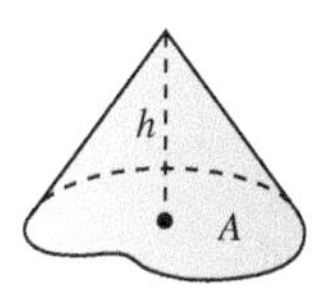

Right Circular Cone

Volume $= \frac{\pi r^2 h}{3}$

Lateral Surface Area $= \pi r\sqrt{r^2 + h^2}$

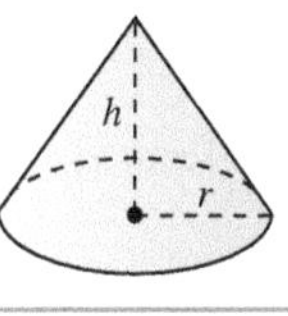

Frustum of Right Circular Cone

Volume $= \frac{\pi(r^2 + rR + R^2)h}{3}$

Lateral Surface Area $= \pi s(R + r)$

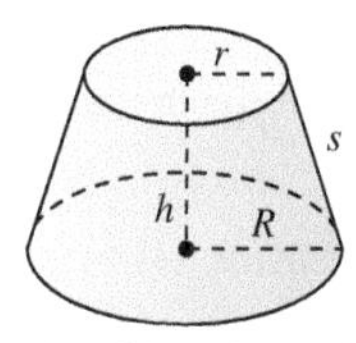

Right Circular Cylinder

Volume $= \pi r^2 h$

Lateral Surface Area $= 2\pi rh$

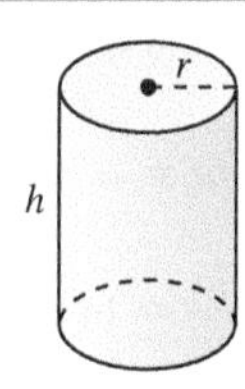

Sphere

Volume $= \frac{4}{3}\pi r^3$

Surface Area $= 4\pi r^2$

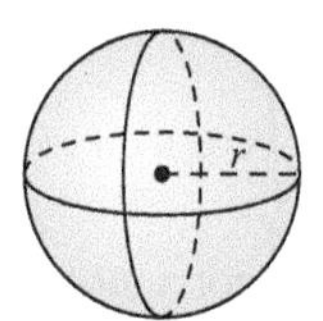

Wedge

(A = area of upper face,

B = area of base)

$A = B \sec \theta$

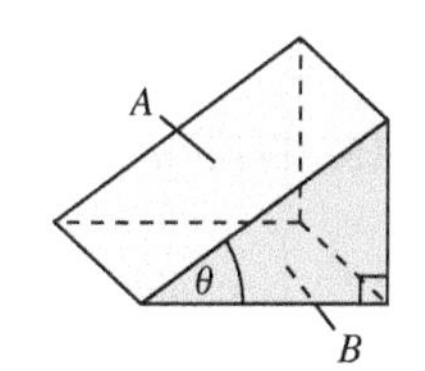

DERIVATIVES AND INTEGRALS

Basic Differentiation Rules

1. $\frac{d}{dx}[cu] = cu'$
2. $\frac{d}{dx}[u \pm v] = u' \pm v'$
3. $\frac{d}{dx}[uv] = uv' + vu'$
4. $\frac{d}{dx}\left[\frac{u}{v}\right] = \frac{vu' - uv'}{v^2}$
5. $\frac{d}{dx}[c] = 0$
6. $\frac{d}{dx}[u^n] = nu^{n-1}u'$
7. $\frac{d}{dx}[x] = 1$
8. $\frac{d}{dx}[|u|] = \frac{u}{|u|}(u'), \quad u \neq 0$
9. $\frac{d}{dx}[\ln u] = \frac{u'}{u}$
10. $\frac{d}{dx}[e^u] = e^u u'$
11. $\frac{d}{dx}[\log_a u] = \frac{u'}{(\ln a)u}$
12. $\frac{d}{dx}[a^u] = (\ln a)a^u u'$
13. $\frac{d}{dx}[\sin u] = (\cos u)u'$
14. $\frac{d}{dx}[\cos u] = -(\sin u)u'$
15. $\frac{d}{dx}[\tan u] = (\sec^2 u)u'$
16. $\frac{d}{dx}[\cot u] = -(\csc^2 u)u'$
17. $\frac{d}{dx}[\sec u] = (\sec u \tan u)u'$
18. $\frac{d}{dx}[\csc u] = -(\csc u \cot u)u'$
19. $\frac{d}{dx}[\arcsin u] = \frac{u'}{\sqrt{1 - u^2}}$
20. $\frac{d}{dx}[\arccos u] = \frac{-u'}{\sqrt{1 - u^2}}$
21. $\frac{d}{dx}[\arctan u] = \frac{u'}{1 + u^2}$
22. $\frac{d}{dx}[\operatorname{arccot} u] = \frac{-u'}{1 + u^2}$
23. $\frac{d}{dx}[\operatorname{arcsec} u] = \frac{u'}{|u|\sqrt{u^2 - 1}}$
24. $\frac{d}{dx}[\operatorname{arccsc} u] = \frac{-u'}{|u|\sqrt{u^2 - 1}}$
25. $\frac{d}{dx}[\sinh u] = (\cosh u)u'$
26. $\frac{d}{dx}[\cosh u] = (\sinh u)u'$
27. $\frac{d}{dx}[\tanh u] = (\operatorname{sech}^2 u)u'$
28. $\frac{d}{dx}[\coth u] = -(\operatorname{csch}^2 u)u'$
29. $\frac{d}{dx}[\operatorname{sech} u] = -(\operatorname{sech} u \tanh u)u'$
30. $\frac{d}{dx}[\operatorname{csch} u] = -(\operatorname{csch} u \coth u)u'$
31. $\frac{d}{dx}[\sinh^{-1} u] = \frac{u'}{\sqrt{u^2 + 1}}$
32. $\frac{d}{dx}[\cosh^{-1} u] = \frac{u'}{\sqrt{u^2 - 1}}$
33. $\frac{d}{dx}[\tanh^{-1} u] = \frac{u'}{1 - u^2}$
34. $\frac{d}{dx}[\coth^{-1} u] = \frac{u'}{1 - u^2}$
35. $\frac{d}{dx}[\operatorname{sech}^{-1} u] = \frac{-u'}{u\sqrt{1 - u^2}}$
36. $\frac{d}{dx}[\operatorname{csch}^{-1} u] = \frac{-u'}{|u|\sqrt{1 + u^2}}$

Basic Integration Formulas

1. $\int kf(u)\,du = k\int f(u)\,du$
2. $\int [f(u) \pm g(u)]\,du = \int f(u)\,du \pm \int g(u)\,du$
3. $\int du = u + C$
4. $\int u^n\,du = \frac{u^{n+1}}{n+1} + C, \quad n \neq -1$
5. $\int \frac{du}{u} = \ln|u| + C$
6. $\int e^u\,du = e^u + C$
7. $\int a^u\,du = \left(\frac{1}{\ln a}\right)a^u + C$
8. $\int \sin u\,du = -\cos u + C$
9. $\int \cos u\,du = \sin u + C$
10. $\int \tan u\,du = -\ln|\cos u| + C$
11. $\int \cot u\,du = \ln|\sin u| + C$
12. $\int \sec u\,du = \ln|\sec u + \tan u| + C$
13. $\int \csc u\,du = -\ln|\csc u + \cot u| + C$
14. $\int \sec^2 u\,du = \tan u + C$
15. $\int \csc^2 u\,du = -\cot u + C$
16. $\int \sec u \tan u\,du = \sec u + C$
17. $\int \csc u \cot u\,du = -\csc u + C$
18. $\int \frac{du}{\sqrt{a^2 - u^2}} = \arcsin\frac{u}{a} + C$
19. $\int \frac{du}{a^2 + u^2} = \frac{1}{a}\arctan\frac{u}{a} + C$
20. $\int \frac{du}{u\sqrt{u^2 - a^2}} = \frac{1}{a}\operatorname{arcsec}\frac{|u|}{a} + C$

Definition of the Six Trigonometric Functions

Right triangle definitions, where $0 < \theta < \pi/2$.

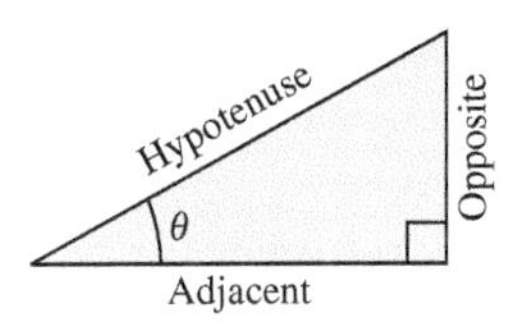

$$\sin\theta = \frac{\text{opp}}{\text{hyp}} \qquad \csc\theta = \frac{\text{hyp}}{\text{opp}}$$

$$\cos\theta = \frac{\text{adj}}{\text{hyp}} \qquad \sec\theta = \frac{\text{hyp}}{\text{adj}}$$

$$\tan\theta = \frac{\text{opp}}{\text{adj}} \qquad \cot\theta = \frac{\text{adj}}{\text{opp}}$$

Circular function definitions, where θ is any angle.

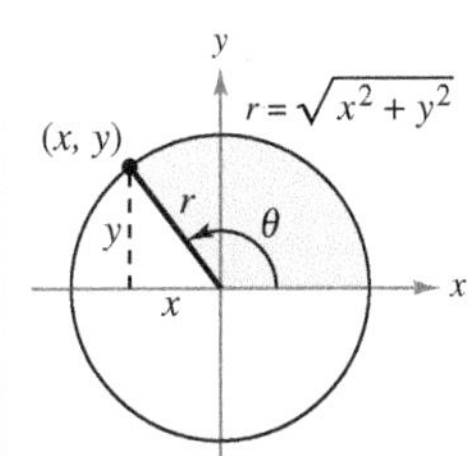

$$\sin\theta = \frac{y}{r} \qquad \csc\theta = \frac{r}{y}$$

$$\cos\theta = \frac{x}{r} \qquad \sec\theta = \frac{r}{x}$$

$$\tan\theta = \frac{y}{x} \qquad \cot\theta = \frac{x}{y}$$

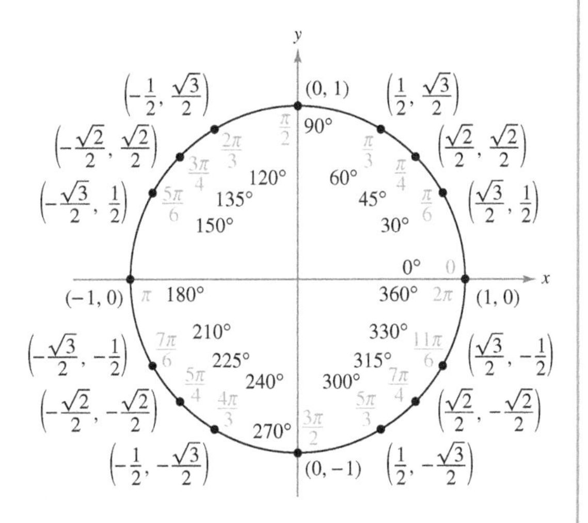

Reciprocal Identities

$$\sin x = \frac{1}{\csc x} \qquad \sec x = \frac{1}{\cos x} \qquad \tan x = \frac{1}{\cot x}$$

$$\csc x = \frac{1}{\sin x} \qquad \cos x = \frac{1}{\sec x} \qquad \cot x = \frac{1}{\tan x}$$

Quotient Identities

$$\tan x = \frac{\sin x}{\cos x} \qquad \cot x = \frac{\cos x}{\sin x}$$

Pythagorean Identities

$$\sin^2 x + \cos^2 x = 1$$

$$1 + \tan^2 x = \sec^2 x \qquad 1 + \cot^2 x = \csc^2 x$$

Cofunction Identities

$$\sin\left(\frac{\pi}{2} - x\right) = \cos x \qquad \cos\left(\frac{\pi}{2} - x\right) = \sin x$$

$$\csc\left(\frac{\pi}{2} - x\right) = \sec x \qquad \tan\left(\frac{\pi}{2} - x\right) = \cot x$$

$$\sec\left(\frac{\pi}{2} - x\right) = \csc x \qquad \cot\left(\frac{\pi}{2} - x\right) = \tan x$$

Even/Odd Identities

$$\sin(-x) = -\sin x \qquad \cos(-x) = \cos x$$

$$\csc(-x) = -\csc x \qquad \tan(-x) = -\tan x$$

$$\sec(-x) = \sec x \qquad \cot(-x) = -\cot x$$

Sum and Difference Formulas

$$\sin(u \pm v) = \sin u \cos v \pm \cos u \sin v$$

$$\cos(u \pm v) = \cos u \cos v \mp \sin u \sin v$$

$$\tan(u \pm v) = \frac{\tan u \pm \tan v}{1 \mp \tan u \tan v}$$

Double-Angle Formulas

$$\sin 2u = 2 \sin u \cos u$$

$$\cos 2u = \cos^2 u - \sin^2 u = 2\cos^2 u - 1 = 1 - 2\sin^2 u$$

$$\tan 2u = \frac{2 \tan u}{1 - \tan^2 u}$$

Power-Reducing Formulas

$$\sin^2 u = \frac{1 - \cos 2u}{2}$$

$$\cos^2 u = \frac{1 + \cos 2u}{2}$$

$$\tan^2 u = \frac{1 - \cos 2u}{1 + \cos 2u}$$

Sum-to-Product Formulas

$$\sin u + \sin v = 2\sin\left(\frac{u + v}{2}\right)\cos\left(\frac{u - v}{2}\right)$$

$$\sin u - \sin v = 2\cos\left(\frac{u + v}{2}\right)\sin\left(\frac{u - v}{2}\right)$$

$$\cos u + \cos v = 2\cos\left(\frac{u + v}{2}\right)\cos\left(\frac{u - v}{2}\right)$$

$$\cos u - \cos v = -2\sin\left(\frac{u + v}{2}\right)\sin\left(\frac{u - v}{2}\right)$$

Product-to-Sum Formulas

$$\sin u \sin v = \frac{1}{2}[\cos(u - v) - \cos(u + v)]$$

$$\cos u \cos v = \frac{1}{2}[\cos(u - v) + \cos(u + v)]$$

$$\sin u \cos v = \frac{1}{2}[\sin(u + v) + \sin(u - v)]$$

$$\cos u \sin v = \frac{1}{2}[\sin(u + v) - \sin(u - v)]$$